CONSTRUCTIVISM IN MATHEMATICS

VOLUME II

STUDIES IN LOGIC

AND

THE FOUNDATIONS OF MATHEMATICS

VOLUME 123

ELSEVIER

AMSTERDAM • BOSTON • HEIDELBERG • LONDON • NEW YORK • OXFORD
PARIS • SAN DIEGO • SAN FRANCISCO • SINGAPORE • SYDNEY • TOKYO

CONSTRUCTIVISM IN MATHEMATICS

AN INTRODUCTION

VOLUME II

A.S. TROELSTRA
*Universiteit van Amsterdam,
The Netherlands*

D. VAN DALEN
*Rijksuniversiteit Utrecht,
The Netherlands*

ELSEVIER

AMSTERDAM • BOSTON • HEIDELBERG • LONDON • NEW YORK • OXFORD
PARIS • SAN DIEGO • SAN FRANCISCO • SINGAPORE • SYDNEY • TOKYO

ELSEVIER B.V.	ELSEVIER Inc.	ELSEVIER Ltd	ELSEVIER Ltd
Sara Burgerhartstraat 25	525 B Street, Suite 1900	The Boulevard, Langford Lane	84 Theobalds Road
P.O. Box 211, 1000 AE	San Diego, CA 92101-4495	Kidlington, Oxford OX5 1GB	London WC1X 8RR
Amsterdam, The Netherlands	USA	UK	UK

First edition 1988
Second impression 2004

Library of Congress Cataloging in Publication Data

Troelstra, A.S. (Anne Sjerp)
Constructivism in mathematics/A.S. Troelstra, D. van Dalen.
p. cm. – (Studies in logic and the foundations of mathematics; v. 121, 123)
Bibliography: p
Includes indexes.
ISBN 0-444-70266-0 (v. 1)
ISBN 0-444-70358-6 (v. 2)
1. Constructive mathematics. I. Dalen, D. van (Dirk), 1932- II. Title. III. Series.
QA9.56.T74 1988
511.3—dc19 88-5240
CIP

British Library Cataloguing in Publication Data
A catalogue record is available from the British Library.

ISBN: 0-444-70358-6

Transferred to digital printing 2005
Printed and bound by Antony Rowe Ltd, Eastbourne

This book is dedicated
to the memory of our teacher
Arend Heyting

CONSTRUCTIVISM IN MATHEMATICS: CONTENTS

Volume I

Volume II

CONTENTS OF VOLUME II

PRELIMINARIES (FROM VOL. 1)

1. *Internal references.* Within chapter n, "$k.m$" refers to the subsection numbered $k.m$ in chapter n; "section k" refers to section k of chapter n; "section $k.m$" refers to section m of chapter k ($k \neq n$), and "$k.l.m$" refers to subsection $l.m$ of chapter k ($k \neq n$). "Exercise $k.l.m$" or "E$k.l.m$" refers to the exercise numbered $k.l.m$ at the end of chapter k. An exercise numbered $k.l.m$ should be regarded as belonging to section $k.l$.

2. *Bibliographical references* are given as author's name followed by year of publication, possibly followed by a letter in the case of more than one publication in the same year by the same author, e.g. "as shown by Brouwer (1927)...", or "...in Brouwer (1927A) it was proved that...", or "the bar theorem (Brouwer 1954)...". References to works by two authors appear as "Kleene and Vesley (1965)".

In the case of three or more authors we use the abbreviation "et al.", e.g. "Constable et al. (1986)". The bibliography at the end of the book contains only items which are actually referred to in the text; by the completion of the logic bibliography Müller (1987) has relieved us of the task of providing something approaching a complete bibliography of constructivism.

In the rest of this section some general notational conventions are brought together, to be consulted when needed. There may be local deviations from these conventions. At the end of the book there is an index of notations of more than purely local use.

3. *Definitions.* $\equiv$ indicates *literal identity*, modulo renaming bound variables. $:=$ is used as the *definition symbol*, the defining expression appears on the right-hand side.

4. *Variables, substitution.* The concept of free and bound variable is defined as usual; for the sets of free and bound variables of the expression $\mathfrak{a}$ we use $\mathrm{FV}(\mathfrak{a})$ and $\mathrm{BV}(\mathfrak{a})$, respectively. As a rule we regard expressions which differ only in the names of bound variables as isomorphic; that is to say, bound variables are used as position markers only.

For the result of *simultaneous substitution* of $t_1, \ldots, t_n$ for the variables $x_1, \ldots, x_n$ in the expression $\mathfrak{a}$ we write $\mathfrak{a}[x_1, \ldots, x_n/t_1, \ldots, t_n]$. We shall often use a looser notation: once $\mathfrak{a}(x_1, \ldots, x_n)$ has appeared in a context, we write $\mathfrak{a}(t_1, \ldots, t_n)$ for $\mathfrak{a}[x_1, \ldots, x_n/t_1, \ldots, t_n]$. Using vector notation as an abbreviation we can also write $\mathfrak{a}[\vec{x}/\vec{t}\,]$ for $\mathfrak{a}[x_1, \ldots, x_n/t_1, \ldots, t_n]$.

In using the substitution notation we shall as a rule tacitly assume the terms to be free for the variables in the expression considered (or we assume that a suitable renaming of bound variables is carried out). In our description of logic in chapter 2 we are more explicit about these matters.

We shall frequently economize on parentheses by writing Ax or At (A a formula) instead of $A(x)$ or $A(t)$.

5. *Logical symbols.* As logical symbols we use

$$\bot, \neg, \wedge, \vee, \to, \leftrightarrow, \forall, \exists, \exists!$$

both formally and informally. In bracketing we adopt the usual convention that $\neg$, $\forall$, $\exists$ bind stronger than any of the binary operators, and that $\wedge$, $\vee$ bind stronger than $\to$, $\leftrightarrow$. Occasionally dots are used as separating symbols instead of parentheses. In discussing formal systems it will be usually clear from the context whether the symbol is used as part of the formalism or on the metalevel. Where it is necessary to avoid confusion we use $\Rightarrow$, $\Leftrightarrow$ on the metalevel; some times a comma serves as a conjunction, and "iff" abbreviates "if and only if". Unless stated otherwise, $\neg$, $\leftrightarrow$, $\exists!$ are regarded as abbreviations defined by

$$\neg A := A \to \bot;$$

$$A \leftrightarrow B := (A \to B) \wedge (B \to A);$$

$$\exists! x A := \exists x(A \wedge \forall y(x = y \leftrightarrow A[x/y])).$$

For repeated equivalences we sometimes write $A \leftrightarrow B \leftrightarrow C \leftrightarrow \cdots$ meaning $(A \leftrightarrow B) \wedge (B \leftrightarrow C) \wedge \cdots$. For iterated finite conjunctions and disjunctions we use

$$\bigwedge, \bigvee.$$

Notation for restricted quantifiers:

$$\forall x \in X, \qquad \exists x \in X.$$

For iterated quantifiers we use the abbreviations

$$\forall x_1 x_2 \ldots x_n := \forall x_1 \forall x_2 \ldots \forall x_n, \qquad \exists x_1 x_2 \ldots x_n := \exists x_1 \exists x_2 \ldots \exists x_n,$$

and for iterated restricted quantifiers

$$\forall x, y \in A := \forall x \in A \, \forall y \in A \quad \text{etcetera.}$$

For possibly undefined expressions $\mathfrak{a}$, $E\mathfrak{a}$ means "$\mathfrak{a}$ exists", or "$\mathfrak{a}$ is well-defined" (cf. the use of $t\downarrow$ in recursion theory).

6. *Set-theoretic notation.* Our standard set-theoretic symbols are

$$\emptyset, \in, \notin, \subset, \supset, \setminus, \cap, \cup, \bigcap, \bigcup.$$

Here $\subset$ and $\supset$ are used for not necessarily proper *inclusion*, i.e. $X \subset Y := \forall x(x \in X \to x \in Y)$, etc. We also use the standard notations

$$\{x_1, x_2, \ldots, x_n\} \quad \text{(for finite sets)},$$

$$\{x : A(x)\}, \quad \{f(x) : A(x)\}, \quad \{x \in B : A(x)\} \quad (f \text{ a function}).$$

For *complements* of sets relative to some fixed set we often use ^{c}. The fixed superset involved will be clear from the context. Thus if X is a subset of $\mathbb{N}$, we write X^c for $\mathbb{N} \setminus X$.

For finite *cartesian products* we use $\times$, for arbitrary cartesian products Π.

If X is a set and $\sim$ an equivalence on X, we write $X/\sim$ for the set of equivalence classes of X modulo $\sim$. For the *equivalence* class of an $x \in X$ we write $x/\sim$, $x_\sim$, $(x)_\sim$ or $[x]_\sim$.

The *set of functions* from X to Y is written as $X \to Y$ or Y^X. *Restrictions* are indicated by $\upharpoonright$ or $|$. $\mathrm{P}(X)$ is the *powerset* of X.

"f is a function from X to Y" is written as $f \in X \to Y$ (and only occasionally as $f\colon X \to Y$); the use of this notation must not be regarded as a commitment to the set of all functions from X to Y as a well-defined totality. If t is a term, we can also introduce "t regarded as a function of the parameter (variable) x" by one of the following notations:

$$\lambda x.t \quad \text{or} \quad f : x \mapsto t.$$

Notations (in diagrams) for *injections*, *surjections*, *bijections*, and *embeddings* are $\rightarrowtail$, $\twoheadrightarrow$, $\rightarrowtail\!\!\!\!\!\twoheadrightarrow$, $\hookrightarrow$, respectively.

For the *characteristic function* of a relation R we use χ_R, where $\chi_R(t) = 0 \leftrightarrow R(t)$.

For the function f applied to the argument t we usually write $f(t)$, or even ft, if no confusion can arise; in general we drop parentheses whenever we can safely do so. In certain chapters we use $t(t')$ or tt' for term t applied to term t' (e.g. in chapter 9); in such cases we use square brackets instead of parentheses to refer to occurrences in a term, writing $t[x]$ instead of $t(x)$, etc.

Pairs, or *n-tuples* for fixed length, are usually also indicated by means of parentheses (,); in case of finite sequences of variable length we use $\langle\ ,\ \rangle$: $\langle x_1, x_2, \ldots, x_n\rangle$ is a sequence of length n (notation $\mathrm{lth}\langle x_1, x_2, \ldots, x_n\rangle = n$). For concatenation we use $*$. A sequence of arguments may also be indicated by vector notation; thus we write $f(\vec{t})$ or $f\vec{t}$ for $f(t_1, \ldots, t_n)$, where $\vec{t} \equiv (t_1, \ldots, t_n)$; $\vec{t} = \vec{s}$ indicates equality of vectors.

We use the abbreviation $\lambda x_1 x_2 \ldots x_n.t$ for $\lambda x_1(\lambda x_2 \ldots (\lambda x_n.t)\ldots)$. If we wish to regard t as a function in n arguments $x_1, x_2, \ldots, x_n$ we use comma's: $\lambda x_1, x_2, \ldots, x_n.t$.

For the *graph* of the function f we write graph(f), dom(f) is its *domain*, range (f) its *range*.

7. *Mathematical constants.* The following constants:

$$\mathbb{N}, \mathbb{Z}, \mathbb{Q}, \mathbb{R}, \mathbb{B}, \mathbb{C},$$

denote the natural numbers, the integers, the rationals, the reals, Baire space, and the complex numbers, respectively. Throughout the book n, m, i, j, k, unless indicated otherwise, are supposed to range over $\mathbb{N}$, and $\alpha, \beta, \gamma, \delta$ over $\mathbb{B}$ or a subtree of $\mathbb{B}$. In metamathematical work we use $\bar{n}$, $\bar{m}$ for numerals.

For an infinite sequence $\mathfrak{a}_0, \mathfrak{a}_1, \mathfrak{a}_2, \ldots$ (i.e. a function with domain $\mathbb{N}$) we use the notation $\langle \mathfrak{a}_n \rangle_n$. A notation such as $\lim \langle x_n \rangle_n$ is self-explanatory.

The notation for arithmetical operations is standard; the multiplication dot $\cdot$ is often omitted.

8. *Formal systems and axioms.* Formal systems are designated by combinations of roman boldface capitals, e.g. **HA**, **IQC**, $\mathbf{EM}_0\restriction$, etc.

If **H** is a system based on intuitionistic logic, $\mathbf{H}^c$ is used for the corresponding system based on classical logic.

For the language of a formal system **H** we often write $\mathcal{L}(\mathbf{H})$. If $\mathbf{H}'$ extends **H**, and $\mathcal{L}(\mathbf{H}')$ extends $\mathcal{L}(\mathbf{H})$, while $\mathbf{H}' \cap \mathcal{L}(\mathbf{H}) = \mathbf{H}$, then $\mathbf{H}'$ is said to be a *conservative extension* of **H** ($\mathbf{H}'$ is *conservative over* **H**). A *definitional extension* is a special case of a conservative extension, where the extra symbols of $\mathscr{L}(\mathbf{H}')$ can be replaced by explicit definitions (cf. 2.7.1).

Axiom schemas and rules are usually designated by combinations of roman capitals: REFL, BI, $\mathrm{BI_D}$, FAN, WC-N, $\mathrm{CT_0}$, etc.

$\mathbf{H} \vdash A$ means "A is derivable in the system $\mathbf{H}$", and $\mathbf{H} + \mathrm{XYZ} \vdash A$ "A is derivable in $\mathbf{H}$ with the axiom schema or rule XYZ added". Occasionally we use subscript notation: $\vdash_{\mathbf{H}} A$ instead of $\mathbf{H} \vdash A$.

9. *Validity.* For validity of a sentence A in a model $\mathscr{M}$ we use the notation $\mathscr{M} \models A$. It is to be noted that if the relations, functions, and constants of the model $\mathscr{M}$ are defined constructively, then $\mathscr{M} \models A$ also makes sense constructively (cf. 13.1.7).

CHAPTER 7

THE TOPOLOGY OF METRIC SPACES

This chapter deals with the topology of metric spaces; a few elementary facts are needed in chapter 15, but otherwise the material is not used in the rest of the book. Familiarity with the corresponding parts of the classical theory is not necessary, but helpful. As a standard reference for classical point-set topology we use Engelking (1968).

The contents of the chapter may be seen as a continuation of chapter 6. Of particular interest is the notion of a metrically located set (ML-set); section 3 contains a definition of topologically located set (L-set) which coincides with the notion of an ML-set for complete totally bounded sets if we assume FAN (section 4).

Assuming FAN, we can also prove that complete, locally totally bounded spaces coincide with complete, locally compact metric spaces; each locally compact complete metric space can be given a topologically equivalent metric such that the notions of L-set and ML-set coincide for this metric (section 5).

Among other results in this chapter we mention a version of the Lindelöf covering theorem in Constructive Recursive Mathematics (section 2) and the existence of "one-point" compactifications for locally compact complete metric spaces (section 5).

1. Basic definitions

1.1. In 4.1.5 we have already defined the notions of topology, basis, topological space, point of a space, open set, interior of a set X (*notation*: $\mathrm{Int}(X)$), continuous mapping, homeomorphism, and in 5.3.12–13 metric space, accumulation point and the *canonical topology* for a metric space with as basis the open balls $U(a, x)$ with radius a and center x.

Convention. We shall in this chapter reserve U, V, W, possibly with sub-

or superscripts, for open sets, X, Y, Z for sets of points in general.

Sometimes we use U_x, V_x for arbitrary opens containing the point x (*neighbourhoods* of x). For $U(\varepsilon, y)$ we sometimes write $U_\varepsilon(y)$. □

1.2. DEFINITION. $\mathscr{B} \subset \mathscr{T}$ is a *subbasis* for the topological space $\Gamma \equiv (X, \mathscr{T})$ if the family of finitely indexed intersections $\{V_0 \cap \cdots \cap V_n : \forall i \leq n(V_i \in \mathscr{B})\}$ is a basis for Γ. □

1.3. EXAMPLES *of metric spaces.*

Example 1. $\mathbb{R}, \mathbb{Q}$ (cf. 5.3.15).

Example 2. Let T be a spread. T can be given a metric as follows. For $\alpha, \beta \in T$, and $n \in \mathbb{N}$, put

$$\rho^*(\bar{\alpha}n, \bar{\beta}n) := \begin{cases} 2^{-n} & \text{if } \bar{\alpha}n = \bar{\beta}n, \\ 2^{-k} & \text{if } \alpha k \neq \beta k \quad \text{and} \quad \bar{\alpha}k = \bar{\beta}k \quad \text{for some } k \leq n. \end{cases}$$

Then

$$\rho(\alpha, \beta) := \lim\langle \rho^*(\bar{\alpha}n, \bar{\beta}n)\rangle_n$$

defines a metric on T. The reader should verify that the topology induced by this metric is the same one as we defined in subsection 4.1.5.

Example 3. *Separable Hilbert space* l^2 consists of all sequences $\langle x_n \rangle_n \subset \mathbb{R}$ such that $\Sigma\{x_n^2 : n \in \mathbb{N}\}$ converges, with a metric

$$\rho(\langle x_n \rangle_n, \langle y_n \rangle_n) := \left(\sum_{n=0}^{\infty} (x_n - y_n)^2 \right)^{1/2}.$$

Example 4. Let $M_k \equiv (X_k, \rho_k)$, $k \leq n$, be metric spaces. The product $\Pi\{M_k : k \leq n\}$ can be given a metric

$$\rho(\vec{x}, \vec{y}) := \max\{\rho_k(x_k, y_k) : k \leq n\},$$

where $\vec{x} \equiv (x_0, \ldots, x_n)$, $\vec{y} \equiv (y_0, \ldots, y_n)$.

Example 5. Let $M \equiv (X, \rho)$ be a metric space; then $M' \equiv (X, \rho')$, where ρ' is defined from ρ by $\rho'(x, y) := \min\{\rho(x, y), 1\}$, is a metric space with a bounded metric, since $\forall xy(\rho'(x, y) \leq 1)$. Observe that the topologies induced by M and M' are the same.

Example 6. Let $M_k \equiv (X, \rho_k)$, $k \in \mathbb{N}$, be an infinite sequence of metric spaces. The product $M := \prod\{M_k : k \in \mathbb{N}\}$ can be metrized by

$$\rho(\langle x_n\rangle_n, \langle y_n\rangle_n) := \sup\{2^{-n}\cdot \rho'_n(x_n, y_n) : n \in \mathbb{N}\},$$

where

$$\rho'_n(x_n, y_n) := \min\{1, \rho_n(x_n, y_n)\}.$$

1.4. EXAMPLES *of topological spaces.*

Example 1. Metric spaces with their canonical topology (5.3.13).

Example 2. On each set X there is a smallest (or "coarsest") topology, the *trivial* topology, with basis $\mathscr{B} \equiv \{X\}$. Classically, this means that the topology $\mathscr{T}$ is the collection $\{\emptyset, X\}$; constructively there are many "doubtful" intermediates: for any proposition P there is an open set $\hat{P} := \bigcup\{X : P\}$.

At the other extreme we can provide each set X with the *discrete* topology $\mathscr{T} \equiv \mathscr{P}(X)$ consisting of all subsets of X (cf. 5.3.15).

Example 3. Let $(X, <)$ be a set with a strict partial order (i.e. $\forall x \neg(x < x)$, $\forall xyz(x < y \land y < z \to x < z)$). The *interval topology* is obtained if we take the open intervals $(x, y) := \{z : x < z < y\}$ as a subbasis. Both $\mathbb{R}$ and $\mathbb{R}^{\mathrm{be}}$ yield special cases of the interval topology.

Example 4. A partially ordered set (poset) $(K, \leq)$ can be made into a topological space $\Gamma \equiv (K, \mathscr{T})$ with as opens the upwards monotone sets of K, i.e.

$$W \in \mathscr{T} := \forall k, k' \in K(k \in W \land k' \geq k \to k' \in W).$$

This topology is often called the *Alexandrov topology*.

Example 5. Let $\Gamma \equiv (X, \mathscr{T})$ be a topological space; then for $Y \subset X$, $\Gamma' \equiv (Y, \mathscr{T}')$ with $\mathscr{T}' \equiv \{Y \cap V : V \in \mathscr{T}\}$, is also a topological space; $\mathscr{T}'$ is the *relative topology on Y induced by* $\mathscr{T}$.

Example 6. Let X be a set with apartness $\#$ (i.e. $\#$ satisfies AP1–3 of 5.2.9 or 8.1.2). The sets of the form $x^* := \{y : y \# x\}$ can be taken as a subbasis for a topology.

Example 7. Let $\Gamma_i \equiv (X_i, \mathscr{T}_i)$, $i \in I$, be a collection of topological spaces, and define the *topological product* $(X, \mathscr{T})$ as the space with set of points

$\{f : \forall i \in I(f(i) \in X_i)\}$, and as basis for the topology all sets of the form

$$\left\{f \in X : \exists n \forall k \leq n\left(f(i_k) \in V_{i_k} \in \mathcal{T}_i\right)\right\}.$$

We leave it as an exercise to show that for $I = \mathbb{N}$ the topology induced by the metric in example 6 of 1.3 coincides with the product topology.

1.5. Definition. Let $\Gamma = (X, \mathcal{O})$ be a topological space, $Y \subset X$. x is a *closure point* of Y iff $\forall W_x \exists y(y \in Y \cap W_x)$. If Y contains all its closure points, Y is said to be *closed*. For any Y, Y^- is the set of all closure points of Y; Y^- is closed. □

1.6. Proposition. *Let $\Gamma = (X, \mathcal{O})$ be a topological space. Then for arbitrary subsets Y, Y_1, Y_2 of X, and arbitrary $V \in \mathcal{O}$:*
(i) $\emptyset^- = \emptyset$, $X^- = X$, $Y \subset Y^- = Y^{--}$;
(ii) $(Y_1 \cap Y_2)^- \subset Y_1^- \cap Y_2^-$;
(iii) $(\bigcap\{Y_i : i \in I\})^- \subset \bigcap\{Y_i^- : i \in I\}$;
(iv) $Y_1 \subset Y_2 \to Y_1^- \subset Y_2^-$;
(v) $(Y_1^- \cup Y_2^-)^- = (Y_1 \cup Y_2)^-$;
(vi) $V^{c-} = V^c$ (*the complement of an open set is closed*).

Proof. Exercise. □

Weak counterexamples can be given in $\mathbb{R}$ to the following classically valid assertions:
(vii) each closed set is the complement of an open set; and
(viii) $Y_1^- \cup Y_2^- = (Y_1 \cup Y_2)^-$.
We shall leave these to the reader. □

1.7. Proposition. *Let $\Gamma \equiv (X, \mathcal{T})$ be a topological space and let $Y, Y' \subset X$. Then*:
(i) $\text{Int}(\text{Int}(Y)) = \text{Int}(Y)$, $\text{Int}(\emptyset) = \emptyset$, $\text{Int}(X) = X$;
(ii) $Y \subset Y' \to \text{Int}(Y) \subset \text{Int}(Y')$;
(iii) $\text{Int}(Y \cap Y') = \text{Int}(Y) \cap \text{Int}(Y')$.

Proof. Trivial. □

1.8. Definition. Let $M_i \equiv (X_i, \rho_i)$, $i \in \{0,1\}$ be metric spaces. $f \in M_0 \to M_1$ is *continuous* iff

$$\forall k \forall x \in X_0 \exists m \forall y \in X_0\left(\rho_0(x, y) < 2^{-m} \to \rho_1(f(x), f(y)) < 2^{-k}\right),$$

and *uniformly continuous* iff

$$\forall k \exists m \forall x, y \in X_0\big(\rho_0(x, y) < 2^{-m} \to \rho_1(f(x), f(y)) < 2^{-k}\big);$$

f is *sequentially continuous* iff for each $\langle x_n \rangle_n \subset X$ with $\lim \langle x_n \rangle_n = x$ (i.e. $\lim \langle \rho_0(x_n, x) \rangle_n = 0$), one has $\lim \langle f(x_n) \rangle_n = f(x)$. □

EXAMPLE. For any metric space (M, ρ) the mapping $\rho \in M^2 \to \mathbb{R}$ is uniformly continuous.

1.9. PROPOSITION. *For M_i as above and $f \in M_0 \to M_1$ we have: f uniformly continuous $\to$ f continuous $\to$ f sequentially continuous.* □

1.10. *Digression on "weakly continuous mappings".* In classical point-set topology, the definition of a topology in terms of closed sets is equivalent to the definition in terms of open sets. As noted in 1.6(vii) this is not obviously the case constructively, since not every closed set is the complement of an open set.

Let us call a mapping f from a topological space Γ_0 to a topological space Γ_1 to be *weakly continuous* if the original of a closed set is always closed. It is easy to verify that:
(i) f is weakly continuous iff $\forall Y \subset X_0(f[Y^-] \subset f[Y]^-)$;
(ii) f continuous $\to$ f weakly continuous.

On the other hand, "f weakly continuous" does not imply "f continuous", as may be seen from the following weak counterexample.

Let $\mathbb{N}, \mathbb{Q}$ be regarded as topological spaces with the topology induced by the usual metric. We define $f \in \mathbb{N} \to \mathbb{Q}$ as follows. Let $\alpha \in \mathbb{N} \to \mathbb{N}$ be any sequence such that $\exists n(\alpha n = 0)$ is unknown, and put

$$f(n) := \begin{cases} n & \text{if } \neg(\alpha n = 0 \wedge \forall m < n(\alpha m \neq 0)), \\ 1 + n^{-1} & \text{if } (\alpha n = 0 \wedge \forall m < n(\alpha m \neq 0)). \end{cases}$$

Now observe that:
(iii) f is injective, i.e. f is a bijection from $\mathbb{N}$ to $f[\mathbb{N}]$, in the strong sense that

$$n \neq m \to f(n) \# f(m).$$

(iv) f is continuous since $\mathbb{N}$ has the discrete topology, i.e. each subset of $\mathbb{N}$ is open.
(v) f^{-1} is weakly continuous on $f[\mathbb{N}]$, where $f[\mathbb{N}]$ is provided with the (relative) topology induced by the metric on $\mathbb{Q}$.
(vi) f^{-1} is not continuous on $f[\mathbb{N}]$.

To prove (v) it suffices to show that for each $X \subset \mathbb{N}$, $f[X]$ is closed in $f[\mathbb{N}]$. Let $y \in f[X]^-$; since $y \in f[\mathbb{N}]$, we know that either $y = n$ for some n, and $\neg(\alpha n = 0 \wedge \forall m < n(\alpha m \neq 0))$, or $y = 1 + n^{-1}$ for some $n \in \mathbb{N}$, and $\alpha n = 0 \wedge \forall m < n(\alpha m \neq 0)$. In the first case, $U(\frac{1}{2}, y)$ must contain a $z \in f[X]$. If $y = n > 1$, z must be n, and then $y \in f[X]$; or if $y = n = 1$, it may happen that $z = f(m) = 1 + m^{-1}$. Then consider $U((2m)^{-1}, y)$; for some z' and m', $z' = f(m') \in U((2m)^{-1}, y)$, and this can only happen if $z' = y = m' = 1$, so again $y \in f[X]$.

Similarly in the second case where $y = 1 + n^{-1}$: again consider $U((2n)^{-1}, y)$, etc.

To prove (vi), consider $\langle x_n \rangle_n \subset f[\mathbb{N}]$ defined by

$$x_n = 1 \quad \text{if } \neg(\alpha n = 0 \wedge \forall m < n(\alpha m \neq 0)),$$

$$x_n = 1 + n^{-1} \quad \text{if } (\alpha n = 0 \wedge \forall m < n(\alpha m \neq 0)).$$

Then obviously $\langle x_n \rangle_n$ converges to 1, but we cannot prove convergence of $\langle f^{-1}(x_n) \rangle_n$.

This example may be taken as an indication that the notion of a topology defined by means of closed sets is less satisfactory than topology defined by means of open sets. □

1.11. Definition. A *separated space* $\Gamma \equiv (X, \mathcal{T})$ is a space with apartness relation # (i.e. # satisfies AP1–3 of 5.2.9 or 8.1.2) such that:
(i) $\forall x(\{y : y \# x\} \in \mathcal{T})$;
(ii) $\forall U \forall xy(x \in U \rightarrow y \in U \vee x \# y)$. □

Note that in a separated space

$$x \# y \leftrightarrow \exists U(x \in U \wedge y \notin U),$$

and that each metric space is separated. Observe also that a homeomorphism f between separated spaces Γ_0 and Γ_1 is a *strong injection*, that is to say $\forall x, y \in \Gamma_0(x \# y \rightarrow fx \# fy)$.

2. Complete separable metric spaces

2.1. Definition. Let $M \equiv (X, \rho)$ be a metric space. $\langle x_n \rangle_n \subset X$ is a *Cauchy sequence* (relative to ρ) with *modulus* α iff

$$\forall kmm'(\rho(x_{\alpha k+m}, x_{\alpha k+m'}) < 2^{-k}).$$

In the presence of AC_{00} this is of course equivalent to

$$\forall k \exists n \forall m m' (\rho(x_{n+m}, x_{n+m'}) < 2^{-k}).$$

M is (metrically) *complete* if every Cauchy sequence is convergent, i.e. for each Cauchy sequence $\langle x_n \rangle_n$ there is an x such that $\lim \langle x_n \rangle_n = x$. □

2.2. DEFINITION. A metric space $M \equiv (X, \rho)$ is *separable* if there is a sequence $\langle p_n \rangle_n \subset X$ (a *basis sequence* for the space) such that $\{p_n : n \in \mathbb{N}\}^- = X$, or in terms of the metric

$$\forall x \in X \forall k \exists n (\rho(p_n, x) < 2^{-k}).$$

We call a complete separable metric space, for short, a *CSM-space*. □

In the rest of this chapter we shall assume the axiom of countable choice AC_0 (even if not needed for every proposition), with the exception of section 3.9, where we shall discuss to what extent we can do without it. We first introduce a standard representation of CSM-spaces.

Since all points of a CSM-space figure as closure points of the p_i's, we can represent them by sequences of p_i's. By carefully keeping track of the convergence moduli we obtain a tree in which the nodes stand for p_i's, and such that the (infinite) branches yield precisely all the points of X.

2.3. *Standard representation of* CSM-*spaces.* Let $\langle r_n \rangle_n$ be a standard enumeration of $\mathbb{Q}$, and let $M \equiv (X, \rho)$ be a CSM-space, with basis sequence $\langle p_n \rangle_n$. By AC_{00}, there is an $\alpha \in \mathbb{N}^3 \to \mathbb{N}$ such that

$$|\rho(p_i, p_j) - r_{\alpha(i,j,k)}| < 2^{-k}.$$

Again by AC_{00}, we can find, for each $x \in X$, sequences $\beta, \gamma \in \mathbb{N}^{\mathbb{N}}$ such that $\forall k \forall m (\rho(p_{\beta(\gamma k+m)}, x) < 2^{-k-1})$. That is to say, $\langle p_{\beta k} \rangle_k$ is a Cauchy sequence with modulus γ.

DEFINITION. A *representation* of a CSM-space consists of a spread T and an assignment $\varphi \in (T \setminus \{\langle\ \rangle\}) \to \langle p_n \rangle_n$ such that:
(i) $\forall \alpha \in T(\langle \varphi(\bar{\alpha}(n+1)) \rangle_n$ is a Cauchy sequence);
(ii) $\forall x \in X \exists \alpha \in T(x = \lim \langle \varphi(\bar{\alpha}(n+1)) \rangle_n)$. □

N.B. Any representation can be transformed into a representation over the universal spread T_U, using the continuous mapping described in 4.1.4.

Construction of the standard representation. Let us write $\alpha_Q(i, j, k)$ for $r_{\alpha(i,j,k)}$. The set

$$\{j : \alpha_Q(i, j, k+1) < 2^{-k+1}\}$$

is obviously decidable, and is moreover inhabited since it contains i (observe that $|\rho(p_i, p_i) - \alpha_Q(i, i, k+1)| < 2^{-k-1}$, hence $\alpha_Q(i, i, k+1) < 2^{-k+1}$). We now define the tree T by specifying its branches:

$$\beta \in T := \forall k\big(\alpha_Q(\beta k, \beta(k+1), k+1) < 2^{-k+1}\big),$$

which can easily be reformulated as a decidable condition on finite sequences of natural numbers; let $\varphi\langle n_0, \ldots, n_k\rangle := p_{n_k}$.

2.4. PROPOSITION. *(T, φ) is a representation of (X, ρ) with the following properties:*

(i) *For all $\beta \in T$, $\langle p_{\beta n}\rangle_n$ is a Cauchy sequence with modulus $\lambda k.k + 3$.*

(ii) *For any $\beta \in T$ let $x_\beta \in X$ be $\lim\langle p_{\beta n}\rangle_n$, then $\forall k(\rho(x_\beta, p_{\beta k}) < 5 \cdot 2^{-k} < 2^{-k+3})$.*

(iii) $\forall n(\rho(x, p_{\beta n}) < 2^{-n-1}) \to \beta \in T$.

(iv) $\forall n(\rho(p_{\beta n}, p_{\beta(n+1)}) < 2^{-n}) \to \beta \in T$.

(v) *Let $V'_n := \{x_\beta : \beta \in T \wedge \beta \in n\}$. If $\beta \in T$, then $\gamma = \lambda x.\beta(x+4) \in T$, $x_\gamma = x_\beta$, and $\forall k(U(2^{-k-1}, x_\gamma) \subset V'_{\bar{\gamma}(k+1)} \subset U(5 \cdot 2^{-k}, p_{\gamma k}))$.*

PROOF. (i), (ii) are left to the reader.

(iii) If $\forall n(\rho(x, p_{\beta n}) < 2^{-n-1})$, then $\rho(p_{\beta n}, p_{\beta(n+1)}) < 2^{-n-1} + 2^{-n-2}$, hence $\alpha_Q(\beta n, \beta(n+1), n+1) < 2^{-n} + 2^{-n-2} < 2^{-n+1}$.

(iv) Similarly.

(v) If $\beta \in T$, then so does $\lambda x.\beta(x+1) = \delta$, since $\alpha_Q(\delta n, \delta(n+2), n+2) = \alpha_Q(\beta(n+1), \beta(n+2), n+2) < 2^{-n-1}$; hence $\rho(p_{\delta n}, p_{\delta(n+1)}) < 2^{-n-1} + 2^{-n-2}$, and $\alpha_Q(\delta(n), \delta(n+1), n+1) < \rho(p_{\delta n}, p_{\delta(n+1)}) + 2^{-n-1} < 2^{-n} + 2^{-n-2} < 2^{-n+1}$.

Let $\gamma = \lambda x.\beta(x+4)$; then obviously $x_\gamma = x_\beta$, and $\rho(x_\gamma, p_{\gamma n}) = \rho(x_\beta, p_{\beta(n+4)}) < 2^{-n-1}$. Now choose y, δ such that $\rho(x_\gamma, y) < 2^{-k-1}$, $\forall n(\rho(y, p_{\delta n}) < 2^{-n})$; then $\rho(p_{\gamma k}, p_{\delta(k+1)}) \le \rho(x_\gamma, p_{\gamma k}) + \rho(x_\gamma, y) + \rho(y, p_{\delta(k+1)}) < 2^{-k-1} + 2^{-k-1} + 2^{-k-1} = 3 \cdot 2^{-k-1}$, and therefore $\alpha_Q(\gamma k, \delta(k+1), k+1)) < 3 \cdot 2^{-k-1} + 2^{-k-1} = 2^{-k+1}$. From this we conclude that $\varepsilon = \bar{\gamma}(k+1) * \lambda x.\delta(k+1+x) \in T$; so $y = x_\varepsilon \in V'_{\bar{\gamma}(k+1)}$ and therefore $U(2^{-k-1}, x_\gamma) \subset V'_{\bar{\gamma}(k+1)}$. The rest is easy and left to the reader. □

In the terminology of classical mathematics the theorem shows that (X, ρ) is a *quotient* of the tree space T with topology inherited from Baire space (and thus in fact a quotient from Baire space itself). Cf. Engelking (1968, 2.4).

The essential result might have been described more concisely, but we also wanted to show in detail which sort of closure operations on the sequences of T are necessary for the argument; we see that closure under "shift" $\beta \mapsto \lambda x.\beta(x+1)$, and replacement of initial segments by other initial segments suffice.

2.5. COROLLARY. *A* CSM-*space is the continuous image of Baire-space.*

PROOF. Immediate. □

We shall now give some consequences of nonclassical axioms, which to some extent are parallel to our proofs for the special case of the reals.

2.6. THEOREM (WC-N). *Let* $M \equiv (X, \rho)$ *be a* CSM-*space, and let* $\{X_i : i \in I\}$, *with* $I \subset \mathbb{N}$, *be a* cover *of* M, *i.e.* $\bigcup\{X_i : i \in I\} \supset X$. *Then* $\{\mathrm{Int}(X_i) : i \in I\}$ *also covers* M.

PROOF. Similar to 6.3.3, now using the standard representation for CSM-spaces. By the assumptions of the theorem we have

$$\forall x \in X\, \exists n \in \mathbb{N}(n \in I \wedge x \in X_n),$$

and thus, for α ranging over T, (T, φ) the standard representation of 2.3,

$$\forall \alpha \exists n(n \in I \wedge x_\alpha \in X_n),$$

and therefore with WC-N

$$\forall \alpha \exists m \exists n \forall \beta \in \bar{\alpha}m(n \in I \wedge x_\beta \in X_n), \text{ i.e.} \tag{1}$$

$$\forall \alpha \exists m \exists n(V'_{\bar{\alpha}m} \subset X_n).$$

Combining this with 2.4(v), we see that for any x we can find a γ such that

$$x = x_\gamma, \quad \forall k\left(U(2^{-k-1}, x_\gamma) \subset V'_{\bar{\gamma}(k+1)}\right);$$

applying (1) to this γ, we find for certain k and n

$$U(2^{-k-1}, x_\gamma) \subset V'_{\bar{\gamma}(k+1)} \subset X_n,$$

so $x \in \mathrm{Int}(X_n)$. □

2.7. THEOREM (WC-N). *Let M be a CSM-space, M' a separable metric space. Then all mappings from M to M' are continuous.*

PROOF. From 2.6, similar to the derivation of 6.3.4 from 6.3.3; this is left to the reader. □

2.8. THEOREM (WC-N). *Let M be a separable metric space, M' any metric space. Then any sequentially continuous $f \in M \to M'$ is continuous.*

PROOF. Let $\langle p_n \rangle_n$ be a basis sequence for $M \equiv (X, \rho)$ and let $x \in X$. For each i we construct a sequence $\langle p_{i,n} \rangle_n$ enumerating $\{ p_n : \rho(p_n, x) < 2^{-i} \}$, possibly with repetitions. Now for any α in $\mathbb{N}^{\mathbb{N}}$ the sequence $\langle p_{m,\alpha m} \rangle_m$ converges to x. By the assumption of sequential continuity we find for each k

$$\forall \alpha \exists m \forall n \geq m \big(\rho'\big(f(p_{n,\alpha n}), f(x)\big) < 2^{-k} \big),$$

and therefore with WC-N

$$\forall \alpha \exists m m' \forall \beta \in \bar{\alpha} m' \forall n \geq m \big(\rho'\big(f(p_{n,\beta n}), f(x)\big) < 2^{-k} \big).$$

Take for α any sequence, e.g. $\lambda x.0$; without loss of generality we may assume $m = m'$, and so

$$\forall \beta \in \bar{\alpha} m \, \forall n \geq m \big(\rho'\big(f(p_{n,\beta n}), f(x)\big) < 2^{-k} \big).$$

If we apply this to $\beta = \alpha m * \langle i \rangle$, $n = m$, we see that $\forall i (\rho'(f(p_{n,i}), f(x)) < 2^{-k})$. Thus

$$\forall n \big(\rho(p_n, x) < 2^{-m} \to \rho'\big(f(p_n), f(x)\big) < 2^{-k} \big).$$

Let $\rho(y, x) < 2^{-m}$; then there is a $\langle p_{\beta n} \rangle_n$ such that $\forall n (\rho(p_{\beta n}, y) < 2^{-n})$; if $\rho(y, x) < 2^{-m} - 2^{-n}$, then for $t \geq n$ $\rho(p_{\beta t}, x) \leq \rho(p_{\beta t}, y) + \rho(y, x) < 2^{-m}$, so $\rho'(f(p_{\beta t}), f(x)) < 2^{-k}$; by sequential continuity at y, we can find a $t \geq n$ such that $\rho'(f(p_{\beta t}), f(y)) < 2^{-k}$, so $\rho'(f(x), f(y)) < 2^{-k+1}$. □

COROLLARY *to the proof of the theorem* (WC-N). *Let $M \equiv (X, \rho)$ be an enumerable space, i.e. $X = \langle p_n \rangle_n$, and let M' be any metric space. Then for any $x \in X$, if $f \in M \to M'$ is sequentially continuous at x, then f is also continuous at x.* □

2.9. *Discussion.* The usual classical proof proceeds by contraposition and uses the axiom of choice. This is unavoidable for the "local" version: sequential continuity at x implies continuity at x; but the global version, as in the theorem, can be proved without the axiom of choice! Namely, take any x and suppose f not to be continuous in x. Then for each k there are y_k with $\rho(y_k, x) < 2^{-k}$, $\rho'(fy_k, fx) > 2^{-n}$(n fixed); for the y_k one can take the first p_i with the required properties (this is possible since we consider the global version).

2.10. PROPOSITION (*Lindelöf's theorem under the assumption of* C-N). *Let* $M \equiv (X, \rho)$ *be a* CSM-*space, and let* $\mathscr{A}$ *be an open covering of* X, *i.e. all elements of* $\mathscr{A}$ *are open and* $\bigcup\mathscr{A} \supset X$. *Then there is an enumerable subcover* $\mathscr{B} = \langle V_n \rangle_n \subset \mathscr{A}$.

PROOF. We again make use of the standard representation (T, φ) of M. By the assumptions of the theorem

$$\forall x \exists k \exists V \in \mathscr{A}\big(U(2^{-k}, x) \subset V\big),$$

hence also

$$\forall \alpha \exists k \exists V \in \mathscr{A}\big(U(2^{-k}, x_\alpha) \subset V\big).$$

Therefore, if $\langle p_n \rangle_n$ is a basis sequence for M,

$$\forall \alpha \exists k m \exists V \in \mathscr{A}\big(U(2^{-k}, p_m) \subset V \wedge \rho(x_\alpha, p_m) < 2^{-k-1}\big).$$

Now apply C-N; we find a neighbourhood function $\gamma \in K_0$ such that

$$\forall n\Big(\gamma n \neq 0 \to \forall \alpha \in n \, \exists V \in \mathscr{A}\Big(U\big(2^{-j_1(\gamma n \dot{-} 1)}, p_{j_2(\gamma n \dot{-} 1)}\big) \subset V \wedge \rho\big(x_\alpha, p_{j_2(\gamma n \dot{-} 1)}\big) < 2^{-j_1(\gamma n - 1) - 1}\Big)\Big).$$

We can enumerate the m of the form $\gamma n \dot{-} 1$ for $\gamma n \neq 0$; let δ be an enumerating function, then $\{U(2^{-j_1(\delta n)}, p_{j_2(\delta n)}) : n \in \mathbb{N}\}$ is a covering, and

$$\forall n \exists V \in \mathscr{A}\Big(U\big(2^{-j_1(\delta n)}, p_{j_2(\delta n)}\big) \subset V\Big).$$

By an application of countable choice we find $\langle V_m \rangle_m \subset \mathscr{A}$ such that

$$\forall n\Big(U\big(2^{-j_1(\delta n)}, p_{j_2(\delta n)}\big) \subset V_n\Big). \quad \square$$

In Constructive Recursive Mathematics we can obtain similar results. First of all, we have the following theorem.

2.11. THEOREM (CRM). *Let $M \equiv (X, \rho)$ be a* CSM-*space, $M' \equiv (X', \delta')$ a metric space. Then any mapping $f \in M \to M'$ is continuous.*

PROOF. A proof can be given by generalizing the argument in 6.4.12. Another proof is outlined in the exercises 7.2.2–5. □

For the discussion of separable metric spaces in CRM we shall introduce some notations.

2.12. NOTATIONS. Let $M \equiv (X, \rho)$ be a separable metric space with basis $\langle p_n \rangle_n$. Let $\{x\}$ be a total recursive function such that for some $p \in X$

$$A_M(x) \equiv \forall n\big(\rho(p_{\{x\}(n)}, p_{\{x\}(n+1)}) < 2^{-n-1}\big); \tag{1}$$

we write $[x]_M$ for $\lim\langle p_{\{x\}(n)}\rangle_n$. Observe that $A_M(x)$ is in fact (equivalent to) an almost negative predicate in x.

If $\{u\}$ enumerates a sequence $\langle q_n \rangle_n$, $q_n = [\{u\}(n)]_M$ with Cauchy modulus $\lambda n.n$, then there is an index c_M such that $[\{c_M\}(u)]_M$ is $\lim\langle q_n\rangle_n$ in the metric completion of M.

Because of ECT_0 and the fact that $A_M(x)$ is an almost negative condition on x, we find that for a CSM-space M and a separable metric space $M' \equiv (X', \rho')$, any $f \in M \to M'$ can be represented by a z such that $\{z\}(x)$ is defined for all x satisfying (1), and $[x]_M = [y]_M \to [\{z\}(x)]_{M'} = [\{z\}(y)]_{M'}$.

If $M_0 := \{y \in \mathbb{R} : y > 0\}$ with metric inherited from $\mathbb{R}$, we let

$$S_n := U([j_1 n]_{M_0}, [j_2 n]_M).$$

Here and in the sequel, the use of the notation $[x]_M$ is tacitly taken to imply that x satisfies (1). □

2.13. DEFINITION. A partial recursive α is said to be an *effective covering of open spheres of M* if

$$\forall [x]_M (E(\alpha x) \wedge [x]_M \in S_{\alpha x}). \quad \square$$

2.14. PROPOSITION (CRM). *Let $M \equiv (X, \rho)$ be a separable metric space, and let α be an effective covering of M by open spheres. Then there is an enumerable subcovering of α.*

Proof. Let $\langle p_n \rangle_n$ be a basis for M, and let M^* be the metric completion of M; $\langle p_n \rangle_n$ is also a basis for M^*. If $[x]_{M^*}$ is a point of M^*, then

$$\rho(\{x\}(n), [x]_{M^*}) < 2^{-n}.$$

We put

$$\gamma(x, y, n) \simeq \begin{cases} \{x\}(n) & \text{if } \neg\exists z \leq n\, Tyyz, \\ \{x\}(n') & \text{if } (n' < n) \wedge Tyyn'. \end{cases}$$

Therefore $\lim\langle p_{\gamma(x,y,n)} \rangle_n = [x]_{M^*}$ if $\neg\exists z Tyyz$, and if $Tyyz$, then

$$\lim\langle p_{\gamma(x,y,n)} \rangle_n = p_{\{x\}(z)}.$$

Let $A_M(x)$ be as in 2.12; we readily see that

$$A_M(x) \rightarrow \forall y A(\Lambda n.\gamma(x, y, n)).$$

α is an effective covering, and so

$$A_M(x) \rightarrow E(\alpha x) \wedge [x]_M \in S_{\alpha x}.$$

We put

$$\gamma_1(x, y) := \alpha(\Lambda n.\gamma(x, y, n)).$$

Let R be the r.e. predicate defined by

$$R(x, y) := \exists z\Big[Tyyz \wedge \forall i \leq z E\gamma(x, y, i) \wedge$$
$$\forall i < z\big(\rho\big(p_{\gamma(x,y,i)}, p_{\gamma(x,y,i+1)}\big) < 2^{-i-1}\big)\Big].$$

So if $R(x, y)$, then $A_M(\Lambda n.\gamma(x, y, n))$ and $\lim\langle p_{\gamma(x,y,n)} \rangle_n = p_{\gamma(x,z)}$, where $Tyyz$ holds. Let γ_2 enumerate $\{j(x, y) : R(x, y)\}$, and put

$$\gamma_3 m := \Lambda n.\gamma(j_1(\gamma_2 m), j_2(\gamma_2 m), n).$$

If $(x, y) \in R$, then $[\gamma_3(j(x, y))]_M$ exists as a point of M. Let σ be partial recursive such that

$$E\sigma(n, x) \leftrightarrow [x]_{M^*} \in S_n,$$

and put

$$\delta_1(x, y) := \sigma(\gamma_1(x, y), x),$$

$$\delta_2(x) := \Lambda y.\delta_1(x, y),$$

$$\delta(x) := \Lambda n.\gamma(x, \delta_2(x), n).$$

If $[x]_{M^*}$ exists as a point of M, i.e. $[x]_{M^*} = [x]_M$, then $[\Lambda n.\gamma(x, y, n)]_{M^*}$ is a point of M for all y, so $\delta_1(x, y)$ is defined for all y and $\delta_2(x)$ is the index of a total recursive function; hence $\exists z T(\delta_2 x, \delta_2 x, z)$ and therefore we have

$$\boldsymbol{E}\sigma(\gamma_1(x, \delta_2 x), x),$$

i.e. $\boldsymbol{E}\delta_1(x, \delta_2 x)$. On the other hand $(x, \delta_2 x) \in R$, so $\boldsymbol{E}(\delta x)$, with δx in the range of γ_3. Also

$$\delta_1(x, \delta_2 x) \simeq \sigma(\alpha(\Lambda n.\gamma(x, \delta_2 x, n)), x) \simeq \sigma(\alpha(\delta x), x),$$

hence $[x]_M \in S_{\alpha(\delta x)}$. □

2.15. THEOREM (*Lindelöf covering theorem for a* CSM-*space in* CRM). *Any open cover of a* CSM-*space has an enumerable subcover.*

PROOF. Let $M \equiv (X, \rho)$ be a CSM-space with basis $\langle p_n \rangle_n$. Put

$$x \in A := \forall n\left(\rho\left(p_{\{x\}(n)}, p_{\{x\}(n+1)}\right) < 2^{-n-1}\right),$$

and let $\{W_i : i \in I\}$ be an open cover of M. Then

$$\forall x \in A\, \exists n \exists i \in I([x]_M \in S_n \subset W_i).$$

By ECT_0 we find a partial recursive α such that

$$\forall x \in A(\boldsymbol{E}(\alpha x) \wedge \exists i \in I([x]_M \in S_{\alpha n} \subset W_i);$$

by the preceding proposition we can find an enumerable subcovering of the $S_{\alpha n}$, enumerated by γ, say; then

$$\forall n \exists i \in I(S_{\gamma n} \subset W_i),$$

and a final application of AC_0 gives the desired result. □

3. Located sets

We shall now introduce the notion of locatedness, which is of obvious constructive interest. The most satisfactory theory for this notion is obtained in the case of complete totally bounded metric spaces (CTB-spaces), to be discussed in the next section.

3.1. Definition. Let $M \equiv (X, \rho)$ be a metric space and let $Y \subset X$. Y is said to be *metrically located* (notation $\mathrm{ML}_M(Y)$, or simply $\mathrm{ML}(Y)$) if there is a function $\varphi \in X \to \mathbb{R}$ such that:
(i) $\forall y \in Y(\rho(y, x) \geq \varphi(x))$,
(ii) $\forall k \exists y \in Y(\rho(y, x) < \varphi(x) + 2^{-k})$. □

These conditions imply that $\varphi(x)$ is the greatest lower bound of $\{\rho(y, x) : y \in Y\}$, and this justifies the notation $\rho(Y, x)$ for $\varphi(x)$. One might consider weakening (ii) to
(ii)′ $\forall k \neg\neg \exists y \in Y(\rho(y, x) < \varphi(x) + 2^{-k})$, but this leads to a less satisfactory notion.

Being metrically located is a global property. There is a corresponding local property which can be defined in arbitrary topological spaces.

3.2. Definition. Let $\Gamma \equiv (X, \mathscr{T})$ be a topological space. $Y \subset X$ is *topologically located* (notation: $\mathrm{L}(Y)$ or $\mathrm{L}_\Gamma(Y)$) if Y is inhabited and the following holds:

$$\forall x \in X \forall U_x(\exists y(y \in U_x \cap Y) \vee \exists V_x(V_x \cap Y = \emptyset)).$$

For metric spaces this condition is equivalent to

$$\forall x \in X \forall \varepsilon > 0(\exists y \in X(y \in U(\varepsilon, x) \cap Y) \vee \exists \delta > 0(U(\delta, x) \cap Y = \emptyset)). \quad \square$$

The following proposition is easy.

3.3. Proposition. *Let* $M \equiv (X, \rho)$ *be a metric space provided with the canonical topology for metric spaces, then, for any subset* Y *of* X, $\mathrm{ML}(Y) \to \mathrm{L}(Y)$.

Proof. Immediate. □

The converse is not generally true. We give a counterexample.

3.4. *Counterexample to* $\mathrm{L}(Y) \to \mathrm{ML}(Y)$. Take $M \equiv \mathbb{N}^{\mathbb{N}}$ with the standard metric introduced in 1.3. For any closed inhabited $X \subset \mathbb{N}^{\mathbb{N}}$ define

$$\Phi(X) := \{n : \exists \alpha \in X(\alpha \in n)\}.$$

$\Phi(X)$ satisfies $\langle\, \rangle \in \Phi(X)$, $n \in \Phi(X) \leftrightarrow \exists k(n * \langle k \rangle \in \Phi(X))$. Conversely, if Y is a subset of $\mathbb{N}$ such that

$$\langle\ \rangle \in Y \wedge \forall n(n \in Y \leftrightarrow \exists k(n * \langle k \rangle \in Y)), \tag{1}$$

then

$$\Psi(Y) := \{\alpha : \forall n(\bar{\alpha}n \in Y)\}$$

is inhabited and closed. The operations Φ and Ψ are inverse to each other (exercise). For Y satisfying (1) one can show

$$\mathrm{ML}(\Psi(Y)) \leftrightarrow \forall n(n \in Y \vee n \notin Y) \tag{2}$$

(exercise); also for Y satisfying (1)

$$n \in Y \vee n \notin Y \leftrightarrow n \in Y \vee \forall k(n * \langle k \rangle \notin Y), \tag{3}$$

$$\mathrm{L}(\Psi(Y)) \leftrightarrow \forall \alpha \forall n(\bar{\alpha}n \in Y \vee \exists m(\bar{\alpha}m \notin Y)) \tag{4}$$

(exercise). The right-hand side of (4) is implied by

$$\forall nk(n \in Y \vee n * \langle k \rangle \notin Y). \tag{5}$$

If $Y_\beta \subset \mathbb{N}$ is given by

$$\langle\ \rangle \in Y_\beta, \quad \forall n(\langle k+1 \rangle * n \in Y_\beta),$$

$$\forall nk(\langle 0, k \rangle * n \in Y_\beta \leftrightarrow \beta k = 0),$$

then $\exists k(\beta k = 0) \leftrightarrow \langle 0 \rangle \in Y_\beta$, and (5) holds, so $\mathrm{L}(\Psi(Y_\beta))$; on the other hand, from (3) and (2) we see that $\mathrm{ML}(\Psi(Y_\beta))$ would imply $\langle 0 \rangle \in Y_\beta \vee \forall k(\langle 0, k \rangle \notin Y_\beta)$, i.e. $\exists k(\beta k = 0) \vee \neg\exists k(\beta k = 0)$. □

For a counterexample in $\mathbb{R}^2$, see E7.3.2. We collect some very easy properties in the following proposition.

3.5. PROPOSITION. *For any subsets Y, Y_i of a metric space $M \equiv (X, \rho)$:*
(i) $\mathrm{L}(Y) \to \mathrm{L}(Y^-)$;
(ii) $\mathrm{ML}(Y) \to \mathrm{ML}(Y^-)$, *in fact*, $\rho(x, Y) = \rho(x, Y^-)$ *for all* x;
(iii) *located and metrically located sets are closed under finite unions; moreover, for metrically located* Y_i, $i \leq n$, *and* $Y := \bigcup\{Y_i : i \leq n\}$, $\rho(x, Y) = \min\{\rho(x, Y_i) : i \leq n\}$.

PROOF. Exercise. □

On the other hand, neither the located nor the metrically located sets are closed under finite intersections or complements (exercise).

There are but few general theorems on located and metrically located sets in separable metric spaces. An interesting example is 3.7, with its corollary 3.8.

3.6. DEFINITION. Let $\Gamma \equiv (X, \mathscr{T})$ be a topological space, $Y \subset Z$, $Z \subset X$. Y is said to be *strongly contained* in Z (notation $Y \subset\subset Z$) iff

$$\forall x \in X \exists V_x (V_x \cap Y = \emptyset \vee V_x \subset Z). \quad \square$$

3.7. THEOREM. *Let* $M \equiv (X, \rho)$ *be a complete metric space, let* $Y, Z \subset X$ *and assume* $\mathrm{L}(Y)$. *Then*

$$Y^{-} \subset \mathrm{Int}(Z) \leftrightarrow Y \subset\subset Z.$$

PROOF. The implication from the right to the left is trivial. For the converse, we may without loss of generality suppose Z to be open, say $Z = W$. Let x be any point of X; Y is located, hence

$$\forall n \big(\exists y (y \in U(2^{-n}, x) \cap Y) \vee \exists k \big(U(2^{-k}, x) \cap Y = \emptyset\big)\big).$$

With countable choice we find a $\gamma \in \mathbb{N} \to \{0,1\}$ such that

$$\forall n \big[(\gamma n = 1 \to \exists y (y \in U(2^{-n}, x) \cap Y)) \wedge \\ \big(\gamma n = 0 \to \exists k \big(U(2^{-k}, x) \cap Y = \emptyset\big)\big)\big].$$

Without loss of generality we may assume

$$\forall n (\gamma n \geq \gamma(n+1)).$$

We construct a sequence $\langle y_n \rangle_n \subset Y$ such that:

$$\gamma n = 1 \to y_n \in U(2^{-n}, x) \cap Y;$$

$$\gamma 0 = 0 \to y_0 \in Y;$$

$$\gamma(n+1) = 0 \to y_{n+1} = y_n \wedge \exists k \big(U(2^{-k}, x) \cap Y = \emptyset\big).$$

$\langle y_n \rangle_n$ clearly is a Cauchy sequence, hence $y = \lim \langle y_n \rangle_n \in Y^{-}$; therefore we have $U(2^{-m}, y) \subset W$ for some m. Now consider y_{m+2}; if $\gamma(m+2) = 1$, we have $\rho(y_{m+2}, x) < 2^{-m-2}$, and

$$\rho(y_{m+2}, y) < \sum_{n=m+2}^{\infty} \rho(y_{n+1}, y_n) < 3 \left(\sum_{n=m+2}^{\infty} 2^{-n-1} \right) = 3 \cdot 2^{-m-2};$$

so $\rho(y, x) < 2^{-m}$; hence also $\rho(y, x) < 2^{-m} - 2^{-k}$, i.e. $U(2^{-k}, x) \subset W$.

On the other hand, if $\gamma(m+2)=0$, $U(2^{-k}, x) \cap Y = \emptyset$ for some k. □

3.8. COROLLARY. *Let* $M \equiv (X, \rho)$ *be a complete metric space*, Y *closed, located, and let* x *satisfy* $\forall y \in Y(\rho(x, y) > 0)$. *Then*:
(i) $\exists \varepsilon > 0(U(\varepsilon, x) \cap Y = \emptyset)$;
(ii) *if, moreover*, Y *is metrically located, then* $\rho(Y, x) > 0$.

PROOF. (i) Apply 3.7 with $Z := \{y : \rho(y, x) \# 0\}$. Then $Y^- \subset Z$, and for x we find a V_x with $V_x \cap Y = \emptyset$.
(ii) Immediate from (i). □

3.9. *Eliminating* AC; *explicit versions.* So far we have freely used the axiom of countable choice. In many instances this can be avoided by giving more explicit versions of the definitions. For example, instead of defining a CSM-space as in 2.2, and subsequently deducing the facts in 2.4, we can start from 2.4, giving rise to an alternative definition.

DEFINITION. Let $\langle p_n \rangle_n$ be a countable sequence of objects, on which a metric ρ is given; ρ is specified by a function $\alpha_Q \in \mathbb{N}^3 \to \mathbb{Q}$ such that

$$\rho(p_i, p_j) = \lim \langle \alpha_Q(i, j, k) \rangle_k,$$

$$|\rho(p_i, p_j) - \alpha_Q(i, j, k)| < 2^{-k}.$$

Now the CSM-space given by $\langle p_n \rangle_n$ and ρ is found as the metric completion of $(\langle p_n \rangle_n, \rho)$. □

Let us now see how for example corollary 3.8(ii) can be reformulated without AC. Assume the distance function for Y to be specified by a function $\beta_Q \in \mathbb{N}^2 \to \mathbb{Q}$, $\beta_Q(i, k) = r_{\beta(i,k)}$ for some standard enumeration $\langle r_n \rangle_n$ of $\mathbb{Q}$, such that

$$|\rho(p_n, Y) - \beta_Q(n, k)| < 2^{-k},$$

where $\langle p_n \rangle_n$ is the generating sequence for the CSM-space (X, ρ); and let $\langle p_{\alpha n} \rangle_n$ determine $x \in X$ with fixed rate of convergence, i.e. $\rho(p_{\alpha n}, x) < 2^{-n}$. Define γ by

$$\gamma n = 0 \quad \text{if } \beta_Q(\alpha(n+2), n+2) > 2^{-n} + 2^{-n-1} \quad \text{or} \quad \gamma(n-1) = 0,$$

$$\gamma n = 1 \quad \text{if } \beta_Q(\alpha(n+2), n+2) \le 2^{-n} + 2^{-n-1}.$$

Clearly $\gamma \in \mathbb{N} \to \{0,1\}$, $\forall n(\gamma(n+1) \leq \gamma(n))$, and

$$\gamma n = 0 \to \rho(x, Y) > 2^{-n} + 2^{-n-1} - 2^{-n-2} - 2^{-n-2} = 2^{-n},$$

$$\gamma n = 1 \to \rho(x, Y) < 2^{-n} + 2^{-n-1} + 2 \cdot 2^{-n-2} = 2^{-n+1}.$$

On the other hand, the explicit form of locatedness requires the existence of a δ such that

$$\rho\left(\lim\langle p_{\delta(k,n,m)}\rangle_n, p_m\right) < \rho(Y, p_m) + 2^{-k};$$

if $x_{k,m} = \lim\langle p_{\delta(k,n,m)}\rangle_n \in Y$ for all k and m, then $\rho(x_{k,m}, p_{\delta(k,n,m)}) < 2^{-n}$. We define y_n by

$$\gamma(n) = 1 \to y_n = \lim\langle p_{\delta(n+2,m,\alpha(n+2))}\rangle_m,$$

$$\gamma(n) = 0 \to y_n = y_{n-1}.$$

Then $y = \lim\langle y_n\rangle_n \in Y$. We now compute $\rho(y, x)$, etc. We leave the completion of the argument as an exercise.

4. Complete totally bounded spaces

We now turn to complete, separable, totally bounded metric spaces, which in constructive mathematics play the role of compact metric spaces.

4.1. Definition. Let $M \equiv (X, \rho)$ be a metric space; a finitely indexed $Y \subset X$ is an *ε-net* for M iff $\forall x \in X \exists y \in Y(\rho(x, y) < \varepsilon)$.

M is *totally bounded* if for each k there is a 2^{-k}-net for M, and M is a *CTB-space* if M is totally bounded and complete. □

N.B. A totally bounded space is obviously separable.

4.2. *Standard representation of* CTB-*spaces*. Let (X, ρ) be a CTB-space, let $\{p_{k,0}, \ldots, p_{k,n(k)}\}$ be a 2^{-k}-net for all $k \in \mathbb{N}$, and let $\langle p_n\rangle_n$ be a suitable enumeration of the union of the 2^{-k}-nets for all k. Then we can describe (X, ρ) by functions $\alpha_Q \in \mathbb{N}^3 \to \mathbb{Q}$, $\beta \in \mathbb{N} \to \mathbb{N}$ such that

$$\left|\rho(p_i, p_j) - \alpha_Q(i, j, k)\right| < 2^{-k}$$

and

$$\forall ki \exists j \exists m \leq \beta k\left(\alpha_Q(i, m, j) < 2^{-k} - 2^{-j}\right).$$

This second condition is in fact equivalent to

$$\forall ki \exists m \leq \beta k \big(\rho(p_i, p_m) < 2^{-k} \big).$$

To see this, we note that, given k, i, j, m such that $\alpha_Q(i, m, j) < 2^{-k} - 2^{-j}$, it follows that $\rho(p_i, p_m) < 2^{-k}$, and conversely, if $\rho(p_i, p_m) < 2^{-k}$, then for some j, $\rho(p_i, p_m) < 2^{-k} - 2^{-j+1}$, hence $\alpha_Q(i, m, j) < 2^{-k} - 2^{-j}$.

It will be obvious how, in the absence of AC_0, the functions α and β can be used to define CTB-spaces. Note that $\{p_0, \ldots, p_{\beta m}\}$ is a 2^{-m+1}-net for each m.

We now construct a decidable, finitely branching tree (a fan) as follows. For each j, k

$$\big\{ i : \alpha_Q(j, i, k+1) < 2^{-k-1} \wedge i \leq \beta k \big\}$$

is inhabited and finite. Put

$$\gamma \in T \equiv \forall k \big(\gamma k \leq \beta(k+2) \wedge \alpha_Q(\gamma k, \gamma(k+1), k+1) < 2^{-k+1} \big);$$

by our earlier observations T will be a fan.

4.3. Proposition. *Let* $x_\gamma \equiv \lim \langle p_{\gamma k} \rangle_k$, *and let T be as above,* $\varphi\langle n_0, \ldots, n_k \rangle = p_{n_k}$. *Then* (T, φ) *is a representation for the* CTB-*space such that*:

(i) $\forall \gamma \in T(\langle p_{\gamma k} \rangle_k$ *is a Cauchy-sequence with modulus* $\lambda k.k + 3)$;
(ii) $\forall k(\rho(x_\gamma, p_{\gamma n}) < 5 \cdot 2^{-k} < 2^{-k+3}$;
(iii) $\forall n(\rho(x, p_{\gamma n}) < 2^{-n} \wedge \gamma n \leq \beta(n+2)) \rightarrow \gamma \in T$;
(iv) $\forall n(\rho(p_{\gamma n}, p_{\gamma(n+1)}) < 2^{-n} \wedge \gamma n \leq \beta(n+2)) \rightarrow \gamma \in T$;
(v) $\forall \gamma \in T \forall k(V'_{\bar{\gamma}(k+1)} \subset U(2^{-k+3}, p_{\gamma k}) \subset U(2^{-k+4}, x_\gamma))$ *where* V'_n *is defined as in* 2.4(v);
(vi) $\forall x \in X \exists \gamma \in T(x = x_\gamma \wedge \forall k(U(2^{-k-1}, x) \subset V'_{\bar{\gamma} k}))$.

Proof. (i)–(iv) are routine, (v) is a direct consequence of (ii); (vi) can be proved quite similarly to 2.4(v) (exercise). □

4.4. Corollary. *Each* CTB-*space is a uniformly continuous image of Cantor-space.*

Proof. Each fan (with standard topology) can be homeomorphically embedded into $2^{\mathbb{N}}$, and each subtree of $2^{\mathbb{N}}$ is a continuous image of $2^{\mathbb{N}}$ (cf. 4.7.5).

4.5. PROPOSITION. *Let $M \equiv (X, \rho)$ be a totally bounded metric space, $M' \equiv (X', \rho')$ an arbitrary metric space, and let $f \in M \to M'$ be uniformly continuous. Then $f[X]$ is totally bounded in M'. If $M' \equiv \mathbb{R}$, then f has a supremum and an infimum on M.*

PROOF. Exercise. □

On the other hand, the uniformly continuous image of a CTB-space need not be closed; for example $[0,3] \subset \mathbb{R}$ is a CTB-space, which is mapped onto $[0,1] \cup [1,2]$ by the f such that $f(x) = x$ for $x \in [0,1]$, $f(x) = 1$ for $x \in [1,2]$, and $f(x) = x - 1$ for $x \in [2,3]$. □

4.6. In classical mathematics, closed, totally bounded sets are compact, i.e. each open cover has a finite subcover, and compactness is an important property of topological spaces. It is almost immediate that also in constructive mathematics compactness is preserved by continuous mappings. However, in BCM we can show for very few spaces compactness in this sense: finite, discrete spaces are compact, and also, for example, a sequence of points converging to a limit, together with this limit (exercise). But we cannot show in BCM that $[0,1]$ or $2^{\mathbb{N}}$ are compact, since the principles we adopted for BCM are compatible with CRM, while $[0,1]$ and $2^{\mathbb{N}}$ are not compact in CRM (cf. 4.7.6, 6.4.2). In fact, it is easy to show that we cannot distinguish between the topology of $2^{\mathbb{N}}$ and $\mathbb{N}^{\mathbb{N}}$ in CRM, see the next theorem.

4.7. THEOREM (CRM). *$2^{\mathbb{N}}$ and $\mathbb{N}^{\mathbb{N}}$ are homeomorphic.*

PROOF. See E4.7.7. □

4.8. *Discussion.* As the preceding theorem shows, the usual notion of compactness is not very useful in the context of BCM or CRM; complete totally bounded sets are the obvious substitute (these are the "compact" sets in much of the literature on BCM). The CTB-spaces have the important property that uniformly continuous functions have a supremum, and as we saw, total boundedness is also preserved by mappings uniformly continuous on totally bounded sets.

However, this indicates that we also have to adapt the notion of homeomorphism, for pointwise continuous mappings need not be uniformly continuous on CTB-sets. Bridges (1979) introduced the following notion of a strongly continuous mapping: for metric spaces $M_i \equiv (X_i, \rho_i)$, $i \in \{0,1\}$,

$f \in M_0 \to M_1$ is said to be *strongly continuous*, iff for all CTB-images $Y \subset M_0$

$$\forall \varepsilon > 0 \exists \delta > 0 \forall x \in M_0 \forall y \in Y(\rho(x, y) < \delta \to \rho(fx, fy) < \varepsilon).$$

Here Y is a *CTB-image* if $Y = g[Z]$ for some uniformly continuous g, Z a CTB-space.

This notion seems to be adequate in the context of functional analysis, but it remains to be seen whether it is satisfactory in the context of point-set topology. For a complete M_0 we may take CTB-sets Y instead of CTB-images as in the definition above.

A very different route, which is perhaps more appropriate for the development of constructive topology, is to take the notion of covering as a primitive.

This is the route taken in so-called "locale" theory (see e.g. Fourman and Grayson 1982). A brief sketch of this approach is given in 6.7.

If we assume FAN as an axiom, we get a feasible theory of compact spaces. So let us introduce officially the following definition.

4.9. Definition. A topological space is *compact* iff each open cover contains a finite subcover. □

As shown in 6.3.5 segments $[a, b]$ with $a < b$ are compact on the assumption of FAN. We can generalize the results of 6.3.5–8 to the following theorem.

4.10. Theorem (FAN). *Let M be a* CTB-*space, M' a separable metric space. Then*:

(i) *Each open cover of M has a finite subcover.*

(ii) *Each continuous $f \in M \to M'$ is uniformly continuous.*

(iii) *Each continuous $f \in M \to \mathbb{R}$ has a supremum and an infimum on M, and if $f(x) > 0$ for all $x \in M$, then* $\inf(f) > 0$.

Moreover, if we adopt the stronger principle FAN* (*the combination of* FAN *with continuity*) *we have*

(iv) *Each $f \in M \to \mathbb{R}$ is uniformly continuous and has a supremum.*

(v) *For each $f \in M \to \mathbb{R}$, $\forall x \in M(f(x) > 0) \to$* $\inf(f) > 0$.

Proof. Very similar to the proofs in section 6.3, using the standard representation for CTB-spaces (exercise). □

We note also that on the assumption of FAN a continuous mapping is strongly continuous in the sense of Bridges (1979) (exercise).

We shall now turn to located sets in CTB-spaces.

4.11. PROPOSITION

(i) *A* CTB-*subspace of a metric space is metrically located.*

(ii) *A closed, metrically located subset of a* CTB-*space is a* CTB-*subspace.*

PROOF. (i) Let $M \equiv (X, \rho)$, $Y \subset X$ a CTB-space. Then Y has a basis $\langle q_n \rangle_n$ such that for some $\beta \in \mathbb{N}^{\mathbb{N}}$, for each k $\{q_0, \ldots, q_{\beta k}\}$ is a 2^{-k}-net for Y. Then

$$\rho(x, Y) = \lim \langle (\min\{\rho(x, q_i) : i \leq \beta k\}) \rangle_k.$$

(ii) Let $M \equiv (X, \rho)$ be a CTB-space with basis $\langle p_n \rangle_n$, $\beta \in \mathbb{N} \to \mathbb{N}$ such that for all k $\{p_0, \ldots, p_{\beta k}\}$ is a 2^{-k}-net for M. Let $Y \subset X$ be closed and metrically located.

We construct a basis $\langle q_n \rangle_n$ for Y, and a $\gamma \in \mathbb{N}^{\mathbb{N}}$ such that $\{q_0, \ldots, q_{\gamma k}\}$ is a 2^{-k}-net for Y, as follows. We know that Y is inhabited. Suppose $\{q_0, \ldots, q_{\gamma k}\}$ to have been constructed; $\{0, 1, \ldots, \beta(k+3)\}$ contains a finite, inhabited subset Z such that

$$i \notin Z \to \rho(p_i, Y) > 2^{-k-3},$$

$$i \in Z \to \rho(p_i, Y) < 2^{-k-2}.$$

Choose for each $i \in Z$ a $q_i' \in Y$ such that $\rho(p_i, p_i') < 2^{-k-2}$. Then $\{q_i' : i \in Z\}$ is a 2^{-k-1}-net for Y; enumerate $\{q_i' : i \in Z\}$ as $q_{\gamma k+1}, \ldots, q_{\gamma(k+1)}$. □

The next proposition is an immediate corollary to the standard representation for CTB-spaces.

4.12. DEFINITION. An *ε-cover* $\mathscr{V}$ of a space $M \equiv (X, \rho)$ is a cover by sets of diameter $\leq \varepsilon$, i.e.

$$\forall Y \in \mathscr{V} \forall x, y \in Y (\rho(x, y) < \varepsilon).$$

The *diameter* diam(Y) of a bounded located set is $\sup\{\rho(y, y') : y, y' \in Y\}$. □

4.13. PROPOSITION. *Let M be a* CTB-*space. M has a finite 2^{-k}-cover by closed, metrically located sets for each k.*

PROOF. Take the $V'_{\bar{\gamma}k}$ of 4.3(vi). □

A very convenient tool in the construction of located sets is the following proposition.

4.14. PROPOSITION. *Let $M \equiv \langle X, \rho \rangle$ be a* CTB-*space and let $f \in M \to \mathbb{R}$ be uniformly continuous. There is an enumerable $Y \subset \mathbb{R}$ such that for all $a \in \mathbb{R}$ avoiding Y (i.e. $\forall y \in Y(a \# y)$) the sets*

$$X_a := \{x \in X : f(x) < a\}, \qquad X'_a := \{x \in X : f(x) > a\}$$

are either empty or metrically located; moreover, they are each others metric complement, i.e.

$$X'_a = \{y : \rho(y, X_a) > 0\}, \qquad X_a = \{y : \rho(y, X'_a) > 0\}.$$

PROOF. For each k, let $\{X_{k,1}, \ldots, X_{k,N(k)}\}$ be a cover of X consisting of CTB-sets of diameter $\leq 2^{-k}$, and let

$$c_{k,i} := \inf\{f(x) : x \in X_{k,i}\}, \qquad c'_{k,i} := \sup\{f(x) : x \in X_{k,i}\}.$$

Let Y be the collection of $c_{k,i}$ and $c'_{k,i}$, and let a avoid Y. Then $c_{k,i} < a$ or $c_{k,i} > a$, and $c'_{k,i} < a$ or $c'_{k,i} > a$, for all k and i. Note that X_a is empty iff no $c_{k,i} < a$ exists, and that X'_a is empty iff no $c'_{k,i} > a$ exists.

For each $c_{k,i} < a$ choose $x_{k,i} \in X_{k,i}$ with $f(x_{k,i}) < a$, and for each $c'_{k,i} > a$ choose $x'_{k,i} \in X'_{k,i}$ with $f(x'_{k,i}) > a$.

Now, if $x \in X_a$ and $x' \in X'_a$, then $x \in X_{k,i}$ and $x' \in X_{k,j}$ for certain i and j, hence

$$c_{k,i} \leq f(x) < a, \qquad c'_{k,j} \geq f(x') > a,$$

and so $c_{k,i} < a$, $c'_{k,j} > a$, $\rho(x, x_{k,i}) < 2^{-k}$, $\rho(x', x'_{k,j}) < 2^{-k}$. Thus $\{x_{k,i} : c_{k,i} < a\}$ is a 2^{-k}-net for X_a, and $\{x_{k,j} : c'_{k,j} > a\}$ a 2^{-k}-net for X'_a. This shows by 4.11 that X_a and X'_a are metrically located.

The final assertion of the theorem we leave as an exercise. □

The following corollary shows that there are many "well-behaved" ε-neighbourhoods, even if they are not always located.

4.15. COROLLARY. *Let M be a* CTB-*space, and let Y be a metrically located subset of M; then for each δ and ε such that $0 < \delta < \varepsilon$ we can find an ε', $\delta < \varepsilon' < \varepsilon$, such that $U(\varepsilon', Y)$ is open and metrically located.*

PROOF. Apply 4.14 to $f(x) \equiv \rho(x, Y)$; the choice of ε' is made possible by 6.4.10(ii). □

In the presence of FAN, we have the following pleasant equivalence.

4.16. THEOREM (FAN). *Let* $M \equiv (X, \rho)$ *be a* CTB-*space and let* Y *be a subset of* X; *then* $\mathrm{ML}(Y) \leftrightarrow \mathrm{L}(Y)$.

PROOF. Assume $\mathrm{L}(Y)$, then for any fixed k,

$$\forall x \in X\big(\exists y\big(y \in Y \cap U(2^{-k}, x)\big) \vee \exists n\big(U(2^{-n}, x) \cap Y = \emptyset\big)\big),$$

hence

$$\mathscr{V} := \big\{U(2^{-k}, x) : \exists y\big(y \in Y \cap U(2^{-k}, x)\big)\big\} \cup \{Y^*\}, \quad \text{where}$$

$$Y^* := \big\{z : \exists m\big(U(2^{-m}, z) \cap Y = \emptyset\big)\big\},$$

is an open cover of M; by FAN, M is compact, so let $\{Y^*, U_0, \ldots, U_n\}$ with $U_i \equiv U(2^{-k}, x_i)$ be a finite subcover of $\mathscr{V}$. For each $i, i \le n$, let $y_i \in U(2^{-k}, x_i) \cap Y$, then $\{y_0, \ldots, y_n\}$ is a 2^{-k+1}-net for Y, since $\{U_0, \ldots, U_n\}$ covers Y. □

Classically, there is a well-known theorem stating that each continuous bijection from a compact space to a compact space is a homeomorphism. In BCM, one can formulate a substitute definition.

4.17. DEFINITION. Let $M \equiv (X, \rho)$, $M' \equiv (X, \rho)$. $f \in M \to M'$ is a *strong injection* (cf. 1.11) iff

$$\forall x, y \in X\big(\rho(x, y) > 0 \to \rho'(fx, fy) > 0\big),$$

and f is *hyperinjective* iff for any pair Z, Z' of CTB-subsets of X

$$\forall \varepsilon > 0\big[\forall z \in Z\, \forall z' \in Z'(\varepsilon < \rho(z, z')) \to$$
$$\exists \delta > 0\, \forall z \in Z\, \forall z' \in Z'\big(\delta < \rho(fz, fz')\big)\big]. \quad \square$$

The substitute for the classical theorem now becomes as follows.

4.18. THEOREM. *Let* M *be a* CTB-*space,* M' *a metric space and* $f \in M \to M'$ *uniformly continuous and hyperinjective; then* f^{-1} *is uniformly continuous and hyperinjective on* $f[X]$, *and* $f[X]$ *is a* CTB-*space.*

PROOF. Let $M \equiv (X, \rho)$, $M' \equiv (X', \rho')$. We have to show that for any given $\varepsilon > 0$ we can find a $\delta > 0$ such that

$$\forall x', y' \in f[X]\big(\rho'(x', y') < \delta \to \rho\big(f^{-1}(x'), f^{-1}(y')\big) < \varepsilon\big).$$

Choose $\varepsilon > 0$, and let $X_0, \ldots, X_n$ be a finite $\frac{1}{3}\varepsilon$-cover of M consisting of CTB-subspaces. Then either:

$$\sup\{\rho(x, X_j) : x \in X\} < \tfrac{1}{3}\varepsilon \quad \text{for some } j \leq n, \quad \text{or} \tag{1}$$

$$\sup\{\rho(x, X_j) : x \in X\} > 0 \quad \text{for all } j \leq n. \tag{2}$$

If (1) holds, then for all $x, y \in f[X]$,

$$\rho\big(f^{-1}(x'), f^{-1}(y')\big) \leq \operatorname{diam}(X) \leq \varepsilon,$$

since $\rho(f^{-1}(x'), X_j) < \frac{1}{3}\varepsilon$, $\rho(f^{-1}(y'), X_j) < \frac{1}{3}\varepsilon$, $\operatorname{diam}(X_j) < \frac{1}{3}\varepsilon$, etc. In this case we are done.

Now assume (2). Then we can find (4.14) an η such that $0 < \eta < \frac{1}{3}\varepsilon$ and such that

$$Z_j = \{x: \rho(x, X_j) > \eta\}^-$$

is a CTB-subset of X for $j \leq n$. Let

$$\delta_j := \inf\{\rho'(x, y): x \in f[Z_j] \wedge y \in f[X_j]\},$$

$$\delta := \min\{\delta_0, \ldots, \delta_n\}.$$

$0 < \eta \leq \inf\{\rho(x, y): x \in Z_j \wedge y \in X_j\}$, and hence $\delta > 0$ by the hyperinjectivity of f.

Now, if $\rho'(fx, fy) < \frac{1}{2}\delta$, and $y \in X_j$, then $\rho'(fx, f[X_j]) < \delta_j$, and hence $fx \notin f[Z_j]$, so $x \notin Z_j$; also $\rho(x, X_j) \leq \eta$, hence $\rho(x, z) < 2\eta$ for some $z \in X_j$. Therefore

$$\rho(x, y) \leq \rho(x, z) + \operatorname{diam}(X_j) < 2\eta + \tfrac{1}{3}\varepsilon < \varepsilon.$$

It is easy to see that f^{-1} is also hyperinjective. By the uniform continuity of f, $f[X]$ is totally bounded, and if $\langle x_n \rangle_n \subset X$, $\langle f(x_n) \rangle_n$ Cauchy in $f[X]$, then by the uniform continuity of f^{-1}, $\langle x_n \rangle_n$ is a Cauchy sequence in X, with limit x, say. By the continuity of f, $\langle f(x_n) \rangle_n \to fx$, so $f[X]$ is complete. □

If FAN is assumed, we have the following proposition.

4.19. PROPOSITION (FAN). *A strong injection from a* CTB-*space into a metric space is a hyperinjection.*

PROOF. Exercise. □

4.20. COROLLARY (FAN). *If* $M \equiv (X, \rho)$ *is a* CTB-*space,* M' *a metric space, and* $f \in M \to M'$ *a continuous strong injection, then* f^{-1} *exists on* $f[X]$, *and* $f \in X \to f[X]$ *is a (strongly injective) homeomorphism.* □

5. Locally compact spaces

5.1. DEFINITION. A CSM-space M is said to be *locally totally bounded* if M has an enumerable open cover of totally bounded sets. M is said to be *locally compact* if M has an enumerable open cover of sets with compact closure. □

On the assumption of FAN, the notions of locally compact space and locally totally bounded space coincide.

Assuming WC-N, the intuitionistic version of the Lindelöf covering theorem is available, and so, in this case, a locally totally bounded (locally compact) space might be defined as a CSM-space in which each point has an open, totally bounded neighbourhood (open neighbourhood with compact closure).

Without assuming extra axioms, the notion of a locally totally bounded space does not seem to be very manageable, and accordingly, in the literature on BCM attention is restricted to spaces in which every bounded set is contained in a CTB-set (such as $\mathbb{R}^n$ for example). Clearly, this notion (called "locally compact" in the literature on BCM) is not invariant under homeomorphisms: the open disc $U(1, (0,0))$ in $\mathbb{R}^2$ is not locally compact in this sense.

In this section we shall assume FAN throughout, and prove some facts about locally compact spaces as defined above.

5.2. NOTATION. We use the following abbreviation:

$$U_0 \underline{\cup} \cdots \underline{\cup} U_{n-1} := (U_0 \cup \cdots \cup U_{n-1})^- . \quad \square$$

5.3. PROPOSITION (FAN). *Let* M *be a* CSM-*space such that every bounded subset of* M *is contained in a* CTB-*subspace of* M; *then for any subset* Y $\mathrm{ML}(Y) \overset{?}{\leftrightarrow} \mathrm{L}(Y)$.

PROOF. Let $M \equiv (X, \rho)$, $Y \subset X$, $\mathrm{L}(Y)$, $p_0 \in Y$, $x \in X$, $\rho(x, p_0) < n_0 \in \mathbb{N}$. We will show that $\rho(x, Y)$ exists. Let Z be a CTB-subspace (and hence compact) such that $U(n_0, x) \subset Z$; choose a CTB-space Z' such that $U(n_0 + 1, x) \cup U(1, Z) \subset Z'$; then $Z \subset\subset Z'$ and $y \notin Z \to \rho(x, y) \geq n_0$, $y \notin Z' \to \rho(x, y) \geq n_0 + 1$. By L($Y$) it follows that for each $k \in \mathbb{N}$

$$\forall z \in Z'\big(\exists p\big(p \in U(2^{-k}, z) \cap Y\big) \vee \exists\delta > 0\big(U(\delta, y) \cap Y = \emptyset\big)\big).$$

By the compactness of Z' the cover

$$\mathscr{U} := \{U(2^{-k}, z) : \exists p\big(p \in U(2^{-k}, z) \cap Y\big) \wedge z \in Z'\} \cup \{Y^*\},$$

where

$$Y^* := \{z \in Z' : \exists\delta > 0\big(U(\delta, z) \cap Y = \emptyset\big)\}$$

has a finite subcover

$$\{U\big(2^{-k}, z_{k,0}\big), \ldots, U\big(2^{-k}, z_{k,n(k)}\big)\} \cup \{Y^*\}.$$

We determine points $q_{k,i}$ such that

$$q_{k,i} \in U\big(2^{-k}, z_{k,i}\big) \cap Y.$$

Now let

$$d_0 := \inf\big(\{\rho(x, q_{0,i}) : i \leq n(0)\} \cup \{\rho(x, p_0)\}\big),$$

$$d_{k+1} := \inf\big(\{\rho(x, q_{k+1,i}) : i \leq n(k+1)\} \cup \{d_k\}\big).$$

By definition $\forall k(d_{k+1} \leq d_k)$. Note that $\forall k(d_k \leq n_0)$.

For any $y \in Y$ either $y \notin Z$ or $y \in Z'$, hence either $\rho(x, y) \geq n_0$, and then $d_k \leq \rho(x, y)$, or $y \in U(2^{-k}, z_{k,i})$, hence $d_k \leq \rho(q_{k,i}, x) \leq \rho(x, y) + 2^{-k}$, so $d_k - 2^{-k} \leq \rho(x, y)$. Therefore $\forall y \in Y(d_k - 2^{-k} \leq \rho(x, y))$. On the other hand

$$\forall k \exists y \in Y\big(d_k > \rho(x, y) - 2^{-k}\big)$$

(choose a suitable $q_{k,i}$ for y). It remains to be shown that $\langle d_k \rangle_k$ is a Cauchy sequence.

Let $\rho(x, q_{k+1,i}) - d_{k+1} < 2^{-k-1}$, then $q_{k+1,i} \in Z'$ (for if not, then $\rho(x, q_{k+1,i}) \geq n_0 + 1$ while $\rho(x, q_{k+1,i}) \leq 2^{-k-1} + d_{k+1} \leq n_0 + 2^{-k-1}$, hence $\neg\neg q_{k+1,i} \in Z'$, which implies $q_{k+1,i} \in Z'$, since Z' is closed).

$Z' \cap Y$ is covered by $\{U(2^{-k}, z_{k,j}) : j \leq n(k)\}$ with $q_{k,j} \in U(2^{-k}, z_{k,j})$ for $j \leq n(k)$. Let $q_{k+1,i} \in U(2^{-k}, z_{k,j})$, then $\rho(q_{k+1,i}, q_{k,j}) < 2^{-k+1}$.

Therefore

$$d_{k+1} > \rho(x, q_{k+1,i}) - 2^{-k-1} \geq \rho(x, q_{k,j}) - 2^{-k-1} - 2^{-k-1}$$
$$\geq d_k - 3\cdot 2^{-k},$$

and thus

$$d_k \geq d_{k+1} \geq d_k - 3\cdot 2^{-k}.$$

As a result, $\lim\langle d_n\rangle_n = d = \rho(x, Y)$. □

5.4. THEOREM (FAN). *Let* $M \equiv (X, \rho)$ *be a locally compact space. Then there is an* $M' \equiv (X, \rho')$, *homeomorphic to* M, *such that in* M' *each bounded inhabited subset is contained in a compact subset, and such that, for subsets* $Y \subset X$, $\mathrm{ML}(Y) \leftrightarrow \mathrm{L}(Y)$ *in* M'.

PROOF. Let $\langle U_n\rangle_n$ be an open cover of M such that each U_n^- is a CTB-subset. From this cover we can construct a cover $\langle X_n\rangle_n$ such that $X = \bigcup\{X_n : n \in \mathbb{N}\}$, each X_n a CTB-subset, and such that for all n $X_n \subset\subset X_{n+1}$, as follows. Put

$$X_0 := U_0^-.$$

Suppose X_k to have been constructed as $U_0 \underline{\cup}\, U_1 \underline{\cup} \cdots \underline{\cup}\, U_{n(k)}$, and let $U_{i_1}, \ldots, U_{i_p}$ be a finite subcover of $\langle U_n\rangle_n$ covering X_k. Then we put $n(k+1) = \max\{n(k), i_1, \ldots, i_p\} + 1$, and we put

$$X_{k+1} := U_0 \underline{\cup} \cdots \underline{\cup}\, U_{n(k+1)}.$$

Clearly $X_k \subset\subset X_{k+1}$; so we find (exercise) an $\varepsilon_k > 0$ such that $U(\varepsilon_k, X_k) \subset X_{k+1}$. The function

$$f_k(x) := \sup\{1 - \rho(x, X_k)\cdot \varepsilon_k^{-1}, 0\}$$

satisfies

$$f_k \text{ is a uniformly continuous function in } X \to [0,1],$$
$$x \in X_k \to f_k(x) = 1, \qquad x \notin X_{k+1} \to f_k(x) = 0.$$

Let $\rho''(x, y) := \min\{1, \rho(x, y)\}$, then (X, ρ'') is homeomorphic with M. If we define ρ' by

$$\rho'(x, y) := \rho''(x, y) + \sum_{k=0}^{\infty} |f_k(x) - f_k(y)|,$$

then (X, ρ') is homeomorphic to (X, ρ). For let $\rho'(x, y) < \varepsilon$, then $\rho''(x, y) < \varepsilon_n$ for a suitable n. Conversely, if $x \in X_n$ and $\rho''(x, y) < \varepsilon_n$, then $y \in X_{n+1}$. Also, because of the uniform continuity of $f_0, \ldots, f_{n+1}$, we can find an $\eta > 0$ such that

$$\forall xy\big(\rho(x, y) < \eta \to |f_i(x) - f_i(y)| < \varepsilon/(n+2)\big).$$

Take $\delta := \min(\frac{1}{2}\varepsilon n, \varepsilon/(n+2))$, then

$$\rho(x, y) < \delta \to \rho'(x, y) < 2\varepsilon.$$

Clearly, on each X_n the identity mapping is a uniformly continuous mapping from (X_n, ρ) to (X_n, ρ'). Finally, note that if $\rho(x, y) < n \in \mathbb{N}$ and $x \in X_m$, then $y \in X_{n+m+2}$, and thus each bounded inhabited subset is contained in a CTB-subspace. The remaining assertion of the theorem follows by the preceding proposition. □

Next we shall construct the "*one-point*" *compactification* of a locally compact space; i.e. we show how a locally compact M can be homeomorphically embedded in a CTB-space such that M appears as the metric complement of a single point.

5.5. THEOREM.
(i) *Let $M \equiv (X, \rho)$ be a* CSM-*space such that each bounded inhabited subset is contained in a* CTB-*subset; then there is a* CTB-*space (X', ρ') and $X'' \subset X'$ such that $(X'', \rho' \restriction X'')$ is homeomorphic to M, and X'' is in (X', ρ') the metric complement of a single point.*
(ii) (FAN). *Any locally compact space can be homeomorphically embedded in a* CTB-*space as the metric complement of a single point.*

PROOF. It is sufficient to prove (i); (ii) is a corollary. Let $\langle p_n \rangle_n$ be a basis sequence for M, and let f_n be given by

$$f_n(x) := \min(1, \rho(x, p_n)).$$

Let $M^* \equiv (X^*, \rho^*)$ be the product space $\prod\{[0,1]: n \in \mathbb{N}\}$ with metric

$$\rho^*(\langle x_n \rangle_n, \langle y_n \rangle_n) := \sum_{n=0}^{\infty} 2^{-n}|x_n - y_n|;$$

X is embedded in M^* by $\varphi : x \mapsto \langle f_n(x)\rangle_n$. We note that

$$\begin{aligned}\rho^*(\varphi x, \varphi y) &= \sum_{n=0}^{\infty} 2^{-n}|f_n(x) - f_n(y)| \\ &= \sum_{n=0}^{\infty} 2^{-n}|\min(1, \rho(x, p_n)) - \min(1, \rho(y, p_n))| \\ &\le \sum_{n=0}^{\infty} 2^{-n}|\rho(x, p_n) - \rho(y, p_n)| \\ &\le \sum_{n=0}^{\infty} 2^{-n}\rho(x, y) = 2\rho(x, y),\end{aligned}$$

that is to say, φ is uniformly continuous.

Now we shall show that $\varphi[X]$ is totally bounded in the metric ρ^*. For a given sequence $\langle a_k\rangle_k$ we put

$$X_k := \{x \in X : \rho(x, p_1) < a_k\}, \quad Y_k := \{x \in X : \rho(x, p_1) > k - 1\}.$$

By 4.14 we can choose $a_k \in (k-1, k)$ such that X_k^- becomes totally bounded and closed. $X = X_k \cup Y_k$. The diameter of $\varphi[Y_k]$ converges to 0 for increasing k, and X_k is totally bounded; therefore $\varphi[X]$ is totally bounded with respect to ρ^*.

We let Y be the ρ^*-closure of $\varphi[X] \cup \{\omega\}$, where $\omega := \langle 1\rangle_n$; Y is also totally bounded in ρ^*. For each totally bounded $Z \subset X$ one has

$$\rho^*(\varphi[Z], \omega) > 0. \tag{1}$$

This is proved as follows. Let $\{p_0, \ldots, p_n\}$ be a $\frac{1}{2}$-net for Z. Then for each $x \in Z$ we can find some p_i, $i \le n$, such that $\rho(p_i, x) < \frac{1}{2}$, hence $|1 - \rho(p_i, x)| > \frac{1}{2}$, and therefore $\rho^*(\varphi(x), \omega) > 2^{-n-1}$; thus $\rho^*(\varphi[Z], \omega) \ge 2^{-n-1}$. Next we define

$$\rho_0(x, y) := \rho^*(\varphi(x), \varphi(y)) + \left|\rho^*(\varphi(x), \omega)^{-1} - \rho^*(\varphi(y), \omega)^{-1}\right|.$$

Then (X, ρ) and (X, ρ_0) are homeomorphic, which can be established as follows.

(a) If $Z \subset X$, Z a CTB-subset of (X, ρ), then $\rho^*(\varphi[Z], \omega) > 0$, and thus the identity mapping between $(Z, \rho \restriction Z)$ and $(Z, \rho_0 \restriction Z)$ is uniformly continuous.

(b) If $Z' \subset X$ is bounded in ρ_0 by k, say, then $\inf\{\rho^*(\varphi(x), \omega) : x \in Z\} > 0$. For if $x \in Z'$, $x_0 \in Z'$, then $\rho_0(x_0, x) \le k$ for $x \in Z'$, and

$\rho^*(\varphi(x_0), \omega) > 2^{-p}$ for some $p \in \mathbb{N}$. Then $\rho^*(\varphi(x), \omega)^{-1} \le \rho_0(x_0, x) + \rho^*(\varphi(x_0), \omega)^{-1} < k + 2^{-p}$, hence $\rho^*(\varphi(x), \omega) > (k + 2^{-p})^{-1}$. Therefore for some m, and all $x \in Z'$, $\rho^*(\varphi(x), \omega) > 2^{-m}$. Hence, if $x \in Z'$, it follows that $\rho(x, p_i) < 1$ for some $i \le m$, therefore Z' is also bounded with respect to ρ.

(c) The identity map from (X, ρ_0) to (X, ρ) is uniformly continuous on each ρ_0-bounded $Z' \subset X$: by (b), Z' is also ρ-bounded, hence contained in a CTB-subset Z of (X, ρ_0), and Z is also a CTB-subset with respect to ρ_0. If $\{p_0, \ldots, p_k\}$ is a $\frac{1}{3}\varepsilon$-net for Z with respect to ρ_0, $\varepsilon < 1$, then for all $x, y \in Z$:

$$\rho_0(x, y) < \tfrac{1}{3}\varepsilon \cdot 2^{-k} \to \rho(x, y) < \varepsilon.$$

Since the premiss implies $|\min(1, \rho(x, p_n)) - \min(1, \rho(y, p_n))| < \frac{1}{3}\varepsilon$ for all $n \le k$, it follows that if $\rho(x, p_m) < \frac{1}{3}\varepsilon$, then $\rho(y, p_m) < \frac{1}{3}\varepsilon$, and hence $\rho(x, y) \le \rho(x, p_m) + \rho(y, p_m) < \varepsilon$.

Finally, we show that $\varphi[X]$ is the metric complement $\{\omega\}^*$ of $\{\omega\}$ in $Y \equiv (\varphi[X] \cup \omega)^-$ with respect to ρ^*. From (1) we see that $\varphi[X] \subset \{\omega\}^*$, so it remains to be shown that $\{\omega\}^* \subset \varphi[X]$.

Let $y \in \{\omega\}^*$, i.e. $\rho^*(y, \omega) > 0$. Since y belongs to Y, there must be a sequence $\langle x_n \rangle_n \subset X$ such that $\varphi(x_n)$ converges to y in the metric ρ^*. However, $\langle x_n \rangle_n$ is contained in CTB-subset Z of X, since there is a positive lower bound to $\{\rho^*(\varphi(x_n), \omega) : n \in \mathbb{N}\}$ (cf. (b) above), and $\langle x_n \rangle_n$ is a Cauchy sequence in ρ_0, hence also in ρ, with a limit $x \in Z$; clearly $\varphi(x) = \lim \langle \varphi(x_n) \rangle_n = y$ with respect to ρ^*, and so $y \in \varphi[X]$. Thus we have proved $\{\omega\}^* \subset \varphi[X]$.

We can now take in the theorem $X' := Y$ as above, $\rho' := \rho^* \upharpoonright Y$, $X'' := \varphi[X]$. □

REMARK. Part (i) of the theorem has been established *without* assuming FAN. If we assume FAN throughout, some steps can be shortened; assuming C-N we can achieve further shortcuts (exercise). We leave it to the reader to show that any two one-point compactifications are homeomorphic (E7.5.3). □

6. Notes

6.1. The example in 1.10 is taken from Troelstra (1966), 2.1.8. (Grayson, 1983) shows that in intuitionistic higher-order logic the existence of two distinct topological spaces with the same collection of closed sets is consistent. The definition of separated space in 1.1.11 is taken from Grayson (1981).

6.2. As mentioned in 6.5.2, standard representations for compact spaces were already given by Brouwer (standard representations are special cases of spreads in Brouwer's sense); for arbitrary CSM-spaces they are given, for example, in Troelstra (1966, proof of II 2.5). As for the theorems 2.6 and 2.7, see 6.5.2. For the continuity theorem in CRM (2.11) see 6.5.5.

Theorem 2.8 is from Troelstra (1967). A corresponding result for constructive recursive mathematics is in Orevkov (1971).

The intuitionistic version of Lindelöf's theorem (2.10) is in Troelstra (1966, II 2.5; 1969, 13.2.1). The proof of Lindelöf's theorem in constructive recursive mathematics (2.15) is taken from Kushner (1973, theorem 1 in section 9.3); the theorem generalizes results of Moschovakis and Tseitin.

6.3. The notion of a metrically located set in a compact metric space is due to Brouwer (1919 for $\mathbb{R}^2$; 1926B for compact metric spaces). A notion of "topologically located" for compact spaces is given by Freudenthal (1937A). The general definition of "topologically located subset" used in this book is taken from Troelstra (1966, I 4.2, cf. also Troelstra 1969, 12.5.3); similar definitions have been proposed, independently, by van Dalen and Schultz. The counterexample of 3.4 is due to Grayson.

Theorem 3.7 is proved in Troelstra (1966, 1969) under the redundant assumption of WC-N; the corollary 7.3.8 appears in Bishop (1967, ch. 6, lemma 7, p. 177).

6.4. As noted above, Brouwer already gave standard representations of CTB-spaces, and theorem 4.4 is in essence due to him.

Theorem 4.10 is part of the intuitionistic tradition. As to 4.10(i) see 6.5.2, as to 4.10(ii), see Brouwer (1923A, 1924, 1924A, 1927, 1954). Theorem 4.14 is essentially theorem 8 of Chapter 4 of Bishop (1967, p. 102), 4.16 is taken from Troelstra (1966, II 3.7), and 4.18 from Bridges (1979, 2.6.4).

6.5. Theorem 5.4 is found in Troelstra (1966, IV 4.1; also in 1968C) with a more complicated proof than the one presented here. Theorem 5.5 is proved in Bishop (1967) in the context of BCM, and in Troelstra (1968C) in the context of intuitionistic mathematics. See also E7.5.1–3.

6.6. The exercises 7.2.6–7 and 7.4.10–11 deal with connectedness. Troelstra (1967A) deals with connectedness in intuitionistic mathematics (i.e. assuming WC-N and FAN); Bridges (1978, 1978A) studied connectedness on the basis of BCM. Grayson (1981, subsection 5.3, section 7; and 1982, section 4) deals with connectedness in a constructive context, and establishes certain independence results involving connectedness with the help of sheaf-semantics.

Dimension theory is treated constructively by Brouwer (1926B), Berg et al. (1976, 1977), and the Jordan curve theorem by Brouwer (1925A), and Berg et al. (1975).

6.7. *Pointless topology; the theory of locales.* In the pioneering paper by Freudenthal (1937A) CTB-spaces are given by specifying the intersection relations between the elements of a sequence of closed basis elements, that is to say, for each finite collection $A_{n_0}, \ldots, A_{n_p}$ of the sequence we know whether their intersection is inhabited or empty; points are defined entities. Troelstra (1966) represents an attempt to generalize this approach to CSM-spaces. A different version of "pointless" axiomatics for intuitionistic topology, based on strong inclusion as a primitive notion, is described in Troelstra (1968B).

A more recent and promising approach to constructive topology is the theory of *locales* or *generalized spaces*. We cannot go into details here. The key to this generalization is the observation that many topological properties can be expressed entirely in terms of the lattice of open sets, without explicit reference to the points of the space. In terms of notions to be introduced in sections 13.4 and 14.3 we can describe the essential idea as follows.

To each continuous mapping f from a space X to a space Y corresponds a mapping f^- from the opens of Y to the opens of X (namely $f^-(Z) = \{y : fy \in Z\}$) which preserves $\cap, \cup$, and maps Y to X. The idea is now to look at the category Frm of *frames* which has as objects complete Heyting algebras $\Omega, \Omega', \ldots$, and as morphisms the mappings preserving $\wedge, \vee, \top$. The opposite category of *locales* Loc $\equiv$ (Frm)^{op} may then be viewed as a generalization of Top, the category of topological spaces. A point t in a space gives rise to a frame morphism t^* from the opens of the space to P({0}) $\equiv \mathbb{1}$, the singleton with the discrete topology given by

$$t^*(V) := \{0 : t \in V\}.$$

Quite generally, we define a *point* in a locale X as a frame morphism from X into P({0}); but there are many examples of locales without any points (cf. Fourman and Scott 1979, section 3.5).

To show how we may transcend in locale theory certain difficulties arising in pointset topology, we consider the notion of compactness.

As is clear from theorem 4.7, the notion of compactness (in the sense of "every open cover has a finite subcover") loses all interest if we use the usual pointset approach to topology and assume Church's thesis. On the other hand there is a possibility to describe the topology of Cantor space and $[0,1]$ via the covering relations for the elements of a suitable basis.

Take the case of Cantor space with basis $\{V_n : n$ code of a finite 01-sequence$\}$. A notion of *cover of* V_n by basis elements is *defined* by the following clauses:

(i) $\{V_n\}$ covers V_n.

(ii) If $\mathscr{W}$ covers $V_{n * \langle 0 \rangle}$ and $\mathscr{W}'$ covers $V_{n * \langle 1 \rangle}$, then $\mathscr{W} \cup \mathscr{W}'$ covers V_n.

(iii) Let $\mathscr{W}$ be a cover of V_n and let $\mathscr{W}'$ be an arbitrary family of basis elements; then $\mathscr{W} \cup \mathscr{W}'$ covers V_n.

(iv) All covers of V_n are generated by (i)–(iii).

This notion of cover permits us to define a complete Heyting algebra Ω, (13.4.9), as follows. Let $\mathscr{X}$ be the family of collections $\{V_n : n \in X\}$ such that:

(v) $n \in X \wedge m \preccurlyeq n \rightarrow m \in X$;

(vi) if W covers V_n and $\forall m \in X(V_m \in \mathscr{W})$, then $n \in X$.

(Alternatively we may consider arbitrary collections $\{V_n : n \in X\}$ modulo an equivalence relation $\cong$ defined by $\{V_n : n \in X\} \cong \{V_m : m \in Y\}$ iff each V_n, $n \in X$, is covered by $\{V_m : m \in Y\}$, and vice versa.)

X is made into a frame with $\cap$ for $\wedge$, for the join one takes the closure under (v) and (vi) of the union. V_n corresponds to closure under (v) and (vi) of $\{V_n\}$.

If CT_0 holds, we cannot show that Ω corresponds to the pointset topology of Cantor space, but Ω is nevertheless a well-defined complete Heyting algebra which is compact: the definition of cover ensures that all covers of any V_n by sets of elements of the basis have a finite subcover, *without* assuming FAN.

An introduction to the topic of locales may be found in Johnstone (1982, ch. 2); see also Fourman and Grayson (1982).

Exercises

7.1.1. Prove the proposition in 1.6, and give weak counterexamples to (vii) and (viii) in 1.6.

7.1.2. Prove the properties (i) and (ii) in 1.10.

7.1.3. Show that for any separable metric space $M \equiv (X, \rho)$ with basis sequence $\langle p_n \rangle_n$, we can find a sequence $\langle q_n \rangle_n$ such that:
(i) $\langle q_n \rangle_n$ is a basis sequence for M;
(ii) $\langle q_n \rangle_n$ is discrete, i.e. $\forall nm(q_n = q_m \vee q_n \# q_m)$;
(iii) $\forall n \exists m(q_n = p_m)$ (i.e. $\{q_n : n \in \mathbb{N}\} \subset \{p_n : n \in \mathbb{N}\}$) (Troelstra 1966, 2.2.1; 1969, 12.2).

7.1.4. If $\Gamma \equiv (X, \mathcal{T})$ is a topological space, and $Y \subset X$, then Y is said to be *nowhere dense* in Γ iff $\forall V \exists W \exists x(x \in W \subset V \wedge W \cap Y = \emptyset)$ (V, W ranging over $\mathcal{T}$). Let $M \equiv (X, \rho)$ be a CSM-space, and $\langle Y_n \rangle_n$ a sequence of subsets of X, Y_n nowhere dense in M for all n. Prove that $(\bigcup_n Y_n)^c$ is dense in M.

7.1.5. Prove the statement at the end of 1.11.

7.2.1. Complete the proof of 2.4.

7.2.2. We use the notations of 2.12. Let $M \equiv (X, \rho)$ be a complete separable metric space. $Y \subset X$ is said to be *traceable with index t* if for all $[x]_M$, $k \in \mathbb{N}$,

$$Y \cap U(2^{-k}, [x]_M) \neq \emptyset \rightarrow [\{t\}(x, k)]_M \in Y \cap U(2^{-k}, [x]_M).$$

Let $L_m := \{[x]_M : E\{m\}(x) \wedge A_M(x)\}$; the L_m are called the *listable* sets. A *completely listable* set is a set of the form L_m satisfying $[x]_M = [y]_M \wedge E\{m\}(x) \rightarrow E\{m\}(y)$. Show:
(i) $U(2^{-k}, [x]_M)$ is completely listable;
(ii) an enumerable set of points is traceable.

7.2.3. Prove the following lemma (notation as in the preceding exercise): if $[x]_M \in L_m$ and L_n is completely listable, Y a traceable set with index t, and $Y \cap L_n = \emptyset$, then for some k $U(2^{-k}, [x]_M) \cap Y = \emptyset$. *Hint.* Let $P(u, v, n) := T(c_M, v, j_1 u) \wedge T(n, U(j_1 u), j_2 u)$ and construct m with the recursion theorem such that

$$\{m\}(y) \simeq \begin{cases} x & \text{if } \forall u < y \neg P(u, m, n), \\ \{t\}(x, \min_y P(y, m, n)) & \text{otherwise.} \end{cases}$$

Show, using Markov's principle, that $\exists u P(u, m, n)$, and that we may take $\min_u P(u, m, n)$ for k.

7.2.4. With the same notation as in the preceding two exercises:
(i) a completely listable, nonempty L_n in M contains a basis point (point of the basis sequence, 2.2);
(ii) the L_n above is traceable;
(iii) if L_n, L_m are completely listable, $L_n \cap L_m = \emptyset$, then we can find a k, recursively in x, n and m, such that for $[x]_M \in L_n$, $U(2^{-k}, [x]_M) \cap L_m = \emptyset$. *Hint.* Use (ii) and E7.2.3.

7.2.5. Prove theorem 2.11. *Hint.* Consider, for any $[x]_M$, the sets

$$L^1(x) := \{[y]_M : \rho'(f[x]_M, f[y]_M) < 2^{-k-1}\},$$

$$L^2(x) := \{[y]_M : \rho'(f[x]_M, f[y]_M > 2^{-k-1}\};$$

both sets are completely listable and do not intersect. Apply (iii) of exercise 7.2.4.

7.2.6. A topological space $\Gamma \equiv (X, \mathscr{T})$ is *connected* iff $\forall V W(V, W$ inhabited $\wedge V \cup W = X \to \exists x(x \in V \cap W))$. Show that this is equivalent to: each finite open cover $\mathscr{U} \equiv \{U_0, \ldots, U_n\}$ of X, consisting of inhabited sets, can be arranged in a chain, i.e. there exists a surjection $f \in \{i : i \le m\} \to \{i : 1 \le n\}$ such that $\forall i \le m\, \exists x(x \in U_{f(i)} \cap U_{f(i+1)})$.

7.2.7. $\Gamma \equiv (X, \mathscr{T})$ is *chain-connected* iff for each open cover $\mathscr{U}$, and all $x, y \in X$, there is an $f \in \{i : i \le m\} \to \mathscr{U}$ such that $\forall i < m\, \exists z(z \in f(i) \cap f(i+1)) \wedge x \in f(0) \wedge y \in f(m))$. A chain-connected space is obviously connected. Prove the following three properties to be equivalent (Moerdijk 1986):

(i) Γ is chain connected.
(ii) For all sets Y, all continuous $f \in \Gamma \to Y_d$ are constant (here Y_d is the set Y provided with the discrete topology).
(iii) Each continuous $f \in \Gamma \to (\mathrm{P}(\{0\}))_d$ is constant.

7.3.1. Supply proofs to the following assertions in 3.4:
(i) Φ and Ψ are inverse to each other;
(ii) formula (1) implies formula (4).

7.3.2. Give a (weak) counterexample to "located implies metrically located" in $\mathbb{N}$ and $\mathbb{R}^2$. *Hint.* Let φ be a mapping from $\mathbb{N}$ onto $\mathbb{N}$ defined by: $\varphi(n) = 1$ if $\beta n = 0 \wedge \forall k < n \neg (\beta k = 0)$, $\varphi(n) = n + 2$ otherwise. $X := \{n \ge 3 : n \in \mathbb{N}\}$ is metrically located in $\mathbb{N}$, but $\varphi[X]$ is metrically located in $\varphi[\mathbb{N}]$ iff $\exists n(\beta n = 0) \vee \neg\exists n(\beta n = 0)$ (Troelstra 1966, 2.1.9).

7.3.3. Show that $\mathbb{R}$ can be homeomorphically embedded into $\mathbb{R}^2$ such that the image is not located in $\mathbb{R}^2$.

7.3.4. Prove 3.5.

7.3.5. Complete the argument in 3.9.

7.3.6. Show that a metrically located subset of a CSM-space is separable.

7.4.1. Prove proposition 4.3.

7.4.2. Prove 4.5.

7.4.3. Show that $\{x_n : n \in \mathbb{N}\} \cup \{x\}$, with $\lim \langle x_n \rangle_n = x$, is compact.

7.4.4. Prove 4.10.

7.4.5. Assuming FAN, show that a continuous mapping between metric spaces is strongly continuous in the sense of 4.8.

7.4.6. Show that X_a and X'_a in 4.14 are each others metric complement.

7.4.7. Prove 4.19.

7.4.8. Show that each strong homeomorphism from a metric space M onto a metric space M' is hyperinjective.

7.4.9. Let Y, Z be subsets of a CSM-space $(X, \rho) \equiv M$, Y a CTB-subset such that $Y \subset \mathrm{Int}(Z)$. Assuming FAN, show $\exists \varepsilon > 0(U(\varepsilon, Y) \subset Z)$.

7.4.10. A topological space is *locally connected* if every point has a connected neighbourhood. Prove the following version of the Hahn–Mazurkiewicz theorem: every compact, closed, connected, and locally connected space is the continuous image of $[0,1]$, where connected is defined as in E7.2.6 (Grayson 1981).

7.4.11. Show in CRM that $[0,1]$ is connected, but not chain-connected; for the definitions of connected and chain-connected see E7.2.6.

7.5.1. Assuming FAN and C-N, indicate some simplifications in the proof of 5.5(i).

7.5.2. Prove the statement in the proof of 5.4: $\exists \varepsilon_k > 0(U(\varepsilon_k, X_k) \subset X_{k+1})$. *Hint.* Cf. the proof of 5.3.

7.5.3. Show that the one-point compactification of a CTB-space is unique up to homeomorphism (Troelstra 1968C, 4.12; Bridges 1979, exercise 2.4).

7.5.4. Suppose $M \equiv (X, \rho)$ to be a locally compact space. $M' \equiv (X', \rho')$ is said to be a *minimal compactification* of M, if M is homeomorphic to (X'', ρ'') with $X'' \subset X'$, $\rho'' = \rho' \upharpoonright X''$, $(X'')^- = X'$ in M' and $\rho'(x, y) > 0 \to x \in X'' \vee y \in Y''$. Give an example of a locally compact space which cannot be shown to have a minimal compactification (Troelstra 1968C, 4.13).

CHAPTER 8

ALGEBRA

Intuitionistic algebra is more complicated than classical algebra in various ways; to begin with, algebraic structures as a rule do not carry a decidable equality relation. This difficulty is partly met by the introduction of a strong inequality relation, the so-called *apartness relation*, special cases of which we have met before (5.2.7, 7.1.11). Furthermore there is the awkward abundance of all kinds of substructures, and hence of quotient structures.

In section 1 we quickly go over the basic properties of equality, apartness, and order.

Section 2 deals with group theory; we consider groups with apartness, and consequently also introduce the positive "dual" of a subgroup, the *antisubgroup*. By way of illustration some properties of free abelian groups are discussed. In the same spirit rings and ideals are covered in section 3.

Section 4 contains the basic facts of linear algebra over fields; this material is needed in section 5 which deals with polynomial rings. In particular the properties of principal ideals (f) and their *anti-ideals* A_f are related to the properties of their polynomials.

Section 6 contains some material concerning local rings, fields, and their Kripke models. Finally, an elementary proof of the fundamental theorem of algebra (i.e. the algebraic closedness of $\mathbb{C}$) is given in section 7.

1. Identity, apartness and order

1.1. *Identity.* In classical mathematics the identity relation is completely neutral, it does not influence, and is not influenced by mathematics proper. Similarly, its role in classical logic is modest: all one can do is add cardinality conditions. The situation is completely different in intuitionistic mathematics; there the nature and construction of the objects determine specific properties of the identity relation. E.g. the identity relation on the natural numbers is, by the very construction of natural numbers, decidable, and the method of construction of the real numbers entails the stability of the identity relation on $\mathbb{R}$ (cf. 5.2.10). So the mathematics influences the properties of the identity relation. Conversely, as the reader will have observed, we essentially use in some proofs and constructions certain nontrivial properties of the identity relation.

In the sequel we will use "identity" and "equality" as synonymous. The "minimal" theory of identity has the axioms (cf. 2.1.8):

REFL $\quad x = x,$

SYM $\quad x = y \to y = x,$

TRANS $\quad x = y \wedge y = z \to x = z,$

and in the presence of functions and other relations besides $=$ we require the schema of replacement for all properties A:

REPL $\quad x = y \to (A[z/x] \to A[z/y]).$

SYM and TRANS follow from REFL in the presence of REPL (cf. 2.1.8). The following properties are the most familiar ones in mathematical practice:

$$\neg\neg x = y \to x = y \quad (\textit{stability}),$$

$$x = y \vee x \neq y \quad (\textit{decidability}).$$

Equality on $\mathbb{N}$, $\mathbb{Z}$, and $\mathbb{Q}$ is decidable, and it is stable on $\mathbb{R}$ and $\mathbb{C}$ (5.2.10).

1.2. *Apartness.* The second most important relation in mathematics is the *apartness relation*, a strong notion of inequality. The basic properties of the apartness relation are (cf. 5.2.9, 5.6.8 for the case of $\mathbb{R}$)

AP1 $\quad \neg x \# y \leftrightarrow x = y,$

AP2 $\quad x \# y \to y \# x,$

AP3 $\quad x \# y \to x \# z \vee y \# z.$

A set with decidable equality automatically carries an apartness relation, namely the inequality (exercise 8.1.1). On the other hand the apartness relation influences the identity relation:

$$\neg\neg x = y \leftrightarrow \neg\neg\neg x \# y \leftrightarrow \neg x \# y \leftrightarrow x = y.$$

Hence, a structure with an apartness relation has a stable equality.

In certain circumstances there is a use for a weaker kind of apartness relation, one with AP1 replaced by $\neg x \# x$. We will call this notion *preapartness*.

A *warning on notation*: some authors use the symbol " $\neq$ " for the apartness relation, and other authors use "apartness" and "tight apartness" where we use "preapartness" and "apartness".

As observed by Fourman, a set with an apartness relation usually carries more than one: let A be some proposition such that $\neg\neg A$ holds, and put $x \#_A y := (x \# y \wedge A)$. Then $\#_A$ is an apartness relation as well. The last two properties are obvious. For the first one, assume $\neg x \#_A y$, i.e. $\neg(x \# y \wedge A)$. Now if $x \# y$, then $\neg\neg x \# y$ and since $\neg\neg A$ was given, $\neg\neg(x \# y \wedge A)$. Contradiction. So $\neg x \# y$, and therefore $x = y$.

There is also a *natural* example of a set with two distinct apartness relations: the set of lawless sequences with the negation of identity, and the canonical apartness $\alpha \# \beta := \exists x(\alpha x \neq \beta x)$, cf. 12.2.4.

Apartness relations on sets X and Y canonically determine apartness relations on the Cartesian product and the exponent.

PROPOSITION.
(i) *If X and Y are sets with apartness relations $\#_X$, $\#_Y$, then $X \times Y$ has the canonical apartness relation $\#_{X\times Y}$ given by*

$$(x, y) \#_{X\times Y}(x', y') := x \#_X x' \vee y \#_Y y'.$$

(ii) *If X has an apartness relation $\#$, then X^Y has the canonical apartness relation $\#'$ given by*

$$f \#' g := \exists x(fx \# gx).$$

PROOF. Routine. □

1.3. There is a natural condition on the identity relation: identical objects have the same properties (Leibniz); is there also a natural condition with respect to the apartness relation? The first one that comes to mind concerns functions: if a function takes two objects to two objects that are apart, then they should have been apart in the first place. The principle seems quite plausible. If one does not know whether two objects are apart, then one cannot—extensionally speaking—sharply distinguish them. But then one could not hope to distinguish sharply between their images.

A property (or set) A is sometimes called *extensional* if $x = y \rightarrow (Ax \rightarrow Ay)$ (i.e. A satisfies REPL). In practice we are always dealing with such entities.

In classical mathematics one can reformulate the replacement condition as follows: $Bx \rightarrow x \neq y \vee By$. This formulation suggests the following intuitionistic condition: $Bx \rightarrow x \# y \vee By$. In words, if $x \in B$ and you pick an element y, then $y \in B$ or x and y are apart.

Not every set has this property. Consider e.g. $B = \{0\} \subset \mathbb{R}$; we do not have $\forall y(0 \# y \vee 0 = y)$.

1.4. Definition.
(i) X is *strongly extensional* if $\forall xy(x \in X \rightarrow x \# y \vee y \in X)$, .
(ii) $f \in X \rightarrow Y$ is *strongly extensional* if $\{(x, y) | f(x) \# y\}$ is so. □

Thus a strongly extensional set (or relation) has rather exclusive characteristics: either you are in it, or you are positively distinct from all its members.

1.5. Examples.

(1) $X = \{x \in \mathbb{R} : x \# 0\}$ is strongly extensional. Let $a \in X$, i.e. $a \# 0$, then for any b, $b \# a \vee b \# 0$. So $b \# a \vee b \in X$.

(2) $\#$ on X as a property of pairs is strongly extensional. To see this, put $A := \{(x, y) \in X^2 : x \# y\}$, and let $x \# y$. Then, for any (u, v), $u \# x \vee u \# y$, and hence $u \# x \vee u \# v \vee v \# y$, which amounts to $(u, v) \in X \vee (u, v) \# (x, y)$.

(3) Addition on $\mathbb{R}$ is strongly extensional. To see this, put $P := \{(x, y, z) \in \mathbb{R}^3 : x + y \# z\}$, and let $x + y \# z$. Then, for any (u, v, w), $(u + v, w) \# (x + y, z)$ or $u + v \# w$ (apply (2)). Now $u + v \# x + y \rightarrow u \# x \vee v \# y$ (by 5.3.10). So $(u + v, w) \# (x + y, z)$ is equivalent to $(u, v, w) \# (x, y, z)$. This shows $(u, v, w) \# (x, y, z) \vee (u, v, w) \in P$.

Strong extensionality for functions can be formulated in a more appealing way.

1.6. Proposition. *$f \in X \rightarrow Y$ is strongly extensional* $\Leftrightarrow \forall xx'(f(x) \# f(x') \rightarrow x \# x')$.

Proof. $\Rightarrow$: Let $F = \{(x, y): f(x) \# y\}$, and $f(x) \# f(x')$. $(x, f(x')) \in F$, so $(x', f(x')) \in F \vee (x, f(x')) \# (x', f(x'))$. From the definition of F it immediately follows that $x \# x'$.

$\Leftarrow$: Take the same F, and let $(x, y) \in F$. Now for any (x', y') we get $f(x') \# f(x) \vee f(x') \# y$ from $f(x) \# y$. This in turn implies $x \# x' \vee f(x') \# y' \vee y' \# y$, hence $(x', y') \in F \vee (x, y) \# (x', y')$. □

1.7. *Order.* The treatment of the various notions of order differs from the classical one in that in classical mathematics the reflexive and irreflexive orders are interdefinable. Here we have to treat them separately.

DEFINITION. A *partial order* $\leq$ satisfies the following axioms:

$$x \leq y \wedge y \leq z \rightarrow x \leq z,$$

$$x \leq y \wedge y \leq x \leftrightarrow x = y. \quad \square$$

In order to introduce an irreflexive positive partial order one would have to introduce an apartness relation and then put $x < y := x \leq y \wedge x \# y$.

In general one cannot always introduce such a positive order relation. Consider e.g. the partial order of inclusion of subsets of a given set; the subsets do not allow an apartness relation, let alone a positive partial ordering.

The theory of linear order is formulated in terms of $<$ rather than $\leq$. Since we want to consider linear orders not only on decidable sets, we drop the trichotomy and replace it by comparability.

DEFINITION. The axioms of *linear order* are:

$$x < y \wedge y < z \rightarrow x < z \qquad (\textit{transitivity}),$$

$$\neg(x < y \vee y < x) \leftrightarrow x = y \qquad (\textit{asymmetry}),$$

$$x < y \rightarrow x < z \vee z < y \qquad (\textit{comparability}). \quad \square$$

From the axioms one immediately deduces the following.

PROPOSITION. *For any linear order* $<$:
(i) $\neg x < x$,
(ii) $x < y \vee y < x$ *is an apartness relation.*

PROOF. Immediate. □

In the context of partial elements and the existence predicate we have to reformulate the properties of identity, apartness and order in E-logic with equality, cf. 2.2.4. Each of the basic relations, $=$, $\#$, $<$ is assumed to be strict, i.e.

$$s = t \to Es \wedge Et,$$

$$s \# t \to Es \wedge Et,$$

$$s < t \to Es \wedge Et.$$

The axioms of apartness and order have to be supplemented in the obvious way. For completeness we list them below:

$$t \# s \to s \# t,$$

$$Es \wedge t \# t' \to s \# t \vee s \# t',$$

$$Es \wedge Et \wedge \neg s \# t \leftrightarrow s = t,$$

$$t < t' \wedge t' < t'' \to t < t'',$$

$$Es \wedge Et \wedge \neg(s < t \vee t < s) \leftrightarrow s = t,$$

$$Es \wedge t < t' \to t < s \vee s < t'.$$

2. Groups

2.1. The basic operations of a group, multiplication and inverse, are both total, i.e. $Es \wedge Et \to Est$ and $Es \to Es^{-1}$. So, if we start with existing elements, multiplication and inverse will again yield existing elements. Thus there is no immediate objection to handle group theory in a plain language without E-predicate. However, groups with "partial elements" are perfectly legitimate; nonetheless, we will for the time being drop the existence predicate.

DEFINITION. A *group* is a structure $(A, \cdot, {}^{-1}, e)$ satisfying the following axioms:

$$x \cdot e = e \cdot x = x,$$

$$x \cdot (y \cdot z) = (x \cdot y) \cdot z,$$

$$x \cdot x^{-1} = x^{-1} \cdot x = e.$$

The group is *abelian* if it satisfies

$$x \cdot y = y \cdot x.$$

A *group with apartness relation* has strictly extensional $\cdot$ and $^{-1}$, i.e.

$$xy \# x'y' \to x \# x' \vee y \# y',$$

$$x^{-1} \# y^{-1} \to x \# y.$$

As usual, we mostly write xy for $x \cdot y$, etc. □

2.2. PROPOSITION. *In a group with apartness relation we have*:
(i) $xy \# e \to x \# e \vee y \# e$,
(ii) $x \# x' \to yx \# yx' \wedge xy \# x'y$,
(iii) $x \# y \to x^{-1} \# y^{-1}$.

PROOF. (i) Apply strong extensionality to $xy \# ee$.
(ii) Consider $y^{-1}(yx) \# y^{-1}(yx')$, etc.
(iii) Obvious. □

The "computational" part of group theory, i.e. the verification of relations between words, does not present any new features when compared with classical group theory. Mathematically speaking (as opposed to logically) constructive group theory starts to diverge considerably from its classical counterpart when we take into account notions as "subgroup", "factorgroup", etc.

Since this chapter is mainly devoted to the mathematical aspects of algebra we will not go into group theory as a first-order theory.

A major reason for the divergence of intuitionistic mathematics from classical mathematics is the unusual behaviour of sets of objects in the absence of the principle of the excluded middle and in the peculiarities of choice sequences. In algebra the first phenomenon is of considerable importance. The presence of strange subgroups, even in finite groups, gives rise to unexpected difficulties.

One might be tempted to eliminate this second-order aspect altogether in favour of first-order concepts, but that is a rather chimerical project in view of the importance of homomorphisms, kernels, etc. Moreover, a considerable part of the traditional notions and techniques can be salvaged.

2.3. EXAMPLES.

(1) Consider the additive group $\mathbb{Z}^2$. We define the subgroup $G := \{(a, b): b = 0 \vee A\}$, where A is some undecided statement. As it is, G is a perfectly good subgroup, but we do not know whether $G = \mathbb{Z}^2$ or $G \cong \mathbb{Z}$. Note that we do not have $\forall x \in \mathbb{Z}^2(x \in G \vee x \notin G)$, i.e. G is not a *removable* subset of $\mathbb{Z}^2$.

(2) Consider $\mathbb{Z}/2\mathbb{Z}$ (i.e. $\{0,1\}$ with the obvious addition). $G_A := \{x = 0 \vee A\}$, where A is some proposition. Roughly speaking there are as many subgroups G_A as there are inequivalent propositions A (E8.2.2). In general the subgroups G_A are not removable.

(3) To add a restriction of removability would go too far, as the following example shows: $\{(x,0) : x \in \mathbb{R}\}$ is a subgroup of $\mathbb{R}^2$, but it is not removable. Nonetheless it is one of the most natural subgroups one can think of.

Antisubgroups, subgroups (2.4–7). The machinery of subgroups below is presented in a kind of "dual" terms in analogy to the relation apartness/equality. We present a positive notion that allows for some refinements not to be found in the traditional theory.

2.4. DEFINITION. An *antisubgroup* A of a group G is subset of G with the properties:
(i) $\neg e \in A$,
(ii) $xy \in A \rightarrow x \in A \vee y \in A$,
(iii) $x^{-1} \in A \rightarrow x \in A$.
The antisubgroup A is *normal* if it satisfies
(iv) $xy \in A \rightarrow yx \in A$.
The antisubgroup A is *compatible with the apartness* of G if
(v) $a \in A \Rightarrow a \# e$. □

2.5. EXAMPLES.

(1) $A := \{x \in \mathbb{Z} : \neg 2|x\}$ is an antisubgroup of $\mathbb{Z}$. Evidently $0 \notin A$. Since $\neg 2|(x + y) \rightarrow \neg 2|x \vee \neg 2|y$, (ii) is satisfied. Finally $2|x \leftrightarrow 2|-x$.

(2) $A = \{x + iy : y \# 0\}$ is an antisubgroup of the additive group of $\mathbb{C}$, the set of complex numbers.

(3) The subset of the group of 2×2 matrices of the form

$$\begin{pmatrix} a & b \\ c & d \end{pmatrix},$$

with $c \# 0 \vee b \# 0 \vee a \# d$ is a normal antisubgroup of $\mathrm{GL}_2(\mathbb{R})$ (E8.2.3).

2.6. PROPOSITION. *Let A be an antisubgroup of G. A^c (the complement of A) satisfies*:
(i) $e \in A^c$,
(ii) $x \in A^c \wedge y \in A^c \rightarrow xy \in A^c$,
(iii) $x \in A^c \rightarrow x^{-1} \in A^c$,
(iv) $\neg\neg x \in A^c \rightarrow x \in A^c$.

PROOF. Immediate. □

Observe that the properties (i), (ii), (iii) are exactly those defining a subgroup. We say that A^c is the *subgroup determined by A*.

2.7. PROPOSITION.
(i) *A is a normal antisubgroup* $\Leftrightarrow (yxy^{-1} \in A \rightarrow x \in A) \Leftrightarrow (x \in A \rightarrow yxy^{-1} \in A)$.
(ii) *If A is a normal antisubgroup, then A^c is a normal subgroup, i.e.* $x \in A^c \rightarrow yxy^{-1} \in A^c$.

PROOF. Immediate. □

We do not intend to develop group theory in detail and we will assume that the reader is familiar with the traditional facts and methods of group theory, such as can be found in any textbook. We will freely use unproblematic results and concepts, and comment when necessary on constructive aspects.

Factor groups, homomorphisms (2.8–10)

2.8. Let us recall the construction of factor groups. Let D be a normal subgroup, D defines an equivalence relation $\sim$ on G: $a \sim b := a^{-1}b \in D$. The equivalence class of a is the coset $aD = \{ax : x \in D\}$. Multiplication of cosets is defined via their representing elements: $(aD)\cdot(bD) = abD$. This multiplication obviously is well-defined. The cosets form a group G/D with unit D and inverse $(aD)^{-1} = a^{-1}D$.

In case D is the normal subgroup determined by a normal antisubgroup A we know more.

PROPOSITION. *Let A be a normal antisubgroup, and $D = A^c$, then G/D is a group with a canonical apartness relation defined by* $aD \# bD := ab^{-1} \in A$.

PROOF. Define $aD \# bD := ab^{-1} \in A$. We check the properties:
(i) $\neg(aD \# bD) \Leftrightarrow \neg ab^{-1} \in A \Leftrightarrow ab^{-1} \in D \Leftrightarrow aD = bD$.
(ii) Symmetry is obvious.
(iii) $aD \# bD \Leftrightarrow ab^{-1} \in A \Leftrightarrow (ac^{-1})\cdot(cb^{-1}) \in A \Rightarrow ac^{-1} \in A \vee cb^{-1} \in A \Leftrightarrow aD \# cD \vee cD \# bD$.
(iv) $a_1b_1D \# a_2b_2D \Leftrightarrow b_1^{-1}a_1^{-1}a_2b_2 \in A \Leftrightarrow a_1^{-1}a_2b_1^{-1}b_2 \in A \Rightarrow a_1^{-1}a_2 \in A \vee b_1^{-1}b_2 \in A \Leftrightarrow a_1D \# a_2D \vee b_1D \# b_2D$.
(v) $a^{-1}D \# b^{-1}D \Leftrightarrow a^{-1}b \in A \Leftrightarrow aD \# bD$. □

2.9. DEFINITION. A *homomorphism* is a mapping σ from a group G_1 to a group G_2 such that $\sigma(xy) = \sigma x \cdot \sigma y$ (notation: $\sigma : G_1 \to G_2$ or $G_1 \to_\sigma G_2$ or $G_1 \to_\sigma G_2$).

$\sigma : G_1 \to G_2$ is an *isomorphism* if $\sigma x = \sigma y \to x = y$, and σ is a *strong isomorphism* if moreover $x \# y \to \sigma x \# \sigma y$.

If G_1 and G_2 are groups with apartness we shall require homomorphisms to be strongly extensional. □

REMARKS. (i) Note that the preservation of products does not imply strong extensionality (E8.2.5). Observe that homomorphisms preserve the unit element and inverse operation.

(ii) The canonical mapping from G to G/D given by $a \mapsto aD$, for a normal subgroup D, is a homomorphism, and if the antisubgroup A is compatible with the apartness of G, then the canonical mapping $G \to G/A^c$ is strongly extensional.

2.10. THEOREM. *If G_2 is a group with apartness and $\sigma : G_1 \to G_2$, then $A := \{x \in G_1 | \sigma x \# e_2\}$ is a normal antisubgroup and there is a unique σ^* such that the diagram in fig.* 8.1 *commutes.* (*As usual in diagrams of this type, the dotted arrow indicates the mapping the existence of which is being asserted, and ! indicates uniqueness of the solution.*)

Moreover σ^* is a strong embedding (i.e. a strong isomorphism onto a subgroup of G_2).

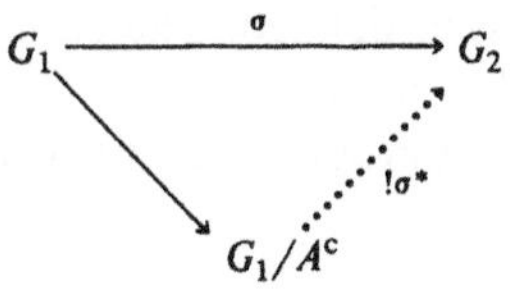

Fig. 8.1.

PROOF. We check the properties of a normal antisubgroup.

(i) $\sigma(e_1) = e_2$, so $e_1 \notin A$.

(ii) $ab \in A \Leftrightarrow \sigma(ab) \# e_2 \Leftrightarrow \sigma(a) \cdot \sigma(b) \# e_2 \Rightarrow \sigma(a) \# e_2 \vee \sigma(b) \# e_2 \Leftrightarrow a \in A \vee b \in A$.

(iii) and (iv) are obvious.

Define $\sigma^*(aA^c) := \sigma a$, then σ^* is a homomorphism. σ^* is also a strong embedding: let $aA^c \# bA^c$, then $ab^{-1} \in A$. So $\sigma(ab^{-1}) \# e_2$ and hence $\sigma^*(aA^c) = \sigma(a) \# \sigma(b) = \sigma^*(bA^c)$. □

2.11. *Models of groups.* Kripke models of group theory (*Kripke groups* for short) have classical groups at their nodes, as one can easily conclude from the form of the axioms not involving apartness (or one may quote the properties of models of geometric theories, E2.6.14).

We shall now characterize the apartness relation on a Kripke group G in terms of the groups G_k at the nodes k.

The set $N_k := \{a \in G_k : k \nVdash a \# e\}$ is a normal subgroup of G_k: $a, b \in N_k \Rightarrow k \nVdash a \# e \wedge k \nVdash b \# e \Rightarrow k \nVdash ab \# e \Rightarrow ab \in N_k$, and $a \in N_k \Rightarrow k \nVdash a \# e \Rightarrow k \nVdash b^{-1}ab \# e \Rightarrow b^{-1}ab \in N_k$.

So there are normal subgroups N_k, such that $k \Vdash a \# b \Leftrightarrow ab^{-1} \notin N_k$. The axiom $\neg a \# b \leftrightarrow a = b$ implies the following condition on the normal subgroups N_k:

$$\forall k' \geq k(\varphi_{kk'}(a) \in N_{k'}) \Rightarrow a = e). \tag{1}$$

Furthermore, the Kripke semantics requires $k \Vdash a \# e \Rightarrow k' \Vdash a \# e$ for $k' \geq k$, so

$$\varphi_{kk'}(a) \in N_{k'} \Rightarrow a \in N_k. \tag{2}$$

Conversely, a Kripke group with (classical) groups and normal subgroups G_k, N_k associated to the nodes such that (1) and (2) hold, carries an apartness relation defined by $k \Vdash a \# b := ab^{-1} \notin N_k$.

We check the axioms:

(i) $\#$ is obviously symmetric.

(ii) By (1) $\#$ satisfies $\neg a \# b \rightarrow a = b$.

(iii) $k \Vdash a \# b \Leftrightarrow ab^{-1} \notin N_k \Leftrightarrow ac^{-1}cb^{-1} \notin N_k \Rightarrow ac^{-1} \notin N_k \vee cb^{-1} \notin N_k \Leftrightarrow k \Vdash a \# c \vee c \# b$.

(iv) $k \Vdash a \# b \Leftrightarrow ab^{-1} \notin N_k \Rightarrow ac(bc)^{-1} \notin N_k = k \Vdash ac \# bc$, likewise for $ca \# bc$. Now, $k \Vdash ab \# e \Leftrightarrow ab \notin N_k \Rightarrow a \notin N_k \vee b \notin N_k \Rightarrow k \Vdash a \# e \vee b \# e$.

From these two facts one easily concludes the strong extensionality of $\cdot$ and $^{-1}$ (cf. E8.2.6). □

The family $\{N_k : k \in K\}$ does not necessarily form a Kripke group since $\varphi_{kk'}$ need not take N_k into $N_{k'}$. However, by condition (2) the complements $A_k := N_k^c$ do form a monotone family, so they form a (Kripke-) normal antisubgroup:

(i) $k \Vdash e \in A \Leftrightarrow e \in A_k \Leftrightarrow e \notin N_k$. Contradiction. So no k forces $e \in A$, hence for all k, $k \Vdash \neg e \in A$.

(ii) Let $a, b \in G_k$, $k \Vdash ab \in A \Leftrightarrow ab \in A_k \Leftrightarrow ab \notin N_k \Rightarrow$ (classically) $a \notin N_k \vee b \notin N_k \Rightarrow k \Vdash a \in A \vee b \in A$.

(iii) Left to the reader.

The internal normal antisubgroup A determines an internal normal subgroup A^c, which, however, is trivial. For,

$$k \Vdash a \in A^c \Leftrightarrow k \Vdash \neg a \in A \Leftrightarrow \forall k' \geq k(\varphi_{kk'}(a) \in N_{k'}) \Leftrightarrow a = e.$$

EXAMPLES. We indicate at each node k the group G_k and the normal subgroup N_k.

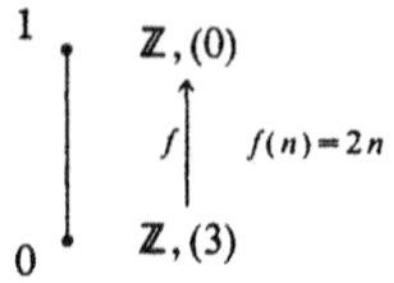

Fig. 8.2.

Example 1. See fig. 8.2. Observe that in this model N_0 is not mapped onto N_1. The apartness is given by $3n + i \# 0$ for $i = 1, 2$.

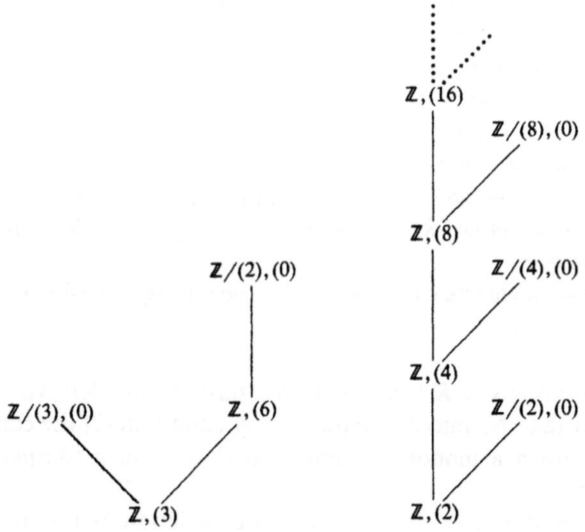

Fig. 8.3.

Examples 2 and 3. See fig. 8.3. At each node the group G_k and the normal subgroup N_k are indicated. The homomorphisms $\varphi_{kk'}$ are canonical.

2.12. *Free abelian groups.* In order to demonstrate group theory in action we will now digress somewhat and discuss a special topic. The topic is chosen for no reason but the occasion to illustrate the extra care needed in a classically trivial subject.

Let us consider the notion of a free abelian group. We want a free group F on the generator set S to satisfy the usual universal property: there is a canonical mapping from S into F such that for each mapping φ from S into a group G there is a *unique* homomorphism $\sigma : F \to G$ such that the diagram in fig. 8.4 commutes.

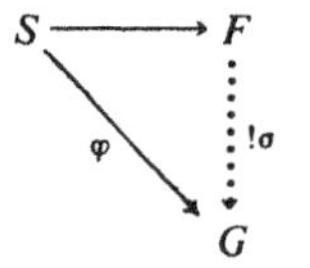

Fig. 8.4.

Let us for a moment assume that S is finite: $S = \{1, \ldots, k\}$. Could we take for F the usual direct sum $\mathbb{Z} \oplus \mathbb{Z} \oplus \cdots \oplus \mathbb{Z}$? We recall that the direct sum $A \oplus B$ of two abelian groups A and B is defined as the Cartesian product with coordinatewise operations.

Indeed, the standard definition of σ works here:

$$\sigma((n_1, \ldots, n_k)) = n_1\varphi(1) + \cdots + n_k\varphi(k).$$

What to do, however, when S is not such a nice set?

Consider, for example, the set $\{a, b\}$ where a and b are real numbers for which it is unknown if they are equal or not. Clearly, we cannot define a canonical mapping from $\{a, b\}$ to $\mathbb{Z} \oplus \mathbb{Z}$, nor from $\{a, b\}$ to $\mathbb{Z}$, such that the diagram in fig. 8.4 can be obtained. For the canonical mapping would preempt the question of the equality of a and b.

In this case we have to revert to the old construction of a free (not necessarily abelian) group.

Let A be a set. We form the set A^* of all finite sequences of elements of A; this is an unproblematic step from the constructive point of view. We include the empty sequence ε ($:= \langle\ \rangle$) in A^* (e.g. as the empty function from $\emptyset$ to A). In case A is empty this is even the only element of A^*. As usual $x * y$ stands for the concatenation of x and y. Intuitively we think of the finite sequences as elements of the abelian monoid over A. However, instead of the usual additive notation we use the sequences themselves; normally we regroup the elements in a word, e.g. *aabcabbac*, into groups of equals (*aaaabbbcc*), but since we cannot decide equality, this procedure does not work.

Since commutativity (plus associativity) allows us to neglect the order in words, the following definition captures the notion "abelian".

2.13. DEFINITION. Let $\vec{a}$ and $\vec{b}$ be words of length n and m. $\vec{a} \sim \vec{b}$ iff $n = m$ and there exists a permutation σ of $\{1, \ldots, n\}$ such that $b_i = a_{\sigma i}$ $(i \leq n)$. □

Observe that $\varepsilon \sim \vec{a}$ iff $\vec{a} \equiv \varepsilon$.

2.14. LEMMA. $\sim$ *is a congruence relation with respect to concatenation.*

PROOF. Immediate. □

2.15. Recall that a *monoid* is a structure $(A, \cdot, e)$ with an associative operation (multiplication) $\cdot$ and a neutral element e. As usual we denote the operation in an abelian monoid by $+$.

LEMMA. $A^*/\sim$ *is a free abelian monoid over* A.

PROOF. Addition on $A^*/\sim$ is defined by $(\vec{a}/\sim) + (\vec{b}/\sim) := (\vec{a} * \vec{b}/\sim)$. The neutral element is $\varepsilon/\sim$. By definition of $\sim$, addition is commutative. Define i by $i(a) = a/\sim$ and let f be a mapping from A into an abelian monoid M, as shown in fig. 8.5. Now put $g(a/\sim) := f(a)$ for all $a \in A$. We extend g canonically to arbitrary elements of $A^*/\sim$ by

$$g(\langle a_1, \ldots, a_n\rangle/\sim) := g(a_1/\sim) \cdot \ldots \cdot g(a_n/\sim) \quad \text{and}$$

$$g(\varepsilon/\sim) := 0.$$

The diagram commutes by definition. For the uniqueness of g, suppose that

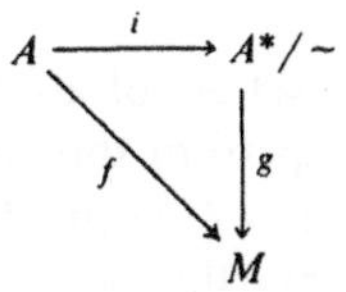

Fig. 8.5.

h also makes the diagram commute, then

$$h(i(a)) = f(a) \quad \text{for all } a \in A, \quad \text{so } h(a/\sim) = f(a) = g(a/\sim).$$

Therefore $h = g$, since h is a homomorphism. □

Note that the proof does not assume anything about A; A can be any set. E.g. if $A = \emptyset$, then $A^*/\sim\, = \{\varepsilon/\sim\} = \{0\}$.

The next step is to embed the free abelian monoid in a free abelian group. The procedure is similar to that of embedding $\mathbb{N}$ in $\mathbb{Z}$.

2.16. DEFINITION. For an abelian monoid M we define

$$(u, v) \sim (u', v') := u + v' = u' + v. \quad \square$$

2.17. LEMMA. *$\sim$ is an equivalence relation on M^2.*

PROOF. Immediate. □

2.18. DEFINITION. Let M be an abelian monoid. $G_M := \langle M^2/\sim, +, -, 0\rangle$, with

$$(u, v)/\sim \; + \; (u', v')/\sim \; := (u + u', v + v')/\sim,$$
$$-(u, v)/\sim \; := (v, u)/\sim,$$
$$0 := (0,0)/\sim. \quad \square$$

2.19. LEMMA. *G_M is an abelian group.*

PROOF. Immediate. □

2.20. LEMMA. *Let $j(m) := (m, 0)/\sim$, then, for each f from the abelian monoid M to the abelian group G, there is a unique $g: G_M \to G$ such that the diagram in fig.* 8.6 *commutes.*

PROOF. Define $g((u, v)/\sim) := f(u) - f(v)$. The remainder of the proof is routine. □

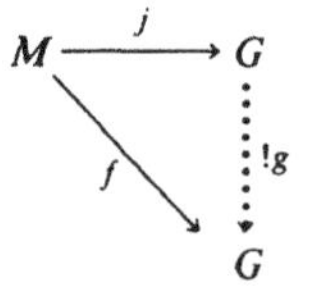

Fig. 8.6.

We now combine the lemmas 2.15 and 2.20 to obtain the desired free abelian group over A.

2.21. THEOREM. *Every set A generates a free abelian group F_A.*

PROOF. The diagram in fig. 8.7 is the result of applying lemmas 2.15 and 2.20.

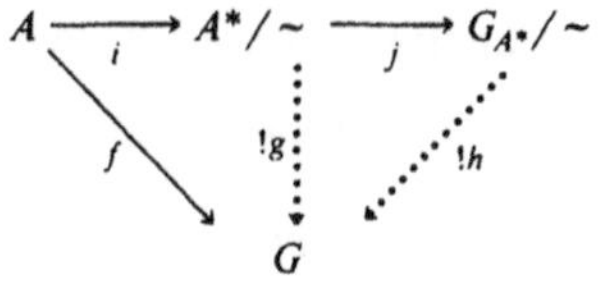

Fig. 8.7.

2.22. EXAMPLES. We will give some illustrations, using Kripke models over the poset $(\{0, 1\}, <)$.

Example 1. The embedding of the generating set is given by two mappings φ and ψ such that the diagram in fig. 8.8 commutes, so we get $g(1) = \psi(a) = (1, 0)$. We leave it to the reader to show that the Kripke group at the right-hand side is indeed the free abelian group over the given set.

The construction of free abelian Kripke groups is simplified by the fact that the groups at the nodes are free abelian groups themselves (E8.2.8).

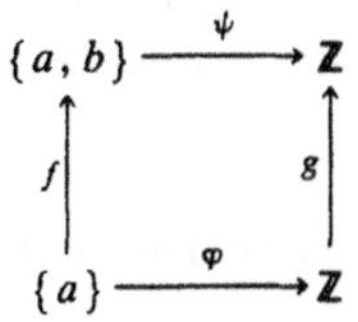

$$
\begin{array}{ccc}
\{a, b\} & \xrightarrow{\ \psi\ } & \mathbf{Z} \\
f \uparrow & & \uparrow g \\
\{a\} & \xrightarrow{\ \varphi\ } & \mathbf{Z}
\end{array}
\qquad
\begin{array}{l}
f(a) = a \\
\varphi(a) = 1 \\
\psi(a) = (1, 0) \\
\psi(b) = (0, 1)
\end{array}
$$

Fig. 8.8.

Example 2. See fig. 8.9. Now $g(\varphi(a)) := \psi(a)$ and $g(\varphi(b)) := \psi(a)$ yields $g(1, 0) = 1 = g(0, 1)$ and $g(x, y) = x + y$. Again the reader should check the freeness. This example illustrates the usefulness of the equivalence relation of definition 1. The bottom node thinks that a and b are distinct (and hence inequivalent) whereas the top node thinks that a and b are identical (and hence equivalent).

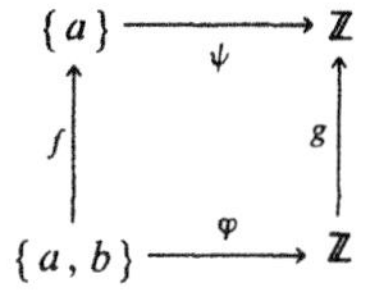

$$f(a) = f(b) = a$$
$$\varphi(a) = (1,0)$$
$$\varphi(b) = (0,1)$$
$$\psi(a) = 1$$

Fig. 8.9.

Example 3. We will now show that just putting free abelian groups at the nodes does not suffice to construct a free abelian group, see fig. 8.10. The right-hand group obviously is free abelian. In order to make the middle group a subgroup of this free group we put $\rho(n) = n$ and hence $g(n) = n$. Let us now try to determine φ and ψ such that the middle group is free over the left-hand set (note that in each node there can only be one generator, and it does not matter if we take the same element plus the identity mapping). Now there are a few choices for φ and ψ (depending on our choice of generators of $\mathbb{Z}$ and $2\mathbb{Z}$). Consider $\psi(a) = 1$, $\varphi(a) = 2$. Then $g(\varphi(a)) = g(2) = 2 \neq 1 = \psi(a)$. We cannot choose φ and ψ such that the diagram commutes. Hence the middle group is not free!

The example also shows that *a subgroup of a free abelian group need not be free*.

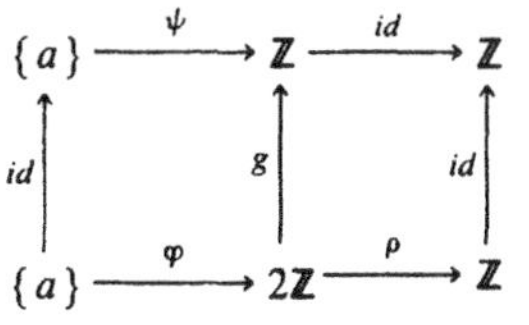

Fig. 8.10.

3. Rings and modules

We will not treat ring theory in full generality; for a demonstration of the methods and problems of constructive ring theory commutative rings with unity and an apartness relation will do very well. For the same reason we restrict the role of E^+-logic to a few remarks.

Rings, integral domains, fields (3.1–3)

3.1. DEFINITION. A (commutative) *ring with apartness* is a structure $\boldsymbol{R} = (R, \#, +, \cdot, -, 0, 1)$ satisfying the apartness axioms AP1–3, the ring identities for $+$ and $\cdot$:

$$x + 0 = x, \qquad x \cdot 1 = x,$$

$$x + y = y + x, \qquad xy = yx,$$

$$x + (y + z) = (x + y) + z, \qquad x(yz) = (xy)z,$$

$$x + (-x) = 0, \qquad x(y + z) = xy + xz,$$

strong extensionality of $+$ and $\cdot$, nontriviality:

$$x + y \# x' + y' \to x \# x' \vee y \# y',$$

$$xy \# x'y' \to x \# x' \vee y \# y',$$

$$0 \# 1.$$

A ring is an *integral domain* if it satisfies

$$x \# 0 \wedge y \# 0 \to xy \# 0.$$

A ring is a *field* if it satisfies

$$x \# 0 \to \exists y(xy = 1).$$

If we leave out the axioms involving $\#$, and add $0 \neq 1$, we obtain the notion of a *ring*. Usually, it will be clear from the context what kind of ring is considered. □

If no confusion can arise we shall abuse our notation and denote the ring $\boldsymbol{R}$ simply by R.

3.2. EXAMPLES.

(1) $\mathbb{Z}, \mathbb{Q}, \mathbb{R}$ are rings with an apartness relation (even integral domains, respectively fields). Most standard constructions from ring theory also yield rings (matrix rings, etc.) with apartness relation.

(2) The boolean ring of subsets of an inhabited set is a commutative ring without apartness relation.

(3) $\mathbb{R}^{\mathrm{be}}$ is a commutative ring without apartness relation (5.6.8).

We list a number of immediate consequences of the definition.

3.3. Proposition. *If R is a ring, then:*
(i) $x + y \# 0 \to x \# 0 \vee y \# 0$,
(ii) $xy \# 0 \to x \# 0 \wedge y \# 0$.
If R is an integral domain, then:
(iii) $x \neq 0 \wedge xy = 0 \to y = 0$,
(iv) $x \# y \wedge z \# 0 \to xz \# yz$.
If R is a field, then:
(v) *R is an integral domain,*
(vi) $x \# 0 \to \exists! y(xy = 1)$.

Proof. (i) follows immediately from the strong extensionality of $+$.
(ii) Apply strong extensionality to $x \cdot y \# x \cdot 0$ and $x \cdot y \# 0 \cdot y$.
(iii) By logic we get from the definition $x \neq 0 \wedge y \neq 0 \to xy \neq 0$. Now let $x \neq 0$ and $xy = 0$; assume $y \neq 0$, then $xy \neq 0$. Contradiction. Hence $\neg\neg y = 0$, and by the stability of $=$, $y = 0$.
(iv) Immediate.
(v) Let $x_1 \# 0$, $x_2 \# 0$, then there exist y_1, y_2 such that $x_1 y_2 = 1 = x_2 y_2$. Hence $x_1 x_2 y_1 y_2 \# 0$. Now apply (ii).
(vi) Immediate. □

Since in a field, for an element a apart from zero, there is a unique b with $ab = 1$, we will introduce the notation $b = a^{-1}$. We will return to the inverse as a partial operation in the framework of E^+-logic (3.14, section 6).

The notions of *homomorphism*, *isomorphism*, and *strong isomorphism* are straightforward generalizations of 2.9. Homomorphisms between rings with apartness are required to be strongly extensional.

Modules and vector spaces (3.4–6)

3.4. Definition. A *module* (with apartness) is a (two-sorted) structure $(R, M, \cdot)$, where R is a ring and M is an abelian group (with apartness), and where $\cdot$ is a function from $R \times M$ to M, scalar multiplication, such that the following identities hold (where $r,s \in R$ and $a,b \in M$):

$$(rs) \cdot a = r \cdot (s \cdot a),$$

$$1 \cdot a = a,$$

$$(r + s) \cdot a = r \cdot a + s \cdot a,$$

$$r \cdot (a + b) = r \cdot a + r \cdot b;$$

and also strong extensionality:

$$r \cdot a \# s \cdot b \to r \# s \vee a \# b,$$

$$r \cdot a \# 0 \to r \# 0 \wedge a \# 0.$$

A *vector space* is a module over a field. □

For convenience we have denoted the apartness relations on R and M by the same symbol; in general no confusion arises. By dropping the clauses involving # we obtain a module without apartness. The context will always make clear what kind of module we are considering. As is customary in algebra texts we have also made no notational distinction between ring multiplication and scalar multiplication.

For vector spaces we can show the following convenient properties of the apartness relation.

3.5. PROPOSITION. *In a vector space the following hold*:
(i) $r \# 0 \wedge a \# 0 \to ra \# 0$,
(ii) $r \# 0 \wedge a \# a' \to r \cdot a \# r \cdot a'$,
(iii) $r \# r' \wedge a \# 0 \to r \cdot a \# r' \cdot a$.

PROOF. (i) Since $r \# 0$, r^{-1} exists. So $a \# 0 \to r^{-1} \cdot (r \cdot a) \# 0 \Rightarrow r \cdot a \# 0$.
(ii) and (iii) follow immediately. □

Ideals, anti-ideals (3.6–9)

3.6. DEFINITION. An *anti-ideal* of a ring R is a subset A satisfying:
(i) $0 \notin A$,
(ii) $x + y \in A \to x \in A \vee y \in A$,
(iii) $xy \in A \to x \in A \wedge y \in A$.
An *ideal* I is a subset of R satisfying:
(iv) $0 \in I$,
(v) $x, y \in I \to x - y \in I$,
(vi) $x \in R,\ y \in I \to xy \in I$. □

3.7. EXAMPLES.
(1) $\{k \in \mathbb{Z} : \neg n | k\}$, for $n \neq 0$, is an anti-ideal in $\mathbb{Z}$ (the complement of $n\mathbb{Z}$).
(2) $\{a \in \mathbb{R} : a \# 0\}$ is an anti-ideal in $\mathbb{R}$.

(3) $\{f \in \mathscr{C}(\mathbb{R},\mathbb{R}) : |f(0)| > 0\}$ is an anti-ideal in the ring $\mathscr{C}(\mathbb{R},\mathbb{R})$ of continuous functions from $\mathbb{R}$ to $\mathbb{R}$.

The definition of an anti-ideal does not require it to contain any elements at all, so $\emptyset$ is a perfectly good anti-ideal. And $A := \{n \in \mathbb{Z} : (\neg 4|n \wedge F) \vee (\neg 9|n \wedge \neg F)\}$, where F is an undecided statement (e.g. Fermat's conjecture), is just as well an anti-ideal; however, we cannot show A to be inhabited. All we know in this particular case is that A is not empty.

Inhabited anti-ideals are characterized by the fact that they contain 1. For suppose that $a \in A$, then by (iii) of the definition $1 \in A$.

As usual, ideals may be used to define factor rings, and as in the case of groups, anti-ideals are the tools for introducing an apartness relation on factor rings.

3.8. PROPOSITION.

(i) *If A is an anti-ideal, then A^c is an ideal, and A^c is proper if A is inhabited.*

(ii) *Let I be an ideal of R, then $R/I := \{a + I : a \in R\}$ with operations*

$$(a + I) + (b + I) := a + b + I,$$

$$(a + I) \cdot (b + I) := ab + I,$$

is a ring (the so-called factor ring of R over I), provided $1 \notin I$ (i.e. I is a proper ideal).

(iii) *If A is an inhabited anti-ideal, then R/A^c is a ring with apartness relation, given by $a + A^c \# b + A^c$ iff $a - b \in A$.*

PROOF. Routine. Observe that we have to require A in (iii) to be inhabited because we consider here rings with unity only. □

From now on we will tacitly assume ideals to be proper. The relation between homomorphisms and (anti-)ideals is given by the following theorem.

3.9. THEOREM.

(i) *For an inhabited anti-ideal A the canonical mapping $j \in R \to R/A^c$ is a homomorphism, and if R has an apartness compatible with A, then j is strongly extensional (as required).*

(ii) *If R_2 is a ring with apartness and $\sigma \in R_1 \to R_2$ a homomorphism, then $A_\sigma = \{a \in R_1 : \sigma a \# 0\}$ is an inhabited anti-ideal. If R_1 is also a ring* *with apartness, then A_σ is compatible with #. There is a unique strong embedding σ^* such that the diagram in fig. 8.11 commutes.*

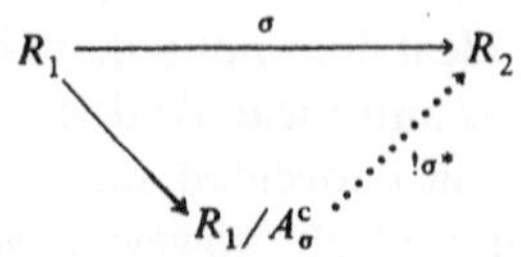

Fig. 8.11.

PROOF. Left to the reader. Observe that A_σ is inhabited, because $\sigma(1) = 1$. Cf. 2.10. □

Prime (anti-)ideals, minimal anti-ideals, and maximal ideals (3.10–12)

3.10. Prime ideals play an important role in traditional algebra, but their usual definitions

$$I \text{ is a prime ideal} \Leftrightarrow xy \in I \to x \in I \vee y \in I,$$

$$I \text{ is a prime ideal} \Leftrightarrow x \notin I \wedge xy \in I \to y \in I,$$

are either too strong or too weak. Under the first definition not even (X) is a prime ideal of $R[X]$: pick a, b such that $ab = 0$, but $a = 0$ or $b = 0$ is undecided and consider $(X + a)(X + b)$. The second definition is not strong enough to make the factor ring over a prime ideal an integral domain. We will approach the matter from the side of the anti-ideals; we could, however, use 3.11(i) below as definition.

For exactly the same reasons we have modified the traditional definition of integral domain. The field $\mathbb{R}$ does not satisfy $xy = 0 \to x = 0 \vee y = 0$, so the traditional definition is too strong.

DEFINITION. Let R be a ring and $A \subset R$:
(i) A is a *prime anti-ideal* of R if

$$1 \in A, \quad x \in A \wedge y \in A \to xy \in A.$$

(ii) A is a *minimal anti-ideal* of R if

$$1 \in A, \quad \forall x \in A \, \exists y \in R(1 - xy \notin A). \quad \square$$

We have to use the notion of minimal anti-ideal instead of maximal ideal since we are aiming for the familiar result that the factor ring modulo a (complement of a) minimal anti-ideal is a field.

3.11. PROPOSITION.
(i) *A is prime $\Leftrightarrow$ R/A^c is an integral domain.*
(ii) *A is minimal $\Leftrightarrow$ R/A^c is a field.*
(iii) *A minimal anti-ideal is prime.*
(iv) *If A is minimal and B is an inhabited anti-ideal such that $B \subseteq A$, then $B = A$.*

PROOF. (i), (ii), and (iii) are obvious. For (iv) let $a \in A$, then $1 - ab \notin A$ for some $b \in R$. Now $1 = (1 - ab) + ab \in B$, so $1 - ab \in B$ or $ab \in B$. Hence $ab \in B$, and therefore $a \in B$. □

We can imitate the traditional construction of the field of fractions of an integral domain. Indeed, the only novel point is the introduction of the apartness relation.

3.12. THEOREM. *For each integral domain R there exists a field F_R such that R is strongly embedded in F_R and for each strong embedding $\varphi \in R \to F$ into a field there is a unique strong embedding σ such that the diagram in fig.* 8.12 *commutes.*

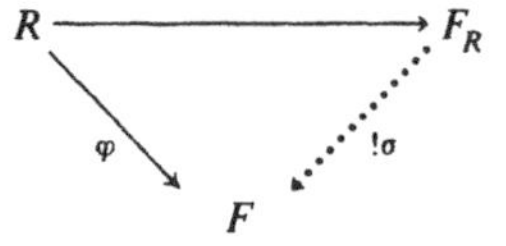

Fig. 8.12.

PROOF. Define the relation $\sim$ on $A = \{(a, b) : a \in R,\ b \# 0\}$ by

$$(a, b) \sim (a', b') := ab' = a'b.$$

It is a matter of routine to check that $\sim$ is an equivalence relation. Next we define the ring operations and $\#$ on $A/\sim$, writing $[t]$ for $[t]_\sim$:

$$[(a, b)] + [(c, d)] := [(ad + bc, bd)],$$
$$[(a, b)] - [(c, d)] := [(ad - bc, bd)],$$
$$[(a, b)] \cdot [(c, d)] := [(ac, bd)],$$
$$[(a, b)] \# [(a, d)] := ad \# bc.$$

Again one easily checks that the operations are well-defined, that $[(0, 1)]$ and $[(1, 1)]$ are the zero and the unit, and that the resulting structure is a field, F_R.

Following the tradition we write a/b for $[(a, b)]$. The inverse of a/b, for $a \# 0$, is b/a. The strong embedding $R \to F_R$ is given by $a \mapsto a/1$.

Now let $\varphi \in R \to F'$ be a strong embedding. Define $\sigma(a/b) = \varphi(a) \cdot \varphi(b)^{-1}$. Clearly σ is a strong embedding and the diagram commutes. If τ makes the diagram commute, then $\tau(a/1) = \varphi(a)$, $\tau(b/1) = \varphi(b)$ and hence $\tau(a/b) = \tau((a/1)\cdot(1/b)) = \tau((a/1)(b/1)^{-1}) = \tau(a/1)\cdot\tau(b/1)^{-1} = \varphi(a)\cdot\varphi(b)^{-1} = \sigma(a/b)$. Hence σ is unique. □

For localization in general see E 8.6.4.

3.13. *Rings with partial elements.* We will now reformulate the theory of rings in the language of logic with the existence predicate. For our formulation we choose E^+-logic.

The logical axioms and rules are taken from 2.2.3, the equality axioms from 2.2.4, and the description operator from 2.2.9. The axioms for apartness now are

$$t \# s \to s \# t,$$

$$Et \wedge Es \wedge \neg t \# s \leftrightarrow t = s,$$

$$t_1 \# t_2 \wedge Es \to t_1 \# s \vee t_2 \# s.$$

The apartness relation is strict, i.e.

$$t \# s \to Et \wedge Es.$$

The axioms for commutative rings with apartness are

$$t + 0 \simeq t, \qquad t \cdot 1 \simeq t,$$

$$t + s \simeq s + t, \qquad t \cdot s \simeq s \cdot t,$$

$$t + (s + r) \simeq (t + s) + r, \qquad t(sr) \simeq (ts)r,$$

$$Et \to (t + (-t) = 0), \qquad t(s + r) \simeq ts + tr,$$

$$0 \# 1.$$

Furthermore, we stipulate that the operations are strict and strongly extensional.

The axiom for an integral domain can be taken from 3.1. The field axiom allows an alternative formulation:

3.14. Proposition. *In a ring the following two statements are equivalent*:
(i) $t \# 0 \to \exists y(ty = 1)$,
(ii) $t \# 0 \to E(Iy.(xy = 1))$.
If we denote $Iy.(ty = 1)$ *by* t^{-1}, *then the latter statement reads* $t \# 0 \to Et^{-1}$.

PROOF. Left to the reader. □

Anti-ideals and ideals can be introduced as in 3.6.

4. Linear algebra

In this section we shall treat the traditional theory of vector spaces, i.e. modules over a field. We assume familiarity with the standard notions and techniques of linear algebra, such as matrix, vector, matrix multiplication, etc. We will adopt in the present section the convention to denote scalars by greek lower case letters. The contents of the section will be formulated in the ordinary language of logic, the role of E^+-logic will be restricted to a few occasional remarks.

Notions of dependence (4.1–2). The fundamental notion in constructive linear algebra is that of a positive analogue of independent, *free*.

4.1. DEFINITION. Let V be a vector space over a field F, $\{a_1, \ldots, a_n, b\} \subset V$, and write $\Sigma\alpha_i a_i$ for $\alpha_1 a_1 + \cdots + \alpha_n a_n$.
(i) $\{a_1, \ldots, a_n\}$ is *free* if $\forall\vec{\alpha} \# 0(\Sigma\alpha_i a_i \# 0)$.
(ii) $\{a_1, \ldots, a_n\}$ is *dependent* if $\exists\vec{\alpha} \# 0(\Sigma\alpha_i a_i = 0)$.
(iii) $\{a_1, \ldots, a_n\}$ is *weakly independent* if $\forall\vec{\alpha} \neq 0(\Sigma\alpha_i a_i \neq 0)$.
(iv) $\{a_1, \ldots, a_n\}$ is *independent* if $\forall\vec{\alpha}(\Sigma\alpha_i a_i = 0 \to \vec{\alpha} = 0)$.
(v) b is *free* from $\{a_1, \ldots, a_n\}$ if $\forall\vec{\alpha}(b \# \Sigma\alpha_i a_i)$.
(vi) b *depends on* $\{a_1, \ldots, a_n\}$ if $\exists\vec{\alpha}(b = \Sigma\alpha_i a_i)$.
(vii) b is *independent* from $\{a_1, \ldots, a_n\}$ if $\forall\vec{\alpha}(b \neq \Sigma\alpha_i a_i)$.
Here we have used 0 for the zero-vector. □

For convenience we will most of the time use phrases such as "the vectors $a_1, \ldots, a_n$ are free" instead of "$\{a_1, \ldots, a_n\}$ is free", etc.

The notions introduced in definition 4.1 relate in various ways.

4.2. PROPOSITION.
(i) $\{a_1, \ldots, a_n\}$ *is free* $\Rightarrow$ $\{a_1, \ldots, a_n\}$ *is independent*;
(ii) $\{a_1, \ldots, a_n\}$ *is independent* $\Rightarrow$ $\{a_1, \ldots, a_n\}$ *is weakly independent*;
(iii) b *is free from* $\{a_1, \ldots, a_n\}$ $\Rightarrow$ b *is independent of* $\{a_1, \ldots, a_n\}$;
(iv) a_0 *is free from* $\{a_1, \ldots, a_n\}$ *and* $\{a_1, \ldots, a_n\}$ *free* $\Leftrightarrow$ $\{a_0, \ldots, a_n\}$ *free*.

PROOF. (i), (ii), (iii) are obvious. The only case to be considered is the implication from left to right in (iv). So let a_0 be free from $a_1, \ldots, a_n$, and let $\vec{\alpha} \# 0$. Write $\Sigma\alpha_i a_i$ for $\alpha_1 a_1 + \cdots + \alpha_n a_n$, $\Sigma'\alpha_i a_i$ for $\alpha_0 a_0 + \cdots + \alpha_n a_n$. We must show $\Sigma'\alpha_i a_i \# 0$. Now $\alpha_0 \# 0 \vee \mathsf{V}\{\alpha_i : i > 0\} \# 0$. If $\alpha_0 \# 0$, we consider $\alpha_0^{-1}(\Sigma'\alpha_i a_i) = a_0 + \Sigma'\alpha_0^{-1}\alpha_i a_i$. But this is apart from 0, since a_0 is free from $a_1, \ldots, a_n$. Hence $\Sigma'\alpha_i a_i \# 0$.

If $\mathsf{V}\{\alpha_i : i < 0\} \# 0$, then $\Sigma\alpha_i a_i \# 0$ (since $a_1, \ldots, a_n$ are free). So $\Sigma'\alpha_i a_i \# 0$ or $\Sigma'\alpha_i a_i \# \Sigma\alpha_i a_i$. The second disjunct implies $\alpha_0 \# 0$ and brings us back to the case above. □

One easily sees that the set of (finite) linear combinations of vectors from a given set A is a vector space, called the subspace generated by A.

We say that two sets of vectors are equivalent if they generate the same subspace.

Basis and dimension (4.3–7). We will now proceed to introduce the basic notions of vector spaces, *basis* and *dimension*. The basic theorem here is the *Austauschsatz* ("exchange theorem").

4.3. Lemma. *If $b \# 0$ and b depends on $a_1, \ldots, a_n$, then there exists an i such that $a_1, \ldots, a_{i-1}, b, a_{i+1}, \ldots, a_n$ and $a_1, \ldots, a_n$ are equivalent.*

Proof. Since $b \# 0$ and $b = \Sigma\alpha_i a_i$, there is an i such that $\alpha_i \# 0$ (+ is strongly extensional). Now one sees that a_i depends on $a_1, \ldots, a_{i-1}, b, a_{i+1}, \ldots, a_n$. From this one easily concludes the required equivalence. □

4.4. Theorem. ("*Austauschsatz*"). *If $x_1, \ldots, x_m$ are free and $x_1, \ldots, x_m$ depend on $y_1, \ldots, y_n$, then $m \leq n$ and there is a subset $z_1, \ldots, z_{n-m}$ of $y_1, \ldots, y_n$ such that $x_1, \ldots, x_m, z_1, \ldots, z_{n-m}$ is equivalent with $y_1, \ldots, y_n$.*

Proof. We apply the lemma above to the vectors $x_1, \ldots, x_m$, one by one. The details are left to the reader. □

Corollary. *If in the Austauschsatz $m = n$, then $x_1, \ldots, x_n$ is equivalent to $y_1, \ldots, y_n$ and $y_1, \ldots, y_n$ are free.* □

The corollary tells us that if a vector space is generated by n free vectors, then all finite free generating sets have n elements. Let us call an arbitrary set of vectors *free* if all its finite subsets are free.

4.5. DEFINITIONS.

(i) If a vector space V is generated by a set B of free vectors, then B is called a *basis* of V.

(ii) If V has a basis of n elements, then V has *dimension* n. By the above the dimension of V, $\dim(V)$, is well-defined. □

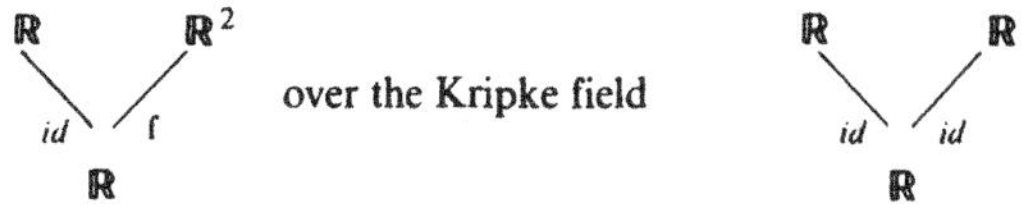

Fig. 8.13.

Observe that not all vector spaces have a dimension, e.g. $\{a \in \mathbb{R} : A \vee a = 0\}$, where A is an undecided statement, is a vector space to which we cannot assign a dimension. We can also present an example by means of a Kripke model (with restrictions) of a vector space as shown in fig. 8.13. Here $f(\alpha) := (\alpha, 0)$. This model does not force $\exists x_1 x_2(\{x_1, x_2\}$ is a basis), nor $\exists x(\{x\}$ is a basis). It does force "there are at most two free generators".

Observe that there is a difference between "minimal set of generators" and "basis". E.g. in the vector space model in fig. 8.14, where $f(\alpha, \beta) := \alpha$, the vectors $(0, 1)$, $(1, 0)$ form a minimal set of generators, but they are not free. They are, however, independent.

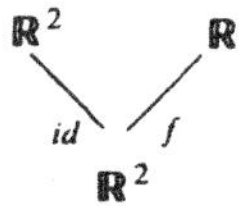

Fig. 8.14.

If a vector space V and a subspace W have bases, then we have the traditional result $\dim(V/W) = \dim V - \dim W$. In general, however, there is not much of the classical dimension theory that remains true.

For convenience we introduce a few more notions.

4.6. DEFINITION. Let $a_1, \ldots, a_n$ be vectors in V.

(i) V has *dimension at most* n ($\dim(V) \leq n$) if $a_1, \ldots, a_n$ generate V.

(ii) V has *dimension at least* n ($\dim(V) \geq n$) if $a_1, \ldots, a_n$ are free.

(iii) V has the *weak dimension* n ($\dim_w(V) = n$) if $a_1, \ldots, a_n$ are independent and generate V. □

The term "dimension" is only used as a "façon de parler" in the above definition. V does not necessarily *have* a dimension.

Next we consider the decomposition of a vector space into the direct sum of two subspaces.

4.7. DEFINITION.
(i) $W = U \oplus V$ if $\forall w \in W \exists u \in U \exists v \in V(w = u + v) \wedge \forall u \in U \forall v \in V(u \# 0 \vee v \# 0 \rightarrow u + v \# 0)$.
(ii) $W = U \oplus_{\mathrm{w}} V$ if $\forall w \in W \exists u \in U \exists v \in V(w = u + v) \wedge \forall u \in U \forall v \in V(u + v = 0 \rightarrow u = 0 \wedge v = 0)$. □

From the definition of (weak) direct sum we immediately conclude

$$W = U \oplus V \Rightarrow W = U \oplus_{\mathrm{w}} V,$$

$$W = U \oplus V \Leftrightarrow \forall u \in U \forall v \in V(u \# 0 \vee v \# 0 \rightarrow u \# v).$$

Rank of a matrix (4.8–14)

4.8. For practical purposes the vector spaces F^n over a field F play an important role. In order to study them, and their linear mappings, we will have to investigate the properties of linear equations. All notions and conventions from traditional linear algebra will be adopted. In particular, a linear mapping $\varphi \in F^n \rightarrow F^m$ will be represented by a matrix $A = (\alpha_{ij})_{1 \leq i \leq m, 1 \leq j \leq n}$ such that $\varphi(\beta_1, \ldots, \beta_n) = (\Sigma_j \alpha_{1j}\beta_j, \ldots, \Sigma_j \alpha_{mj}\beta_j)$ (Σ_j abbreviates in an obvious way summation over j). In matrix notation: $A\vec{b}^{\mathrm{T}}$.

A linear mapping φ into F is called a *linear form*, and φ is represented by a vector $\vec{a}$ such that $\varphi(\vec{b}) = \vec{a}\vec{b}^{\mathrm{T}}$.

The linear forms form the dual vector space of F^n with the canonical apartness relation, and $\varphi \# \psi \leftrightarrow \vec{a} \# \vec{a}'$, where $\vec{a}$ and $\vec{a}'$ are the representing vectors of φ and ψ.

DEFINITION. The *row* (*column*) *rank* of a matrix A is the dimension of the subspaces of F^n (F^m) generated by the row (column) vectors of A. Notation: $\mathrm{rank}_{\mathrm{r}}(A)$, $\mathrm{rank}_{\mathrm{c}}(A)$ stand for the row rank, column rank of A. When the subscript is irrelevant we will just write $\mathrm{rank}(A)$. □

What we said about the dimension of vector spaces holds automatically for the ranks of matrices; in particular, not every matrix has a row (or column) rank.

We can draw a number of straightforward conclusions from the definition.

4.9. PROPOSITION. *Let* $A = (\alpha_{ij})_{1 \le i \le m, 1 \le j \le n}$ *be a matrix.*

(i) rank $A = 0 \Rightarrow \forall ij(\alpha_{ij} = 0)$.

(ii) rank $A > 0 \Rightarrow \exists ij(\alpha_{ij} \# 0)$.

(iii) *If A' is obtained from A by a permutation of rows (columns), then* $\mathrm{rank}_r(A) = \mathrm{rank}_r(A')$, $(\mathrm{rank}_c(A) = \mathrm{rank}_c(A'))$

(iv) *If A' is obtained from A by a multiplication with $\alpha \# 0$, then* $\mathrm{rank}(A) = \mathrm{rank}(A')$.

PROOF. Routine. □

4.10. THEOREM. *If the row or column rank of A exists, then* $\mathrm{rank}_r(A) = \mathrm{rank}_c(A)$.

PROOF. Induction on $\max(m, n)$ of the matrix A. For $\max(m, n) = 1$ the theorem is trivial. So let the theorem hold for $\max(m, n) = k$, and let $\max(m, n) = k + 1$ for A. The case $\mathrm{rank}_c(A) = 0$ is obvious; consider therefore $\mathrm{rank}_c(A) > 0$, then $\alpha_{ij} \# 0$, some i, j. It is no restriction to suppose $i = j = 1$. Now we sweep the first row and column, this yields a matrix A' with the same ranks:

$$A' = \left(\begin{array}{c|ccc} \alpha & 0 & \cdots & 0 \\ \hline 0 & & & \\ \vdots & & B & \\ 0 & & & \end{array}\right)$$

One immediately sees that B has ranks one less than A. Apply the induction hypothesis to B: $\mathrm{rank}_c B = \mathrm{rank}_r B$. But $\mathrm{rank}_c A = \mathrm{rank}_c A' = \mathrm{rank}_c B + 1 = \mathrm{rank}_r B + 1 = \mathrm{rank}_r A' = \mathrm{rank}_r A$. □

From now on we may, by the theorem, use the *rank* of the matrix, which can be taken to be the row or column rank ad libitum.

Determinants (4.11–15)

4.11. We will now introduce determinants. In classical mathematics the use of determinants can largely be replaced by suitable dimension arguments (as long as one is interested, say, in the *existence* of solutions rather than their particular form). We can, however, use dimension arguments only under special circumstances. Moreover, the use of determinants may reduce the logical complexity of certain problems. E.g. "A is invertible", which is of the form $\exists B(AB = I)$, reduces to a quantifier free statement: $\det A \# 0$.

For our purpose the elegant definitions of determinant, such as that of Weierstrass: "$\det A$ is an alternating, multilinear mapping of the row vectors with $\det I = 1$", are not suitable since the accompanying existence proof invariably uses "sweeping techniques" which assume PEM. Hence, we stick to the old explicit definition.

DEFINITION. "det" is a function from square matrices to F given by

$$\det A = \sum_{\sigma} \operatorname{sign}(\sigma) \alpha_{1,\sigma(1)} \alpha_{2,\sigma(2)} \cdots \alpha_{n,\sigma(n)},$$

where σ ranges over all permutations of $\{1, \ldots, n\}$ and $\operatorname{sign}(\sigma)$ is the sign of σ. □

The following facts follow immediately from the definition.

4.12. LEMMA. *Let A be a matrix as before. Then*:
(i) $\det A$ *is linear in the rows and columns.*
(ii) $\det A$ *is alternating in its rows and columns.*
(iii) $\det I = 1$.
(iv) $\det A = \Sigma_i (-1)^{i+j} \alpha_{ij} \det A_{ij}$ *for a fixed j (and for a fixed i, $\Sigma_j (-1)^{i+j} \alpha_{ij} \det A_{ij}$), where A_{ij} is the submatrix with entries $\alpha_{p,q}$, $p \neq i$, $q \neq j$.*
(v) *If A contains two identical rows (columns), then* $\det A = 0$.
(vi) $\det A = \det A^{\mathrm{T}}$. □

Cramer's rule explicitly expresses solutions of equation in terms of determinants.

4.13. PROPOSITION. *If* $\det A \# 0$, *then there exists for each $\vec{b}$ a unique $\vec{x}$ such that $A\vec{x}^{\mathrm{T}} = \vec{b}^{\mathrm{T}}$ and*

$$x_i = (\det A)^{-1} \det A(i) \quad (1 \leq i \leq n),$$

where $A(i)$ is obtained from A by replacing the ith column by $\vec{b}^{\mathrm{T}}$.

PROOF. Routine. □

From Cramer's rule it follows that if det $A \# 0$, the linear transformation φ represented by A has an inverse φ^{-1} with a matrix A^{-1} (a routine calculation). Hence φ^{-1} is also linear. Since φ^{-1} is represented by a matrix it is obviously strongly extensional (cf. E8.4.3). Since both φ and φ^{-1} are strongly extensional, φ is a strong isomorphism.

The proof of the product theorem for determinants, $\det(A) \cdot \det(B) = \det(AB)$ presents no special intuitionistic difficulties, the reader may consult a suitable text on linear algebra (cf. Lang 1965, ch. XIII, section 4).

We can now collect our results.

4.14. THEOREM. *The following are equivalent*:
(i) det $A \# 0$,
(ii) *A is invertible (i.e. there is a (unique) B with $AB = BA = I$),*
(iii) $\vec{x} \# 0 \Rightarrow A\vec{x} \# 0$,
(iv) *A is (represents) a strong isomorphism,*
(v) *the antikernel of A is $\{\vec{x} : \vec{x} \# 0\}$.*

PROOF. Left to the reader. □

The following proposition provides an alternative characterization of the rank of a matrix.

4.15. PROPOSITION. $\operatorname{rank}(A) = k \Leftrightarrow$ *A has a maximal submatrix with determinant $\# 0$ of size $k \times k$.*

PROOF. We first show that a matrix A with a rank, that has an $m \times m$ submatrix B with det $B \# 0$, has rank $\geq m$.

Consider the submatrix B and the row vectors it determines in A, $\vec{a}_1, \ldots, \vec{a}_m$, say. If $\vec{x} \# 0$, $\vec{x} \in F^k$, then $B\vec{x} \# 0$, since $\vec{x}$ belongs to the antikernel of B (by the above theorem (v)), so the restrictions of $\vec{a}_1, \ldots, \vec{a}_n$ to B are free, and therefore they are free themselves. Now, by the Austauschsatz, rank $A \geq m$. It is a routine matter to show $\operatorname{rank}(A) = k \Rightarrow$ the determinants of all $m \times m$ submatrices with $m > k$ are zero and at least one $k \times k$ submatrix has determinant $\# 0$. The proposition now follows. □

Equations and solution spaces (4.16–18). The theory of linear equations for systems with (a matrix with) rank is similar to its classical version, we list some properties.

4.16. PROPOSITION. *Let A be an $n \times m$ matrix*

$$\begin{pmatrix} \alpha_{11} & \cdots & \alpha_{1m} \\ \vdots & & \vdots \\ \alpha_{n1} & \cdots & \alpha_{nm} \end{pmatrix};$$

(i) *If* rank $A = k$, *then the solution space of $A\vec{x}^{\mathrm{T}} = 0$ has dimension $n - k$.*
(ii) *If* rank A = rank$(A, \vec{b}^{\mathrm{T}}) = k$, *then the solutions of $A\vec{x}^{\mathrm{T}} = \vec{b}^{\mathrm{T}}$ form a coset modulo an $(n-k)$-dimensional subspace.*
(iii) rank $A = k \Leftrightarrow$ *there is a maximal submatrix B of size $k \times k$ with* det $B \# 0$, *and the column vectors "through" B generate the range of A in F^m. The unit vectors $\vec{e}_i$ (with entries $1 \dot{-} |i - j|$, $j = 1, \ldots, n$) corresponding to the row indices of B, generate a subspace V of F^n on which A acts as a strong isomorphism.*
(iv) *Let W be the kernel of A and V as in* (iii), *then $F^n = V \oplus W$ (cf.* 4.7).
(v) *The antikernel of A consists of sums of vectors from the kernel and vectors apart from $\vec{0}$ in V.*

PROOF. We only consider (iv). Let $\vec{x} \in W$ and $\vec{y} \in V$. If $\vec{y} \# 0$, then $A\vec{y}^{\mathrm{T}} \# 0 = A(-\vec{x}^{\mathrm{T}})$, so $\vec{x} + \vec{y} \# 0$. If $\vec{x} \# 0$, then $\vec{x} \# -y$ or $\vec{y} \# 0$. In the latter case we again get $\vec{x} + \vec{y} \# 0$, so we are done. Now we pick bases of V and W (which exist!); from the preceding lines we immediately conclude that they form together a basis of F^n. Hence $F^n = V \oplus W$. □

If we drop the existence of the rank of A, there is little we can say in general about equations, kernels, etc.

4.17. EXAMPLE. Consider a field in which $ab = 0$, but neither $a = 0$ or $b = 0$ is known (we have seen an example in $\mathbb{R}$, more examples will follow later). Let

$$A = \begin{pmatrix} a & 0 \\ 0 & b \end{pmatrix}$$

Clearly det $A = 0$, but we cannot indicate an element of the antikernel. For, $A\vec{x}^{\mathrm{T}} = ax_1 + bx_2 \# 0 \to a \# 0 \vee b \# 0$. Neither can we indicate an element of the kernel which is apart from 0: $A\vec{x}^{\mathrm{T}} = 0 = ax_1 + bx_2$. Suppose $(x_1, x_2) \# 0$, then $a = 0 \vee b = 0$. Observe that A does not have a rank. However,

$$B = \begin{pmatrix} 1 & 0 \\ 0 & a \end{pmatrix}$$

does not have a rank either, but we can at least say that the kernel contains the subspace generated by (0, 3), and the antikernel contains at least the vectors (x_1, x_2) with $x_1 \# 0$. In general we have the following result.

4.18. PROPOSITION. *If A contains a $k \times k$ submatrix B with* det $B \# 0$, *then there is a k-dimensional subspace V of F^n such that the maximal antisubspace of V (i.e. $\vec{x} \in V: x \# 0\}$) is contained in the antikernel of A.*

PROOF. Obvious. □

As the above example shows, there is no immediate dual to this proposition, i.e. the fact that all $(k+1)$-sized submatrices have determinant 0 does not tell us much about the dimension of the solution space.

5. Polynomial rings

5.1. The theory of polynomial rings distinguishes itself from its classical counterpart mainly by the absence of a good notion of degree of a polynomial. E.g., consider the polynomial $aX^3 + bX^2 + 2X + 5$ over $\mathbb{R}$, where it is unknown if $a = 0$, $b = 0$; we cannot assign a unique degree to it, all we know is that "the degree is at least one and at most three".

DEFINITION.
(i) The ring of (*formal*) *power-series* $R[\![X]\!]$ over the ring R is the set $R^{\mathbb{N}}$ with the operations $+, -, \cdot$ given by

$$(f+g)(n) := f(n) + g(n),$$

$$(f-g)(n) := f(n) - g(n),$$

$$(f \cdot g)(n) := \sum_{i=0}^{n} f(i)g(n-i).$$

The zero $\mathbf{0}$ and unit $\mathbf{1}$ are given by $\mathbf{0}(n) := 0$, $\mathbf{1}(n) := 1 \dot{-} n$. The apartness relation on $R[\![X]\!]$ is the canonical apartness relation induced by R:

$$f \# g := \exists n (f(n) \# g(n)).$$

(ii) The ring of *polynomials* $R[X]$ is the subset of $R[\![X]\!]$ defined by $\exists n \forall m f(n+m) = 0$.

We shall adopt the traditional notation for power series and write

$$\sum_{k=0}^{\infty} a_k X^k \quad \text{for } f \in R[\![X]\!] \quad \text{with } a_k = f(k), \quad \text{and}$$

$$\sum_{k=0}^{n} a_k X^k \quad \text{for } f \in R[X], \quad \text{where } f(k) = 0 \quad \text{for } k > n. \quad \square$$

The reader can easily show that $R[\![X]\!]$ and $R[X]$ are rings. Note that in addition to the usual equations, one has to check extra axioms concerning apartness.

X is called the "unknown" and f a "polynomial in the unknown X". The notion can be extended to polynomial rings in finitely many unknowns:

$$R[X_1, \ldots, X_{n+1}] := (R[X_1, \ldots, X_n])[X_{n+1}].$$

Since a permutation of the unknowns yields an isomorphism of the polynomial rings we will disregard the order of the unknowns.

There is a canonical embedding of R into $R[X]$, given by $a \mapsto f$ with $f(0) = a$, $f(k) = 0$, $k > 0$. In practice we will consider R as a subring of $R[X]$. If we write polynomials in the standard form, X appears as an element of $R[X]$, and we will henceforth treat it as such.

5.1. Proposition. *If $\varphi : R_1 \to R_2$ and $a \in R_2$, then there is a unique σ such that $\sigma(X) = a$ and the diagram in fig.* 8.15 *commutes.*

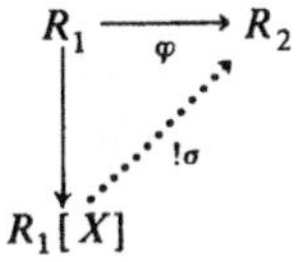

Fig. 8.15.

Proof. Obvious. $\square$

5.2. Definition. Let $R_1 \subset R_2$ and $a \in R_2$, then we denote the subring of R_2 generated by a over R_1 by $R_1[a]$. Let σ be the homomorphism from the above proposition, associated to $R_1 \hookrightarrow R_2$. If σ is an isomorphism onto $R[a]$, then a is called *transcendental* over R_1. If $\sigma(f) = 0$ for some $f \# 0$, then a is called *algebraic* over R_1. $\square$

Observe that $\sigma : R_1[X] \to R_1[a]$ and that $R_1[a] \cong R_1[X]/I$, where I is the kernel of σ. Now a is transcendental over R_1 iff $I = (0)$ iff for all polynomials $b_nX^n + \cdots + b_0$, $b_na^n + \cdots + b_0 = 0 \to b_0 = \cdots = b_n = 0$. And a is algebraic over R_1 iff I is inhabited iff $b_na^n + \cdots + b_0 = 0$ for some polynomial with $\bigvee\{b_i \# 0 : i \le n\}$.

Polynomials over an integral domain (5.4–5)

5.4. THEOREM. *R is an integral domain $\Rightarrow$ $R[X]$ is an integral domain.*

The proof is based on the following lemma.

5.5. LEMMA. *Let R be an integral domain and $f = a_nX^n + \cdots + a_0$, $g = b_mX^m + \cdots + b_0$, $fg = c_{n+m}X^{n+m} + \cdots + c_0$, then*

$$a_ib_j \# 0 \to \exists k\big(i + j \le k \le n + m \wedge c_k \# 0\big). \tag{1}$$

PROOF. Induction on $n + m - (i + j)$. For $i + j = 0$ (1) is trivial. Now let $a_ib_j \# 0$, then $c_{i+j} - \Sigma\{a_kb_{k'} : k + k' = i + j \wedge k \ne i\} \# 0$, so $c_{i+j} \# 0 \vee \bigvee\{a_kb_{k'} \# 0 : k + k' = i + j,\ k \ne i\}$. If $c_{i+j} \# 0$, we are done. The remaining disjunction implies, with $xy \# 0 \to x \# 0 \wedge y \# 0$, the following:

$$a_{i+j} \# 0 \vee a_{i+j-1} \# 0 \vee \cdots \vee a_{i+1} \# 0 \vee b_{j+1} \# 0 \vee \cdots \vee b_{i+j} \# 0.$$

Using $a_i \# 0 \wedge b_j \# 0$ we get in the integral domain R

$$a_{i+j}b_j \# 0 \vee \cdots \vee a_{i+1}b_j \# 0 \vee a_ib_{j+1} \# 0 \vee \cdots \vee a_ib_{i+j} \# 0.$$

By the induction hypothesis this yields

$$\bigvee\big\{c_k \# 0 : i + j + 1 \le k \le n + m\big\}. \ \square$$

PROOF *of theorem 5.4.* $f \# 0 \wedge g \# 0 \to a_ib_j \# 0$ for some i, j, hence $c_k \# 0$ for some k, and thus $fg \# 0$. $\square$

It now immediately follows that $R[X_1, \ldots, X_n]$ is an integral domain if R is so. The lemma also serves to prove that the invertible elements $R[X]$ for a domain R are exactly the invertible elements of R itself. Suppose $f \cdot g = 1$, then (in the notation of the lemma) $c_i = 0$ for $i > 0$. And hence $a_ib_j = 0$ for all $(i, j) \ne (0, 0)$. Since $a_0b_0 = 1$, we get $b_j = 0$ $(j > 0)$ from $a_ib_j = 0$. Similarly $a_i = 0$ $(i > 0)$.

The remainder theorem (5.6–9)

5.6. DEFINITION. The polynomial $a_nX^n + \cdots + a_0$ is *regular* if $a_n \# 0$. $\square$

Regular polynomials are easier to deal with than polynomials in general; they form, however, a too restricted class. Moreover, even quite simple operations, such as addition, lead outside the class of regular polynomials.

The following lemma is a first step towards the euclidean algorithm.

5.7. LEMMA. *Let* $f(X) = a_n X^n + \cdots + a_0$, $g(X) = b_m X^m + \cdots + b_0$, *then there exist* $k \in \mathbb{N}$, $q, r \in R[X]$ *such that* $(b_m)^k f(X) = q(X)g(X) + r(X)$, *where* $r(X) = c_{m-1} X^{m-1} + \cdots + c_0$.

PROOF. The case $m > n$ is trivial, so consider $m \leq n$. Use induction on $n - m$. We have

$$b_m f(X) - a_n X^{n-m} g(X) = d_{n-1} X^{n-1} + \cdots + d_0 = f_1(X).$$

If $n - 1 = m$ we are done. If not, apply induction to f_1 and g:

$$b_m^{k'} f_1(X) = q'(X)g(X) + r'(X),$$

and hence

$$b_m^{k'+1} f(X) = q'(X)g(X) + r'(X) + a_n b_m^{k'} X^{n-m} g(X).$$

So

$$b_m^{k'+1} f(X) = \left(a_n b_m^{k'} X^{n-m} + q'(X)\right) g(X) + r'(X). \quad \square$$

Observe that we have not said anything about the "degree" of $r(X)$, or about its leading coefficient. We simply do not know! If b_m is invertible, as it is in the case of a regular g over a field, then we can even write

$$f(X) = q(X)g(X) + r(X).$$

Then also q and r are unique (E8.5.2). As a special case we get the following theorem.

5.8. THEOREM (*remainder theorem*). *If* $f(X) \in R[X]$ *and* $a \in R$, *then*

$$\exists! q(X) \in R[X](f(X) = (X - a)q(X) + f(a)),$$

and its corollary

$$(X - a) | f(X) \Leftrightarrow f(a) = 0.$$

This in turn implies that $a_n X^n + \cdots + a_0$ *has at most* n *zero's, or, if* f *has*

$n+1$ zero's (multiple zero's allowed), then $f = 0$ (i.e. $a_0 = a_1 = \cdots = a_n = 0$). □

In classical mathematics it suffices to study the vanishing of a polynomial; the remainder theorem tells us that if there are too many zeros all coefficients disappear. In constructive mathematics the "dual" problem is of importance: when has a polynomial values apart from zero? We have already considered the problem for $\mathbb{R}[X]$ in 6.2.6, 6.2.7; the general case is treated below.

5.9. PROPOSITION. *Let $f = a_n X^n + \cdots + a_n \in F[X]$, F a field.*
(i) $f \# f(0) \Rightarrow \exists a \in F(f(a) \# 0)$,
(ii) $a_k \# 0 \Rightarrow \exists a \in F(f(a) \# 0 \wedge f'(a) \# 0 \wedge \cdots \wedge f^{(k)}(a) \# 0)$,
where F is assumed to have enough elements and $i \neq 0$, $a \# 0 \Rightarrow ia \# 0$.

PROOF. (i) From $f \# f(0)$ it follows that some coefficient other than a_0 is apart from zero, say $a_k \# 0$. Select $n+1$ distinct elements $u_0, \ldots, u_n$ (i.e. $u_i \# u_j$, $i \neq j$) and consider the system of equations

$$\begin{array}{ccc} a_0 + a_1 u_0 + \cdots + a_n u_0^n & = & f(u_0), \\ \vdots & & \vdots \\ a_0 + a_1 u_n + \cdots + a_n u_n^n & = & f(u_n). \end{array}$$

The determinant of the matrix is

$$\prod_{i>j} (u_i - u_j) \quad (\textit{Vandermonde}),$$

so we can solve a_k: $a_k = c_0 f(u_i) + \cdots + c_n f(u_n)$. Now we conclude $\exists i (f(u_i) \# 0)$ from $a_k \# 0$.

(ii) Determine n points $u_0, \ldots, u_{n-1}$ such that $f(u_i) \# 0$ and repeat the argument for $f' = na_n X^{n-1} + \cdots + ka_k X^{k-1} + \cdots + a_1$. The case of the higher derivatives is similar. □

We will now turn to factorization and congruence of polynomials. When not specified otherwise, we are dealing with a field F.

Factorization and congruence of polynomials (5.10–20)

5.10. DEFINITION. Let $f = a_n X^n + \cdots + a_0$.
(i) *f has degree n* if $a_n \# 0$ (notation $\deg f = n$),
(ii) *f has degree at least m* if $a_m \# 0$ (notation $\deg f \geq m$),

(iii) *f has degree at most m* if $a_n = a_{n+1} = \cdots = a_{m+1} = 0$ (notation $\deg f \le n$). □

Recall that f is *regular* if $\deg f = n$, and observe that for an integral domain D we have $\forall f, g \in D[X](fg$ is regular $\leftrightarrow f$ and g are regular). Moreover, if fg is regular, then $\deg(fg) = \deg f + \deg g$.

We shall now investigate divisibility in the general case, i.e. where polynomials are not necessarily regular. So the problem is: given f and g, is there an h such that $g = fh$? Or dually, is $g \# fh$ for all h?

The problem requires a certain amount of preparatory notions and lemmas.

5.11. LEMMA. *Let $f(X) = a_nX^n + \cdots + a_0 \in F[X]$, and $a_r \# 0$ for some $r \le n$, and let A be the $m' \times m$ matrix*

$$A = \begin{pmatrix} a_0 & 0 & \cdots & 0 \\ a_1 & a_0 & \cdots & \vdots \\ \vdots & & & 0 \\ \vdots & & \cdots & a_0 \\ a_n & & \cdots & \vdots \\ 0 & & \cdots & \vdots \\ \vdots & & & \vdots \\ 0 & & \cdots & a_n \end{pmatrix}$$

Then there is an $s \le r$ (respectively $s \ge r$) and an invertible maximal submatrix

$$B = \begin{pmatrix} a_s & & \\ & \ddots & \\ & & a_s \end{pmatrix}$$

of A with $a_s \# 0$.

PROOF. Induction on r. The case $r = 0$ is trivial. So assume the lemma for all $r' < r$, and consider the submatrix C with main diagonal a_r. Now $a_r \# 0$, so $a_r^m \# 0$. Hence $a_r^m \# \det C \vee \det C \# 0$. If $\det C \# 0$, we are done. Consider therefore $a_r^m \# \det C$. Expand the determinant according to definition of 4.11, then one summand is a_r^m, hence one of the remaining summands must be apart from 0. Therefore at least one a_t, $t < r$, is apart from 0. So we may apply the induction hypothesis. Exactly the same argument works for $s \ge r$. □

5.12. *The anti-ideal of a polynomial.* It is clear that for any polynomial $f \in R[X]$ the set (f), i.e. $\{gf : g \in R[X]\}$ is an ideal, the so-called principal ideal generated by f. We now ask ourselves how and when f determines an anti-ideal.

DEFINITION. $A_f := \{g \in F[X] : \forall h \in F[X](g \# fh)\}$. □

EXAMPLE. A_{X^2} is the set of polynomials $a_n X^n + \cdots + a_1 X + a_0$ with $a_1 \# 0 \vee a_0 \# 0$.

Evidently, if $f = 0$, then A_f is an anti-ideal (the maximal one). Before we establish the result for $f \# 0$, we note that one does not have to test all polynomials h by the following lemma.

5.13. LEMMA.
(i) $g \in A_f \Leftrightarrow$ *for all* h *of length at most that of* g, $g \# hf$,
(ii) $g \in (f) \Leftrightarrow$ *there is an* h *of length at most that of* g *such that* $g = hf$.

PROOF. (i) We only have to show $\Leftarrow$. Let $h \in F[X]$, then we can write $h = h_1 + h_2$, where $h_1(X) = c_p X^p + \cdots + c_0$, $h_2 = c_m X^m + \cdots + c_{p+1} X^{p+1}$ and $g(X) = b_p X^p + \cdots + b_0$. By assumption $g \# h_1 f$, so $g \# hf \vee h_1 f \# hf$. In the first case we are done, in the second case $h_2 f \# 0$, so the degree of hf is at least $p + 1$ and thus $g \# hf$.
(ii) Consider $\Rightarrow$. Let $g = hf$; split h as in part (i), assume $g \# hf$ and derive a contradiction. □

We now return to the question: is A_f an anti-ideal if $f \# 0$? Let us first introduce some more notations.

5.14. DEFINITION. Let $f := a_n X^n + \cdots + a_0$, $h := c_m X^m + \cdots + c_0$. The coefficients of hf are given by the following product:

$$\begin{pmatrix} a_0 & & \\ \vdots & \ddots & \\ \boxed{\begin{matrix} a_s & \ddots & \\ \vdots & & a_0 \\ & B & \vdots \\ a_n & & \\ & \ddots & a_s \end{matrix}} \\ & \ddots & \vdots \\ & & a_n \end{pmatrix} \begin{pmatrix} c_0 \\ \vdots \\ c_m \end{pmatrix} (= A\vec{c}^{\mathrm{T}}).$$

B is an invertible $(m+1) \times (m+1)$ submatrix of A, whose existence is guaranteed by lemma 5.11. For $g = b_m X^m + \cdots + b_0$ we define

$$g^* := B^{-1} \begin{pmatrix} b_s \\ \vdots \\ b_m \\ 0 \\ \vdots \\ 0 \end{pmatrix}$$

Observe that g^* depends on f and on the choice of B. $*$ is a linear mapping that provides a *test for membership of* A_f as follows.

5.15. PROPOSITION. *Let f and g be as given above, then*

$$g \in A_f \Leftrightarrow g \# fg^*.$$

PROOF. It suffices to show $\Leftarrow$. Let $g \# fg^*$. We want to compare f and fh, where we may suppose h to be of the form $c_m X^m + \cdots + c_0$. From $g \# fg^*$ it follows that $g \# fh \vee fh \# fg^*$. In the first case we are done, in the second case we conclude $h \# g^*$, and hence, by strong extensionality of B^{-1}, $Bh \# Bg^*$.

We consider the product fh in matrix form, Ah:

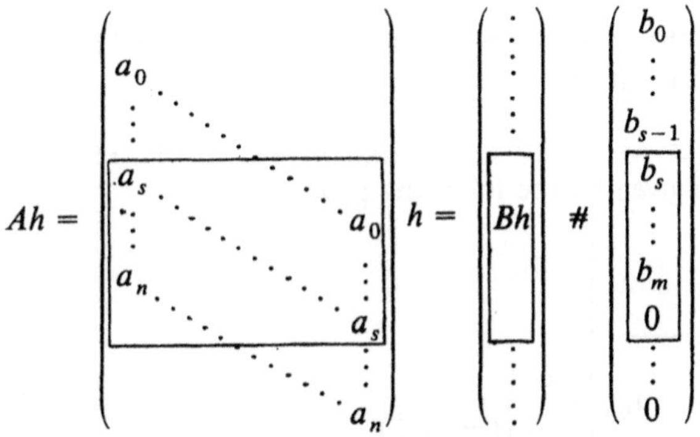

since the vectors in the boxes, i.e. Bh and Bg^*, are apart. This shows $g \# fh$ for all h of degree at most that of g. Hence $g \in A_f$. $\square$

We can now show that A_f is an anti-ideal if $f \# 0$.

5.16. THEOREM. *If $f \# 0$, then A_f is an anti-ideal compatible with $\#$. If, moreover, $f \# f(0)$, then A_f is proper.*

PROOF. (i) Clearly $0 \notin A_f$.

(ii) If $g_1 g_2 \in A_f$, then for arbitrary h $g_1 g_2 \# h g_2 f$. Hence $g_1 \# hf$. So $g_1 \in A_f$. Similarly $g_2 \in A_f$.

(iii) If $g_1 + g_2 \in A_f$, then by the above lemma $g_1 + g_2 \# (g_1 + g_2)^* f = g_1^* f + g_2^* f$ (*is linear). So $g_1 \# g_1^* f \vee g_2 \# g_2^* f$, thus $g_1 \in A_f \vee g_2 \in A_f$.

(iv) We now show that A_f is compatible with $\#$. Let $g \in A_f$, then $g \# f \cdot 0 = 0$.

(v) If $f \# f(0)$, then $a_r \# 0$ for some $r > 0$. So we can determine by 5.14 the submatrix B with $s \leq r$, such that $1^* \# 0$ and hence $1 \# 0 = f \cdot 1^*$. So $1 \in A_f$, i.e. A_f is proper. □

Another conclusion from the criterion is the following proposition.

5.17. PROPOSITION. *If $f \# 0$, then $A_f^c = (f)$.*

PROOF. $g \notin A_f \Leftrightarrow \neg g \# g^* f \Leftrightarrow g = g^* f \Rightarrow g \in (f)$.
Conversely, $\exists h(g = hf) \Rightarrow \neg g \in A_f$. □

5.18. *The factor ring over a polynomial.* Having investigated A_f for $f \# 0$, we are interested in the quotient structure $F[X]/(f)$. This is a ring (with apartness relation) if A_f is proper (or inhabited), and this is guaranteed by $f \# f(0)$ (5.16).

We will adopt the following notation for the factor ring:

$$F[\alpha] := F[X]/(f) \quad \text{with } \alpha := X + (f).$$

It follows from the definition that

$$g \in (f) \Leftrightarrow g(\alpha) = 0, \qquad g \in A_f \Leftrightarrow g(\alpha) \# 0.$$

It is a well-known fact that for g with degree at most that of f, $g(\alpha) = 0 \Leftrightarrow g = 0$. We shall prove a dual, but stronger, theorem.

5.19. PROPOSITION. *Let $f = a_n X^n + \cdots + a_0$, $g = b_m X^m + \cdots + b_0$, $f \# f(0)$, and $\forall i(b_i \# 0 \to \exists j > i(a_j \# 0))$, then (in the notation above):*

$$g \# 0 \Leftrightarrow g(\alpha) \# 0.$$

PROOF. $\Leftarrow$ is obvious.

$\Rightarrow$ Let $b_i \# 0$. We show $g(\alpha) \# 0$ by induction on $m - i$.

$m - i = 0$. There is a $j > i$ such that $a_j \# 0$. We now apply lemma 5.11 to obtain an $s \geq j$ and a corresponding submatrix which defines g^*. Since $s > m$ we have $g^* = 0$, and hence $g \# fg^*$. This proves $g \in A_f$, so $g(\alpha) \# 0$.

$m - i > 0$. Let $a_j \# 0$ with $j > i$. $g \# 0 \Rightarrow g \# fg^* \vee fg^* \# 0$. In the first case we are done:

$$\begin{pmatrix} a_0 & & 0 & \\ \vdots & \ddots & & \\ a_j & & \ddots & \\ \vdots & & & a_0 \\ & & & \vdots \\ \hline \vdots & & & \vdots \\ a_s & & & \vdots \\ \vdots & \ddots & & \vdots \\ a_n & & \ddots & \vdots \\ & \ddots & & a_s \\ \hline & 0 & \ddots & \vdots \\ & & & a_n \end{pmatrix} (g^*) = \begin{pmatrix} \vdots \\ \hline b_s \\ \vdots \\ b_m \\ 0 \\ \vdots \\ 0 \\ \hline \vdots \end{pmatrix} \# 0.$$

The horizontal lines in the left and the right matrix correspond

In the second case we find a coefficient a_s of fg^* apart from 0. By lemma 5.11 we may assume $s \geq j > i$, so $b_k \# 0$ for some $k > i$. Therefore $m - k < m - i$ and we may apply the induction hypothesis. □

Our next step is to investigate the notions of "prime" and "relatively prime" for polynomials and to relate them to (anti-)ideals. We first establish some useful facts.

5.20. Lemma. *Let* $f, g, h, k \in F[X]$, $h \# h(0)$, *then*:

(i) $f \# 0 \rightarrow (f \# hk \vee k \in A_f)$,

(ii) $f \# 0 \rightarrow (k \in A_f \rightarrow f \# f(0))$,

(iii) $f \# 0 \wedge g \# 0 \rightarrow (k \in A_f \rightarrow k \in A_g \vee f \# g)$.

Proof. (i) $f \# 0 \rightarrow f \# hk \vee hk \# 0$. If $hk \# 0$, then $k \# 0$. So let $k = \Sigma d_i X^i$, and $d_i \# 0$ for some i. Thus hk has degree at least $i + 1$. By looking at the coefficients we see that $hk \# f \vee \exists j > i(a_j \# 0)$; thus we have shown $d_i \# 0 \rightarrow \exists j > i(a_j \# 0)$, so by proposition 5.19, $hk \# 0 \rightarrow k \in A_f$. Thus $f \# hk \vee k \in A_f$.

(ii) From $f \# 0$ we conclude $f \# f(0) \vee f(0) \# 0$. So let $f(0) \# 0$. Now $k \in A_f \to k \# k \cdot f(0)^{-1} \cdot f$ and $k \# k \cdot f(0)^{-1} \cdot f \to f \# f(0)$ by strong extensionality of multiplication.

(iii) We shall use the membership test of proposition 5.15 for A_f and A_g; denote the * maps by k_f^* and k_g^*. Recall that the map * was defined by an invertible submatrix B of A. So determine this B for the matrix A belonging to f (N.B. it is no restriction to assume that f and g have the same length), consider the submatrix D of the matrix C belonging to g that has the same position as B in A. Since $\det B \# 0$, we have $\det D \# 0 \vee \det B \# \det D$.

In the latter case $f \# g$. So we assume now $\det D \# 0$. Recall that the * map operates on a vector $k^{\sim}$ consisting of a tail piece of k supplemented with zero's (cf. definition 5.14).

Now, $k \in A_f$ implies $k \# k_f^* f$. So $k \# k_g^* g \vee k_g^* g \# k_f^* f$. In the first case we have $k \in A_g$, in the second case $k_g^* \# k_f^* \vee g \# f$. If $g \# f$ we are done, so consider $k_g^* \# k_f^*$; by definition (and the above assumptions) $B^{-1}k^{\sim} \# D^{-1}k^{\sim}$. So $B \# D$ and hence $f \# g$. □

REMARK. We can reformulate (b) as follows: $f \# 0 \wedge A_f$ inhabited $\Leftrightarrow f \# f(0)$ (use theorem 5.16).

Primeness and relative primeness for polynomials (5.21–28)

5.21. DEFINITION.
(i) f and g in $F[X]$ are *relatively prime* if

$$\forall h h_1 h_2 \big(h \# h(0) \to f \# h_1 h \vee g \# h_2 h\big),$$

(ii) f in $F[X]$ is *prime* if $f \# f(0) \wedge \forall gh(g \# g(0) \wedge h \# h(0) \to f \# gh)$. □

Observe that $f \# 0$ or $g \# 0$ if f and g are relatively prime. When f is prime the elements of the anti-ideal of f should be relatively prime with f. Actually we have the following proposition.

5.22. PROPOSITION. *Let* $f \# 0$; *then*

$$f \textit{ is prime} \Leftrightarrow f \# f(0) \wedge \forall g \in A_f (f \textit{ and } g \textit{ are relatively prime}).$$

PROOF. ⇒ Let $g \in A_f$ and $h, h_1, h_2 \in F[X]$ such that h $\# h(0)$. From $f \# 0$ we conclude $f \# h_1 h \vee h_1 h \# 0$. If the first disjunct holds we are done, so consider the second disjunct.

We get $h \# 0 \wedge h_1 \# 0$; and hence $ch \# 0$ for all $c \# 0$ in F. We now apply 5.20(iii): $f \# 0 \wedge ch \# 0 \to (g \in A_f \to g \in A_h \vee f \# ch)$. So let $g \in$

A_h, then $g \# h_2 h$, which is of the required form. If $f \# ch$, then $f \# h_1 h \vee h_1 h \# ch$. We again consider the second disjunct and conclude $h_1 \# c$. From $h_1 \# 0$ we conclude $h_1 \# h_1(0) \vee h_1(0) \# 0$; in the second case we get $h_1 \# h_1(0)$ by putting $c = h_1(0)$. Now we use the fact that f is prime, so $f \# h_1 h$.

$\Leftarrow$ Let $g \# g(0) \wedge h \# h(0)$, then by 5.20(i) $f \# gh \vee g \in A_f$. If $g \in A_f$, then f and g are relatively prime, so $f \# hg \vee g \# 1 \cdot g$. Hence we obtain $f \# hg$. So f is prime. □

We want to characterize the factor rings modulo prime polynomials, but in order to do so we have to establish a number of auxiliary properties of "prime" and "relatively prime".

5.23. LEMMA. *The following are equivalent*:

$$\mathrm{RP}_1(f, g) \qquad \left[g \# 0 \wedge \forall hk\left(h \in A_g \to hf + kg \# 0\right)\right] \vee \left[f \# 0 \wedge \forall hk\left(k \in A_f \to hf + kg \# 0\right]\right.,$$

$$\mathrm{RP}_2(f, g) \qquad (f \# 0 \vee g \# 0) \wedge \forall hk\left(h \in A_g \vee k \in A_f \to hf + kg \# 0\right).$$

PROOF. $\mathrm{RP}_2(f, g) \Rightarrow \mathrm{RP}_1(f, g)$ is a matter of logic.

$\mathrm{RP}_1(f, g) \Rightarrow \mathrm{RP}_2(f, g)$: assume the first disjunct of $\mathrm{RP}_1(f, g)$, then $f \# 0 \vee g \# 0$. Now let $h \in A_g \vee k \in A_f$, it suffices to consider $k \in A_f$. Let $*$ be the map connected with the membership test for A_g (5.14), then $hf + kg = (h^*g + d)f + kg = (h^*f + k)g + df$, where $d = h - h^*g$. From $k \in A_f$ we conclude $k \# -h^*f$; so by $g \# 0$, we get $(k + h^*f)g \# 0$ (recall that $F[X]$ is an integral domain, cf. 5.4). This yields $hf + kg \# 0 \vee df \# 0$. From the latter disjunct we conclude $d \# 0$ and hence (by the criterion) $h \in A_g$, which in turn implies $hf + kg \# 0$. This establishes the desired implication. □

5.24. LEMMA. $\mathrm{RP}_2(f, g) \Rightarrow$ *f and g are relatively prime.*

PROOF. Let $h \# h(0)$. By $\mathrm{RP}_2(f, g)$, $f \# 0 \vee g \# 0$, hence by 5.20(i) $f \# hk_1 \vee k_1 \in A_f \vee g \# hk_2 \vee k_2 \in A_g$. Applying $\mathrm{RP}_2(f, g)$ we get $f \# hk_1 \vee g \# hk_2 \vee k_1 g - k_2 f \# 0$.

It remains to consider $k_1 g \# k_2 f$. We see that $k_1 g \# k_1 k_2 h \vee k_1 k_2 h \# k_2 f$ so $g \# k_2 h \vee f \# k_1 h$. This shows that f and g are relatively pri—

We will show that $RP_1(f, g)$, $RP_2(f, g)$, and "f and g are relatively prime" are equivalent, but first we need a lemma that shows that relative primeness is "preserved under the euclidean algorithm".

5.25. LEMMA. *f and g are relatively prime* $\Rightarrow RP_1(f, g)$.

PROOF. We first note that for a pair of relatively prime polynomials f and g we have $f \# 0 \vee g \# 0$ (cf. E8.5.3).

In the proof the polynomials f, g, h, and k will occur, we fix the notation

$$f = a_n X^n + \cdots + a_0, \qquad g = b_m X^m + \cdots + b_0,$$

$$h = c_p X^p + \cdots + c_0, \qquad k = d_l X^l + \cdots + d_0.$$

We start by proving an auxiliary fact,

$$(h \# 0 \vee k \# 0) \wedge \exists i > 0\big(a_i \# 0 \wedge \forall j \geq i(c_j = 0)\big) \rightarrow hf + kg \# 0. \tag{1}$$

We use induction on l.

$l = 0$. Let $d_0 \# 0$, then from the relative primeness of f and g it follows that $g \# -d_0^{-1}hf$ and hence $hf + kg \# 0$.

Next we show that if (1) holds for all regular k with degree ≤ 1, then it holds for k of the form $d_l X^l + \cdots + d_0$. Let such a k be given; we write $k = k' + k''$, where k' is a regular initial segment of k, or $k' = 0$. Now $hf + k'g \# 0$, and so $hf + kg \# 0 \vee k \# k'$.

In the latter case we can find another regular k' of higher degree and repeat the argument. Induction on l-degree (k') yields the statement. In particular, this finishes the case $l = 0$.

So let (1) be given for all k with $l < l^*$. It suffices to consider a regular k with degree l^*. Now $f = qk + r$, with r of degree at most $l^* - 1$, and consequently $q \# q(0)$. Since f and g are relatively prime we have

$$f \# gk \vee g \# -qh \quad \text{so } r \# 0 \vee g + qh \# 0.$$

Now a simple computation yields $f(g + qh) - rg = q(hf + kg)$. From the assumption of (1) for $g + qh$ and r we conclude that the left-hand side is apart from 0, hence $hf + kg \# 0$. This finishes the proof of (1). We now return to the proof of $RP_1(f, g)$.

Given that f and g are relatively prime, we assume $f \# 0$. Furthermore, let $k \in A_f$, then $f \# f(0)$ (5.20(ii)). Now put $f = f' + f''$, where f' is a regular initial segment of f. Observe that $g \in A_f$, since $f \# f(0)$ and f and g are relatively prime. Using the regularity of f' we determine q and r such that $g = qf' + r$.

By 5.20(iii) we get $g \in A_{f'} \vee f \# f'$. If the latter holds we may repeat the argument with an initial segment f' of higher degree. If the first disjunct holds we get $r \# 0$. We may then apply (1) to the left-hand side of the following equation:

$$(h + qg)f + rg = (hf + kg) - qgf''.$$

So $hf + kg \# 0 \vee qgf'' \# 0$. In the first case we are done, in the second case we again may raise the degree of f' and repeat the argument. As before this leads in a finite number of steps to $\mathrm{RP}_1(f, g)$. □

COROLLARY. $\mathrm{RP}_1(f, g) \Leftrightarrow \mathrm{RP}_2(f, g) \Leftrightarrow$ *f and g are relatively prime.* □

5.26. DEFINITION. An element a of a ring R is *zero-divisor free* if

$$\forall x \in R(x \# 0 \to xa \# 0). \quad □$$

5.27. LEMMA. *Let $f \# 0$ and $F[\alpha] = F[X]/(f)$, then*

$$f \textit{ and } g \textit{ are relatively prime} \Leftrightarrow g(\alpha) \textit{ is zero-divisor free}.$$

PROOF. $\Rightarrow$. Let $k(\alpha) \# 0$, i.e. $k \in A_f$. Since $f \# 0$, we have $\forall h \forall k \in A_f$ $(hf + kg \# 0)$, so $kg \in A_f$ and therefore $k(\alpha)g(\alpha) \# 0$.

$\Leftarrow$ is left to the reader. □

5.28. LEMMA. *If f, g_1 and f, g_2 are relatively prime, then so are f, $g_1 g_2$.*

PROOF. By definition we get $f \# 0 \vee (g_1 \# 0 \wedge g_2 \# 0)$. First consider $f \# 0$. Put $F[\alpha] = F[X]/(f)$. $g_1(\alpha)$ and $g_2(\alpha)$ are zero-divisor free by 5.27. Hence, by definition, $g_1(\alpha)g_2(\alpha)$ is zero-divisor free, so f and $g_1 g_2$ are relatively prime.

Next consider $g_1 \# 0$ and $g_2 \# 0$. We have $g_1 \# g_1(0) \vee g_2 \# g_2(0) \vee g_1(0) \# 0 \vee g_2(0) \# 0$. If $g_1 \# g_1(0)$ or $g_2 \# g_2(0)$, then by the definition of relatively primeness $f \# 0$, so we may repeat the above argument. Let therefore $g_2(0) \# 0$. We must show $h \in A_{g_1 g_2} \Rightarrow hf + kg_1 g_2 \# 0$. Observe that $h \in A_{g_2(0) g_1} \Leftrightarrow h \in A_{g_1}$. We now apply 5.20(iii); since $g_1 g_2 \# 0$ and $g_1 \# 0$ we have $h \in A_{g_1 g_2} \Rightarrow h \in A_{g_1} \vee g_1 g_2 \# g_1 g_2(0)$. So $h \in A_{g_1}$ or $g_2 \# g_2(0)$. In the latter case we are done. So let $h \in A_{g_1}$, then $hf + kg_1 g_2 \# 0$. □

Prime polynomials, their ideals and factor rings (5.29–31).

We now get to the characterization of a polynomial ring over a field modulo a prime polynomial. Whereas in classical algebra the factor ring is a field, we only get an integral domain. The reason is that we lack information on the degree of a prime polynomial.

5.29. THEOREM. *If $f \# 0$, then*

$$f \text{ is prime} \Leftrightarrow F[X]/(f) \text{ is an integral domain.}$$

PROOF. $\Rightarrow$ By definition $f \# f(0)$, so $1 \in A_f$ and hence $1 \# 0$ in $F[\alpha]$. Let $g_1(\alpha) \# 0$, $g_2(\alpha) \# 0$. Then f, g_1 and f, g_2 are relatively prime (5.22), so by 5.28 f and $g_1 g_2$ are relatively prime. By 5.27 $g_1(\alpha)g_2(\alpha)$ is zero-divisor free, in particular $g_1(\alpha)g_2(a) \# 0$.

$\Leftarrow$ $1 \# 0$ in $K[\alpha]$, so $f \# f(0)$ (5.20(ii)). We will show f prime by applying proposition 5.22. Let $g \in A_f$, then $g(\alpha) \# 0$. Now since $F[\alpha]$ is an integral domain, $g(\alpha)$ is zero-divisor free and so f and g are relatively prime. This shows that f is prime. □

COROLLARY. *f is prime $\Leftrightarrow$ (f) is a prime ideal.*

PROOF. Combine the theorem above and proposition 3.11. □

If moreover f is a regular prime polynomial, then $F[\alpha]$ is a field. We first establish a useful lemma.

5.30. LEMMA. *Let f and g be relatively prime and let f be regular, then there are unique h and k such that $hf + kg = 1$. Moreover, the degree of k is less than the degree of f.*

PROOF. The case that f is a constant is trivial. So consider f of degree $n > 0$. Let $x = x_{n-1}X^{n-1} + \cdots + x_0 \in F(x_0, \ldots, x_{n-1})[X]$, where $F(x_0, \ldots, x_{n-1})$ is the field of rational functions over F. We apply the division algorithm in $F(x_0, \ldots, x_{n-1})[X]$:

$$xg = qf + r \quad \text{where } r \text{ has degree at most } n - 1.$$

The coefficients of r are linear in the x_j:

$$r_i = a_{i+1,1}x_0 + \cdots + a_{i+1,n}x_{n-1}.$$

Now consider the ring $F[X]/(f)(= F[\alpha])$. If we substitute elements of F for the x_j in x and r, we obtain elements x' and r' of $F[X]$, and hence elements $x'(\alpha)$, $r'(\alpha)$ of $F[\alpha]$. Say we substitute $c_0, \ldots, c_{n-1}$ for $x_0, \ldots, x_{n-1}$, where $c_j \# 0$ for some j. We may apply proposition 5.19 to x': $x'(\alpha) \# 0$. By 5.27 $g(\alpha)$ is zero-divisor free, so $x'(\alpha) \cdot g(\alpha) \# 0$. Hence $r'(\alpha) \# 0$. This shows $x' \# 0 \to r' \# 0$ (again using 5.19). Now we apply 4.14, and conclude that the matrix A belonging to r is invertible. Therefore we can find a polynomial $k = k_{n-1}X^{n-1} + \cdots + k_0$, such that $kg = -hf + 1$. Since f is regular, h is also uniquely determined. □

5.31. THEOREM. *If f is a regular prime polynomial, then $F[X]/(f)$ is a field.*

PROOF. Apply the preceding lemma. □

6. Fields and local rings

From a constructive point of view fields and local rings are next of kin. Since, constructively, 0 is not necessarily the only noninvertible element in a field, the similarity is even greater than in classical mathematics. For the moment we do not require rings to have an apartness relation. We recall that when we use the E-predicate, it will be in the framework of E^+-logic.

6.1. DEFINITION. A *local ring* is a commutative ring with a unit element such that

$$\forall x[\exists y(xy = 1) \vee \exists y((1 - x)y = 1)],$$

or in the language with E-predicate,

$$Et \to [\exists y(ty = 1) \vee \exists y((1 - t)y = 1)]. \quad \square$$

REMARK. The latter condition has the following elegant reformulation: $Et \to Et^{-1} \vee E(1 - t)^{-1}$ (cf. E8.6.1).

EXAMPLE. $\{nm^{-1} \in \mathbb{Q} : \neg 2|m\}$ is a local ring, but not a field.

6.2. LEMMA. $A = \{a \in R : Ea^{-1}\}$ *is a minimal anti-ideal of the local ring* R.

PROOF. (i) It is clear that $0 \notin A$; furthermore $1 \in A$, so A is proper.

(ii) $ab \in A \Rightarrow E(ab)^{-1} \Rightarrow E(a^{-1}b^{-1}) \Rightarrow Ea^{-1} \wedge Eb^{-1} \Rightarrow a \in A \wedge b \in A$.

(iii) $a + b \in A \Rightarrow E(a + b)^{-1} \Rightarrow \exists z((a + b)z = 1) \Rightarrow \exists z(az + bz = 1)$. So let c satisfy Ec and $ac + bc = 1$. By $E(ac)^{-1} \vee E(1 - ac)^{-1}$, we get

$E(ac)^{-1} \vee E(bc)^{-1}$. Hence, by (ii), we get $Ea^{-1} \vee Eb^{-1}$, and thus $a \in A \vee b \in A$.

(iv) (minimality, 3.10) Let $a \in A$, then Ea^{-1}. Now $Ea^{-1} \Rightarrow \exists z(az = 1) \Rightarrow \exists z(1 - az = 0) \Rightarrow \exists z(1 - az \notin A)$. □

The following theorem states that a local ring carries a canonical preapartness relation.

6.3. Theorem. *The relation $a \# b := E(a-b)^{-1}$ on a local ring R is a preapartness relation.*

Proof. (i) $\#$ is obviously symmetric.

(ii) $\#$ is strict: $a \# b \Rightarrow E(a-b)^{-1} \Rightarrow E(a-b) \Rightarrow Ea \wedge Eb$ (note that the inverse, although introduced by the description operator, is strict).

(iii) $a \# b \Rightarrow E(a-b)^{-1}$. Now, if Ec, then by the totality of $+$ and $-$, $(a-b) = ((a-c)+(c-b))$. Hence $E((a-c)+(c-b))^{-1}$, so $(a-c)+(c-b) \in A$, and therefore $(a-c) \in A \vee (b-c) \in A$; i.e. $a \# c \vee b \# c$.

(iv) Addition is strongly extensional: $a + b \# a' + b' \Rightarrow (a - a' + b - b') \in A \Rightarrow (a - a') \in A \vee (b - b') \in A \Rightarrow a \# a' \vee b \# b'$.

(v) Multiplication is strongly extensional: $ab \# a'b' \Rightarrow ab - a'b' \in A \Rightarrow ab - a'b + a'b - a'b' \in A \Rightarrow (a-a')b \in A \vee a'(b-b') \in A \Rightarrow (a-a') \in A \vee (b-b') \in A \Rightarrow a \# a' \vee b \# b'$.

(vi) $0 \# 1$, since $1 \in A$. □

6.4. Remarks. Observe that $\#$ need not be an apartness relation. Consider the local ring $\mathbb{Z}_{(p)}$, which is the localization of $\mathbb{Z}$ with respect to (p) (where p is prime), i.e. the set nm^{-1} where m does not contain a factor p. Since p is not invertible, $\neg p \# 0$. But $p \neq 0$.

The minimal anti-ideal A of invertible elements is by definition compatible with $\#$. So R/A^c is a field with an induced relation $\# : x + A^c \# 0 := x \in A$. Now $\neg(x + A^c \# 0) \Leftrightarrow x \notin A \Leftrightarrow x \in A^c$, so R/A^c is a field with apartness relation.

We can now characterize fields as special local rings.

6.5. Proposition. *Let R be a local ring with canonical $\#$. Then*

$$R \text{ is a field} \Leftrightarrow \# \text{ is an apartness relation} \Leftrightarrow \forall x(\neg Ex^{-1} \rightarrow x = 0).$$

The proof is left to the reader. □

6.6. *Kripke rings and fields.* Consider a Kripke model R of ring theory (a *Kripke ring*) in the language of total elements; one can easily check that the structures at the nodes are (classical) rings (or one may use the fact that the theory of rings is axiomatized by geometric sentences, as far as the # -free fragment is concerned, cf. E2.6.14).

In the ring R_k at node k the set $I_k := \{a \in R_k : k \not\Vdash a \# 0\}$ is an ideal.

Let $\varphi_{kk'} \in R_k \to R_{k'}$ be the homomorphism belonging to the nodes k, k' (with $k \leq k'$), then we infer from $k' \not\Vdash \varphi_{kk'}(a) \# 0 \to k \not\Vdash a \# 0$,

$$\varphi_{kk'}(a) \in I_{k'} \Rightarrow a \in I_k. \tag{α}$$

Moreover, $k \Vdash \neg a \# 0 \Rightarrow k \Vdash a = 0$ yields the following:

$$\forall k' \geq k(\varphi_{kk'}(a) \in I_{k'}) \Rightarrow a = 0. \tag{β}$$

Conversely, when proper ideals I_k in the rings R_k are given, we may define a relation # on R by $k \Vdash a \# b := a - b \notin I_k$.

PROPOSITION.
(i) *If the ideals I_k satisfy (α), then # is a preapartness relation on R and $+, \cdot, -$ are strongly extensional. If, moreover, (β) holds, then # is an apartness relation.*
(ii) *R is an integral domain $\Leftrightarrow \forall k(I_k$ is prime$)$.*
(iii) *R is a local ring $\Leftrightarrow \forall k(R_k$ is a local ring$)$.*
(iv) *R is a field $\Leftrightarrow$ R is a local ring and (β) holds.*

PROOF. (i) is immediate.

(ii) R is an integral domain if $ab \# 0 \Rightarrow a \# 0 \wedge b \# 0$. So $ab \notin I_k \Leftrightarrow k \Vdash ab \# 0 \Leftrightarrow k \Vdash a \# 0 \wedge b \# 0 \Leftrightarrow a \notin I_k$ and $b \notin I_k$. This shows that I_k is prime.

Conversely, if the I_k are prime, by the same argument R is an integral domain.

(iii) One may check the equivalence for the formulation of local ring in the language of total elements, but one may also use the fact that in E^+-logic the theory of local rings is geometric (E2.6.14).

(iv) $\Rightarrow$ is immediate. For $\Leftarrow$, consider R with its canonical preapartness relation. Obviously, (α) holds: the ideals I_k consist of the noninvertible elements and are maximal. Since an element that is invertible in R_k remains so in $R_{k'}$ for $k' \geq k$, we have $\varphi_{kk'}^{-1}(I_{k'}) \subset I_k$ (and also $\varphi_{kk'}(I_k^c) \subset I_{k'}^c$). (β) tells us that # is an apartness relation, so by 6.5, R is a field. □

6.7. Examples.

(1) In this model $R = \{nm^{-1} : \neg 3|m\}$, $I = 3R$, and the homomorphism is the embedding of R in $\mathbb{Q}$. Both $\mathbb{Q}$ and R are local rings, (α) and (β) hold, so the model is a field. We see that, e.g. $2 \# 0$, $3 \neq 0$, but not $3 \# 0$. Observe that $\forall x(x = 0 \vee x \neq 0)$ holds, but not $\forall x(x = 0 \vee x \# 0)$. See fig. 8.16.

Fig. 8.16.

(2) Consider the (classical) reals $\mathbb{R}$ and let $^*\mathbb{R}$ be a nonstandard extension with finite part $^*\mathbb{R}_f$. The $^*\mathbb{R}_f$ is a local ring with maximal ideal I consisting of all infinitesimals (noninvertibles). The homomorphisms in the diagram in fig. 8.17 are the canonical injection i and projection p. The conditions (α) and (β) hold, so R is a field. We have listed a few properties in fig. 8.17.

$$R \Vdash \forall x(x \neq 0 \rightarrow Ex^{-1}),$$
$$R \nVdash \forall xy(x = y \vee x \neq y),$$
$$R \Vdash \forall xy(x \neq y \leftrightarrow x \# y).$$

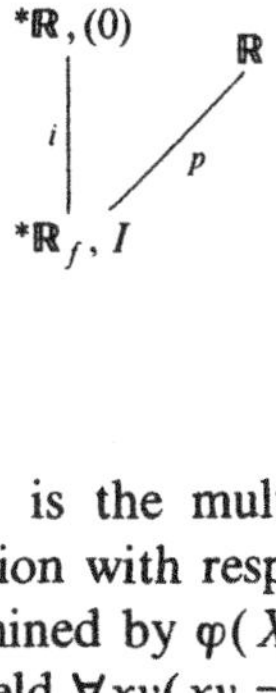

Fig. 8.17.

(3) Here F is a (classical) field and S is the multiplicatively closed complement of (X, Y). S^{-1} is the localization with respect to S (cf. Lang 1965, ch. 2, section 3). φ and ψ are determined by $\varphi(X) = X$, $\varphi(Y) = 0$, $\psi(X) = 0$, $\psi(Y) = Y$. In this model of a field $\forall xy(xy = 0 \rightarrow x = 0 \vee y = 0)$ does not hold, see fig. 8.18.

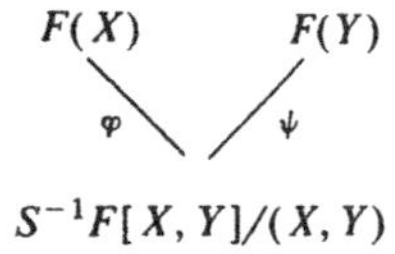

Fig. 8.18.

7. The fundamental theorem of algebra

We have seen that the field of real numbers is real closed (6.2.7), i.e. every $f \in \mathbb{R}[x]$ of odd degree has a zero in $\mathbb{R}$. We will now show that $\mathbb{C}$ is algebraically closed, i.e. that every $f \in \mathbb{C}[X]$, with $f \# (0)$, has a zero. A number of constructive proofs have been given; the proof below is an intuitionistic version of proofs by H. Kneser (1940) and M. Kneser (1981).

7.1. THEOREM (*fundamental theorem of algebra*). *Let* $f \in \mathbb{C}[X]$ *be such that* $f(X) = b_n X^n + \cdots + b_0$ *with* $b_k \# 0$ *for some* $k > 0$, *then* $\exists z \in \mathbb{C}$ $(f(z) = 0)$.

The theorem is established by means of the following lemma, which is used in an approximation of the desired zero.

7.2. LEMMA. *Let* $f = X^n + b_{n-1}X^{n-1} + \cdots + b_0 \in \mathbb{C}[X]$ $(n > 0)$, *then*

$$\exists q \in (0,1)\, \forall c > |b_0|\, \exists z \in \mathbb{C}\big(|z| < c^{1/n} \wedge |f(z)| < qc\big).$$

PROOF. Let a c be given such that $|b_0| < c$. We first approximate the coefficients of f by suitable numbers $a_0, \ldots, a_n$ apart from zero, and to simplify the argument we choose the a_i such that they have a rational modulus. So determine $a_0, \ldots, a_n$ such that the following conditions are met:

$$|a_i| > 0, \quad |a_0| < c, \quad |a_i - b_i| < \alpha, \quad a_n = 1, \quad |a_i| \in \mathbb{Q},$$
$$a_i \neq a_j \quad \text{for } i \neq j.$$

We will specify α later on. We want to find a k and a $z \in \mathbb{C}$ such that $a_0^{-1} a_k z^k < 0$. Since both a_0 and a_k are apart from zero, this can be done. The number z has also to meet the following requirements

$$|a_k| \cdot r^k < |a_0|, \qquad r^n < |a_0|, \quad \text{where } r = |z|. \tag{1}$$

Let us use Σ for summation over i ranging from 0 to n, and Σ^* for sumation over $\{1, \ldots, k-1, k+1, \ldots, n\}$.

Now $|f(z)| = |\Sigma b_i z^i| \leq |\Sigma a_i z^i| + |\Sigma(b_i - a_i)z^i| < \Sigma a_i z^i + \alpha\Sigma|z|^i$. Put $g(X) = \Sigma a_i X^i$. By the conditions on z, $|a_0 + a_k z^k| = |a_0| - |a_k z^k|$, so

$$|g(z)| \leq |a_0 + a_k z^k| + |\Sigma^* a_i z^i| = |a_0| - |a_k z^k| + |\Sigma^* a_i z_i|. \tag{2}$$

Our strategy is to make $|\Sigma^* a_i z^i|$ small with respect to $|a_k z^k|$, while controlling the difference between $|a_0|$ and $|a_k z^k|$. In order to construct k, r and to keep an eye on the various absolute values in (2), we define an auxiliary real-valued function m on $[0,\infty)$:

$$m(s) = \max\{|a_i|s^i : 1 \le i \le n\}.$$

Observe that m is continuous, strictly monotone, tends to infinity, and piecewise coincides with the functions $|a_i|s^i$. To be precise, there is a finite number of segments I_i such that

$$m(s) = |a_i|s^i \quad \text{for } s \in I_i. \tag{3}$$

We remark that not all the $|a_i|s^i$ need actually occur, so there are at most n segments and the rightmost segment is unbounded. By the intermediate value theorem (6.1.5) there is a t such that

$$m(t) = |a_0|. \tag{4}$$

Hence, by the monotonicity of m

$$r \le t \to |a_k|r^k \le |a_0| \quad \text{for all } k. \tag{5}$$

Since we have chosen the values $|a_i|$ to be rational, we can actually decide the position of t with respect to the segments I_i (E8.7.1). The same applies to the points $3t, t, 3^{-1}t, 3^{-2}t, 3^{-3}t, \dots$.

Similarly, we can exactly determine which segments I_i play a role in (3). By the above we determine numbers k_i such that $3^{-i}t \in I_k$. As the sequence $3t, t, 3^{-1}t, \dots$ descends, we have $n \gtrsim k_{-1} \ge k_0 \ge k_1 \ge \cdots$. Therefore there is a j such that $k_{j-1} = k_j = k_{j+1}$. We pick the first such j. Since before j we can at most have coinciding couples, we have $k_{2i-1} \le n - i$ for $2i - 1 \le j$. Hence we get

$$j \le 2(n - k_j). \tag{6}$$

We now fix k and r:

$$k = k_j, \tag{7}$$

$$r = 3^{-j}t. \tag{8}$$

We choose a z such that (8) holds (z plays no further role). By (8) $r \le t$, so $|a_k|r^k \le |a_k|t^k \le |a_0|$ (by (5)). Given r and k we have to estimate the terms in (2); we will use the function m for this purpose.

The decrease of m from $3^{-i}t$ to $3^{-i-1}t$ is estimated by majorizing it by the decrease of $|a_k|s^k$, where $3^{-i}t \in I_k$. For, by definition $m(s)$ majorizes $|a_k|s^k$ on $[0, 3^{-i}t]$. The estimates on the differences of the function m and the relevant functions $|a_k|s^k$ will eventually enable us to majorize the values of $|a_i|r^i$ in terms of $|a_k|r^k$.

From t to $3^{-j}t$ (i.e. r) it takes j steps. For the first step we have $m(t) = |a_{k_0}|t^{k_0}$, since $t \in I_{k_0}$. So the next value, $m(3^{-1}t)$, majorizes $|a_{k_0}| \cdot 3^{-k_0} \cdot t^{k_0}$, thus $m(3^{-1}t) \geq 3^{-k_0} \cdot m(t)$. By repeating this step we get $m(3^{-j}t) \geq 3^{-k_0-k_1-\cdots-k_{j-1}} \cdot m(t)$.

Now consider the sequence $k_0, k_1, k_2, \ldots, k_{j-1}$ and recall that no more than two consecutive terms may coincide, that the sequence is descending, and that $n \geq k_0$, $k_{j-1} = k_j = k$. So $k_0 + (k_1 + k_2) + (k_3 + k_4) + \cdots + k_{j-1} \leq n + (k_1 + k_1) + (k_3 + k_3) + \cdots + k \leq n + 2(n-1) + 2(n-2) + \cdots + 2(k+1) + k = n^2 - k^2$. Hence

$$m(r) = m(3^{-j}t) \geq 3^{-k^2-n^2} \cdot m(t). \tag{9}$$

By $k_{j+1} = k_j = k$ we have $3^{-1} \cdot r \in I_k$. Therefore $|a_i|r^i = 3^i|a_i|(3^{-1} \cdot r)^i \leq 3^i \cdot m(3^{-1} \cdot r) = 3^i \cdot |a_k|(3^{-1} \cdot r)^k = 3^{i-k}|a_k|r^k$, and hence

$$\sum_{1 \leq i < k} |a_i|r^i \leq \sum_{1 \leq i < k} 3^{i-k}|a_k|r^k = \tfrac{1}{2}(1 - 3^{1-k})|a_k|r^k. \tag{10}$$

Similarly, $3r \in I_{k_{j-1}} = I_k$, so $m(3r) = |a_k|(3r)^k$ and hence

$$\sum_{k < i \leq n} |a_i|r^i \leq \sum_{k < i \leq n} 3^{k-i}|a_k|r^k = \tfrac{1}{2}(1 - 3^{k-n})|a_k|r^k. \tag{11}$$

From (2), (10), (11) it follows that for z with modulus r and argument determined by $a_0^{-1}a_kz^k < 0$ we have

$$\begin{aligned} |g(z)| &\leq |a_0| - |a_k|r^k + \tfrac{1}{2}(1 - 3^{1-k})|a_k|r^k + \tfrac{1}{2}(1 - 3^{k-n})|a_k|r^k \\ &< |a_0| - \tfrac{1}{2} \cdot 3^{1-k}|a_k|r^k = |a_0| - \tfrac{1}{2} \cdot 3^{1-k} \cdot m(r) \\ &\leq |a_0|\left(1 - \tfrac{1}{2} \cdot 3^{1-k+k^2-n^2}\right) \\ &\leq |a_0|\left(1 - \tfrac{1}{2} \cdot 3^{1-n^2}\right) \end{aligned}$$

(by $m(t) = |a_0|$ and (9)). So $|f(z)| \leq (1 - \frac{1}{2} \cdot 3^{1-n^2})|a_0| + \alpha\Sigma|r|^i$. We now choose α such that

$$\begin{aligned} \alpha\Sigma|r|^i < \tfrac{1}{4} \cdot 3^{1-n^2}, \quad \text{then } |f(z)| &\leq \left(1 - \tfrac{1}{4} \cdot 3^{1-n^2}\right)|a_0| \\ &< \left(1 - \tfrac{1}{4} \cdot 3^{1-n^2}\right)c. \end{aligned}$$

Now put $q = (1 - \frac{1}{4} \cdot 3^{1-n^2})$, then the lemma holds, since $|z| < c^{1/n}$ holds by (5). □

7.3. *Proof of a restricted version of the fundamental theorem.* We approximate a zero of $X^n + \cdots + a_0$ using the above lemma. Put $z_0 := 0$, and assume that we have found z_m with the property $|f(z_m)| < q^m c$.

Now apply the lemma to $f(z_m + X)$, which has leading coefficient 1 and constant $f(z_m)$:

$$\exists z\left(|z| < (q^m c)^{1/n} \wedge |f(z_m + z)| < q^{m+1}c\right).$$

Pick such a z and put $z_{m+1} := z_m + z$, then

$$|z_{m+1} - z_m| < (q^m c)^{1/n} \quad \text{and} \quad f(z_{m+1}) < q^{m+1}c.$$

Clearly $z^* = \lim\langle z_m \rangle_m$ exists and $|f(z^*)| = 0$. So regular polynomials have zero's in $\mathbb{C}$. □

For the remainder of the proof we recall the fact from 5.9. Let $f \in \mathbb{C}[X]$, then

$$a_k \# 0 \to \exists z \bigwedge_{i \le k} \left(f^{(i)}(z) \# 0\right).$$

7.4. Proof *of the fundamental theorem.* Let $f(X) = a_n X^n + \cdots + a_0$ with $a_k \# 0\,(k > 0)$. Put $g(Y) = Y^n f(c + Y^{-1})$, where $c \in \mathbb{C}$ such that $f(c) \# 0$, $f'(c) \# 0, \ldots, f^{(k)}(c) \# 0$. Then g is a regular polynomial in Y:

$$g(Y) = f(c)Y^n + f'(c)Y^{n-1} + \cdots + (1/n!)f^{(n)}(c).$$

We now apply the restricted fundamental theorem:

$$g(Y) = f(c)\prod_{j=1}^{n}(Y - \alpha_j), \quad \alpha_j \in \mathbb{C}.$$

Using a simple property of the symmetric functions of the zero's of polynomials we see that

$$f^{(k)}(c) = k! \sum_{(i_1 \ldots i_k)} \alpha_{i_1} \cdots \alpha_{i_k}.$$

Since $f^{(k)}(c) \# 0$, also $\alpha_{i_1} \# 0, \ldots, \alpha_{i_k} \# 0$ for some k-tuple $i_1, \ldots, i_k$. Now substitute $Y = (X - c)^{-1}$ in g, then

$$f(X) = f(c)\prod_{j=1}^{n}(1 + \alpha_j c - \alpha_j X).$$

Conclusion: f has k zero's and a factorization of the form $d(X - \gamma_1)\cdots(X - \gamma_k)\cdots(\beta_{k+1}X - \gamma_{k+1})\cdots(\beta_n X - \gamma_n)$. □

The fundamental theorem of algebra essentially rests on the axiom of countable choice (cf. 6.1.5, 15.3.6–7). We have made use of AC-NN in the determination of t in (4) in the proof of 7.2 and in the repeated choice of the $a_0, \ldots, a_n$ of 7.2 in the process of 7.3.

8. Notes

8.1. *Apartness and ordering.* Brouwer (1919) introduced the notion of apartness (örtlich verschieden, Entfernung). In a later paper (1923) he considered the relationships between equality, apartness, and inequality. The axioms of the theory of apartness were formulated by Heyting (1925). Brouwer considered various notions of order, cf. (1925). In Brouwer (1927A) he presented a quite unusual argument of a metamathematical nature to identify the so-called "virtual order" and the "unextendable order". The notion of "linear order" in 1.7 is the most useful one for practical purposes.

8.2. *General constructive algebra.* The study of algebraic structures in an intuitionistic setting was undertaken by Heyting (1941). He formulated all the familiar notions: groups, rings, integral domains, fields, polynomial rings, etc. Whereas constructive algebra in the tradition of Kronecker had dealt with discrete structures (i.e. with decidable equality), Heyting considered structures in full generality, equipped with an apartness relation. The notion of an antisubstructure, implicit in Heyting's treatment of ideals in polynomial rings, was formulated explicitly by D.S. Scott (1979). (N.B. the first draft of this paper contains a good deal more than the published version). Ruitenburg (1982, 1982A) deals with intuitionistic algebra in the spirit of Heyting and Scott. Our presentation partly follows his. Scott (1979) also contains a treatment of structures with partial elements and a formulation of the main properties in terms of E-logic.

8.3. *Linear algebra* was studied by Heyting (1927); he considered skew fields with an apartness relation, and a noncommutative version of determinants.

8.4. *Polynomial ring theory* as formulated here is mostly a reformulation of Heyting (1941) in more general terms. In particular the facts, that the factor ring of a polynomial ring over a prime polynomial (ideal) is but an integral domain, and that the prime has to be regular to obtain a field, are noteworthy results of Heyting (1941).

8.5. *Fields and local rings.* The close relation between fields and local rings was brought out by the sheaf semantics, but was already foreshadowed by the insight that in fields there are more elements than zero and those apart from zero. The characterization of Kripke fields and local rings are specializations of familiar facts from sheaf models.

8.6. *Fundamental theorem.* Since the fundamental theorem of algebra had definitely been settled by Gauss, one had realized that the existence of zeros of polynomials was established in a noneffective way. Brouwer and de Loor (1924) and Weyl (1924) gave satisfactory constructive proofs, and Brouwer (1924B) extended the theorem to its general form.

8.7. *Other work.* Constructive algebra comprises considerably more than the material presented here; we have restricted ourselves to an exposition of the basic facts of the theory on the basis of a general constructivistic programme.

There are certain schools in constructive algebra with particular features, we will mention some of the work that has been done and is still going on.

The Kronecker style of algebra, roughly speaking the algebra of discrete structures (i.e. with decidable equality), has found its adherents mainly in the school of Bishop; in the intervening period there were some isolated contributions to "discrete constructive algebra", in particular Hermann (1926) and Van der Waerden (1930) should be mentioned. For information on this subject see Mines et al. (1988), and Seidenberg (1974, 1978).

Constructive algebra in the sense of "recursive algebra" was introduced by Fröhlich and Shepherdson (1955); one may describe this as the study of algebraic structures over $\mathbb{N}$ with recursive operations and relations (including equality). There exists a considerable amount of literature on the subject, cf. Crossley (1981), and Metakides and Nerode (1975, 1982). Ershov's treatment of algebra in the framework of the theory of enumerations (1977) also belongs to this area.

Recent work in topos theory, such as Hyland (1982), McCarty (1984), and Rosolini (1986), seems to confirm D.S. Scott's opinion that a good deal of recursive algebra finds its natural home in one of the recursive (respectively realizability, effective) toposes.

In general, topos theory provides an ideal setting for testing constructive algebraic theories, albeit for reasons that have no underlying constructive philosophy, cf. Wraith (1979), Borceux and van den Bossche (1983), and Kock (1976).

Exercises

8.1.1. Show that $\neq$ is an apartness relation if $\forall xy(x = y \vee x \neq y)$.

8.1.2. Prove proposition 1.2.

8.1.3. (i) Show that the class of strongly extensional sets is closed under union and intersection.

(ii) Let $f \in A \to B$ be strongly extensional. If $X \subset B$ is strongly extensional, then so is $f^{-1}[X]$.

8.1.4. Give a Kripke model that does not carry an apartness relation.

8.1.5. Show that the power set of an inhabited set does not admit an apartness relation.

8.1.6. The theories of apartness and linear order have been presented in a Hilbert-type formulation. Show that the natural deduction formulations below yield equivalent systems.

Equality.

$$\frac{}{x = x} \qquad \frac{x = y}{y = x} \qquad \frac{x = y \quad y = z}{x = z}$$

Apartness. The rules of equality plus

$$\frac{\begin{array}{c}[x \# y]^{(1)} \\ \vdots \\ \bot\end{array}}{x = y}\,(1) \qquad \frac{x \# y \quad x = y}{\bot}$$

$$\frac{x \# y \quad \begin{array}{c}{}^{(1)}[x \# z] \\ \vdots \\ A\end{array} \quad \begin{array}{c}[y \# z]^{(1)} \\ \vdots \\ A\end{array}}{A}\,(1)$$

$$\frac{x = x' \quad x \# y}{x' \# y}$$

Linear order. The rules of equality, apartness plus

$$\frac{x < y \quad y < z}{x < z} \qquad \frac{\neg x < y \quad \neg y < x}{x = y}$$

$$\frac{x < y \quad x = y}{\bot} \qquad \frac{x < y \quad \begin{array}{c}{}^{(1)}[z < y] \\ \vdots \\ A\end{array} \quad \begin{array}{c}[x < z]^{(1)} \\ \vdots \\ A\end{array}}{A}\,(1)$$

$$\frac{x < y}{x \# y} \qquad \frac{x \# y \quad \begin{array}{c}{}^{(1)}[x < y] \\ \vdots \\ A\end{array} \quad \begin{array}{c}[y < x]^{(1)} \\ \vdots \\ A\end{array}}{A}\,(1)$$

$$\frac{x = x' \quad x < y}{x' < y} \qquad \frac{y = y' \quad x < y}{x < y'}$$

8.2.1. Consider the group of $n \times n$ matrices over $\mathbb{R}$ with det # 0. Show that this is a group with apartness relation, for $A \# B := \exists \vec{x}(A\vec{x} \# B\vec{x})$.

8.2.2. Show that in 2.3(2) $(A \leftrightarrow B) \Leftrightarrow G_A = G_B$.

8.2.3. Show that in 2.5(3) a normal antisubgroup is defined.

8.2.4. Give an example of a group with antisubgroup A and apartness #, such that $a \in A \to a \# e$ does not hold.

8.2.5. Give an example of a (group) isomorphism that is not strongly extensional.

8.2.6. Check the strong extensionality of $\cdot$ and $^{-1}$ in a Kripke group with # defined by normal subgroups at the nodes.

8.2.7. Give an example of a (Kripke) group such that neither $\forall xy(xy = yx)$ nor $\exists xy(xy \# yx)$ holds.

8.2.8. Let G be a free abelian Kripke group. Show that the groups at the nodes are all free.

8.3.1. Show that in a ring $x \# y \to x + z \# y + z$.

8.3.2. Show that for a decidable ideal I (i.e. $\forall x(x \in I \vee x \notin I)$) R/I has a decidable equality. Similarly, if I is *stable* (i.e. $\forall x(\neg\neg x \in I \to x \in I)$), then R/I has a stable equality.

8.3.3. If A is a minimal anti-ideal and I is an ideal such that $1 \notin I$ and $A^c \subset I$, then $A^c = I$. This shows that A^c is a maximal ideal.

8.3.4. The following notion of maximal ideal is too strong: $1 \notin I$ and $\forall J(I \subset J \to I = J \vee J = R)$. Show that the resulting factor ring has decidable equality. Show that the existence of such ideals in $\mathbb{Z}$ is equivalent with PEM.

8.4.1. Show that "weakly dependent" does not imply "dependent", and that "independent" does not imply "free from" (cf. 4.1).

8.4.2. Give an example of a subspace of $\mathbb{R}^2$ without a dimension.

8.4.3. Consider the linear mapping φ represented by a matrix A. Show that φ is strongly extensional.

8.5.1. Let $f \in F[X]$ be of degree at most n. If for some $a_0, \ldots, a_n \in F$ $\bigwedge\{a_i \# a_j : i < j \le n\}$ and $\bigwedge\{f(a_i) = 0 : i \le n\}$, then $f = 0$.

8.5.2. If $f, g \in F[X]$ with g regular, then there are unique q and r such that $f = qg + r$, where r is a shorter polynomial than g.

8.5.3. If f and g are relatively prime, then $f(0) \# 0 \vee g(0) \# 0$.

8.6.1. Show that for a ring the statements $\forall x(\exists y(xy = 1) \vee \exists y((1 - x)y = 1))$ and $\forall x(Ex^{-1} \vee E(1 - x)^{-1})$ are equivalent (recall that $x^{-1} = Iy.xy = 1$).

8.6.2. Let D be a Kripke integral domain with ideals I_k determined by the apartness relation. Show that I_k contains all zerodivisors in D_k, and that D_k is nilpotent free.

8.6.3. Give a Kripke field in which $2 = 0 \vee 3 = 0$, but in which $2 = 0$ and $3 = 0$ fail. This shows that fields do not necessarily have a characteristic.

8.6.4. Let S be a multiplicatively closed subset of a ring R with $\#$, not containing 0 (i.e. $a, b \in S \leftrightarrow ab \in S, 0 \notin S$). The ring of fractions R_S is a generalization of the field of fractions (cf. 3.12); it is defined as follows. Define $(a, b) \sim (c, d) := \exists s \in S(s(ad - bc) = 0)$. Show that $\sim$ is an equivalence relation. R_S is the set of equivalence classes, with addition and multiplication defined as in 3.12.
(i) Show that R_S is a ring.
In the following take for S a prime anti-ideal A compatible with $\#$.
(ii) Define $(a, b) \# (c, d) := \exists s \in A(s(ad - bc) \# 0)$. Show that this relation induces a preapartness relation on R_A such that $+$ and $\cdot$ are strongly extensional.
(iii) Show that $\#$ need not be an apartness relation.
(iv) Show that R_A is a local ring.
(v) Show that the canonical preapartness relation on R_A is included in the above relation $\#$, but not necessarily vice versa.
(vi) Show that $A^{-1}A$ (i.e. $\{a/b \in R_A : a \in A\}$) is a minimal anti-ideal in R_A.

8.6.5. A set is weakly finite if each infinite sequence of its elements has repetitions. Show that a weakly finite integral domain is a field with the property $\forall x(x = 0 \vee x \# 0)$ (Ruitenburg 1982).

8.6.6. Let R be a local ring such that $\forall x(\neg Ex^{-1} \to x = 0)$, then R is a field under the canonical apartness relation.

8.7.1. Give an effective method to determine i such that $t \in I_i$ in the proof of lemma 7.2 (cf. (4) and the following lines).

CHAPTER 9

FINITE-TYPE ARITHMETIC AND THEORIES OF OPERATORS

The first two sections of this chapter are devoted to the metamathematics of intuitionistic finite-type arithmetic $\mathbf{HA}^\omega$. In constructive mathematics there is a considerable difference between function types and set types, since functions, even choice sequences, carry more effective information than sets: each input from the domain gives an output. Accordingly there are three kinds of formal constructive type theories: theories based on function types, theories based on set types, and combinations of both.

$\mathbf{HA}^\omega$ is a "neutral" (minimal) theory of the first kind: it permits extensional as well as nonextensional interpretations of equality between functionals, and does not incorporate any principles obtained from a closer analysis of the constructive function concept, such as Church's thesis, or continuity principles for choice sequences.

$\mathbf{HA}^\omega$ represents an obvious way of extending the expressive power of **HA** and **EL** without increasing proof-theoretic strength. However, due to the absence of set types and its limited range of function types, $\mathbf{HA}^\omega$ is not particularly convenient for formalizing the practice of constructive mathematics. On the other hand $\mathbf{HA}^\omega$ lends itself very well to metamathematical research and presents us with interesting examples of metamathematical techniques and results for constructive functional type theories.

$\mathbf{HA}^\omega$ can be embedded as a subsystem in many other formalisms for constructive mathematics, such as the operator-part **APP** of Feferman's theories of operators and classes, Martin-Löf's type theories (chapter 11), and in various **ZF**-like intuitionistic set theories (section 11.8).

In sections 3–5 we will consider in some detail the theory **APP**. This theory codifies the notion of "untyped operation" (such as in the untyped lambda calculus) with a *partial* application operation. In **APP** we have the abstract version of the main tools of recursion theory (combinatory logic), so that we can formulate an abstract version of realizability, similar in all respects to Kleene's numerical realizability (cf. section 4.4), except that the operations are not assumed to be partial recursive.

As a system for formalizing constructive mathematics **APP** has certain advantages over **HA**, though in this respect the extension of **APP** with elementary classes, the system $\mathbf{EM}_0$ of 8.2, is still better.

Section 6 discusses choice principles and extensionality for $\mathbf{HA}^\omega$ and **APP**.

Section 7 demonstrates several metamathematical techniques applied to **APP**. As typical results we obtain the DP (disjunction property), EDN (explicit definability for numbers), a general form of Markov's rule, the recursive continuity rule, and the fan rule. By suitable adaptations the techniques used apply to $\mathbf{HA}^\omega$ and many other intuitionistic systems as well, but we shall not discuss the necessary adaptations in this book; **APP** just represents a typical and not too complicated case.

Section 8 briefly discusses the extension of **APP** to theories with operators and classes.

1. Intuitionistic finite-type arithmetic

In this section we describe intuitionistic finite-type arithmetic (in several variants) and some models of intrinsic interest for this theory: the model of the hereditarily recursive operations HRO and its extensional counterpart, the model of the hereditarily effective operations HEO, as well as the intensional continuous functionals ICF.

1.1. DEFINITION (*type symbols*). The set of *finite-type* symbols $\mathscr{T}$ is inductively generated by the clauses:
(i) $0 \in \mathscr{T}$,
(ii) $\sigma, \tau \in \mathscr{T} \Rightarrow (\sigma \times \tau) \in \mathscr{T}$ (*product types*),
(iii) $\sigma, \tau \in \mathscr{T} \Rightarrow (\sigma \to \tau) \in \mathscr{T}$ (*function types*).
As an alternative notation for $(\sigma \to \tau)$ we use $(\sigma\tau)$. We write 1 for (00), $n+1$ for $(n0)$. Outer parentheses in type symbols are usually omitted. In the alternative notation we also save on parentheses by association to the *right*, i.e. $\sigma_0\sigma_1\sigma_2\sigma_3$ is short for $(\sigma_0(\sigma_1(\sigma_2(\sigma_3))))$. $\sigma_1 \times \cdots \times \sigma_n$ abbreviates $(\cdots((\sigma_1 \times \sigma_2) \times \sigma_3) \cdots \times \sigma_n)$.

The subset $\mathscr{T}_\to$ of the *linear types* is the set of type symbols generated by (i), (iii) alone. □

Another notation, which is rather often used in the literature, is $(\sigma)\tau$ for $(\sigma \to \tau)$. In this notation $\sigma_0\sigma_1\sigma_2\sigma_3$ becomes $(\sigma_0)(\sigma_1)(\sigma_2)\sigma_3$.

1.2. DEFINITION. A *finite-type structure* consists of a collection $\langle(M_\sigma, =_\sigma)\rangle_{\sigma\in\mathcal{T}}$, M_σ a set, $=_\sigma$ an equivalence relation on M_σ which we call the *equality on* M_σ; $M_{\sigma\to\tau}$ is a collection of mappings from M_σ to M_τ respecting the equalities, i.e. if $f \in M_{\sigma\to\tau}$, then $\forall x, y \in M_\sigma(x =_\sigma y \to fx =_\tau fy)$, and $M_{\sigma\times\tau}$ is $M_\sigma \times M_\tau$ with

$$(x, y) =_{\sigma\times\tau} (x', y') \Leftrightarrow x =_\sigma x' \wedge y =_\tau y'. \quad \Box$$

N.B. We have chosen to represent the sort of a type structure as sets with an equivalence relation, instead of sets with their given equality; the reason is that this permits a more flexible, less cumbersome description of many examples. If we wish, we can always form equivalence classes.

In this chapter, we shall only consider *standard* type structures in which $(M_0, =_0)$ is (isomorphic to) $\mathbb{N}$ with the usual equality.

1.3. *Examples of type structures.*
(A) *The full type structure* $\langle(F_\sigma, =_\sigma)\rangle_{\sigma\in\mathcal{T}}$ is defined by:

$F_0 := \mathbb{N}$, $\quad =_0$ is equality on $\mathbb{N}$;

$F_{\sigma\times\tau} := \{(x, y) : x \in F_\sigma, y \in F_\tau\}$ with the induced equality;

$F_{\sigma\to\tau} :=$ collection of all functions from F_σ to F_τ;

$$f =_{\sigma\to\tau} g := \forall x \in F_\sigma(fx =_\tau gx).$$

(B) *The structure of the hereditarily recursive operations.*
$\mathrm{HRO} := \langle(\mathrm{HRO}_\sigma, =_\sigma)\rangle_{\sigma\in\mathcal{T}}$ is inductively defined by:

$$\mathrm{HRO}_0 := \mathbb{N},$$

$$\mathrm{HRO}_{\sigma\to\tau} := \{x \in \mathbb{N} : \forall y \in \mathrm{HRO}_\sigma(\{x\}(y) \in \mathrm{HRO}_\tau)\},$$

$$\mathrm{HRO}_{\sigma\times\tau} := \{x \in \mathbb{N} : j_1x \in \mathrm{HRO}_\sigma \wedge j_2x \in \mathrm{HRO}_\tau)\},$$

(using standard Kleene bracket notation) and for all $\sigma \in \mathcal{T}$, $x, y \in \mathrm{HRO}_\sigma$,

$$(x =_\sigma y) := (x = y).$$

Thus $\mathrm{HRO}_{0\to0}$ consists of all codes of total recursive functions, and equality is equality of codes (and not extensional equality between the functions represented by the codes).

(C) *The structure of the hereditarily effective operations.* HEO $:= \langle(\mathrm{HEO}_\sigma, =_\sigma)\rangle_{\sigma\in\mathcal{T}}$ is inductively defined by:

$$\mathrm{HEO}_0 := \mathbb{N},$$

$$x =_0 y := x = y,$$

$$\mathrm{HEO}_{\sigma\to\tau} := \{x \in \mathbb{N} : \forall yy'(y =_\sigma y' \to \{x\}(y) =_\tau \{x\}(y'))\},$$

$$x =_{\sigma\to\tau} y := x \in \mathrm{HEO}_{\sigma\to\tau} \land y \in \mathrm{HEO}_{\sigma\to\tau} \land$$
$$\forall z \in \mathrm{HEO}_\sigma(\{x\}(z) =_\tau \{y\}(z)),$$

$$\mathrm{HEO}_{\sigma\times\tau} := \{x \in \mathbb{N} : j_1 x \in \mathrm{HEO}_\sigma \land j_2 x \in \mathrm{HEO}_\tau\},$$

$$x =_{\sigma\times\tau} y := (j_1 x =_\sigma j_1 y) \land (j_2 x =_\tau j_2 y).$$

$\mathrm{HEO}_{0\to 0} = \mathrm{HRO}_{0\to 0}$, but equality between elements of $\mathrm{HEO}_{0\to 0}$ is equality of the corresponding functions. Note that the domains for the types and equality are defined simultaneously.

REMARKS. (i) The concept of the full type structure makes sense not only classically, but also constructively. However, its extent is largely determined by the general principles one is willing to accept. Thus, if one accepts ECT_0 (cf. 4.4.8), the full type structure coincides with HEO. On the other hand, when viewed classically, the full type structure is uncountable, whereas HRO and HEO are countable.

(ii) HRO_2 contains an operation assigning to each (code of a) total recursive function its code (namely the identity operation), but this operation is not represented in HEO_2!

1.4. *Primitive recursive functionals.* We shall in particular be interested in type structures closed under finite-type recursion. That is to say, the type structure should satisfy the following closure conditions:

(i) 0 (zero) and S (successor) are in the type structure;

(ii) closure under explicit definition;

(iii) closure under recursion: if f, g are of types σ, $\sigma \times 0 \to \sigma$ respectively, then there is an h of type $0 \to \sigma$ such that

$$h(0) = f, \qquad h(Sz) = g(hz, z).$$

It is intuitively immediate that the full type structure (example (A) in 1.3) is closed under finite-type recursion, although a formal proof in Zermelo–Fraenkel set theory **ZF** requires a certain amount of work.

Examples (B) and (C) are also closed under finite-type recursion, as we shall see below in 1.9.

At first sight it might seem as if the closure conditions generate at type 1 only primitive recursive functions. This, however, is not true: the fact that the values in the recursion schema may be of type σ, instead of type 0 only, permits us to define functions which are not primitive recursive. For example, the well-known Ackermann function (cf. Kleene 1952, section 55) given by

$$\begin{cases} f(0,x) &= Sx, \\ f(Sz,0) &= f(z,1), \\ f(Sz,Sx) &= f(z,f(Sz,x)), \end{cases}$$

can be introduced by finite-type recursion, defining $f_z = \lambda x.f(z,x)$ as a function of z (exercise).

We shall now describe a language for axiomatizing the theory of finite-type structures closed under finite-type recursion.

1.5. *The language* $\mathscr{L}$(**HA**$^\omega$). For each $\sigma \in \mathscr{T}$ there is a countably infinite supply of variables of type σ; we shall use $x^\sigma, y^\sigma, z^\sigma, u^\sigma, v^\sigma, w^\sigma$ for such variables. For each $\sigma \in \mathscr{T}$ there is a binary predicate $=_\sigma$ for equality at type σ; and for all $\sigma, \tau \in \mathscr{T}$ there is an application operator $\mathrm{Ap}^{\sigma,\tau}$.

Furthermore, the language contains the following constants, for all $\sigma, \tau, \rho \in \mathscr{T}$, with "$c$ a constant of type σ" indicated by "$c \in \sigma$":

$$0 \in 0 \quad (\textit{zero});$$

$$S \in 0 \to 0 \quad (\textit{successor});$$

$$\mathbf{p}^{\sigma,\tau} \in \sigma \to (\tau \to (\sigma \times \tau)) \quad (\textit{pairing operator});$$

$$\mathbf{p}_0^{\sigma,\tau} \in (\sigma \times \tau) \to \sigma, \qquad \mathbf{p}_1^{\sigma,\tau} \in (\sigma \times \tau) \to \tau \quad (\textit{unpairing operators});$$

$$\mathbf{k}^{\sigma,\tau} \in \sigma \to (\tau \to \sigma),$$

$$\mathbf{s}^{\rho,\sigma,\tau} \in (\rho \to (\sigma \to \tau)) \to ((\rho \to \sigma) \to (\rho \to \tau)) \quad (\textit{combinators});$$

$$\mathbf{r}^\sigma \in \sigma \to ((\sigma \to (0 \to \sigma)) \to (0 \to \sigma)) \quad (\textit{recursor}).$$

Terms:

(i) variables and constants of type σ are terms of type σ;

(ii) if t is a term of type $\sigma \to \tau$, t' a term of type σ, then $\mathrm{Ap}^{\sigma,\tau}(t, t')$ is a term of type τ.

Formulas:

(iii) prime formulas are expressions of the form $t =_\sigma s$, where t and s are terms of type σ;

(iv) a prime formula is a formula; arbitrary formulas are built from prime formulas with the help of the logical operators $\rightarrow, \wedge, \vee, \forall x^\sigma, \exists x^\sigma$.

1.6. NOTATION. We use $t, t', t'', t_i, s, s', \ldots$ for terms; we sometimes write t^σ, s^σ to indicate that t, s are terms of type σ; this can also be expressed by $t \in \sigma$, $s \in \sigma$. $\mathrm{Ap}^{\sigma,\tau}(t, t')$ will be written as (tt'); outer parentheses are mostly dropped, and $t_0 t_1 \ldots t_n$ is short for $((t_0 t_1) t_2) \ldots t_n$ (association to the *left*). (N.B. This deviates from our notation in the case of **EL**, where $\varphi(t)$ is the standard notation for φ applied to t.) Occasionally we insert parentheses for readability.

If we wish to indicate variables occurring free in a compound term we use square brackets: $t[x]$, $t[x, y]$, and so on; also $t[t']$ as shorthand for $t[x/t']$, etc.

Type superscripts of variables, constants, and terms will be mostly omitted; types will be assumed to be fitting, i.e. if we write tt', t is assumed to be of type $\sigma \rightarrow \tau$, t' of type σ for certain $\sigma, \tau \in \mathscr{T}$.

We shall use $\vec{u}, \vec{v}, \vec{w}, \vec{x}, \vec{y}, \vec{z}, \vec{U}, \ldots, \vec{Z}$ for finite sequences (possibly empty) of variables of finite type, and $\vec{t}, \vec{s}, \vec{S}$ for finite sequences of terms. Let $\vec{x} \equiv (x_1, \ldots, x_n)$, $\vec{y} \equiv (y_1, \ldots, y_m)$; $\forall \vec{x}, \exists \vec{x}$ abbreviate $\forall x_1 \forall x_2 \ldots \forall x_n$, $\exists x_1 \exists x_2 \ldots \exists x_n$, respectively, $\forall \vec{x}\vec{y}$ is $\forall \vec{x} \forall \vec{y}$, etc. For an empty sequence $\vec{x}$, $\forall \vec{x}$ and $\exists \vec{x}$ are empty strings of quantifiers.

If $\vec{t} \equiv (t_1, \ldots, t_n)$, $\vec{s} \equiv (s_1, \ldots, s_m)$, then $\vec{t}\vec{s}$ in a formula $A(\vec{t}\vec{s})$ stands for a finite sequence of terms $(t_1 s_1 \ldots s_m, t_2 s_1 \ldots s_m, \ldots, t_n s_1 \ldots s_m)$.

For $\boldsymbol{p}ts$ we also write (t, s). □

1.7. *Axioms of* $\mathbf{HA}^\omega$. The logical basis of $\mathbf{HA}^\omega$ is many-sorted **IQC** with equality. That is to say, for equality at all types we assume

$$x = x, \qquad x = y \rightarrow y = x, \qquad x = y \wedge y = z \rightarrow x = z,$$

$$y = z \rightarrow xy = xz, \qquad x = y \rightarrow xz = yz.$$

The defining equations for the constants are

$$\boldsymbol{k}xy = x, \qquad \boldsymbol{s}xyz = xz(yz),$$

$$\boldsymbol{p}_0(\boldsymbol{p}xy) = x, \qquad \boldsymbol{p}_1(\boldsymbol{p}xy) = y, \qquad \boldsymbol{p}(\boldsymbol{p}_0 z)(\boldsymbol{p}_1 z) = z,$$

$$\boldsymbol{r}xy0 = x, \qquad \boldsymbol{r}xy(Sz) = y(\boldsymbol{r}xyz)z.$$

Finally we have the arithmetical axioms

$$Sx = Sy \to x = y, \qquad \neg 0 = Sx,$$

$$A(0) \wedge \forall x(A(x) \to A(Sx)) \to \forall y A(y).$$

Here $\neg A := A \to S0 = 0$.

N.B. Elsewhere in the literature $\mathbf{N\text{-}HA}^\omega$ is often used instead of $\mathbf{HA}^\omega$.

The combinators $\boldsymbol{k}, \boldsymbol{s}$ permit *explicit definition* as shown by the next proposition.

1.8. PROPOSITION. *Let $t[x]$ be any term of* $\mathbf{HA}^\omega$, *then there is a term, which we will denote by $\lambda x.t[x]$, such that*:
(i) $(\lambda x.t[x])(t') = t[t']$;
(ii) $\lambda x.(tx) = t$ *if* $x \notin \mathrm{FV}(t)$;
(iii) $x \notin \mathrm{FV}(t') \cup \mathrm{FV}(t'') \Rightarrow t' = t'' \to \lambda x.t[y/t'] = \lambda x.t[y/t'']$.

PROOF. $\lambda x^\sigma.t$ is defined by induction on the complexity of t:
(a) if $x \notin \mathrm{FV}(t)$, we put $\lambda x.t := \boldsymbol{k}t$;
(b) $\lambda x^\sigma.x^\sigma := \boldsymbol{s}^{\sigma,0\sigma,\sigma}\boldsymbol{k}^{\sigma,0\sigma}\boldsymbol{k}^{\sigma,0}$;
(c) if $x \notin \mathrm{FV}(t)$, $\lambda x.tx := t$;
(d) if $x \in \mathrm{FV}(t)$, or $x \in \mathrm{FV}(t')$ and $t' \not\equiv x$, we put $\lambda x.tt' := \boldsymbol{s}(\lambda x.t)(\lambda x.t')$.

Now (ii) is immediate from this definition, (i) and (iii) are proved simultaneously. For example, in the induction step for (i), if $x \in \mathrm{FV}(t)$: $(\lambda x.tt')t'' \equiv \boldsymbol{s}(\lambda x.t)(\lambda x.t')t'' = (\lambda x.t)t''((\lambda x.t')t'') = t[x/t''](t'[x/t'']) = tt'[x/t'']$, etc. □

N.B. For a definition of $\lambda x.t[x]$ with clause (c) and the restriction "$x \in \mathrm{FV}(t)$" in (d) dropped, (ii) of the theorem does not generally hold.

CONVENTION. $\lambda \vec{x}.(t_1, \ldots, t_n) := (\lambda \vec{x}.t_1, \ldots, \lambda \vec{x}.t_n)$. □

1.9. *The models* HRO, HEO. We shall now show that the type structures HRO and HEO are models of $\mathbf{HA}^\omega$, or, more precisely, can be made into models for $\mathbf{HA}^\omega$ by choosing appropriate interpretations of the constants.

In both models, $\mathrm{Ap}^{\sigma,\tau}$ is uniformly interpreted by partial recursive function application, i.e. if the terms t, t' are interpreted as $[\![t]\!]$ and $[\![t']\!]$, respectively, then $\mathrm{Ap}^{\sigma,\tau}(t, t')$ is interpreted by $\{[\![t]\!]\}([\![t']\!])$. As to the other constants, we can take in HRO:

$$[\![0]\!]_{\mathrm{HRO}} := 0,$$

$$[\![S]\!]_{\mathrm{HRO}} := \Lambda x.x + 1,$$

$$[\![k^{\sigma,\tau}]\!]_{\mathrm{HRO}} := \Lambda xy.x,$$

$$[\![s^{\rho,\sigma,\tau}]\!]_{\mathrm{HRO}} := \Lambda xyz.\{\{x\}(z)\}(\{y\}(z)),$$

$$[\![p^{\sigma,\tau}]\!]_{\mathrm{HRO}} := \Lambda xy.j(x, y),$$

$$[\![p_i^{\sigma,\tau}]\!]_{\mathrm{HRO}} := \Lambda x.j_{i+1}x \quad (i \le 1),$$

$[\![r^{\sigma}]\!]_{\mathrm{HRO}}$ is a number r such that

$$\begin{cases} \{r\}(x, y, 0) \simeq x, \\ \{r\}(x, y, Sz) \simeq \{y\}(\{r\}(x, y, z), z), \end{cases}$$

where $\{r\}(t_1, t_2, t_3)$ abbreviates $\{\{\{r\}(t_1)\}(t_2)\}(t_3)$. r can be constructed with the help of the recursion theorem.

For HEO, we can use exactly the same interpretation for the constants, that is to say for each constant c we have $[\![c]\!]_{\mathrm{HEO}} := [\![c]\!]_{\mathrm{HRO}}$. The only difference is in the interpretation of equality. The reader can easily verify the axioms of $\mathbf{HA}^{\omega}$ in HRO and HEO.

N.B. Strictly speaking it is not correct to speak about *the* model HRO or HEO for $\mathbf{HA}^{\omega}$, since we have a lot of freedom in the exact choice of the interpretation of the constants (cf. E9.2.6). However, we shall neglect this point when the exact version does not matter.

1.10. *Embedding* **HA** *in* $\mathbf{HA}^{\omega}$. It is easy to see that **HA** is contained as a subsystem in $\mathbf{HA}^{\omega}$ under a suitable embedding. In particular, for each n-ary primitive recursive function symbol φ we can find a $t_\varphi \in (0 \times \cdots \times 0) \to 0$ satisfying the same recursion equations; for example, if φ is defined from ψ, χ by recursion: $\varphi(0, \vec{x}) = \psi(\vec{x})$, $\varphi(Sz, \vec{x}) = \chi(\varphi(z, \vec{x}), z, \vec{x})$, we must have

$$t_\varphi(0, \vec{x}) = t_\psi(\vec{x}), \qquad t_\varphi(Sz, \vec{x}) = t_\chi(t_\varphi(z, \vec{x}), \vec{x}, z). \tag{1}$$

In $\mathbf{HA}^{\omega}$ we code n-tuples $(t_1, \ldots, t_n)$ by

$$(t_1) := t_1,$$

$$(t_1, t_2) := \boldsymbol{p}t_1t_2,$$

$$(t_1, \ldots, t_{n+1}) := ((t_1, \ldots, t_n), t_{n+1});$$

with help of $\boldsymbol{p}_0$, $\boldsymbol{p}_1$ one readily defines inverse operations $\boldsymbol{p}_{n,i}$ such that

$$\boldsymbol{p}_{n,i}(t_0,\ldots,t_{n-1}) = t_i \quad (i < n).$$

Now $t_\psi(t_0,\ldots,t_{n-1})$ is taken to mean t_ψ applied to the tuple $(t_0,\ldots,t_{n-1})$, etc.

Returning to (1) above, we can define t_φ as follows, assuming t_ψ, t_χ to have been defined:

$$t^* := \lambda z\vec{x}.\boldsymbol{r}(t_\psi \vec{x})(\lambda uz.t_\chi(u, \vec{x}, z))z,$$

$$t := \lambda u.t^*(\boldsymbol{p}_{n+1,0}u)(\boldsymbol{p}_{n+1,1}u)\ldots(\boldsymbol{p}_{n+1,n}u).$$

Interpreting each φ of **HA** by t_φ, and interpreting variables of **HA** by type 0 variables of $\mathbf{HA}^\omega$, we obtain the required embedding of **HA** in $\mathbf{HA}^\omega$.

1.11. *The systems* $\mathbf{I\text{-}HA}^\omega$, $\mathbf{E\text{-}HA}^\omega$. The system $\mathbf{HA}^\omega$ is noncommittal as to the exact nature of equality between objects of higher type, as is clearly seen by comparing the models HRO and HEO.

So we can strengthen $\mathbf{HA}^\omega$ by being more explicit about equality. One obvious possibility is to require equality to be extensional, i.e. we add for all suitable types of y, z, and x:

EXT $\quad \forall yz(\forall x(yx = zx) \to y = z).$

The resulting system we call $\mathbf{E\text{-}HA}^\omega$ ("E" from "extensional"). If we define *extensional equality* $=_{e,\sigma}$ as follows:

$$=_{e,0} \text{ is } =_0,$$

$$x =_{e,\sigma\times\tau} y := \boldsymbol{p}_0 x =_{e,\sigma} \boldsymbol{p}_0 y \wedge \boldsymbol{p}_1 x =_{e,\tau} \boldsymbol{p}_1 y,$$

$$x =_{e,\sigma\tau} y := \forall z^\sigma(xz =_{e,\tau} yz),$$

then it is easy to see that, relative to $\mathbf{HA}^\omega$, EXT is equivalent to

$$\forall xy(x =_\sigma y \leftrightarrow x =_{e,\sigma} y) \tag{1}$$

for all types σ. This makes it possible to give an alternative formulation of $\mathbf{E\text{-}HA}^\omega$ in the sublanguage of $\mathscr{L}(\mathbf{HA}^\omega)$ with equality of type 0 only (cf. E9.1.4).

We can also think of our objects of type σ as rules or algorithms, and interpret $=_\sigma$ as equality of rules; then it is natural to require

$$x^\sigma =_\sigma y^\sigma \vee \neg x^\sigma =_\sigma y^\sigma.$$

An even stronger way of expressing the decidability is to require the existence of an equality functional $e^\sigma \in \sigma \to (\sigma \to 0)$ satisfying

$$e^\sigma xy \le 1, \qquad e^\sigma xy = 0 \leftrightarrow x =_\sigma y.$$

The resulting system we call **I-HA**$^\omega$ ("I" from "intensional"). HEO is a model of **E-HA**$^\omega$, HRO a model of **I-HA**$^\omega$ (take $[\![e^\sigma]\!]_{\mathrm{HRO}} := \Lambda xy.\mathrm{sg}|x - y|$).

1.12. *The system* **HA**$^\omega_0$. Let $\mathscr{L}_0$ be the sublanguage of $\mathscr{L}$(**HA**$^\omega$) with equality of type 0 only, which therefore has the pleasant property that all prime formulas are decidable. **HA**$^\omega_0$ is a subsystem of **HA**$^\omega$ in the sublanguage $\mathscr{L}_0$, which is axiomatized as **HA**$^\omega$, except that the equality axioms are now postulated in the form:

$$u(\boldsymbol{k}xy) = ux, \quad u(\boldsymbol{s}xyz) = u(xz(yz)),$$

$$u(\boldsymbol{p}_0(\boldsymbol{p}xy) = ux, \quad u(\boldsymbol{p}_1(\boldsymbol{p}xy)) = uy,$$

$$u(\boldsymbol{p}(\boldsymbol{p}_0z)(\boldsymbol{p}_1z)) = uz,$$

$$u(\boldsymbol{r}xy0) = ux,$$

$$u(\boldsymbol{r}xy(Sz)) = u(y(\boldsymbol{r}xyz)z).$$

Note that the embedding of **HA** in **HA**$^\omega$ described in 1.10 actually embeds **HA** in **HA**$^\omega_0$.

The relationship between **HA**$^\omega_0$ and **HA**$^\omega$ is simple: **HA**$^\omega_0$ is the $\mathscr{L}_0$-fragment of **HA**$^\omega$, i.e. the following theorem holds.

1.13. Theorem (*Rath 1978*). **HA**$^\omega$ *is conservative over* **HA**$^\omega_0$.

Proof. We define a mapping $^\circ$ on formulas of **HA**$^\omega$ by:
(i) $(t =_0 s)^\circ := (t =_0 s)$,
(ii) $(t =_\sigma s)^\circ := \forall y^{\sigma 0}(yt =_0 ys)$ for $\sigma \neq 0$ (y a "fresh" variable, not occurring in the formulas to be translated),
(iii) $^\circ$ is a homomorphism with respect to $\wedge, \to, \vee, \forall, \exists$.
For A in $\mathscr{L}_0$, $A^\circ \equiv A$, and always $\mathrm{FV}(A) = \mathrm{FV}(A^\circ)$. By induction on the length of derivations

$$\mathbf{HA}^\omega + \Gamma \vdash A \Rightarrow \mathbf{HA}^\omega_0 + \Gamma^\circ \vdash A^\circ.$$

All the logical axioms and rules, as well as induction are trivially preserved under $^\circ$ and the defining equations of the constants are translated by $^\circ$ into their **HA**$^\omega_0$-versions. Only the equality axioms need some care. Consider e.g.

$x^{\sigma\to\tau} = y^{\sigma\to\tau} \to x^{\sigma\to\tau}z^{\sigma} = y^{\sigma\to\tau}z^{\sigma}$. This becomes under $^{\circ}$:

$$\forall u^{(\sigma\to\tau)\to 0}(ux = uy) \to \forall v^{\tau\to 0}(v(xz) = v(yz)).$$

Let $t := \lambda w.v(wz)\ (w \in \sigma \to \tau)$; so from $\forall u(ux = uy)$ we have $tx = ty$, i.e. $v(xz) = v(yz)$, etc. □

1.14. PROPOSITION. **I-HA**$^{\omega}$ *and* **E-HA**$^{\omega}$ *are conservative over* **HA**.

PROOF. In order to show that **I-HA**$^{\omega}$ is conservative over **HA**, we interpret **I-HA**$^{\omega}$ in **HA**; with each sentence A of **I-HA**$^{\omega}$ there corresponds in the obvious way an interpretation $[\![A]\!]_{\mathrm{HRO}}$ which is arithmetical (more precisely, a formula in the definitional extension of **HA** with Kleene-brackets for partial recursive function application). It is indeed routine to verify that

$$\mathbf{I\text{-}HA}^{\omega} \vdash A \Rightarrow \mathbf{HA} \vdash [\![A]\!]_{\mathrm{HRO}}.$$

It remains to be shown that for **HA** (as embedded in **HA**$^{\omega}$) we have, for each formula A in the language of **HA**:

$$\mathbf{HA} \vdash A \leftrightarrow [\![A]\!]_{\mathrm{HRO}}.$$

This is done by induction on the complexity of A; the only difficulty is in the basis case, where we have to show

$$\mathbf{HA} \vdash t[x_1, \ldots, x_n] = [\![t[x_1, \ldots, x_n]]\!]_{\mathrm{HRO}}$$

for all terms of **HA**. But this in turn is proved by induction on the term complexity, showing first for all primitive recursive function symbols φ:

$$\mathbf{HA} \vdash \varphi(x_1, \ldots, x_n) = [\![t_{\varphi}(x_1, \ldots, x_n)]\!]_{\mathrm{HRO}}.$$

Using HEO instead of HRO, we similarly obtain that **E-HA**$^{\omega}$ is conservative over **HA**. □

1.15. *The intensional continuous functionals.* We shall now describe two variants of an interesting model of **HA**$^{\omega}$ which is neither a model of **I-HA**$^{\omega}$ nor of **E-HA**$^{\omega}$: a "crude" version ICF* and a "refined" version ICF. The type structure $\langle(\mathrm{ICF}_{\sigma}^{*}, =_{\sigma})\rangle_{\sigma\in\mathcal{T}}$ is defined as follows. All sets $\mathrm{ICF}_{\sigma}^{*}$ are subsets of $\mathbb{N}^{\mathbb{N}}$, and equality is extensional equality between functions:

$$\alpha =_{\sigma} \beta := \alpha \in \mathrm{ICF}_{\sigma}^{*} \wedge \beta \in \mathrm{ICF}_{\sigma}^{*} \wedge \forall x(\alpha x = \beta x).$$

The application operation is partial continuous application, i.e. for

$\alpha \in \mathrm{ICF}^*_{\sigma\to\tau}$, $\beta \in \mathrm{ICF}^*_\sigma$, application of α to β is $\alpha|\beta$, where

$$\alpha|\beta = \gamma := \forall x \exists y\big(\alpha(\langle x\rangle * \bar{\beta}y) = \gamma x + 1 \wedge \forall z < y(\alpha(\langle x\rangle * \bar{\beta}z) = 0)\big).$$

Now we put

$$\mathrm{ICF}^*_0 := \{\alpha : \forall x(\alpha x = \alpha 0)\},$$

$$\mathrm{ICF}^*_{\sigma\to\tau} := \{\alpha : \forall \beta \in \mathrm{ICF}^*_\sigma \exists \gamma \in \mathrm{ICF}^*_\tau(\alpha|\beta = \gamma)\},$$

$$\mathrm{ICF}^*_{\sigma\times\tau} := \{\alpha : j_1\alpha \in \mathrm{ICF}^*_\sigma \wedge j_2\alpha \in \mathrm{ICF}^*_\tau\}.$$

Each $n \in \mathbb{N}$ has a unique representative $\lambda x.n$ in ICF^*_0. However, as one can easily check, in ICF^*_1 each $\alpha \in \mathbb{N}^\mathbb{N}$ has many representatives. We can avoid this in the structure ICF, as follows. Let Φ be the following partial operation on functions

$$(\Phi\alpha)\langle\ \rangle = 0, \qquad (\Phi\alpha)\langle x\rangle = 0,$$

$$(\Phi\alpha)(\langle x, y\rangle * n) = z + 1 \quad \text{iff } \alpha(\langle x\rangle * \lambda u.y) = z + 1.$$

$\Phi\alpha$ is total if $\alpha|\lambda u.y$ is defined for all y. In particular $\Phi\alpha$ is total for all $\alpha \in \mathrm{ICF}^*_1$. Now it follows from the definition that all α that represent a given $\alpha^* \in \mathbb{N}^\mathbb{N}$ yield the same $\Phi\alpha$, which is also a representative of α^*. Note that $\Phi(\Phi(\alpha)) = \Phi\alpha$.

We define $\langle(\mathrm{ICF}_\sigma, =_\sigma)\rangle_{\sigma\in\mathcal{T}}$ almost as ICF*, except that we put

$$\mathrm{ICF}_0 := \{\alpha : \forall x(\alpha x = \alpha 0)\},$$

$$\mathrm{ICF}_{0\to 0} := \{\Phi\alpha : \forall \beta \in \mathrm{ICF}_0 \exists \gamma \in \mathrm{ICF}_0(\alpha|\beta = \gamma)\},$$

$$\mathrm{ICF}_{\sigma\to\tau} := \{\alpha : \forall \beta \in \mathrm{ICF}_\sigma \exists \gamma \in \mathrm{ICF}_\tau(\alpha|\beta = \gamma)\} \text{ for } \sigma \to \tau \not\equiv 0 \to 0.$$

We have

$$\mathrm{ICF}^* \models \forall z^2 \in \mathrm{EXT}\, \exists u^2 \forall \alpha^1 \forall \beta^1 \in \bar{\alpha}(u\alpha)(z^2\beta = z^2\alpha),$$

where

$$z^2 \in \mathrm{EXT} := \forall \alpha^1 \beta^1(\forall x(\alpha x = \beta x) \to z^2\alpha = z^2\beta).$$

In ICF we may drop the condition "$z^2 \in \mathrm{EXT}$", i.e.

$$\mathrm{ICF} \models \forall z^2 \exists u^2 \forall \alpha^1 \forall \beta^1 \in \bar{\alpha}(u\alpha)(z^2\beta = z^2\alpha).$$

We leave the verification of these facts as exercises.

1.16. ICF *and* IFC* *as models for* $\mathbf{HA}^{\omega}$. We have already indicated how $=_{\sigma}$, $\mathrm{Ap}^{\sigma,\tau}$ are to be interpreted; it remains to define the interpretations of the constants. For ICF* this is entirely straightforward ("$\Lambda\alpha$" is "$\Lambda^1\alpha$" of 3.7.15):

$$[\![0]\!]_{\mathrm{ICF}^*} := \lambda x.0,$$

$$[\![S]\!]_{\mathrm{ICF}^*} := \Lambda\alpha.\lambda x.S(\alpha 0),$$

$$[\![k]\!]_{\mathrm{ICF}^*} := \Lambda\alpha\Lambda\beta.\alpha,$$

$$[\![s]\!]_{\mathrm{ICF}^*} := \Lambda\alpha\Lambda\beta\Lambda\gamma.(\alpha|\gamma)|(\beta|\gamma),$$

$$[\![p]\!]_{\mathrm{ICF}^*} := \Lambda\alpha\Lambda\beta.j(\alpha;\beta),$$

$$[\![p_0]\!]_{\mathrm{ICF}^*} := \Lambda\alpha.j_1\alpha,$$

$$[\![p_1]\!]_{\mathrm{ICF}^*} := \Lambda\alpha.j_2\alpha.$$

As to $[\![r]\!]_{\mathrm{ICF}^*}$, we have to appeal to the recursion theorem for partial continuous function application (cf. 3.7.14). We can find an ε such that

$$\varepsilon|(\delta,\alpha,\beta,\gamma) \simeq \begin{cases} \alpha & \text{if } \gamma 0 = 0, \\ \beta|(\delta|(\alpha,\beta,\gamma,\lambda z.\gamma 0 \dot- 1),\lambda z.\gamma 0 \dot- 1) & \text{if } \gamma 0 > 0. \end{cases}$$

By the recursion theorem we can find a δ_1 such that

$$\delta_1|(\alpha,\beta,\gamma) \simeq \begin{cases} \alpha & \text{if } \gamma 0 = 0, \\ \beta|(\delta_1|(\alpha,\beta,\lambda z.\gamma 0 \dot- 1),\lambda z.\gamma 0 \dot- 1) & \text{if } \gamma 0 > 0. \end{cases}$$

Now take

$$[\![r]\!]_{\mathrm{ICF}^*} := \Lambda\alpha\Lambda\beta\Lambda\gamma.\delta_1|(\alpha,\beta,\gamma).$$

For ICF we have to be more careful: we want to make certain that type 1 objects α are functions of the form $\Phi(\beta)$, in other words they should satisfy $\Phi\alpha = \alpha$. We write $[\![\ ]\!]$ for $[\![\ ]\!]_{\mathrm{ICF}}$.

We cannot be assured that the canonical choice of a function expressing successor, indicated by the operator "$\Lambda\alpha$" in $\Lambda\alpha.\lambda x.S(\alpha x)$, yields an element $[\![S]\!]$ satisfying $\Phi([\![S]\!]) = [\![S]\!]$. The solution is to take $[\![S]\!] = \Phi([\![S]\!]_{\mathrm{ICF}^*})$.

We can do this systematically as follows. We use canonical smn-operators $\Lambda^{\sigma}\alpha$ for all $\sigma \in \mathcal{T}$ such that

$$\Lambda^{\sigma}\alpha.t[\alpha] := \Lambda\alpha.t[\alpha] \quad \text{for } \sigma \neq 0,$$

$$\Lambda^{0}\alpha.t[\alpha] := \Phi(\Lambda\alpha.t[\alpha]),$$

and we put

$$[\![S]\!] := \Lambda^0\alpha.\lambda x.S(\alpha 0),$$

$$[\![k^{\sigma,\tau}]\!] := \Lambda^\sigma\alpha\Lambda^\tau\beta.\alpha,$$

$$[\![s^{\rho,\sigma,\tau}]\!] := \Lambda^{\sigma\to\tau}\alpha\,\Lambda^{\rho\to\sigma}\beta\,\Lambda^\rho\gamma.(\alpha|\gamma)\,|(\beta|\gamma), \quad \text{etc.}$$

The verification of the axioms of $\mathbf{HA}^\omega$ is left to the reader.

N.B. The definition of $\Lambda^\sigma\alpha$ from $\Lambda\alpha$ is ad hoc; in particular, $\Lambda^0\alpha$ is in general different from $\Lambda^0\alpha$ in 3.7.15.

1.17. Remarks.

(i) We can also define an extensional type structure ECF* (the *extensional continuous functionals*), which is related to ICF* in the same way as HEO is related to HRO; for example,

$$\mathrm{ECF}^*_{\sigma\to\tau} := \{\alpha : \forall\beta\beta'(\beta =_\sigma \beta' \to \alpha|\beta =_\tau \alpha|\beta'\},$$

$$\beta =_{\sigma\to\tau} \gamma := \forall\alpha\in\mathrm{ECF}^*_\sigma(\beta|\alpha =_\tau \gamma|\alpha), \quad \text{etc.}$$

Also there is a structure ECF similarly related to ICF; however, ECF and ECF* are isomorphic (exercise).

The resulting structure ECF (or ECF*) is known as the *countable functionals* (Kleene 1959) or *continuous functionals* (Kreisel 1959), and has been extensively investigated, especially its recursion theory (see e.g. Bergstra 1976, Hyland 1978, and Normann 1980). More information is found in Troelstra (1973, ch. II, section 6).

(ii) In fact the whole definition of ICF, ICF*, and ECF can be given relative to any universe $\mathscr{U}'$ of functions which is a model of **EL** (3.6.2). In particular, we may take the set of total recursive functions $\mathscr{R} \equiv$ TREC for $\mathscr{U}$. Writing ICF($\mathscr{U}$), ECF($\mathscr{U}$), etc. for the versions relativized to $\mathscr{U}$, one has: ECF($\mathscr{R}$), for $\mathscr{R} \equiv$ TREC, is isomorphic to HEO; this fact can be proved in **HA** + MP (but not in **HA** alone; cf. Beeson 1975).

1.18. *Reduction to linear types*: $\mathbf{HA}^\omega_\to$. In the literature $\mathbf{HA}^\omega$ and its variants are often formulated for linear types only. To be specific, $\mathbf{HA}^\omega_\to$ is defined as $\mathbf{HA}^\omega$, but without $\boldsymbol{p}, \boldsymbol{p}_0, \boldsymbol{p}_1$, with linear types only; and with sequences of constants for simultaneous recursion instead of single recursors. That is to say, for each sequence $\boldsymbol{\sigma}^*$ of types $(\sigma_1, \ldots, \sigma_n)$ we have a sequence $(\vec{\boldsymbol{r}}^{\,\sigma})^* \equiv (\boldsymbol{r}_1^{\sigma_1}, \ldots, \boldsymbol{r}_n^{\sigma_n})$ of constants such that

$$\vec{\boldsymbol{r}}\vec{x}\vec{y}0 = \vec{x}, \quad \vec{\boldsymbol{r}}\vec{x}\vec{y}(Sz) = \vec{y}(\vec{\boldsymbol{r}}\vec{x}\vec{y}z)z,$$

where $\vec{x} \equiv (x_1, \ldots, x_n)$, $\vec{y} \equiv (y_1, \ldots, y_n)$, $x_i \in \sigma_i$, $y_i \in \sigma_1 \cdots \sigma_n 0 \sigma_i$.

In fact, the sequences $(\boldsymbol{r}^\sigma)^*$ can be defined explicitly from the single recursors $\boldsymbol{r}^\sigma$, as shown by Schütte (1977, theorem 17.12), but we have included them for simplicity.

It is not possible to obtain $\mathbf{HA}^\omega$ from $\mathbf{HA}^\omega_\rightarrow$ by definitional extension (Barendregt 1974), but we have a result almost as good: $\mathbf{HA}^\omega$ is "almost definitional" over $\mathbf{HA}^\omega_\rightarrow$ (as appears from the *proof* of proposition 1.19).

1.19. PROPOSITION. $\mathbf{HA}^\omega$ *is conservative over* $\mathbf{HA}^\omega_\rightarrow$.

PROOF. To each $\sigma \in \mathcal{T}$ we assign a sequence $\Gamma\sigma$ of types in $\mathcal{T}_\rightarrow$ as follows. Let $\Gamma\sigma \equiv (\sigma_1, \ldots, \sigma_m)$, $\Gamma\tau \equiv (\tau_1, \ldots, \tau_n)$, and put:

(i) $\Gamma 0 := (0)$;

(ii) $\Gamma(\sigma \times \tau) := (\sigma_1, \ldots, \sigma_m, \tau_1, \ldots, \tau_n)$;

(iii) $\Gamma(\sigma \rightarrow \tau) := (\sigma_1 \cdots \sigma_m\tau_1, \ldots, \sigma_1 \cdots \sigma_m\tau_n)$.

We can define sequences $\vec{\boldsymbol{k}}^{\Gamma\sigma, \Gamma\tau}$, $\vec{\boldsymbol{s}}^{\Gamma\rho, \Gamma\sigma, \Gamma\tau}$ such that

$$\vec{\boldsymbol{k}}\vec{x}\vec{y} = \vec{x}, \qquad \vec{\boldsymbol{s}}\vec{x}\vec{y}\vec{z} = \vec{x}\vec{z}(\vec{y}\vec{z}).$$

Thus, for example, we want $\boldsymbol{k}'$, $\boldsymbol{k}''$ such that

$$\boldsymbol{k}'x_1x_2y_1y_2 = x_1, \qquad \boldsymbol{k}''x_1x_2y_1y_2 = x_2;$$

for these we can take

$$\boldsymbol{k}' := \lambda x_1x_2y_1y_2.x_1, \qquad \boldsymbol{k}'' := \lambda x_1x_2y_1y_2.x_2 \quad \text{etc.}$$

Recursion sequences $\vec{\boldsymbol{r}}$ have already been postulated in the definition of $\mathbf{HA}^\omega_\rightarrow$.

Now we define a mapping Γ on terms and formulas of $\mathbf{HA}^\omega$ as follows:

(iv) To each x^σ a sequence $\Gamma x^\sigma \equiv (x_1^{\sigma_1}, \ldots, x_n^{\sigma_n})$ is assigned, where $(\sigma_1, \ldots, \sigma_n) \equiv \Gamma\sigma$. If x^σ, y^τ are distinct, we shall assume $\Gamma x^\sigma \cap \Gamma y^\tau = \emptyset$.

(v) $\Gamma\boldsymbol{k}^{\sigma,\tau} := \vec{\boldsymbol{k}}^{\Gamma\sigma,\Gamma\tau}$; $\Gamma\boldsymbol{s}^{\rho,\sigma,\tau} := \vec{\boldsymbol{s}}^{\Gamma\rho,\Gamma\sigma,\Gamma\tau}$; $\Gamma 0 := 0$; $\Gamma(\mathrm{S}) := \mathrm{S}$; $\Gamma\boldsymbol{r}^\sigma := \vec{\boldsymbol{r}}^{\Gamma\sigma}$.

(vi) Let $\Gamma x^\sigma \equiv \vec{x}$, $\Gamma y^\tau \equiv \vec{y}$, then $\Gamma\boldsymbol{p}^{\sigma,\tau} := (\lambda\vec{x}\vec{y}.\vec{x}) * (\lambda\vec{x}\vec{y}.\vec{y})$ ($*$ denotes concatenation of finite sequences), and $\Gamma\boldsymbol{p}_0^{\sigma,\tau} := \lambda\vec{x}\vec{y}.\vec{x}$, $\Gamma\boldsymbol{p}_1^{\sigma,\tau} := \lambda\vec{x}\vec{y}.\vec{y}$.

(vii) $\Gamma(tt') := (\Gamma t)(\Gamma t')$.

(viii) $\Gamma(t = s) := (\Gamma t = \Gamma s)$.

(ix) Γ is a homomorphism with respect to $\wedge, \vee, \rightarrow, \perp$; $\Gamma(Qx^\sigma A) := (Q\Gamma x^\sigma)(\Gamma A)$ for $Q \in \{\forall, \exists\}$.

We leave it as an exercise to show

$$\mathbf{HA}^\omega + \Delta \vdash A \Rightarrow \mathbf{HA}^\omega_\rightarrow + \Gamma(\Delta) \vdash \Gamma(A),$$

where $\Gamma(\Delta) = \{\Gamma A : A \in \Delta\}$; moreover $\Gamma(A) \equiv A$ for $A \in \mathbf{HA}^\omega_\rightarrow$. □

1.20. *Pairing in* **E-HA**$^\omega$. In an extensional context, pairing is unproblematic; in fact, if we define **E-HA**$^\omega_{\rightarrow}$ similarly to **HA**$^\omega_{\rightarrow}$, we have the following proposition.

PROPOSITION. **E-HA**$^\omega$ *is a definitional extension of* **E-HA**$^\omega_{\rightarrow}$.

PROOF. Exercise. □

1.21. REMARKS

(i) There is a parallel between the types of $\mathcal{T}$ and the formulas in the $\rightarrow$, $\wedge$ fragment of **IPC**; the operation Γ and types of $\mathcal{T}$ correspond to the replacement of an $\rightarrow\wedge$-formula A by a set of $\rightarrow$-formulas $\{A_0, \ldots, A_{n-1}\}$ such that $A \leftrightarrow \bigwedge\{A_i : i < n\}$. The types of the combinators $\boldsymbol{s}$ and $\boldsymbol{k}$ have the form of two well-known axioms for the $\rightarrow$-fragment of **IPC** (if we write function types with $\rightarrow$), and the types of $\boldsymbol{p}$, $\boldsymbol{p}_0$, $\boldsymbol{p}_1$ correspond to axioms for conjunction if we write $\wedge$ for $\times$.

(ii) In **E-HA**$^\omega$ we can even achieve a reduction to types n (*pure types*), see e.g. Troelstra (1973, 1.8.5–8) (the idea is due to Kleene).

2. Normalization, and a term model for HA$^\omega$

2.1. Intuitively speaking, each closed term of type 0 of **HA**$^\omega$ denotes a natural number. The question is, do the axioms of **HA**$^\omega$ also permit us to evaluate each closed $t \in 0$, i.e. can we always find a numeral $\bar{n}$ such that **HA**$^\omega \vdash t = \bar{n}$?

This question is not entirely trivial already for a simple system such as primitive recursive arithmetic **PRA**, and the answer is even less obvious for the terms of **HA**$^\omega$, where a closed term of type 0 can be introduced via terms of arbitrarily complex types.

We shall reach a positive answer by studying the process of *reduction* of terms. The idea of reduction of terms is suggested by the fact that in evaluating a term in, say, **PRA** from defining equations, there is an obvious direction of computation associated with each defining equation. Thus to $x \cdot Sy = x \cdot y + x$ corresponds the step of replacing the left-hand side by the right-hand side in a computation.

The study of reduction, moreover, is not only of interest as a tool in showing that terms can be evaluated, but is also needed for the introduction of syntactically defined "term models" for **HA**$^\omega$, and as a preparation for the study of normalization in natural deduction systems in the next chapter.

The term models are important tools in the proof of certain metamathematical results of section 7. The conversions below do not completely mimic the axioms of $\mathbf{HA}^{\omega}$, cf. 2.6.

2.2. DEFINITION (*conversion and reduction*). The relation of *conversion*, conv (we say that "*t converts to t′*" if $t \operatorname{conv} t'$), holds between terms t and t' of $\mathbf{HA}^{\omega}$, if there are t_1, t_2, t_3 such that:

Con1	$t \equiv \boldsymbol{k}t_1t_2$,	$t' \equiv t_1$, or
Con2	$t \equiv \boldsymbol{s}t_1t_2t_3$,	$t' \equiv t_1t_3(t_2t_3)$, or
Con3	$t \equiv \boldsymbol{p}_i(\boldsymbol{p}t_1t_2)$,	$t' \equiv t_{i+1} \quad (i \leq 1)$, or
Con4	$t \equiv \boldsymbol{r}t_1t_20$,	$t' \equiv t_1$, or
Con5	$t \equiv \boldsymbol{r}t_1t_2(St_3)$,	$t' \equiv t_2(\boldsymbol{r}t_1t_2t_3)t_3$.

A t of one of the above forms is called a *redex*. The relation of reduction, written as $\succcurlyeq$, is inductively generated by the following stipulations:

$$t \succcurlyeq t;$$

$$t \operatorname{conv} t' \Rightarrow t \succcurlyeq t';$$

$$t \succcurlyeq t' \Rightarrow t''t \succcurlyeq t''t';$$

$$t \succcurlyeq t' \Rightarrow tt'' \succcurlyeq t't'';$$

$$t \succcurlyeq t', t' \succcurlyeq t'' \Rightarrow t \succcurlyeq t''.$$

If $t \succcurlyeq t'$, we shall say that t *reduces* to t'. We write $t \succ_1 t'$ if t' is obtained from t by converting a single redex in t. Observe that $\succcurlyeq$ is the transitive and reflexive closure of $\succ_1$.

A *reduction sequence* is sequence of terms $t_1 \succ_1 t_2 \succ_1 t_3 \succ_1 \cdots$. □

2.3. DEFINITION. t is in *normal form*, or t is *normal*, if t does not contain a redex.

NF is the set of all terms in normal form, and CNF the set of all *closed* terms in normal form. NF_{σ}, CNF_{σ} are the sets of elements of type σ in NF and CNF respectively.

A reduction sequence *terminates* if it ends in a normal term. □

EXAMPLE. By our definition $\lambda xy.Sx \equiv \boldsymbol{s}(\boldsymbol{kk})S$; if we put

$$t \equiv \boldsymbol{r}(S0)(\lambda xy.Sx)(S0), \qquad t' \equiv \boldsymbol{r}(S0)(\lambda xy.Sx)0,$$

then

$$\begin{aligned}&\boldsymbol{r}(S0)(\lambda xy.Sx)(SS0) \succ_1 (\lambda xy.Sx)t(S0) \equiv \boldsymbol{s}(\boldsymbol{kk})St(S0) \succ_1\\ &\boldsymbol{kk}t(St)(S0) \succ_1 \boldsymbol{k}(St)(S0) \succ_1 St \succ_1 S((\lambda xy.Sx)t'0) \equiv\\ &S(\boldsymbol{s}(\boldsymbol{kk})St'0) \succ_1 S(\boldsymbol{kk}t'(St')0) \succ_1 S(\boldsymbol{k}(St')0) \succ_1 S(St') \succ_1\\ &SSS0 \equiv 3\end{aligned}$$

is a terminating reduction sequence.

An inductive characterization of NF is given by the following proposition.

2.4. PROPOSITION. $s \in \mathrm{NF}$ *iff s has one of the following forms*:

$$\begin{aligned}&0,\ St_1,\ \boldsymbol{k},\ \boldsymbol{k}t_1,\ \boldsymbol{s},\ \boldsymbol{s}t_1,\ \boldsymbol{s}t_1t_2,\\ &\boldsymbol{r},\ \boldsymbol{r}t_1,\ \boldsymbol{r}t_1t_2,\ \boldsymbol{r}t_1t_2tt_3\ldots t_n,\\ &\boldsymbol{p}_i,\ \boldsymbol{p}_it_0t_1\ldots t_n,\ \boldsymbol{p},\ \boldsymbol{p}t_1,\ \boldsymbol{p}t't'',\\ &x^\sigma,\ x^\sigma t_1\ldots t_n,\end{aligned}$$

where $t_1,\ldots,t_n \in \mathrm{NF}$; $t_0 \in \mathrm{NF}$ *and not of the form* $\boldsymbol{p}s_1s_2$; $t \in \mathrm{NF}$, $t \neq 0$ *and not of the form* Ss'; t' *and* $t'' \in \mathrm{NF}$.

PROOF. Straightforward from the definition. □

2.5. PROPOSITION.
(i) *If* $t \in \mathrm{CNF}_0$, *then t is a numeral* (*i.e.* $t \equiv S^n0$ *for some* $n \in \mathbb{N}$).
(ii) *If* $t \in \mathrm{CNF}_{\sigma\times\tau}$, *then* $t = \boldsymbol{p}t_1t_2$ *for suitable* t_1, t_2.

PROOF. (i) and (ii) are proved simultaneously by induction on the complexity of t.

Let $t \in \mathrm{CNF}_0$, then $t \equiv 0$, $t \equiv St^*$, $t \equiv \boldsymbol{r}t_1t_2t^*t_3\ldots t_n$, or $t \equiv \boldsymbol{p}_it_1\ldots t_n$ for appropriate $t^*, t_1^*, t_1,\ldots,t_n \in \mathrm{CNF}$. Since t^* is a closed term of type 0 in normal form, it is by the induction hypothesis a numeral; this rules the third case out.

Similarly, in the fourth case $t_1 \in \sigma\times\tau$, but then by the induction hypothesis, $t_1 \equiv \boldsymbol{p}t't''$ and $\boldsymbol{p}_it_1$ is a redex; so this case is also excluded. We leave the case $t \in \sigma\times\tau$ to the reader. □

REMARK. It follows that elements of CNF_0 can be evaluated. Since on the other hand obviously $t \succcurlyeq t' \Rightarrow \mathbf{HA}^\omega \vdash t = t'$, it suffices to show that all type 0 terms can be reduced to normal form in order to reach the result that all closed terms of type 0 can be evaluated in $\mathbf{HA}^\omega$. □

2.6. DEFINITION. In a few places we shall add to the conversion relation an additional clause:

Con6 $\quad t \equiv \boldsymbol{p}(\boldsymbol{p}_0 t)(\boldsymbol{p}_1 t), \qquad t' \equiv t,$

with a correspondingly enlarged notion of reduction (*sp-reduction*). "sp" abbreviates "surjective pairing", since $t = \boldsymbol{p}(\boldsymbol{p}_0 t)(\boldsymbol{p}_1 t)$ expresses surjectivity of $\boldsymbol{p}$. □

In the sequel the use of sp-reduction (sp-normal form, etc.) will always be mentioned explicitly.

2.7. PROPOSITION.
(i) *Proposition* 2.4 *also holds with respect to sp-normal form provided we require that* t' *and* t'' *are not simultaneously of the form* $\boldsymbol{p}_0 t_1'$ *and* $\boldsymbol{p}_1 t_1''$, *respectively.*
(ii) *Proposition* 2.5 *also holds for closed terms in sp-normal form.*

PROOF. Easy. □

2.8. DEFINITION. The *reduction tree* T_t of a term t is a (finitely branching) tree, each node labelled with a term, such that t is the root of the tree, and the immediate successors of a node labelled with t' are nodes labelled with terms t'' such that $t' \succ_1 t''$.

The branches of T_t are the reduction sequences starting from t. $h(t)$, the *height* of t, is the height of T_t. □

EXAMPLE. The reduction tree of $t \equiv \boldsymbol{s}xy(\boldsymbol{k}uv)$ is shown in fig. 9.1 below; in this case $h(t) = 4$.

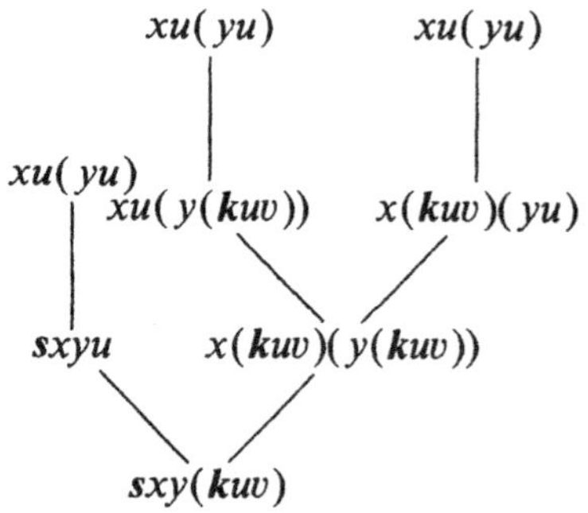

Fig. 9.1.

2.9. Definition. t is said to be *strongly normalizable* (notation: SN(t)) iff T_t is finite. For strongly normalizable terms the height is defined. A sequence $t_1, \ldots, t_n$ is *strongly normalizable* (SN($t_1, \ldots, t_n$)) iff SN(t_1) $\wedge \cdots \wedge$ SN(t_n). □

Our ultimate goal is to prove strong normalizability for all terms. As a warming-up exercise we shall first establish an easier result.

2.10. Proposition. *Each term of* $\mathbf{HA}^{\omega}_{\rightarrow}$ *reduces to normal form (by some reduction sequence).*

Proof. The proof uses the method of *computability* in its simplest form (Tait 1967). We define "computability predicates" Comp_σ for all $\sigma \in \mathscr{T}_{\rightarrow}$ as follows:
(i) $\mathrm{Comp}_0(t) :\equiv t \in 0$ and t reduces to normal form,
(ii) $\mathrm{Comp}_{\sigma\rightarrow\tau}(t) :\equiv \forall t' \in \mathrm{Comp}_\sigma(tt' \in \mathrm{Comp}_\tau)$ and t reduces to normal form.
Observe now that it is sufficient to show that each constant and each variable is computable, for if $t_1, \ldots, t_n$ are computable, so is each applicative combination of these terms.

We first mention a convenient property of computability of $\mathbf{HA}^{\omega}_{\rightarrow}$. One easily sees that each type σ is uniquely of the form $\sigma_1\sigma_2\ldots\sigma_n 0$. It can readily be shown that

$$t \in \mathrm{Comp}_\sigma \Leftrightarrow \forall t_1 \in \mathrm{Comp}_{\sigma_1} \ldots \forall t_n \in \mathrm{Comp}_{\sigma_n}(t_1 \ldots t_n t \in \mathrm{Comp}_0). \tag{1}$$

$0 \in \mathrm{Comp}_0$, $S \in \mathrm{Comp}_1$ are almost immediate and left to the reader. To see that $\boldsymbol{k}^{\sigma,\tau} \in \mathrm{Comp}_{\sigma\tau\sigma}$, let $t \in \mathrm{Comp}_\sigma$, $t' \in \mathrm{Comp}_\tau$, then $\boldsymbol{k}tt' \equiv t \in \mathrm{Comp}_\tau$, hence $\forall t \in \mathrm{Comp}_\sigma(\boldsymbol{k}t \in \mathrm{Comp}_{\tau\sigma})$, therefore $\boldsymbol{k} \in \mathrm{Comp}_{\sigma\tau\sigma}$. Similarly we prove $\boldsymbol{s}^{\rho,\sigma,\tau}$ to be computable. For variables x^σ we use the fact that $x^\sigma t_1^{\sigma_1} \ldots t_n^{\sigma_n} \in \mathrm{Comp}_0$ for $t_i^{\sigma_i} \in \mathrm{Comp}_{\sigma_i}$ in combination with (1).

The case of $\boldsymbol{r}^\sigma$ is more complicated. For terms t in Comp_0 we define

$$t \in D_k :\equiv t \text{ has a normal form } S^k t',\ t' \text{ normal},\ t' \not\equiv 0,$$
$$t' \text{ not of the form } St''.$$

Clearly $\mathrm{Comp}_0 = \bigcup\{D_k : k \in \mathbb{N}\}$. We shall show by induction on k

$$\forall t_1 t_2 t_3 (t_1 \in \mathrm{Comp}_\sigma \wedge t_2 \in \mathrm{Comp}_{\sigma 0 \sigma} \wedge t_3 \in D_k \rightarrow \boldsymbol{r} t_1 t_2 t_3 \in \mathrm{Comp}_\sigma).$$

So assume $t_1 \in \mathrm{Comp}_\sigma$, $t_2 \in \mathrm{Comp}_{\sigma 0 \sigma}$, $t_3 \in D_k$.

Case 1. $k = 0$, t_3 reduces to 0. Then $rt_1t_2t_3 \succcurlyeq t_1$ and t_1 has a normal form by the assumption $t_1 \in \mathrm{Comp}_\sigma$.

Case 2. $k = 0$, t_3 reduces to $t_3' \not\equiv 0$, then $rt_1t_2t_3$ has a normal form $rt_1't_2't_3'$ with t_1', t_2' normal forms of t_1, t_2, respectively. Also $rt_1t_2t_3 \in \mathrm{Comp}_\sigma$, since if $\sigma \equiv \sigma_1\sigma_2 \ldots \sigma_n 0$ and $s_i \in \mathrm{Comp}_{\sigma_i}$ with normal form s_i' $(1 \le i \le n)$, then $rt_1t_2t_3s_1 \ldots s_n$ has a normal form $rt_1't_2't_3's_1' \ldots s_n'$.

Case 3 (*induction step*). Let t_3 have a normal form $S^{k+1}t_3'$, $t_3' \not\equiv 0$, t_3' not beginning with S. Then $rt_1t_2t_3 \succcurlyeq rt_1t_2(S^{k+1}t_3') \succcurlyeq t_2(rt_1t_2(S^kt_3'))(S^kt_3') \equiv t^*$. Since $S^kt_3' \in D_k$, we have $rt_1t_2(S^kt_3') \in \mathrm{Comp}_\sigma$ by the induction hypothesis. So $t^* \in C_\sigma$. □

2.11. *Strict reductions.* There are many variants to the definition of computability (cf. Troelstra 1973, II, section 2). One such variant is obtained by restricting attention to certain uniquely prescribed reduction sequences, as follows.

DEFINITION. A *minimal redex* of t is a redex not containing another redex as a subterm. The *leftmost minimal redex* (*l.m.r.*) is a minimal redex such that no other redex occurs to the left of it. Similarly "rightmost minimal redex". A *strict reduction sequence* is a reduction sequence such that the next term in the sequence is obtained by contracting the l.m.r. □

We may correspondingly adapt the notion of computability by replacing "t reduces to normal form" by "t strictly reduces to normal form". We leave the details to the reader. The normal form obtained via a strict reduction sequence is obviously unique.

From 2.10 we obtain the following corollary.

2.12. COROLLARY. *Each* $t \in \mathrm{CNF}_0$ *of* $\mathbf{HA}^\omega_\rightarrow$ *can be evaluated, i.e.* $\mathbf{HA}^\omega_\rightarrow \vdash t = \bar{n}$ *for some numeral* $\bar{n}$. □

We now turn to the proof of strong normalization. We introduce the notion of strong computability as a tool.

2.13. DEFINITION (*strong computability* SC).
(i) $\mathrm{SC}_0(t) := \mathrm{SN}(t) \wedge t \in 0$,
(ii) $\mathrm{SC}_{\sigma\rightarrow\tau}(t) := \mathrm{SN}(t) \wedge \forall t' \preccurlyeq t \forall t''(\mathrm{SC}_\sigma(t'') \rightarrow \mathrm{SC}_\tau(t't''))$.
We call a finite sequence $\vec{t}$ *of terms strongly computable* (*notation* $\mathrm{SC}(\vec{t})$) *if each element of the sequence is strongly computable. The type index is dropped when no confusion arises.* □

2.14. LEMMA.
(i) $\mathrm{SC}(t) \to \mathrm{SN}(t)$, $\mathrm{SC}_0(t) \leftrightarrow \mathrm{SN}(t)$;
(ii) $\mathrm{SC}(t) \wedge t \succcurlyeq t' \to \mathrm{SC}(t')$;
(iii) $\mathrm{SC}(t) \wedge \mathrm{SC}(s) \to \mathrm{SC}(ts)$;
(iv) $\mathrm{SN}(xt_0t_1 \ldots t_{n-1}) \leftrightarrow \mathrm{SN}(t_0) \wedge \ldots \wedge \mathrm{SN}(t_{n-1})$;
(v) $\mathrm{SN}(xt_0t_1 \ldots t_{n-1}) \to \mathrm{SC}(xt_0 \ldots t_{n-1})$;
(vi) $\mathrm{SC}(x^\sigma)$, *and hence* SC_σ *is inhabited for all* $\sigma \in \mathcal{T}$;
(vii) $\forall t' \in \mathrm{SC}_\sigma(tt' \in \mathrm{SC}_\tau) \to t \in \mathrm{SC}_{\sigma \to \tau}$;
(viii) $\mathrm{SC}(t) \leftrightarrow \forall s \prec_1 t(\mathrm{SC}(s))$.

PROOF.
(i) Immediate by the definition.
(ii) Holds since $\succcurlyeq$ is transitive, and subtrees of finite trees are finite.
(iii) Immediate since $\preccurlyeq$ is reflexive.
(iv) Immediate.
(v) Let $t \equiv xt_0 \ldots t_{n-1}$, $t \in \tau$, $\mathrm{SN}(t)$. We apply induction on τ. If $\tau = 0$, apply (i). If $\tau = \sigma \to \rho$, let $\mathrm{SC}_\sigma(s)$; then $s, t_0, \ldots, t_{n-1}$ are all SN by (i), (iv), and if $t \succcurlyeq t'$, $t' \equiv xt'_0 \ldots t'_{n-1}$ with $t'_0 \ldots t'_{n-1} \in \mathrm{SN}$. Therefore $\mathrm{SC}_\rho(t's)$ by the induction hypothesis, hence $t \in \mathrm{SC}$.
(vi) follows from (v).
(vii) Assume $\forall t' \in \mathrm{SC}_\sigma(tt' \in \mathrm{SC}_\tau)$. If $t \succcurlyeq t''$, then $t''t' \preccurlyeq tt'$, hence by (ii) $\mathrm{SC}(t''t')$ for all $t' \in \mathrm{SC}_\sigma$; and since $x^\sigma \in \mathrm{SC}_\sigma$, in particular $\mathrm{SC}(tx^\sigma)$, and so $\mathrm{SN}(tx_\sigma)$ and therefore also $\mathrm{SN}(t)$. We conclude that $\mathrm{SC}(t)$ holds.
(viii) Left to the reader. □

2.15. THEOREM. $\mathrm{SC}(t)$, *hence* $\mathrm{SN}(t)$ *for all* $t \in \mathbf{HA}^\omega_\to$.

PROOF. By (iii) of the lemma, it suffices to prove that all atomic terms (variables and constants) are strongly computable.

For any term $s \in \mathrm{SN}$, let $h(s)$ be the height of the reduction tree of s, and put $h(s_1, \ldots, s_n) = h(s_1) + \cdots + h(s_n)$.

Case 1. $\mathrm{SC}(0)$, $\mathrm{SC}(S)$ are immediate.

Case 2. $\mathrm{SC}(\boldsymbol{k}^{\sigma,\tau})$: let $t_1, \ldots, t_n$ be terms in SC such that $t' \equiv \boldsymbol{k}t_1 \ldots t_n \in 0$; we show $\mathrm{SC}(t')$, i.e. $\mathrm{SN}(t')$ by induction on $h(t_1) + \cdots + h(t_n)$. If $\boldsymbol{k}t_1 \ldots t_n \succ_1 s$, then $s \equiv t_1t_3 \ldots t_n$ or $s \equiv \boldsymbol{k}t'_1 \ldots t'_n$ with $t_i \succ_1 t'_i$ for some i, $t_j \equiv t'_j$ for $j \not\equiv i$. In the first case $\mathrm{SC}(s)$ by (iii) of the lemma, in the second case $\mathrm{SC}(s)$ by the induction hypothesis, since $h(t'_1) + \cdots + h(t'_n) < h(t_1) + \cdots + h(t_n)$. Thus $\forall t_1, \ldots, t_n \in \mathrm{SC}\,[\mathrm{SC}_0(\boldsymbol{k}t_1 \ldots t_n)]$, and by (vii) $\mathrm{SC}(\boldsymbol{k})$.

Case 3. We leave it to the reader to prove $s^{\rho,\sigma,\tau} \in \mathrm{SC}$.

Case 4. $\mathrm{SC}(x^\sigma)$ has been proved.

Case 5. Let $\vec{r} \equiv (r_1, \ldots, r_m)$ be a sequence of simultaneous recursion constants. We prove $\mathrm{SC}(\vec{r})$ by showing, simultaneously for $1 \leq i \leq m$,

$$\mathrm{SC}(\vec{t}_1) \wedge \mathrm{SC}(\vec{t}_2) \wedge \mathrm{SC}(t_3) \rightarrow \mathrm{SC}(r_i \vec{t}_1 \vec{t}_2 t_3).$$

So let $\mathrm{SC}(\vec{t}_1)$, $\mathrm{SC}(\vec{t}_2)$, $\mathrm{SC}(t_3)$. We use induction with respect to the pair $(\nu(t_3), h(\vec{t}_1) + h(\vec{t}_2) + h(t_3))$, where $\nu(t) \equiv \max\{n : \exists t'(t \succcurlyeq S^n t')\}$. If $r_i \vec{t}_1 \vec{t}_2 t_3 \succ_1 s$, then one of the following subcases occurs.

Subcase 5a. $s \equiv r_i \vec{t}_1{}' \vec{t}_2{}' t_3'$ with either $\vec{t}_1{}' \prec_1 \vec{t}_1$ or $\vec{t}_2{}' \prec_1 \vec{t}_2$, or $t_3' \prec_1 t_3$. Then $\mathrm{SC}(s)$ by the induction hypothesis.

Subcase 5b. $t_3 = 0$, $s = t_{1,i}$, where $\vec{t}_1 \equiv t_{1,0}, \ldots, t_{1,p-1}$. Then $\mathrm{SC}(s)$ is immediate.

Subcase 5c. $t_3 = St_3''$, $s \equiv t_{2,i}(\vec{r}\vec{t}_1\vec{t}_2 t_3'')t_3''$, where $\vec{t}_2 \equiv t_{2,0}, t_{2,1}, \ldots, t_{2,q-1}$. Now $h(t_3'') = h(t_3)$, $\nu(t_3'') = \nu(t_3) - 1$, and so $\mathrm{SC}(s)$ by the induction hypothesis. □

We now extend the proof of strong normalization to $\mathbf{HA}^\omega$.

2.16. THEOREM. *For all terms t of* $\mathbf{HA}^\omega$ $\mathrm{SN}(t)$ *holds, also with respect to sp-reduction.*

PROOF. For a sequence of terms $\vec{t}$ of $\mathbf{HA}^\omega_\rightarrow$, $\mathrm{SN}(\vec{t})$ is a consequence of the preceding theorem.

Let t^σ be any term of $\mathbf{HA}^\omega$. Under the mapping Γ as defined in 1.19, t^σ is represented by a sequence $\Gamma t \equiv (t_1, \ldots, t_n)$ with types in a sequence $\Gamma\sigma$ and the t_i terms of $\mathbf{HA}^\omega_\rightarrow$.

Now we observe that each single step reduction of t corresponds under Γ to a finite sequence (of length ≥ 1) of reductions applied to Γt.

In particular, let us consider the case of the pairing and unpairing constants. $\mathbf{p}_1(\mathbf{p}t^\sigma s^\tau) \succ_1 t$; $\mathbf{p}_1(\mathbf{p}ts)$ becomes under Γ a sequence of terms with reductions

$$(\lambda\vec{x}\vec{y}.\vec{x})((\lambda\vec{x}\vec{y}.\vec{x}) * (\lambda\vec{x}\vec{y}.\vec{y})(\Gamma t, \Gamma s)) \succcurlyeq (\lambda\vec{x}\vec{y}.\vec{x})(\Gamma t, \Gamma s) \succcurlyeq \Gamma t,$$

where $\vec{x} \in \Gamma\sigma$, $\vec{y} \in \Gamma\tau$. $\mathbf{p}(\mathbf{p}_0 t)(\mathbf{p}_1 t)$ is transformed into

$$(\lambda\vec{x}\vec{y}.\vec{x} * \lambda\vec{x}\vec{y}.\vec{y})((\lambda\vec{x}\vec{y}.\vec{x})(\Gamma t) * (\lambda\vec{x}\vec{y}.\vec{y})(\Gamma t)) \succcurlyeq$$

$$(\lambda\vec{x}\vec{y}.\vec{x} * \lambda\vec{x}\vec{y}.\vec{y})(\vec{t}_1^* * \vec{t}_2^*) \succcurlyeq (\vec{t}_1^* * \vec{t}_2^*),$$

where $\Gamma t \equiv t'_1, \ldots, t'_n, t^*_1 \equiv (t'_1, \ldots, t'_n), t^*_2 \equiv (t'_{n+1}, \ldots, t'_{n+m})$. Further details are left to the reader. □

REMARK. One can give a *direct* proof (de Vrijer 1987) of strong normalization for $\mathbf{HA}^\omega$, by extending SC to product types:

$$\mathrm{SC}_{\sigma\times\tau}(t) := \mathrm{SN}(t) \wedge \forall t't''(t \succcurlyeq \mathbf{p}t't'' \to \mathrm{SC}_\sigma(t') \wedge \mathrm{SC}_\tau(t''))$$

(exercise).

COROLLARY. *For all* $t \in \mathrm{CNF}_0$, *t can be evaluated in* $\mathbf{HA}^\omega$: *for some* $\bar{n}$ $\mathbf{HA}^\omega \vdash t = \bar{n}$. □

We now turn to versions of $\mathbf{HA}^\omega$ with the λ-abstraction operator as a primitive.

2.17. Instead of using combinators, one may also consider formulations of $\mathbf{HA}^\omega$ with λ-abstraction as a primitive notion with an axiom schema:

CON $\quad (\lambda x^\sigma.t)(t') = t[x/t']$,

with the tacit convention that no variable free in t' is to become bound by substitution in t; if necessary we rename the variables bound in t. The combinators then become definable:

$$\mathbf{k}^{\sigma,\tau} := \lambda x^\sigma y^\tau . x^\sigma, \qquad \mathbf{s}^{\rho,\sigma,\tau} := \lambda x^{\rho(\sigma\tau)} y^{\rho\sigma} z^\rho . xz(yz).$$

A delicate point is the formulation of the equality rules in the presence of λ-abstraction. It seems natural to request

$$\forall x(t = t') \to \lambda x.t = \lambda x.t',$$

but this introduces extensionality (cf. section 6) and therefore we shall not postulate it. However, we are more interested in the system of terms with a primitive notion of λ-abstraction than in the corresponding theory λ-$\mathbf{HA}^\omega$.

So we shall describe first the natural notion of conversion, and then let λ-$\mathbf{HA}^\omega$ be the least strengthening of $\mathbf{HA}^\omega$ respecting the equality relation generated by λ-conversion.

2.18. DEFINITION (*λ-conversion*). In the presence of the λ-operator as a primitive, $t \operatorname{conv} t'$ holds if there are t_1, t_2, t_3 such that:

Con1′ $\quad t \equiv (\lambda x.t_1)(t_2), \qquad t' \equiv t_1[x/t_2]$ (*β-conversion*), or

Con2′ $\quad t \equiv \lambda x.t_1 x, \qquad t' \equiv t_1$ where $x \notin \mathrm{FV}(t_1)$ (*η-conversion*), or Con3, or Con4, or Con5 as in 2.2. The notion of *redex* is widened accordingly. The corresponding *reduction* relation $\succcurlyeq$ is now generated by

$$t \succcurlyeq t,$$
$$t \operatorname{conv} t' \Rightarrow t \succcurlyeq t',$$
$$t \succcurlyeq t' \Rightarrow \lambda x.t \succcurlyeq \lambda x.t',$$
$$t \succcurlyeq t' \Rightarrow tt'' \succcurlyeq t't'',$$
$$t \succcurlyeq t' \Rightarrow t''t \succcurlyeq t''t',$$
$$t \succcurlyeq t', t' \succcurlyeq t'' \Rightarrow t \succcurlyeq t''.$$

$\succ_1$ and *reduction sequence* are defined as before. □

2.19. DEFINITION. λ-$\mathbf{HA}^\omega$ is defined completely similarly to $\mathbf{HA}^\omega$, but now λ is a primitive, and we require as a schema

$$t \succcurlyeq t' \Rightarrow \lambda\text{-}\mathbf{HA}^\omega \vdash t = t'.$$

λ-$\mathbf{HA}^\omega_\rightarrow$ is λ-$\mathbf{HA}^\omega$ without product types. □

N.B. λ-$\mathbf{HA}^\omega$ identifies more terms than the combinatory version, see 2.26.

REMARK. We shall have little occasion to study λ-$\mathbf{HA}^\omega$ beyond the strong normalizability of its set of terms. The notion of "model of λ-$\mathbf{HA}^\omega$" is a delicate one; the problems are similar to the ones encountered in formulating the notion of "model of the untyped λ-calculus". A precise treatment of the latter notion has been given by Koymans (1982, 1984).

If we drop the conversion rule

$$t \succcurlyeq t' \Rightarrow \lambda x.t \succcurlyeq \lambda x.t',$$

while retaining λ as a primitive, we obtain a theory λ'-$\mathbf{HA}^\omega$ which is very close to $\mathbf{HA}^\omega$: the defined λ in $\mathbf{HA}^\omega$ models the λ of λ'-$\mathbf{HA}^\omega$, and in λ'-$\mathbf{HA}^\omega$ we can define combinators $\boldsymbol{s}$ and $\boldsymbol{k}$. However, the definition of λ in $\mathbf{HA}^\omega$, followed by the construction of combinators from this λ in the obvious way, does not give us back the original $\boldsymbol{s}$ and $\boldsymbol{k}$ (cf. 11.2.18).

For λ-terms we have to adapt our proof of strong normalizability, by establishing a property stronger than strong normalization.

2.20. DEFINITION. t is *strongly computable under substitution* (notation: $\mathrm{SC}^*(t)$) if, for each substitution of strongly computable terms for the free variables of t, the resulting term is strongly computable (where strongly computable is taken in the sense of 2.13 and 2.16, remark). □

2.21. THEOREM. $SC^*(t)$ *for each term of* λ-$\mathbf{HA}^{\omega}_{\rightarrow}$.

PROOF. We leave most of the details to the reader, and shall consider only the case of terms introduced by λ-abstraction.

Suppose $SC^*(t[x])$; we want to show that $\lambda x^{\sigma}.t[x] \in SC^*$. Let $t[x] \in \tau$, and let $t^*[x]$ be the result of substituting strongly computable terms for variables of t other than x; then $t^* \in SC^*$ by the induction hypothesis. (N.B. $SC^*(x^{\sigma})$ can be proved as before.) Now $SC^*(\lambda x^{\sigma}.t[x])$ follows from $SC(\lambda x^{\sigma}.t^*[x])$ for all possible $t^*[x]$. Let $\vec{t} \equiv (t_1, \ldots, t_n)$ be such that $SC(\vec{t})$, $(\lambda x.t^*)\vec{t} \in 0$. We intend to show $(\lambda x.t^*)\vec{t} \in SC$ by induction on $h(t^*) + h(t_1) + \cdots + h(t_n)$. By lemma 2.14 (viii) it suffices to show $SC(s)$ for all s such that $(\lambda x.t^*)\vec{t} \succ_1 s$.

Case 1. $s \equiv t^*[t_1]t_2 \ldots t_n$ (β-conversion applied to $(\lambda x.t^*)t_1$). Then $t^*[t_1]$ is SC since $t \in SC^*$, and therefore $s \in SC$.

Case 2. $t^*[x] \equiv t''x$, $x \notin FV(t'')$, $s \equiv t''\vec{t}$ (η-conversion applied to $\lambda x.t''x$). Now $s \equiv ((t''x)[x/t_1])t_2 \ldots t_n$, and this can be handled as in case 1.

Case 3. $s \equiv \lambda x.t''[x]t_1' \ldots t_n'$, where either $t^* \succ_1 t''$ and $t_i' \equiv t_i$, or $t'' \equiv t^*$ and $t_i' \prec_1 t_i$ for some i, $t_j' \equiv t_j$ for $j \neq i$. Then $SC(s)$ follows from our induction hypothesis. $\Box$

COROLLARY. $SN(t)$ *for each* t *of* λ-$\mathbf{HA}^{\omega}$. $\Box$

PROOF. Strong normalization for λ-$\mathbf{HA}^{\omega}$ may be reduced to the corresponding assertion for λ-$\mathbf{HA}^{\omega}_{\rightarrow}$ as before. $\Box$

An application of this result is found in 10.8.4.

Next we describe the construction of term models; this material is not essential to the rest of the chapter except for the details of the proof of the Church–Rosser theorem (2.25 below) which are used again in 4.19. The techniques used are familiar from combinatory logic and the lambda calculus.

Just as HRO, the term model of the closed terms in normal form (also called CNF) described below is countable, and satisfies CT. In fact, one can show that the type 1 objects of CNF are precisely the provably recursive functions of first-order arithmetic. From this it follows that, in contrast, the principle

$$\forall x(\forall y \exists z\, Txyz \rightarrow \exists \alpha \forall y(\alpha y = \{x\}(y))),$$

stating that all total recursive functions occur as type 1 objects, is valid in HRO, but not in CNF.

2.22. *The term model* CNF. We wish to construct a model for $\mathbf{HA}^\omega$ from the closed terms in normal form, such that the objects of type σ are the elements of CNF_σ, and application of $t^{\sigma\to\tau}$ to t_1^σ is the (unique) normal term t' such that $tt_1 \succcurlyeq t'$; constants are interpreted by themselves, equality at all types is equality of normal form. The result will be a standard model, since the set of numerals representing type 0 is isomorphic to $\mathbb{N}$.

To see that this indeed describes a model, we have to show that the normal form of a term is unique. We can achieve this by restricting attention to normal forms reached by strict reduction sequences. For example, the equation $sxyz = xz(yz)$ may be verified in CNF as follows. Let $t_1, t_2, t_3 \in \mathrm{CNF}$. A strict reduction sequence for $st_1t_2t_3$ starts with $st_1t_2t_3 \succ_1 t_1t_3(t_2t_3) \succ_1 \cdots$ and after the second term it becomes identical with the strict reduction sequence starting from $t_1t_3(t_2t_3)$, so $st_1t_2t_3$ and $t_1t_3(t_2t_3)$ denote the same term in the model.

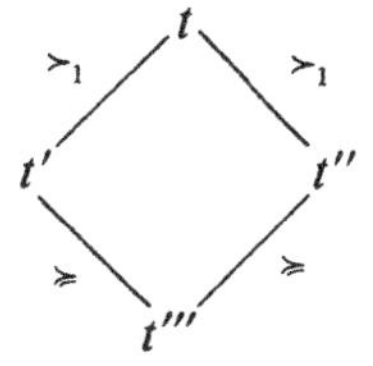

Fig. 9.2.

More natural perhaps is to prove that all reduction sequences for t terminate in the same normal form. This could be done by proving the *weak Church–Rosser property* (WCR): if $t \succ_1 t'$, $t \succ_1 t''$, then for some t''', $t' \succcurlyeq t'''$ and $t'' \succcurlyeq t'''$. See fig. 9.2. With the help of WCR one easily proves by induction on $h(t)$ that each t has a unique normal form (the property UN(t)). For let t be any term with $h(t) = k + 1$, and consider two reduction sequences starting from t, say

$$t \succ_1 t_1 \succ_1 t_2 \succ_1 \cdots \succ_1 t^*, \qquad t \succ_1 t_1' \succ_1 t_2' \succ \cdots \succ_1 t^{**}.$$

If $t_1 \equiv t_1'$, we are done: the induction hypothesis applies to t_1. If $t_1 \not\equiv t_1'$, we find by WCR t''' with $t_1 \succcurlyeq t'''$, $t_1' \succcurlyeq t'''$, and $h(t_1) \le k$, $h(t_1') \le k$, hence the unique normal forms of t_1 and of t_1' which exist by the induction hypothesis must coincide.

For purposes of generalization however it is more advantageous to give a direct proof of the *Church–Rosser property* (CR): if $t \succcurlyeq t'$, $t \succcurlyeq t''$, there is a t''' such that $t' \succcurlyeq t'''$, $t'' \succcurlyeq t'''$. As a first step we define the following.

2.23. DEFINITION. The relation $\succcurlyeq^*$ between terms is inductively generated by the clauses:

(i) $t \succcurlyeq^* t$,

(ii) $t \operatorname{conv} t' \Rightarrow t \succcurlyeq^* t'$,

(iii) $t_1 \succcurlyeq^* t_1', t_2 \succcurlyeq^* t_2' \Rightarrow t_1t_2 \succcurlyeq^* t_1't_2'$. □

Let us write $t \succcurlyeq^*_n t'$ iff there is a sequence $t \equiv t_0 \succcurlyeq^* t_1 \succcurlyeq^* t_2 \succcurlyeq^* \cdots \succcurlyeq^* t_n$. □

REMARK. $\succcurlyeq^*$ is not transitive; $t \succcurlyeq^* t'$ holds iff either $t \equiv t'$ or t' is obtained from t by simultaneously converting a finite set of pairwise disjoint redexes. $\succcurlyeq$ is the transitive closure of $\succcurlyeq^*$.

2.24. LEMMA (*diamond lemma*). *If $t \succcurlyeq^* t'$, $t \succcurlyeq^* t''$, then for some t''', $t' \succcurlyeq^* t'''$, $t'' \succcurlyeq^* t'''$. See fig. 9.3.*

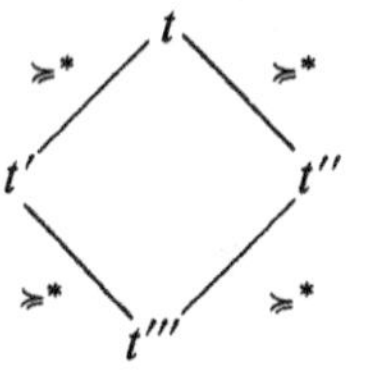

Fig. 9.3.

PROOF. By definition $t \succcurlyeq^* t'$ can be derived by repeated application of (i)–(iii) in 2.23; the steps may be arranged linearly, e.g. if $t \succcurlyeq^* t'$, there is a finite sequence $t_0 \succcurlyeq^* t_0'$, $t_1 \succcurlyeq^* t_1', \ldots, t_n \succcurlyeq^* t_n'$ with $t_n \equiv t$, $t_n' \equiv t'$ such that for each $i \leq n$ either $t_i \equiv t_i'$ or $t_i \operatorname{conv} t_i'$ or $\exists k, m < i(t_i \equiv t_k t_m \wedge t_i' \equiv t_k' t_m')$. In this case we can say that $t \succcurlyeq^* t'$ has a proof of length n.

Now let $t \succcurlyeq^* t'$, $t \succcurlyeq^* t''$ by proofs of length n and m, respectively; we apply induction on $n + m$.

Case 1. $t \succcurlyeq^* t'$ because $t \operatorname{conv} t'$; we can inspect the possible cases. E.g. let $t \equiv st_1t_2t_3$, $t' \equiv t_1t_3(t_2t_3)$, then $t'' \equiv st_1't_2't_3'$ with $t_i \succcurlyeq^* t_i'$ $(i \in \{1,2,3\})$. Clearly we can take $t''' \equiv t_1't_3'(t_2't_3')$. The other possible subcases we leave to the reader.

Case 2. $t \succcurlyeq^* t''$ because of $t \operatorname{conv} t''$, we argue as in case 1.

Case 3. If neither case 1 nor case 2 applies, it must be the case that $t \equiv t_1t_2$, $t' \equiv t_1't_2'$, $t'' \equiv t_1''t_2''$ and $t_1t_2 \succcurlyeq^* t_1't_2'$, $t_1t_2 \succcurlyeq^* t_1''t_2''$ obtained from $t_1 \succcurlyeq^* t_1'$, $t_1 \succcurlyeq^* t_1''$ and $t_2 \succcurlyeq^* t_2'$, $t_2 \succcurlyeq^* t_2''$; by the induction hypothesis we find t_1''', t_2''', with $t_1' \succcurlyeq^* t_1'''$, $t_1'' \succcurlyeq^* t_1'''$, $t_2' \succcurlyeq^* t_{2'''}$, $t_2'' \succcurlyeq^* t_2'''$, so $t_1't_2' \succcurlyeq^* t_1'''t_2'''$, etc. □

2.25. PROPOSITION (*CR-property*). *If* $t \succcurlyeq_n^* t'$, $t \succcurlyeq_m^* t''$, *then for some* t''' $t' \succcurlyeq_m^* t'''$, $t'' \succcurlyeq_n^* t'''$. *See fig.* 9.4.

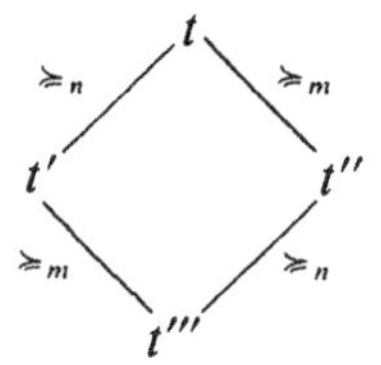

Fig. 9.4.

PROOF. By induction on $n + m$; we leave the details to the reader. □

REMARK. We cannot extend the proof of CR so as to include sp-conversions (J.W. Klop, see Barendregt 1981, 15.3.15). However, WCR can be established also with respect to sp-conversions (exercise), so that by the argument outlined in 2.22 we get uniqueness of normal form also in this case.

COROLLARY. UN(t) *for all* t *of* $\mathbf{HA}^\omega$. □

2.26. REMARK (*on the equality generated by* $\succcurlyeq$ *for* λ-*terms*). We can now substantiate our warning in 2.19: the conversion rules for a primitive λ-operator induce a stronger notion of equality than for the defined one.

For example, if we put $\boldsymbol{k} := \lambda xy.x$, $\boldsymbol{s} := \lambda xyz.xz(yz)$, we find $\boldsymbol{skk} \succcurlyeq \lambda x.x$, $\boldsymbol{sks} \succcurlyeq \lambda x.x$, but $\boldsymbol{skk}$ and $\boldsymbol{sks}$ are distinct terms in normal form which therefore cannot be identified in $\mathbf{HA}^\omega$. □

2.27. *Continuous evaluation of definable type-two objects.* If t^2 is a closed type 2 term of $\mathbf{HA}^\omega$, $t^2\alpha$ can be evaluated using only finitely many values of

α. This is already made plausible by the fact that ICF is a model for $\mathbf{HA}^\omega$; a straightforward proof can be given by extending the notion of reduction.

In this extension we assign to a specific type 1 variable x^* (to be regarded as a constant in this connection) a function α by adding to the conversion clauses of 2.2:

Con7 $\quad t \equiv x^*(S^n0), \qquad t' \equiv S^{\alpha n}0,$

where $S^0 0 := 0$, $S^{n+1}0 := S(S^n 0)$.

The notion of reduction, strict reduction, etc. is accordingly extended, and we can easily show strong normalization, uniqueness of normal form, etc.

By this method we see that each type 2 object definable in $\mathbf{HA}^\omega$ must be continuous, since it is representable by a t^2 in closed normal form, and $t^2 x^*$ is reducible to a numeral for any α assigned to x^*, and for each α at most finitely many values are used in evaluating t^2 for the argument α; therefore we conclude to the truth of

$$\forall\alpha\exists n\forall\beta \in \bar{\alpha}n\left(t^2\alpha = t^2\beta\right).$$

Formalizing this argument leads to *rules* of continuity; we shall treat this in more detail for the case of **APP** in section 7.

For each α the construction also yields a term model CNF^α, which can be defined completely similarly to CNF. □

3. The theory APP

3.1. *Theories of operations.* We now wish to consider type-free theories, where all objects may be thought of as operations (rules) which can be freely applied to each other; self-application is meaningful, but need not be total. The principal example of such a domain will be that of the partial recursive functions, which via their codes may be identified with natural numbers; application is partial recursive function application.

Though in such a theory an operation always describes a partial function on the universe, the converse need not be the case, i.e. we must distinguish between partial functions and operations (cf. 3.14 below).

Below we shall describe a sort of minimal axiomatic basis **APP** for theories of operators. **APP** contains arithmetic and the means for introducing all recursive functions. It was Feferman (1975, 1979) who introduced **APP**-based theories for the formalization of constructive mathematics.

An important reason for our interest in **APP** is the smooth and elegant abstract version of Kleene's realizability it permits; this will be treated in section 5; section 4 will be devoted to some special models of **APP**.

3.2. *The language* $\mathscr{L}$(**APP**). There are variables x, y, z, u, v, w for operations, and constant symbols for certain specific operations: $\boldsymbol{k}, \boldsymbol{s}$ (combinators), $\boldsymbol{p}, \boldsymbol{p}_0, \boldsymbol{p}_1$ (pairing and unpairing operators), 0 (zero), S (successor), P (predecessor), and $\boldsymbol{d}$ (definition by numerical cases).

There is also a binary partial operator Ap for term formation: if t, s are terms, so is Ap(t, s); we write ts for Ap(t, s) and $t_1 \ldots t_n$ abbreviates $(\ldots((t_1 t_2)t_3)\ldots)t_n$, as before. There are three predicate constants: $=$ for equality, N for the natural numbers, and $\boldsymbol{E}$ for existence.

Prime formulas are of the forms $t = s$, $\boldsymbol{E}t$, Nt (also written $t \in N$); formulas are built from prime formulas by $\rightarrow, \wedge, \vee, \forall, \exists$. We use some standard abbreviations:

$$\bot := (0 = S0), \qquad \top := (0 = 0),$$

$$t \simeq s := \boldsymbol{E}t \vee \boldsymbol{E}s \rightarrow t = s.$$

We may regard $\vee$ as defined by

$$A \vee B := \exists x \in N((x = 0 \rightarrow A) \wedge (x \neq 0 \rightarrow B)).$$

Furthermore the numerals are introduced as usual, and we sometimes write (s, t) for $\boldsymbol{p}st, (t)_i$ for $\boldsymbol{p}_i t$.

Sometimes we use n, m as variables ranging over N, i.e. $\forall n$ and $\exists n$ abbreviate $\forall n(n \in N \rightarrow \cdots)$ and $\exists n(n \in N \wedge \cdots)$, respectively. □

3.3. *The system* **APP**. The system is based on intuitionistic predicate logic with existence predicate and equality; all primitive functions and relations are assumed to be strict. We can choose to present the system either with the convention that free variables are schematic, or with free variables existing. For **APP** it is slightly easier and more natural to use E^+-logic, i.e. the version with "existing variables" (cf. 2.2.3).

Actually, by stating axioms schematically with arbitrary terms instead of variables, we can give an axiomatization which works for both variants. Strictness amounts to

$$t = s \rightarrow \boldsymbol{E}t \wedge \boldsymbol{E}s, \qquad Nt \rightarrow \boldsymbol{E}t;$$

$$\boldsymbol{E}(ts) \rightarrow \boldsymbol{E}t \wedge \boldsymbol{E}s.$$

The relation between $\boldsymbol{E}$ and $\exists$ is given by

$$\boldsymbol{E}t \leftrightarrow \exists x(x = t),$$

and replacement is guaranteed by

$$Nt \wedge t = s \rightarrow Ns;$$

$$E(rs) \wedge t = s \rightarrow rt = rs;$$

$$E(tr) \wedge t = s \rightarrow tr = sr.$$

The nonlogical axioms are

$$Et \rightarrow kst \simeq s;$$

$$Et \wedge Et' \rightarrow E(stt'),\ stt't'' \simeq tt''(t't'');$$

$$Et' \rightarrow p_0(ptt') \simeq t;\ Et' \rightarrow p_1(pt't) \simeq t;$$

$$0 \in N, \quad t \in N \rightarrow St \in N, \quad St \neq 0;$$

$$P0 = 0, \quad t \in N \rightarrow Pt \in N, \quad P(St) \simeq t;$$

$$Et_1 \wedge Et_2 \wedge t \in N \wedge t' \in N \wedge t \neq t' \rightarrow dt_1t_2tt = t_1 \wedge dt_1t_2tt' = t_2;$$

$$A(0) \wedge \forall x \in N(Ax \rightarrow A(Sx)) \rightarrow \forall x \in N\, A(x) \quad (\textit{induction}).$$

REMARKS. (i) Since N is strict all natural numbers exist. Moreover, from the axioms one can easily derive the (conditional) existence of a number of terms, e.g. Ek, Es, Ep, Ep_i, $Et \rightarrow E(kt)$, etc.

(ii) It is easy to see how the axioms can be reformulated with variables, e.g. in E-logic,

$$Ex \wedge Ey \rightarrow Esxy, \qquad Ey \rightarrow kxy \simeq x,$$

and in E^+-logic $Esxy$, $kxy \simeq x$, etc. □

3.4. *The model of the partial recursive operations* (PRO). We now introduce the "standard" model of the system **APP**. The objects are the (indices of) the partial recursive functions PRO, equality is numerical equality, N is $\mathbb{N}$, Et means "t is defined", application is partial recursive function application, i.e. $[\![tt']\!]_{\mathrm{PRO}} = \{[\![t]\!]_{\mathrm{PRO}}\}[\![t']\!]_{\mathrm{PRO}}$; for the constants we put (compare the interpretation of $\mathbf{HA}^\omega$ in HRO):

$$[\![0]\!]_{\mathrm{PRO}} := 0,$$

$$[\![S]\!]_{\mathrm{PRO}} := \Lambda x.Sx,$$

$$[\![P]\!]_{\mathrm{PRO}} := \Lambda x.prd(x),$$

$$[\![p]\!]_{\mathrm{PRO}} := \Lambda xy.j(x, y), \qquad [\![p_i]\!]_{\mathrm{PRO}} := \Lambda x.j_{i+1}x \quad (i \leq 1),$$

$$[\![k]\!]_{\mathrm{PRO}} := \Lambda xy.x, \qquad [\![s]\!]_{\mathrm{PRO}} := \Lambda xyz.\{\{x\}(z)\}(\{y\}(z)),$$

$$[\![d]\!]_{\mathrm{PRO}} := \Lambda xyuv.(x \cdot \mathrm{sg}'|u - v| + y \cdot \mathrm{sg}|u - v|).$$

3.5. PROPOSITION (*lambda abstraction or explicit definition in* **APP**). *For each term* $t[x]$ *there is a term* $\lambda x.t[x]$ *such that*

$$\mathbf{APP} \vdash E(\lambda x.t[x]) \wedge (Es \to (\lambda x.t[x])(s) \simeq t[s]);$$

moreover $\mathrm{FV}(\lambda x.t[x]) \equiv \mathrm{FV}(t[x]) \setminus \{x\}$.

PROOF. Similar to the case of $\mathbf{HA}^\omega$ we put:
(a) $\lambda x.x := \boldsymbol{skk}$;
(b) $\lambda x.y := \boldsymbol{k}y$ for $y \not\equiv x$, $\lambda x.c := \boldsymbol{k}c$ for individual constants c;
(c) $\lambda x.tt' := \boldsymbol{s}(\lambda x.t)(\lambda x.t')$.
We leave the verification to the reader. □

REMARKS. (i) Let **TAPP** be the extension of **APP** with total application, i.e. **APP** + TOT, where

TOT $\quad Exy$.

In **TAPP** we can copy the definition of $\lambda x.t$ as given for $\mathbf{HA}^\omega$ in 1.8 (dropping all reference to types). In **APP**, the adoption of the definition of 1.8 would result in a λ-operator such that

$$Es \to (\lambda x.t[x](s) \simeq t[s]),$$

but in general we cannot prove $E(\lambda x.t[x])$ for this definition. (The definition of 1.8 gives us $\lambda x.yz \equiv \boldsymbol{k}(yz)$ which is not always defined in a model in which application is not total.)

(ii) The "$Es \to$ " in the statement of the proposition cannot be dropped: think of the case where $x \notin \mathrm{FV}(t)$.

(iii) In a system with E-logic the statement of the proposition has to be modified to

$$\mathbf{APP} \vdash Ey_1 \wedge \cdots \wedge Ey_n \to E(\lambda x.t[x]) \wedge (Es \to (\lambda x.t[x])(s) \simeq t[s]),$$

where $\mathrm{FV}(t[x]) \setminus \{x\} \subset \{y_1, \ldots, y_n\}$, and TOT must be formulated as $\forall xy(Exy)$.

Recursive functions in **APP** (3.7–3.11). We wish to show that **HA** can be embedded in **APP**; for that purpose we have to develop the rudiments of recursion theory in **APP**, so that we are able to introduce all primitive recursive functions.

3.7. LEMMA (*fixed point operator*). *There is a term* fix *satisfying*

$$\mathbf{APP} \vdash E(\mathrm{fix}(x)) \wedge (\mathrm{fix}(x))y \simeq x(\mathrm{fix}(x))y.$$

PROOF. Put $\chi := \lambda zy.x(zz)y$, fix $:= \lambda x.\chi\chi$. By the preceding proposition $E\chi$, and $\text{fix}(x) \simeq \chi\chi \simeq (\lambda zy.x(zz)y)\chi \simeq \lambda y.x(\chi\chi)y$, so $E(\text{fix}(x))$; also $(\text{fix}(x))y \simeq x(\chi\chi)y \simeq x(\text{fix}(x))y$. □

REMARK. In **TAPP** we have an operator Fix such that

$$x(\text{Fix}(x)) = \text{Fix}(x);$$

take $\lambda x.\chi'\chi'$ for $\chi' \equiv \lambda z.x(zz)$. In **APP** we cannot prove $E(\text{Fix}(x))$ (consider in PRO an x with $\{x\}$ nowhere defined). □

3.8. LEMMA (*existence of recursor*). *In* **APP** *there is a term* $\boldsymbol{r}$ *satisfying*

$$Et' \to \boldsymbol{r}tt'0 \simeq t,$$

$$Nt'' \wedge t'' \neq 0 \to \boldsymbol{r}tt't'' \simeq t'(\boldsymbol{r}tt'(Pt''))t''.$$

PROOF. For $\boldsymbol{r}xy$ we can take fix(ρ) for a ρ such that

$$\rho f0 = x, \qquad \rho f(Sn) = y(fn)n.$$

Then, writing φ for fix

$$\varphi\rho n = \rho(\varphi\rho)n = x \quad \text{if } n = 0, \qquad \varphi\rho n = y(\varphi\rho m) \quad \text{if } n = Sm.$$

So put

$$\rho := \lambda fn.\boldsymbol{d}(\boldsymbol{k}x)(\lambda z.y(f(Pz))z)n0n,$$

$$\boldsymbol{r} := \lambda xy.\varphi\rho.$$

Now, assuming Et',

$$\boldsymbol{r}tt'0 \simeq \varphi\rho 0 \simeq \rho(\varphi\rho)0 \simeq \boldsymbol{d}(\boldsymbol{k}t)(\lambda z.t'(\varphi\rho(Pz))z)000 \simeq \boldsymbol{k}t0 \simeq t.$$

If $n \in N$, $n \neq 0$, then

$$\boldsymbol{r}tt'n \simeq \varphi\rho n \simeq \rho(\varphi\rho)n \simeq \boldsymbol{d}(\boldsymbol{k}t)(\lambda z.t'(\varphi\rho(Pz))z)n0n$$

$$\simeq (\lambda z.t'(\varphi\rho(Pz))z)n \simeq t'(\varphi\rho(Pn))n \simeq t'(\boldsymbol{r}tt'(Pn))n. \quad □$$

REMARK. In **TAPP**, a simpler definition is possible:

$$\rho := \lambda fn.\boldsymbol{d}t(t'(f(Pn))n)n0.$$

But since we cannot assume $t'(f(P0))0$ to exist, we cannot show in **APP** $Et' \rightarrow \boldsymbol{r}tt'0 \simeq t$.

3.9. DEFINITION.

$$t + t' := \boldsymbol{r}t(\boldsymbol{s}(\boldsymbol{kk})S)t',$$

$$t < t' := t \in N \wedge t' \in N \wedge \exists n \in N(t + Sn = t'),$$

$$t > t' := t' < t, \text{ etc.}$$

$$\mathrm{Adm}(f) := \forall n(fn \in N) \vee \exists m(fm = 0 \wedge \forall m' < m(fm' \in N)). \quad \Box$$

It is routine to verify the well-known elementary properties of $\neq$, $<$ on $\mathbb{N}$. (Cf. section 3.2.)

3.10. LEMMA (*minimum operator*). *In* **APP** *there exists a term* μ *such that*

$$\mathrm{Adm}(f) \rightarrow (\mu(f) = n \leftrightarrow fn = 0 \wedge \forall m < n(fm > 0)). \tag{1}$$

PROOF. Let us again write φ for fix, and put

$$f^+ := \lambda x.f(Sx),$$

$$M := \lambda xf.\boldsymbol{d}(\boldsymbol{k}0)(\lambda g.S(xg))(f0)0f^+,$$

$$\mu := \varphi M.$$

Then

$$\mu f \simeq \varphi Mf \simeq M(\varphi M)f \simeq M\mu f \simeq \boldsymbol{d}(\boldsymbol{k}0)(\lambda g.S(\mu g))(f0)0f^+,$$

and this latter expression is equal to

$$\boldsymbol{k}0f^+ = 0 \quad \text{if } f0 = 0, \text{ and}$$

$$(\lambda g.S(\mu g))f^+ \simeq S(\mu f^+) \quad \text{if } f0 > 0.$$

The required property (1) of μ now follows by induction over n, assuming Adm(f). Assume (1) to have been established up to n, then we have for $n + 1$

$$(\mathrm{Adm}(f^+) \wedge f0 \in N) \vee f0 = 0,$$

and by the induction hypothesis

$$\mathrm{Adm}(f^+) \rightarrow (\mu f^+ \simeq n \leftrightarrow f^+n = 0 \wedge \forall m < n(f^+m > 0)).$$

Then

$$\begin{aligned}\mu f = n+1 &\leftrightarrow \mu f \simeq \mu f^{+} + 1 = n+1 \wedge f0 > 0\\ &\leftrightarrow \mu f^{+} = n \wedge f0 > 0\\ &\leftrightarrow f^{+}n = 0 \wedge \forall m < n(f^{+}m > 0) \wedge f0 > 0\\ &\leftrightarrow f(n+1) = 0 \wedge \forall m < n+1(fm > 0).\end{aligned}$$

The first equivalence uses the fact that no St can be 0. □

REMARK. For **TAPP** again a simplification is possible (exercise).

3.11. THEOREM. *All $f \in$ TREC are definable in* **APP**, *i.e. for each k-argument recursive f there is a term t_f such that*

$$f(m_1, \dots, m_k) \simeq n \Leftrightarrow \mathbf{APP} \vdash t_f \overline{m}_1 \cdots \overline{m}_k = \overline{n},$$

where $\overline{m}$ is the m-th numeral $S^m 0$. For primitive recursive φ we can show uniformly in the arguments that t_φ satisfies the same recursion equations as φ, that is to say if e.g. φ is defined from ψ, χ by $\varphi(\vec{x}, 0) = \psi(\vec{x})$, $\varphi(\vec{x}, Sz) = \chi(\varphi(\vec{x}, \vec{z}), \vec{x}, \vec{z})$, then in **APP** $\vdash \vec{x} \in N \to t_\varphi \vec{x} 0 = t_\psi \vec{x}$, $\vdash \vec{x} \in N \wedge z \in N \to t_\varphi \vec{x}(Sz) = t_\chi(t_\varphi \vec{x} z)\vec{x} z$, *etc.*

PROOF. Straightforward by combination of 3.8, 3.10. For the case of $\boldsymbol{r}$ and μ we have to see that the cases with parameters are also accounted for. E.g., if we want to find h such that $h\vec{x}0 \simeq f\vec{x}$, $h\vec{x}(Sn) \simeq g\vec{x}n(h\vec{x}n)$, take $h := \lambda \vec{x}.\boldsymbol{r}(f\vec{x})(\lambda m.g\vec{x}(Pm))$. □

COROLLARY. **HA** *and* **EL** *are subsystems of* **APP**. □

3.12. PROPOSITION. **APP** *is conservative over* **HA**.

PROOF. Use the fact that the interpretation of **APP** in PRO can be defined in **HA** (cf. the corresponding interpretation of $\mathbf{HA}^\omega$ in HRO in 1.9) and check that for arithmetical sentences A in **APP**

$$\mathbf{HA} \vdash A \leftrightarrow [\![A]\!]_{\mathrm{PRO}}.\ \square$$

3.13. REMARK. PRO in fact models a stronger theory, satisfying besides the axioms of **APP** also:

SP $\quad \boldsymbol{p}(\boldsymbol{p}_0 t)(\boldsymbol{p}_1 t) \simeq t$ *(surjective pairing)*,

$\mathrm{D_V}$ $\quad \forall xyuv(\boldsymbol{d}xyuu = x \wedge (u \neq v \to \boldsymbol{d}xyuv = y))$

(full definition by cases). □

3.14. *The distinction between operations and functions.* If D is any subdomain of the universe, and we are given a predicate A such that

$$\forall x \in D\, \exists ! y A(x, y),$$

this defines, intuitively, a function on D. However, there need not be an *operation* f such that

$$\forall x \in D(E(fx) \wedge A(x, fx)).$$

To see this, let

$$Dx := \neg Exx \vee \exists z(xx = z),$$

and put

$$A(x, y) := (\neg Exx \wedge y = 0) \vee$$
$$(xx = 0 \wedge y = 1) \vee (xx \neq 0 \wedge y = 0).$$

Observe that $\neg\neg Dx$ for all x. Now suppose there is an operation $f \in D$ such that

$$\forall x \in D\, A(x, fx). \tag{1}$$

If $\neg E(ff)$, we have $ff = 0$; contradiction. If $E(ff)$, and $ff = 0$, we find $ff = 1$; contradiction. Finally, if $E(ff), ff \neq 0$, we find $ff = 0$; again a contradiction, hence $f \notin D$. Thus we have derived a contradiction from (1).

Translating this counterexample into PRO, we see that this is an abstract variant of the example in 4.4.11. We shall return to this matter in 5.13.

Now we shall consider the embedding of type structures in **APP**.

3.15. *The structure* IT *within* **APP**. We define the (abstract) *intensional type structure* $\langle(\mathrm{IT}_\sigma, =)\rangle_{\sigma \in \mathcal{T}}$ by:

$$x \in \mathrm{IT}_0 := x \in N,$$

$$x \in \mathrm{IT}_{\sigma \times \tau} := \boldsymbol{p}_0 x \in \mathrm{IT}_\sigma \wedge \boldsymbol{p}_1 x \in \mathrm{IT}_\tau \wedge \boldsymbol{p}(\boldsymbol{p}_0 x)(\boldsymbol{p}_1 x) = x,$$

$$x \in \mathrm{IT}_{\sigma \to \tau} := \forall y \in \mathrm{IT}_\sigma(xy \in \mathrm{IT}_\tau).$$

Equality in IT is at each type the restriction of equality in **APP**.

We can now interpret $\mathbf{HA}^\omega$ in **APP**, letting the variables of type σ range over IT_σ. $\mathrm{Ap}^{\sigma,\tau}$ becomes Ap restricted to $\mathrm{IT}_{\sigma\to\tau} \times \mathrm{IT}_\sigma$; for the constants $0, S, P, \boldsymbol{k}^{\sigma,\tau}, \boldsymbol{s}^{\rho,\sigma,\tau}, \boldsymbol{r}^\sigma, \boldsymbol{p}, \boldsymbol{p}_0, \boldsymbol{p}_1$ we take the corresponding (defined or primitive) constants of **APP** restricted to the appropriate domains. □

Proposition. $\mathbf{HA}^\omega \vdash A \Rightarrow \mathbf{APP} \vdash [\![A]\!]_{\mathrm{IT}}$ *for sentences* A.

Proof. Straightforward. □

3.16. *The structure* ET *within* ***APP***. We define the *extensional type structure* $\mathrm{ET} := \langle(\mathrm{ET}_\sigma, =_\sigma)\rangle_{\sigma\in\mathcal{T}}$ in **APP** by:

$$x \in \mathrm{ET}_0 := x \in N,$$

$$x =_0 y := x \in \mathrm{ET}_0 \wedge y \in \mathrm{ET}_0 \wedge x = y,$$

$$x \in \mathrm{ET}_{\sigma\times\tau} := \boldsymbol{p}_0 x \in \mathrm{ET}_\sigma \wedge \boldsymbol{p}_1 x \in \mathrm{ET}_\tau \wedge \boldsymbol{p}(\boldsymbol{p}_0 x)(\boldsymbol{p}_1 x) = x,$$

$$x =_{\sigma\times\tau} y := (\boldsymbol{p}_0 x =_\sigma \boldsymbol{p}_0 y) \wedge (\boldsymbol{p}_1 x =_\tau \boldsymbol{p}_1 y),$$

$$x \in \mathrm{ET}_{\sigma\to\tau} := \forall yz(y =_\sigma z \to xy =_\tau xz),$$

$$x =_{\sigma\to\tau} y := x \in \mathrm{ET}_{\sigma\to\tau} \wedge y \in \mathrm{ET}_{\sigma\to\tau} \wedge \forall z \in \mathrm{ET}_\sigma(xz =_\tau yz).$$

$\mathbf{E\text{-}HA}^\omega$ is now interpreted in **APP** via ET in the same way as $\mathbf{HA}^\omega$ was interpreted in **APP** via IT. We have to show that $\boldsymbol{k}$, $\boldsymbol{s}$, etc. belong to the corresponding sets, e.g. $\boldsymbol{s}^{\sigma,\tau,\rho} \in \mathrm{ET}_{(\sigma\tau\rho)(\sigma\rho)\tau\rho}$, etc. □

Proposition. $\mathbf{E\text{-}HA}^\omega \vdash A \Rightarrow \mathbf{APP} \vdash [\![A]\!]_{\mathrm{ET}}$ *for sentences* A.

Proof. Straightforward. □

Observe that if we interpret **APP** in PRO, IT becomes a (version of) HRO, and ET a version of HEO (cf. 1.9). □

Convention. If in the sequel we speak about a *type-σ object* in (a model of) **APP**, we mean an element of IT_σ, and if we talk about an *extensional type-σ object*, we mean an element of ET_σ.

Since $\mathrm{IT}_0 = \mathrm{ET}_0 = N$, we can also use the more suggestive notation $N \to N$ or N^N for $\mathrm{IT}_1 = \mathrm{ET}_1$. □

4. Models for APP

In this section we shall review some of the most interesting models of **APP**, to show the wide range of possibilities. First we give some more mathematical, "recursion-theoretic" models; then we describe two term models which are useful for metamathematical results concerning **APP** (cf. section 7). We have already encountered one model of **APP**, our standard example PRO, the model of the partial recursive operations.

PRO may be regarded as the "type-free" version of HRO (or HRO as the typed version of PRO). Naturally, one should like to have a type-free version of ICF* and ICF as well, such that the typed objects of the type-free model are precisely the objects of ICF*, respectively ICF. For ICF* a corresponding type-free model is easily described.

4.1. *The model* PCO *of the partial continuous operations.* The objects of PCO are elements of $\mathbb{N}^{\mathbb{N}}$, and application of α to β, written as $\alpha|\beta$, is defined, as in 3.7.9 by

$$\alpha|\beta = \gamma := \forall x \exists y \big(\alpha(\langle x\rangle * \bar{\beta}y) = \gamma x + 1 \wedge \forall z < y\big(\alpha(\langle x\rangle * \bar{\beta}z) = 0\big)\big).$$

We may also express this by first defining

$$\alpha(\beta) = x := \exists y\big(\alpha(\bar{\beta}y) = x + 1 \wedge \forall z < y\big(\alpha(\bar{\beta}z) = 0\big)\big),$$

and then

$$\alpha|\beta = \gamma := \forall x\big(\alpha_{\langle x\rangle}(\beta) = \gamma(x)\big),$$

where

$$\alpha_{\langle x\rangle} := \lambda n.\alpha(\langle x\rangle * n).$$

For the natural numbers we take $N = \{\lambda x.y \colon y \in \mathbb{N}\}$; $=$ is interpreted as (extensional) equality between functions, and the interpretation $[\![c]\!] \equiv [\![c]\!]_{\mathrm{PCO}}$ of the various constants c of **APP** is given by:

$$[\![0]\!] := \lambda x.0, \qquad [\![S]\!] := \Lambda\alpha(\lambda x.S(\alpha x)), \qquad [\![P]\!] := \Lambda\alpha.\lambda x.\mathrm{prd}(\alpha x),$$

$$[\![\boldsymbol{k}]\!] := \Lambda\alpha\Lambda\beta.\alpha, \qquad [\![\boldsymbol{s}]\!] := \Lambda\alpha\Lambda\beta\Lambda\gamma.(\alpha|\gamma)|(\beta|\gamma),$$

$$[\![\boldsymbol{p}]\!] := \Lambda\alpha\Lambda\beta.j(\alpha,\beta), \qquad [\![\boldsymbol{p}_i]\!] := \Lambda\alpha.j_{i+1}\alpha \quad (i \leq 1),$$

$$[\![\boldsymbol{d}]\!] := \Lambda\alpha\Lambda\beta\Lambda\gamma\Lambda\delta\big(\lambda x(\alpha x \cdot \mathrm{sg}'|\gamma 0 - \delta 0|) + \lambda x(\beta x \cdot \mathrm{sg}|\gamma 0 - \delta 0|)\big).$$

Here $\Lambda\alpha$ is defined as in 3.7.15. This model can be used to show that **APP** is conservative over elementary analysis **EL**.

Extensional type 2 objects (cf. 3.16, convention) are continuous (to be precise, continuous with respect to the number-theoretic functions represented by the type 1 objects); in fact, there is even a type 3 object which assigns to extensional type 2 objects the modulus at $[\![\lambda x.0]\!]$ (E9.4.2).

The models PRO and PCO suggest that a minimum of recursion theory suffices to construct a model of **APP**. We shall now show that this is indeed generally true: any partial combinatory algebra can be extended to a model for **APP**.

4.2. DEFINITION. A *partial combinatory algebra* (pca) is a structure $(A, \cdot, k, s)$, with A a set, $\cdot$ a partial binary operation ("application"), and k, s two elements of A such that

$$\forall xy(kxy = x), \qquad \forall xyz(sxyz \simeq xz(yz)),$$
$$\forall xy\, \boldsymbol{E}(sxy), \qquad \exists xy(x \neq y).$$

Here we have written "xy" for "$x \cdot y$" and used as before the convention that $t_1 t_2 \ldots t_{n-1} t_n$ stands for $(\ldots(t_1 t_2)\ldots t_{n-1})t_n$. The condition on a pca of nontriviality, i.e. $\exists xy(x \neq y)$, is in fact equivalent to $k \neq s$ (exercise). □

REMARK. Any model of **APP** is in fact also a pca.

4.3. PROPOSITION. *Any* pca *can be expanded to a model of* **APP** *by suitable interpretations of* $\boldsymbol{p}$, $\boldsymbol{p}_0$, $\boldsymbol{p}_1$, $\boldsymbol{d}$, 0, S, P, *and* N.

PROOF. First we observe that for any given t constructed by application from constants k, s and variables we can define, as in 3.5, $\lambda x.t$ such that

$$\boldsymbol{E}(\lambda x.t) \quad \text{and} \quad (\lambda x.t)(t') \simeq t[x/t']$$

for all terms t' free for x in t. We abbreviate (dropping the application dot $\cdot$):

$$i := skk, \qquad \top := k, \qquad \bot := ki \equiv \lambda xy.y;$$
$$p := \lambda xy(\lambda z.zxy);$$
$$p_0 := \lambda x.xk \equiv \lambda x.x\top, \qquad p_1 := \lambda x.x(ki) \equiv \lambda x.x\bot.$$

We readily see that

$$p_0(pxy) = x, \qquad p_1(pxy) = y.$$

We introduce the *combinatory numerals*, and the successor S^*:

$$[\![0]\!] \equiv 0^* := i, \qquad [\![n+1]\!] \equiv (n+1)^* := p \bot n^*,$$
$$[\![S]\!] \equiv S^* := \lambda x.p \bot x.$$

Obviously $S^* n^* = (n+1)^*$.

As a first approximation to the predecessor we use

$$P' := \lambda x . p_1 x;$$

then

$$P'(S^* x) = x, \qquad P'0^* = \perp .$$

For $tt't''$ we may use the suggestive notation

if t then t' else t'',

for if $t \equiv \top$ (true), then $tt't'' \simeq t'$, and if $t \equiv \perp$ (false), then $tt't'' \simeq t''$. Note that $p_0 0^* = \top$, $p_0(S^* x) = \perp$, hence, if we put

$$[\![P]\!] \equiv P^* := \lambda x.(p_0 x)0^*(P'x),$$

we have indeed

$$P^*(S^* x) = x, \qquad P^* 0^* = 0^*.$$

Any pca contains a fixed point operator fix (as in 3.7), and thus similar to 3.8 we can put

$$\rho := \lambda fn. (\text{if } p_0 n \text{ then } kxn \text{ else } \lambda z.yz(f(P^* z))n)$$

and

$$r := \lambda xy.\text{fix}(\rho);$$

r behaves as a recursor. So on the set $N^* := \{n^* : n \in N\}$ of combinatory numerals we have explicit definition (via λx), and primitive recursion; $Z^* := k0^*$ satisfies $Z^* x = 0^*$, so Z^* represents the zero-function; for the projections we can take $\lambda x_1 \ldots x_n . x_i$. Thus we have all primitive recursive functions available, and we can construct a term t such that $tn^* m^* = |n - m|^*$; if we now take

$$d := \lambda xyuv. (\text{if } p_0(tuv) \text{ then } x \text{ else } y),$$

we have a definition-by-cases operator.

The model for **APP** is then obtained by taking N^*, interpreting 0, S, P by 0^*, S^*, P^*, respectively, and interpreting all other $\boldsymbol{c}$ by the corresponding c as defined above. □

REMARK. In view of the fact that e.g. the partial recursive operations form a pca, one might ask why we define PRO separately, and why we have explicitly listed as primitives in **APP** certain operations such as $\boldsymbol{p}$, $\boldsymbol{p}_0$, $\boldsymbol{p}_1$, $\boldsymbol{d}$ which are in fact explicitly definable in any pca. The reason is that the definition of p, etc., as outlined above, introduces a very special relationship between these constants and k and s, which is mostly irrelevant and restricts us in our choice of model. In particular, if we define in PCO $\boldsymbol{p}, \boldsymbol{p}_0, \boldsymbol{p}_1, N$ as above, N corresponds only to a subset of the domain $\mathbb{N}$, and $\boldsymbol{p}$ cannot be assumed to be surjective. □

4.4. COROLLARY. *Each model of combinatory logic or the λ-calculus can be made into a model of* **TAPP**.

PROOF. A model of combinatory logic is nothing but a combinatory algebra, i.e. a pca with total application. As to the definition of a model of the λ-calculus, see e.g. Barendregt (1981, ch. 5 of the *second* edition). In any case, any model of the λ-calculus is also a combinatory algebra. □

4.5. *The graph model* $\mathrm{P}(\omega)$. This model for the λ-calculus was discovered by Plotkin (1972) and, independently, by Scott (1975, 1975A). It has been extensively studied in the context of models of the λ-calculus (see Barendregt 1981, sections 18.2, 19.2). We shall first present the *graph model* straightforwardly as a model of **TAPP**, that is to say as a combinatory algebra with extra operators.

The domain of the model consists of all subsets of $\mathbb{N}$ (this suggests the notation $\mathrm{P}(\mathbb{N})$ for the model, but we follow the tradition and use $\mathrm{P}(\omega)$). In the description below we shall choose a particular coding of finite subsets of $\mathbb{N}$ and a specific pairing function on $\mathbb{N}$; for the pairing function we can take our usual j; for $j(x, y)$ we also use the abbreviation (x, y).

The finite set $\{k_0, \dots, k_p\}$ with $k_0 < k_1 < \cdots < k_p$ is coded by

$$[k_0, \dots, k_p] := \sum_{i=0}^{p} 2^{k_i}, \qquad [\emptyset] := 0,$$

and we shall write

$$e_n := \{k_0, \dots, k_p\} \quad \text{if } n = [k_0, \dots, k_p].$$

We use X, Y, Z for arbitrary subsets of $\mathbb{N}$. Since the coding of finite sets is onto $\mathbb{N}$, we can use variables ranging over $\mathbb{N}$ also for finite sets. We shall often neglect the distinction between finite sets and their codes in our notation, and thus write, e.g.,

$$x \subset V := e_x \subset V, \qquad z \in x := z \in e_x, \quad \text{etc.}$$

On $\mathrm{P}(\omega)$ we can define a total application operation $\cdot$ by

$$X \cdot Y := \{z : \exists y \subset Y((y, z) \in X)\}.$$

$P(\omega)$ can be made into a combinatory algebra putting

$$[\![k]\!]_{P(\omega)} := \{(x,(y,z)) : z \in x\},$$

$$[\![s]\!]_{P(\omega)} := \{(x,(y,(z,w))) : \exists z' \subset z((z',(u,w)) \in x) \wedge \forall u' \in u \exists z' \subset z((z',u') \in y)\}.$$

We leave it as an exercise to show that indeed, dropping the subscript $P(\omega)$,

$$([\![k]\!] \cdot X) \cdot Y = X, \qquad (([\![s]\!] \cdot X) \cdot Y) \cdot Z = (X \cdot Z) \cdot (Y \cdot Z).$$

As interpretation of N we take the set $\{\{n\} : n \in \mathbb{N}\}$, and we put

$$[\![0]\!] := \{0\}, \qquad [\![S]\!] := \{(2^n, n+1) : n \in \mathbb{N}\}$$

(recall that 2^n codes the singleton $\{n\}$, and that indeed on this definition $[\![S]\!] \cdot \{n\} = \{n+1\}$). Similarly

$$[\![P]\!] := \{(2^n, n \dot{-} 1) : n \in \mathbb{N}\}.$$

For the pairing operators we take

$$[\![p]\!] := \{(2^n,(0,2n)) : n \in \mathbb{N}\} \cup \{(0,(2^m,2m+1)) : m \in \mathbb{N}\},$$

$$[\![p_0]\!] := \{(2^{2n}, n) : n \in \mathbb{N}\}, \qquad [\![p_1]\!] := \{(2^{2n+1}, n) : n \in \mathbb{N}\}.$$

Finally, for the definition-by-cases constant take

$$[\![d]\!] := \{(y,(x,(2^m,(2^n,z)))) : (m = n \wedge z \in x) \vee (m \neq n \wedge z \in y)\}.$$

Equality is simply equality of sets in $P(\omega)$, and we leave it to the reader to verify that thus $P(\omega)$ indeed becomes a model of **TAPP**.

4.6. *The topology of* $P(\omega)$. Some more insight into the structure of $P(\omega)$ is obtained by introducing an appropriate topology on $P(\omega)$. We shall summarize the principal definitions and facts below, without proofs.

If we put $O_{e_n} := \{X : n \subset X\}$, we see that $O_{e_n} \cap O_{e_m} = \{X : n \subset X \wedge m \subset X\} = \{X : n \cup m \subset X\} = O_{e_n \cup e_m}$; therefore the O_{e_n} are a basis for a topology on $P(\omega)$.

This topology is different from the well-known (Cantor-set) product topology on the characteristic functions of subsets of $\mathbb{N}$ (4.1.5). The present topology has basis elements which only tell which numbers are *in* the set; for this reason it is sometimes called the positive information topology. The reader can easily check that it is only T_0.

Taking the continuous mappings (in the sense of this topology) from $P(\omega)$ to $P(\omega)$ as operations, and encoding these mappings in $P(\omega)$ itself, we obtain a model of the lambda calculus. The basic facts concerning this model essentially are exercises in topology, we leave the proofs to the reader. Details can be found in Barendregt (1981, section 18.2).

A characterization of functions continuous in the sense of this topology is given by the following proposition.

PROPOSITION. *Let* $(\mathrm{P}(\omega))^k$ *be given the product topology. Then*:
(i) $f \in \mathrm{P}(\omega) \to \mathrm{P}(\omega)$ *is continuous iff* $f(X) = \bigcup\{f(e_n) : n \subset X\}$ *for all* X.
(ii) $f \in (\mathrm{P}(\omega))^k \to \mathrm{P}(\omega)$ *is continuous iff it is continuous in each of its variables separately.* □

COROLLARY. *A continuous* $f \in \mathrm{P}(\omega) \to \mathrm{P}(\omega)$ *is monotone.* □

DEFINITION. If $f \in \mathrm{P}(\omega) \to \mathrm{P}(\omega)$ is continuous, define

$$\mathrm{graph}(f) := \{(n, m) : m \in f(e_n)\},$$

and if $X \in \mathrm{P}(\omega)$, let fun(X) be the function given by

$$\mathrm{fun}(X)(Y) := \{m : \exists n \subset Y((n, m) \in X)\}. \quad \square$$

N.B. The application we defined before satisfies

$$X \cdot Y := \mathrm{fun}(X)(Y).$$

PROPOSITION.
(i) fun(graph(f)) $= f$, *for continuous* f;
(ii) $\forall X$(fun(X) *continuous*);
(iii) *application is continuous*;
(iv) *if* $f \in (\mathrm{P}(\omega))^{k+2} \to \mathrm{P}(\omega)$ *is continuous, then the function* g *given by* $g(\vec{Y}) = \mathrm{graph}(\lambda X.f(X, \vec{Y}))$ *is continuous.* □

We can now interpret the λ-operator in the model as follows.

4.7. DEFINITION. The set of P(ω)-λ-*terms* is constructed from variables and constants for the elements of P(ω), by means of application and λ-abstraction. We interpret the closed P(ω)-λ-terms by

$$[\![X]\!] := X \quad \text{for } X \in \mathrm{P}(\omega),$$

$$[\![tt']\!] := [\![t]\!] \cdot [\![t']\!],$$

$$[\![\lambda x.t[x]]\!] := \mathrm{graph}(\lambda X.[\![t[X]]\!]). \quad \square$$

4.8. PROPOSITION. $\mathrm{P}(\omega)$ *is weakly extensional, i.e.*

$$\forall X([\![t[X]]\!] = [\![s[X]]\!]) \rightarrow [\![\lambda x.t[x]]\!] = [\![\lambda x.s[x]]\!]. \quad \square$$

4.9. *The finite-type objects in* $\mathrm{P}(\omega)$. Any function α of $\mathbb{N}^{\mathbb{N}}$ can be canonically represented in $\mathrm{P}(\omega)$ by a type 1 object $\hat{\alpha} \equiv \{(2^n, \alpha n) : n \in \mathbb{N}\}$, since by the definition of application $\hat{\alpha} \cdot \{n\} = \{\alpha n\}$. However, this representation of α is by no means unique. For example, if α is constant, it is also represented by $\{(0, \alpha 0)\}$, and it is also irrelevant what pairs of the form $(2^n + 2^m, k)$ (for $n \neq m$) are contained in any representation of α. But note that if X_1 and X_2 both represent α, i.e. $\forall n(X_1 \cdot \{n\} = X_2 \cdot \{n\} = \{\alpha n\})$, then $X_1 \cup X_2$ also represents α, since application is continuous and hence monotone. But from this it follows that any type 2 object must be extensional, i.e. if Z is a type 2 object, and X_1, X_2 represent α, then $Z \cdot X_1 = Z \cdot X_2$; for $Z \cdot X_1 \subset Z \cdot (X_1 \cup X_2)$, $Z \cdot X_2 \subset Z \cdot (X_1 \cup X_2)$ and since the results of these applications are all singletons, we have in fact $Z \cdot X_1 = Z \cdot (X_1 \cup X_2) = Z \cdot X_2$.

As we shall show below, this holds for the finite type objects in $\mathrm{P}(\omega)$ in general.

4.10. DEFINITION. A *partial equivalence relation* on a set A is a binary relation $\sim$ such that for all $x, y, z \in A$

$$x \sim y \rightarrow y \sim x, \qquad x \sim y \wedge y \sim z \rightarrow x \sim z.$$

The restriction of $\sim$ to $\{x : x \sim x\}$ is an equivalence relation.

We assign to each type symbol σ a partial equivalence relation $=_\sigma$ on $\mathrm{P}(\omega)$ as follows. We put

$$\mathrm{I}_\sigma := \{X : X =_\sigma X\},$$

$$X =_0 Y := \exists n(X = \{n\} = Y),$$

$$X =_{\sigma \rightarrow \tau} Y := \forall Z \in \mathrm{I}_\sigma(X \cdot Z =_\tau Y \cdot Z),$$

$$X =_{\sigma \times \tau} Y := P_1 \cdot X =_\sigma P_1 \cdot Y \wedge P_2 \cdot X =_\tau P_2 \cdot Y,$$

where

$$P_0 = [\![\mathbf{p}_0]\!], \qquad P_1 = [\![\mathbf{p}_1]\!].$$

Note that $Z \in \mathrm{I}_{\sigma \rightarrow \tau}$ iff $\forall X \in \mathrm{I}_\sigma(Z \cdot X \in \mathrm{I}_\tau)$, and that the relations $=_{\sigma \rightarrow \tau}$, $=_{\sigma \times \tau}$ are partial equivalence relations if $=_\sigma$ and $=_\tau$ are so. $\square$

4.11. LEMMA. *Let* $=_\tau$ *be a partial equivalence relation such that*

$$\forall XX'(X =_\tau X' \to X \cup X' =_\tau X) \quad (\textit{the join-property}),$$

then $=_{\sigma\tau}$ *also has the join-property.*

PROOF. Assume $Z =_{\sigma\to\tau} Z'$, then $\forall X \in \mathrm{I}_\sigma(Z \cdot X =_\tau Z' \cdot X)$, and by the definition of application $(Z \cup Z') \cdot X = (Z \cdot X) \cup (Z' \cdot X) =_\tau Z \cdot X$, therefore $Z \cup Z' =_{\sigma\tau} Z$. □

4.12. LEMMA. *Let* $X =_0 Y := \exists n(X = Y \cdot \{n\})$. *Then*:
(i) *if* $=_\sigma$ *has the join-property, then* $\forall Z \in \mathrm{I}_{\sigma 0} \forall XX'(X =_\sigma X' \to Z \cdot X =_0 Z \cdot X')$;
(ii) $\forall Z \in \mathrm{I}_{\sigma\tau} \forall XX'(X =_\sigma X' \to Z \cdot X =_\tau Z \cdot X')$.

PROOF. We leave (i) to the reader. We prove (ii) by induction on the complexity of τ; (i) covers the basis case $\tau \equiv 0$.

If $\tau \equiv \rho\rho'$, $Z \in \mathrm{I}_{\sigma\tau}$, $X =_\sigma X'$, we have to show $Z \cdot X =_\tau Z \cdot X'$, i.e. $\forall Y \in \mathrm{I}_\rho(Z \cdot X \cdot Y =_{\rho'} Z \cdot X' \cdot Y)$, but this means showing $\forall Y(T \cdot X =_{\rho'} T \cdot X')$, where $T :=$ graph $\lambda X.(Z \cdot X \cdot Y)$ (or we may take for T the interpretation of $\lambda x.(Z \cdot x \cdot Y)$ for the combinatorially defined λx); this holds since $T \in \mathrm{I}_{\sigma\rho'}$, for which the relevant property holds by induction hypothesis.

We leave the case $\tau \equiv \rho \times \rho'$ to the reader. □

4.13. PROPOSITION. *In* P(ω) IT *and* ET (3.15, 3.16) *coincide.*

PROOF. Immediate by the preceding lemma. □

4.14. PROPOSITION. *Each type* 2 *operation in* P(ω) *is continuous* (*and has a neighbourhood function*).

PROOF. Let X be a type 2 object in P(ω) and let $\hat{\alpha} = \{(2^n, \alpha n) : n \in \mathbb{N}\}$, then $X \cdot \hat{\alpha} = \{m\}$ for some m; hence $\hat{\alpha}$ contains a finite subset with code a say such that $(a, m) \in X$. Let x be the maximal argument of α occurring in a, i.e.

$$x := \max\{i : (2^i, \alpha i) \in a\};$$

then

$$\mathrm{P}(\omega) \models \bar{\alpha}(x+1) = \bar{\beta}(x+1) \to X\alpha = X\beta. \quad \Box$$

REMARK. Appealing to AC-NN we can find a neighbourhood function for X, but there is no type (2)1 object in P(ω) which assigns to any type 2 object its associate, since such an operation cannot exist in an extensional finite type structure, as we shall see in 6.11.

The argument above can be extended to prove that the finite type objects of P(ω) coincide with the extensional continuous functionals ECF (1.17). □

4.15. *The submodels* $\Sigma(\omega)$ *and* RE(ω). Inspection shows that restriction to enumerable subsets of $\mathbb{N}$, i.e. sets of the form $\{n : \exists m\alpha(n, m)\}$, where α ranges over a universe of functions satisfying the axioms of **EL** (i.e. closed under "recursive in") constitute a submodel $\Sigma(\omega)$: all the interpretations in P(ω) of the constants $\boldsymbol{k}$, $\boldsymbol{s}$, etc. are in fact enumerable sets, and the abstraction operator as defined also goes from enumerable sets to enumerable sets.

We may restrict attention still further to the r.e. sets, giving rise to the submodel RE(ω), which can be described in arithmetic. This provides a method for showing that **TAPP** is conservative over **EL** and **HA**.

4.16. *Term models for* **APP**. Many different term models may be constructed (see e.g. Beeson 1985, VI, Section 6), but we shall here concentrate on two models, needed for some of our metamathematical results in section 7: the model CNFS of closed terms in normal form, and the model CTS of all closed terms (the "S" in CNFS and CTS refers to "strict reduction", see below). For this purpose we have to define contractions and reductions for our type-free terms.

4.17. DEFINITION. $t \operatorname{conv} t'$ (*t converts to t'*) is a binary relation between terms which holds in the following cases: there are t_0, t_1, t_2 such that one of the following applies:

(1) $t \equiv \boldsymbol{k}t_0t_1,\qquad t' \equiv t_0;$

(2) $t \equiv \boldsymbol{s}t_0t_1t_2,\qquad t' \equiv t_0t_2(t_1t_2);$

(3) $t \equiv \boldsymbol{p}_i(\boldsymbol{p}t_0t_1),\qquad t' \equiv t_i(i \in \{0,1\});$

(4) $t \equiv P(St),\qquad t' \equiv t;$

(5) $t \equiv P0,\qquad t' \equiv 0;$

(6) $t \equiv \boldsymbol{d}t_0t_1\bar{n}\bar{m}\qquad t' \equiv t_0$ if $n = m$ and $t' \equiv t_1$ if $n \neq m$.

A possible strengthening of this rule is

(6′) $t \equiv \boldsymbol{d}t_0t_1t_2t_3$, t_2, t_3 closed and in normal form; if $t_2 \equiv t_3$, then $t' \equiv t_0$, and if $t_2 \not\equiv t_3$, then $t' \equiv t_1$.

4.18. Definition. The reduction relations $\succcurlyeq$, $\succ_1$, $\succcurlyeq^*$ are defined relative to conv exactly as in 2.2. We shall say that t, t' are *r-equal* if there is a chain $t \equiv t_0, t_1, t_2, \ldots, t_n \equiv t'$ such that for all $i < n$ we have $t_i \preccurlyeq t_{i+1}$ or $t_{i+1} \preccurlyeq t_i$. *Reduction sequences* are defined as before; in a *strict reduction sequence* each t_{i+1} is obtained from t_i by contraction of the leftmost minimal redex. *Normal form* is also defined as before (2.3). □

Observe that $t \equiv \boldsymbol{kk}((\lambda x.xx)(\lambda x.xx))$ can be reduced to $\boldsymbol{k}$, but t strictly reduces only to itself. So t has a normal form, but cannot be strictly reduced to this normal form.

4.19. Theorem (*Church–Rosser property*). *If $t \succcurlyeq t'$, $t \succcurlyeq t''$, there is a t''' such that $t' \succcurlyeq t'''$ and $t'' \succcurlyeq t'''$.*

Proof. As in the typed case. $\boldsymbol{P}$ and the contraction rule for $\boldsymbol{d}$ introduce some extra cases which are easily dealt with. □

4.20. *The models* CNFS, CNFS$^\alpha$. From the Church–Rosser theorem it follows that the normal form of a term, if existing, is unique; this suggests to define the model CNFS as consisting of all closed terms in normal form, with Ap(t, t') being the uniquely determined term t'' in normal form such that $tt' \succcurlyeq t''$, if existing. However, thus we do not obtain a model for **APP**, since on this definition the condition of *strictness* is not satisfied; if t_1 is in normal form, and t_2 any term without normal form, e.g. $(\lambda x.xx)(\lambda x.xx)$, then $\boldsymbol{k}t_1t_2$ ought to be undefined, but t_1 exists in the model and is defined; so we have in the model $\boldsymbol{k}t_1t_2 = t_1$, which conflicts with $\boldsymbol{E}\boldsymbol{k}t_1t_2 \to \boldsymbol{E}t_2$.

Therefore (as in Beeson 1985), we interpret Ap(t, t') as the unique normal t'' such that tt' strictly reduces to t'', if existing. For N we take $\{S^n0 : n \in \mathbb{N}\}$, and the constants are interpreted by themselves; equality is equality of normal form. So this model has a recursive domain with decidable (recursive) equality (but the relation $tt' = t''$ is not decidable; why?). The verification that CNFS is indeed a model is straightforward; e.g. we verify $\boldsymbol{s}t_1t_2t_3 = t_1t_3(t_2t_3)$ in the model by showing that the strict reduction sequence of the left-hand side coincides with the strict reduction sequence of $t_1t_3(t_2t_3)$ after a single step.

Without difficulty we can in fact incorporate the stronger rule (6′) for conversions in our notion of reduction. This gives definition by cases for the universe, in the model CNFS.

It is also easy to give a variant CNFS$^{\alpha}$, depending on the function $\alpha \in \mathbb{N} \to \mathbb{N}$: choose a designated variable x^*, which is to play the role of a fixed type 1 object, and add to the conversions

$$x^*(S^n 0) \operatorname{conv} S^{\alpha n} 0 \quad \text{for all } n.$$

4.21. *The model* CTS. Now we take as domain all closed terms; two closed terms are equal in CTS iff they are r-equal. The Church–Rosser theorem guarantees that we obtain a consistent notion of equality for which distinct terms in normal form cannot be identified (exercise); N is interpreted as before, the constants are interpreted by themselves and application by term composition. More customary would be to take as domain the equivalence classes with respect to r-equality of the closed terms, but this is of course essentially equivalent.

Note that, in fact, CTS is a model with *total* application, i.e. a model of **TAPP**. Formalization of CTS in **HA** can be used to obtain conservativeness of **TAPP** over **HA**. □

5. Abstract realizability in APP

5.1. We proceed to present an abstract version of Kleene's realizability, formulated in **APP**. This version is essentially due to Feferman (1975).

Definition. To each formula A of **APP** we assign a new formula $x\,\mathbf{r}\,A$, with $x \notin \mathrm{FV}(A)$; $\mathrm{FV}(x\,\mathbf{r}\,A) = \{x\} \cup \mathrm{FV}(A)$ where $x\,\mathbf{r}\,A$ is defined by induction on A:

(i) $x\,\mathbf{r}\,P := Ex \wedge P$ for P prime;

(ii) $x\,\mathbf{r}(A \wedge B) := (\mathbf{p}_0 x\,\mathbf{r}\,A) \wedge (\mathbf{p}_1 x\,\mathbf{r}\,B)$;

(iii) $x\,\mathbf{r}(A \to B) := Ex \wedge \forall y(y\,\mathbf{r}\,A \to xy\,\mathbf{r}\,B)$;

(iv) $x\,\mathbf{r}(\forall y A) := \forall y(xy\,\mathbf{r}\,A)$;

(v) $x\,\mathbf{r}(\exists y A) := E(\mathbf{p}_0 x) \wedge (\mathbf{p}_1 x\,\mathbf{r}\,A)[y/\mathbf{p}_0 x]$.

Assuming $\vee$ to be defined, we do not need a separate clause for $\vee$; if $\vee$ is taken as a primitive, we put

(vi) $x\,\mathbf{r}(A \vee B) := (\mathbf{p}_0 x = 0 \wedge \mathbf{p}_1 x\,\mathbf{r}\,A) \vee (\mathbf{p}_0 x = 1 \wedge \mathbf{p}_1 x\,\mathbf{r}\,B)$. □

N.B. For free variables x the "$Ex \wedge$" in (i) and (iii) is redundant, but not when arbitrary terms t are substituted for x; cf. lemma 5.3(iii) below.

5.2. DEFINITION. A formula is *∃-free* (*existence-free*) if it does not contain ∃ (nor ∨, if ∨ is included as a primitive). □

5.3. LEMMA
(i) $t\mathbf{r}A$ *is ∃-free for all* A;
(ii) $(t\mathbf{r}A)[x/s] \equiv t[x/s]\mathbf{r}A[x/s]$ *if* x *is not bound in* A *or in* $t\mathbf{r}A$;
(iii) $\mathbf{APP} \vdash t\mathbf{r}A \to Et$.

PROOF. By induction on A. □

5.4. THEOREM (*soundness of abstract realizability*). *For* A *in* $\mathscr{L}(\mathbf{APP})$

$$\mathbf{APP} \vdash A \Rightarrow \mathbf{APP} \vdash t\mathbf{r}A$$

for some term t. *More generally, if* Γ *is a set of sentences, then*

$$\mathbf{APP} + \Gamma \vdash A \Rightarrow \mathbf{APP} + \Gamma^* \vdash \exists x(x\mathbf{r}A),$$

where $\Gamma^* \equiv \{\exists x(x\mathbf{r}B) : B \in \Gamma\}$.

PROOF. By induction on the length of derivations in **APP**; cf. 4.4.10(ii). □

5.5. DEFINITION. Let A be any formula without ∨ or ∃ in its strictly positive parts (i.e. a Rasiowa–Harrop formula, cf. 2.3.23, 3.5.14). We define a term τ_A for such A inductively:

$$\tau_A := 0 \quad \text{for } A \text{ prime},$$

$$\tau_{A \wedge B} := \mathbf{p}\tau_A\tau_B,$$

$$\tau_{A \to B} := \tau_{\forall x B} := \lambda x.\tau_B$$

(note that we do not need clauses for ∨ and ∃). □

5.6. LEMMA. *For ∃-free* A

$$\mathbf{APP} \vdash A \leftrightarrow \exists x(x\mathbf{r}A) \leftrightarrow \tau_A\mathbf{r}A.$$

PROOF. Straightforward, by induction on A (cf. 4.4.5). □

5.7. DEFINITION. Let the *extended axiom of choice* be the following schema:

EAC $\quad \forall x(Ax \to \exists yBxy) \to \exists f\forall x(Ax \to B(x, fx))$

for ∃-free A. □

5.8. LEMMA. *Let* EAC(A, B) *be an instance of* EAC *with premiss* $\forall x(Ax \to \exists y Bxy)$. *Then there is a term* σ_A *(depending on* A *only) such that*

$$\mathbf{APP} \vdash \sigma_A \,\mathbf{r}\, \mathrm{EAC}(A, B).$$

PROOF. Take

$$\sigma_A := \lambda z.\boldsymbol{p}(\lambda x.\boldsymbol{p}_0(zx\tau_A))(\lambda xv.\boldsymbol{p}_1(zx\tau_A)),$$

and assume

$$z \,\mathbf{r}\, \forall x(Ax \to \exists y Bxy).$$

Then $zx\,\mathbf{r}(Ax \to \exists yBxy)$, and so

$$u\,\mathbf{r}\,Ax \to zxu\,\mathbf{r}\,\exists y Bxy.$$

Since $u\,\mathbf{r}\,Ax \to \tau_A\,\mathbf{r}\,Ax$, we have $u\,\mathbf{r}\,Ax \to zx\tau_A\,\mathbf{r}\,\exists yBxy$, i.e.

$$u\,\mathbf{r}\,Ax \to E\boldsymbol{p}_0(zx\tau_A) \wedge \boldsymbol{p}_1(zx\tau_A)\mathbf{r}\,B(x, \boldsymbol{p}_0(zx\tau_A)),$$

and from this we see that

$$\sigma_A z\,\mathbf{r}\,\exists f \forall x(Ax \to B(x, fx)). \quad \square$$

5.9. THEOREM (*axiomatization of abstract realizability*).
(i) **APP** + EAC $\vdash A \leftrightarrow \exists x(x\,\mathbf{r}\,A)$,
(ii) **APP** + EAC $\vdash A \Leftrightarrow$ **APP** $\vdash \exists x(x\,\mathbf{r}\,A)$.

PROOF. (i) is shown by induction on A (cf. 4.4.10(i)). (ii) follows from the soundness theorem together with lemma 5.8. □

5.10. DEFINITION. Let Γ be the class of formulas A such that all B occurring negatively in A are $\exists$-free.

This is equivalent to the following inductive definition of Γ: Γ is the least class $\mathscr{X}$ such that:
(i) prime formulas are in $\mathscr{X}$,
(ii) $\mathscr{X}$ is closed under $\wedge$, $\vee$, $\forall$, $\exists$;
(iii) if A is $\exists$-free and $B \in \mathscr{X}$, then $A \to B \in \mathscr{X}$. □

5.11. LEMMA. $A \in \Gamma \Rightarrow$ **APP** $\vdash \exists x(x\,\mathbf{r}\,A) \to A$.

PROOF. By induction on A. Assume $A \in \Gamma$. Let us consider two of the induction steps.

Case 1. $A \equiv B \to C$. Then B is $\exists$-free, $C \in \Gamma$. Assume $\exists x(x \mathbf{r} (B \to C))$ and B, then by 5.6 $\tau_B \mathbf{r} B$. Let $x \mathbf{r} (B \to C)$ ($\exists E$), then $x \tau_B \mathbf{r} C$, and so, by induction hypothesis, C. This shows $B \to C$.

Case 2. $A \equiv \exists y B$. $A \in \Gamma$, hence $B \in \Gamma$. Assume $\exists x(x \mathbf{r} \exists y By)$, so $\exists x(\mathbf{p}_1 x \mathbf{r} B[y/\mathbf{p}_0 x])$, i.e. $\exists yx(x \mathbf{r} By)$. With the induction hypothesis $\exists y B$ follows. □

5.12. THEOREM. **APP** + EAC *is conservative over* **APP** *with respect to formulas in* Γ.

PROOF. Obviously, **APP** + EAC is conservative over **APP** with respect to $\{B : \mathbf{APP} \vdash \exists x(x \mathbf{r} B) \to B\}$ by 5.9; combine this with the lemma. □

5.13. REMARK. The strengthening of EAC to EAC$^+$, where the condition "A $\exists$-free" has been dropped, is inconsistent; this follows from the example in 3.14 where it was shown that for certain D and A

$$\forall x(Dx \to \exists! y A(x, y)) \to \exists z \forall x(Dx \to E(zx) \wedge A(x, zx)) \qquad (1)$$

is inconsistent with **APP**.

At the same time this shows that EAC is inconsistent with classical logic, since the predicate D of 3.14 can be rewritten in classical logic as an $\exists$-free formula, and then (1) becomes an instance of EAC itself. □

5.14. *Recursive realizability and abstract realizability.* By means of the model PRO we can express the exact relationship between recursive realizability as introduced in 4.4.2 and our abstract realizability notion for **APP**.

Closed terms t and sentences A of **APP** become closed terms $[\![t]\!]_{\mathrm{PRO}}$ and sentences $[\![A]\!]_{\mathrm{PRO}}$ of **HA*** under the interpretation of **APP** in PRO, where **HA*** is the definitional extension of **HA** obtained by permitting Kleene-brackets $\{\cdot\}$ as a primitive in the language (cf. 2.7.6, section 3.7). Below we shall systematically disregard the difference between **HA** and **HA***; also we drop the subscript PRO. The interpretation $[\![\ ]\!]$ is readily extended to arbitrary terms and formulas of **APP**, letting free variables in t or A correspond to the same free variables in $[\![t]\!]$ or $[\![A]\!]$; in other words, the obvious definition of $[\![t]\!]$ and $[\![A]\!]$ by recursion on the complexity of $[\![t]\!]$ and $[\![A]\!]$ is extended with a clause $[\![x]\!] := x$ for variables x.

Let $\mathbf{r}$ denote abstract realizability in **APP** as before, and let $\mathbf{r}'$ be recursive realizability. Then we have the following theorem.

5.15. THEOREM. *For all A in* $\mathscr{L}$(**APP**)

$$\mathbf{HA} \vdash [\![\exists x(x\mathbf{r}A)]\!] \leftrightarrow \exists x(x\mathbf{r}'[\![A]\!]).$$

For the proof, we need the following lemma.

5.16. LEMMA.
(i) *If A is* $\exists$*-free in* **APP**, *then* $[\![A]\!]$ *is almost negative* (4.4.4) *in* **HA**.
(ii) *If F is an instance of* EAC, *then* $[\![F]\!]$ *is an instance of* ECT_0.
(iii) *If A belongs to* Γ *as defined in* 5.10, *then* $[\![A]\!]$ *belongs to the corresponding class* Γ_0 *of* **HA**; Γ_0 *is defined as* Γ *except that "*$\exists$*-free" is replaced by almost negative.*
(iv) **HA** + ECT_0 *is conservative over* **HA** *with respect to formulas of* Γ_0.

PROOF. (i) This is proved by induction on the complexity of A. The only nontrivial case is the case of prime formulas. Let $P \equiv (t = s)$ be prime in **APP**; by repeated use of equivalences $t_1 = t_2 \leftrightarrow \exists x(t_1 = x \wedge t_2 = x)$, $t_1t_2 = x \leftrightarrow \exists yz(t_1 = y \wedge t_2 = z \wedge yz = x)$, $\exists x(A \wedge \exists yB) \leftrightarrow \exists xy(A \wedge B)$ ($y \notin \mathrm{FV}(A)$), we find **APP** $\vdash P \leftrightarrow \exists x(P_1 \wedge \cdots \wedge P_n)$, where each P_i has one of the forms $x = y$, $xy = z$, or $x = c$ (c a constant).

Under the interpretation in PRO we can find a term $t[w]$ of **HA** such that **HA** $\vdash [\![P]\!] \leftrightarrow \exists w(t[w] = 0)$, by repeated use of $\{x\}(y) = z \leftrightarrow \exists u(Txyu \wedge Uu = z)$, $\exists x(A \wedge \exists yB) \leftrightarrow \exists xy(A \wedge B)$ ($y \notin \mathrm{FV}(A)$) and $\exists v_1v_2A(v_1, v_2) \leftrightarrow \exists vA(j_1v, j_2v)$.

(ii) Immediate by (i).

(iii) Immediate by (i).

(iv) is proved as in 5.12. □

5.17. *Proof of theorem* 5.15. **APP** + EAC $\vdash A \leftrightarrow \exists x(x\mathbf{r}A)$, so **HA** + $\mathrm{ECT}_0 \vdash [\![A]\!] \leftrightarrow [\![\exists x(x\mathbf{r}A)]\!]$ (by (ii) of the lemma); also **HA** + $\mathrm{ECT}_0 \vdash [\![A]\!] \leftrightarrow \exists x(x\mathbf{r}'[\![A]\!])$; hence **HA** + $\mathrm{ECT}_0 \vdash [\![\exists x(xrA) \leftrightarrow \exists x(x\mathbf{r}'[\![A]\!])$ since both sides of the equivalence belong to Γ_0 (lemma 5.16(iii)); since **HA** + ECT_0 is conservative over **HA** with respect to Γ_0, we find the statement of the theorem. □

6. Extensionality and choice in APP and $\mathbf{HA}^\omega$

6.1. We recall the intuitive justification of choice principles (4.2.1): let $\forall x\exists yA(x, y)$ be proved, then a constructive proof *must* provide us with an operation ("recipe") which constructs y for each given x. However, if x

ranges over some domain with a specified equality relation, we cannot expect that in general the operation provided by the proof will respect this equality relation—especially not if the equality relation is standard extensional equality (cf. our examples in 4.2.1).

As we shall see from the results below, a formal conflict arises in particular when additional nonclassical axioms are present which enforce the nonextensional character of operations.

We shall first present a few results for **APP**. With respect to **APP**, let

Ext $\quad \forall yz(\forall x(yx \simeq zx) \to y = z)$,

or equivalently

$$\forall yz(\forall uv(yu = v \leftrightarrow zu = v) \to y = z),$$

and let $\mathrm{D_V}$ be the axiom of full definition by cases (3.13):

$\mathrm{D_V} \quad \forall xyuv(\boldsymbol{d}uvxx = u \wedge (x \neq y \to \boldsymbol{d}uvxy = v))$.

The axiom of choice $\mathrm{AC_V}$ for the universe is the schema

$\mathrm{AC_V} \quad \forall x \exists y A(x, y) \to \exists z \forall x(Ezx \wedge A(x, zx))$.

6.2. Proposition (*Feferman* 1975). *Assume* $\boldsymbol{e}$ *to be an operator for deciding equality*:

$$\forall xy[(\boldsymbol{e}yx = 1 \wedge x = y) \vee (\boldsymbol{e}xy = 0 \wedge x \neq y)]. \tag{1}$$

Then **APP** + Ext + (1) $\vdash \perp$.

Proof. (Beeson 1985, p. 235). The existence of a fixed point operator (3.7) guarantees the existence of a g such that $gx \simeq \boldsymbol{d}10(\boldsymbol{e}g(\lambda x.0))1$, i.e.

$$gx \simeq 1 \quad \text{if } g = \lambda x.0, \qquad gx \simeq 0 \quad \text{if } g \neq \lambda x.0.$$

Clearly $g \neq \lambda x.0$ by the first clause, hence $\forall x(gx = 0)$; but then by Ext $g = \lambda x.0$; contradication. □

N.B. The proof does not even use induction!

Corollary. **APP** + Ext + $\mathrm{D_V}$ + $\forall xy(x = y \vee x \neq y) \vdash \perp$. □

Remark. The argument for the corollary also goes through if $\mathrm{D_V}$ is weakened to

$\mathrm{D'_V} \quad \exists f \forall xyuv(fuvxx = u \wedge (x \neq y \to fuvxy = v))$, that is to say, the f required by D_V need not coincide with the $\boldsymbol{d}$ of numerial definition by cases. Clearly **APP** + $D_V + \Gamma$ is conservative over **APP** + $D'_V + \Gamma$ whenever Γ is a set of sentences not containing $\boldsymbol{d}$.

6.3. COROLLARY. **APP** + Ext + $AC_V + \forall xy(x = y \vee x \neq y) \vdash \bot$.

PROOF. It suffices to show

$$\mathbf{APP} + AC_V + \forall xy(x = y \vee x \neq y) \vdash D'_V.$$

To see this, apply AC_V to

$$\forall uvxy \exists z((z = u \wedge x = y) \vee (z = v \wedge x \neq y)). \quad \square$$

On the positive side, we have the following proposition.

6.4. PROPOSITION. **TAPP** + Ext + EAC *is consistent.*

PROOF (sketch). Suppose **APP** + Ext + EAC + TOT $\vdash 1 = 0$, then by realizability

$$\mathbf{APP} + \exists x(x\,\mathbf{r}\,\mathrm{Ext}) + \exists y(y\,\mathbf{r}\,\mathrm{TOT}) \vdash 1 = 0.$$

However, Ext and TOT ($\equiv \forall xy(\boldsymbol{E}xy)$) are self-realizing ($\exists$-free) hence

$$\mathbf{TAPP} + \mathrm{Ext} \vdash 1 = 0.$$

Now if we take a model for **APP** constructed from (as in 4.3) some fully extensional model for the λ-calculus such as D_{∞} (cf. Barendregt 1981, 18.3; Bethke 1986) the consistency of **TAPP** + Ext + EAC is immediate. $\square$

6.5. COROLLARY. *Let the axiom of choice for finite type objects be the schema, for all* $\sigma \in \mathcal{T}$,

$$\mathrm{AC}_{\mathrm{IT}} \qquad \forall x \in \mathrm{IT}_{\sigma}\, \exists y A(x, y) \rightarrow \exists z \forall x \in \mathrm{IT}_{\sigma}(\boldsymbol{E}zx \wedge A(x, zx)),$$

where IT_{σ} *is defined as in* 3.15. *Then* **TAPP** + *Ext* + AC_{IT} *is consistent.*

PROOF. $x \in \mathrm{IT}_{\sigma}$ is equivalent to an $\exists$-free formula, hence $\mathrm{AC}_{\mathrm{IT}}$ is a special case of EAC. $\square$

6.6. We have already commented on the fact that AC! in general is false (see 3.14). On the other hand, the schema

$$\mathrm{AC}_{\mathrm{ET}}! \quad \forall x \in \mathrm{ET}_\sigma \exists y \in \mathrm{ET}_\tau A(x, y) \wedge \forall x, x' \in \mathrm{ET}_\sigma \forall y, y' \in \mathrm{ET}_\tau$$
$$(A(x, y) \wedge A(x', y') \wedge x =_\sigma x' \to y =_\tau y')$$
$$\to \exists z \in \mathrm{ET}_{\sigma\to\tau} \forall x \in \mathrm{ET}_\sigma A(x, zx),$$

(ET_σ is defined as in 3.16), is derivable in **APP** + EAC (see E9.6.4).

Extensionality and choice in **HA**$^\omega$ (6.7–12)

6.7. Definition. Let $=_e$ be extensional equality as defined in 1.11 (we drop the type subscript):

$$\mathrm{AC}_{\sigma,\tau} \qquad \forall x^\sigma \exists y^\tau A(x, y) \to \exists z^{\sigma\to\tau} \forall x^\sigma A(x, zx),$$
$$\mathrm{AC}_{\sigma,\tau}! \qquad \forall x^\sigma \exists! y^\tau A(x, y) \to \exists x^{\sigma\to\tau} \forall x^\sigma A(x, zx),$$
$$\mathrm{EXT}_{\sigma,\tau} \qquad \forall x^\sigma y^\sigma z^{\sigma\to\tau}(x =_e y \to zx =_e zy).$$

Furthermore, AC_σ is $\mathrm{AC}_{\sigma,\tau}$ for all $\tau \in \mathscr{T}$, $\mathrm{AC}_\sigma!$ is $\mathrm{AC}_{\sigma,\tau}!$ for all $\tau \in \mathscr{T}$, etc. □

Note that $\bigcup\{\mathrm{EXT}_{\sigma,\tau} : \sigma, \tau \in \mathscr{T}\}$ is equivalent to EXT as defined in 1.11.

6.8. Church's thesis is not compatible with extensionality and choice in **HA**$^\omega$; to what extent choice, extensionality, and Church's thesis are compatible is indicated by the following theorem.

Theorem.
(i) **HA**$^\omega$ + CT + $\mathrm{EXT}_{1,0}$ + $\mathrm{AC}_{1,0} \vdash 1 = 0$,
(ii) **E-HA**$^\omega$ + ECT_0 + AC_0 + AC! $\nvdash 1 = 0$,
(iii) **I-HA**$^\omega$ + ECT_0 + AC $\nvdash 1 = 0$.
N.B. ECT_0 *is here taken with respect to the language of arithmetic.*

Proof (i). Applying $\mathrm{AC}_{1,0}$ to CT

$$\forall \alpha \exists x \forall y \exists v(Txyv \wedge Uv = \alpha y),$$

we find a z^2 such that

$$\forall \alpha y \exists v(T(z^2\alpha, y, v) \wedge Uv = \alpha y).$$

Thus z^2 picks an index of a total recursive function which extensionally coincides with α. Let $\varphi_x(y) = 0$ if $\neg Txxy$, 1 otherwise, then

$$\forall xy \exists v \big(T(z^2 \varphi_x, y, v) \wedge Uv = \varphi_x y \big).$$

So determine $z^2(\lambda x.0) = k$, then

$$k = z^2 \varphi_x \leftrightarrow \varphi_x =_e \lambda x.0 \leftrightarrow \neg \exists y Txxy,$$

since, by $\mathrm{EXT}_{1,0}$ $\alpha =_e \varphi_x \leftrightarrow z^2\alpha = z^2\varphi_x$. Moreover, by CT, $\lambda x.z^2\varphi_x$ is recursive, and thus $\{x : \exists y Txxy\}$ would be recursive. Contradiction.

(ii) Interpret **HA**$^\omega$ in HEO. Then instances of ECT_0 are translated into other instances of ECT_0; EXT is automatically satisfied, instances of AC_0 translate into instances of ECT_0, and the translation of AC! is easily provable from ECT_0.

To verify the latter fact, let us consider an instance of AC!:

$$\forall x^\sigma \exists ! y^\tau A(x, y) \rightarrow \exists z^{\sigma \rightarrow \tau} \forall x^\sigma A(x, zx).$$

The premiss translates under the interpretation in HEO into

$$\forall x \in \mathrm{HEO}_\sigma \exists y \in \mathrm{HEO}_\tau [\![A(x, y)]\!] \wedge$$
$$\forall xyuv ([\![A(x, u)]\!]_{\mathrm{HEO}} \wedge [\![A(y, v)]\!]_{\mathrm{HEO}} \wedge x =_\sigma y \rightarrow u =_\tau v).$$

Applying ECT_0 to the first half yields a z such that

$$\forall x \in \mathrm{HEO}_\sigma (\{z\}(x) \in \mathrm{HEO}_\tau \wedge A(x, \{z\}(x));$$

the second part shows that z *must* satisfy

$$x =_\sigma y \rightarrow \{z\}(x) =_\tau \{z\}(y),$$

and thus $z \in \mathrm{HEO}_{\sigma \rightarrow \tau}$. So $[\![\mathrm{AC!}]\!]_{\mathrm{HEO}}$ is derivable from ECT_0. We have shown that the interpretation of **HA**$^\omega$ + AC! + AC_0 + EXT in HEO is provably true in **HA** + ECT_0, hence consistent.

(iii) For this result interpret **HA**$^\omega$ similarly in the model HRO. □

6.9. Definition. Let us call a formula of **EL** *almost negative* iff it is constructed from formulae $t =_0 s$, $\exists x^0(t =_0 s)$, $\exists \alpha(t =_0 s)$ by means of $\wedge, \forall, \rightarrow$. Now the schema of *generalized continuity* is

$$\mathrm{GC} \quad \forall \alpha [A\alpha \rightarrow \exists \beta B(\alpha, \beta)] \rightarrow \exists \gamma \forall \alpha [A\alpha \rightarrow E(\gamma | \alpha) \wedge B(\alpha, \gamma | \alpha))]$$

for almost negative A. (N.B. $\exists \alpha(t = s) \leftrightarrow \exists n(t[\alpha/r_n] = s[\alpha/r_n])$ for a sequence $\langle r_n \rangle_n$ of primitive recursive functions dense in $\mathbb{N}^{\mathbb{N}}$, since all terms of **EL** are continuous in their function variables.) □

6.10. Completely parallel to theorem 6.8 we now have a theorem indicating to what extent extensionality, choice, and continuity are compatible.

THEOREM. *Taking* GC *with respect to the language of* **EL**, *we have*
(i) $\mathbf{HA}^\omega + \text{WC-N} + \text{EXT}_{2,0} + \text{AC}_{2,0} \vdash 1 = 0$,
(ii) $\mathbf{HA}^\omega + \text{EXT} + \text{GC} + \text{AC}_1 + \text{AC!} \nvdash 1 = 0$,
(iii) $\mathbf{HA}^\omega + \text{GC} + \text{EXT}_{1,0} + \text{AC} \nvdash 1 = 0$.

PROOF of (i). Clearly, $\forall z^2 \forall \alpha \exists x (z^2\alpha = x)$, and by WC-N

$$\forall z^2 \forall \alpha \exists x \forall \beta \in \bar{\alpha}x\big(z^2\beta = z^2\alpha\big).$$

Applying this to $\alpha \equiv \lambda u.0$ we find

$$\forall z^2 \exists x \forall \beta \in \overline{\lambda u.0}\,(x)\big(z^2\beta = z^2(\lambda u.0)\big);$$

and with $\text{AC}_{2,0}$ we find $\Phi \in 2 \to 0$ such that

$$\forall z^2 \forall \beta \in \overline{\lambda u.0}'(\Phi z^2)\big(z^2\beta = z^2(\lambda u.0)\big).$$

In other words, Φz^2 is a modulus of continuity of z^2 at $\lambda u.0$. Now define

$$\Phi^* = \lambda z^2.\big(\lambda u.0(\Phi z^2)\big);$$

this is again a functional of type $2 \to 0$.

Suppose $\Phi^*(\lambda\alpha.0) = n_0$, and let γ_0 be any neighbourhood function for $\lambda\alpha.0$ such that $\gamma_0 n_0 = 1$, $\gamma_0 m = 0$ for $m \prec n_0$, $\gamma_0 m = 1$ for all m such that $\neg m \preccurlyeq n_0$.

For each β we construct a neighbourhood function γ_β as follows:

$$\gamma_\beta m = 0 \quad \text{for } m \preccurlyeq n_0;$$

$$\gamma_\beta\big(n_0 * \langle x\rangle\big) = \begin{cases} 2 & \text{if } \beta x = 0 \wedge \forall y < x(\beta y = 0), \\ 1 & \text{otherwise;} \end{cases}$$

$$\gamma_\beta m = 1 \quad \text{if } \neg n_0 \preccurlyeq m \wedge \neg m \preccurlyeq n_0.$$

For each β the γ_β is clearly a neighbourhood function, so

$$\forall \alpha\beta \exists yx\big(\gamma_\beta(\bar{\alpha}x) = y + 1\big),$$

and there is a primitive recursive Ψ such that

$$\forall \alpha\beta \exists x\big(\gamma_\beta(\bar{\alpha}x) = \Psi j(\alpha, \beta) + 1\big).$$

Put $\Psi_\beta := \lambda\alpha.\Psi(j(\alpha, \beta))$, then by $EXT_{2,0}$

$$\forall x(\beta x = 0) \leftrightarrow \Psi_\beta = \lambda\alpha.0 \rightarrow \Phi^*(\Psi_\beta) = \Phi^*(\lambda\alpha.0) = n_0,$$

$$\exists x(\beta x \neq 0) \rightarrow \Phi^*(\Psi_\beta) > n_0.$$

Hence the value of Ψ_β cannot be computed continuously in β at $\lambda x.0$, while WC-N requires continuity in β.

We omit the proof of (ii) and (iii). See Troelstra (1977C). □

6.11. COROLLARY (*to the proof of* 6.10(i)). *An extensional functional of type* 3 *assigning to each extensional* z^2 *its modulus at* $\lambda x.0$ *does not exist, that is*

$$\mathbf{HA}^\omega + \mathrm{MOD} + \mathrm{EXT}_{2,0} \vdash 1 = 0,$$

where

$$\mathrm{MOD} \qquad \forall\alpha^1 \in \overline{\lambda x.0}\,(\varphi_m z^2)(z^2\alpha = z^2(\lambda x.0)).$$

PROOF. The only application of $AC_{2,0}$ in the proof of 6.10(i) occurs in the construction of φ_m (i.e. Φ). The final appeal to WC-N can be replaced by renewed appeal to Φ_m: apply φ_m to $\lambda\beta.\Phi^*(\Psi_\beta)$, and with MOD we obtain a contradiction as before. □

6.12. On the other hand, the following result can be easily proved by several different methods.

PROPOSITION. $\mathbf{E\text{-}HA}^\omega$ + AC $\nvdash 1 = 0$.

PROOF. See e.g. E9.6.7, where an even stronger result is indicated. □

7. Some metamathematical applications

7.1. In this section we shall show how, in the case of **APP**, models and realizability can be used to obtain several metamathematical properties of **APP**, such as the existence property, and closure under certain rules, such as Church's rule and a continuity rule. We shall also introduce a notion of uniform forcing, in order to obtain a fan-rule (or rule of uniform continuity). Later we shall encounter this notion of uniform forcing in other guises, first in the form of "elimination of lawless sequences" (12.3.11) and again as forcing in sheaves over a special site (section 15.6).

The techniques we develop here can also be applied to many other systems, such as $\mathbf{HA}^{\omega}$ and extensions with set variables, but we shall not describe these extensions and modifications here (one may consult e.g. Beeson 1985).

We shall presuppose familiarity with the methods of coding syntax into arithmetic. Also we must assume the reader to be familiar with the contents of section 3.6, where it is shown how finitary inductive definitions can be replaced by explicit definitions within **HA** or **EL**.

Our first aim will be to obtain the numerical existence property and a derived rule of choice; for this purpose we need a variant of the notion of realizability introduced in section 5.

Throughout this section we write $\bar{n}$ for the numeral S^n0.

7.2. DEFINITION (**q**-*realizability*). We define $x\mathbf{q}A$ by the same clauses as for $x\mathbf{r}A$, replacing everywhere $\mathbf{r}$ by $\mathbf{q}$, except for the implication clause, which becomes

$$x\mathbf{q}(A \to B) = Ex \wedge \forall y(y\mathbf{q}A \to xy\mathbf{q}B) \wedge (A \to B). \quad \square$$

7.3. PROPOSITION (*properties of* **q**-*realizability*).

(i) $\mathbf{APP} \vdash t\mathbf{q}A \to A$;

(ii) *For* $\exists$-*free* A *there is a term* τ_A *such that* $\mathbf{APP} \vdash \exists x(x\mathbf{q}A) \to \tau_A\mathbf{q}A$, $\mathbf{APP} \vdash A \to \exists x(x\mathbf{q}A)$;

(iii) $\mathbf{APP} \vdash A \Rightarrow \mathbf{APP} \vdash t\mathbf{q}A$ *for some* t, *with* $\mathrm{FV}(t) \subset \mathrm{FV}(A)$ (*soundness*).

PROOF. (i) is proved by induction on A; (ii) as for **r**-realizability, in fact, the same terms τ_A may be chosen as in 5.5; (iii) requires only very slight modifications of the soundness proof for **r**-realizability (exercise). $\square$

7.4. COROLLARY. *Let* A *be* $\exists$-*free and* IT_σ *as defined in* 3.15. *Then*:

(i) $\mathbf{APP} \vdash \forall x(Ax \to \exists yB(x, y)) \Rightarrow \mathbf{APP} \vdash \forall x(Ax \to E(tx) \wedge B(x, tx))$ *for a term* t *with* $\mathrm{FV}(t) \subset (\mathrm{FV}(A) \cup \mathrm{FV}(B)) \setminus \{x\}$;

(ii) $\mathbf{APP} \vdash \forall x \in \mathrm{IT}_\sigma \exists yB(x, y) \Rightarrow \mathbf{APP} \vdash \forall x \in \mathrm{IT}_\sigma(E(tx) \wedge B(x, tx))$ *for a term* t *with* $\mathrm{FV}(t) \subset \mathrm{FV}(B) \setminus \{x\}$;

(iii) $\mathbf{APP} \vdash \exists xBx \Rightarrow \mathbf{APP} \vdash Bt \wedge Et$ *with* $\mathrm{FV}(t) \subset \mathrm{FV}(B) \setminus \{x\}$.

PROOF. (i) Assume $\vdash \forall x(Ax \to \exists yB(x, y))$, then by soundness for **q**-realizablity, for some t $\vdash t\mathbf{q}\forall x(Ax \to \exists yB(x, y))$, hence

$$\vdash \forall x(Etx) \wedge \forall u(u\mathbf{q}Ax \to txu\mathbf{q}\exists yB(x, y)));$$

and since $\vdash Ax \to \tau_A \mathbf{q} Ax$ we obtain

$$\vdash \forall x(Ax \to tx\tau_A \mathbf{q} \exists y B(x, y)).$$

We put $t^* \equiv \lambda x.\mathbf{p}_0(tx\tau_A)$, then we see that

$$\vdash \forall x(Ax \to \mathbf{E}(t^*x) \wedge \mathbf{p}_1(tx\tau_A)\mathbf{q} B(x, t^*x))$$

and thus $\vdash \forall x(Ax \to \mathbf{E}(t^*x) \wedge B(x, t^*x))$.

(ii) is a special case of (i).

(iii) can also be obtained as a special case of (i), taking $A'x \equiv (0 = 0)$, $B'(x, y) \equiv B(y)$; or directly by a simplification of the argument for (i). □

7.5. Corollary. *Let* $\exists x \in N C(x)$ *and* $A \vee B$ *be closed.*
(i) $\mathbf{APP} \vdash \exists x \in N C(x) \Rightarrow \mathbf{APP} \vdash C\bar{n}$ *for some numeral* $\bar{n}$ (EDN).
(ii) $\mathbf{APP} \vdash A \vee B \Rightarrow \mathbf{APP} \vdash A$ *or* $\mathbf{APP} \vdash B$ (DP).

Proof. (i) Let $\mathbf{APP} \vdash \exists x \in N C(x)$. By (iii) of the preceding corollary we find a closed t such that $\mathbf{APP} \vdash t \in N \wedge C(t)$. Now $t \in N$ must also hold in the term model CNFS, and therefore there exists a strict reduction sequence from t to a numeral $\bar{n}$, i.e. $\mathbf{APP} \vdash t = \bar{n}$; therefore $\mathbf{APP} \vdash C(\bar{n})$.

(ii) Immediate from (i) using $A \vee B \leftrightarrow \exists x \in N((x = 0 \wedge A) \vee (x = 1 \wedge B))$. □

Remark. If we use E-logic instead of E$^+$-logic, we can even drop the condition that $\exists x \in N C(x)$ and $A \vee B$ are closed; for if $\vdash t \in N \wedge C(t)$, t must exist and this can, by strictness, only happen if all its components exist; thus t cannot contain free variables.

Our next aim is to obtain Church's rule and a derived rule of continuity. This requires more delicate methods.

7.6. Definition (*graph of the valuation function for terms*). We consider terms containing at most the variable x^* free; for such terms we define inductively $\mathrm{VAL}^\alpha(\ulcorner t \urcorner, x)$, expressing "$x$ is the value of t under assignment of $\alpha \in \mathbb{N} \to \mathbb{N}$ to x^*".

We use $\ulcorner t \urcorner$ for the gödelnumber of t; to be quite definite, we can assign codes to terms e.g. as follows. We assign number $c^\sim$ to the constants c and a number $x^{*\sim}$ to x^*: $0^\sim := 0$, $S^\sim := 1$, $P^\sim := 2$, $\mathbf{p}^\sim := 3$, $\mathbf{p}_0^\sim := 4$, $\mathbf{p}_1^\sim := 5$, $\mathbf{k}^\sim := 6$, $\mathbf{s}^\sim := 7$, $\mathbf{d}^\sim := 8$, $\mathbf{x}^{*\sim} := 9$. Then we assign the gödelnumbers $\ulcorner t \urcorner$

by

$$\ulcorner c \urcorner = j(0, c^{\sim}), \qquad \ulcorner x^* \urcorner = j(0, x^{*\sim}),$$
$$\ulcorner t_1 t_2 \urcorner = j(1, j(\ulcorner t_1 \urcorner, \ulcorner t_2 \urcorner)).$$

We now define $\mathrm{VAL}^{\alpha}(\ulcorner t \urcorner, x)$ as the least relation closed under

$$\mathrm{VAL}^{\alpha}(\ulcorner c \urcorner, c); \qquad \mathrm{VAL}^{\alpha}(\ulcorner x^* \urcorner, \alpha);$$
$$Exy \wedge \mathrm{VAL}^{\alpha}(\ulcorner t_1 \urcorner, x) \wedge \mathrm{VAL}^{\alpha}(\ulcorner t_2 \urcorner, y) \to \mathrm{VAL}^{\alpha}(\ulcorner t_1 t_2 \urcorner, xy).$$

As explained in detail in section 3.6, such an inductive definition can in fact be replaced by an explicit definition in **APP** (the details of section 3.6 concern explicit definitions in **HA** and **EL**, but, obviously, apply without change to **APP**).

If x^* does not occur in t, we can simply write $\mathrm{VAL}(\ulcorner t \urcorner, x)$ instead of $\mathrm{VAL}^{\alpha}(\ulcorner t \urcorner, x)$. □

REMARK. We can easily extend VAL to terms containing variables from a list $\langle v_n \rangle_n$ (not containing x^*), modulo an assignment to the variables; the assignment of gödelnumbers is extended by adding $v_n^{\sim} := n + 10$, $\ulcorner v_n \urcorner := j(0, v_n^{\sim})$, and VAL^{α} now becomes $\mathrm{VAL}^{\alpha}_{w}$, defined as VAL but with an additional clause

$$E(wn) \to \mathrm{VAL}^{\alpha}_{w}(\ulcorner v_n \urcorner, wn).$$

However, we shall not need this additional generality. □

7.7. PROPOSITION. *Let* $\mathrm{FV}(t[v_0, \ldots, v_p, x^*]) \subset \{v_0, \ldots, v_p, x^*\}$, *then*:
(i) $\mathrm{VAL}^{\alpha}(\ulcorner t_1 t_2 \urcorner, z) \leftrightarrow \exists xy(\mathrm{VAL}^{\alpha}(\ulcorner t_1 \urcorner, x) \wedge \mathrm{VAL}^{\alpha}(\ulcorner t_2 \urcorner, y) \wedge xy = z)$;
(ii) $\mathrm{VAL}^{\alpha}(\ulcorner t \urcorner, x) \wedge \mathrm{VAL}^{\alpha}(\ulcorner t \urcorner, y) \to x = y$;
(iii) $\mathbf{APP} \vdash Et[n_0, \ldots, n_p, \alpha] \to \mathrm{VAL}^{\alpha}(\ulcorner t[\bar{n}_0, \ldots, \bar{n}_p, x^*] \urcorner, t[n_0, \ldots, n_p, \alpha])$;
(iv) $\mathbf{APP} \vdash \forall n_0 \ldots n_p \in N\, \forall \alpha \in N^N\, \forall y(\mathrm{VAL}^{\alpha}(\ulcorner t[\bar{n}_0, \ldots, \bar{n}_p, x^*] \urcorner, y) \to t[n_0, \ldots, n_p, \alpha] = y)$; ($t_1$, t_2 *in* (i) *and* (ii) *closed*).

PROOF. (i) Define

$$E(x, y) := [(x = j(0,0) \wedge y = 0) \vee (x = j(0,1) \wedge y = S) \vee \cdots \vee (x = j(0,8) \wedge y = \boldsymbol{d}) \vee (x = j(0,9) \wedge y = \alpha) \vee \exists x_1 x_2 y_1 y_2 (x = j(1, j(x_1, x_2)) \wedge \mathrm{VAL}^{\alpha}(x_1, y_1) \wedge \mathrm{VAL}^{\alpha}(x_2, y_2) \wedge y_1 y_2 = y)].$$

Then one shows $E(x, y) \to \mathrm{VAL}^\alpha(x, y)$, and by induction over VAL, $\mathrm{VAL}^\alpha(x, y) \to E(x, y)$. Now the statement immediately follows.

(ii) $\mathrm{VAL}^\alpha(x, x') \to \forall y(\mathrm{VAL}^\alpha(x, y) \to x' = y)$ is proved by induction over VAL with the help of (i).

(iii) By induction on t; as to the basis of the induction, $\mathrm{VAL}(\ulcorner c \urcorner, c)$ is immediate and $\mathrm{VAL}(\ulcorner \bar{n} \urcorner, n)$ follows by induction.

(iv) follows by combining (iii) and (ii). □

7.8. *Validity in the model* CNFS. Let $\mathrm{SRED}(x, y)$ be the arithmetical predicate expressing "the term with gödelnumber x strictly reduces to the term with gödelnumber y". With the help of this predicate validity in the models CNFS and CNFS^α for any sentence of **APP** can be defined in arithmetic. Specifically, for sentences we have

$$[\![t_1 = t_2]\!] := \exists t \in \mathrm{CNFS}(\mathrm{SRED}(\ulcorner t_1 \urcorner, \ulcorner t \urcorner) \wedge \mathrm{SRED}(\ulcorner t_2 \urcorner, \ulcorner t \urcorner)),$$

$$[\![t \in N]\!] := \exists n \in N(\mathrm{SRED}(\ulcorner t \urcorner, \ulcorner \bar{n} \urcorner)),$$

$$[\![Et]\!] := \exists t' \in \mathrm{CNFS}(\mathrm{SRED}(\ulcorner t \urcorner, \ulcorner t' \urcorner)),$$

and for compound sentences

$$[\![A \circ B]\!] := [\![A]\!] \circ [\![B]\!] \quad \text{for } \circ \in \{\wedge, \vee, \to\},$$

$$[\![(Qv_i)A(v_i)]\!] := (Qt \in \mathrm{CNFS})[\![A[v_i/t]]\!] \quad \text{for } Q \in \{\forall, \exists\}.$$

(As they stand the clauses do not literally yield sentences in **HA**, but this is easily remedied.) □

Formalizing the soundness of the model CNFS for **APP** yields the following.

7.9. PROPOSITION. *For sentences* A:

$$\mathbf{APP} \vdash A \Rightarrow \mathbf{HA} \vdash [\![A]\!].$$

PROOF. By induction on the length of deductions we show that when $\mathbf{APP} \vdash A(v_0, \ldots, v_n)$ with $\mathrm{FV}(A) \subset \{v_0, \ldots, v_n\}$, then also $\mathbf{HA} \vdash [\![\forall v_0 \ldots v_n A(v_0, \ldots, v_n)]\!]$. □

Quite similarly we can define validity in any model CNFS^α, replacing "SRED" by "SRED^α", etc. Validity in CNFS^α is definable in elementary analysis. We then get the following proposition.

7.10. Proposition. *For formulas A containing at most x^* free*

$$\mathbf{APP} \vdash A \Rightarrow \mathbf{EL} \vdash [\![A]\!]^\alpha,$$

where the open brackets $[\![\ ]\!]$ *now refer to validity in the model* CNFS$^\alpha$. □

7.11. Lemma. *In* **APP** *we can prove*

$$\mathrm{VAL}(\ulcorner t \urcorner, x) \wedge \mathrm{SRED}(\ulcorner t \urcorner, \ulcorner t' \urcorner) \rightarrow \mathrm{VAL}(\ulcorner t' \urcorner, x),$$

and similarly with VAL$^\alpha$, SRED$^\alpha$.

Proof. We apply induction on the length of the reduction sequences from $\ulcorner t \urcorner$ to $\ulcorner t' \urcorner$. For the basis case of a single-step reduction, we apply a subinduction with respect to the depth of the redex being converted.

Case 1. t itself is a redex. For example, let $t \equiv \mathbf{k}t_1t_2$, $t' \equiv t_1$. Then, if $\mathrm{VAL}(\ulcorner \mathbf{k}t_1t_1 \urcorner, x)$, we can (by 7.7(i)) find y, z, u such that $\mathrm{VAL}(\ulcorner \mathbf{k}t_1 \urcorner, y)$, $\mathrm{VAL}(\ulcorner t_2 \urcorner, z)$, $\mathrm{VAL}(\ulcorner t_1 \urcorner, u)$, and since $\mathrm{VAL}(\ulcorner \mathbf{k} \urcorner, \mathbf{k})$, it follows that $\mathbf{k}u = y$, $yz = x$. Then $x = \mathbf{k}uz = u$, hence $\mathrm{VAL}(\ulcorner t_1 \urcorner, x)$.

Case 2 (*subinduction step*). Let $t \equiv t_1t_2$, $t' \equiv t_1t_2'$, t_2 reducing in a single step to t_2'; assume $\mathrm{VAL}(\ulcorner t \urcorner, x)$. Then there are y, z such that $\mathrm{VAL}(\ulcorner t_1 \urcorner, y)$, $\mathrm{VAL}(\ulcorner t_2 \urcorner, z)$, $yz = x$; by induction hypothesis $\mathrm{VAL}(\ulcorner t_2' \urcorner, z)$ and therefore also $\mathrm{VAL}(\ulcorner t' \urcorner, yz)$, i.e. $\mathrm{VAL}(\ulcorner t' \urcorner, x)$, etc. □

7.12. Corollary. *For terms* $t[x_0, \ldots, x_p, x^*]$, $t'[x_0, \ldots, x_p, x^*]$ *with* $\mathrm{FV}(t), \mathrm{FV}(t') \subset \{x_0, \ldots, x_p, x^*\}$ *we have in* **APP**

$$\forall n_0 \ldots n_p \forall \alpha \in N^N$$

$$\mathrm{SRED}^\alpha\left(\ulcorner t\left[\bar{n}_0, \ldots, \bar{n}_p, x^*\right]\urcorner, \ulcorner t'\left[\bar{n}_0, \ldots, \bar{n}_p, x^*\right]\urcorner\right)$$
$$\rightarrow t\left[n_0, \ldots, n_p, \alpha\right] = t'\left[n_0, \ldots, n_p, \alpha\right]\Big).$$

Proof. Combine 7.11 with 7.7. □

7.13. In the rest of this section we shall write $T^\alpha(m, x_0, x_1, \ldots, u)$ for $T(m, \langle\alpha\rangle, \langle x_0, x_1, \ldots\rangle, u)$ as defined in 3.7.8; T is Kleene's T-predicate for partial continuous functionals.

We shall also use the suggestive notation N^N for the type 1 objects in **APP** (IT_1 of 3.15).

Lemma. $\mathrm{SRED}^\alpha(\ulcorner t \urcorner, \ulcorner t' \urcorner)$ *is r.e. in* α, *i.e. we can find a numeral* $\bar{n}$ *such that in* **EL**

$$\vdash \mathrm{SRED}^\alpha(x, y) \leftrightarrow \exists u T^\alpha(\bar{n}, x, y, u).$$

In particular we can also find a primitive recursive φ *such that for all numerals* $\bar{n}$

$$\vdash \mathrm{SRED}^{\alpha}(x, \ulcorner \bar{n} \urcorner) \leftrightarrow \exists u(T^{\alpha}(\varphi(x), u) \wedge Uu = n).$$

PROOF. Straightforward. □

7.14. THEOREM (*Church's rule and the continuity rule for* **APP**).
(i) *Let* $\mathrm{FV}(A) \subset \{x, y\}$, *then* $\mathbf{APP} \vdash \forall x \in N \exists y \in NA(x, y) \Rightarrow \mathbf{APP} \vdash \forall n \in N \exists u \in N(T(\bar{m}, x, u) \wedge A(x, Uu))$ *for some numeral* $\bar{m}$.
(ii) *Let* $\mathrm{FV}(A) \subset \{x, \alpha\}$, $\mathbf{APP} \vdash \forall \alpha \in N^N \exists x \in NA(\alpha, x)$. *Then for some numeral* $\bar{m}$ $\mathbf{APP} \vdash \forall \alpha \in N^N(\exists u T^{\alpha}(\bar{m}, u) \wedge A(\alpha, Uu))$.

PROOF. (i) can be obtained as a special case of (ii); so we shall prove (ii). Assume $\mathbf{APP} \vdash \forall \alpha \in N^N \exists x \in NA(\alpha, x)$. By 7.4 we can find a closed t^* such that

$$\mathbf{APP} \vdash \forall \alpha \in N^N(t^*\alpha \in N \wedge A(\alpha, t^*\alpha)).$$

Choose the fixed variable x^*, then

$$\mathbf{APP} \vdash x^* \in N^N \to t^*x^* \in N.$$

Then, for an arbitrary term model CNFS^{α} (α assigned to x^*), we can show in **EL**, which is contained in **APP**,

$$\vdash \forall \alpha \in N^N \exists n \in N\, \mathrm{SRED}^{\alpha}(\ulcorner t^*x^* \urcorner, \ulcorner \bar{n} \urcorner)$$

(by the soundness of the term models of CNFS^{α}). By lemma 7.13, we find a numeral $\bar{m}$ (namely $m = \varphi(\ulcorner t^*x^* \urcorner)$) such that

$$\mathbf{EL} \vdash \exists u(T^{\alpha}(\bar{m}, u) \wedge \mathrm{SRED}^{\alpha}(\ulcorner t^*x^* \urcorner, \ulcorner \overline{Uu} \urcorner)),$$

and since $\mathrm{VAL}^{\alpha}(\ulcorner t^*x^* \urcorner, t^*\alpha)$ we find

$$\mathbf{APP} \vdash \exists u(T^{\alpha}(\bar{m}, u) \wedge t^*\alpha = Uu)).$$

This immediately yields the desired conclusion. □

7.15. COROLLARIES.
(i) **HA** *is closed under Church's rule.*
(ii) **HA** *and* **APP** *are closed under the general form of Markov's rule*: *if* $\vdash \forall x \in N(A(x) \vee \neg A(x))$ *and* $\vdash \neg\neg \exists x \in NA(x)$, *then* $\vdash \exists x \in NA(x)$.

(iii) **APP** $\vdash \forall\alpha \in N^N \exists x \in N A(\alpha, x) \Rightarrow$ **APP** $\vdash \exists\gamma \in K_0 \forall\alpha \in N^N$ $A(\alpha, \gamma(\alpha))$, *where* K_0 *is the set of neighbourhood functions defined as usual* (4.6.8).

Proof. (i) Use 7.14(i) and conservativity of **APP** over **HA**.

(ii) We give the proof for **HA**; for **APP** the argument is only notationally different. We use (i) and 7.14(i) to reduce Markov's rule to Markov's rule for primitive recursive predicates. Assume $\vdash \forall x \in N(A(x, \vec{y}) \vee \neg A(x, \vec{y}))$ and $\vdash \neg\neg\exists x A(x, \vec{y})$, $\mathrm{FV}(A) \subset \{x, \vec{y}\}$. By closure under Church's rule for some numeral $\bar{n}$,

$$\vdash \exists u(T(\bar{n}, x, \vec{y}, u) \wedge Uu = 0) \leftrightarrow A(x, \vec{y}),$$

that is to say for some primitive recursive φ, $A(x, \vec{y}) \leftrightarrow \exists x(\varphi(x, \vec{y}) = 0)$. Hence $\vdash \neg\neg\exists x(\varphi(x, \vec{y}) = 0)$. Now appeal to 3.5.4 (closure of **HA** under Markov's rule for primitive recursive predicates) and conclude $\vdash \exists x(\varphi(x, \vec{y}) = 0)$, hence $\vdash \exists x A(x)$.

(iii) is left to the reader. □

7.16. *The fan rule.* We shall now show **APP** to be closed under the fan rule with the help of a notion of forcing.

Below, $\boldsymbol{T}$ is a fixed fan (4.1.2) definable in **APP**, the important special case being the binary tree ($\boldsymbol{T} \equiv T_{01}$). We shall let p, q, r range over nodes of $\boldsymbol{T}$, and α, β over infinite branches of $\boldsymbol{T}$. As usual $p \preccurlyeq q$ indicates that p is an initial segment of q. We introduce the abbreviations

$$p \succcurlyeq_n q := p \succcurlyeq q \wedge (\mathrm{lth}(p) - \mathrm{lth}(q) = n),$$

$$\forall p \succcurlyeq q(A) := \forall p(p \succcurlyeq q \rightarrow A),$$

$$\forall p \succcurlyeq_n q(A) := \forall p(p \succcurlyeq_n q \rightarrow A).$$

7.17. Definition (*of uniform forcing*). We select a special variable α which is treated as a constant for an arbitrary infinite branch of $\boldsymbol{T}$. We shall define a relation $\Vdash$ between nodes of $\boldsymbol{T}$ and formulas of A (in which α is not being quantified). "$p \Vdash A$" is pronounced as "*p forces A* or "*A* is true at p".

To each term $t[\alpha]$ containing at most α free, there will be canonically associated a term $\hat{t}$ describing t as a function of α, i.e. $\hat{t}\alpha = t[\alpha]$.

A variable x may be thought of as an *arbitrary* operation applied to α. This suggests the following definition of $\hat{t}$ for each term t of **APP**:

(i) $\hat{x}$ is a variable assigned to x, such that $\hat{x} \equiv \hat{y} \Rightarrow x \equiv y$ (we may in fact take $\hat{x} \equiv x$);

(ii) $\hat{\alpha} := \lambda x.x$;

(iii) $\hat{c} := \lambda x.c$ for constants of **APP**;
(iv) $(t_1t_2)\hat{} := s\hat{t}_1\hat{t}_2$.
Clearly, for terms t not containing variables besides α, $\hat{t} = \lambda\alpha.t[\alpha]$ (for our defined λ-operator); and by a slight modification of the definition we can even arrange $(t\alpha)\hat{} \equiv t$ if t is closed.

If $\mathrm{FV}(t) \subset \{x_1, \dots, x_n, \alpha\}$, we have in **APP**:

$$\mathbf{APP} \vdash \forall\alpha \in N \to N\big((\hat{t}[x_1, \dots, x_n, \alpha])\alpha \simeq t[\hat{x}_1\alpha, \dots, \hat{x}_n\alpha, \alpha]\big).$$

In the definition below we tacitly avoid clashes of variables.

Forcing for *atomic A* is now defined by

$$p \Vdash \mathbf{E}t := \forall\alpha \in p\big(\mathbf{E}(\hat{t}\alpha)\big), \qquad p \Vdash \bot := \bot\ ;$$

$$p \Vdash t = s := \forall\alpha \in p(\hat{t}\alpha = \hat{s}\alpha);$$

$$p \Vdash Nt := \exists m \forall q \succcurlyeq_m p\, \exists x \in N\, \forall\alpha \in q(\hat{t}\alpha = x).$$

For *compound A* we define:

$$p \Vdash A \wedge B := (p \Vdash A) \wedge (p \Vdash B);$$

$$p \Vdash A \to B := \forall q \succcurlyeq p((q \Vdash A) \to (q \Vdash B));$$

$$p \Vdash \forall x A x := \forall\hat{x}\forall q \succcurlyeq p((q \Vdash \mathbf{E}x) \to (q \Vdash Ax));$$

$$p \Vdash \exists x A x := \exists m \forall q \succcurlyeq_m p\, \exists\hat{x}(q \Vdash Ax).$$

We can treat $\vee$ as defined; if not, we must add a clause

$$p \Vdash A \vee B := \exists m \forall q \succcurlyeq_m p((q \Vdash A) \vee (q \Vdash B)).$$

Intuitively, we may think of $p \Vdash A$ as: A is (uniformly) true on the basis of the information that $\alpha \in p$. □

REMARK. Observe that

$$p \Vdash \mathbf{E}t \leftrightarrow \exists m \forall q \succcurlyeq_m p\, \forall\alpha \in q\mathbf{E}(\hat{t}\alpha),$$

$$p \Vdash t = s \leftrightarrow \exists m \forall q \succcurlyeq_m p\, \forall\alpha \in q(\hat{t}\alpha = \hat{s}\alpha).$$

Furthermore, as we shall see below,

$$(p \Vdash \exists x A x) \leftrightarrow \exists\hat{x}(p \Vdash \mathbf{E}x \wedge Ax),$$

which might have been used as an alternative clause for $p \Vdash \exists x Ax$.

7.18. LEMMA. *In* **APP**, *for all formulas A*:
(i) $p \Vdash A[x/t] \equiv (p \Vdash A(x))[\hat{x}/\hat{t}\,]$,

(ii) *and if* $\alpha \notin \mathrm{FV}(t) \cup \mathrm{FV}(s)$, *then for all* p

$$\mathbf{APP} \vdash \forall\alpha \in p(t\alpha = s\alpha) \to ((p \Vdash A)[\hat{x}/\hat{t}] \to (p \Vdash A)[\hat{x}/\hat{s}]).$$

PROOF. By induction on A. □

7.19. LEMMA. *In* **APP** *for all formulas* A:
(i) $p \Vdash A \wedge q \succcurlyeq p \Rightarrow q \Vdash A$ (*monotonicity*);
(ii) $\exists m \forall q \succcurlyeq_m p(q \Vdash A) \Rightarrow p \Vdash A$ (*covering property*);
(iii) $p \Vdash \exists x A x \leftrightarrow \exists \hat{x}(p \Vdash Ax \wedge \boldsymbol{E}x)$.

PROOF. (i) and (ii) are proved by formula induction.
(i) Entirely straightforward.
(ii) We check two cases.

Case 1. Suppose for some m $\forall q \succcurlyeq_m p(q \Vdash B \to C)$; we have to show $p \Vdash B \to C$. Let $q' \succcurlyeq p$, $q' \Vdash B$. Then either $q' \succcurlyeq q$ for some $q \succcurlyeq_m p$, and then $q' \Vdash B \to C$ ((i) applied to $q \Vdash B \to C$), hence $q' \Vdash C$; or $p \preccurlyeq q' \prec q$ for some $q \succcurlyeq_m p$. In the latter case, we have $\forall q \succcurlyeq_{m'} q'(q \Vdash B)$ for $m' = m - (\mathrm{lth}(q') - \mathrm{lth}(p))$, and hence $\forall q \succcurlyeq_{m'} q(q \Vdash C)$; hence by induction hypothesis $q' \Vdash C$.

Case 2. Suppose for some $m \forall q \succcurlyeq_m p(q \Vdash \exists x A x)$, then $\forall q \succcurlyeq_m p \exists n \forall r \succcurlyeq_n q$ $\exists \hat{x}(r \Vdash Ax \wedge \boldsymbol{E}x)$. From this we find an m' such that $\forall r \succcurlyeq_{m'} p \exists \hat{x}$ $(r \Vdash Ax \wedge \boldsymbol{E}x)$ (applying (i)), i.e. $p \Vdash \exists x A x$. □

(iii) The implication from the right to the left is immediate. Suppose conversely $p \Vdash \exists x A x$; then for some m $\forall q \succcurlyeq_m p \exists \hat{x}(q \Vdash \boldsymbol{E}x \wedge Ax)$. Let $q_1, \ldots, q_n$ $(n = 2^m)$ enumerate $\{q : q \succcurlyeq_m p\}$, and suppose $\hat{x}_i$ to satisfy $q_i \Vdash \boldsymbol{E}x_i \wedge Ax_i$. By repeated use of definition by cases we construct a term t such that

$$\forall \alpha \in q_i(\hat{x}_i \alpha = t),$$

then

$$\forall \alpha \in q_i(\hat{x}_i \alpha = \hat{t}\alpha).$$

As a result (with 7.18) $\forall q \succcurlyeq_m p(q \Vdash A(t) \wedge \boldsymbol{E}t)$, and with the covering property $p \Vdash A(t) \wedge \boldsymbol{E}(t)$, hence $\exists \hat{x}(p \Vdash A(x) \wedge \boldsymbol{E}(x))$. □

7.20. LEMMA.

$$\mathbf{APP} \vdash (p \Vdash \forall x \in NA(x)) \leftrightarrow \forall y \in N((p \Vdash A)[\hat{x}/\lambda\alpha.y]).$$

PROOF. Suppose first $p \Vdash \forall x \in NA(x)$, then $\forall x(p \Vdash x \in N \to A(x))$, hence in particular for all x $(p \Vdash Nx) \to (p \Vdash A(x))$. Now $(p \Vdash Nx) \leftrightarrow$

$\exists n \forall q \succcurlyeq_n p \exists y \in N \forall \alpha \in q(\hat{x}\alpha = y)$. If we take $\hat{x} = \lambda\alpha.y$, we find with $n = 0$ $\forall \alpha \in p(\hat{x}\alpha = y)$, so $p \Vdash Nx$, hence $p \Vdash A(x)$, i.e. $(p \Vdash A)[\hat{x}/\lambda\alpha.y]$.

Conversely, suppose $\forall y \in N((p \Vdash A)[\hat{x}/\lambda\alpha.y])$, $q \succcurlyeq p$, $q \Vdash Nx$. Then for some n $\forall q' \succcurlyeq_n q \exists y \in N \forall \alpha \in q'(\hat{x}\alpha = y)$, and by assumption $\exists q' \succcurlyeq_n q$ $(q' \Vdash A(x))$, and hence by the covering property $q \Vdash A(x)$. This holds for all $q \succcurlyeq p$, and so $p \Vdash \forall x \in N A(x)$. □

7.21. LEMMA. *In* **APP**

$$\langle\ \rangle \Vdash A(0) \wedge \forall x \in N(A(x) \to A(Sx)) \to \forall x \in N A(x).$$

PROOF. Let $p \Vdash A(0)$, $p \Vdash \forall x \in N(A(x) \to A(Sx))$. It follows that $(p \Vdash A(x))[\hat{x}/\lambda\alpha.0]$, and $\forall y \in N(p \Vdash A(x) \to A(Sx))[\hat{x}/\lambda\alpha.y]$, i.e.

$$\forall y \in N((p \Vdash A(x))[\hat{x}/\lambda\alpha.y] \to (p \Vdash A(x))[\hat{x}/\lambda\alpha.Sy]);$$

hence $\forall y \in N((p \Vdash A)[\hat{x}/\lambda\alpha.y])$, and thus by the preceding lemma $p \Vdash \forall x \in N A(x)$. □

7.22. THEOREM (*soundness of uniform forcing*).

$$\mathbf{APP} + \alpha \in \mathbf{T} \vdash A \Rightarrow \mathbf{APP} \vdash (\langle\ \rangle \Vdash A).$$

PROOF. By induction on the length of derivations. The only nontrivial case turns out to be induction, which is taken care of by the preceding lemma. We leave the verification of all other axioms and rules to the reader. □

7.23. THEOREM (*fan rule for* **APP**).

$$\mathbf{APP} \vdash \forall \alpha \in \mathbf{T}\, \exists x \in N A(\alpha, x) \Rightarrow$$
$$\mathbf{APP} \vdash \exists z \in N \forall \alpha \in \mathbf{T}\, \exists x \in N \forall \beta \in \bar{\alpha} z A(\alpha, x).$$

PROOF. Suppose $\mathbf{APP} \vdash \forall \alpha \in \mathbf{T}\, \exists x \in N A(\alpha, x)$. By $\mathbf{q}$-realizability we find a term t^* such that

$$\mathbf{APP} + \alpha \in \mathbf{T} \vdash t^*\alpha \in N, \qquad \mathbf{APP} + \alpha \in \mathbf{T} \vdash A(\alpha, t^*\alpha).$$

The soundness of forcing shows that $\mathbf{APP} \vdash (\langle\ \rangle \Vdash t^*\alpha \in N)$, i.e. $\mathbf{APP} \vdash \exists n \in N \forall q \succcurlyeq_n \langle\ \rangle \exists x \in N \forall \alpha \in q(t^*\alpha = x)$. We can take this n for z and find $\mathbf{APP} \vdash \exists n \in N \forall \alpha \in \mathbf{T}\, \forall \beta \in \bar{\alpha} n A(\beta, t^*\alpha)$. □

The next lemma plays no role here, but is useful later and of interest in itself.

7.24. LEMMA. *Let* $A(x_1, \ldots, x_n)$ *be a formula of arithmetic as embedded in* **APP**, *and* $\mathrm{FV}(A) \subset \{x_1, \ldots, x_n\}$. *Then*

$$\mathbf{APP} \vdash \forall \hat{x}_1 \ldots \hat{x}_n \forall x_1, \ldots, x_n \in N\big(\forall \alpha \in p(\hat{x}_1 \alpha = \hat{x}_1) \wedge \cdots \wedge \\ \forall \alpha \in p(\hat{x}_n \alpha = \hat{x}_n) \to (A(x_1, \ldots, x_n) \leftrightarrow p \Vdash A)\big)$$

(α *ranging over* $\boldsymbol{T}$).

PROOF. By induction on the complexity of A. (N.B. The interpretation $\hat{t}$ of an arithmetical t is a constant function.) □

8. Theories of operators and classes

8.1. We shall now consider extensions of **APP** in which we do not only consider operators but also classes (or sets). That is to say, the classes form a subdomain of the universe of operators; this may be taken care of in the language either by adding an extra predicate Cl, where $\mathrm{Cl}(x)$ expresses "x is a set", or by having a distinct set of *class variables* $X, Y, Z, \ldots$, which are assumed to range over a subdomain; Cl is then defined by

$$\mathrm{Cl}(x) := \exists x(x = X).$$

If we think of the individuals as ranging over algorithms, classes may be thought of as *codes* or *names* for certain definable sets of individuals; the codes are themselves individuals, and so a (code for a) class can play a double role, namely as an algorithm and as a set (and an algorithm has again a double role: argument and operation). Self-application thus makes sense also for classes, and algorithms may be applied to classes.

The basic extension $\mathbf{EM}_0$ we shall consider here is rather close in spirit to $\mathbf{HAS}_0$, intuitionistic second-order arithmetic with arithmetical comprehension (3.8.11).

8.2. *Description of* $\mathbf{EM}_0$ *and* $\mathbf{EM}_0\!\restriction$. We choose the language with a separate countable supply of class variables, since this slightly facilitates the description. New operation symbols c_n for the formation of classes are added to the language. The logic is two-sorted intuitionistic predicate logic.

DEFINITION. A formula $A(x_1, \ldots, x_n, X_1, \ldots, X_m)$ is said to be *stratified with respect to classifications*, iff all its atomic formulas are of the form

$t_1 = t_2$, $t_1 t_2 = t_3$, $t_1 \in X_i$ for t_1, t_2, t_3 constants or individual variables. A is *elementary* if A is stratified and does not contain bound class variables. □

N.B. A stratified formula may be thought of as a formula which is "second-order"; the classes play the role of sets of individuals throughout.

We shall assume each formula A to have a gödelnumber $\ulcorner A \urcorner$ associated with it.

The specific axioms for classes are now:

(I) $\forall X \exists x (X = x)$; $t \in X \to Et \wedge EX$ (the classes form a subdomain of the universe and $\in$ is strict).

(II) $\exists X(N = X)$ (or $\mathrm{Cl}(N)$) (the natural numbers form a class).

(III) Let $A \equiv A(y, x_1, \ldots, x_n, X_1, \ldots, X_m)$ be any elementary formula. Then

$$\text{ECA} \quad \exists Y\big(c_{\ulcorner A \urcorner}\vec{x}\vec{X} = Y \wedge \forall y\big[y \in Y \leftrightarrow A(y, \vec{x}, \vec{X})\big]\big)$$

(the axiom schema of *elementary comprehension*).

For better readability we sometimes write $c_{\ulcorner A \urcorner}(\vec{x}, \vec{X})$ for $c_{\ulcorner A \urcorner}\vec{x}\vec{X}$.

This completes the description of $\mathbf{EM}_0$. We shall also have reason to consider the weaker system in which induction is restricted to elementary formulas; this we denote by $\mathbf{EM}_0\upharpoonright$. $\mathbf{EM}_0\upharpoonright$ is quite similar to the system $\mathbf{HAS}_0$ we have encountered before in 3.8.11.

8.3. *Some examples of classes.* We will use $\{y : A(y, \vec{x}, \vec{X})\}$ as a suggestive notation for $c_{\ulcorner A \urcorner}(\vec{x}, \vec{X})$, where A is elementary. Some examples of classes and class operations in $\mathbf{EM}_0$ and $\mathbf{EM}_0\upharpoonright$ are in the following list:

$$V := \{x : x = x\},$$

$$\emptyset := \{x : \bot\},$$

$$\{t, s\} := \{x : x = t \vee x = s\},$$

$$\neg X := \{x : \neg x \in X\},$$

$$X \cup Y := \{x : x \in X \vee x \in Y\},$$

$$X \times Y := \{z : \exists xy(z = pxy \wedge x \in X \wedge y \in Y\} \quad \text{or}$$

$$\{z : \exists xyu(u = px \wedge z = uy \wedge x \in X \wedge y \in Y)\},$$

$$X^Y := \{f : \forall y(y \in Y \to fy \in X)\} \quad \text{or}$$

$$\{f : \forall y(y \in Y \to \exists x(x = fy \wedge y \in X)\}, \quad \text{etc.}$$

If $A(y, \vec{x})$ is an arithmetical formula, i.e. A is the translation of a formula of **HA** (3.12), then there is a class operation $c_{\ulcorner A \urcorner}$ in $\mathbf{EM}_0$, since $A(y, \vec{x})$ is elementary.

A little more is true: if we define the translation of a formula $A(y, \vec{x}, \vec{X})$ without bound set-quantifiers in the same way, we also find an elementary formula of $\mathbf{EM}_0$. Thus we see that $\mathbf{HAS}_0$ is contained in $\mathbf{EM}_0\restriction$.

The preference for the terminology "class" over "set" is due to the fact that, in the context of set theory, the universe V is a proper class, not a set. But observe that the Russell-paradox is blocked, since $\neg x \in x$ is not elementary. It is quite easy to construct models for $\mathbf{EM}_0$ and $\mathbf{EM}_0\restriction$, as we shall see in 8.5 below.

8.4. *Nonextensionality for classes.* Extensionality for clases is expressed by

$$\forall XY(\forall x(x \in X \leftrightarrow x \in Y) \to X = Y).$$

PROPOSITION. *In* $\mathbf{EM}_0\restriction$ *extensionality for classes is refutable.*

PROOF (Gordeev, in Beeson 1985). Define $\emptyset := \{x : x \neq x\}$. Using the comprehension constants we find g such that $gzf = \{x \in \{\emptyset\} : x = fz\}$; the fixed point operator yields an f such that $fz \simeq gzf$. g is total, hence f is total as well. $ff = gff = \{x \in \{\emptyset\} : x = ff\}$; $ff = \emptyset$ would yield $\emptyset \in \emptyset$, a contradiction. Now $x \in ff \to ff = \emptyset$, which is again contradictory. Thus $\forall x(x \notin ff)$; by extensionality for classes $ff = \emptyset$, and we have a contradiction. $\square$

8.5. *The standard model construction (Feferman* 1975).

Let $\mathscr{M} := \langle M, N, 0, S, P, k, s, p, p_0, p_1, d\rangle$ be any model of **APP**. We show how to define a model $\mathscr{M}^*$ of $\mathbf{EM}_0$ by defining $\mathrm{Cl} := \bigcup\{\mathrm{Cl}_n : n \in \mathbb{N}\}$, $\in := \bigcup\{\in_n : n \in \mathbb{N}\}$, where the predicate Cl_n and relation $\in_n$ are defined by induction over n. We begin with choosing codes for the classes, more or less arbitrarily:

$$\mathrm{code}(N) := \langle 0, 0\rangle;$$

and for $c_{\ulcorner A \urcorner}$ we take

$$\mathrm{code}(c_{\ulcorner A \urcorner}) := \lambda y.\langle 1, \ulcorner A \urcorner, y\rangle,$$

so that we can give $c_{\ulcorner A \urcorner}\vec{x}\vec{X}$ a code,

$$\mathrm{code}(c_{\ulcorner A \urcorner}\vec{x}\vec{X}) := \langle 1, \ulcorner A \urcorner, \langle \vec{x}, \vec{X}\rangle\rangle.$$

From now on we shall not bother to distinguish between classes and their codes in our notation. We define

$$\mathrm{Cl}_0 := \{N\}, \qquad x \in_0 N := N(x).$$

Suppose Cl_n, $\in_n$ to have been defined already, such that $\in_{n-1} \subset \in_n$, $\mathrm{Cl}_{n-1} \subset \mathrm{Cl}_n$ Let $x_1, \ldots, x_n$ be arbitrary, $a_1, \ldots, a_m \in \mathrm{Cl}_n$, $p = \ulcorner A(y, \vec{x}, \vec{X})\urcorner$, A elementary, then

$$c_p\vec{x}\vec{a} \in \mathrm{Cl}_{n+1},$$

$$y \in_{n+1} c_p\vec{x}\vec{a} \ \text{ iff } (A, \mathrm{Cl}_n, \in_n) \models A(y, \vec{x}, \vec{a}).$$

We put

$$\mathrm{Cl} = \bigcup\{\mathrm{Cl}_n : n \in \mathbb{N}\}, \qquad \in = \bigcup\{\in_n : n \in \mathbb{N}\},$$

and so for $a \in \mathrm{Cl}_n$ we have $x \in a$ iff $x \in_n a$.

It is now straightforward to show that this yields a model of $\mathbf{EM}_0$. Let us check ECA for example. Suppose $p \equiv \ulcorner A(y, \vec{x}, \vec{X})\urcorner$, $\vec{X} \equiv (X_1, \ldots, X_n)$, $X_i \in \mathrm{Cl}_n$. Then $y \in c_p\vec{x}\vec{X}$ iff $\langle \mathcal{M}, \mathrm{Cl}_n, \in_n\rangle \models A(y, \vec{x}, \vec{X})$, the definition of which is fixed, so $\langle \mathcal{M}, \mathrm{Cl}, \in\rangle \models A(y, \vec{x}, \vec{X})$. □

8.6. Theorem. $\mathbf{EM}_0{\restriction}$ *is conservative over* $\mathbf{HAS}_0$.

Proof. Not only can we specialize the preceding argument to the case of the recursive model built over any nonstandard model of arithmetic, but we can also apply the same idea with respect to Kripke models. Thus, given any Kripke model $\mathcal{K} \equiv (\mathrm{K}, \leq, \mathrm{D}_1, \mathrm{D}_2, \Vdash)$, for $\mathbf{HAS}_0$, with $(\mathrm{K}, \leq)$ a countable tree, $\bigcup\{\mathrm{D}_1(k) : k \in \mathrm{K}\}$, $\bigcup\{\mathrm{D}_2(k) : k \in \mathrm{K}\}$ both countable, we can construct a model $\mathcal{K}^*$ for $\mathbf{EM}_0{\restriction}$ from this, such that $\mathcal{K}^*$ validates the same sentences of $\mathbf{HAS}_0$ as $\mathcal{K}$ does.

Because of $\forall nm(n = m \vee n \neq m)$, equality between the elements of $\mathrm{D}_1(k)$ at each node may be interpreted as proper (ordinary) equality on $\mathrm{D}_1(k)$. The sets at node k (i.e. $\mathrm{D}_2(k)$) are families $\mathcal{S} \equiv \{S_{k'} : k' \geq k\}$ satisfying $k'' \geq k' \to S_{k''} \supset S_{k'}$.

The model $\mathcal{K}^*$ for $\mathbf{EM}_0{\restriction}$ is now constructed as follows. Application is defined by

$$\{d\}(d') = d'' \leftrightarrow \exists e(Tdd'e \wedge Ue = d'').$$

T and U of course exist, since $\mathcal{K}$ is a model of $\mathbf{HAS}_0$ and a fortiori of $\mathbf{HA}$. At each node we now construct $\in_n$, Cl_n by a variant of the recipe in 8.5:

$$\mathrm{Cl}_0(k) := \{\mathrm{D}_1(k)\} \cup \mathrm{D}_2(k).$$

We assume that $D_1(k)$, the interpretation of N, has the code 0 for all k; the elements of $D_2(k)$ get assigned an even number as a code (according to some enumeration of $\bigcup\{D_2(k): k \in K\}$ say); if $\mathscr{S} \in D_2(k)$, it retains the same code at all higher nodes.

Suppose $\mathrm{Cl}_n(k)$ to have been defined, and let $p \equiv \ulcorner A(y, \vec{x}, \vec{X})\urcorner$. We assign to c_p the odd-valued code

$$c_p = \Lambda x.(2\cdot\langle 1, p, x\rangle + 1).$$

Let $\mathscr{S}_1, \ldots, \mathscr{S}_q$ be families (representing sets) with nodes assigned in Cl_n at node k; if $\vec{x}$ is a sequence of elements of $D_1(k)$, $\vec{z}$ a sequence of codes for $\mathscr{S}_1, \ldots, \mathscr{S}_q$, then $\mathscr{S}$ has code $2\cdot\langle 1, p, \langle \vec{x}, \vec{z}\rangle\rangle + 1$. We put

$$k \Vdash y \in_{n+1} \mathscr{S} := k \Vdash A(y, \vec{x}, \mathscr{S}_1, \ldots, \mathscr{S}_k).$$

The proof is completed by establishing the following lemma, the proof of which is left to the reader.

LEMMA. *Each $\mathscr{S}$ introduced in the model construction is extensionally equal to an $\mathscr{S}_{2n}$ of the basic model $\mathscr{K}$, i.e. $k \Vdash d \in \mathscr{S}_{2n}$ iff $k \Vdash d \in \mathscr{S}$ ($2n$ being the code of $\mathscr{S}_{2n}$), since for $\mathscr{S}$, $\mathscr{S}'$ which are extensionally equal, and for elementary A,*

$$k \Vdash A(y, \vec{x}, \mathscr{S}) \Leftrightarrow k \Vdash A(y, \vec{x}, \mathscr{S}').$$

Thus we may think of Cl *as being the collection of arithmetically definable sets of $\mathscr{K}^*$ interpreted in the Kripke model $\mathscr{K}$; induction for elementary formulas automatically follows from induction already valid in* $\mathbf{HAS}_0$. □

N.B. The metamathematics of the proof is at several places classical.

8.7. *The system* $\mathbf{EM}_0 + \Sigma\mathrm{AX}$. $\mathbf{EM}_0$ can be reinforced in several directions. A very natural extension is the addition of a *disjoint-sum* or *join axiom*, stating the existence of an operation Σ such that

$$\Sigma\mathrm{AX} \qquad \forall x \in Y(fx \in \mathrm{Cl}) \rightarrow \exists X(\Sigma(f, y) = X \wedge \forall z(z \in \Sigma(f, y) \leftrightarrow \exists x \in Y \exists x' \in fx(z = \mathbf{p}xx'))).$$

In the presence of Σ we can also form general products, i.e. there is a Π such that

$$\forall x \in X(fx \in \mathrm{Cl}) \rightarrow \forall z(z \in \Pi fx \leftrightarrow \exists x \in X(zx \in fx)).$$

To see this, note that if $\forall x \in X(fx \in \mathrm{Cl})$, then

$$z \in \Sigma fX \leftrightarrow \exists xx'(x \in X \wedge x' \in fx \wedge z = pxx').$$

We want

$$z \in \Pi fX \leftrightarrow \forall x \in X(px(zx) \in \Sigma fX),$$

and since $\forall x(x \in X \to px(zx) \in Y)$ is logically equivalent to an elementary formula $A(z, X, Y)$ we can take $c_{\ulcorner A \urcorner}(X, \Sigma fX)$ for ΠfX.

8.8. *Extending the standard model construction to* $\mathbf{EM}_0 + \Sigma\mathrm{AX}$. Assuming the classical notion of ordinal, we can easily extend the model construction of 8.3 to cover $\Sigma\mathrm{AX}$ as well. Let α, β range over ordinals; we shall define Cl and $\in$ as

$$\mathrm{Cl} := \bigcup\{\mathrm{Cl}_\alpha : \alpha \text{ ordinal}\}, \qquad \in := \bigcup\{\in_\alpha : \alpha \text{ ordinal}\}.$$

The specification of indices is extended by giving $\Sigma(f, Y)$ a code $\langle 2, f, Y\rangle$.

Suppose Cl_α, $\in_\alpha$ to have been defined. To validate ECA we put, for $a_1, \ldots, a_m \in \mathrm{Cl}_\alpha$, $p = \ulcorner A(y, \vec{x}, \vec{X})\urcorner$,

$$c_p(\vec{x}, \vec{a}) \in \mathrm{Cl}_{\alpha+1},$$

$$y \in_{\alpha+1} c_p(\vec{x}, \vec{a}) := \langle \mathscr{M}, \mathrm{Cl}_\alpha, \in_\alpha \rangle \Vdash A(y, \vec{x}, \vec{a}).$$

$\Sigma\mathrm{AX}$ is made valid by taking, if $a \in \mathrm{Cl}_\alpha$, $\forall x \in_\alpha a(fx \in_\alpha \mathrm{Cl}_\alpha)$,

$$\langle 2, f, y\rangle \in \mathrm{Cl}_{\alpha+1}, \qquad y \in_{\alpha+1} \Sigma fa := \exists x \in_\alpha a\, \exists x' \in_\alpha fx(pxx' = y).$$

For limit ordinals one takes $\mathrm{Cl}_\lambda := \bigcup\{\mathrm{Cl}_\alpha : \alpha < \lambda\}$, $\in_\lambda := \bigcup\{\in_\alpha : \alpha < \lambda\}$. $\square$

9. Notes

9.1. *Finite-type arithmetic.* The primitive recursive functionals of finite-type have been the subject of extensive investigations since Gödel's Dialectica paper (1958); these researches also led to the study of $\mathbf{HA}^\omega$ and its variants. (In the traditional intuitionistic literature objects of higher type scarcely play a role. Brouwer (e.g. 1918) mentions, but does not actually use, a hierarchy of sets, or species in Brouwer's terminology. (The title of Brouwer (1942) is particularly misleading: in this paper he shows that choice sequences which take choice sequences as values can be reduced to ordinary choice sequences, etc.)

Most of the material in section 1 and 2 is contained in Troelstra (1973), where also references to the earlier literature may be found (see in particular 2.2.35 and section 8 of chapter 2). For the research inspired by Gödel (1958) see the Introductory Note to *Gödel 1958, 1972* in volume 2 of Gödel's Collected Works (1987).

ICF* as defined in this chapter is not quite the same as ICF* in Troelstra (1973, 2.6.23); ICF has also been modified (cf. Troelstra 1973, 2.6.2), so as to permit a more uniform definition of the interpretation of $\boldsymbol{k}^{\sigma,\tau}$, $\boldsymbol{s}^{\rho,\sigma,\tau}$, $\boldsymbol{r}^{\sigma}$.

The main novelty in section 2 is the extension of strong normalization to $\mathbf{HA}^{\omega}$ with Cartesian products, using the reduction to strong normalization for $\mathbf{HA}^{\omega}$. The first normalization proof for the typed lambda calculus is due to Turing (see Gandy 1980A).

9.2. *Theories of operations with self-application.* The computational basis of all theories considered is a slight extension (called "APP" in Feferman (1979)) of the notion of a partial combinatory algebra. This basis is similar to the axioms for Friedman's BRFT (1971) and uniform reflexive structures (Strong 1968, Wagner 1969).

APP as defined here corresponds to "**EON**" in Beeson (1985). Feferman (1975, 1979) introduced theories of operators and classes; the axioms for operators plus arithmetic correspond to **APP**. The metamathematics of Feferman's systems has been extensively investigated by Beeson in a number of papers. These results are brought together, with a number of additions and corrections in proofs, in Beeson (1985).

In our exposition in sections 3 and 5 in particular we have made liberal use of Renardel de Lavalette (1974A).

The model PRO is already described in Feferman (1975). Beeson (1977) introduced a model for **APP** based on partial functions from $\mathbb{N}$ to $\mathbb{N}$ with partial continuous application; the model PCO with total functions, also used in Beeson (1985), is technically simpler. The name "Kleene's second model" used by Beeson seems to us historically not very appropriate, since Kleene (1959, and in Kleene and Vesley 1965) introduced partial continuous function application in a typed context only.

$\mathrm{P}(\omega)$ as a model for **APP** is also discussed in Beeson (1985), but it is to be noted that theorem 8.2.1 on page 138 is incorrect. In fact from the result in 6.11 we know that an extensional type structure cannot contain a modulus functional, while on the other hand by 4.13 IT is extensional in $\mathrm{P}(\omega)$. (The proof of this fact presented in section 4 is due to I. Bethke.)

In this connection we want to mention an interesting open problem: IT in PCO corresponds exactly to ICF*, but we do not know of a model of **APP** in which IT corresponds to ICF. More generally, does there exist a standard model for **APP** in which all type two objects of IT correspond to continuous functionals (that is, continuous with respect to the number-theoretic functions represented by the type 1 objects), and in which IT contains a modulus of continuity for the type two objects?

Beeson (1977) already considers the model of closed terms in normal form, but without the restriction to strict reductions; a corrected version is used in Beeson (1985) (this is essentially CNFS as in 4.20). A slightly different method for introducing CNFS has been given by Renardel de Lavalette (1984A): he considers conversion rules under conditions. For example, the conversion rule for $\boldsymbol{k}$ becomes

$$t' \text{ in normal form} \Rightarrow \boldsymbol{k}tt' \text{ conv } t.$$

Renardel de Lavalette also established that **TAPP** + EAC is conservative over **HA**; the proof is based on a combination of realizability and a special kind of forcing, originally used by Goodman (1978) and Beeson (1979) for $\mathbf{HA}^{\omega}$ (cf. also Beeson 1985, XV, section 2). Similar results by purely proof-theoretic means have been obtained by Gordeev (1988).

The proof of 6.10(i) was obtained by analyzing a counterexample of Kreisel (1962A; cf. also Troelstra 1973, 2.6.7) showing that there is no extensional continuous modulus functional of type 3 in the extensional continuous functionals (the countable functionals of Kleene (1959)).

9.3. *Metamathematics.* **APP** has been chosen as a typical example for illustrating a number of metamathematical techniques and results. For **APP** the results are due to Beeson; for $\mathbf{HA}^{\omega}$ the corresponding results had been obtained before. (See e.g. Troelstra 1973, 2.7.8, 3.7; Troelstra 1977B).

The method of proving a fan rule (uniform continuity rule) for $\mathbf{HA}^{\omega}$ in Troelstra (1977B; preprint 1974) is quite different from the method used in this book: a special notion of "fan-computability" is used. The fan rule may also be obtained as a corollary to closure under bar induction rules and continuity rules.

The methods used here are in principle the same as in Beeson (1985), but there are some expository differences. In particular: (1) in formalizing metamathematics we have made liberal use of elementary inductive definitions, relying on the results in section 3.6 showing them to be equivalent to explicit ones; (2) we used a definition of uniform forcing which clearly reveals its connection with the definition of validity in Beth models over finitely branching trees and the elimination of lawless sequences (12.3.11);

(3) **q**-realizability is obtained by changing the $\rightarrow$-clause of **r**-realizability instead of the $\vee$- and $\exists$-clauses. This is similar to the switch from Kleene slash to Aczel slash (cf. 3.5.7, E3.5.3).

As to closure under rules of bar induction, see Hayashi (1982) and the references given there.

9.4. *Theories of operators and classes.* The results in this section are mostly due to Feferman (1975, 1979). The nonextensionality for classes (8.4) was proved by Feferman (1975) in the presence of D_V (cf. 6.2). Beeson (1980) gave a Kripke model argument for $\mathbf{EM}_0\upharpoonright$ conservative over **HA** (cf. Beeson 1985, p. 322); theorem 8.6 represents a slight strengthening of this result.

9.5. *Further reading.* The principal sources for further information are Beeson (1985) (in particular for Feferman-style theories) and Troelstra (1973) (for $\mathbf{HA}^\omega$ and its extensions).

Exercises

9.1.1. Show that the Ackermann function can be defined with the help of finite-type recursion.

9.1.2. Construct by means of finite-type recursion an enumerating two-place function for all unary functions in PRIM.

9.1.3. Give the proof of 1.10 in detail.

9.1.4. Show that $\mathbf{HA}^\omega$ + (1) is equivalent to $\mathbf{E}\text{-}\mathbf{HA}^\omega$ + EXT, where (1) is the schema equating defined extensional equality and primitive equality (1.11). Show that $\mathbf{E}\text{-}\mathbf{HA}^\omega$ is conservative over the system **S** formulated in the sublanguage $\mathscr{L}_0$ with equality of type 0 only, and with all axioms of $\mathbf{HA}^\omega$ stated for defined extensional equality.

9.1.5. Complete the proof of 1.13.

9.1.6. Check that in ICF*:

$$\forall z^2 \in \mathrm{EXT}\,\exists u^2 \forall \alpha^1 \forall \beta^1 \in \bar{\alpha}(u\alpha)(z^2\beta = z^2\alpha), \tag{1}$$

and in ICF:

$$\forall z^2 \exists u^2 \forall \alpha^1 \forall \beta^1 \in \bar{\alpha}(u\alpha)(z^2\beta = z^2\alpha). \tag{2}$$

Show that (2) is false for ICF*.

9.1.7. Show that in ICF* we have a $\Psi \in 1 \to 1$ such that

$$\begin{cases} \forall \alpha^1 \forall x^0 (\alpha x = (\Psi\alpha)x) \quad \text{and} \\ \forall \alpha^1 \beta^1 (\forall x(\alpha x = \beta x) \to \Psi\alpha = \Psi\beta). \end{cases} \tag{1}$$

Show that in ICF there is a z^3 such that

$$\forall u^2 \forall \alpha^1 \in \overline{\lambda u.0}(z^3 u^2)\big(u^2\alpha^1 = u^2(\lambda u.0)\big), \tag{2}$$

and deduce from this the existence of a w^{210} in ICF such that

$$\forall u^2 \forall \alpha^1 \forall \beta^1 \in \bar{\alpha}(wu\alpha)[u\alpha = u\beta]. \tag{3}$$

Show that a similar assertion holds in ICF* if we replace "$\forall u^2$" by "$\forall u^2 \in \mathrm{EXT}$" (Troelstra 1973, 2.6.3).

9.1.8. Use the model HRO to show that

$$\textbf{I-HA}^{\omega} + \exists z^2 \forall \alpha^1 \forall x^0 \exists u^0 \big(T(z^2\alpha, x, u) \wedge Uu = \alpha x\big)$$

is conservative over **HA**, and use HEO to show that

$$\textbf{E-HA}^{\omega} + \forall \alpha^1 \exists z^0 \forall x^0 \exists u^0 (T(z, x, u) \wedge Uu = \alpha x)$$

is conservative over **HA**.

9.1.9. ICF* or ICF can be used to show that $\mathbf{HA}^{\omega}$ is conservative over **EL** (cf. 1.14). Can you obtain still stronger conclusions from this proof? (cf. (3) in E9.1.7.)

9.1.10. Show that in $\mathbf{HA}^{\omega}$ constants for simultaneous recursion can be straightforwardly defined with the help of $\boldsymbol{p}, \boldsymbol{p}_0, \boldsymbol{p}_1$.

9.1.11. Give a complete formulation of the extensional versions ECF, ECF* of ICF and, respectively, ICF* and prove that ECF and ECF* are isomorphic (1.17).

9.1.12. Complete the proof of 1.20.

9.1.13. Show how to obtain $\mathbf{E\text{-}HA}^{\omega}$ as a definitional extension of $\mathbf{E\text{-}HA}^{\omega}_{\to}$ (cf. Troelstra 1973, 1.6.17).

9.1.14. Let $\alpha \le \beta := \forall x(\alpha x \le \beta x)$. A fan functional $\varphi_{uc} \in 210$ is a functional satisfying

$$\forall z^2 \forall \beta \forall \alpha \le \beta \forall \alpha' \le \beta\big(\bar{\alpha}(\varphi_{\mathrm{uc}} z\beta) = \bar{\alpha}'(\varphi_{\mathrm{uc}} z\beta) \to z\alpha = z\alpha'\big).$$

Assuming FAN for $\mathscr{U}$, show that φ_{uc} exists in ECF($\mathscr{U}$) and ICF($\mathscr{U}$) (notation of 1.17(ii)). Show also that φ_{uc} does not exist in ECF($\mathscr{R}$) and ICF($\mathscr{R}$) (Troelstra 1973, 2.6.4, 2.6.6, 2.6.9–10), where $\mathscr{R} \equiv$ TREC.

9.1.15. Is in ICF($\mathscr{U}$) the existence of a φ^*_{uc} satisfying

$$\forall z^2 \forall \beta \forall \alpha \le \beta\big(z^2\alpha \le \varphi^*_{\mathrm{uc}} z^2\beta\big)$$

equivalent to the existence of φ_{uc}?

9.1.16. Show how to define φ_{uc} from $\lambda z^2.\varphi_{uc}z^2(\lambda x.0)$.

9.2.1. Complete the proofs of 2.5 and 2.10.

9.2.2. Show (2.11) that the proof of computability also works if we consider strict reductions only.

9.2.3. Check lemma 2.14, (viii).

9.2.4. Treat the remaining cases in the proofs of 2.15 and 2.16.

9.2.5. Give a direct proof of strong computability for terms of **HA**$^\omega$ (cf. remark after 2.16).

9.2.6. Show that it is possible to choose the interpretations of $\boldsymbol{k}, \boldsymbol{s}, \boldsymbol{r}, \boldsymbol{p}, \boldsymbol{p}_0, \boldsymbol{p}_1, 0, S, P$ in HRO in such a way that the model CNF is isomorphically embedded in HRO. *Hint:* use the fact that for each partial recursive function the set of codes contains an infinite recursive set.

Describe another HRO-variant (i.e. a particular choice for the interpretation of the constants) in which certain terms with distinct normal forms are identified (Troelstra 1973, 2.5.9).

9.2.7. Use the first half of E9.2.6 to obtain an alternative proof of UN(t) for closed t: each closed t has at least *some* normal form.

9.2.8. The term-evaluation property for closed terms of type 0 fails in **I-HA**$^\omega$; why? Can you strengthen **I-HA**$^\omega$ consistently so as to restore this property? *Hint:* look at the next exercise.

9.2.9. Extend the conversion rules by

$$e^\sigma ts \text{ conv } 0 \quad \text{if } t, s \in \text{CNF} \quad \text{and} \quad t \equiv s,$$

$$e^\sigma ts \text{ conv } 1 \quad \text{if } t, s \in \text{CNF} \quad \text{and} \quad t \not\equiv s.$$

Show that the Church–Rosser property holds for the corresponding notion of reduction, as well as strong normalizability. Use this to construct a term model for **I-HA**$^\omega$ (Troelstra 1973, 2.3.1–5).

9.2.10. Complete the proof of 2.21.

9.2.11. Complete the proof of 2.25.

9.2.12. Verify the properties claimed for the generalized notion of reduction in 2.27.

9.3.1. Elaborate the remarks at the end of 3.3.

9.3.2. Complete the proof of 3.5.

9.3.3. Give a simpler definition of $\boldsymbol{\mu}$ in **TAPP** (remark at the end of 3.10).

9.3.4. Describe in more detail how to establish the second half of 3.11.

9.3.5. Show that in a first-order theory with total application and constants $\boldsymbol{k}, \boldsymbol{s}, \boldsymbol{d}$ such that

$$\boldsymbol{k}xy = x, \qquad \boldsymbol{s}xyz = xz(yz),$$

$$x = y \rightarrow \boldsymbol{d}xy = \boldsymbol{k}, \qquad x \neq y \rightarrow \boldsymbol{d}xy = \boldsymbol{s},$$

we can derive $\forall xy \neg\neg(x = y)$.

9.3.6. Let α, β, γ be variables ranging over type 0 objects; show that **APP** + $\forall \alpha \exists x \in N \forall y \in N \exists u \in N(Txyu \wedge Uu = \alpha x)$ is conservative over **HA**.

9.4.1. Show that in a pca $k \neq s$.

9.4.2. Show (4.2) that PCO contains a type 3 object which is a modulus for extensional type 2 objects at $[\![\lambda x.0]\!]$ *Hint:* See E9.1.7.

9.4.3. Show that $[\![\boldsymbol{k}]\!], [\![\boldsymbol{s}]\!], [\![\boldsymbol{p}]\!], [\![\boldsymbol{p}_i]\!] (i \leq 1)$, and $[\![\boldsymbol{d}]\!]$ as defined in 4.5 satisfy the equations for $\boldsymbol{k}, \boldsymbol{s}, \boldsymbol{p}, \boldsymbol{p}, \boldsymbol{p}_i, \boldsymbol{d}$ in **APP**. (Beeson 1985).

9.4.4. Complete the proof of 4.12.

9.4.5. Complete the proof of 4.19, and show in detail that CNFS is indeed a model of **APP**.

9.4.6. Show that equality in CTS is not recursive. If we extend the definition of reduction and r-equality so as to include the conversion rule (6′) of 4.17, does CTS then remain a model of **TAPP**? (cf. Barendregt 1981, section 6.6).

9.4.7. Show that in CTS no two terms in distinct normal form can be identified.

9.4.8. Which functions of $\mathbb{N}^{\mathbb{N}}$ are represented in CTS and in CNFS?

9.4.9. Give an example where graph(fun(X)) $\neq X$ in P(ω). *Hint:* consider $\{n\}$.

9.5.1. Establish the soundness theorem 5.4. Use e.g. the Hilbert-type formalization of E^+-logic in section 2.4.

9.5.2. Prove 5.6 and 5.9.

9.5.3. Derive a contradiction from the instance of EAC^+ ((1) in 5.13), where $A(x) := \exists y(xx \neq y)$, $B(x, y) := (xx \neq y \wedge Ey)$.

9.5.4. Let $\mathrm{AC}^{\vee}$ be the special case of choice for disjunctions.

$$\forall x(Ax \vee Bx) \rightarrow \exists f \forall x((fx = 0 \wedge Ax) \vee (fx = 1 \wedge Bx)),$$

and let DNS (*Double Negation Shift*) be the schema

$$\forall x \neg\neg A \rightarrow \neg\neg \forall x A.$$

Show that **APP** + $\mathrm{AC}^{\vee}$+ DNS is inconsistent (Renardel de Lavalette 1984A, III2.9(i), 2.12(iv)).

9.6.1. Show with the help of P(ω) or the term model CTS that **TAPP** + EAC is consistent.

9.6.2. Show the consistency of **TAPP** + Ext + EAC + $\mathrm{AC}_{\mathrm{IT}}$.

9.6.3. Show that **TAPP** + Ext + $\forall z^2 \forall u^1 \exists x^0 \forall v^1 \in \bar{u}x(zv = zu)$, where $\forall z^\sigma$ stands for $\forall z \in \mathrm{IT}_\sigma$, is conservative over **EL**.

9.6.4. Show that $\mathrm{AC}_{\mathrm{ET}}!$ is derivable in **APP** + EAC.

9.6.5. Associate to each formula A of $\mathbf{HA}^\omega$ another formula $A_{\mathrm{mr}}(\vec{x})$ ($\vec{x}$ *modified realizes* A or $\vec{x}$ **mr**-*realizes* A, alternative notation: $\vec{x}\,\mathbf{mr}\,A$) such that $\vec{x} \cap \mathrm{FV}(A) = \emptyset$, $\mathrm{FV}(\exists \vec{x} A_{\mathrm{mr}}(\vec{x})) = \mathrm{FV}(A)$, as follows:
(i) $A_{\mathrm{mr}} := A$ for A prime, $\vec{x}$ empty,
(ii) $(A \wedge B)_{\mathrm{mr}}(\vec{x}, \vec{y}) := A_{\mathrm{mr}}(\vec{x}) \wedge B_{\mathrm{mr}}(\vec{y})$ ($\vec{x}$, $\vec{y}$ disjoint),
(iii) $(A \to B)_{\mathrm{mr}}(\vec{x}) := \forall \vec{y}(A_{\mathrm{mr}}(\vec{y}) \to B_{\mathrm{mr}}(\vec{x}\vec{y}))$,
(iv) $(\forall z A(z))_{\mathrm{mr}}(\vec{x}) := \forall z\, A_{\mathrm{mr}}(\vec{x}z)$,
(v) $(\exists z A)_{\mathrm{mr}}(z, \vec{x}) := A_{\mathrm{mr}}(z, \vec{x})$.
Prove that for $\exists$-free A $A_{\mathrm{mr}} \equiv A$, and that for all A, if $\mathbf{HA}^\omega \vdash A$, then $\mathbf{HA}^\omega \vdash A_{\mathrm{mr}}(\vec{t})$ for a suitable sequence of terms $\vec{t}$. More precisely, if $\mathrm{FV}(A) = \vec{x}$, then $\mathbf{HA}^\omega \vdash A \Rightarrow \mathbf{HA}^\omega \vdash A_{\mathrm{mr}}(\vec{t}\vec{x})$ for a sequence of closed terms $\vec{t}$ (Troelstra 1973, 3.4.5).

9.6.6. Find a simple axiomatic characterization of the set $\{A\colon \mathbf{HA}^\omega \vdash \exists \vec{x} A_{\mathrm{mr}}(\vec{x})\}$ (Troelstra 1973, 3.4.7–3.4.8; see also the correction in Troelstra (1974)).

9.6.7. Show $\mathbf{HA}^\omega$ + AC + EXT to be consistent. *Hint:* use E9.6.5, Troelstra (1973, 3.6.6(ii)). Strengthen the result with E9.6.6.

9.7.1. Verify proposition 7.3.

9.7.2. Prove 7.7 in full.

9.7.3. Give the proof of 7.9 and 7.10 in detail.

9.7.4. Prove the remaining cases of 7.11.

9.7.5. Prove 7.14(i) and 7.15(iii).

9.7.6. Prove 7.18.

9.7.7. Prove 7.22.

9.7.8. Prove 7.24.

9.7.9. Give versions of 7.14(i) and (ii) where extra numerical parameters and (for (ii)) function parameters are present.

9.7.10. Generalize the fan rule to the case where arbitrary function parameters are present.

9.7.11. Modify A_{mr} of E9.6.5 to A_{mq}, defined as A_{mr} except for clause (iii)

$$(A \to B)_{\text{mq}}(\vec{x}) := \forall \vec{y}(A_{\text{mq}}(\vec{y}) \to B_{\text{mq}}(\vec{x}\vec{y})) \wedge (A \to B).$$

Show $\mathbf{HA}^\omega \vdash A \Rightarrow \mathbf{HA}^\omega \vdash A_{\text{mq}}(\vec{t})$ for suitable $\vec{t}$ with $\text{FV}(\vec{t}) \subset \text{FV}(A)$, and also $\mathbf{HA}^\omega \vdash A_{\text{mq}}(\vec{x}) \to A$ (*modified* **q**-*realizability* or **mq**-*realizability*: Troelstra 1973, 3.4.5).

9.7.12. Give proofs of the continuity rule and the fan rule for $\mathbf{HA}^\omega$. *Hint:* **mq**-realizability of E9.7.11 can play the role of **q**-realizability.

9.8.1. The system $\mathbf{EM}_0 + \Sigma\text{AX}$ can be still further extended by adding the *schema of inductive generation*, consisting of the principle of *closure* and an *induction schema*:

$$\text{IG} \qquad \begin{cases} \exists X(i(Y,R) = X \wedge \forall x \in Y \forall y(pyx \in R \to y \in X) \to x \in X) \\ \forall x \in Y(\forall y(pyx \in R \to A(y)) \to A(x)) \to \forall x \in i(Y,R) A(x) \end{cases}$$

Show how to extend the standard model construction of $\mathbf{EM}_0 + \Sigma\text{AX} + \text{IG}$ (cf. 8.8 and Feferman 1979, III.1).

9.8.2. Show that N and induction on N can in fact be obtained from IG without assuming them (Feferman 1975; 1979, II.6).

9.8.3. Show that if we only want to prove $\mathbf{EM}_0\!\upharpoonright$ to be conservative over **HA**, we can in fact simplify the argument of 8.6 so that the only appeal to classical metamathematics in the argument is in our reliance on the completeness theorem for Kripke semantics.

CHAPTER 10

PROOF THEORY OF INTUITIONISTIC LOGIC

In this chapter we study the structure of deductions in certain systems of intuitionistic predicate logic, in particular N-**IQC** and N-**IQCE** for languages without equality.

The greater part of this chapter presupposes sections 2.1 and 2.2 only; section 8 also requires familiarity with 9.1 and 9.2. Sections 2 and 3 discuss normalization and the structure of normal derivations; sections 4–6 contain applications. Section 7 compares the system of natural deduction with sequent calculi with left- and right-introduction rules, such as Gentzen's well-known systems LJ and LK.

Section 8 describes N-**IQC** as a calculus of terms. The "crude discharge convention" of earlier sections is relaxed here, thus better bringing out the parallel with normalization and reduction in $\mathbf{HA}^{\omega}$. This section also serves to set the stage for the type theories in the next chapter.

1. Preliminaries

1.1. In sections 1–3 we deal with N-**IQC** and N-**IQCE** simultaneously.

Where definitions are given for N-**IQCE**, the same definitions apply to N-**IQC**, possibly dropping some clauses; we shall not always bother to indicate this explicitly. Our version of N-**IQCE** does not contain equality.

In **IQCE** we shall regard ***Et*** as subformula of any formula containing quantifiers.

We also note, for future use, that $\bot_i$ may be restricted to instances with atomic conclusion only; the general case of $\bot_i$ is easily seen to be derivable from these instances. The restriction to atomic instances of $\bot_i$ is not always an advantage, however.

We first have to describe certain conventions in the notation and description of proof trees. A completely rigorous treatment avoiding all notational ambiguity easily degenerates into pedantry, and we shall not attempt such a treatment here, but we have to supplement at certain points the description given in section 2.1.

In describing proof trees we are primarily concerned with *formula occurrences* (f.o.'s for short) instead of just formulas; an f.o. may be specified by a formula plus data determining its position in a proof tree. Occurrences of formulas which are identical modulo renaming of bound variables are said to be isomorphic.

1.2. *Discharge of assumptions.* Under the crude discharge convention (CDC, cf. 2.1.4(ii)) all assumptions are discharged as early as possible.

If we do not accept the CDC, the numbering of assumption classes is really necessary, for example in distinguishing the derivations below:

$$\rightarrow\mathrm{I}\frac{\rightarrow\mathrm{I}\dfrac{A^{(1)}}{A \rightarrow A}\ (1)}{A \rightarrow (A \rightarrow A)} \qquad\qquad \rightarrow\mathrm{I}\frac{\rightarrow\mathrm{I}\dfrac{A^{(1)}}{A \rightarrow A}}{A \rightarrow (A \rightarrow A)}\ (1)$$

The deduction on the left obeys the CDC: the second application of $\rightarrow$I discharges an empty set of assumptions. In the deduction on the right the discharge of the f.o. marked (1) is postponed until the second application of $\rightarrow$I. Without numbering the two deductions would be indistinguishable.

1.3. Notation. We slightly extend our notations for proof trees introduced in section 2.1. Let

$$\begin{array}{c}[A_1][A_2]\cdots[A_n]\\ \mathcal{D}\end{array}$$

be a derivation with certain assumption classes containing f.o.'s of the form $A_1, A_2, \ldots, A_n$, respectively. Given derivations

$$\begin{array}{ccc}\mathcal{D}_1 & & \mathcal{D}_n\\ A_1, & \ldots & , A_n\end{array}$$

we shall write

$$\begin{array}{c}\mathcal{D}_1\ \mathcal{D}_2 \qquad \mathcal{D}_n\\ [A_1][A_2]\cdots[A_n]\\ \mathcal{D}\end{array}$$

for the derivation $\mathcal{D}'$ obtained by replacing each occurrence in $[A_i]$ of $\mathcal{D}$ by $\mathcal{D}_i$.

1.4. Definition. The leftmost premiss in an application α of an E-rule is called the *major* premiss of α; the other premisses (if any) are the *minor* premisses of α. □

N.B. The definition refers to the rules as presented in chapter 2; for a consistent application of the terminology we must not change the order of the premisses!

1.5. *Conditions on variables.* From now on we shall assume that the collection of variables occurring bound in some formula occurrence of a deduction is disjoint from the collection of variables occurring free. We can always arrange this by renaming bound variables. Without this assumption the process of normalization as described in the next section cannot always be carried out (cf. E10.3.1).

In fact it will be convenient to introduce further restrictions.

DEFINITION. x is said to be the *proper parameter* of an application of ∀I:

$$\frac{\begin{array}{c}\mathcal{D}\\ A\end{array}}{\forall y A[x/y]}$$

if x actually occurs in A. x is said to be a *proper parameter* of an application of ∃E:

$$\frac{\begin{array}{c}\\ \mathcal{D}\\ \exists y A[x/y]\end{array} \quad \begin{array}{c}[A][Ex]\\ \mathcal{D}'\\ C\end{array}}{C}$$

if $[Ex]$ is inhabited or if x actually occurs in A and the set of assumptions $[A]$ is inhabited.

x is said to be a *proper parameter* of $\mathcal{D}$ if it is the proper parameter of some application of ∀I or ∃E in $\mathcal{D}$. □

We now prove the following lemma.

LEMMA. *Let $\Gamma \vdash A$ in* **N-IQCE** *or* **N-IQC**, *and let $\mathcal{V}$ be a countably infinite set of variables; suppose* $\mathrm{FV}(\{A\} \cup \Gamma) \cap \mathrm{BV}(\{A\} \cup \Gamma) = \emptyset$. *Then we can find a deduction $\mathcal{D}$ of A from Γ such that:*

(i) *the sets of bound variables and free variables in $\mathcal{D}$ are disjoint and all proper parameters of $\mathcal{D}$ belong to $\mathcal{V}$;*

(ii) *each proper parameter of $\mathcal{D}$ is proper parameter of a single application of* ∀I *or* ∃E;

(iii) *the proper parameter of an* ∀I*-application α in $\mathcal{D}$ occurs only above α;*

(iv) *the proper parameter of an* ∃E*-application α in $\mathcal{D}$ occurs only above the minor premiss of α.*

PROOF. By induction on $\mathscr{D}$. Observe that this boils down to a systematic renaming of proper parameters starting with the uppermost applications of $\forall$I and $\exists$E. We leave the details as an exercise. $\square$

DEFINITION. A deduction satisfies *the variable condition* if it has the properties (i)–(iv) above. $\square$

In combining proof trees as for example in the substitution notation of 1.3 we shall tacitly assume that the resulting proof trees again satisfy the variable condition, if necessary by renaming proper parameters.

1.6. *Normalization as the removal of detours.* An introduction immediately followed by an elimination, as for example in

$$\frac{\dfrac{\begin{matrix}\mathscr{D}_1\\ A\end{matrix}\qquad\begin{matrix}\mathscr{D}_2\\ B\end{matrix}}{A\wedge B}}{A}$$

results in a "local maximum" of complexity at the formula occurrence $A \wedge B$. The process of normalization systematically removes such detours, in this case by replacing the derivation above by $\mathscr{D}_1$. But local maxima may also occur in more complex situations, for example in

$$\frac{\dfrac{\begin{matrix}\mathscr{D}\\ A\vee B\end{matrix}\qquad \dfrac{\begin{matrix}\mathscr{D}_1\\ C\end{matrix}\quad\begin{matrix}\mathscr{D}_1'\\ D\end{matrix}}{C\wedge D}\qquad \dfrac{\begin{matrix}\mathscr{D}_2\\ C\end{matrix}\quad\begin{matrix}\mathscr{D}_2'\\ D\end{matrix}}{C\wedge D}}{C\wedge D}}{C}$$

the occurrences of $C \wedge D$ also represent local maxima: they are first introduced and two steps later eliminated again.

In order to discuss such local maxima we need some terminology.

1.7. DEFINITION. A *branch* of a derivation $\mathscr{D}$ is a sequence $A_1, \dots, A_n$ of f.o.'s such that A_1 is a top formula, A_n the root of $\mathscr{D}$, and each A_{i+1} occurs immediately below A_i for $1 \leq i < n$. $\square$

1.8. DEFINITION. A *segment* $A_1, \dots, A_n$ of a derivation $\mathscr{D}$ is a sequence of isomorphic f.o.'s such that:

(i) A_1 is not the conclusion of an $\vee$E- or $\exists$E-application;

(ii) A_n is not a minor premiss of an $\vee$E- or $\exists$E-application;

(iii) A_i $(1 \leq i < n)$ is a minor premiss of an $\vee$E- or $\exists$E-application with conclusion A_{i+1}.

We shall use $\sigma, \sigma', \ldots$ for arbitrary segments. □

Informally speaking, a nontrivial (i.e. of length > 1) segment consists of f.o.s that are repeated because of applications of the rules $\exists$E or $\vee$E.

EXAMPLE.

$$\dfrac{A \vee (B \vee C) \qquad \dfrac{\dfrac{A}{A \vee B}}{(\mathrm{d})(A \vee B) \vee C} \qquad \dfrac{B \vee C \qquad \dfrac{\dfrac{B}{A \vee B}}{(\mathrm{a})(A \vee B) \vee C} \qquad \dfrac{C}{(\mathrm{b})(A \vee B) \vee C}}{(\mathrm{c})(A \vee B) \vee C}}{(\mathrm{e})(A \vee B) \vee C}$$

In this tree the following sequences of f.o.'s are examples of segments: (d), (e); (a), (c), (e); (b), (c), (e). Note that e.g. the single f.o. A is also a segment according to the definition.

1.9. DEFINITION. Let $\sigma \equiv A_1, \ldots, A_n$ be a segment in $\mathscr{D}$. We say that σ is the *premiss* of a rule-application α if A_n is the premiss of α; σ is the *conclusion of* α if A_1 is the conclusion of α.

Similarly we define *minor* and *major premiss* for segments, and *proper parameter* of a segment.

If a segment σ consists of f.o.'s of the form A, A is said to be *the formula of* σ. □

1.10. DEFINITION. The *degree* $d(\sigma)$ of a segment σ is the number of occurrences of logical operators in the formula of σ. σ is a *cut* in $\mathscr{D}$ if σ is the conclusion of an application of an I-rule or $\bot_{\mathrm{i}}$ and the premiss of an application of an E-rule. (N.B. if $\bot_{\mathrm{i}}$ is restricted to atomic formulas then the "or $\bot_{\mathrm{i}}$" may be dropped.)

σ is *maximal* in $\mathscr{D}$ if σ is a cut and $d(\sigma) = \max\{d(\sigma') : \sigma' \text{ is a cut in } \mathscr{D}\}$.

A *cut formula* is a cut of length 1, a *maximal formula* is a maximal segment of length 1. □

N.B. Any f.o. which is neither a minor premiss nor the conclusion of an $\vee$E- or $\exists$E-application is itself a segment of length 1.

REMARK. A more liberal definition of cut is the following: a cut segment σ is premiss of an E-rule application, and σ is either of length 1 and conclusion of an application of an I-rule or $\bot_i$, or σ is of length > 1. This permits a local definition of "cut formula" (lowest f.o. in a cut): a cut formula is premiss of an E-rule and conclusion of an application of $\bot_i$, $\exists$E, $\vee$E, or an I-rule.

1.11. DEFINITION (*redundant* $\vee$E, $\exists$E; *normal derivation*). An application α of $\vee$E

$$\dfrac{\begin{matrix} & {}^{(n)}[A] & [B]^{(n)} \\ \mathcal{D} & \mathcal{D}' & \mathcal{D}'' \\ A \vee B & C & C \end{matrix}}{C}\,(n)$$

is *redundant* if $[A] = \emptyset$ or $[B] = \emptyset$. An application α of $\exists$E

$$\dfrac{\begin{matrix} & {}^{(n)}[A][Ex]^{(n)} \\ \mathcal{D} & \mathcal{D}' \\ \exists y A[x/y] & C \end{matrix}}{C}\,(n)$$

is *redundant* if $[A]$ and $[Ex]$ are empty.

A derivation $\mathcal{D}$ is *normal* if $\mathcal{D}$ contains no cuts and no redundant applications of $\vee$E or $\exists$E. □

2. Normalization

2.1. Below we shall define the transformation steps of various kinds which serve to transform a derivation into a normal one. We shall always assume the result of the transformation to be a proof satisfying the variable condition, if necessary by renaming parameters.

2.2. *Detour-conversions.* We first show how to remove cuts of length 1. For derivations with the cuts of length 1 as major premiss of the conclusion we define *conversions* as follows, writing "conv" for "converts to":

$\wedge$*-conversion*:

$$\dfrac{\dfrac{\begin{matrix} \mathcal{D}_1 & \mathcal{D}_2 \\ A_1 & A_2 \end{matrix}}{A_1 \wedge A_2}}{A_i} \quad \text{conv} \quad \begin{matrix} \mathcal{D}_i \\ A_i \end{matrix} \quad \text{for } i \in \{1,2\}.$$

$\vee$*-conversion*:

$$\dfrac{\dfrac{\begin{matrix}\mathcal{D}\\ A_i\end{matrix}}{A_1\vee A_2}\quad \begin{matrix}{}^{(n)}[A_1]\\ \mathcal{D}_1\\ C\end{matrix}\quad \begin{matrix}[A_2]^{(n)}\\ \mathcal{D}_2\\ C\end{matrix}}{C}\,{\scriptstyle (n)} \qquad \text{conv}\qquad \begin{matrix}\mathcal{D}\\ [A_i]\\ \mathcal{D}_i\\ C\end{matrix} \quad \text{for } i\in\{1,2\}.$$

$\to$ *-conversion*:

$$\dfrac{\dfrac{\begin{matrix}[A]^{(n)}\\ \mathcal{D}\\ B\end{matrix}}{A\to B}\quad \begin{matrix}\mathcal{D}_1\\ A\end{matrix}}{B}\,{\scriptstyle (n)} \qquad \text{conv}\qquad \begin{matrix}\mathcal{D}_1\\ [A]\\ \mathcal{D}\\ B\end{matrix}$$

$\forall$*-conversion*:

$$\dfrac{\dfrac{\begin{matrix}[Ex]^{(n)}\\ \mathcal{D}\\ A\end{matrix}}{\forall y A[x/y]}\,{\scriptstyle (n)}\quad \begin{matrix}\mathcal{D}_1\\ Et\end{matrix}}{A[x/y][y/t]} \qquad \text{conv}\qquad \begin{matrix}\mathcal{D}_1\\ [Et]\\ \mathcal{D}[x/t]\\ A[x/t]\end{matrix}$$

$\exists$*-conversion*:

$$\dfrac{\dfrac{\begin{matrix}\mathcal{D}\\ A[y/t]\end{matrix}\quad \begin{matrix}\mathcal{D}'\\ Et\end{matrix}}{\exists x A[y/x]}\quad \begin{matrix}{}^{(n)}[A][Ey]^{(n)}\\ \mathcal{D}''\\ C\end{matrix}}{C}\,{\scriptstyle (n)} \qquad \text{conv}\qquad \begin{matrix}\mathcal{D}\qquad \mathcal{D}'\\ [A[y/t]][Et]\\ \mathcal{D}''[y/t]\\ C\end{matrix}$$

We have to verify that the derivations obtained by applying these conversions are again correct.

Consider e.g. $\forall$-conversion. The proper parameters of $\mathcal{D}(x)$ are distinct from x (by 1.5(ii)) and from t (by 1.5(iii), (iv)), and $A[x/y][y/t] \equiv A[x/t]$ since y does not occur free in A. The assumptions in $\mathcal{D}$ on which $\forall y A[x/y]$ depends do not contain any occurrence of x and are thus not changed by the substitution of t for x in $\mathcal{D}$. So the conversion yields a correct deduction of $\Gamma \vdash A[x/t]$.

As to $\exists$-conversion, let $y \in \mathrm{FV}(A)$; then t does not contain y free (1.5(iv)), no assumptions in $\mathcal{D}''$ except $[A]$, $[Ey]$ contain y free (by the conditions on $\exists$E). The proper parameters of $\mathcal{D}$ and $\mathcal{D}'$ are distinct from the proper parameters of $\mathcal{D}''$ and from y, and do not occur in t (1.5(iv)).

The reader can now easily check the correctness of the results of the other conversions.

2.3. *Permutation conversions.* Cut segments of length greater than one cannot be removed by detour conversions alone; for example in the tree of 1.6

$$\dfrac{\dfrac{\begin{matrix}\mathscr{D}\\ A\vee B\end{matrix}\qquad \dfrac{\begin{matrix}\mathscr{D}_1\\ C\end{matrix}\qquad\begin{matrix}\mathscr{D}_1'\\ D\end{matrix}}{C\wedge D}\qquad \dfrac{\begin{matrix}\mathscr{D}_2\\ C\end{matrix}\qquad\begin{matrix}\mathscr{D}_2'\\ D\end{matrix}}{C\wedge D}}{C\wedge D}}{C}$$

we should like to make the transformation to

$$\dfrac{\begin{matrix}\mathscr{D}\\ A\vee B\end{matrix}\quad\begin{matrix}\mathscr{D}_1\\ C\end{matrix}\quad\begin{matrix}\mathscr{D}_2\\ C\end{matrix}}{C}$$

but this transformation cannot be obtained by a finite number of detour conversions. We may now extend our notion of conversion, so as to permit transformations as above, but that would introduce a kind of nonlocal conversions, as the segments involved may be arbitrarily long. It is more convenient to introduce purely local conversions which reduce the length of a cut segment. This can be achieved by the so-called *permutation conversions*.

$\vee$*-perm conversion*:

$$\text{E-rule}\ \dfrac{\dfrac{\begin{matrix}\mathscr{D}\\ A\vee B\end{matrix}\quad\begin{matrix}\mathscr{D}_1\\ C\end{matrix}\qquad\begin{matrix}\mathscr{D}_2\\ C\end{matrix}}{C}\quad \mathscr{D}'}{D}\quad \text{conv}\quad \dfrac{\begin{matrix}\mathscr{D}\\ A\vee B\end{matrix}\quad \dfrac{\begin{matrix}\mathscr{D}_1\\ C\end{matrix}\ \ \mathscr{D}'}{D}\quad \dfrac{\begin{matrix}\mathscr{D}_2\\ C\end{matrix}\ \ \mathscr{D}'}{D}}{D}$$

$\exists$*-perm conversion*:

$$\text{E-rule}\ \dfrac{\dfrac{\begin{matrix}\mathscr{D}\\ \exists x Ax\end{matrix}\quad\begin{matrix}\mathscr{D}'\\ C\end{matrix}}{C}\quad \mathscr{D}''}{D}\quad \text{conv}\quad \dfrac{\begin{matrix}\mathscr{D}\\ \exists x Ax\end{matrix}\quad \dfrac{\begin{matrix}\mathscr{D}'\\ C\end{matrix}\ \ \mathscr{D}''}{D}}{D}$$

Here $\mathscr{D}''$ stands for the derivation or pair of derivations of the minor premisses of the E-rule. Observe that, by definition, C is a major premiss, and hence the lower occurrence of C is the end of the segment under consideration. The effect of the permutation conversions is to "push upwards" the E-rule application that is eligible for a detour conversion; in finitely many steps it reaches the corresponding E-rule. Note that while we shorten the segment of occurrences of C, we pay by increasing the length of the segment with occurrences of D. We can now achieve the transformation in the example above by first applying an $\vee$E-permutation and then two $\wedge$-contractions to the resulting subdeductions of the form

$$\dfrac{\dfrac{\begin{matrix}[A_i]\\ \mathscr{D}_i\\ C\end{matrix} \qquad \begin{matrix}\mathscr{D}'_i\\ D\end{matrix}}{C \wedge D}}{C} \qquad i \in \{1,2\}.$$

REMARK. Permutation conversions also apply to the more liberal notion of cut segment in 1.10, remark.

2.4. *Immediate simplifications.* Redundant applications of $\vee$E or $\exists$E can be removed by the following conversions:

$$\mathscr{D}' \equiv \left\{ \dfrac{\begin{matrix}\mathscr{D}\\ A_1 \vee A_2\end{matrix} \quad \begin{matrix}\mathscr{D}_1\\ C\end{matrix} \quad \begin{matrix}\mathscr{D}_2\\ C\end{matrix}}{C} \right. \quad \text{conv} \quad \begin{matrix}\mathscr{D}_i\\ C\end{matrix} \qquad (i \in \{1,2\})$$

if *no* assumptions of the subderivation $\mathscr{D}_i$ are discharged in $\mathscr{D}'$, and

$$\mathscr{D}' \equiv \left\{ \dfrac{\begin{matrix}\mathscr{D}\\ \exists x A x\end{matrix} \quad \begin{matrix}\mathscr{D}''\\ C\end{matrix}}{C} \right. \quad \text{conv} \quad \begin{matrix}\mathscr{D}''\\ C\end{matrix}$$

if *no* assumptions of the subderivation $\mathscr{D}''$ are discharged in $\mathscr{D}'$.

2.5. *$\perp$-conversions.* If we do not restrict $\perp_i$ to the case of atomic conclusions, we need extra conversions following exactly the same pattern as the detour reductions:

$$\dfrac{\dfrac{\begin{matrix}\mathcal{D}\\ \bot\end{matrix}}{A_1\wedge A_2}}{A_i} \quad \text{conv} \quad \dfrac{\begin{matrix}\mathcal{D}\\ \bot\end{matrix}}{A_i} \qquad\qquad \dfrac{\dfrac{\begin{matrix}\mathcal{D}\\ \bot\end{matrix}}{A\to B}\quad \begin{matrix}\mathcal{D}'\\ A\end{matrix}}{B} \quad \text{conv} \quad \dfrac{\begin{matrix}\mathcal{D}\\ \bot\end{matrix}}{B}$$

$$\dfrac{\dfrac{\begin{matrix}\mathcal{D}\\ \bot\end{matrix}}{A_1\vee A_2}\quad \begin{matrix}{}^{(n)}[A_1]\\ \mathcal{D}_1\\ C\end{matrix}\quad \begin{matrix}[A_2]^{(n)}\\ \mathcal{D}_2\\ C\end{matrix}}{C}\,(n) \quad \text{conv} \quad \begin{matrix}\dfrac{\begin{matrix}\mathcal{D}\\ \bot\end{matrix}}{[A_i]}\\ \mathcal{D}_i\\ C\end{matrix} \qquad\qquad \dfrac{\dfrac{\begin{matrix}\mathcal{D}\\ \bot\end{matrix}}{\forall x Ax}\quad \begin{matrix}\mathcal{D}'\\ Et\end{matrix}}{At} \quad \text{conv} \quad \dfrac{\begin{matrix}\mathcal{D}\\ \bot\end{matrix}}{At}$$

$$\dfrac{\dfrac{\begin{matrix}\mathcal{D}\\ \bot\end{matrix}}{\exists x A(x)}\quad \begin{matrix}[A(y)]^{(n)}\\ \mathcal{D}'\\ C\end{matrix}}{C}\,(n) \quad \text{conv} \quad \begin{matrix}\dfrac{\begin{matrix}\mathcal{D}\\ \bot\end{matrix}}{[A(y)]}\\ \mathcal{D}'\\ C\end{matrix}$$

In the case of $\vee$E and $\exists$E we can also use the following shortcuts:

$$\dfrac{\dfrac{\begin{matrix}\mathcal{D}\\ \bot\end{matrix}}{A_1\vee A_2}\quad \begin{matrix}\mathcal{D}_1\\ C\end{matrix}\quad \begin{matrix}\mathcal{D}_2\\ C\end{matrix}}{C} \quad \text{conv} \quad \dfrac{\begin{matrix}\mathcal{D}\\ \bot\end{matrix}}{C} \qquad\qquad \dfrac{\dfrac{\begin{matrix}\mathcal{D}\\ \bot\end{matrix}}{\exists x A}\quad \begin{matrix}\mathcal{D}'\\ C\end{matrix}}{C} \quad \text{conv} \quad \dfrac{\begin{matrix}\mathcal{D}\\ \bot\end{matrix}}{C}$$

2.6. DEFINITION. We say that $\mathcal{D}$ *reduces in one step to* $\mathcal{D}'$ (notation $\mathcal{D} \succ_1 \mathcal{D}'$) if $\mathcal{D}'$ is obtained from $\mathcal{D}$ by an application of a conversion to a subderivation of $\mathcal{D}$. We shall say that $\mathcal{D}$ reduces to $\mathcal{D}'$ ($\mathcal{D} \succcurlyeq \mathcal{D}'$) if there is a finite sequence $\mathcal{D} \equiv \mathcal{D}_1 \succ_1 \mathcal{D}_2 \succ_1 \mathcal{D}_3 \cdots \succ_1 \mathcal{D}_n \equiv \mathcal{D}'$ (so $\succcurlyeq$ is the reflexive and transitive closure of $\succ_1$). □

N.B. A conversion applied to a subdeduction of $\mathcal{D}$ may remove a cut, or reduce the length of a cut, while introducing new cuts.

2.7. For certain results we need to distinguish between several possibilities for the set of conversions with respect to which $\succ_1$, $\succcurlyeq$ are defined.

CONVENTION. Let us use "d" for "detour" (referring to detour- and $\bot$-conversions), "s" for "immediate simplification", and "p" for "permutation". Then "dp-normal" means "normal with respect to detour-, $\perp$-, and permutation conversions", etc. Thus "normal" as defined before coincides with "dps-normal".

2.8. THEOREM (*normalization for* N-IQCE). *Each derivation* $\mathscr{D}$ *in* N-IQCE *reduces to a normal derivation.*

PROOF. Let $d := \max\{d(\sigma) : \sigma \text{ a cut in } \mathscr{D}\}$ and let n be the sum of the lengths of all maximal cuts in $\mathscr{D}$.

The *induction value* $i(\mathscr{D})$ is then defined as $\omega \cdot d + n$ (i.e. an ordinal $< \omega^2$); $\mathscr{D}$ is normal if $i(\mathscr{D}) = 0$. The theorem is proved if we can indicate a suitable conversion that lowers $i(\mathscr{D})$.

By a suitable choice of the maximal segment to which we apply a conversion we can achieve a lowering of the induction value. Let us call σ a *t.m.d. segment* (top segment of maximal degree) in $\mathscr{D}$ if no maximal σ' occurs in a branch of $\mathscr{D}$ above σ. Now apply a conversion to the *rightmost* t.m.d. cut of $\mathscr{D}$; then the resulting $\mathscr{D}'$ has a lower induction value.

For example, if at the rightmost t.m.d. segment consisting of a formula occurrence $A \to B$ we apply a conversion

$$\dfrac{\dfrac{\begin{matrix}[A]\\ \mathscr{D}'\\ B\end{matrix}}{A \to B} \qquad \begin{matrix}\mathscr{D}''\\ A\end{matrix}}{B} \quad \text{conv} \quad \begin{matrix}\mathscr{D}''\\ [A]\\ \mathscr{D}'\\ B\end{matrix}$$

then the repeated substitution of $\mathscr{D}''$ at each f.o. of $[A]$ cannot increase the induction value, since $\mathscr{D}$ does not contain a t.m.d. cut in $\mathscr{D}''$ above the minor premiss A of $\to$E (such a cut would have to occur to the right of $A \to B$, contrary to our assumption). We leave it to the reader to verify the other cases. $\square$

REMARK. The proof works equally well for the more liberal notion of cut segment in 1.10, remark.

One might also consider contractions of I-rule applications followed by E-rule applications, e.g.

$$\dfrac{\dfrac{A \wedge B}{A} \qquad \dfrac{A \wedge B}{B}}{A \wedge B} \quad \text{contr} \quad A \wedge B.$$

Such contractions are not covered by the preceding considerations.

3. The structure of normal derivations of N-IQCE

3.1. Suppose $\pi \equiv A_0, \ldots, A_n$ is part of a branch in a deduction $\mathscr{D}$, i.e. for all $i < n$, A_{i+1} occurs immediately below A_i, and no A_j is minor premiss of an E-rule. Then, for all $i < n$, either A_{i+1} is subformula of A_i or A_i is subformula of A_{i+1} ("subformula property"), unless A_i is major premiss of an $\vee$E- or $\exists$E-application, in which case the subformula property may fail to hold.

The more refined notion of a track has been devised so as to retain also the subformula property when passing through the major premiss of an $\vee$E- or $\exists$E-application. In a track, when arriving at an A_i which is major premiss of an $\vee$E- or $\exists$E-application α we take for A_{i+1} a hypothesis discharged by α. More formally we have the following definition.

DEFINITION. A *track* of a derivation $\mathscr{D}$ is a sequence of f.o.'s $A_0, \ldots, A_n$ such that:

(i) A_0 is a top f.o. in $\mathscr{D}$ not discharged by an application of $\vee$E or $\exists$E;

(ii) A_i for $i < n$ is not a minor premiss of $\to$E or $\forall$E and either:

(a) A_i is not a major premiss of $\vee$E or $\exists$E and A_{i+1} is directly below A_i, or

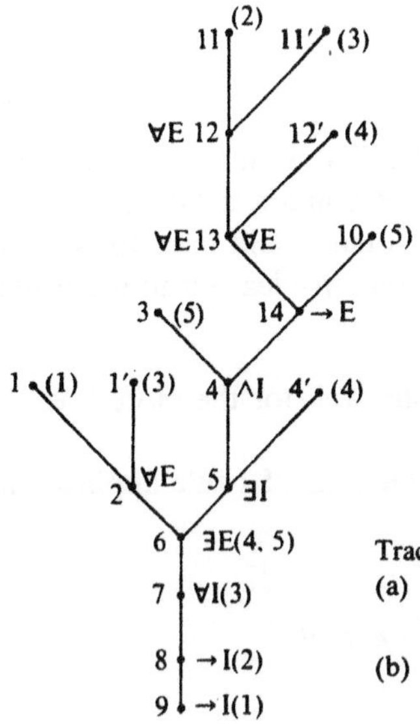

Tracks:
(a) 1–9; 11–14, 4–9;
 1, 2, 4′, 5–9;
(b) 1, 2, 10; 11′; 1′;
 1, 2, 12′.

Fig. 10.1.

(b) A_i is the major premiss of an $\vee$E- or $\exists$E-application α, and A_{i+1} is an assumption discharged in $\mathcal{D}$ by α;

(iii) A_n is either:

(a) minor premiss of $\to$ E or $\forall$E, *or*

(b) the conclusion of $\mathcal{D}$, *or*

(c) major premiss of an $\vee$E- or $\exists$E-application α in case no assumptions are discharged by α. □

N.B. For N-**IQC** one can drop "$\forall$E" under (ii) and (iii).

3.2. Example. Consider the following derivation:

$$\dfrac{\dfrac{\dfrac{\dfrac{{}^{(1)}\forall x\exists y Pxy \quad Eu^{(3)}}{\exists y Puy}\,\forall E \qquad \dfrac{\wedge I\;\dfrac{Puv^{(5)} \qquad \dfrac{\dfrac{\dfrac{{}^{(2)}\forall xy(Pxy\to Pyx)\quad Eu^{(3)}}{\forall y(Puy\to Pyu)}\,\forall E \quad Ev^{(4)}}{Puv\to Pvu}\,\forall E \qquad Puv^{(5)}}{Pvu}\,\to E}{Puv\wedge Pvu} \qquad Ev^{(4)}}{\exists y(Puy\wedge Pyu)}\,\exists I}{\exists y(Puy\wedge Pyu)}\,(4,5)\,\exists E}{\forall x\exists y(Pxy\to Pyx)}\,(3)\,\forall I}{\forall xy(Pxy\to Pyx)\to\forall x\exists y(Pxy\wedge Pyx)}\,(2)\to I$$

$$\dfrac{\forall xy(Pxy\to Pyx)\to\forall x\exists y(Pxy\wedge Pyx)}{\forall x\exists yPxy\to[\forall xy(Pxy\to Pyx)\to\forall x\exists y(Pxy\wedge Pyx)]}\,(1)\to I$$

with its tree representation shown in fig. 10.1. Dropping the accented f.o.'s $(1', 4', 11', 12')$ gives a derivation in N-**IQC**.

Each track can be viewed as a sequence of segments $\sigma_1, \ldots, \sigma_n$.

3.3. Proposition. *Let $\mathcal{D}$ be a normal derivation in* **IQCE**, *and let $\pi \equiv \sigma_0, \ldots, \sigma_n$ be a track in $\mathcal{D}$. Then there is a segment σ_i in π, the* minimum segment *of the track, which separates two (possibly empty) parts of π, called the* E-part *(elimination part) and the* I-part *(introduction part) of π such that*:

(i) *for each σ_j in the* E-*part one has $j < i$, σ_j is the major premiss of an* E-*rule, and the formula of σ_{j+1} is a subformula of the formula of σ_j*;

(ii) *σ_i for $i \neq n$ is a premiss of an* I-*rule or premiss of $\bot_i$ (and then of the form $\bot$)*;

(iii) *for each σ_j in the* I-*part one has $i < j$, and if $j \neq n$, then σ_j is a premiss of an* I-*rule and its formula is a subformula of the formula of σ_{j+1}.*

PROOF. All applications of E-rules in a track of $\mathcal{D}$ have to precede all applications of I-rules, for if not, a maximal segment would occur in the track. σ_i is the first segment in the track which is not premiss of an E-rule anymore. (i)–(iii) follow immediately. □

3.4. DEFINITION. A *track of order* 0, or a *main track*, in a dsp-normal derivation $\mathcal{D}$ is a track ending in the conclusion of $\mathcal{D}$. A *track of order* $n+1$ is a track ending in the minor premiss of an $\to$E- or $\forall$E-application of which the major premiss belongs to a track of order n. □

EXAMPLE. In the proof tree above the tracks under (a) are of order 0, the tracks under (b) of order 1. □

3.5. PROPOSITION. *In a* dsp-*normal derivation each f.o. belongs to some track.*

PROOF. By induction on the height of normal deductions. For example, suppose $\mathcal{D}$ ends with an $\vee$E-application:

$$\mathcal{D} \equiv \left\{ \dfrac{\begin{array}{ccc} & {}^{(n)}[A_1] & [A_2]^{(n)} \\ \mathcal{D}_1 & \mathcal{D}_2 & \mathcal{D}_3 \\ A_1 \vee A_2 & C & C \end{array}}{C}\,(n) \right.$$

The tracks of $\mathcal{D}$ are:

(a) tracks of order >0 in $\mathcal{D}_1$, or tracks of order >0 in $\mathcal{D}_{i+1}$ not beginning in $[A_i]$, or

(b) of the form π_1, π_2 with π_1 a track of order 0 in $\mathcal{D}_1$ and π_2 a track of order >0 in $\mathcal{D}_{i+1}$ beginning in $[A_i]$, or

(c) of the form π_1, π_2, C, where π_1 is a track of order 0 in $\mathcal{D}_1$, π_2 a track of order 0 in $\mathcal{D}_{i+1}$ beginning in $[A_i]$, or

(d) of the form π_i, C, where π_i is a track of order 0 in $\mathcal{D}_{i+1}$ not beginning in $[A_i]$.

Now apply the induction hypothesis for the $\mathcal{D}_i$. □

REMARK. The proposition also holds for dp-normal derivations, if we appropriately adapt the notion of a track (exercise).

Propositions 3.3 and 3.5 together yield the following important theorem.

3.6. THEOREM (*subformula property*). *Let $\mathscr{D}$ be a normal deduction in* **IQCE** *for $\Gamma \vdash A$. Then each formula in $\mathscr{D}$ is a subformula of a formula in $\Gamma \cup \{A\}$. (Et counts as subformula of any quantified formula.)* □

N.B. The theorem also holds for dp-normal derivations, cf. the remark above.

Certain applications of the normalization theorem can be obtained by looking not at tracks, but at the simpler notion of a main branch.

3.7. DEFINITION. A *main branch* (or *spine*) of a normal derivation $\mathscr{D}$ is a branch $A_0, \ldots, A_n$ which does not pass through the minor premiss of any E-rule application, ends in the conclusion of $\mathscr{D}$, and begins in a top node of $\mathscr{D}$. □

3.8. PROPOSITION. *Let $A_0, \ldots, A_n$ be a main branch in a* d-*normal derivation $\mathscr{D}$. Then there is an A_i such that:*
(i) *A_j for $j < i$ is a major premiss of an* E-*rule application*;
(ii) *A_i, if $i \neq n$ is a premiss of an* I-*rule or of $\perp_{\mathrm{i}}$*;
(iii) *A_j, for $i < j < n$ is a premiss of an* I-*rule and a subformula of A_{j+1}.*

PROOF. By the observation that no I-rule application can precede any E-rule application in the main branch of a normal proof. □

REMARK. For d-normal derivations we do not have the subformula property. Several applications of normalization however depend on 3.8 alone, and 3.8 can be proved much more easily than 3.3. The fact that for each derivation there exists a d-normal derivation can be derived from strong normalization for terms of $\mathbf{HA}^{\omega}$ (see section 8).

4. The decidability of IPC

4.1. *Reformulation of* **IPC** *with the CDC.* Preparatory to our proof of the decidability of **IPC** we give a reformulation of the formalism.

For any proof tree in **IPC** we can add at each node labelled with a f.o. A the finite set Γ of assumptions open above the node where A occurs (i.e. the assumptions of the subdeduction ending in A); we indicate this by writing at the node $\Gamma \vdash' A$. The applications of the various rules now take the form

$$\frac{\Gamma_1 \vdash' A_1 \quad \Gamma_2 \vdash' A_2}{\Gamma_1 \cup \Gamma_2 \vdash' A_1 \wedge A_2} \qquad \frac{\Gamma \vdash' A_1 \wedge A_2}{\Gamma \vdash' A_i} \quad (i \in \{1,2\})$$

$$\frac{\Gamma \vdash' A_i}{\Gamma \vdash' A_1 \vee A_2} \quad (i \in \{1,2\}) \qquad \frac{\Gamma_0 \vdash' A \vee B \qquad \Gamma_1 \vdash' C \qquad \Gamma_2 \vdash' C}{\Gamma_0 \cup (\Gamma_1 \setminus \{A\}) \cup (\Gamma_2 \setminus \{B\}) \vdash' C}$$

$$\frac{\Gamma \vdash' B}{\Gamma \setminus \{A\} \vdash A \to B} \qquad \frac{\Gamma_1 \vdash' A \to B \quad \Gamma_2 \vdash' A}{\Gamma_1 \cup \Gamma_2 \vdash' B}$$

$$\frac{\Gamma \vdash' \bot}{\Gamma \vdash' A}$$

At the top nodes of the deduction tree rewritten in this style, we find $A \vdash' A$. The meaning of $\Gamma \vdash' A$ is "there is a deduction of A depending on assumptions listed in Γ".

One easily convinces oneself that $\Gamma \vdash A$, "A has a deduction with open assumptions from Γ" (that is to say, not all formulas of Γ actually have to occur as open assumptions in the deduction) may correspondingly be specified by the following rules and axioms:

Axioms:

$\Gamma \cup \{A\} \vdash A$;

Rules:

$$\frac{\Gamma \vdash A_1 \quad \Gamma \vdash A_2}{\Gamma \vdash A_1 \wedge A_2} \qquad \frac{\Gamma \vdash A_1 \wedge A_2}{\Gamma \vdash A_i} \quad (i \in \{1,2\})$$

$$\frac{\Gamma \vdash A_i}{\Gamma \vdash A_1 \vee A_2} \quad (i \in \{1,2\}) \qquad \frac{\Gamma \vdash A \vee B \quad \Gamma \cup \{A\} \vdash C \quad \Gamma \cup \{B\} \vdash C}{\Gamma \vdash C}$$

$$\frac{\Gamma \cup \{A\} \vdash B}{\Gamma \vdash A \to B} \qquad \frac{\Gamma \vdash A \to B \quad \Gamma \vdash A}{\Gamma \vdash B}$$

$$\frac{\Gamma \vdash \bot}{\Gamma \vdash A}$$

4.2. LEMMA. *From a deduction tree $\mathscr{D}'$ for $\Gamma \vdash' A$ we can construct an order-isomorphic deduction tree $\mathscr{D}$ for $\Gamma \vdash A$ such that to each node $\Gamma_i \vdash' A_i$ in $\mathscr{D}'$ there corresponds in $\mathscr{D}$ a node $\Gamma_i^* \vdash A_i$ with $\Gamma_i^* \supset \Gamma_i$, Γ_i^* containing formulas occurring in $\mathscr{D}'$ only.*

In particular, if we have a normal deduction of $\Gamma \vdash' A$, then, by the subformula property, each node $\Gamma_i \vdash' A_i$ of $\mathcal{D}'$ contains only subformulas of $\Gamma \cup \{A\}$; therefore the same is true for the tree $\mathcal{D}$.

PROOF. Left as an exercise. □

4.3. THEOREM. **IPC** *is decidable.*

PROOF. We shall call expressions $\Gamma \vdash A$ *sequents*. If $\Gamma \vdash A$ can be deduced, there is also a normal deduction for $\Gamma \vdash A$ which, by the subformula property, involves sequents $\Gamma_i \vdash A_i$ with $\Gamma_i \cup \{A_i\}$ subformulas of $\Gamma \cup A$ only.

We define a *problem* as a finite set of sequents $\{\Gamma_1 \vdash A_1, \ldots, \Gamma_n \vdash A_n\}$. Let S be any finite set of formulas closed under taking subformulas. An *S-sequent* is a sequent with all its formulas in S; an *S-problem* is a problem consisting of S-sequents.

For any S-sequent $\Gamma \vdash A$ we define the notion of an S-successor set as follows. $\theta \equiv \{\Gamma_1 \vdash A_1, \ldots, \Gamma_i \vdash A_i\}$ $(i \leq 3)$ is an S-successor of $\Gamma \vdash A$ if θ is an S-problem, and

$$\frac{\Gamma_1 \vdash A_1 \quad \cdots \quad \Gamma_i \vdash A_i}{\Gamma \vdash A}$$

is an instance of an **IPC**-rule, or $\theta \equiv \{\Gamma \vdash A\}$ and $\Gamma \vdash A$ is an axiom.

An *S-successor* of a problem θ is obtained by taking an S-successor set for each sequent in θ, and then taking their union.

Observe that, for any given finite S, there are only finitely many S-problems, and that each S-problem has only finitely many S-successors.

We construct a tree τ associated with a given problem $\theta \equiv \theta_{\langle\rangle}$ as follows. Let S be the set of subformulas occurring in θ. If θ_n has already been constructed, then the problems $\theta_{n * \langle i \rangle}$ attached to the nodes $n * \langle i \rangle$ are the S-successors of θ_n. Let $m(S)$ be the upper bound on the number of S-problems. We construct the tree of S-problems up to depth $m(S)$; then each branch terminates in a problem which either consists of axioms or in a problem, not consisting of axioms only, which repeats a problem lower down along the branch.

Clearly, if $\theta_{\langle\rangle} = \{\Gamma \vdash A\}$, then $\Gamma \vdash A$ is derivable in **IPC** iff the problem tree τ contains a branch terminating in a problem consisting of axioms only. □

4.4. *Illustration of the method.* The problem tree for $\{\emptyset \vdash ((P \to Q) \to P) \to P\}$ takes the form, using "PQP" as an abbreviation for $P \to (Q \to P)$:

$$\begin{array}{c} \cdots \\ | \\ \{PQP \vdash PQP;\ PQP, P \vdash P \to Q;\ PQP, P \vdash Q\} \\ | \\ \{PQP \vdash PQP;\ PQP, P \vdash Q\} \\ | \\ \{PQP \vdash PQP;\ PQP \vdash P \to Q\} \\ | \\ \{PQP \vdash P\} \\ | \\ \{\emptyset \vdash PQP \to P\}. \end{array}$$

Continuing the tree upwards leads to repetition of the same problem, hence Peirce's law $PQP \to P$ is not derivable in **IPC**.

5. Other applications of normalization

5.1. An easy corollary of the subformula property is the following theorem.

THEOREM (*separation theorem*). *If* $\Gamma \vdash A$ *in* **IQC**, *then there is a deduction of* A *from* Γ *in* **IQC** *using only the rules for the logical operators actually occurring in* $\Gamma \cup \{A\}$; *here* $\perp$ *is regarded as a zero-place operator.* □

By means of the separation theorem we can settle the relationship between intuitionistic and minimal logic.

5.2. THEOREM. *Let* **N-MQCE**, **N-MQC** *be the systems of minimal logic obtained from* **N-IQCE** *and* **N-IQC**, *respectively, by dropping the rule* $\perp_i$. *Let* P *be a fixed proposition letter, and assume that* P *does not occur in* $\Gamma \cup \{A\}$. *For any* B *define* $B^* := B[\perp/P]$, *i.e.* $\perp$ *is everywhere replaced by* P, *and let* $\Gamma^* := \{B^* : B \in \Gamma\}$. *Let us use* $\vdash_m$ *for derivability in* **N-MQCE**, *and* $\perp_i$ *for derivability in* **N-IQCE**. *Then*

$$\Gamma \vdash_m A \Leftrightarrow \Gamma^* \vdash_i A^*,$$

and similarly for **N-MQC** *and* **N-IQC**.

PROOF. Assume $\Gamma \vdash_m A$, then also $\Gamma^* \vdash_m A^*$, hence $\Gamma^* \vdash_i A^*$, since no rule in **N-MQCE** involves $\perp$ (i.e. $\perp$ behaves as an arbitrary proposition letter).

Conversely, if $\Gamma^* \vdash_{\mathrm{i}} A^*$, then there is a normal proof in **N-IQCE** deriving A^* from Γ^*; such a proof does not involve $\bot_{\mathrm{i}}$ by the separation theorem, hence also $\Gamma^* \vdash_{\mathrm{m}} A^*$ and therefore $\Gamma \vdash_{\mathrm{m}} A$. □

5.3. PROPOSITION (*disjunction under hypotheses*). *Assume that no s.p.p. of a formula of* Γ *has* $\vee$ *or* $\exists$ *as a principal logical operator, and let* $\vdash$ *denote deducibility in* **IQC** *or* **IQCE**. *Then* $\Gamma \vdash A \vee B \Rightarrow \Gamma \vdash A$ *or* $\Gamma \vdash B$.

PROOF. Let $\mathcal{D}$ be a normal deduction of $\Gamma \vdash A \vee B$, and let π be a main branch in $\mathcal{D}$. If the last step in $\mathcal{D}$ is an application of $\vee$I, we are done: deleting this final application from $\mathcal{D}$ yields a proof of $\Gamma \vdash A$ or of $\Gamma \vdash B$.

If the deduction $\mathcal{D}$ is of the form

$$\dfrac{\begin{matrix}\mathcal{D}'\\ \bot\end{matrix}}{A \vee B}\,\bot_{\mathrm{i}},$$

we can take the deduction

$$\dfrac{\begin{matrix}\mathcal{D}'\\ \bot\end{matrix}}{A}\,\bot_{\mathrm{i}}$$

and we are also done. Therefore we may assume that the I-part of π is empty and that π ends with an application of an E-rule. The first formula A_1 of π must belong to Γ, since it cannot be an assumption discharged by $\to$I or $\forall$I, nor can it be an assumption discharged by $\vee$E or $\exists$E since π does not pass through a minor premiss of these rules.

But as long as π does not pass through a major premiss of $\vee$E or $\exists$E, each A_{j+1} in the E-part $A_0, A_1, \ldots, A_i$ of π is an s.p.p. of A_j, hence of A_0; and since A_0 does not contain an s.p.p. with $\vee$ or $\exists$ as a principal operator, π does not contain a major premiss of $\vee$E or $\exists$E, and then the conclusion of $\mathcal{D}$ cannot contain $\vee$ or $\exists$, contradicting our initial assumptions. □

COROLLARY. $\vdash \neg C \to A \vee B \;\Rightarrow\; \vdash \neg C \to A$ *or* $\vdash \neg C \to B$. □

The preceding argument did not use the notion of track and did not depend on normalization with respect to permutative reductions. The following refinement of the preceding theorem however does need normalization also with respect to permutative reductions.

5.4. PROPOSITION (*disjunction under hypotheses, refinement*). *Let Γ be such that no formula of Γ has an s.p.p. with $\vee$ as principal logical operator. Then, in* **IQC**, $\Gamma \vdash B_1 \vee B_2 \Rightarrow \Gamma \vdash B_1$ *or* $\Gamma \vdash B_2$.

PROOF. Let $\mathcal{D}$ be a normal deduction of $\Gamma \vdash B_1 \vee B_2$. There is only a single endsegment; for assume there would be at least two, say σ_1, σ_2, then there is a lowest occurrence α of $\vee$E with major premiss of the form $C \vee D$ such that σ_1 and σ_2 pass through the minor premisses of α. Let π be the track passing through $C \vee D$. Then the first f.o. A_0 in π cannot be discharged since $C \vee D$ occurs in the E-part of π and there is no $\rightarrow$I-application below $C \vee D$, since the f.o.'s below $C \vee D$ all belong to end segments. Therefore $A_0 \in \Gamma$; but this is impossible since it would force $C \vee D$ to be an s.p.p. of A_0 which is excluded. Therefore $\mathcal{D}$ must necessarily have the form

$$\mathcal{D} \equiv \left\{ \dfrac{\begin{matrix}\mathcal{D}_1\\ \exists x_1\, C_1(x_1)\end{matrix} \qquad \dfrac{\begin{matrix}\mathcal{D}_2\\ \exists x_2\, C_2(x_2)\end{matrix} \qquad \begin{matrix} \dfrac{\begin{matrix}\mathcal{D}_n\\ \exists x_n\, C_n(x_n)\end{matrix} \quad \dfrac{\begin{matrix}\mathcal{D}'_n\\ B_i\end{matrix}}{B_1 \vee B_2}}{\vdots} \\ B_1 \vee B_2\end{matrix}}{B_1 \vee B_2}}{B_1 \vee B_2} \right.$$

From this deduction we readily extract a correct derivation of $\Gamma \vdash B_i$:

$$\dfrac{\begin{matrix}\mathcal{D}_1\\ \exists x_1\, C_1(x_1)\end{matrix} \qquad \dfrac{\begin{matrix}\mathcal{D}_2\\ \exists x_2\, C_2(x_2)\end{matrix} \qquad \begin{matrix} \dfrac{\begin{matrix}\mathcal{D}_n\\ \exists x_n\, C_n(x_n)\end{matrix} \quad \begin{matrix}\mathcal{D}'_n\\ B_i\end{matrix}}{\vdots} \\ B_i\end{matrix}}{B_i}}{B_i}$$

If π does end with an application of an I-rule or the $\bot_i$-rule, the desired conclusion is immediate. □

As a counterpart for existential statements we have the following proposition.

5.5. PROPOSITION (*existence under hypotheses in* **IQC**). *Suppose that $\Gamma \cup \{\exists x A\}$ does not contain function symbols and that no s.p.p. of Γ has $\exists$ as a*

principal operator. Let $\{y_1, \ldots, y_n\}$ *be a complete list of the free variables in* $\Gamma \cup \{\exists x A\}$. *Then:*
(i) *for* $n > 0$, *if* $\Gamma \vdash \exists x A$, *then* $\Gamma \vdash A(y_1) \vee \cdots \vee A(y_n)$;
(ii) *for* $n > 0$, *if* Γ *does not have a disjunction as s.p.p., and if* $\Gamma \vdash \exists x A$, *then* $\Gamma \vdash A(y_i)$ *for some* i, $1 \le i \le n$;
(iii) *if* $n = 0$, *then* $\Gamma \vdash \exists x A \Rightarrow \Gamma \vdash \forall x A$.

PROOF. Cf. E10.5.1. □

As a final application we discuss the relation between E-logic and ordinary logic.

5.6. THEOREM. *Let* **IQCE*** *be* E-*logic with function symbols and* $\simeq$ *as a primitive* (*cf.* 2.2.7) *and with strictness for function symbols and predicate letters. Let* **IQC*** *be ordinary predicate logic with* $=$ *and a distinguished unary predicate letter E in the language, with strictness axioms for E. Let* ^{E} *be the mapping as defined in* 2.2.8, *i.e.*

$$(Rt_0 \ldots t_{n-1})^E := Rt_0 \ldots t_{n-1}; \qquad \bot^E := \bot;$$

$$(t_0 \simeq t_1)^E := Et_0 \wedge Et_1 \to t_0 = t_1;$$

$$(A \circ B)^E := A^E \circ B^E \quad \text{for } \circ \in \{\wedge, \vee, \to\};$$

$$(\forall x A)^E := \forall x[Ex \to A^E];$$

$$(\exists x A)^E := \exists x[Ex \wedge A^E].$$

Then

$$\mathbf{IQCE^*} \vdash A \Leftrightarrow \mathbf{IQC^*} \vdash A^E.$$

PROOF. We recall the axioms for $\simeq$ and E in **IQCE***: $t \simeq t$; $s \simeq t \wedge A(s) \to A(t)$ (A prime); $(Es \vee Et \to s \simeq t) \to s \simeq t$; $E(ft_1 \ldots t_n) \to Et_1 \wedge \cdots \wedge Et_n$ (f a function symbol); $Rt_1 \ldots t_n \to Et_1 \wedge \cdots \wedge Et_n$ (R a relation symbol).

For **IQC*** we assume $t = t$; $s = t \wedge A(s) \to A(t)$ (A atomic), and strictness with respect to E, as for **IQCE***. Let Form(S) be the set of formulas of a system **S**.

Let us call a formula of **IQC*** *bounded* if all quantified subformulas of A are of the form $\forall x(Ex \to B)$ or $\exists x(Ex \wedge B)$. Let $\mathscr{BF}$ be the set of bounded formulas. We define a mapping * from Form(**IQC***) $\cap\, \mathscr{BF}$ to Form(**IQCE***) by the following inductive clauses (P an arbitrary predicate variable):

$$(Et)^* := Et, \qquad (s = t)^* := s \simeq t,$$

$$(Pt_1 \ldots t_n)^* := Pt_1 \ldots t_n, \qquad \bot^* := \bot\,;$$

$$(\forall x(Ex \to A))^* := \forall x A^*;$$

$$(\exists x(Ex \wedge A))^* := \exists x A^*;$$

$$(A \circ B)^* := A^* \circ B^* \quad \text{for } \circ \in \{\wedge, \vee, \to\}.$$

As to the proof of the theorem, the direction from left to right is straightforward, by induction on the length of deductions; for the converse it suffices to show for all Form(**IQC***) $\cap\, \mathscr{BF}$ that

$$\mathbf{IQC}^* \vdash A \Rightarrow \mathbf{IQCE}^* \vdash A^*, \tag{1}$$

since it is easy to show that

$$\mathbf{IQCE}^* \vdash A \leftrightarrow (A^E)^*.$$

(1) can be proved by induction over the length of normal proofs of A from the "nonlogical axioms" for **IQCE*** listed above; we leave the details as an exercise. □

6. Conservative addition of predicative classes

6.1. In this section we provide a proof-theoretic argument for the conservativity of the addition of predicative classes to intuitionistic first-order theories.

Let **T** be a first-order theory (formalized on the basis of **IQCE**, axiomatized by means of some (individual) first-order axioms and a set of axiom schemata $\mathscr{F}_i$.

An axiom schema $\mathscr{F}_i$ is a formula in the language of **T** extended with predicate variables, say

$$\mathscr{F}_i\left(P_1^{n(i,1)}, \ldots, P_{r(i)}^{n(i,r(i))}\right),$$

where P_j is $n(i,j)$-ary. Each substitution of formulas

$$\lambda x_1, \ldots, x_{n(i,j)}.A_j(x_1, \ldots, x_{n(i,j)}) \quad (1 \leq j \leq r(i))$$

for the P_j yields an axiom of **T**.

6.2. DEFINITION. The *weak second-order extension* $\mathbf{T}^*$ (extension by predicative classes) of **T** is defined as follows. We add to the language of **T** relation variables $X^n, Y^n, Z^n, \ldots$, for n-ary relations, for each n, and the corresponding quantifiers $\forall X^n, \exists X^n$. If $t_1, \ldots, t_n$ are individual terms, then $X^n t_1 \ldots t_n$ is a prime formula.

We add the *elementary comprehension schema*

$$\text{ECA} \qquad \exists X^n \forall x_1 \ldots x_n [X^n(x_1, \ldots, x_n) \leftrightarrow A(x_1, \ldots, x_n)]$$

for each A of $\mathbf{T}^*$ not containing bound relation variables. The axiom schemas $\mathscr{F}_i$ of **T** are replaced by corresponding axioms

$$\forall X_1^{n(i,1)} \ldots \forall X_{r(i)}^{n(i,r(i))} \mathscr{F}_i(X_1, \ldots, X_{r(i)}).$$

Then we have the following theorem.

6.3. THEOREM. $\mathbf{T}^*$ *is a conservative extension of* **T**.

PROOF. We make use of a slight extension of the normalization theorem for N-**IQCE**. The complexity of a formula is now to be measured as $\omega \cdot n + m$, where n is the number of relation quantifiers occurring in A, and m is the number of logical symbols. The axiom schema ECA can be dispensed with at the cost of introducing second-order quantifier rules:

$$\forall\text{I}_2 \frac{A(Y^n)}{\forall X^n A(X^n)} \qquad \forall\text{E}_2 \frac{\forall X^n A(X^n)}{A(\lambda x_1 \ldots x_n.B)}$$

$$\exists\text{I}_2 \frac{A(\lambda x_1 \ldots x_n.B)}{\exists X^n A(X^n)} \qquad \exists\text{E}_2 \frac{\exists X^n A(X^n) \quad \begin{matrix}[A(Y^n)]^{(1)} \\ \vdots \\ C\end{matrix}}{C} (1)\,.$$

Here $A(\lambda x_1 \ldots x_n.B)$ is the formula obtained from $A(X^n)$ by replacing each prime formula of the form $X^n(t_1, t_2, \ldots, t_n)$ by $B[x_1, x_2, \ldots, x_n/t_1, t_2, \ldots, t_n]$.

In $\forall I_2$ Y^n does not occur in assumptions on which the f.o. $A(Y^n)$ depends; in $\exists E_2$ Y^n does not occur free in assumptions on which C depends outside the assumption class $[A(Y^n)]$, nor does Y^n occur free in C. B does not contain bound predicate variables.

As before we can assign to a derivation an induction value, now measured as $\omega \cdot c + d$, where c is the maximal complexity of cut formulas and d is the total number of f.o.'s in cuts of complexity c. Thus the induction value is now an ordinal $< \omega^3$.

The notion of subformula is extended by stipulating that $A(\lambda x_1 \ldots x_n.B)$ is a subformula of $\exists X^n A(X^n), \forall X^n A(X^n)$ for any B not containing bound relation variables. With this notion of subformula we prove normalization and give an analysis of the structure of tracks as before, and obtain the subformula property.

Now let A be a formula in the language of **T** such that $\mathbf{T}^* \vdash A$ and let $\mathcal{D}$ be a normal derivation in $\mathbf{T}^*$ with conclusion A. By the subformula property all formulas in $\mathcal{D}$ are either subformulas of A, or of axioms of $\mathbf{T}^*$. The second-order axioms can only occur at top nodes and they cannot appear as subformula of other f.o.'s in $\mathcal{D}$; therefore they occur as first formula of a track followed by $\forall E_2$-applications, until a first-order formula has been reached; this first-order formula is then an instance of an axiom schema in **T**. □

Corollary. $\mathbf{HAS}_0$ (*as described in* 3.8.11) *is conservative over* **HA**.

Proof. Take in the preceding theorem **HA** for **T**, and $\mathbf{HAS}_0$ for $\mathbf{T}^*$. □

7. Sequent calculi

Gentzen not only devised natural deduction calculi in tree form, but also so-called sequent calculi, which have the following convenient feature: whereas in natural deduction one sees at each step only locally the premisses and the conclusion, and has to inspect the whole deduction up to this particular step in order to ascertain the open assumptions, in a sequent calculus the assumptions are shown at each line of the deduction tree. There are many forms of sequent calculi; we already used a sequent version of natural deduction in section 4. In the sequent calculi introduced by Gentzen there is also a more fundamental difference with the natural deduction calculus: the role of introduction and elimination rules is taken over by "left" and "right" rules for the various logical operators (corresponding to introduction (better "treatment") of a logical operator in the antecedent and the consequent of a sequent, respectively).

There is also a special structural rule, the *cut rule*. Normalization for natural deduction corresponds to elimination of uses of the cut rule for Gentzen-type sequent calculi. Below, "sequent calculus" will always mean a Gentzen-style sequent calculus. Kleene (1952, sections 77–80) contains a systematic treatment of sequent calculi.

7.1. *A variant of Kleene's calculus* G3. We axiomatize "A is deducible from Γ" now in a different style, with left-introduction and right-introduction rules as follows. In $\Gamma \vdash A$, Γ is a finite set of formulas.

Axioms:

$$\Gamma \vdash A \quad \text{for } A \in \Gamma.$$

Rules:

L$\wedge$ $\dfrac{\Gamma, A \wedge B, A, B \vdash C}{\Gamma, A \wedge B \vdash C}$ R$\wedge$ $\dfrac{\Gamma \vdash A \quad \Gamma \vdash B}{\Gamma \vdash A \wedge B}$

L$\vee$ $\dfrac{\Gamma, A \vee B, A \vdash C \quad \Gamma, A \vee B, B \vdash C}{\Gamma, A \vee B \vdash C}$ R$\vee$ $\dfrac{\Gamma \vdash A_i}{\Gamma \vdash A_1 \vee A_2}$ $(i \in \{1,2\})$

L$\to$ $\dfrac{\Gamma, A \to B \vdash A \quad \Gamma, A \to B, B \vdash C}{\Gamma, A \to B \vdash C}$ R$\to$ $\dfrac{\Gamma, A \vdash B}{\Gamma \vdash A \to B}$

L$\exists$ $\dfrac{\Gamma, \exists x Ax, Ay \vdash C}{\Gamma, \exists x Ax \vdash C}$ R$\exists$ $\dfrac{\Gamma \vdash At}{\Gamma \vdash \exists x Ax}$

(y not free in $\Gamma \cup \{C\}$)

L$\forall$ $\dfrac{\Gamma, \forall x Ax, At \vdash C}{\Gamma, \forall x Ax \vdash C}$ R$\forall$ $\dfrac{\Gamma \vdash Ay}{\Gamma \vdash \forall x Ax}$

$\perp$-rule $\dfrac{\Gamma \vdash \perp}{\Gamma \vdash A}$. ($y$ not free in Γ)

REMARK. If we interpret Γ as a finite set with multiplicity (i.e. repetitions are permitted, or Γ is a finite sequence modulo the order), we obtain an equivalent system; the verification is left to the reader. The system as given above is virtually identical with the system G3 as given by Kleene (1952, section 80).

7.2. Let us write $\vdash_{\mathrm{K}}$ for deducibility in the system of 7.1, and let $\vdash_{\mathrm{N}}$ be deducibility in the natural deduction system for **IQC**. Then we have the following theorem.

THEOREM. $\Gamma \vdash_K A \Leftrightarrow \Gamma \vdash_N A$.

PROOF. $\Rightarrow$ is proved by induction on the length of deductions in the system of 7.1; this is left as a exercise.

For the converse we apply induction on the length of normal derivations; for each normal derivation of $\Gamma \vdash_N A$ we construct a derivation in the sequent calculus where it is now convenient to take the variant of the sequent calculus (7.1, remark) where the Γ are finite sets with repetitions (the repetitions enter as a result of multiple occurrences of the same open assumption in a natural deduction).

Case 1. Let $\mathcal{D}$ be a normal derivation for $\Gamma \vdash_N A$, and suppose that the final step in $\mathcal{D}$ is an I-rule application. Let $\mathcal{E}$ be the deduction(s) in the sequent calculus corresponding (by induction hypothesis) to the immediate subdeduction(s) of $\mathcal{D}$; apply to $\mathcal{E}$ the corresponding R-rule. For example if $\mathcal{D}$ ends with

$$\dfrac{\begin{matrix}[A]\\ \mathcal{D}'\\ B\end{matrix}}{A \to B},$$

we have by the induction hypothesis a deduction for $\Gamma, A \vdash_K B$, hence by $R\!\to$ we have $\Gamma \vdash_K A \to B$.

Case 2. Let the final step in $\mathcal{D}$ be an application of $\bot_i$; apply the corresponding $\bot$-rule to the derivation given by the induction hypothesis.

Case 3. The conclusion of $\mathcal{D}$ is the consequence of an E-rule application. Let τ be a main branch of $\mathcal{D}$. τ is unique, since the I-part of $\mathcal{D}$ is empty (multiple main branches can only occur as a result of $\wedge$I). τ does not contain a minor premiss, hence no assumption can be discharged along τ.

Thus the first f.o. C of τ belongs to Γ and is major premiss of an E-rule. Suppose e.g. $C \equiv C_1 \to C_2$, then $\mathcal{D}$ has the form

$$\begin{matrix}\dfrac{C_1 \to C_2 \quad \begin{matrix}\mathcal{D}'\\ C_1\end{matrix}}{(C_2)}\\ \mathcal{D}''\\ A\end{matrix}$$

where (C_2) refers to a single occurrence in $\mathcal{D}''$. The f.o. C_1 cannot depend on other assumptions besides the ones on which A depends, since no assumptions are discharged in τ, which passes through the f.o. C_2.

Thus, if $\mathscr{D}$ establishes $\Gamma, C_1 \to C_2 \vdash A$, then $\mathscr{D}'$ establishes $\Gamma \vdash_{\mathrm{N}} C_1$, and $\mathscr{D}''$ shows $\Gamma, C_2 \vdash_{\mathrm{N}} A$. By the induction hypothesis

$$\Gamma \vdash_{\mathrm{K}} C_1, \qquad \Gamma, C_2 \vdash_{\mathrm{K}} A,$$

and therefore by $\to$L, noting that $C_1 \to C_2 \in \Gamma$:

$$\Gamma \vdash_{\mathrm{K}} A.$$

To consider yet another subcase, suppose now that $C \equiv C_1 \vee C_2$, then $\mathscr{D}$ has the form

$$\begin{array}{c} \dfrac{C_1 \vee C_2 \quad \begin{array}{c}[C_1]\\ \mathscr{D}_1\\ B\end{array} \quad \begin{array}{c}[C_2]\\ \mathscr{D}_2\\ B\end{array}}{(B)} \\ \mathscr{D}' \\ A \end{array}$$

Then

$$\begin{array}{c}[C_1]\\ \mathscr{D}_1\\ (B)\\ \mathscr{D}'\\ A\end{array} \quad \text{and} \quad \begin{array}{c}[C_2]\\ \mathscr{D}_2\\ (B)\\ \mathscr{D}'\\ A\end{array}$$

are correct derivations with fewer formulas and therefore by the induction hypothesis

$$\Gamma, C_1 \vdash_{\mathrm{K}} A, \qquad \Gamma, C_2 \vdash_{\mathrm{K}} A.$$

with $C_1 \vee C_2 \in \Gamma$. Then the $\vee$L-rule gives us a deduction of $\Gamma \vdash_{\mathrm{K}} A$. The other cases are left to the reader.

REMARK. The construction of a natural deduction tree from a proof of $\Gamma \vdash_{\mathrm{K}} A$ may actually be carried out so as to obtain a *normal* derivation tree. We leave this as an exercise.

7.3. COROLLARY. *Kleene's system is closed under the cut rule* (“*closed under cut*”), *that is to say, the following rule holds as a derived rule*:

$$\Gamma \vdash_{\mathrm{K}} A \quad \textit{and} \quad \Gamma, A \vdash_{\mathrm{K}} B \Rightarrow \Gamma \vdash_{\mathrm{K}} B.$$

In other words, adding a rule

$$\text{Cut} \qquad \frac{\Gamma \vdash A \quad \Gamma, A \vdash B}{\Gamma \vdash B}$$

does not increase the number of derivable sequents.

PROOF. Suppose $\Gamma \vdash_K A$ and $\Gamma, A \vdash_K B$, then also $\Gamma \vdash_N A$ and $\Gamma, A \vdash_N B$ and thus by an application of $\rightarrow$ E in the natural deduction system

$$\frac{\Gamma \vdash A \quad \dfrac{\Gamma, A \vdash B}{\Gamma \vdash A \rightarrow B}}{\Gamma \vdash B}$$

so there is a derivation for $\Gamma \vdash_K B$. □

7.4. *The Beth-tableau system.* We describe yet another type of sequent calculus, originally devised so as to permit a very direct model construction in a Gödel-type completeness proof for Beth and Kripke semantics. Let Γ, Δ be finite sets of formulas.

Axioms:

$\Gamma \vdash \Delta$ *with* $\Gamma \cap \Delta$ *inhabited or* $\bot \in \Gamma$.

Rules:

$$\text{L}\wedge \quad \frac{\Gamma, A \wedge B, A, B \vdash \Delta}{\Gamma, A \wedge B \vdash \Delta} \qquad \text{R}\wedge \quad \frac{\Gamma \vdash A, \Delta \quad \Gamma \vdash B, \Delta}{\Gamma \vdash A \wedge B, \Delta}$$

$$\text{L}\vee \quad \frac{\Gamma, A \vee B, A \vdash \Delta \quad \Gamma, A \vee B, B \vdash \Delta}{\Gamma, A \vee B \vdash \Delta} \qquad \text{R}\vee \quad \frac{\Gamma \vdash A, B, \Delta}{\Gamma \vdash A \vee B, \Delta}$$

$$\text{L}\rightarrow \quad \frac{\Gamma, A \rightarrow B \vdash A, \Delta \quad \Gamma, A \rightarrow B, B \vdash \Delta}{\Gamma, A \rightarrow B \vdash A} \qquad \text{R}\rightarrow \quad \frac{\Gamma, A \vdash B}{\Gamma \vdash A \rightarrow B, \Delta}$$

$$\text{L}\exists \quad \frac{\Gamma, \exists x A(x), A(y) \vdash \Delta}{\Gamma, \exists x A(x) \vdash \Delta} \qquad \text{R}\exists \quad \frac{\Gamma \vdash A(t), \Delta}{\Gamma \vdash \exists x A(x), \Delta}$$

(y not free in $\Gamma \cup \Delta$)

$$\text{L}\forall \quad \frac{\Gamma, \forall x A(x), A(t) \vdash \Delta}{\Gamma, \forall x A(x) \vdash \Delta} \qquad \text{R}\forall \quad \frac{\Gamma \vdash A(y)}{\Gamma \vdash \forall x A(x), \Delta}$$

(y not free in $\Gamma \cup \Delta$).

Intuitively, one may interpret $A_1, \ldots, A_n \vdash B_1, \ldots, B_m$ as "$A_1 \wedge \cdots \wedge A_n \rightarrow B_1 \vee \cdots \vee B_m$ is intuitionistically true". Observe that:

(i) going from conclusion to premiss the left-hand side of the sequent (the *antecedent* of the sequent) never decreases;
(ii) going from premiss to conclusion, the number of formulas on the right-hand side of the sequent (the *succedent* of the sequent) is only increased (possibly) at an application of R $\to$ or R$\forall$.

Let $\vdash_B$ be used for deducibility in the sequent calculus just described; we wish to show $\Gamma \vdash_K A \Leftrightarrow \Gamma \vdash_B A$, in fact we shall establish the following.

7.5. Theorem. *Let* $\bigvee\Delta$ *indicate the disjunction of the formulas in* Δ, *then*:
(i) $\Gamma \vdash_K A \Rightarrow \Gamma \vdash_B A$;
(ii) $\Gamma \vdash_B \Delta \Rightarrow \Gamma \vdash_K \bigvee\Delta$.

Proof. (i) is proved by induction on the length of deductions. We need an easy *thinning lemma*, the proof of which is left as an exercise: a derivation showing $\Gamma \vdash_B \Delta$ can be transformed into a deduction of $\Gamma, \Delta' \vdash_B \Delta, \Delta''$.

We consider some typical cases of the induction step.

Case 1. If $\Gamma \vdash A$ is an axiom in Kleene's system, then $\Gamma \vdash A$ is also an axiom in the Beth system.

Case 2. The final step in the derivation of $\Gamma \vdash_K A$ is L$\wedge$, R$\wedge$, L$\vee$, R$\to$, L$\exists$, R$\exists$, L$\forall$, R$\forall$; then we can use the corresponding rules of the Beth system applied to the deductions existing by the induction hypothesis.

Case 3. The final step in the deduction for $\Gamma \vdash_K A$ is R$\vee$, i.e.

$$\frac{\Gamma \vdash B_i}{\Gamma \vdash B_1 \vee B_2}$$

with $B_1 \vee B_2 \equiv A$. By induction hypothesis $\Gamma \vdash_B B_i$ and by the thinning lemma $\Gamma \vdash_B B_1, B_2$; now apply R$\vee$ in the Beth system.

Case 4. Suppose the final step in the derivation was an application of L$\to$:

$$\frac{\Gamma, A \to B \vdash A \quad \Gamma, A \to B, B \vdash C}{\Gamma, A \to B \vdash C}$$

By the induction hypothesis there are deductions for $\Gamma, A \to B \vdash_B A$ and

$\Gamma, A \to B, B \vdash_{\mathrm{B}} C$; applying thinning and L$\to$ for the Beth system we find

$$\frac{\Gamma, A \to B \vdash A, C \quad \Gamma, A \to B, B \vdash C}{\Gamma, A \to B \vdash C}$$

Case 5. Suppose the final step in the deduction in the Kleene system to be an application of the $\bot$-rule with conclusion $\Gamma \vdash A$. By the induction hypothesis there is a derivation of $\Gamma \vdash \bot$ in the Beth system. Starting at the bottom of this derivation, we systematically replace the occurrences of $\bot$ on the right-hand side by A. By inspection of the rules we see that all inferences remain correct until we arrive at an axiom $\Gamma \vdash \bot, \Delta$, which holds either because $\bot \in \Gamma$ or because $\Gamma \cap \Delta$ is inhabited. Then $\Gamma \vdash A, \Delta$ is again an axiom. Thus we have found a proof showing $\Gamma \vdash_{\mathrm{B}} A$.

(ii) The easiest way is to show by induction on the length of deductions

$$\Gamma \vdash_{\mathrm{B}} \Delta \Rightarrow \Gamma \vdash_{\mathrm{N}} \bigvee \Delta$$

and then appeal to our earlier result that $\Gamma \vdash_{\mathrm{N}} C \Rightarrow \Gamma \vdash_{\mathrm{K}} C$. A direct proof is also possible, see E10.7.4. □

8. N-IQC as a calculus of terms

8.1. Hitherto we have been tacitly using the CDC as a blanket assumption. The particular normalization procedure and its applications we described in sections 2–6 did not depend on the CDC however, and neither did our discussion of the relationship between sequent calculi and natural deduction in section 7.

In this section we shall present natural deduction as a system of terms denoting proofs, which brings out a parallel, in fact an isomorphism, between deductions in **N-IQC** and a suitable typed lambda calculus, provided we do not stick to the CDC.

8.2. *Terms describing natural deduction proof trees*

(1) For every formula A of the predicate calculus we assume a countably infinite supply of variables $v_0^A, v_1^A, v_2^A, \ldots$ of variables of type A to be available. For distinct A the corresponding sets of variables are disjoint, and the set of typed variables is disjoint from the set of individual variables. We use x^A, y^A, z^A for arbitrary variables of type A; we shall drop the type superscript when clear from the context.

(2) Now we associate terms to natural deductions without the crude discharge convention. Assumptions are assumed to be divided into classes, each class consisting of f.o.'s of the same form, and such that all f.o.'s of a

class are discharged simultaneously. The only rule-applications which permit discharge of two assumption classes are instances of $\vee$E. The term is defined by recursion on the depth of the proof tree. Thus in a proof tree each node (f.o.) gets assigned a term as label, the term assigned to the subdeduction corresponding to the node. If t is a label attached to a f.o. A we write $t \in A$.

To a proof tree of depth one, i.e an assumption of the form A say, we associate a variable of type A as a label. (One may think of such a variable as standing for a hypothetical deduction of A.) Thus

$$x^A \in A.$$

To assumptions in the same assumption class we assign the same variable.

(3) For the recursion step we must indicate how the term assigned to the conclusion is constructed from the term label(s) of the premises. This is shown below:

$\wedge$I $\quad \dfrac{t_0 \in A_0 \quad t_1 \in A_1}{p(t_0, t_1) \in A_0 \wedge A_1}$ $\qquad$ $\wedge$E $\quad \dfrac{t \in A_0 \wedge A_1}{p_i t \in A_i} \quad (i \in \{0,1\})$

$\vee$I $\quad \dfrac{t \in A_i}{k_i t \in A_0 \vee A_1} \quad (i \in \{0,1\})$ $\qquad$ $\vee$E $\quad \dfrac{t \in A \vee B \quad t_0[x^A] \in C \quad t_1[y^B] \in C}{D_{u,v}(t, t_0[u], t_1[v]) \in C}$

$\to$I $\quad \dfrac{t[x^A] \in B}{\lambda y^A . t[y^A] \in A \to B}$ $\qquad$ $\to$E $\quad \dfrac{t \in A \to B \quad t' \in A}{tt' \in B}$

$\forall$I $\quad \dfrac{t[x] \in A(x)}{\lambda y . t[y] \in \forall y A(y)}$ $\qquad$ $\forall$E $\quad \dfrac{t \in \forall x A(x)}{tt' \in A(t')}$

$\exists$I $\quad \dfrac{t_1 \in A(t_0)}{p(t_0, t_1) \in \exists x A(x)}$ $\qquad$ $\exists$E $\quad \dfrac{t \in \exists x A(x) \quad t_1[y, z^{A(y)}] \in C}{E_{u,v}(t, t_1[u, v]) \in C}$

$\bot_i \quad \dfrac{t \in \bot}{\bot_A(t) \in A}$

We observe that each term label completely describes the structure of the proof tree above it.

Simultaneously with the definition of the label t, FV(t) is defined, starting from the stipulations FV(x) $= \{x\}$ (x an individual variable), and FV(x^A) $=$ FV(A) $\cup \{x^A\}$. D and E are to be regarded as variable-binding operators; $D_{u,v}$ binds u in t_0 and v in t_1, and $E_{u,v}$ binds u and v in t_1. The λ-operator binds variables as usual.

REMARK. Note the striking *similarity* between $\to$ and $\forall$: in both cases the rules take the form of λ-abstraction and application, respectively.

In exercise 10.8.5 it is shown that $\wedge$ may be treated completely similarly to $\exists$E, replacing $\boldsymbol{p}_0, \boldsymbol{p}_1$ by an appropriate elimination constant $E'_{u,v}$ analogous to $E_{u,v}$. It would be natural to have a version

$$\exists\mathrm{E}'\frac{t \in \exists x A(x)}{\boldsymbol{p}_1 t \in A(\boldsymbol{p}_0 t)}$$

(from a proof of an existential statement we read off a term $\boldsymbol{p}_0 t$ and a proof $\boldsymbol{p}_1 t$ of $A(\boldsymbol{p}_0 t)$) but this would introduce terms for individuals constructed from deduction terms and thus destroy the correlation with N-**IQC**. On the other hand $\exists$E′ plays a crucial role in Martin-Löf's type theories to be discussed in the next chapter.

EXAMPLE.

$$\cfrac{\cfrac{z^{A\vee B} \in A \vee B \quad \cfrac{x^A \in A}{\boldsymbol{k}_1(x^A) \in B \vee A} \quad \cfrac{y^B \in B}{\boldsymbol{k}_0(y^B) \in B \vee A}}{D_{x,y}(z^{A\vee B}, x^A, y^B) \in B \vee A.}}{\lambda z^{A\vee B}.D_{x,y}(z^{A\vee B}, x^A, y^B) \in A \vee B \to B \vee A}$$

8.3. *Detour conversions and reductions.* There is an obvious way in which the detour conversions correspond to conversion rules for the terms for proof trees. For example, a detour conversion

$$\cfrac{\cfrac{\begin{array}{c}[A]\\ \mathscr{D}\\ B\end{array}}{A \to B} \quad \begin{array}{c}\mathscr{D}'\\ A\end{array}}{B} \qquad \text{conv} \qquad \begin{array}{c}\mathscr{D}'\\ [A]\\ \mathscr{D}\\ B\end{array}$$

corresponds to a conversion of terms: if $\lambda x^A.t[x^A] \in A \to B$ and $t' \in A$ represent

$$\cfrac{\begin{array}{c}\mathscr{D}\\ B\end{array}}{A \to B} \quad \text{and} \quad \mathscr{D}'$$

respectively, then the conversion above yields

$$\lambda x^A.t[x^A](t') \operatorname{conv} t[t'].$$

It is to be noted however that for an exact correspondence we have to drop the CDC. For example, suppose we have proof terms

$$\lambda x^A \lambda y^C.t[x^A, y^C] \in A \to B,$$

$$t'[z^C] \in A \quad (z \not\equiv y).$$

So t' represents a proof tree with a class of open assumptions of the form C marked by z and $\lambda y^C.t$ represents a proof tree in which certain assumptions, marked by y, of the form C are bound. β-conversion results in

$$\lambda y^C.t[t'[z^C], y^C] \in B.$$

Thus after β-conversion the assumptions marked by z are not bound, but on the CDC they ought to have been bound, corresponding to a derivation with term

$$\lambda y^C.t[t'[y^C], y^C] \in B,$$

which is not the result of β-conversion.

Interpreting detour conversions in the manner of the λ-calculus automatically tells us how to handle assumption classes when substituting a derivation for an assumption in performing conversions: it corresponds to the usual term substitution.

The other detour conversions yield the following term conversions:

$$\boldsymbol{p}_0\boldsymbol{p}(t_0, t_1) \text{ conv } t_0,$$

$$\boldsymbol{p}_1\boldsymbol{p}(t_0, t_1) \text{ conv } t_1,$$

$$D_{x,y}(\boldsymbol{k}_0 t, t_0[x^A], t_1[y^B]) \text{ conv } t_0[t],$$

$$D_{x,y}(\boldsymbol{k}_1 t, t_0[x^A], t_1[y^B]) \text{ conv } t_1[t],$$

$$E_{y,z}(\boldsymbol{p}(t_0, t_1), t_2[y, z]) \text{ conv } t_2[t_0, t_1].$$

The $\bot$-conversions correspond to:

$$\boldsymbol{p}_0(\bot_{A \wedge B}(t)) \text{ conv } \bot_A(t),$$

$$\boldsymbol{p}_1(\bot_{A \wedge B}(t)) \text{ conv } \bot_B(t),$$

$$(\bot_{A \to B}(t))(t') \text{ conv } \bot_B(t), \quad \text{where } t' \in B,$$

$$(\bot_{\forall x A(x)}(t))(t') \text{ conv } \bot_{A(t')}(t),$$

$$D_{x,y}(\bot_{A \vee B}(t), t_0[x^A], t_1[y^B]) \text{ conv } t_0[\bot_A(t)] \quad \text{and to } t_1[\bot_B(t)],$$

$$E_{y,z}(\bot_{\exists x A(x)}(t), t_2[y, z]) \text{ conv } t_2[y, \bot_{A(y)}(t)].$$

There is also no problem in formulating permutation conversions and immediate simplifications, for example permuting $\rightarrow$ E upwards over $\vee$E becomes

$$D_{x,y}(t, t_1[x], t_2[y])(t') \operatorname{conv} D_{x,y}(t, t_1[x](t'), t_2[y](t')),$$

and permuting $\vee$E over $\vee$E

$$D_{u,v}\big(D_{x,y}(t, t_1, t_2), t_3, t_4\big) \operatorname{conv} D_{x,y}\big(t, D_{u,v}(t_1, t_3, t_4), D_{u,v}(t_2, t_3, t_4)\big)$$

(x, y not free in t_3, t_4 and u, v not free in t_1, t_2, and t), etc. Reductions are generated by these conversions in the usual way.

8.4. *Strong normalization with respect to detour reductions.* We can now prove strong normalization with respect to detour reductions and $\bot$-reductions (in the λ-calculus sense described above), that is to say, every reduction sequence terminates. The proof consists in a reduction to strong normalization for terms of $\mathbf{HA}^{\omega}$ via the method of "collapsing types". The method as described does not cover permutation reductions, and we have left out immediate simplifications for reasons of expository convenience.

First of all we *collapse* the formulas onto the type structure of $\mathbf{HA}^{\omega}$ by a mapping * as follows:

$$\begin{aligned}
(P)^* &:= 0 \quad \text{for } P \text{ atomic (including } P \equiv \bot);\\
(A \wedge B)^* &:= A^* \times B^*;\\
(\exists x A(x))^* &:= 0 \times A^*;\\
(A \rightarrow B)^* &:= A^* \rightarrow B^* \equiv A^*B^*;\\
(\forall x A(x))^* &:= 0 \rightarrow A^* \equiv 0B^*;\\
(A \vee B)^* &:= 0 \times (A^* \times B^*).
\end{aligned}$$

Note that $(A(x_1,\ldots,x_n))^*$ does not have parameters. Now we define the *collapsing map* * for terms.

(i) The collection of all individual and deduction variables is mapped injectively to the variables of $\mathbf{HA}^{\omega}$ such that an individual variable gets assigned a variable x^0, and a variable x^A gets assigned a variable x^{A^*}. Dropping types, we shall mostly write x^* for the variable assigned to x.

(ii) We assign to the constants $\boldsymbol{p}$, $\boldsymbol{p}_0$, $\boldsymbol{p}_1$, $\boldsymbol{k}_0$, $\boldsymbol{k}_1$, $\bot_A$ and the variable-binding operators λx, λx^A, $D_{u,v}$, $E_{u,v}$ (defined or primitive) constants and variable binding operations $\boldsymbol{p}^*$, $\boldsymbol{p}_0^*, \ldots$, $D^*_{u,v}$, $E^*_{u,v}$ of $\mathbf{HA}^{\omega}$. The mapping * is a homomorphism with respect to application, i.e. $(tt')^* \equiv t^*t'^*$.

(iii) $\boldsymbol{p}^*$ is $\boldsymbol{p}$ of the appropriate type in $\mathbf{HA}^\omega$, i.e. $(\boldsymbol{p}(t, t'))^* \equiv \boldsymbol{p}t^*t'^*$. Here the $\boldsymbol{p}$'s on the left-hand side and on the right-hand side may be pairing operators of different types; we shall permit ourselves the same kind of ambiguity in describing the collapsing map for the other constants and operators:

$$(\boldsymbol{p}_0)^* := \boldsymbol{p}_0, \qquad (\boldsymbol{p}_1)^* := \boldsymbol{p}_1, \qquad (\perp_A)^* := \lambda x^0.0^{A^*},$$

$$(\boldsymbol{k}_0)^* := \lambda x^{A^*}(0,(x^{A^*},0^{B^*})), \qquad (\boldsymbol{k}_1)^* := \lambda y^{B^*}(1,(0^{A^*}, y^{B^*})),$$

where we write (t, t') for $\boldsymbol{p}tt'$, and where 0^σ is used for a canonical element of type σ; for definiteness we take

$$0^0 := 0, \qquad 0^{\sigma\times\tau} := \boldsymbol{p}0^\sigma 0^\tau \equiv (0^\sigma,0^\tau), \qquad 0^{\sigma\to\tau} := \lambda x^\sigma.0^\tau,$$

$$(\lambda x.t)^* := \lambda x^*.t^*.$$

For $(D_{x,y})^*$ we need an F such that

$$F(t, t', t'') \succcurlyeq \begin{cases} t'(\boldsymbol{p}_0(\boldsymbol{p}_1 t)) & \text{if } \boldsymbol{p}_0 t \succcurlyeq 0, \\ t''(\boldsymbol{p}_1(\boldsymbol{p}_1 t)) & \text{if } \boldsymbol{p}_0 t \succcurlyeq S0. \end{cases}$$

This can be achieved by means of the recursor, simulating definition by cases:

$$F(x, y, z) := \boldsymbol{r}(y(\boldsymbol{p}_0(\boldsymbol{p}_1 x)))(\lambda uv.z(\boldsymbol{p}_1(\boldsymbol{p}_1 x)))(\boldsymbol{p}_0 x).$$

We now take

$$(D_{x,y}(t, t_1[x], t_2[y]))^* := F(t^*, \lambda x^*.t_1^*[x^*], \lambda y^*.t_2^*[y^*]).$$

Finally we put

$$(E_{u,v}(t, t_1[u, v]))^* := t_1^*[\boldsymbol{p}_0 t^*, \boldsymbol{p}_1 t^*].$$

We leave it to the reader to see that each detour contraction or $\perp$-contraction of a redex in a derivation term corresponds to a finite non-empty sequence of reduction steps for the collapsed terms, and thus the strong normalization theorem for the terms of $\mathbf{HA}^\omega$ implies strong normalization for deductions with respect to detour- and $\perp$-conversions. In particular, we obtain strong normalization for the fragment of **IQC** without $\vee$ and $\exists$.

8.5. Remark. The strategy used in section 2, namely applying a contraction to the rightmost maximal segment without a maximal segment above it, corresponds to contraction of rightmost minimal redexes for terms. Of course, contracting rightmost minimal redexes is just the "mirror image", so to speak, of contracting leftmost minimal redexes.

It should be noted that as a reduction strategy this works equally well for deductions under the CDC as for the λ-calculus variant discussed in this section.

It is certainly also possible to prove strong normalization if we include permutative reductions. A proof may be obtained by refining strong computability for $\mathbf{HA}^{\omega}$ to strong validity as introduced by Prawitz (1971); for a detailed exposition see Troelstra (1973, chapter 4).

8.6. *Application of the collapsing technique to* **HA**. The collapsing method can also be applied to a suitable natural deduction formulation of **HA**. We present a sketch.

We identify $\perp$ with $0 = S0$, and shall assume the equality rules to be applied to atomic formulas only. Further we restrict the function symbols to the Successor S; for the other primitive recursive functions we add predicate symbols representing their graphs. Thus, for example, if "Sum" is a predicate such that $\text{Sum}(x, y, z) \leftrightarrow x + y = z$, we add the rules

$$\frac{}{\text{Sum}(t, 0, t)} \qquad \frac{\text{Sum}(t, t', t'')}{\text{Sum}(t, St', St'')}$$

Induction is formulated as a rule

$$\text{IND} \qquad \frac{\begin{matrix} \\ \mathcal{D}' \\ A(0) \end{matrix} \quad \begin{matrix} [A(x)]^{(\mathrm{n})} \\ \mathcal{D}(x) \\ A(Sx) \end{matrix}}{A(t)}\,(\mathrm{n})$$

Here x is the proper parameter of the induction application. The term notation for deduction trees can now be extended. We leave it to the reader to formulate this for the graph-predicate constants such as Sum; the instance of induction exhibited above becomes in term notation

$$\frac{t' \in A(0) \quad t''[x, y^{A(x)}] \in A(Sx)}{rt'(\lambda xy.t''[x, y])t \in A(t)}$$

Treating r as a recursor leads to two types of induction conversions, which for proof trees take the forms below:

$$\dfrac{\begin{matrix} & [A(x)] \\ \mathscr{D}' & \mathscr{D}(x) \\ A(0) & A(Sx) \end{matrix}}{A(0)} \quad \text{conv} \quad \begin{matrix} \mathscr{D}' \\ A(0) \end{matrix} \qquad \dfrac{\begin{matrix} & [A(x)] \\ \mathscr{D}' & \mathscr{D}(x) \\ A(0) & A(Sx) \end{matrix}}{A(St)} \quad \text{conv} \quad \begin{matrix} \dfrac{\begin{matrix} & [A(x)] \\ \mathscr{D}' & \mathscr{D}(x) \\ A(0) & A(Sx) \end{matrix}}{[A(t)]} \\ \mathscr{D}(t) \\ A(St) \end{matrix}$$

The type and term mapping is now readily extended to this system; we leave the details as an exercise. From a deduction we can also remove all parameters which do not occur in the conclusion and are not used as proper parameter of an induction, ∃E- or ∀I-application, e.g. by replacing them everywhere by 0. Analyzing the form of natural deductions of this kind we can obtain new proofs for the disjunction property and numerical existence property for **HA** (E10.8.8).

9. Notes

9.1. *Normalization.* For sections 1–3 our principal source has been the monograph by Prawitz (1965). Gentzen (1935) introduced natural deduction calculi NJ, NK as well as the sequent calculi LJ and LK (systems of the type described in section 6, with left and right introduction rules), but he studied cut elimination only for the latter. It was Prawitz (1965) who introduced and studied the notion of normalization for the natural deduction systems.

9.2. *Applications of normalization.* The proof of decidability of **IPC** in section 4 is an adaptation to natural deduction of Kleene's proof for the sequent calculus G3 (1952), the crucial ingredient in both cases being the subformula property.

The theorems 5.3–5 are all taken from Prawitz (1965); theorem 5.2 is found in Malmnås and Prawitz (1968). Theorem 5.6 is due to D. S. Scott (1979, section 5.3), but our proof is (essentially) due to Renardel de Lavalette (unpublished). For the separation theorem cf. Prawitz (1965, p. 54).

The conservativity of predicative classes in section 6 is the general form of a well-known result from axiomatic set theory: the von Neumann–Bernays theory **VNB** with predicative classes is conservative over

Zermelo–Fraenkel set theory **ZF** (cf. Levy (1973) and the references given there, and Flannagan (1976)). The general formulation is in Kreisel (1967), with a model-theoretic proof after the pattern of our proof in 3.8.12, except that results just mentioned concern systems based on classical logic.

9.3. *Sequent calculi.* The "Kleene system" of section 7 is almost identical with the intuitionistic version of the calculus G3 in Kleene (1952), except for the following differences:

(i) In G3 L$\wedge$ is replaced by two rules:

$$\frac{\Gamma, A \wedge B, A \vdash C}{\Gamma, A \wedge B \vdash C} \qquad \frac{\Gamma, A \wedge B, B \vdash C}{\Gamma, A \wedge B \vdash C}$$

and

(ii) instead of the $\bot$-rule there are three rules for negation ($\emptyset$ the empty set, we might also have left a blank):

$$\frac{A, \Gamma \vdash \emptyset}{\Gamma \vdash \neg A} \qquad \frac{\neg A, \Gamma \vdash A}{\neg A, \Gamma \vdash \emptyset} \qquad \frac{\neg A, \Gamma \vdash A}{\neg A, \Gamma \vdash B}$$

As to (i), we leave it to the reader to show that Kleene's two rules are equivalent to our L$\wedge$.

If in (ii) we interpret $\emptyset$ to the right of $\vdash$ as $\bot$, and $\neg A$ as $A \to \bot$, then the rules for $\neg$ become

$$\frac{A, \Gamma \vdash \bot}{\Gamma \vdash A \to \bot} \qquad \frac{A \to \bot, \Gamma \vdash A}{A \to \bot, \Gamma \vdash \bot} \qquad \frac{A \to \bot, \Gamma \vdash A}{A \to \bot, \Gamma \vdash B}$$

The first rule is now a special case of R$\to$, the second rule a special case of the third, and the third rule follows by

$$\dfrac{\dfrac{A \to \bot, \Gamma \vdash A \qquad A \to \bot, \Gamma, \bot \vdash \bot}{A \to \bot, \Gamma \vdash \bot}}{A \to \bot, \Gamma \vdash B}$$

For the converse see E10.7.1.

The equivalence proof between $\vdash_{\mathrm{K}}$ and $\vdash_{\mathrm{N}}$ in section 7 follows Prawitz (1965). Gentzen (1935) gives a proof via a Hilbert-type system.

The Beth tableau system in 7.4 is the system of Fitting (1969) which in turn is quite close to the system used by Beth in his completeness proof for **IQC** (1959, section 145); the proof of $\Gamma \vdash_{\mathrm{K}} C \Rightarrow \Gamma \vdash_{\mathrm{B}} C$ follows Fitting (1969).

9.4. *Formulas-as-types.* Section 8 is based on the "formulas-as-types" idea. The first germ of this idea is the observation in Curry and Feys (1958, ch. 9) that the theory of functionality in the lambda calculus and combinatory logic shows a close parallel with intuitionistic implicational logic.

The idea was later extended to predicate logic by several authors independently (e.g. Howard 1980, widely circulated as preprint since 1969; D. S. Scott 1970; de Bruijn 1970; related ideas are exploited in Läuchli 1970). Since then many authors have used this parallel between reduction for term calculi and normalization of deductions.

The combinatorial relationship between normalization and cut-elimination has been investigated by Zucker (1974) and Pottinger (1977). Normalization corresponds to cut-elimination under a homomorphic mapping, provided the basic cut elimination steps are suitably chosen. Recently a still better correspondence has been achieved by Diller and Unterhalt.

Collapsing mappings (8.4) are used e.g. by Gandy (1980).

9.5. *The dialogue interpretation of intuitionistic logic.* This topic, which has a technical connection with sequent calculi, is not discussed in this book. The dialogue interpretation of logic has been introduced by Lorenzen (1960). The spirit is game-theoretic: a formula is proposed by one of the participants in the dialogue, which then is attacked by the other participant. Defenses and attacks are governed by specific logical rules. A logical statement A is valid of there is a winning strategy for P in dialogues starting with initial thesis A.

For a thorough discussion see in particular Krabbe (1985) and Felscher (1985, 1986), where completeness proofs for the dialogue interpretation are given, showing that the existence of a winning strategy is equivalent to the existence of a deduction in a suitable calculus of sequents. Further information is also found in Lorenzen and Lorenz (1978) and Walton (1985).

Exercises

10.1.1. Construct a derivation containing segments of length 4, 3, 2 and 1; indicate all segments.

10.2.1. Complete the proof of the normalization theorem 2.8.

10.2.2. Extend the normalization theorem to N-**IQCE** for a language with equality. *Hint.* Formulate N-**IQCE** with equality rules for *atomic A* only:

$$\frac{At \quad t = s}{As} \qquad \frac{t = t}{Et} \qquad \frac{Et}{t = t} \qquad \frac{t = s}{s = t}.$$

10.2.3. Give an example in propositional logic to show that detour conversions may introduce new redundant applications of $\vee$E.

10.3.1. Show that there exists no normal derivation in **N-IQC** for $\forall x \forall y(By \wedge Ax) \vdash Ay$ (and hence for normalization *some* condition on the variables is necessary: the example shows that the normalization theorem may fail if the sets of bound and free variables are not disjoint).

10.3.2. Adapt the proof of 3.5 to dp-normal derivations and verify that the subformula property also holds for dp-normal derivations.

10.3.3. Consider **N-IQCE** with equality as in exercise 10.2.2. Adapt the definition of subformula by treating $A[x/s]$ as a subformula of $A[x/t]$, for arbitrary terms s. An equality is to be counted as subformula of any formula. Adapt proposition 3.3 to **N-IQCE** with equality by replacing the notion of a minimum segment of a track by a suitably defined notion of a *minimum part* of a track.

10.3.4. Establish a suitable version of the subformula property for **N-IQCE** with equality.

10.3.5. Adapt proposition 3.8 to **N-IQCE** with equality.

10.3.6. Show that the addition of classical predicate logic to **PRA**, resulting in a system $\mathbf{PRA}^+$, is conservative over **PRA**. N.B. The induction rule remains restricted to quantifier-free formulas! *Hint.* First consider $\mathbf{PRA}^0 \equiv \mathbf{PRA}$ with rules for $\forall$ added. Prove a normalization theorem as usual; take a generalized notion of track where the first and last formula may now also be conclusion or premiss of an application of the induction rule. The normal derivations of quantifier-free statements do not contain quantifiers. Full classical logic is then afterwards justified by embedding $\mathbf{PRA}^+$ in $\mathbf{PRA}^0$ via the Gödel–Gentzen translation ^{g} (cf. 2.3.4).

10.4.1. Prove lemma 4.2.

10.4.2. Apply the decision method to show the underivability of $(P \to Q) \vee (Q \to P)$, $\neg P \vee \neg\neg P$.

10.4.3. Show that for pure intuitionistic predicate logic, i.e. without function symbols or equality, derivability of prenex formulas is decidable.

10.4.4. Find a purely implicational schema F such that $\mathbf{IPC} + F = \mathbf{IPC} + (P \to Q) \vee (Q \to P)$.

10.5.1. Prove 5.5. Show that in **IQCE** no statement of the form $\exists x A$ is derivable. (Prawitz 1965, IV, cor. 7).

10.5.2. Let C be a formula of **IQC** not containing $\to$, and let $\Gamma \equiv \{A_1 \to B_1, \ldots, A_n \to B_n\}$. Prove that if $\Gamma \vdash C$, then $\Gamma \vdash A_i$ for some $i \le n$. (Prawitz 1965, IV, cor. 8).

10.5.3. Prove the *interpolation theorem* for **IQC** in the following form: suppose $\Gamma \vdash A$, and let Γ_1, Γ_2 be such that $\Gamma_1 \cap \Gamma_2 = \Gamma$, $\Gamma_1 \cup \Gamma_2 = \emptyset$. Then there is an interpolation formula for

(Γ_1, Γ_2, A), i.e. a formula F such that:

(i) $\Gamma_1 \vdash F$, $\Gamma_2 \cup \{F\} \vdash A$.

(ii) Each predicate letter and free variable or nonlogical constant of F that occurs positively (negatively) in F occurs positively (negatively) in some formula of Γ_1 and negatively (positively) in some formula of $\Gamma_2 \cup \{\neg A\}$ (Prawitz 1965, ch. III, IV, cor. 5; originally due to Schütte 1962).

Hint. Use induction on the length of a normal deduction; F is to be constructed from minimum formulas using logical constants.

10.5.4. Obtain an analogue of the subformula property for the Hilbert-type axiomatization of **IQC** in 2.4.1, by analyzing the proof of equivalence between the Hilbert-type system and **N-IQC**.

10.5.5. Let $\mathscr{X}$ be any subset of $\{\wedge, \vee, \rightarrow, \bot, \forall, \exists\}$ containing $\rightarrow$. The $\mathscr{X}$-fragment of **IQC** contains all theorems of **IQC** using operations from $\mathscr{X}$ only. Show that the $\mathscr{X}$-fragment of **CQC** is generated by the rules of **IQC** involving only the operations of $\mathscr{X}$, and in addition the Peirce axiom schema $((A \rightarrow B) \rightarrow A) \rightarrow A$.

10.5.6. (*Herbrand's theorem for negations of prenex formulas without* $=$ *and function symbols*). Let B be a prenex formula $\forall \vec{x}_0 \exists \vec{y}_0 \forall \vec{x}_1 \exists \vec{y}_1 \ldots A(\vec{x}_0, \vec{y}_0, \vec{x}_1, \vec{y}_1, \ldots)$ and assume $\mathbf{IQC} \vdash \neg B$. Then we can find a finite conjunction C of substitution instances of A such that $\vdash \neg C$. More precisely, C is of the following form (taking for notational simplicity $B \equiv \forall u \exists x \forall y \exists z A(u, x, y, z)$)

$$A(t_0, b_0, t_1, b_1) \wedge A(t_0', b_0', t_1', b_1') \wedge A(t_0'', b_0'', t_1'', b_1'') \wedge \cdots \wedge A\left(t_0^{(n)}, b_0^{(n)}, t_1^{(n)}, b_1^{(n)}\right),$$

where $a_0, b_0, b_0', \ldots, b_0^{(n)}, b_1, b_1', \ldots, b_1^{(n)}$ are variables and

$$t_0 \in \{a_0\}, \quad t_1 \in \{a_0, b_0\},$$

$$t_0' \in \{a_0, b_0, b_1\}, \quad t_1' \in \{a_0, b_0, b_1, b_0'\},$$

$$\ldots$$

$$t_0^{(n)} \in \{a_0\} \cup \left\{b_k^{(j)} : 0 \leq k \leq 1 \wedge 0 \leq j \leq n-1\right\},$$

$$t_1^{(n)} \in \{a_0\} \cup \left\{b_k^{(j)} : 0 \leq k \leq 1 \wedge 0 \leq j \leq n-1\right\} \cup \left\{b_0^{(n)}\right\}.$$

10.5.7. Deduce from the preceding exercise that for A a negation of a prenex formula in the language without $=$, E and function symbols: $\mathbf{CQC} \vdash A$ iff $\mathbf{IQC} \vdash A$.

10.5.8. Complete the proof of 5.6.

10.6.1. Apply the method of section 6 to show conservativity of $\mathbf{EM}_0\!\upharpoonright$ over **APP** (cf. 9.8.6).

10.7.1. Show that the $\bot$-rule in Kleene's system can be deleted if instead we permit as axioms all $\Gamma \vdash A$ with $A \in \Gamma$ or $\bot \in \Gamma$. Use this fact to show that if we take the negation rules as indicated in 9.3 and define $\bot$ as $A \wedge \neg A$, then we can derive the $\bot$-rule.

10.7.2. Show that the construction of a natural deduction tree from a proof showing $\Gamma \vdash_K A$ may actually be carried out so as to obtain a normal deduction tree.

10.7.3. Prove the thinning lemma in the proof of 7.5.

10.7.4. Prove 7.5(ii) directly (Fitting 1969). *Hint.* First establish as a lemma for derivability in Kleene's system: $\Gamma \vdash_K A_1 \vee (A_2 \vee (A_3 \vee \cdots (A_{n-1} \vee A_n) \cdots))$ iff $\Gamma \vdash_K (\cdots ((A_1 \vee A_2) \vee A_3) \vee \cdots \vee A_{n-1} \vee A_n)$.

10.7.5. Describe the decision procedure for **IPC** on the basis of Kleene's system (Kleene 1952, section 80).

10.7.6. Show that if we liberalize the rules R→, R∀ of the Beth system to

$$\text{R}\!\to \frac{\Gamma, A \vdash B, \Delta}{\Gamma \vdash A \to B, \Delta} \qquad \text{R}\forall \frac{\Gamma \vdash A(y), \Delta}{\Gamma \vdash \forall x A(x), \Delta} \quad (y \notin \text{FV}(\Gamma \cup \Delta)),$$

we obtain a system for classical predicate logic.

10.7.7. Let all formulas in $\Gamma \cup \Delta$ be →, ∀-free, and let us use $\vdash$ and $\vdash'$ for deducibility in the Beth system and its classical counterpart (as described in the preceding exercise), respectively. Show that $\Gamma \vdash \Delta$ iff $\Gamma \vdash' \Delta$. Derive from this: if A is a formula such that all its *proper* positive subformulas are ∀, → -free, then $\vdash A$ iff $\vdash' A$. Can you generalize this still further?

10.8.1. Give a counterexample showing that on the CDC strong normalization does not hold, not even for pure →-logic. *Hint.* Consider the following deduction term:

$$t'[v^P, u^Q] \equiv (\lambda v^P.t[u^Q, v^P])t[v^P, u^Q] \in Q,$$

where $t[v^P, u^Q] \equiv (\lambda u^Q.v^P)[(\lambda v^P.u^Q)[(\lambda u^Q.v^P)(u^Q)]] \in P$ (R. Statman, unpublished; Leivant 1979).

10.8.2. Prove the Church–Rosser theorem for deduction terms with respect to dp-normalization (for deductions with atomic $\bot_i$ alone). Give examples to show that uniqueness of normal form may fail with respect to d-normalization if we permit non-atomic instances of $\bot_i$.

10.8.3. Give a complete list of all permutation conversions in terms of deduction terms.

10.8.4. Show that under the collapsing map * deductions which differ only in a permutation past an ∃E-application are mapped onto the same term.

10.8.5. We may treat ∧E completely parallel to ∃E by using a variable-binding operator E' with rule

$$\frac{t \in A \wedge B \quad t'[x^A, y^B] \in C}{E'_{x,y}(t, t'[x, y]) \in C}$$

where $E'_{x,y}$ binds x and y in t, and with a conversion rule

$$E'_{x,y}(p(t_0, t_1), t'[x, y]) = t'[t_0, t_1].$$

Show that this rule is equivalent to the rules given before, in the sense that from E' we can define inverses $\mathbf{p}_0$, $\mathbf{p}_1$ to $\mathbf{p}$, and conversely from $\mathbf{p}_0, \mathbf{p}_1$ we can define an operator E' with the required properties. Show that surjectivity of pairing, $\mathbf{p}(\mathbf{p}_0 t, \mathbf{p}_1 t) = t$, uniquely determines the relation between $E', \mathbf{p}_0, \mathbf{p}_1$.

10.8.6. Show that a Hilbert-type axiomatization of intuitionistic implicational calculus based on modus ponens and the axiom schemas

$$(A \to (B \to A))$$

$$(A \to (B \to C)) \to ((A \to B) \to (A \to C))$$

corresponds to a calculus of deduction terms, isomorphic to typed combinatory logic with primitives S and K of all appropriate types. Extend this so as to include other propositional operators.

10.8.7. Describe in detail a term representation for the version of **HA** sketched in 8.6, and extend the definition of the collapsing mapping to this system.

10.8.8. Let us call a parameter in a proof tree *redundant* if it is not the proper parameter of an IND-, ∀I-, or ∃E-application and does not occur in the conclusion.

Show, by analyzing the form of ds-normal deductions without redundant parameters in **HA**, that **HA** has the disjunction property (cf. section 8.6).

CHAPTER 11

THE THEORY OF TYPES AND CONSTRUCTIVE SET THEORY

1. Towards a theory of types

1.1. In this section we shall attempt to motivate the type theories in the later sections, taking as our starting point the calculus of terms for intuitionistic predicate logic introduced in section 10.8. We will thus, step by step, introduce the formulas-as-types idea until all the basic ingredients for the type theories are available. Martin-Löf has developed the sort of motivation given here into a semantics for his type theories; this semantics is discussed in 5.9–5.11.

In this chapter we shall adhere to the following convention: *occurrences of free variables in expressions are always indicated by square brackets, whereas parentheses will be reserved for pairing and application.*

That is to say, in the notation $F[x]$, "$[x]$" refers to the free occurrences (if any) of x in F; $F(x)$ is used when F is a function and indicates the result of applying F to the argument x.

1.2. In the term-calculus for N-**IQC**, $t \in A$ may be read as "t proves A", since the term t encodes a natural deduction proof tree for A. On this reading we might equally well, or better, have used the notation $t \vdash A$ ("t derives A").

However, we wish to view the terms t as descriptions of "proofs" or "constructions" such as they occur in the BHK-explanations of the intuitionistic logical operators (cf. 1.3.1–2), and the primitive notion of "proof" or "construction" appearing in those explanations cannot be straightforwardly identified with "formal proof". If we read the terms as descriptions of constructions, certain syntactically distinct terms must be assumed to denote the same construction, thereby introducing a notion of equality between terms preventing their identification with deduction trees.

Nonetheless we shall usually treat expressions such as "t proves A", "t witnesses A", "t is a construction for A," and "t is a construction witnessing the truth of A" as synonyms.

We point out that in the term-calculus "$t \in A$" does not occur as a proposition (which is capable of being true or false) but as an *assertion* or *judgment* (possibly modulo assumptions) which is asserted to be true (modulo assumptions) by the rules of the calculus.

1.3. *Implication and universal quantification.* On the BHK-interpretation any construction of $A \to B$ contains a function for transforming constructions of A into constructions of B. In keeping with this, the $\to$I-rule introduces function-terms as proofs of $A \to B$. The fact that, on the BHK-explanation, *all* constructions of $A \to B$ are in fact functions mapping constructions of A to constructions of B is expressed by the elimination rule $\to$E: given $t \in A \to B$ and $t' \in A$, it makes sense to apply t to t', yielding a proof tt' of B.

This also suggests that if we indeed wish to regard our deduction terms t in "$t \in A$" as BHK-constructions for A, we should identify

$$(\lambda y^A.t)(t_1^A) \quad \text{and} \quad t[y/t_1^A],$$

that is to say the two proof-trees coded by these terms denote the same BHK-proof (and the $\to$-detour conversion preserves the interpretation by a BHK-construction).

For the $\forall$-rules the situation is completely similar; the role of the premiss of an implication is now taken over by a domain of individuals. To bring this out more clearly, suppose we have a many-sorted first-order logic with domains $D, D', D'', \ldots$; we can now formulate the $\forall$-rules with specification of domain:

$$\frac{\begin{array}{c}[x^D \in D]\\ \mathcal{D}\\ t \in A\end{array}}{\lambda x^D.t \in \forall x \in D.A} \qquad \frac{t' \in \forall x \in D.A \quad t \in D}{t'(t) \in A[x/t]}$$

If we use $\forall x \in A.B$, where x is not free in B, (loosely paraphrased as "for all proofs x of A we have (a proof of) B") as an alternative notation for $A \to B$, the analogy is complete.

1.4. *Conjunction and existential quantification.* On the BHK-interpretation, a construction of $A \wedge B$ is a pair of constructions, one for A and one for B, and accordingly we have

$$\frac{t \in A \quad t' \in B}{(t, t') \in A \wedge B} \qquad \frac{t \in A \wedge B}{p_0 t \in A} \qquad \frac{t \in A \wedge B}{p_1 t \in B}$$

where the E-rule expresses the fact that if t is to denote a construction for $A \wedge B$, it should have first and second parts $\boldsymbol{p}_0 t$, $\boldsymbol{p}_1 t$, respectively.

Similarly, constructions for $\exists x \in D.A[x]$ are thought of as pairs (d, e) with $d \in D$ and e a construction for $A[d]$. Accordingly, we have ∃I in complete parallel to ∧I:

$$\frac{t \in D \quad t' \in A[t]}{(t, t') \in \exists x \in D.A[x]}$$

The corresponding E-rules ought to be

$$\exists\text{E}^* \qquad \frac{t \in \exists x \in D.A}{\boldsymbol{p}_0 t \in D} \qquad \frac{t \in \exists x \in D.A}{\boldsymbol{p}_1 t \in A[x/\boldsymbol{p}_0 t]}$$

However, the actual rule ∃E for **IQC** is weaker. By ∃E* we find

$$\frac{t \in \exists y \in D.A \quad t_1[y^D, z^A] \in C[z/(y, z)]}{t_1[\boldsymbol{p}_0 t, \boldsymbol{p}_1 t] \in C[z/t]}$$

(using $t = (\boldsymbol{p}_0 t, \boldsymbol{p}_1 t)$), which reduces to ordinary ∃E in case $z \notin \mathrm{FV}(C)$ and we put $E_{u,v}(t, t_1[u, v]) := t_1[\boldsymbol{p}_0 t, \boldsymbol{p}_1 t]$.

The most important difference between ∃E and ∃E* is that ∃E* introduces terms for individuals containing terms for constructions ($\boldsymbol{p}_0 t$ in ∃E*) as subterms, whereas this is not the case for ∃E.

To make the analogy complete, we may introduce $\exists x \in A.B (x \notin \mathrm{FV}(B))$ as an alternative notation for $A \wedge B$.

We shall skip the case of ∨, but turn to the rules for the natural numbers instead.

1.5. *Rules for* ℕ. As already indicated in 10.8.6, we can incorporate arithmetic in our natural deduction calculus by adding a suitable form of the induction rule. The symmetry between the I- and E-rules can be retained by formulating the closure conditions (generating clauses) of ℕ as I-rules:

$$\mathbb{N}\text{I} \qquad 0 \in \mathbb{N} \qquad \frac{t \in \mathbb{N}}{St \in \mathbb{N}}$$

Induction becomes the corresponding E-rule (in a formulation slightly different from the one given in 10.8.6; $R_{x,y}$ is an operator binding x and y in t_1)

$$\mathbb{N}\text{E} \qquad \frac{t \in \mathbb{N} \quad t_0 \in A[0] \quad t_1[x, y^{A[x]}] \in A[Sx]}{R_{x,y}(t, t_0, t_1) \in A[t]}$$

The use of a constant $\boldsymbol{r}$ with $\boldsymbol{r}tt_0(\lambda xy.t_1)$ instead of $R_{x,y}(t, t_0, t_1)$ corresponds intuitively to a version of $\mathbb{N}$E with

$$\lambda xy.t_1[x, y^{A[x]}] \in \forall x \in \mathbb{N}(A[x] \to A[Sx])$$

as a premiss.

As always, the intuitive significance of the elimination rule is that the I-rules describe all possibilities for introducing constructions in (elements of) $\mathbb{N}$. The obvious equality rules for $R_{x,y}$,

$$R_{x,y}(0, t_0, t_1) = t_0,$$

$$R_{x,y}(St, t_0, t_1) = t_1[x, y / t, R_{x,y}(t, t_0, t_1)],$$

correspond exactly to the induction reductions described in 10.8.6.

In the parallel between $\to$I, $\to$E, $\wedge$I, $\wedge$E on the one hand and $\forall$I, $\forall$E, $\exists$I, $\exists$E on the other hand "$t \in D$" for a domain D and "$t \in A$" for a proposition A are treated completely similarly. If we extend this parallel also to $\mathbb{N}$E, we obtain as a special case of $\mathbb{N}$E

$$\frac{t \in \mathbb{N} \quad t' \in \mathbb{N} \quad t''[x^N, y^N] \in \mathbb{N}}{R_{x,y}(t, t', t'') \in \mathbb{N}}$$

and now the equality rules for $R_{x,y}$ show that $R_{x,y}$ serves to define functions by recursion.

Thus the parallel between domains and propositions leads to the recognition that proof by induction and definition by recursion are basically the same thing.

1.6. *The formulas-as-types concept.* The formulas-as-types idea ("propositions as types" would perhaps have been more accurate) is, that we can in fact regard domains or *types* on the one hand, and *propositions* on the other hand, as the same sort of thing; we only have to think of a proposition as determined by the collection of BHK-constructions of the proposition.

Propositions of the form $\forall x \in A.B[x]$ then correspond to (generalized) cartesian products, with as elements functions f such that $f(x) \in B[x]$ for each $x \in A$. Such a product we shall denote by $\Pi x \in A.B[x]$.

Propositions of the form $\exists x \in A.B[x]$ correspond to disjoint unions, where each element is a pair (d, e) with $d \in A$ and $e \in B[d]$. Such disjoint unions are denoted by $\Sigma x \in A.B[x]$.

$A \to B$ corresponds to the special case $\Pi x \in A.B$, where $x \notin \mathrm{FV}(B)$, and $A \wedge B$ to the special case of $\Sigma x \in A.B$ with $x \notin \mathrm{FV}(B)$, which in turn may be identified with the cartesian product $A \times B$.

The strong $\exists$E*-rule is essential to give full expression to the fact that the elements of $\exists x \in A.B$ are pairs. Note also that (apart from the treatment of prime formulas) the BHK-constructions for compound statements behave like "realizing objects" in an abstract form of realizability (such as e.g. realizability for **APP**, discussed in section 9.5); again, it is the adoption of the strong $\exists E^*$ in our term-calculus which brings out the realizability aspect.

Disjunctions $A \vee B$ correspond to disjoint sums of the type $A + B$. We have not discussed $\vee$ above; cf. 10.8.2 and the next section.

We now get a "list of synonyms" according to the formulas-as-types concept:

type, domain, proposition;
product formation, universal quantification;
disjoint union formation, existential quantification;
product of constant factors, implication;
cartesian product of two factors, conjunction;
disjoint sum of two factors, disjunction;
t is an element of A, t proves A, t is a construction of A, t realizes A.

1.7. *Identity types; the analogy with realizability.* In order to interpret mathematical theories on the basis of the formulas-as-types idea, one ingredient is still needed: the notion of the construction of a prime formula. Thus in arithmetic we need to know what is to be regarded as a construction of $t = s$ for numerical t and s; and more generally, given a domain A and a notion of equality $=_A$ between elements of A, we may consider propositions $t =_A s$ ("t and s are equal elements of A") and ask ourselves what is to be counted as a construction of $t =_A s$. In other words, we have to consider, for each t and s in A, the type $I(A, t, s)$ consisting of the collection of all constructions for $t =_A s$.

Boldly continuing the analogy with realizability suggests that $I(A, t, s)$ should contain a single canonical construction $\boldsymbol{e}$ iff t and s are equal as elements of A, and that it should be empty if t and s are not equal.

Adoption of this proposal gives another source of divergence between "proofs of A" in the intuitive sense and "constructions of A": even if $t =_A s$ can be established by several complicated proofs, there is always at most one trivial construction of $t =_A s$. Compare the situation in arithmetic: $\forall x(t[x] = s[x])$ may have a very complicated proof, but it is always realized by $\Lambda x.0$ (4.4.5).

2. The theory $\mathbf{ML}_0^i$

2.1. In this section we shall describe an intensional version $\mathbf{ML}_0^i$ of what might be called the "arithmetical" part of Martin-Löf's theory of types.

This theory is "intensional" in the sense that we impose only minimal requirements on equality at each type; in section 5 we shall describe a corresponding extensional theory $\mathbf{ML}_0$, where equality is interpreted extensionally throughout.

The theories $\mathbf{ML}_0^i, \mathbf{ML}_0$ correspond to those fragments of the theories in Martin-Löf (1975, 1982), respectively in which neither universes nor the formation of well-founded types occur. There are certain deviations from Martin-Löf's own versions, mainly dictated by reasons of expository convenience.

The language of $\mathbf{ML}_0^i$ contains object-terms (terms for elements of a type) and type-terms. *Object-terms* are built from variables ($v_0, v_1, v_2, \ldots$), certain constant symbols (0, S, $\boldsymbol{e}$), and operators (Ap, $\boldsymbol{p}$, $\boldsymbol{p}_0$, $\boldsymbol{p}_1$, λ, E, D, R). *Type-terms* are built from the type constant N and object-terms with help of type-forming operators Π, Σ, $+$, and I. Furthermore we need equality between terms ($=$), the "element of" relation ($\in$) between object-terms or object-term equations and type-terms, the sequent arrow ($\Rightarrow$), and auxiliary symbols (,). The construction rules for the terms are given simultaneously with the deduction rules.

2.2. *The assertions (judgments) of* $\mathbf{ML}_0^i$. As a calculus, $\mathbf{ML}_0^i$ is a many-sorted system of natural deduction, where the conclusion of a deduction is an *assertion* or *judgment*, possibly depending on certain *assumptions*. That is to say, the assertions may be hypothetical. We shall write $\Gamma \Rightarrow \theta$ for an assertion θ which has been derived under the finite sequence of assumptions Γ; more loosely we shall also use "assertion" for $\Gamma \Rightarrow \theta$.

The term "assertion" or "judgment" serves to emphasize that expressions $\Gamma \Rightarrow \theta$ in the theory are not to be regarded as propositions which may be true or false; $\Gamma \Rightarrow \theta$ is to appear only in a deduction when we have concluded θ to be true on the basis of the assumptions Γ. Thus it makes no sense to say that $\Gamma \Rightarrow \theta$ is false in $\mathbf{ML}_0^i$. Of course, in discussing $\mathbf{ML}_0^i$ metamathematically, we freely discuss expressions $\Gamma \Rightarrow \theta$ regardless of their appearance in a correct formal proof of $\mathbf{ML}_0^i$; if in such a discussion we wish to indicate that $\Gamma \Rightarrow \theta$ is actually derivable, we may write $\mathbf{ML}_0^i \vdash \Gamma \Rightarrow \theta$, or simply $\vdash \Gamma \Rightarrow \theta$.

The assertions of the theory have one of the following four forms:

A type	(A is a *type* or *proposition*);
$t \in A$	(t is an *element* of A, t proves, witnesses, realizes A);
$t = s \in A$	(t and s are *equal elements of* A);
$A = B$	(A and B are *equal types* or *propositions*).

CONVENTION. We shall use x, y, z, u, v, w as metavariables for variables ranging over elements of types, t, s as metavariables for terms denoting elements of types, A, B, C, D as metavariables for type-terms (expressions denoting types), and $\theta, \theta', \theta''$ as metavariables for assertions. □

2.3. *Assumptions.* Each assertion is derived under a finite, possibly empty, sequence of assumptions; such a sequence is called a *context*, and has the form

$$x_0 \in A_0, x_1 \in A_1, \ldots, x_n \in A_n, \tag{1}$$

where A_j contains only x_i free with $i < j$, for all $j \le n$. Thus, a context may be written more suggestively as

$$x_0 \in A_0, x_1 \in A_1[x_0], \ldots, x_n \in A_n[x_0, \ldots, x_{n-1}]. \tag{2}$$

Some of the variables may be "dummy": $\mathrm{FV}(A_i) \subset \{x_0, \ldots, x_{i-1}\}$, but not necessarily $\mathrm{FV}(A_i) = \{x_0, \ldots, x_{i-1}\}$. $\Gamma, \Gamma', \Gamma''$ will be used as metavariables for contexts.

The reader will find it more easy to understand $\mathbf{ML}_0^{\mathrm{i}}$ if the parallel with natural deduction for many-sorted **IQC** is continually kept in mind, in particular the version with deduction terms. The use of typed variables x^A, $y^{B[x]}$ in the natural deduction calculus corresponds to the use of contexts containing $x \in A$ and $y \in B[x]$ in the theory $\mathbf{ML}_0^{\mathrm{i}}$.

2.4. *Structural rules of* $\mathbf{ML}_0^{\mathrm{i}}$. There are two general structural rules, the *assumption rule* and the *rule of thinning*:

$$\text{ASS} \qquad \frac{\Gamma \Rightarrow A \text{ type}}{\Gamma, x \in A \Rightarrow x \in A} \qquad (x \notin \mathrm{FV}(\Gamma)),$$

$$\text{THIN} \qquad \frac{\Gamma \Rightarrow A \text{ type} \quad \Gamma, \Gamma' \Rightarrow \theta}{\Gamma, x \in A, \Gamma' \Rightarrow \theta} \qquad (x \notin \mathrm{FV}(\Gamma \cup \Gamma')).$$

As we shall prove later, for the contexts $x_0 \in A_0, \ldots, x_n \in A_n$ appearing in a derivation $\mathcal{D}$ we can find, for each $i \le n$, a derivation of $x_0 \in A_0, \ldots, x_{i-1} \in A_{i-1} \Rightarrow A_i$ type.

2.5. CONVENTION. In stating the remaining rules we shall use, for all rules except REPL and SUB, the following convention by way of abbreviation. The general form of the rules (except SUB and REPL) is

$$\frac{\Gamma, \Delta_0 \Rightarrow \theta_0 \quad \cdots \quad \Gamma, \Delta_n \Rightarrow \theta_n}{\Gamma \Rightarrow \theta} \tag{1}$$

where the Δ_i contain assumptions discharged by the application of the rule. In most cases we shall drop "Γ" and also "$\Delta_i \Rightarrow$" for Δ_i empty in exhibiting the form of the rules (i.e. contexts not changed by the rule-application are not exhibited).

Ordinarily the premisses of an inference appear on a single line above the horizontal line, and the conclusion immediately below the line. Sometimes, if there are many or relatively long premisses, we have to arrange the premisses in several consecutive lines (not separated by black lines) below each other. See for instance the rule +CONV in 2.10.

2.6. *General equality rules of* $\mathbf{ML}_0^{\mathrm{i}}$.

REFL $\dfrac{t \in A}{t = t \in A}$ $\dfrac{t = t \in A}{t \in A}$ $\dfrac{A \text{ type}}{A = A}$

SYM $\dfrac{t = t' \in A}{t' = t \in A}$ $\dfrac{A = B}{B = A}$

TRANS $\dfrac{t = t' \in A \quad t' = t'' \in A}{t = t'' \in A}$ $\dfrac{A = B \quad B = C}{A = C}$

SUB $\dfrac{\Gamma, x \in A, \Gamma' \Rightarrow \theta \qquad \Gamma \Rightarrow t \in A}{\Gamma, \Gamma'[x/t] \Rightarrow \theta[x/t]}$

REPL1 $\dfrac{\Gamma, x \in A, \Gamma' \Rightarrow B \text{ type} \qquad \Gamma \Rightarrow t = t' \in A}{\Gamma, \Gamma'[x/t] \Rightarrow B[x/t] = B[x/t']}$

REPL2 $\dfrac{\Gamma, x \in A, \Gamma' \Rightarrow s \in B \qquad \Gamma \Rightarrow t = t' \in A}{\Gamma, \Gamma'[x/t] \Rightarrow s[x/t] = s[x/t'] \in B[x/t]}$

REPL3 $\dfrac{t \in A \quad A = B}{t \in B}$

2.7. Remarks. (i) The effect of the first two reflexivity rules is that $t \in A$ may be treated as an abbreviation of $t = t \in A$. Note that from REFL, SYMM, and TRANS we also have

$$\frac{t = t' \in A}{t \in A}$$

since

$$\dfrac{\dfrac{t = t' \in A \qquad \dfrac{t = t' \in A}{t' = t \in A}}{t = t \in A}}{t \in A}$$

As we shall see later,

$$\frac{A = A}{A \text{ type}}$$

is a derived rule, so we may think of "A type" as an abbreviation of "$A = A$".

(ii) From $\Gamma, x \in A, \Gamma' \Rightarrow B = B'$ and $\Gamma \Rightarrow t = t' \in A$ with (i), SUB:

$$\Gamma, \Gamma'[x/t] \Rightarrow B[x/t] = B'[x/t]$$

and with REFL and REPL

$$\Gamma, \Gamma'[x/t] \Rightarrow B'[x/t] = B'[x/t'];$$

combining this with TRANS yields

$$\Gamma, \Gamma'[x/t] \Rightarrow B[x/t] = B'[x/t'].$$

Similarly one can justify the generalization of the second replacement rule

$$\frac{\Gamma, x \in A, \Gamma' \Rightarrow s = s' \in B \qquad \Gamma \Rightarrow t = t' \in A}{\Gamma, \Gamma'[x/t] \Rightarrow s[x/t] = s'[x/t'] \in B[x/t]}$$

The equality rules for types are quite weak; nontrivial equalities between types are generated by term substitution and replacement of terms only.

2.8. *Type-introduction rules of* $\mathbf{ML}_0^i$. Whereas in the traditional formalisms type-formation is relegated to the syntax, in the present system type-statements are part of the axiomatization. The following rules provide the types of $\mathbf{ML}_0^i$.

NTYP $\quad N$ type

ΠTYP $\quad \dfrac{x \in A \Rightarrow B \text{ type}}{\Pi x \in A.B \text{ type}}$

ΣTYP $\quad \dfrac{x \in A \Rightarrow B \text{ type}}{\Sigma x \in A.B \text{ type}}$

+TYP $\quad \dfrac{A \text{ type} \quad B \text{ type}}{A + B \text{ type}}$

ITYP $\quad \dfrac{t \in A \quad s \in A \quad A \text{ type}}{I(A, t, s) \text{ type}}.$

2.9. *Introduction and elimination rules of* $\mathbf{ML}_0^{\mathrm{i}}$.

NI $\quad 0 \in N \qquad \dfrac{t \in N}{St \in N}$

NE $\quad \dfrac{t \in N \quad t_0 \in A[x/0] \quad x \in N, y \in A \Rightarrow t_1 \in A[x/Sx] \quad x \in N \Rightarrow A \text{ type}}{R_{x,y}(t, t_0, t_1) \in A[x/t]}$

($R_{x,y}$ binds x and y in t_1),

ΠI $\quad \dfrac{x \in A \Rightarrow t \in B \qquad x \in A \Rightarrow B \text{ type}}{\lambda x.t \in \Pi x \in A.B}$

ΠE $\quad \dfrac{t \in \Pi x \in A.B \qquad t' \in A \qquad x \in A \Rightarrow B \text{ type}}{\mathrm{Ap}(t, t') \in B[x/t']}$

For Ap(t, t') we shall mostly write tt', and occasionally $t(t')$.

ΣI $\quad \dfrac{t \in A \quad t' \in B[x/t] \quad x \in A \Rightarrow B \text{ type}}{\boldsymbol{p}(t, t') \in \Sigma x \in A.B}$

ΣE $\quad \dfrac{t \in \Sigma x \in A.B \quad A \text{ type}}{\boldsymbol{p}_0 t \in A} \qquad \dfrac{t \in \Sigma x \in A.B \quad x \in A \Rightarrow B \text{ type}}{\boldsymbol{p}_1 t \in B[x/\boldsymbol{p}_0 t]}$

$\boldsymbol{p}(t, t')$ will usually be written as (t, t').

+I $\quad \dfrac{t \in A \quad A \text{ type} \quad B \text{ type}}{\boldsymbol{k}_0 t \in A + B} \qquad \dfrac{t \in B \quad A \text{ type} \quad B \text{ type}}{\boldsymbol{k}_1 t \in A + B}$

+E $\quad \dfrac{t \in A+B \quad x \in A \Rightarrow t_0 \in C[z/\boldsymbol{k}_0 x] \quad y \in B \Rightarrow t_1 \in C[z/\boldsymbol{k}_1 y] \quad z \in A+B \Rightarrow C \text{ type}}{D_{x,y}(t, t_0, t_1) \in C[z/t]}$

($D_{x,y}$ binds x in t_0 and y in t_1, $x \notin \mathrm{FV}(t_1)$, $y \notin \mathrm{FV}(t_0)$).

II $\quad \dfrac{t = t' \in A \quad A \text{ type}}{\boldsymbol{e} \in I(A, t, t')}$

IE $\quad \dfrac{t'' \in I(A, t, t') \quad A \text{ type}}{t = t' \in A}$

NOTATION. We introduce the following notations, which bring out the double role of types:

$$\forall x \in A.B := \Pi x \in A.B, \qquad \exists x \in A.B := \Sigma x \in A.B,$$

$$A \vee B := A + B,$$

and if $x \notin \mathrm{FV}(B)$

$$A \to B := \Pi x \in A.B, \qquad A \wedge B := \Sigma x \in A.B.$$

Also

$$\bot := I(N,0,1). \quad \square$$

The hybrid character of A as type or proposition may be confusing at first, but it is definitely convenient in certain respects. We may switch at will from Π to $\forall$ etc., whichever reading is intuitively more appealing.

EXAMPLES. (i) Suppose $\vdash A$ type and $\vdash B$ type, then we can construct a deduction (cf. 10.8.2)

$$\dfrac{z \in A \vee B \quad \dfrac{\dfrac{\dfrac{\mathcal{D}}{A \text{ type}}}{x \in A \Rightarrow x \in A}\,\text{ASS}}{x \in A \Rightarrow \boldsymbol{k}_1(x) \in B \vee A} \quad \dfrac{\dfrac{\dfrac{\mathcal{D}'}{B \text{ type}}}{y \in B \Rightarrow y \in B}\,\text{ASS}}{y \in B \Rightarrow \boldsymbol{k}_0(x) \in B \vee A}\,{+\text{I}}}{\dfrac{z \in A \vee B \Rightarrow D_{x,y}(z, x, y) \in B \vee A}{\lambda z.D_{x,y}(z, x, y) \in A \vee B \to B \vee A}\,\Pi\text{I}}\,{+\text{E}}$$

(ii) Assume $\vdash I(A, t, t') \wedge I(A, t', t'')$ type. We exhibit an abbreviated deduction of the transitivity of equality at type A; the context $z \in I(A, t, t') \wedge I(A, t', t'')$ is not shown; the bottom line has an empty context.

$$\dfrac{\dfrac{\dfrac{\dfrac{\dfrac{\dfrac{\mathcal{D}}{z \in I(A, t, t') \wedge I(A, t', t'')}}{\boldsymbol{p}_0 z \in I(A, t, t')}}{t = t' \in A} \quad \dfrac{\dfrac{\dfrac{\mathcal{D}}{z \in I(A, t, t') \wedge I(A, t', t'')}}{\boldsymbol{p}_1 z \in I(A, t', t'')}}{t' = t'' \in A}}{t = t'' \in A}}{\boldsymbol{e} \in I(A, t, t'')}}{\lambda z.\boldsymbol{e} \in I(A, t, t') \wedge I(A, t', t'') \to I(A, t, t'')}$$

2.10. *Special equality rules* (CONV-*rules*). Let "conv" be a binary relation between terms of the same type which holds in the following cases, for terms of the appropriate types:

$$R_{x,y}(0,t_0,t_1) \operatorname{conv} t_0, \ R_{x,y}(St,t_0,t_1) \operatorname{conv} t_1[x, y / t, R_{x,y}(t, t_0, t_1)];$$

$$\operatorname{Ap}(\lambda x.t, t') \operatorname{conv} t[x/t'];$$

$$\boldsymbol{p}_i(t_0, t_1) \operatorname{conv} t_i (i \in \{0,1\}), \qquad (\boldsymbol{p}_0 t, \boldsymbol{p}_1 t) \operatorname{conv} t;$$

$$D_{x,y}(\boldsymbol{k}_i t, t_0, t_1) \operatorname{conv} t_i[x_i/t] \quad (i \in \{0,1\}).$$

Here, as everywhere, we tacitly assume a suitable renaming of bound variables, so as to avoid clashes of variables. The special equality rules for N, Π, Σ, and $+$ state, roughly, that if $t \operatorname{conv} t'$, then $t = t' \in A$ for the appropriate type A. More precisely,

$$\text{NCONV} \quad \begin{cases} \dfrac{t_0 \in A[0] \quad x \in N, y \in A \Rightarrow t_1 \in A[x/Sx]}{R_{x,y}(0, t_0, t_1) = t_0 \in A[0]} \\ \dfrac{t \in N \quad t_0 \in A[0] \quad x \in N, y \in A \Rightarrow t_1 \in A[x/Sx] \quad x \in N \Rightarrow A \text{ type}}{R_{x,y}(St, t_0, t_1) = t_1[x,y/t,R_{x,y}(t,t_0,t_1)] \in A[St]} \end{cases}$$

$$\Pi\text{CONV} \quad \frac{\lambda x.t \in \Pi x \in A.B \quad t' \in A \quad x \in A \Rightarrow B \text{ type}}{(\lambda x.t)(t') = t[x/t'] \in B[x/t']}$$

$$\Sigma\text{CONV} \quad \begin{cases} \dfrac{(t_0, t_1) \in \Sigma x \in A.B \quad A \text{ type}}{p_0(t_0, t_1) = t_0 \in A} \qquad \dfrac{(t_0, t_1) \in \Sigma x \in A.B \quad x \in A \Rightarrow B \text{ type}}{p_1(t_0, t_1) = t_1 \in B[x/t_0]} \\ \dfrac{t \in \Sigma x \in A.B}{(p_0 t, p_1 t) = t \in \Sigma x \in A.B} \end{cases}$$

$$+\text{CONV} \quad \frac{\begin{array}{c} x_0 \in A_0 \Rightarrow t_0 \in C[z/k_0 x_0] \\ t \in A_i \quad x_1 \in A_1 \Rightarrow t_1 \in C[z/k_1 x_1] \quad z \in A_0 + A_1 \Rightarrow C \text{ type} \end{array}}{D_{x_0,x_1}(k_i t, t_0, t_1) = t_i[x_i/t] \in C[z/k_i t]}$$

For I-types we add

$$\text{ICONV} \quad \frac{t \in I(A, s, s')}{t = e \in I(A, s, s')}$$

Remarks. (i) The special equality rules might also have been "distributed" over the E-rules since for each type-forming operation the corresponding CONV-rule explains the behaviour of the elimination constant with respect to equality.

(ii) It is to be noted that two forms of conversion are missing: we have not postulated an *η-rule* $\lambda x.tx \operatorname{conv} t$ for $x \notin \mathrm{FV}(t)$, or more formally

$$\frac{\lambda x.tx \in \Pi x \in A.B}{\lambda x.tx = t \in \Pi x \in A.B} \quad (x \notin \mathrm{FV}(t)).$$

We also do not have a *ξ-axiom* (*weak extensionality axiom*)

$$\frac{x \in A \Rightarrow t = t' \in B}{\lambda x.t = \lambda x.t' \in \Pi x \in A.B},$$

nor even the more special case corresponding to the ξ-*rule* in the λ-calculus,

$$\frac{t \succcurlyeq t' \in B \quad x \in A \Rightarrow B \text{ type}}{\lambda x.t = \lambda x.t' \in \Pi x \in A.B}$$

where $\succcurlyeq$ is a reduction relation generated by our conversions.

As we shall see later, $\mathbf{ML}_0^{\mathrm{i}}$ permits an alternative formulation based on combinators in which $\lambda x.t$ can be defined in the usual way (2.18).

2.11. *An example.* Below, in 2.13–16, we shall establish some basic properties of the formalism. However, by way of motivation of the formalism, we shall first present an example. At first sight one might expect that whenever a context $\Gamma \equiv \langle x_0 \in A_0, x_1 \in A_1, \ldots, x_n \in A_n \rangle$ is given, and we drop from Γ those $x_i \in A_i$ for which $x_i \notin \mathrm{FV}(A_n)$, we still have a context. However, the following example shows that this is not true. Let $I_0 := I(N \to N, s, t)$, $I_1 := I(I_0, e, e)$, $I_2 := I(I_1, e, e)$, and assume $s \in N \to N$, $t \in N \to N$ to have been established. Then we can construct the following deduction:

$$\dfrac{\dfrac{\dfrac{\dfrac{\dfrac{\dfrac{\dfrac{\dfrac{\dfrac{\begin{matrix}\mathscr{D}_0 \\ s \in N \to N\end{matrix} \quad \begin{matrix}\mathscr{D}_1 \\ t \in N \to N\end{matrix}}{I_0 \ \textit{type}}}{x \in I_0 \Rightarrow x \in I_0}}{x \in I_0 \Rightarrow x = e \in I_0}}{x \in I_0 \Rightarrow e \in I_0} \quad \mathscr{D}}{x \in I_0 \Rightarrow \quad I_1 \text{ type}}}{x \in I_0,\, y \in I_1 \Rightarrow y \in I_1}}{x \in I_0,\, y \in I_1 \Rightarrow y = e \in I_1}}{x \in I_0,\, y \in I_1 \Rightarrow e \in I_1} \quad \mathscr{D}'}{x \in I_0,\, y \in I_1 \Rightarrow I_2 \text{ type}}$$

Suppose s, t to be two primitive recursive functions which are extensionally equal, but which cannot be shown to be equal by our special equality rules. Then the assumption $x \in I_0$ is really needed to show $e \in I_0$; and this in turn is needed to show that I_1 is a type. Only if we know that I_1 is a type, we may introduce $y \in I_1$ in a context, so $y \in I_1$ has to be preceded by $x \in I_0$; we cannot show that $y \in I_1$ itself is a context. (As we shall see later, it is not difficult to find examples of s and t which cannot be proved to be equal in $\mathbf{ML}_0^{\mathrm{i}}$; cf. 6.1.) The example also shows that a closed assertion θ may need an inhabited context.

In presenting a deduction like the one above, there are some obvious possibilities for abbreviating. For example, the repetition of $\mathscr{D}$ and $\mathscr{D}'$ is made necessary by the form of the ITYP rule; however, we may abbreviate

$$\frac{t \in A \quad t \in A}{I(A, t, t) \text{ type}} \quad \text{to} \quad \frac{t \in A}{I(A, t, t) \text{ type}}$$

Also, a sequence of steps such as

$$\cfrac{\cfrac{x \in I_0 \Rightarrow x \in I_0}{x \in I_0 \Rightarrow x = e \in I_0}}{x \in I_0 \Rightarrow e \in I_0}$$

may be rendered in abbreviated form dropping the middle line. In the sequel we shall rather often use such abbreviations in exhibiting deductions.

Note that if we wish to have the property of contexts mentioned at the end of 2.4, we cannot afford to permute assumptions $x \in A$, $y \in B$ in a context, *even* in the case where A and B are closed; this also follows from the example above.

2.12. Another property we should like to have is that whenever a statement $\Gamma \Rightarrow t \in A$ or $\Gamma \Rightarrow t = s \in A$ has been demonstrated, then $\Gamma \Rightarrow A$ type can be demonstrated as well. As an easy way to guarantee this we have included "type premisses" among the premisses of most of the E- and I-rules and special equality rules (such as $x \in A \Rightarrow B$ type in ΠI, ΠE, ΣE). In some cases these premisses are actually redundant (see e.g. 3.1); on the other hand, at least in ΣI the premiss "$x \in A \Rightarrow B$ type" cannot be omitted without strengthening the system and losing some derived rules, as will be shown in 2.16 below.

Our next aim is to obtain some easy derived rules for the system $\mathbf{ML}_0^{\mathrm{i}}$. We first give a definition.

2.13. DEFINITION.
(i) If $\Gamma \equiv \langle x_0 \in A_0, \ldots, x_n \in A_n \rangle$, then
$\Gamma | i \equiv \langle x_0 \in A_0, \ldots, x_{i-1} \in A_{i-1} \rangle$.
(ii) $\Gamma \Rightarrow A$ type is a *type statement* (TS) of $\Gamma \Rightarrow \theta$ if θ has one of the forms, for suitable t, s and B:

$$A \text{ type}; \qquad t \in A; \qquad t = s \in A; \qquad A = B; \qquad B = A.$$

(iii) A *subtype statement* (STS) of a TS of the form $\Gamma \Rightarrow A$ type is:

$$\begin{array}{l} \Gamma, x \in B \Rightarrow C \text{ type}, \quad \text{if } A \equiv \Pi x \in B.C \quad \text{or } \Sigma x \in B.C, \\ \Gamma \Rightarrow B \text{ type}, \quad \text{if } A \equiv I(B, t, s), \\ \Gamma \Rightarrow B \text{ type or } \Gamma \Rightarrow C \text{ type}, \quad \text{if } A \equiv B + C. \ \square \end{array}$$

Some basic derived rules are collected in the following proposition.

2.14. PROPOSITION.
(i) *If* $\vdash \Gamma \Rightarrow \theta$, *then* $\mathrm{FV}(\theta) \subset \mathrm{FV}(\Gamma)$.
(ii) *Let* $\Gamma \equiv \langle x_0 \in A_0, \ldots, x_n \in A_n \rangle$, $\vdash \Gamma \Rightarrow \theta$. *Then for all* $i \le n$ $\Gamma | i \Rightarrow A_i$ type *is also derivable.*
(iii) *If* $\vdash \Gamma \Rightarrow \theta$, *then the TS of* $\Gamma \Rightarrow \theta$ *are also derivable.*
(iv) *If* $\vdash \Gamma \Rightarrow A$ type, *then the STS of* $\Gamma \Rightarrow A$ type *are also derivable.*
(v) *If* $\vdash \Gamma \Rightarrow I(A, t, s)$ type, *then also* $\vdash \Gamma \Rightarrow t \in A$ *and* $\vdash \Gamma \Rightarrow s \in A$.

PROOF.
(i) Straightforward by induction on the deduction tree for $\Gamma \Rightarrow \theta$.
(ii) Induction on the depth of the deduction tree $\mathcal{D}$ of $\Gamma \Rightarrow \theta$. The only interesting cases are those where the rule applied changes the context not just by the deletion of final assumptions: ASS, THIN, SUB, and REPL. The induction step is trivial for ASS.

If the last step of $\mathcal{D}$ is an instance of THIN, e.g.

$$\frac{\Gamma, \Gamma', y \in B, \Gamma'' \Rightarrow \theta \qquad \Gamma \Rightarrow A \text{ type}}{\Gamma, x \in A, \Gamma', y \in B, \Gamma'' \Rightarrow \theta}$$

then by the induction hypotheses $\vdash \Gamma, \Gamma' \Rightarrow B$ type, and hence by THIN $\vdash \Gamma, x \in A, \Gamma' \Rightarrow B$ type.

If the last step of $\mathcal{D}$ is an instance of SUB, e.g.

$$\frac{\Gamma, x \in A, y \in A_1[x], \Gamma' \Rightarrow \theta \qquad \Gamma \Rightarrow t \in A}{\Gamma, y \in A_1[t], \Gamma'[x/t] \Rightarrow \theta[x/t]}$$

then we have to show $\Gamma \Rightarrow A_1[t]$ type; but by the induction hypothesis $\Gamma, x \in A \Rightarrow A_1$ type, hence by SUB $\Gamma, x \in A \Rightarrow A_1[t]$ type. The treatment of REPL is similar.

(iii) Induction on the depth of derivations. For example, let the last step of the derivation be an instance of ΣE:

$$\frac{\Gamma \Rightarrow t \in \Sigma x \in A.B \quad \Gamma, x \in A \Rightarrow B \text{ type}}{\Gamma \Rightarrow \mathrm{p}_1 t \in B[x/\mathrm{p}_0 t]}$$

By (ii) the second premiss implies the existence of a proof of $\Gamma \Rightarrow A$ type, and thus with the other ΣE-rule $\Gamma \Rightarrow \mathbf{p}_0 t \in A$. Then SUB applied to $\Gamma, x \in A \Rightarrow B$ type yields $\Gamma \Rightarrow B[x/\mathbf{p}_0 t]$ type.

As a second example of the induction step, consider the case where the last rule applied is NE, say

$$\frac{t \in N \quad t_0 \in A[x/0] \quad x \in N, y \in A \Rightarrow t_1 \in A[x/Sx] \quad x \in N \Rightarrow A \text{ type}}{R_{x,y}(t, t_0, t_1) \in A[x/t]},$$

then $A[x/t]$ type follows by an application of SUB to $t \in N$ and $x \in N \Rightarrow A$ type.

(iv) Inspection of the rules shows that a derivation with a conclusion $\Gamma \Rightarrow A$ type must have as its final rule either one of the TYP rules (i.e. NTYP, ΠTYP, ΣTYP, ITYP, +TYP), or SUB, or THIN. Let for example $A \equiv \Pi x \in B.C$. Then a derivation of $\Gamma \Rightarrow A$ type must necessarily be of the following sort:

$$\dfrac{\begin{matrix}\mathcal{D}_1 \\ \Gamma_1' \Rightarrow t_1 \in A_1\end{matrix} \qquad \dfrac{\begin{matrix}\mathcal{D}_0 & & \mathcal{D} \\ \Gamma_0' \Rightarrow t_0 \in A_0 & & \Gamma_0 \Rightarrow \Pi x \in B_0.C_0 \text{ type}\end{matrix}}{\Gamma_1 \Rightarrow \Pi x \in B_1.C_1 \text{ type}}}{\Gamma_2 \Rightarrow \Pi x \in B_2.C_2 \text{ type}}$$

$$\begin{matrix}\vdots \\ \overline{\Gamma_n \Rightarrow \Pi x \in B_n.C_n \text{ type}} \quad (\Gamma_n \Rightarrow \Pi x \in B_n.C_n \equiv \Gamma \Rightarrow A)\end{matrix}$$

The kind of type at the bottom is introduced somewhere higher up in the derivation, after that it is only changed by SUB, the applications of which are shown. Furthermore, only applications of THIN can occur, they have been indicated by double lines.

This derivation is readily transformed into

$$\dfrac{\begin{matrix}\mathcal{D}_1 \\ \Gamma_1' \Rightarrow t_1 \in A_1\end{matrix} \qquad \dfrac{\begin{matrix}\mathcal{D}_0 & & \mathcal{D} \\ \Gamma_0' \Rightarrow t_0 \in A_0 & & \Gamma_0, x \in B_0 \Rightarrow C_0 \text{ type}\end{matrix}}{\Gamma_1, x \in B_1 \Rightarrow C_1 \text{ type}}}{\Gamma_2, x \in B_2 \Rightarrow C_2 \text{ type}}$$

$$\begin{matrix}\vdots \\ \overline{\Gamma, x \in B_n \Rightarrow C_n \text{ type}}\end{matrix}$$

where the applications of SUB and THIN correspond step by step with those of the original derivation.

The argument is quite similar for $A \equiv \Sigma x \in B.C$, $A \equiv I(B, s, t)$, $A \equiv B + C$.

(v) is proved by a completely similar argument. □

2.15. COROLLARY.

$$\frac{A = A}{A \text{ type}}$$

is a derived rule, i.e., if $\vdash \Gamma \Rightarrow A = A$, *then* $\vdash \Gamma \Rightarrow A$ *type.* □

2.16. *An example.* The following example shows that dropping the premiss "$x \in A \Rightarrow A$ type" in the ΣI-rule leads to a strengthening of the system for which 2.14 fails.

One readily shows $\boldsymbol{e} \in I(I(N, 0, 0), \boldsymbol{e}, \boldsymbol{e})$; therefore an application of ΣI with the type statement premiss dropped yields

$$\frac{0 \in N \quad (\boldsymbol{e} \in I(I(N, x, 0), \boldsymbol{e}, \boldsymbol{e}))[x/0]}{(0, \boldsymbol{e}) \in \Sigma x \in N.I(I(N, x, 0), \boldsymbol{e}, \boldsymbol{e})}.$$

If 2.14 holds, there must be a derivation of $x \in N \Rightarrow I(I(N, x, 0), \boldsymbol{e}, \boldsymbol{e})$ type and hence also of $x \in N \Rightarrow \boldsymbol{e} \in I(N, x, 0)$; hence $x \in N \Rightarrow x = 0 \in N$, which is clearly underivable. □

In most of the other rules the type statement premisses are in fact redundant; cf. 3.1. In any case, the correct form of the type statement is as a rule obvious from the other premisses, thus we shall adopt the following convention.

2.17. CONVENTION. In exhibiting (parts of) deductions we shall often omit the type statement premisses. Also we shall from now on tacitly adopt the convention that, for a given $C[x_1, \ldots, x_n]$, $C[t_1, \ldots, t_n]$ is in fact short for $C[x_1, \ldots, x_n/t_1, \ldots, t_n]$.

2.18. *Combinators in* $\mathbf{ML}_0^i$. In $\mathbf{ML}_0^i$ we can define combinators $\boldsymbol{k}$, $\boldsymbol{s}$, $\boldsymbol{p}'$, $\boldsymbol{r}$, $\boldsymbol{f}$ of the appropriate types of

$$\boldsymbol{k} := \lambda xy.x,$$

$$\boldsymbol{s} := \lambda xyz.xz(yz),$$

$$\boldsymbol{p}' := \lambda xy.\boldsymbol{p}(x, y),$$

$$\boldsymbol{r} := \lambda uvw.R_{x,y}(u, v, wxy),$$

$$\boldsymbol{f} := \lambda uvw.D_{x,y}(u, vx, wy).$$

The types of these constants are given by

$$\frac{x \in A \Rightarrow B \text{ type}}{\boldsymbol{k} \in \Pi x \in A\, \Pi y \in B.A}$$

$$\frac{x \in A,\ y \in B[x] \Rightarrow C[x, y] \text{ type}}{\boldsymbol{s} \in \Pi u \in (\Pi x \in A \Pi y \in B.C)\, \Pi v \in (\Pi x \in A.B)\, \Pi w \in A.C[w, vw]}$$

$$\frac{x \in A \Rightarrow B \text{ type}}{\boldsymbol{p}' \in \Pi x \in A\, \Pi y \in B.(\Sigma x \in A.B)}$$

$$\frac{x \in N \Rightarrow A \text{ type}}{\boldsymbol{r} \in \Pi u \in N\, \Pi v \in A[0]\, \Pi w \in (\Pi x \in N\, \Pi y \in A.A[x/Sx]).A[x/u]}$$

$$\frac{z \in A_0 + A_1 \rightarrow C[z] \text{ type}}{\boldsymbol{f} \in \Pi z \in A_0 + A_1\, \Pi u \in B_0\, \Pi v \in B_1.C[z]},$$

where $B_i := \Pi x_i \in A_i.C[z/\boldsymbol{k}_i x_i]$.

We can derive special equality rules for these constants corresponding to the conversions $\boldsymbol{k}t_1t_2 \text{ conv } t_1$, $\boldsymbol{s}t_1t_2t_3 \text{ conv } t_1t_3(t_2t_3)$, $\boldsymbol{p}'(\boldsymbol{p}_0t)(\boldsymbol{p}_1t) \text{ conv } t$, $\boldsymbol{p}_i(\boldsymbol{p}'t_0t_1) \text{ conv } t_i$ $(i \in \{0, 1\})$, $\boldsymbol{r}0t_1t_2 \text{ conv } t_1$, $\boldsymbol{r}(St)t_1t_2 \text{ conv } t_2t(\boldsymbol{r}tt_1t_2)$, $\boldsymbol{f}(\boldsymbol{k}_it)t_0t_1 \text{ conv } t_it$ $(i \in \{0,1\})$.

From these combinators we can in turn define λ, $\boldsymbol{p}$, $D_{x,y}$, $R_{x,y}$ by

$$\begin{aligned}
&\lambda x.t := \boldsymbol{k}t, \quad \text{if } x \notin \mathrm{FV}(t),\\
&\lambda x.tx := t, \quad \text{if } x \notin \mathrm{FV}(t),\\
&\lambda x.x := \boldsymbol{skk},\\
&\lambda x.t_1t_2 := \boldsymbol{s}(\lambda x.t_1)(\lambda x.t_2), \quad \text{if } t_2 \not\equiv x \text{ or } x \notin \mathrm{FV}(t_1),\\
&\boldsymbol{p}(t_1, t_2) := \boldsymbol{p}'t_1t_2,\\
&R_{x,y}(t, t_0, t_1) := \boldsymbol{r}tt_0(\lambda xy.t_1),\\
&D_{x,y}(t, t_0, t_1) := \boldsymbol{f}t(\lambda x.t_0)(\lambda y.t_1).
\end{aligned}$$

These operators defined in terms of the combinators can be shown to satisfy the rules; but it is to be noted that first defining the combinators from λ, $R_{x,y}$, $D_{x,y}$, $\boldsymbol{p}$, $\boldsymbol{k}_0$, $\boldsymbol{k}_1$, $\boldsymbol{p}_0$, $\boldsymbol{p}_1$ and then defining operators λ, $R_{x,y}, \ldots$ again does not yield the identity mapping: $\lambda x.x$ is defined from combinators as $\boldsymbol{skk}$, and this combination is defined from λ as

$$(\lambda xyz.xz(yz))\boldsymbol{kk} = \lambda z.\boldsymbol{k}z(\boldsymbol{k}z) = \lambda z.((\lambda xy.x)z)((\lambda xy.x)z),$$

which cannot be simplified to $\lambda z.z$ in the absence of the ξ-rule (i.e. the rule $t \succcurlyeq t' \Rightarrow \lambda x.t \succcurlyeq \lambda x.t'$).

3. Some alternative formulations of $\mathbf{ML}_0^{\mathrm{i}}$

In this section we shall describe a few alternative formulations of $\mathbf{ML}_0^{\mathrm{i}}$; these formulations are not needed in later sections, therefore this section may be skipped.

3.1. *Dropping the type statement premisses.* We obtain a system equivalent to $\mathbf{ML}_0^{\mathrm{i}}$ by dropping the type statement premisses in all I-, E-, and CONV-rules except ΣI, +I, +E.

This can be established by a closer inspection of the proof of 2.14. Let us call the rules with type premiss dropped "strong rules", and let $\mathbf{ML}_0^*$ be the system containing $\mathbf{ML}_0^{\mathrm{i}}$ and in addition the strong rules except ΣI, +I, +E. The *critical depth* of a deduction $\mathscr{D}$ (cd($\mathscr{D}$) or "cd of $\mathscr{D}$") in $\mathbf{ML}_0^*$ is the maximum number of applications of the strong rules in any branch of $\mathscr{D}$. We wish to show by induction on cd($\mathscr{D}$) that any derivation $\mathscr{D}$ in $\mathbf{ML}_0^*$ can be transformed into a derivation $\mathscr{D}^*$ in $\mathbf{ML}_0^{\mathrm{i}}$ by systematically weakening each application of the strong rules. One observes:

(a) 2.14(ii) holds for $\mathbf{ML}_0^*$, and if $\mathscr{D}$ shows $\vdash^* \Gamma \Rightarrow \theta$, then there is a $\mathscr{D}_i$ with $\mathrm{cd}(\mathscr{D}_i) \le \mathrm{cd}(\mathscr{D})$ for $\vdash^* \Gamma | i \Rightarrow A_i$ type. This is so because the inductive argument of the proof shows that the $\mathscr{D}_i$ are constructed from subderivations of $\mathscr{D}$ with the help of SUB and THIN only.

(b) Inspection of the proofs of 2.14(iii)–(iv), shows that we can in fact prove these simultaneously in a sharpened form: if $\mathscr{D}$ shows $\vdash^* \Gamma \Rightarrow \theta$, then for each TS "$\Gamma \Rightarrow A$ type" of $\Gamma \Rightarrow \theta$ there is a $\mathscr{D}'$ showing $\vdash^* \Gamma \Rightarrow A$ type, and for each STS "$\Gamma' \Rightarrow B$ type" of a TS of $\Gamma \Rightarrow \theta$ there is a $\mathscr{D}''$ with $\mathrm{cd}(\mathscr{D}') = \mathrm{cd}(\mathscr{D}'') = 0$ for $\vdash^* \Gamma' \Rightarrow B$ type, and a $\mathscr{D}''$ with $\mathrm{cd}(\mathscr{D}''') = 0$ for $\vdash^* \Gamma \Rightarrow \theta$. The proof proceeds by induction on cd($\mathscr{D}$).

Consider for example a deduction $\mathscr{D}$ with as final step an application of strong ΣE:

$$\cfrac{\begin{array}{c}\mathscr{D}'\\ \Gamma \Rightarrow t \in \Sigma x \in A . B\end{array}}{\Gamma \Rightarrow \mathbf{p}_1 t \in B[x/\mathbf{p}_0 t]}$$

Then $\mathrm{cd}(\mathscr{D}') = \mathrm{cd}(\mathscr{D}) \dot{-} 1$. By the induction hypothesis we have

$$\begin{array}{l}\mathscr{D}_1' \vdash^* \Gamma \Rightarrow t \in \Sigma x \in A . B,\\ \mathscr{D}_2' \vdash^* \Gamma \Rightarrow \Sigma x \in A . B \text{ type},\\ \mathscr{D}_3' \vdash^* \Gamma, x \in A \Rightarrow B \text{ type},\end{array}$$

with $\mathrm{cd}(\mathscr{D}_1') = \mathrm{cd}(\mathscr{D}_2') = \mathrm{cd}(\mathscr{D}_3') = 0$.

From $\mathscr{D}_3'$ we obtain $\mathscr{D}_4 \vdash {}^*\Gamma \Rightarrow A$ type, cd$(\mathscr{D}_4) = 0$, and thus

$$\frac{\begin{matrix}\mathscr{D}_1' & \mathscr{D}_4 \\ \Gamma \Rightarrow t \in \Sigma x \in A.B & \Gamma \Rightarrow A \text{ type}\end{matrix}}{\Gamma \Rightarrow \mathbf{p}_0 t \in A}$$

is a derivation $\mathscr{D}_5$ of $\Gamma \Rightarrow \mathbf{p}_0 t \in A$ with cd$(\mathscr{D}_5) = 0$. Then

$$\frac{\begin{matrix}\mathscr{D}_1' & \mathscr{D}_3' \\ \Gamma \Rightarrow t \in \Sigma x \in A.B & \Gamma, x \in A \Rightarrow B \text{ type}\end{matrix}}{\Gamma \Rightarrow \mathbf{p}_1 t \in B[x/\mathbf{p}_0 t]}$$

is a derivation $\mathscr{D}_6$ with cd$(\mathscr{D}_6) = 0$; and by the substitution rule applied to $\mathscr{D}_3'$ and $\mathscr{D}_5$ we see that $\Gamma \Rightarrow B[x/\mathbf{p}_0 t]$ type by a derivation with cd zero, etc. □

We now turn to a more liberal formulation of $\mathbf{ML}_0^{\mathrm{i}}$ based on the notion of an *amalgam* of contexts.

3.2. DEFINITION. Two contexts

$$\Gamma \equiv \langle x_0 \in A_0, \ldots, x_n \in A_n \rangle, \qquad \Gamma' \equiv \langle x_0' \in A_0', \ldots, x_m' \in A_m' \rangle$$

are said to be *compatible*, if for all i, j such that $x_i \equiv x_j'$ also $A_i \equiv A_j$ (i.e. A_i and A_j' are syntactically identical modulo the renaming of bound variables).

Let Γ, Γ' be two compatible contexts. The context Γ'' is said to be an *amalgam* of Γ and Γ' if Γ' is obtained by ordering $\Gamma \cup \Gamma'$ in such a way that Γ and Γ' are contained in Γ'' as subsequences. "Compatible" and "amalgam" are defined similarly for finitely many contexts. □

EXAMPLE. Let $\Gamma \equiv \langle x_0 \in A_0, x_1 \in A_1, x_2 \in A_2[x_1] \rangle$, $\Gamma' \equiv \langle y_0 \in A_0', x_1 \in A_1, y_2 \in A_2'[y_0, x_1] \rangle$; all variables free in A_i and A_j' are shown and x_0, x_1, x_2, y_0, y_2 are all distinct. Then

$$x_0 \in A_0, y_0 \in A_0', x_1 \in A_1, x_2 \in A_2[x_1], y_2 \in A_2'[y_0, x_1]$$

and

$$y_0 \in A_0', x_0 \in A_0, x_1 \in A_1, x_2 \in A_2'[y_0, x_1], x_2 \in A_2[x_1]$$

are both amalgams of Γ and Γ'. □

3.3. *A generalization of the thinning rule.* Let THIN* be the following generalization of THIN:

$$\text{THIN*}\qquad \frac{\Gamma \Rightarrow A \text{ type} \quad \Gamma' \Rightarrow \theta}{\Gamma'' \Rightarrow \theta},$$

where Γ'' is an amalgam of the compatible contexts Γ, $x \in A$ and Γ' (for some variable x). We have the following proposition.

PROPOSITION. THIN* *holds as a derived rule in* $\mathbf{ML}_0^i$.

PROOF. Tedious but straightforward. We give an example. Let Γ, Γ' be as in the example of 3.2, and suppose we have premisses

$$x_0 \in A_0, x_1 \in A_1 \Rightarrow A_2 \text{ type} \qquad y_0 \in A_0', x_1 \in A_1, y_2 \in A_2' \Rightarrow \theta.$$

Then $\Gamma'' \Rightarrow \theta$ for the first amalgam mentioned in the example of 3.2, by repeated application of THIN

$$\dfrac{\dfrac{\dfrac{A_0 \text{ type} \quad A_0' \text{ type}}{x_0 \in A_0 \Rightarrow A_0' \text{ type}} \quad x_0 \in A_0, x_1 \in A_1 \Rightarrow A_2 \text{ type}}{x_0 \in A_0, y_0 \in A_0', x_1 \in A_1 \Rightarrow A_2 \text{ type}} \qquad \dfrac{A_0 \text{ type} \quad y_0 \in A_0', x_1 \in A_1, y_2 \in A_2' \Rightarrow \theta}{x_0 \in A_0, y_0 \in A_0', x_1 \in A_1, y_2 \in A_2' \Rightarrow \theta}}{x_0 \in A_0, y_0 \in A_0', x_1 \in A_1, x_2 \in A_2, y_2 \in A_2' \Rightarrow \theta}$$

□

3.4. $\mathbf{ML}_0^i$ *based on* THIN*. We now obtain an equivalent formulation of $\mathbf{ML}_0^i$ by generalizing THIN to THIN*, and interpreting all other rules except ASS, REPL not as in 2.5, but as

$$\frac{\Gamma_0, \Delta_0 \Rightarrow \theta_0 \quad \cdots \quad \Gamma_n, \Delta_n \Rightarrow \theta_n}{\Gamma^* \Rightarrow \theta}$$

where the Δ_i are discharged assumptions and Γ^* an amalgam of $\Gamma_1, \ldots, \Gamma_n$.

3.5. *A tree representation of* $\mathbf{ML}_0^i$. Another possibility for presenting derivations in $\mathbf{ML}_0^i$ is to keep as closely as possible to the deduction trees of natural deduction, though with a term at each node. In that case the open assumptions at each node are not shown in the form of a context, but are found by inspection of the tree above the node. Two features of such a representation are to be noted. The correct formulation of a tree representation is not entirely straightforward.

For example, if we formulate ASS as

$$\frac{\begin{array}{c}\vdots\\ A \text{ type}\end{array}}{x \in A}$$

then an open assumption does not necessarily appear at a top node, but may follow a derivation of a type statement. This can be remedied by formulating the rule as

$$\frac{\begin{array}{c}\vdots\\ A \text{ type}\end{array} \quad x \in A}{x \in A}$$

Similarly, the appropriate version of thinning becomes

$$\frac{\begin{array}{c}\vdots\\ A \text{ type}\end{array} \quad x \in A \quad \begin{array}{c}\vdots\\ \theta\end{array}}{\theta}$$

REPL requires a transformation of trees, etc. We leave it to the reader to devise a tree formulation of $\mathbf{ML}_0^{\mathrm{i}}$.

3.6. *A combinatorial variant of* $\mathbf{ML}_0^{\mathrm{i}}$. Finally we note that 2.18 suggests a combinatorial variant of $\mathbf{ML}_0^{\mathrm{i}}$ with $\boldsymbol{k}, \boldsymbol{s}, \boldsymbol{p}', \boldsymbol{r}, \boldsymbol{f}, \boldsymbol{p}_0, \boldsymbol{p}_1, \boldsymbol{k}_0, \boldsymbol{k}_1$ as primitives, with types as specified there and special equality rules as suggested by the conversions indicated for these constants in 2.18. The discussion in 2.18 also shows how to embed such a variant in $\mathbf{ML}_0^{\mathrm{i}}$ and vice versa.

4. The types N_k and reformulation of the E-rules

In this section we shall carry out some exercitions in the system $\mathbf{ML}_0^{\mathrm{i}}$, namely the explicit definition of the k-element types N_k, a proof that $I(N, 0, 1)$ can play the role of falsum in $\mathbf{ML}_0^{\mathrm{i}}$, and a reformulation of the elimination rules establishing a uniform pattern: the elimination rule expresses that the methods for introducing elements given by the I-rules are exhaustive, i.e. all elements of a type are equal to elements of the form specified by the I-rules.

4.1. PROPOSITION (*definition of* $\perp$). *If* $\mathbf{ML}_0^{\mathrm{i}} \vdash \Gamma \Rightarrow A$ type, *then* $\mathbf{ML}_0^{\mathrm{i}} \vdash \Gamma \Rightarrow \lambda x. \perp_A \in I(N, 0, 1) \rightarrow A$ *for a certain term* $\perp_A$.

Proof. We define $\perp_A$ by induction on the complexity of A:

$$
\begin{aligned}
&\perp_N := 0,\\
&\perp_{I(A,s,t)} := \boldsymbol{e},\\
&\perp_{\Pi x\in A.B} := \lambda x.\perp_B\,,\\
&\perp_{\Sigma x\in A.B} := (\perp_A\,,\perp_B),\\
&\perp_{A+B} := \boldsymbol{k}_0\perp_A \quad (\boldsymbol{k}_1\perp_B \text{ works equally well}).
\end{aligned}
$$

Note that $\perp_A$ is closed, and thus does not depend on variables free in A. The assertion of the theorem is proved by induction on A; the proof repeatedly appeals to 2.14. We check two cases.

(a) Let $\vdash \Gamma \Rightarrow A$, $A \equiv I(B,s,t)$ and abbreviate $I(N,t,s)$ as $t =_N s$. We first construct a deduction $\mathscr{D}$ (not exhibiting Γ)

$$
\dfrac{
\dfrac{
\dfrac{
\dfrac{\dfrac{}{0\in N}\,(\mathrm{NI}) \qquad \dfrac{0\in N}{1\in N}\,\mathrm{NI}}{I(N,0,1)\ \text{type}}\,\mathrm{ITYP}
}{u\in 0=_N 1 \Rightarrow u\in 0 =_N 1}\,\mathrm{ASS}
}{u\in 0=_N 1\Rightarrow 0=1\in N}\,\mathrm{IE}
\qquad
\dfrac{\dfrac{N\ \text{type}}{z\in N\Rightarrow z\in N} \quad \begin{matrix}\mathscr{D}'\\ s\in B\end{matrix} \quad \begin{matrix}\mathscr{D}''\\ t\in B\end{matrix}}{z\in N\Rightarrow R_{x,y}(z,s,t)\in B}\,\mathrm{NE}
}{u\in 0=_N 1\Rightarrow R_{x,y}(0,s,t)=R_{x,y}(1,s,t)\in B}\,\mathrm{REPL}
$$

and two deductions $\mathscr{D}_0$, $\mathscr{D}_1$

$$
\dfrac{\dfrac{\begin{matrix}\mathscr{D}'\\ s\in B\end{matrix} \qquad \begin{matrix}\mathscr{D}''\\ t\in B\end{matrix}}{R_{x,y}(0,s,t)=s\in B}}{s=R_{x,y}(0,s,t)\in B}
\qquad
\dfrac{\dfrac{0\in N}{1\in N} \quad \begin{matrix}\mathscr{D}'\\ s\in B\end{matrix} \quad \begin{matrix}\mathscr{D}''\\ t\in B\end{matrix}}{R_{x,y}(1,s,t)=t\in B}
$$

Combining $\mathscr{D}$, $\mathscr{D}_0$, $\mathscr{D}_1$ with TRANS yields a derivation of $u\in 0 =_N 1 \Rightarrow s=t\in B$, hence

$$
\dfrac{\dfrac{u\in 0=_N 1\Rightarrow s=t\in B}{u\in 0=_N 1\Rightarrow \boldsymbol{e}\in I(B,s,t)}}{\lambda u.\boldsymbol{e}\in I(N,0,1)\to I(B,s,t)}
$$

We have also tacitly used THIN, and left out some other obvious steps.

(b) Let $\vdash \Gamma \Rightarrow A$, $A \equiv \Pi x \in B.C$. Then

$$\dfrac{\dfrac{\dfrac{\dfrac{\dfrac{\mathcal{D}}{0 =_N 1 \text{ type}}}{y \in 0 =_N 1 \Rightarrow y \in 0 =_N 1} \qquad \dfrac{\dfrac{\mathcal{D}'}{x \in B \Rightarrow C[x] \text{ type}}}{x \in B \Rightarrow \lambda y.\bot_C \in 0 =_N 1 \to C}}{y \in 0 =_N 1, x \in B \Rightarrow \bot_C \in C}}{y \in 0 =_N 1 \Rightarrow \lambda x.\bot_C \in \Pi x \in B.C}}{\lambda yx.\bot_C \in 0 =_N 1 \to \Pi x \in B.C}$$

The double line indicates a contraction of several steps into one and is motivated by the induction hypothesis. Once again we did not indicate the thinnings occurring at several places. □

4.2. All the finite types of $\mathbf{HA}^\omega$ are present in the type structure of $\mathbf{ML}_0^{\mathrm{i}}$ if we identify type 0 with type N. Let us use σ, τ for arbitrary finite types. The following proposition is easy.

PROPOSITION. *All primitive recursive functionals are represented by terms in* $\mathbf{ML}_0^{\mathrm{i}}$ *satisfying the appropriate recursion equations.*

PROOF. Straightforward; e.g. if $t \in 0$, $t' \in \sigma$, $t'' \in \sigma \to (0 \to \sigma)$, then the term

$$t^* \equiv \lambda uvz.R_{x,y}(z, u, vyx)$$

satisfies

$$t^*t't''0 = t',$$
$$t^*t't''(St) = t''[x, y/t, t^*t't''t],$$

etc. □

4.3. *The k-element types* N_k. We can define types N_k with elements $0_k, \ldots, (k-1)_k$ and introduction rules

$\mathrm{N_kI}$ $\quad 0_k \in N_k, 1_k \in N_k, \ldots, (k-1)_k \in N_k,$

and an elimination rule with an elimination constant R_k satisfying

$$\mathrm{N_kE} \qquad \frac{t \in N_k \quad t_0 \in C[0_k] \cdots t_{k-1} \in C[(k-1)_k] \qquad x \in N_k \Rightarrow C[x] \text{ type}}{R_k(t, t_0, \ldots, t_{k-1}) \in C[t]}$$

and the special equality rules

$$\mathrm{N_kCONV} \qquad \frac{t_0 \in C[0_k] \cdots t_{k-1} \in C[(k-1)_k] \qquad x \in N_k \Rightarrow C[x] \text{ type}}{R_k(i_k, t_0, \ldots, t_{k-1}) = t_i \in C[i_k]}$$

As follows from 4.1, we can take $I(N,0,1)$ for N_0: the elimination rule for N_0 is

$$\frac{t \in N_0}{R_0(t) \in C}$$

and we can take $\lambda x.\perp_C$ for R_0. (Note that our definition of R_0 is not uniform, it depends on C!) N_0 has no introduction rule.

For all other k we can define N_k, i_k, R_k as follows. Let f_k be a term for a primitive recursive function such that

$$f_k(n) = n \quad \text{for } n < k,$$

$$f_k(n) = k-1 \quad \text{for } n \geq k.$$

Then we can take

$$N_k := \Sigma n \in N. I(N, n, f_k(n)),$$

$$i_k := (i, e) \quad \text{where } i \in \{0, 1, \ldots, k-1\}.$$

In particular we may take the signum function sg (cf. 3.1.3) for $k = 2$, i.e. $f_2 \equiv$ sg. For $k = 2$ we define R_2 by

$$R_2(t, t_0, t_1) := R_{x,y}(\mathbf{p}_0 t, t_0, t_1) \quad (x, y \notin \mathrm{FV}(t_1)).$$

We leave it to the reader to show that this R_2 satisfies the appropriate elimination rule N_2E, and N_2CONV, and to provide a definition of R_k for $k > 2$, and a proof of N_kE for these defined constants. □

In Martin-Löf's papers the N_k are primitives.

CONVENTION. We shall use $\mathbf{0}, \mathbf{1}$ for $0_2, 1_2$.

4.4. *An alternative formulation of* ΣE. Instead of formulating ΣE as in section 2, we can also use the following version:

ΣE*

$$\frac{t \in \Sigma x \in A.B \quad x \in A, y \in B[x] \Rightarrow t_1 \in C[z/(x,y)] \qquad z \in \Sigma x \in A.B \Rightarrow C[z] \text{ type}}{E_{x,y}(t, t_1) \in C[z/t]}$$

where $E_{x,y}$ is an operator binding x and y in t_1. We also need a special equality rule for E:

ΣCONV*

$$\frac{(t_0, t_1) \in \Sigma x \in A.B \quad x \in A, y \in B \Rightarrow t \in C[z/(x,y)] \quad z \in \Sigma x \in A.B \Rightarrow C[z] \text{ type}}{E_{x,y}((t_0, t_1), t) = t[x, y/t_0, t_1] \in C[z/(t_0, t_1)]}$$

That is to say, to the conversion rules we have to add a conversion rule for $E_{x,y}$:

$$E_{x,y}((t_0, t_1), t) \operatorname{conv} t[x, y/t_0, t_1].$$

PROPOSITION. ΣE *with* ΣCONV *is equivalent to* ΣE* *with* ΣCONV*.

PROOF. Assume ΣE and ΣCONV, and define

$$E_{x,y}(t, t_1) := t_1[x, y/\mathbf{p}_0 t, \mathbf{p}_1 t].$$

Then

$$\dfrac{\dfrac{t \in \Sigma x \in A.B}{\mathbf{p}_0 t \in A} \qquad \dfrac{t \in \Sigma x \in A.B}{\mathbf{p}_1 t \in B[x/\mathbf{p}_0 t]} \quad x \in A, y \in B \Rightarrow t_1 \in C[z/(x, y)]}{t_1[x, y/\mathbf{p}_0 t, \mathbf{p}_1 t] \in C[z/(\mathbf{p}_0 t, \mathbf{p}_1 t)]}$$

At the last line we applied the substitution rule twice. With surjectivity of pairing then

$$E_{x,y}(t, t_1) \in C[z/t].$$

Obviously, the conversion rule ΣCONV* is also valid.

Let us now assume, conversely, ΣE* and ΣCONV*, and define

$$\mathbf{p}_0 t := E_{x,y}(t, x), \qquad \mathbf{p}_1 t := E_{x,y}(t, y),$$

then

$$\dfrac{t \in \Sigma x \in A.B \qquad x \in A \Rightarrow x \in A}{\mathbf{p}_0 t \equiv E_{x,y}(t, x) \in A}$$

$$\dfrac{t \in \Sigma x \in A.B[x] \qquad x \in A, y \in B[x] \Rightarrow y \in B[\mathbf{p}_0(x, y)]}{\mathbf{p}_1 t \equiv E_{x,y}(t, y) \in B[\mathbf{p}_0 t]}$$

We must also show $E_{x,y}((t_0, t_1), x) = t_0$ and $E_{x,y}((t_0, t_1), y) = t_1$, which follows easily from ΣCONV*:

$$\dfrac{\dfrac{t_0 \in A \quad t_1 \in B[x/t_0]}{(t_0, t_1) \in \Sigma x \in A.B} \qquad x \in A, y \in B \Rightarrow x \in A}{\mathbf{p}_0(t_0, t_1) \equiv E_{x,y}((t_0, t_1), x) = t_0 \in A}$$

and similarly for $B \equiv B[x]$.

Let $t_0 \in A$, $t_1 \in B[t_0]$, $x \in A \Rightarrow B[x]$ type, and put $C[z] := B[x/\boldsymbol{p}_0 z]$. By the preceding derivation

$$x \in A,\ y \in B \Rightarrow \boldsymbol{p}_0(x, y) = x \in A, \tag{1}$$

hence by REPL

$$x \in A,\ y \in B \Rightarrow B[x] = B[\boldsymbol{p}_0(x, y)], \tag{2}$$

and therefore by SUB

$$B[t_0] = B[\boldsymbol{p}_0(t_0, t_1)]. \tag{3}$$

Also

$$\frac{(t_0, t_1) \in \Sigma x \in A.B[x] \qquad \begin{matrix}\mathscr{D}\\ x \in A,\ y \in B \Rightarrow y \in C[z/(x, y)] \equiv B[\boldsymbol{p}_0(x, y)]\end{matrix}}{E_{x,y}((t_0, t_1), y) = t_1 \in C[z/(t_0, t_1)]}$$

where $\mathscr{D}$ derives $x \in A,\ y \in B \Rightarrow y \in B[\boldsymbol{p}_0(x, y)]$ from $x \in A,\ y \in B \Rightarrow y \in B$, (2), and REPL. Applying (3) to the conclusion yields

$$t_1 \in B[t_0].$$

Finally we have to show surjectivity of pairing. The following derivation is made under the context $x \in A$, $y \in B[x]$ (not shown). Let $\Delta \equiv u \in A$, $v \in B[u]$, then

$$\dfrac{\dfrac{\mathscr{D}_1'}{y = \boldsymbol{p}_1(x, y) \in B[x]} \qquad \dfrac{\begin{matrix}\mathscr{D}_0'\\ \boldsymbol{p}_0(x, y) = x \in A\end{matrix} \quad \dfrac{\Delta \Rightarrow u \in A \qquad \Delta \Rightarrow v \in B[u]}{\Delta \Rightarrow (u, v) \in \Sigma x \in A.B[x]}}{v \in B[x] \Rightarrow (x, v) = \boldsymbol{p}_0 x, y), v) \in \Sigma x \in A.B[x]}}{\dfrac{(x, y) = (\boldsymbol{p}_0(x, y), \boldsymbol{p}_1(x, y)) \in \Sigma x \in A.B[x]}{\boldsymbol{e} \in I(\Sigma x \in A.B[x], (x, y), (\boldsymbol{p}_0(x, y), \boldsymbol{p}_1(x, y)))}}$$

Let $C[z] \equiv I(\Sigma x \in A.B[x], z, (\boldsymbol{p}_0 z, \boldsymbol{p}_1 z))$ and let $\mathscr{D}^*$ be the preceding derivation, then

$$\dfrac{\dfrac{t \in \Sigma x \in A.B[x] \qquad \begin{matrix}\mathscr{D}^*\\ x \in A,\ y \in B \Rightarrow \boldsymbol{e} \in C[(x, y)]\end{matrix}}{E_{x,y}(t, \boldsymbol{e}) \in C[t]}}{t = (\boldsymbol{p}_0 t, \boldsymbol{p}_1 t) \in \Sigma x \in A.B[x]}$$

□

It should be observed that the weakening of ΣE* where $z \notin \mathrm{FV}(C)$ corresponds exactly to ∃E in **N-IQC** (cf. 10.8.2).

4.5. IE *with elimination constant.* The rule ICONV can be replaced by

$$\text{IE}^* \qquad \frac{t \in I(A, t_0, t_1) \quad t' \in C[z/e] \quad z \in I(A, t_0, t_1) \Rightarrow C \text{ type}}{J(t, t') \in C[z/t]}$$

with a corresponding conversion rule $J(t, t') \operatorname{conv} t'$, in other words

$$\text{ICONV}^* \qquad \frac{t \in I(A, t_0, t_1) \quad t' \in C[z/e] \quad z \in I(A, t_0, t_1) \Rightarrow C \text{ type}}{J(t, t') = t' \in C[z/t]}$$

More precisely, we have the following proposition.

PROPOSITION. ICONV *is equivalent to the combination of* IE* *and* ICONV*.

PROOF. In one direction simply take $J(t, t') := t'$ and in the other direction use

$$C[z] := I(I(A, t_0, t_1), z, e).$$

We leave the details as an exercise. □

4.6. *Reformulating* ΠE *(sketch).* The reformulations IE*, ΣE* and the rules +E, NE clearly reveal a pattern: the introduction rules for a type tell us how elements of a type can be found (thus ΣI gives elements of a Σ-type as pairs, and elements of a +-type are given by +I as elements of the form $k_i t$) while the corresponding E-rules show that the I-rules are exhaustive, in the following sense: if we know how to "prove" a type/proposition for a value of a parameter which has the form specified by the I-rule, then we also know how to "prove" it for *all* values of the parameter. Thus in ΣE* the premiss

$$x \in A, y \in B \Rightarrow C[z/(x, y)]$$

gives a term in $C[z]$ for z an arbitrary pair; the conclusion $E_{x,y}(t_1, t) \in C[z/t_1]$ produces a term in $C[z/t_1]$ for t_1 an arbitrary element of $\Sigma x \in A.B$.

The only introduction rule for which the E-rule does not yet have this form is ΠI. What should a modified ΠE* look like, so as to conform to the pattern described above? The ΠI-rule gives elements of the form $\lambda x.t$, so we need an elimination rule which deduces from premisses (i) $t_1 \in \Pi x \in A.B$, and (ii) for all $\lambda x.t$ such that $x \in A \Rightarrow t \in B$ we have $t_2[z/\lambda x.t] \in C[z/\lambda x.t]$, a conclusion of the form $F_z(t_1, t_2) \in C[z/t_1]$, where F is an elimination constant.

However, our syntactical apparatus does not permit us to state a premiss of the form (ii), which is of a schematic character, in fact a collection of infinitely many premisses $x \in A \Rightarrow t \in B$, one for each t.

This difficulty can be removed at a considerable expense in syntactical complexity. Since we think it of some interest to see that a rule ΠE* conforming to the general pattern is possible, we shall present below a brief, necessarily incomplete, sketch of the solution.

We introduce two new syntactical categories: *families of objects over a type*, and *families of types over a type*. We shall use φ, φ', φ'' as metavariables for terms denoting families of objects, and Φ, Φ' for terms denoting families of types. We shall also use variables for object families. To the four basic forms of statement we now add

$$\varphi \in \Phi : A$$

(φ is a family of objects over A ("indexed by A") with a corresponding family of types Φ over A), and

$$\Phi : A \text{ ftype}$$

(Φ is a family of types over A). We also have a general notion of abstraction denoted by ( ) *in front of* an expression. N.B. In this chapter occurrences in types and terms are indicated by *square* brackets!

We list some of the obvious rules for families of types and objects:

$$\text{FTYP} \qquad \frac{x \in A \Rightarrow B \text{ type}}{(x)B : A \text{ ftype}}$$

$$\text{FASS} \qquad \frac{\Phi : A \text{ ftype}}{x \in \Phi : A \Rightarrow x \in \Phi : A} \quad (x \text{ a new variable})$$

$$\text{ABSTR} \qquad \frac{x \in A \Rightarrow t \in B}{(x)t \in (x)B : A}$$

The following rules express that abstraction ( ) is inverse to application:

$$\frac{t \in A \quad \varphi \in \Phi : A}{\varphi(t) \in \Phi(t)} \qquad \frac{t \in A \quad \Phi : A \text{ ftype}}{\Phi(t) \text{ type}}$$

ΠTYP and ΠI take the forms:

$$\Pi\text{TYP*} \qquad \frac{\Phi : A \text{ ftype}}{\Pi(A, \Phi) \text{ type}} \qquad\qquad \Pi\text{I*} \qquad \frac{\varphi \in \Phi : A}{\lambda(\varphi) \in \Pi(A, \Phi)}$$

Here λ is not a variable-binding operator, but indicates that the family φ is regarded as an element of $\Pi(A, \Phi)$. Finally, the elimination rule becomes

$$\Pi\text{E*} \qquad \frac{t \in \Pi(A, \Phi) \quad y \in \Phi : A \Rightarrow t' \in C[z/\lambda(y)]}{F_y(t, t') \in C[z/t]}$$

F_y binds the variable y in t', and satisfies the following special equality rule:

$$\Pi\text{CONV}^* \qquad \frac{\varphi \in \Phi : A \quad y \in \Phi : A \Rightarrow t' \in C[z/\lambda(y)]}{F_y(\lambda(\varphi), t') = t'[y/\varphi] \in C[z/\lambda(\varphi)]}$$

In order to obtain equivalence with the earlier formulation, we write ΠE with application explicitly shown as an operator Ap:

$$\Pi\text{E} \qquad \frac{t \in A \quad t' \in \Pi(A, \Phi)}{\text{Ap}(t', t) \in \Phi(t)}$$

$$\Pi\text{CONV} \qquad \frac{t \in A \qquad\qquad \varphi \in \Phi : A}{\text{Ap}(\lambda(\varphi), t) = \varphi(t) \in \Phi(t)}$$

We can define F in terms of Ap by

$$F_y(t, t') := t'[y/(x)\text{Ap}(t, x)] \quad (x \notin \text{FV}(t)),$$

and conversely, given F we can define

$$\text{Ap}(t, t') := F_y(t', y(t)).$$

We leave the proof of the equivalence of ΠE, ΠCONV with ΠE*, ΠCONV* to the reader.

In the analogy with implication in natural deduction ΠE* corresponds to a form of →E where a *hypothetical rule* (Schroeder-Heister 1984) is discharged:

$$\dfrac{A \to B \qquad \begin{matrix}[A \Rightarrow B]^{(n)} \\ \mathcal{D} \\ C\end{matrix}}{C}\,(n)$$

which may be paraphrased as: if we have shown how to deduce C on the assumption that B can be derived from A, we can conclude C from $A \to B$.

4.7. Proposition. *With the notations introduced in* 2.9 *the laws of intuitionistic predicate logic hold in* $\mathbf{ML}_0^i$.

Proof. It has to be shown that, if $A_1, \ldots, A_n \vdash B$ in **N-IQC**, then $\vdash x_1 \in A_1, \ldots, x_n \in A_n \Rightarrow t \in B$ for a suitable t. The details are routine and are left as an exercise. □

4.8. *Embedding* $\mathbf{HA}^\omega$ *in* $\mathbf{ML}_0^i$. As remarked before, the type structure of $\mathbf{HA}^\omega$ is part of the type structure of $\mathbf{ML}_0^i$, if we identify type 0 with type N. $0, S, \boldsymbol{p}, \boldsymbol{p}_0, \boldsymbol{p}_1$ can be interpreted by the corresponding constants of $\mathbf{ML}_0^i$; from 2.18 we see how $\boldsymbol{k}, \boldsymbol{s}, \boldsymbol{r}$ are to be interpreted; variables may be assumed to be interpreted by an injection into the variables of $\mathbf{ML}_0^i$ (without loss of generality we can assume this to be obtained by dropping type superscripts). Thus we have assigned to each term t of $\mathbf{HA}^\omega$ a translated term $t^\wedge$ in $\mathbf{ML}_0^i$. Formulas are now translated as type-expressions:

$$[t =_\sigma s]^\wedge := I(\sigma, t^\wedge, s^\wedge),$$

$$[A \circ B]^\wedge := A^\wedge \circ B^\wedge \quad \text{for } \circ \in \{\wedge, \vee, \to\},$$

$$\bot^\wedge := I(N, S0, 0),$$

$$[(Qx^\sigma)A]^\wedge := (Qx \in \sigma)(A^\wedge) \quad \text{for } Q \in \{\forall, \exists\}.$$

4.9. THEOREM. *For any sentence* A

$$\mathbf{HA}^\omega \vdash A \quad \Rightarrow \quad \mathbf{ML}_0^i \vdash t \in A^\wedge$$

for a suitable closed term t.

PROOF. By induction on the length of deductions in $\mathbf{HA}^\omega$; as in 4.7, a natural deduction formulation is most convenient. For example, one of the equality axioms translates into

$$\vdash \lambda xyzu.\boldsymbol{e} \in \forall x, y \in \sigma\, \forall z \in \sigma \to \tau(I(\sigma, x, y) \to I(\tau, zx, zy))$$

etc. □

On the other hand, $\mathbf{ML}_0^i$ is not conservative over $\mathbf{HA}^\omega$, as follows from the derivability of the following form of the axiom of choice in $\mathbf{ML}_0^i$.

4.10. PROPOSITION (*axiom of choice*). *Suppose* $\vdash x \in A, y \in B[x] \Rightarrow C[x, y]$ type, *then for a suitable closed term* t

$$\mathbf{ML}_0^i \vdash t \in \Pi x \in A\, \Sigma y \in B.C[x, y] \to$$
$$\Sigma z \in (\Pi x \in A.B)\, \Pi x \in A.C[x, zx].$$

PROOF. Abbreviate $D \equiv \Pi x \in A\, \Sigma y \in B.C$ and assume $\vdash x \in A$, $y \in B \Rightarrow C[x, y]$ type. Let $\mathcal{D}_0$ be the following deduction where Δ abbreviates $u \in D, x \in A$:

$$\dfrac{x \in A, y \in B \Rightarrow C \text{ type} \qquad \dfrac{\dfrac{\dfrac{\dfrac{\dfrac{\dfrac{\mathscr{D}'}{D \text{ type}}}{u \in D \Rightarrow u \in D} \qquad \dfrac{\dfrac{\mathscr{D}''}{A \text{ type}}}{x \in A \Rightarrow x \in A}}{\Delta \Rightarrow ux \in \Sigma y \in B.C}}{\Delta \Rightarrow \boldsymbol{p}_0(ux) \in B}}{\Delta \Rightarrow \lambda x.\boldsymbol{p}_0(ux) \in \Pi x \in A.B} \qquad \Delta \Rightarrow x \in A}{\Delta \Rightarrow (\lambda x.\boldsymbol{p}_0(ux))(x) = \boldsymbol{p}_0(ux) \in B}}{\Delta \Rightarrow C[x, \boldsymbol{p}_0(ux)] = C[x, (\lambda x.\boldsymbol{p}_0(ux))(x)]}$$

With help of $\mathscr{D}_0$ we construct, under the constant context $u \in D$ (not shown), the following deduction:

$$\dfrac{\dfrac{\dfrac{\dfrac{u \in D \qquad x \in A \Rightarrow x \in A}{x \in A \Rightarrow ux \in \Sigma y \in B.C}}{x \in A \Rightarrow \boldsymbol{p}_0(ux) \in B}}{\lambda x.\boldsymbol{p}_0(ux) \in \Pi x \in A.B} \qquad \dfrac{\dfrac{\dfrac{\dfrac{x \in A \Rightarrow x \in A \qquad u \in D}{x \in A \Rightarrow ux \in \Sigma y \in B.C}}{\mathscr{D}_0 x \in A \Rightarrow \boldsymbol{p}_1(ux) \in C[x, \boldsymbol{p}_0(ux)]}}{x \in A \Rightarrow \boldsymbol{p}_1(ux) \in C[x, (\lambda x.\boldsymbol{p}_0(ux))x]}}{\lambda x.\boldsymbol{p}_1(ux) \in \Pi x \in A.C[x, (\lambda x.\ \boldsymbol{p}_0(ux))x]}}{(\lambda x.\boldsymbol{p}_0(ux), \lambda x.\ \boldsymbol{p}_1(ux)) \in \Sigma z \in (\Pi x \in A.B)\Pi x \in A.C[x, zx]}$$

and with a final step

$$\lambda u(\lambda x.\boldsymbol{p}_0(ux), \lambda x.\boldsymbol{p}_1(ux)) \in$$
$$D \rightarrow \Sigma z \in (\Pi x \in A.B)\Pi x \in A.C[x, zx]. \quad \square$$

4.11. Corollary. *Theorem* 4.9 *can be reinforced to*

$$\mathbf{HA}^{\omega} + \mathbf{AC} \vdash A \Rightarrow \mathbf{ML}_0^{\mathrm{i}} \vdash t \in A^{\wedge}$$

for a suitable term t. $\square$

We end this section with an easy but important (with respect to applications in section 8) lemma, the proof of which is left as an exercise.

4.12. Lemma. *We have the following equivalences between propositions in* $\mathbf{ML}_0^{\mathrm{i}}$ (*i.e. there exist terms for the equivalences below*):

(i) $(\exists z \in \Sigma x \in A.B)D[z] \leftrightarrow \exists x \in A\, \exists y \in B.D[(x, y)]$,
(ii) $\exists z \in A + B.D[z] \leftrightarrow \exists x \in A.D[\boldsymbol{k}_0 x] \vee \exists y \in B.D[\boldsymbol{k}_1 y]$,
(iii) $\exists z \in N_k.D[z] \leftrightarrow D[0_k] \vee \cdots \vee D[(k-1)_k]$,
$\exists z \in \perp .D[z] \leftrightarrow \perp$, *where* $\perp \equiv I(N, 0, 1)$. $\square$

5. The theory $\mathbf{ML}_0$

We shall now describe a purely extensional version of the (lowest level of the) theory of types: equality at each type, and equality between types is to be interpreted extensionally throughout. From a classical point of view the most obvious model for $\mathbf{ML}_0^{\mathrm{i}}$ is a fragment of the classical set-theoretical hierarchy.

The syntax of $\mathbf{ML}_0$ is the same as that of $\mathbf{ML}_0^{\mathrm{i}}$; it is convenient however to think of "$t \in A$" as an abbreviation of "$t = t \in A$".

5.1. *Structural and general equality rules of* $\mathbf{ML}_0$. We adopt the same rules ASS, THIN, REFL, SYM, TRANS, SUB, and REPL as for $\mathbf{ML}_0^{\mathrm{i}}$. (This results, in combination with the other rules below, in a certain redundancy.)

5.2. *Type-introduction rules of* $\mathbf{ML}_0$. NTYP is as before; the ΠTYP, ΣTYP, +TYP, and ITYP we add, respectively,

$$\frac{A = A' \quad x \in A \Rightarrow B = B' \quad x \in A \Rightarrow B \text{ type}}{\Pi x \in A.B = \Pi x \in A'.B'}$$

$$\frac{A = A' \quad x \in A \Rightarrow B = B' \quad x \in A \Rightarrow B \text{ type}}{\Sigma x \in A.B = \Sigma x \in A'.B'}$$

$$\frac{A = A' \quad B = B' \quad A \text{ type} \quad B \text{ type}}{A + B = A' + B'}$$

$$\frac{A = A' \quad t = t' \in A \quad s = s' \in A \quad A \text{ type}}{I(A, t, s) = I(A', t', s')}$$

If we treat "$A = A$" as an abbreviation of "A type", or (what amounts to the same) add a rule

$$\frac{A = A}{A \text{ type}},$$

then the rules above contain the original ΠTYP, ΣTYP, +TYP, and ITYP as special cases.

5.3. *Introduction and elimination rules of* $\mathbf{ML}_0$. The rules of $\mathbf{ML}_0^{\mathrm{i}}$ are generalized to

NI $\quad 0 = 0 \in N \qquad \dfrac{t = t' \in N}{St = St' \in N}$

NE $$\frac{t = t' \in N \quad t_0 = t_0' \in A[x/0] \quad \begin{array}{c} x \in N \Rightarrow A \text{ type} \\ x \in N,\, y \in A \Rightarrow t_1 = t_1' \in A[x/Sx] \end{array}}{R_{x,y}(t, t_0, t_1) = R_{x,y}(t', t_0', t_1') \in A[x/t]}$$

($R_{x,y}$ binds x and y in t_1 and t_1'; $x, y \notin \mathrm{FV}(t)$, $\mathrm{FV}(t')$, $\mathrm{FV}(t_0)$, $\mathrm{FV}(t_0')$);

ΠI $$\frac{x \in A \Rightarrow t = t' \in B \qquad x \in A \Rightarrow B \text{ type}}{\lambda x.t = \lambda x.t' \in \Pi x{\in}A.B}$$

ΠE $$\frac{t = t' \in \Pi x{\in}A.B \qquad s = s' \in A \qquad x \in A \Rightarrow B \text{ type}}{\mathrm{Ap}(t,s) = \mathrm{Ap}(t',s') \in B[x/s]}$$

ΣI $$\frac{t = t' \in A \qquad s = s' \in B[x/t] \qquad x \in A \Rightarrow B \text{ type}}{(t,s) = (t',s') \in \Sigma x{\in}A.B}$$

ΣE $$\frac{t = t' \in \Sigma x{\in}A.B \quad A \text{ type}}{\mathbf{p}_0 t = \mathbf{p}_0 t' \in A} \qquad \frac{t = t' \in \Sigma x{\in}A.B \quad x \in A \Rightarrow B \text{ type}}{\mathbf{p}_1 t = \mathbf{p}_1 t' \in B[x/\mathbf{p}_0 t]}$$

+I $$\frac{t = t' \in A \quad A \text{ type} \quad B \text{ type}}{\mathbf{k}_0 t = \mathbf{k}_0 t' \in A + B} \qquad \frac{t = t' \in B \quad A \text{ type} \quad B \text{ type}}{\mathbf{k}_1 t = \mathbf{k}_1 t' \in A + B}$$

+E $$\frac{\begin{array}{r} z \in A + B \Rightarrow C \text{ type} \\ t = t' \in A + B \quad x \in A \Rightarrow t_0 = t_0' \in C[z/\mathbf{k}_0 x] \quad y \in B \Rightarrow t_1 = t_1' \in C[z/\mathbf{k}_1 y] \end{array}}{D_{x,y}(t, t_0, t_1) = D_{x,y}(t', t_0', t_1') \in C[z/t]}$$

($D_{x,y}$ binds x in t_0, t_0', and y in t_1, t_1', $x \notin \mathrm{FV}(t_1)$, $\mathrm{FV}(t_1')$, $y \notin \mathrm{FV}(t_0)$, $\mathrm{FV}(t_0')$). The rules II and IE are as before. The generalized rules imply the original ones since $t = t \in A$ corresponds to $t \in A$.

5.4. *Special equality rules for* $\mathbf{ML}_0$. The special equality rules are as before $\mathbf{ML}_0^{\mathrm{i}}$, except that to ΠCONV we add

$$\frac{\lambda x.\mathrm{Ap}(t,x) \in \Pi x{\in}A.B}{\lambda x.\mathrm{Ap}(t,x) = t \in \Pi x{\in}A.B}$$

for t with $x \notin \mathrm{FV}(t)$. That is to say, we have so-called *η-conversion* ($\lambda x.tx \operatorname{conv} t$ for $x \notin \mathrm{FV}(t)$). Without this extra rule we do not get full extensionality of functions in the traditional form. An alternative to the rule above is

$$\frac{t \in \Pi x{\in}A.B \qquad t' \in \Pi x{\in}A.B \qquad x \in A \Rightarrow tx = t'x \in B \qquad x \in A \Rightarrow B \text{ type}}{t = t' \in \Pi x{\in}A.B}$$

5.5. Remark. The type formation operation + is in fact definable in $\mathbf{ML}_0$ by

$$A + B := \Sigma x \in N_2[(x = \mathbf{0} \to A) \wedge (x = \mathbf{1} \to B)],$$

but the proof is laborious and relies rather heavily on the extensional character of equality (cf. exercise 11.5.1). Actually, + is already definable in $\mathbf{ML}_0^{\mathrm{i}}$, but the definition is far more complicated in that case (cf. exercise 11.5.6).

As for $\mathbf{ML}_0^{\mathrm{i}}$, we can drop type assumptions in many rules; also the first two rules under REPL are in fact redundant. More precisely, one has the following proposition.

5.6. PROPOSITION. *Let* $\mathbf{ML}_0'$ *be the variant of* $\mathbf{ML}_0$ *with all type assumptions in the* I-, E-, *and* CONV-*rules dropped except for* ΣI, +I, *and* +E, *and the first two* REPL *rules left out. Then* $\mathbf{ML}_0'$ *is equivalent to* $\mathbf{ML}_0$, *i.e.* $\mathbf{ML}_0 \vdash \Gamma \Rightarrow \theta$ *iff* $\mathbf{ML}_0' \vdash \Gamma \Rightarrow \theta$.

PROOF. The proof of proposition 2.14 also applies to $\mathbf{ML}_0$ with the first two replacement rules left out, as may be seen by inspection. We may also repeat the more refined analysis of 3.1 to see that we can drop the type assumptions everywhere except in ΣI, +I, and +E. Finally we observe that the first two replacement rules can be derived by induction on the complexity of type B, using the fact that 2.14 holds for $\mathbf{ML}_0$ minus those replacement rules, and also using the extensionality of equality. We leave the details as an exercise to the reader. □

5.7. *Embedding* $\mathbf{E\text{-}HA}^\omega$ *in* $\mathbf{ML}_0$. We use again ^ for the embedding; ^ is defined on terms and types exactly as in 4.8. We obtain the corresponding theorem.

THEOREM. *If* A *is a sentence of* $\mathscr{L}(\mathbf{E\text{-}HA}^\omega)$, *and* $\mathbf{E\text{-}HA}^\omega + \mathrm{AC} \vdash A$, *then* $\mathbf{ML}_0 \vdash t \in A^\wedge$ *for a suitable closed term t.* □

COROLLARY. CT *is false in* $\mathbf{ML}_0$, *ie.,* $\mathbf{ML}_0 \vdash t \in \neg\mathrm{CT}$ *for some closed t. We can also refute certain instances of* WC-N *in* $\mathbf{ML}_0$.

PROOF. Immediate from the theorem in combination with 9.6.8(i). □

In the following subsections 5.8–5.12 we shall describe the semantics of $\mathbf{ML}_0$ according to Martin-Löf.

5.8. Our introduction in section 1 suggests an interpretation of the formalism of type theory in which types are certain special sets, and the introduction rules tell us what the elements of a type are: the I-rules list methods for introducing terms in types; moreover, the form of a term t given by an I-rule indicates which I-rule has been applied. The E-rules corresponding to the I-rules for a type tell us in fact that the I-rules exhaust the possibilities for introducing elements of the given type. The terms given by the E-rules do not show in their form by which method the elements denoted by them were introduced, but such terms can be evaluated by conversion rules.

Compare this with arithmetic: numerals are either of the form 0 or St, corresponding to the two methods for introducing elements of $\mathbb{N}$, directly reflecting the generation of $\mathbb{N}$. Terms such as $(2+3)\cdot 5$ on the other hand are not numerals, but denote numbers and can be computed with a numeral as a result.

5.9. Martin-Löf has developed this sort of explanation into an informal semantics for $\mathbf{ML}_0$, i.e. a systematic method of assigning meaning to the assertions of $\mathbf{ML}_0$ and certain extensions of $\mathbf{ML}_0$. We give a brief sketch.

We have to explain the meaning of the four types of assertions occurring in $\mathbf{ML}_0$. In these explanations, assertions can occur hypothetically: assuming we can assert θ, we can also assert θ; schematically we can indicate this by

$$\begin{array}{c}\theta\\ \cdots\\ \theta'\end{array}$$

"$\cdots$" is to be distinguished from "—" as occurring in the rules of $\mathbf{ML}_0$: — indicates a single step in an argument, $\cdots$ may refer to any acceptable argument involving many steps (not necessarily restricted to rules of $\mathbf{ML}_0$). The semantics is based on the following four principles.

(1) A type A is defined by prescribing how a *canonical* element of A is formed, as well as how two *equal canonical* elements of A are formed. Equality between canonical elements should be reflexive, symmetric, and transitive.

(2) Two types A and B are regarded as *equal* if

$$\begin{array}{c}x \in A\\ \cdots\cdots\\ x \in B\end{array},\quad \begin{array}{c}x \in B\\ \cdots\cdots\\ x \in A\end{array},\quad \begin{array}{c}x = y \in A\\ \cdots\cdots\cdots\\ x = y \in B\end{array},\quad\text{and}\quad \begin{array}{c}x = y \in B\\ \cdots\cdots\cdots\\ x = y \in A\end{array}.$$

(3) An *element* of type A is a method or program which, when executed, yields a canonical element as result(value). N.B. The notion of a method appears here as a primitive.

(4) Two arbitrary *elements* a, b of a set A are *equal* if, when executed, a and b yield equal canonical elements of A as results.

N.B. Terms refer to elements (are descriptions of elements); canonical terms refer to canonical elements, arbitrary terms to arbitrary elements.

A difference between Martin-Löf's semantics and the intuitive explanations in section 1 which should be noted is that where the intuitive explanation takes "2^2" to refer to the number 4, here "2^2" refers to a program with as result SSSS0, denoting 4.

This results in a picture for $\mathbb{N}$ different from the usual one: ordinarily one thinks of natural numbers as built up from below, starting with 0 and taking successors; here an element of $\mathbb{N}$ is either 0 or can be evaluated to a canonical element St_0, t_0 an element of $\mathbb{N}$; so t_0 can be evaluated to 0 or to St_1, t_1 can be evaluated to 0 or to St_2, etc., so the evaluation proceeds as suggested by the picture below:

$$\begin{array}{c} \cdots \\ \uparrow \\ S(S(St_2)) \\ \uparrow \\ S(St_1) \\ \uparrow \\ St_0 \\ \uparrow \\ t \end{array}$$

The evaluation procedure must stop, so we end with $S^n 0$ for some $n \in \mathbb{N}$.

5.10. *Some examples of the informal interpretation.* A canonical element of $A \times B$ is a pair, and can be written as (a, b), where it is not required that a and b are themselves canonical elements.

A canonical element of $\Pi x \in A.B$ is a function and is of the form $\lambda x.t[x]$, where $t[x]$ is an element of $B[x]$ for each $x \in A$.

$(a, b) = (c, d) \in A \times B$ iff $a = c \in A$ and $b = d \in B$ (we know already what equality between elements of A and B means).

Suppose t is an arbitrary element of $\Pi x \in A.B$, and t' an element of A. Then $\mathrm{Ap}(t, t') \equiv tt'$ should be an element of B. Suppose t has a canonical value $\lambda x.t''[x]$, and t' a canonical value t'''. Then the value of $(\lambda x.t''[x])(t')$ is $t''[t''']$, and this tells us how to find a value for $\mathrm{Ap}(t, t')$, i.e. $\mathrm{Ap}(t, t')$ is explained as a method for finding a canonical element of B, provided we have recognized that $t \in A$ and $t''[x] \in B$ for each $x \in A$.

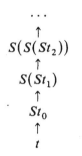

How does one justify the elimination rule NE? Suppose $t \in N$, $t' \in A[0]$, and $x \in N$, $y \in A[x] \Rightarrow t''[x, y] \in A[Sx]$, then $R_{x,y}(t, t', t'')$ must be an element of $A[t]$; in other words we must indicate how to find a value for this element.

The value of t is either 0 or of the form St_0, $t_0 \in N$. In the first case we are done; in the second case $R_{x,y}(St_0, t', t'')$ evaluates to $t''[St_0, R_{x,y}(t_0, t', t'')]$, which has a value provided $R_{x,y}(t_0, t', t'')$ has a value; to convince ourselves that this is a terminating process of evaluation we essentially need the insight that the picture in 5.9 above is well-founded. (The argument is strongly reminiscent of the computability argument in 9.2.10.)

The semantical justification of the rules of $\mathbf{ML}_0$ is now essentially straightforward, thanks to the fact that equality between elements of compound types is completely explained in terms of equality between elements of the constituent types; in particular, it is not necessary to justify the first two replacement rules directly, in view of proposition 5.6.

5.11. For $\mathbf{ML}_0^i$, or its combinatorial variant (3.6) the requirements on equality have been minimalized: equality at each type is regarded as a primitive notion of identity. The application of the semantical principles outlined above is therefore not so straightforward in the case of $\mathbf{ML}_0^i$, and we shall not discuss the problems here in detail. Let us mention only the difficulty in explaining equality of canonical elements of a Π-type in $\mathbf{ML}_0^i$: when is $\lambda x.t[x] = \lambda x.t'[x] \in \Pi x \in A.B$? If we wish to have the effect of equality generated by $\beta\eta$-conversion, the answer should be something like: the programs $(\lambda x.t[x])(x)$ and $(\lambda x.t'[x])(x)$ can be shown to yield the same values uniformly in the argument (for example, if $A \equiv N$, "uniformly" would mean that the proof of "equal values $t[x]$ and $t'[x]$ for all x" does not use a case distinction according to whether x is zero or x is a successor).

On the other hand, $\mathbf{ML}_0^i$, and in particular its combinatorial version, represent subsystems of $\mathbf{ML}_0$ which have more models than $\mathbf{ML}_0$ itself, some of which are particularly easy to describe; and mathematically $\mathbf{ML}_0^i$ seems to be just as powerful as $\mathbf{ML}_0$.

6. Embeddings into APP

In this section we describe interpretations of $\mathbf{ML}_0^i$ and $\mathbf{ML}_0$ in **APP** under which the logical operations in $\mathbf{ML}_0$ correspond to realizability interpretations of logic in **APP**.

6.1. *Interpreting* $\mathbf{ML}_0^i$ *in* **APP**. The types will be interpreted by certain definable sets in **APP**, with the equality for each type interpreted by the equality between elements of **APP**. Constants and variable-binding operators are interpreted by constants and corresponding operators in **APP**.

We describe the interpretation as a syntactic mapping * defined simultaneously for terms and types. We use $\{y : A\}$ as an informal notation in **APP** for A as a predicate in the variable y.

(i) To each variable x we assign a variable x^* of **APP**, such that $x \not\equiv y \Rightarrow x^* \not\equiv y^*$.

(ii) $N^* := N$.

(iii) Suppose the type-expressions A, B to have been interpreted by A^*, B^*, respectively. By clause (i) there corresponds to each variable x free in A, B, a parameter x^*, y^* free in A^* and B^*, respectively, and vice versa. We put

$$(\Pi x \in A.B)^* := \{y : \forall x^* \in A^*(yx^* \in B^*)\},$$

$$(\Sigma x \in A.B)^* := \{\mathbf{p}xy : x \in A^* \wedge y \in B^*)\},$$

$$(A + B)^* := \{\mathbf{p}_0 x : x \in A^*\} \cup \{\mathbf{p}_1 x : x \in B^*\},$$

and if s^* and t^* have been defined,

$$(I(A, s, t))^* := \{0 : s^* = t^*\}.$$

(iv) For terms we put

$$0^* := 0, \qquad (St)^* := S(t^*);$$

$$R_{x,y}(t, t_0, t_1[x, y])^* := \mathbf{r}t_0^*(\lambda y^* x^*.t_1^*[x^*, y^*])t^*;$$

$$(\lambda x.t)^* := \lambda x^*.t^*, \qquad (t_0 t_1)^* := t_0^* t_1^*;$$

$$(t_0, t_1)^* := \mathbf{p}t_0^* t_1^*, \qquad (\mathbf{p}_i t)^* := \mathbf{p}_i t^* \quad (i \in \{0,1\});$$

$$(\mathbf{k}_i t)^* := \mathbf{p}_i t^* \quad (i \in \{0,1\});$$

$$D_{x,y}(t, t_0[x], t_1[y])^* := \mathbf{d}(t_0^*[x^*/\mathbf{p}_1 t^*])(t_1^*[y^*/\mathbf{p}_1 t^*]0(\mathbf{p}_0 t^*)$$

(the λ on the right-hand side is defined in **APP**);

(v) Statements are translated as follows. A context $\langle x_1 \in A_1, \ldots, x_n \in A_n \rangle$ is translated as $x_1^* \in A_1^*, \ldots, x_n^* \in A_n^*$, and

$$(t \in A)^* := t^* \in A^*;$$

$$(A = B)^* := \forall x(x \in A^* \leftrightarrow x \in B^*);$$

$$(t = s \in A)^* := t^* = s^* \wedge t^* \in A^* \wedge s^* \in A^*;$$

$$(A \text{ type})^* := \top \quad (\text{i.e. true}).$$

For this interpretation we have the following theorem.

6.2. THEOREM (*soundness*). *If* $\mathbf{ML}_0^i \vdash \Gamma \Rightarrow \theta$, *then* $\mathbf{APP} \vdash \Gamma^* \to \theta^*$.

PROOF. A straightforward induction on the length of derivations in $\mathbf{ML}_0^i$ (exercise). □

6.3. *Connection with abstract realizability in* **APP**. Assuming $y\,\mathbf{r}\,A \leftrightarrow y \in A^*$, $y\,\mathbf{r}\,B \leftrightarrow y \in B^*$ (where $\mathbf{r}$ is the abstract realizability for **APP** as defined in 9.5.1) we also have

$$y\,\mathbf{r}\,\forall x^* \in A^*.B^*[x^*] \quad\leftrightarrow\quad y \in (\Pi x \in A.B)^*,$$
$$y\,\mathbf{r}\,\exists x^* \in A^*.B^*[x^*] \quad\leftrightarrow\quad y \in (\Sigma x \in A.B)^*,$$
$$y\,\mathbf{r}\,(A^* \vee B^*) \quad\leftrightarrow\quad y \in (A + B)^*.$$

$x\,\mathbf{r}\,A$ can be straightforwardly defined in **APP** for any A of $\mathbf{HA}^\omega$ via the standard embedding of $\mathbf{HA}^\omega$ into **APP**. The types of $\mathbf{HA}^\omega$ are then interpreted by the standard intensional type structure in **APP** (cf. 9.3.15). In fact we have the following proposition.

PROPOSITION. *Let* ^ *be the embedding which maps formulas of* $\mathbf{HA}^\omega$ *to types of* $\mathbf{ML}_0^i$ (*cf.* 4.8) *and let* $^+$ *be the embedding of formulas of* $\mathbf{HA}^\omega$ *into* **APP**, * *as defined in* 6.1, *then there are* φ_B, ψ_B *such that*

$$\mathbf{APP} \vdash (x \in (B^\wedge)^*) \to \varphi_B(x)\,\mathbf{r}\,B^+,$$
$$\mathbf{APP} \vdash (x\,\mathbf{r}\,B^+) \to (\psi_B(x) \in (B^\wedge)^*).$$

PROOF. Exercise. □

6.4. *Interpreting* $\mathbf{ML}_0$ *in* **APP**. The interpretation is similar to the one defined in 6.1, but now we have to define not just predicates representing types in **APP**, but also partial equivalence relations representing the equalities on types.

(i) $N^* := \{(x, y)\colon x = y \wedge x \in N \wedge y \in N\}$, and assuming that A^* and B^* have been defined we put

$$(\Pi x \in A.B)^* := \{(y, y')\colon \forall xx'((x, x') \in A^* \to (yx, y'x') \in B^*[x^*/x])\},$$
$$(\Sigma x \in A.B)^* := \{(\mathbf{p}xy, \mathbf{p}x'y')\colon (x, x') \in A^* \wedge (y, y') \in B^*[x'/x]\},$$
$$(A + B)^* := \{(\mathbf{p}_0x, \mathbf{p}_0x')\colon (x, x') \in A^*\} \cup \{(\mathbf{p}_1x, \mathbf{p}_1x')\colon (x, x^*) \in B^*\},$$

and assuming that t^* and s^* have been defined

$$(I(A, t, s))^* := \{(0,0)\colon (s^*, t^*) \in A^*\}.$$

(ii) On terms * is defined as before.
(iii) On contexts * is defined as before, and for assertions we put

$$(t \in A)^* := (t^*, t^*) \in A^*,$$

$$(A = B)^* := \forall x x'((x, x') \in A^* \leftrightarrow (x, x') \in B^*),$$

$$(t = s \in A)^* := (t^*, s^*) \in A^*,$$

$$(A \text{ type})^* := \top .$$

We again have a soundness theorem.

6.5. Proposition (*soundness*). If $\mathbf{ML}_0 \vdash \Gamma \Rightarrow \theta$, *then* $\mathbf{APP} \vdash \Gamma^* \to \theta^*$, *with * as above.*

Proof. Exercise. □

Remark. The interpretation of logic now does not any longer correspond directly to abstract realizability in **APP**, but rather to a form of "extensional realizability" which cannot be extended to the whole of **APP**. □

7. Extensions of $\mathbf{ML}_0^{\mathrm{i}}$ and $\mathbf{ML}_0$

In this section we consider strengthenings of $\mathbf{ML}_0^{\mathrm{i}}$ and its extensional analogue $\mathbf{ML}_0$. In one direction, there is the possibility of extension by reflection; in another, by transfinite induction.

7.1. *Extension by reflection: universes.* In $\mathbf{ML}_0^{\mathrm{i}}$ types were constructed from the simple type N by means of four type-forming operations Π, Σ, I, and $+$. We can now consider the universe U of all "small" types generated from N by means of these closure operations, and regard this collection itself as a "large" type. That is, we adopt the following introduction rules:

$$\mathrm{UI}\quad \left\{\begin{array}{ll} N \in U \quad \dfrac{A \in U \quad B \in U}{A + B \in U} & \dfrac{A \in U \quad t \in A \quad s \in A}{I(A, t, s) \in U} \\[2ex] \dfrac{A \in U \quad x \in A \Rightarrow B \in U}{\Sigma x \in A . B \in U} & \dfrac{A \in U \quad x \in A \Rightarrow B \in U}{\Pi x \in A . B \in U} \end{array}\right.$$

We also want to express that all elements of U are types, and that U is itself a type:

$$\text{UTYP}\qquad \frac{A \in U}{A\ \text{type}} \qquad \frac{A = B \in U}{A = B} \qquad U\text{type}.$$

We call the resulting system $\mathbf{ML}_1^{\mathrm{i}}$.

If we wanted to express that the rules UI give all possibilities for introducing elements in U, we would have to add a corresponding elimination rule UE. However, if we do not assume that N, $+$, Π, Σ, and I generate *all* small types, we can adopt UI and UTYP only, that is to say we assume that there is a collection of types containing N and which is closed at least under $+$, Σ, Π, and I, which itself can be regarded as a type; this leaves open the possibility of adding further rules to UI, corresponding to further type-forming operations. Innocent as the addition of UI without corresponding elimination rule might seem, it has rather strong consequences, as we shall see below.

7.2. *Remark on* $\bot$. We digress for a moment on $\bot$. In $\mathbf{ML}_0^{\mathrm{i}}$ (4.1) we defined $\bot$ as $I(N,0,1)$, and demonstrated the property $\lambda x.\bot_A \in \bot \to A$, where $\bot_A$ depended on A. This argument is readily extended to $\mathbf{ML}_1^{\mathrm{i}}$, putting $\bot_U := N$.

A perhaps more elegant solution is the addition of a primitive constant $\bot \equiv N_0$ for the empty type, without I-rule (since $\bot$ does not contain an element), but with an elimination rule

$$\frac{t \in N_0}{R_0(t) \in A}$$

for every type A. However, in $\mathbf{ML}_0^{\mathrm{i}}$ (as well as in $\mathbf{ML}_0$) we have no means of proving $I(N,0,1)$ to be "equivalent" to N_0. Of course $\lambda x.R_0(x) \in N_0 \to I(N,0,1)$, but the converse implication would have to be postulated, e.g. by assuming a constant $c \in I(N,0,1) \to N_0$ (cf. Smith A).

In $\mathbf{ML}_1^{\mathrm{i}}$ on the other hand we can prove $I(N,0,1) \to N_0$ without further assumptions, as follows. We have, as an application of NE, with $\top \equiv I(N,0,0)$

$$\frac{t \in N \quad \top \in U \quad \bot \in U}{R_{x,y}(t,\top,\bot) \in U}$$

and $R_{x,y}(0,\top,\bot) = T \in U$, $R_{x,y}(1,\top,\bot) = \bot \in U$ by the NCONV rules. Then, abbreviating $R_{x,y}(z,\top,\bot)$ as $B[z]$ and noting that

$e \in I(N, 0, 0) = B[0]$, we have

$$\dfrac{e \in B[0] \qquad \dfrac{z \in N \Rightarrow B[z] \text{ type} \qquad u \in 0 =_N 1 \Rightarrow u \in 0 =_N 1}{u \in 0 =_N 1 \Rightarrow B[0] = B[1]}}{\dfrac{\dfrac{u \in 0 =_N 1 \Rightarrow e \in B[1] \qquad\qquad B[1] = \bot}{u \in 0 =_N 1 \Rightarrow e \in \bot}}{\lambda u.e \in 0 =_N 1 \to \bot}}$$

If we adopt N_0 as a primitive, we should add

UI′ $N_0 \in U$.

7.3. *The richness of U.* It is tempting to think that the collection of all types which can be *proved* to be types in $\mathbf{ML}_0^{\mathrm{i}}$ would give us a correct interpretation of U. But in fact there are many elements in U which can be proved to exist in $\mathbf{ML}_1^{\mathrm{i}}$ but not in $\mathbf{ML}_0^{\mathrm{i}}$. For example, we can enumerate all "pure types" by means of the following construction:

$$\dfrac{\dfrac{n \in N \qquad N \in U \qquad x \in U \Rightarrow (x \to N) \in U}{R_{x,y}(n, N, x \to N) \in U}}{B \equiv \Sigma n \in N.R_{x,y}(n, N, x \to N) \in U}$$

where B is a type which enumerates all pure types N, $N \to N$, $(N \to N) \to N$, ...; we have no means of proving B type in $\mathbf{ML}_0^{\mathrm{i}}$.

Similarly, we may construct a recursive function not provably recursive in **HA**, whereas the embedding of $\mathbf{ML}_0^{\mathrm{i}}$ in **APP** and hence in **HA** makes it clear that in $\mathbf{ML}_0^{\mathrm{i}}$ we can only prove the existence of recursive functions which are provably recursive in **HA**. Also a truth definition for **HA** can be given along these lines.

7.4. *Iteration of reflection; Girard's paradox.* It is now easy to iterate the idea of collecting small types into a universe; we may postulate a whole sequence $U_1, U_2, U_3, \ldots$ of universes such that for each U_n we have rules $\mathrm{U_nI}$ and $\mathrm{U_nTYP}$ similar to UI and UTYP, and in addition

$U_n \in U_{n+1}$.

On the other hand, we cannot strengthen $\mathbf{ML}_1^{\mathrm{i}}$ by adding

$U \in U$,

for this would mean, intuitively, that U is the universe of all types, from which we can derive a paradox (due to Girard). We give a sketch of the derivation of the paradox, and leave the verification, and the claims made, as an exercise.

Let $A \in U$, $< \in A \to (A \to U)$ (we think of $<$ as a binary relation A), then define

$$P(A,<) := \forall x \in A\, \forall y \in A\, \forall z \in A[x<y \to (y<z \to x<z)] \in U,$$

$$Q(A,\ <) := \forall f \in N \to A[\forall n \in N(f(Sn) < f(n)) \to \bot] \in U,$$

$$\Omega := \ \Sigma A \in U \exists < \in A \to (A \to U)[P(A,\ <) \wedge Q(A,\ <)] \in U.$$

$P(A,\ <)$ expresses transitivity of $<$, $Q(A,\ <)$ well-foundedness; Ω is the collection of all well-founded relations on elements of U. In order to prove $\Omega \in U$ we essentially need $U \in U$.

We can now define $<_\Omega \in \Omega \to (\Omega \to U)$ and prove $P(\Omega,\ <_\Omega) \wedge Q(\Omega,\ <_\Omega)$ as follows. Suppose

$$A^* \equiv (A,\ <_A, p_A, q_A) \in \Omega, \qquad B^* \equiv (B,\ <_B, p_B, q_B) \in \Omega,$$

where (t_1, t_2, t_3, t_4) is short for $(t_1,(t_2,(t_3, t_4)))$, and where

$$p_A \in P(A,\ <_A), \qquad q_A \in Q(A,\ <_A),$$

$$p_B \in P(B,\ <_B), \qquad q_B \in Q(B,\ <_B).$$

Now we define $A^* <_\Omega B^*$ as: there exists an order-preserving map from $(A,\ <_A)$ onto an initial segment $(B,\ <_B)$, or formally

$$A^* <_\Omega B^* := \exists f \in A \to B\, \exists z \in B$$
$$[\forall x \in A\, \forall x \in B(x <_A y \to fx <_B fy) \wedge \forall x \in A(fx <_B z)].$$

For each A^*, $B^* \in \Omega$ this is again a type in U, and we can construct

$$p_\Omega \in P(\Omega,\ <_\Omega), \qquad q_\Omega \in Q(\Omega,\ <_\Omega),$$

and thus, using again $\Omega \in U$,

$$\Omega^* := (\Omega,\ <_\Omega, p_\Omega, q_\Omega) \in \Omega,$$

and we can show for suitable t_{Ω^*}

$$t_{\Omega^*} \in \Omega^* <_\Omega \Omega^*.$$

This yields a contradiction, for let $f \in \lambda n.\Omega^*$, then $\lambda n.t_{\Omega^*} \in \Pi n \in N$ $(f(Sn) <_\Omega f(n))$, and thus $q_\Omega(\lambda n.t_{\Omega^*})(\lambda n.\Omega) \in \bot$.

The argument above may be regarded as a version of the Burali–Forti paradox in the context of the theory of types.

7.5. *Extension of type theory by tree-induction.* We shall now describe an extension in another direction, namely the addition of rules for well-founded tree classes with a form of transfinite induction.

The extension requires a new type-forming operation denoted by W (from "well-founded")

$$\text{WTYP} \qquad \frac{A \text{ type} \quad x \in A \Rightarrow B \text{ type}}{Wx \in A.B \text{ type}}$$

with an introduction rule

$$\text{WI} \qquad \frac{t \in A \qquad t' \in B[x/t] \to Wx \in A.B \qquad x \in A \Rightarrow B \text{ type}}{\sup(t, t') \in Wx \in A.B}$$

Before we state the corresponding elimination rule, let us try to give an intuitive picture of the elements of $Wx \in A.B$. We may think of an element of $Wx \in A.B$ as a well-founded tree, where at each node the immediate predecessors are indexed by $B[x/t] \equiv B[t]$ for some $t \in A$; A is thus a collection of indices of "branching-types". Thus, given $t \in A$, $B[t]$ and a collection of trees of $Wx \in A.B$ indexed by $B[t]$, say $\{T_x : x \in B[t]\}$, we obtain a new tree, $\sup(t, \lambda x.T_x)$, the disjoint union of the T_x with a new root, see fig. 11.1.

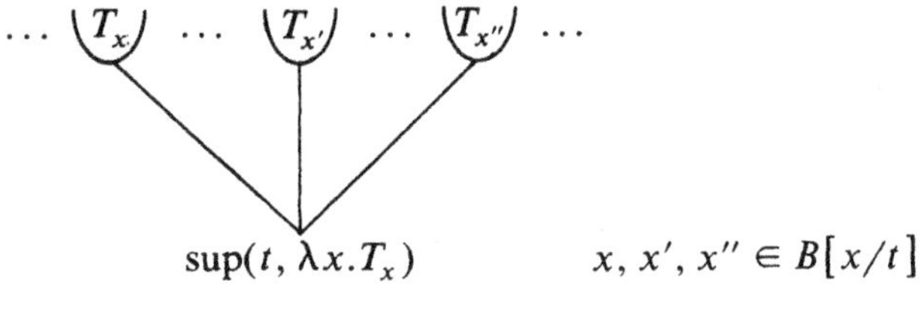

Fig. 11.1.

So we may think of $Wx \in A.B$ as being inductively generated by the construction method just described. The corresponding elimination rule is then a principle of induction over $Wx \in A.B$ stating, intuitively, that $Wx \in A.B$ is the least set generated by the clause of the introduction rule. Formally this becomes

$$\text{WE} \qquad \frac{\begin{array}{l} \Gamma \Rightarrow t \in Wx \in A.B \\ \Gamma, w \in Wx \in A.B \Rightarrow C[w] \text{ type} \\ \Gamma, x \in A, y \in B \to Wx \in A.B, z \in \Pi v \in B.C[y(v)] \Rightarrow t' \in C[\sup(x, y)] \end{array}}{\Gamma \Rightarrow T_{x,y,z}(t, t') \in C[t]}$$

$T_{x,y,z}$ binds x, y, and z in t'; we assume $x, y, z \notin \mathrm{FV}(t)$. There is also a special equality rule WCONV corresponding to an additional conversion

$$T_{x,y,z}(\sup(t,t'),t'')\,\mathrm{conv}\,t''[x,y,z/t,t',\lambda v.T_{x,y,z}(t'(v),t'')].$$

More precisely, we have the following rule:

$$\text{WCONV}\quad \frac{\begin{array}{l}\Gamma \Rightarrow t \in A\\ \Gamma, w \in Wx{\in}A.B \Rightarrow C[w]\ \text{type}\\ \Gamma \Rightarrow t'' \in B[x/t] \to Wx{\in}A.B\\ \Gamma, x \in A, y \in B \to Wx{\in}A.B, z \in \Pi v{\in}B.C[y(v)] \Rightarrow t'' \in C[\sup(x,y)]\end{array}}{T_{x,y,z}(\sup(t,t'),t'') = t''[x,y,z/t,t',\lambda v.T_{x,y,z}(t'(v),t'')] \in C[\sup(t,t')]}$$

This rule expresses recursion over $Wx \in A.B$. We now illustrate the W-types by means of a simple

7.6. EXAMPLE. The "*first tree-class*", or the type of the countably branching trees, may be obtained as follows. For A we take N_2, and we choose $B[x]$ such that $B[\mathbf{0}]$ is empty, and $B[\mathbf{1}]$ an isomorphic copy of N, e.g. by putting

$$B[x] := \Sigma n \in N.I(N_2, \mathbf{1}, x).$$

The elements of $B[\mathbf{1}]$ are of the form $(n, \boldsymbol{e})$ with $n \in N$. Since $B[\mathbf{0}]$ is isomorphic to N_0 or $I(N, 0, 1)$, there is an R such that $\lambda x.R(x) \in B[\mathbf{0}] \to C$ for any type C (cf. 7.2). The first tree-class $\mathrm{Tr} := Wx \in N_2 B[x]$ is not empty; $\lambda x.R(x) \in B[\mathbf{0}] \to \mathrm{Tr}$, hence $\sup(\mathbf{0}, \lambda x.R(x)) \in \mathrm{Tr}$. This corresponds to a tree consisting of a single node; since $B[\mathbf{0}]$ is empty there are no predecessors, see fig. 11.2.

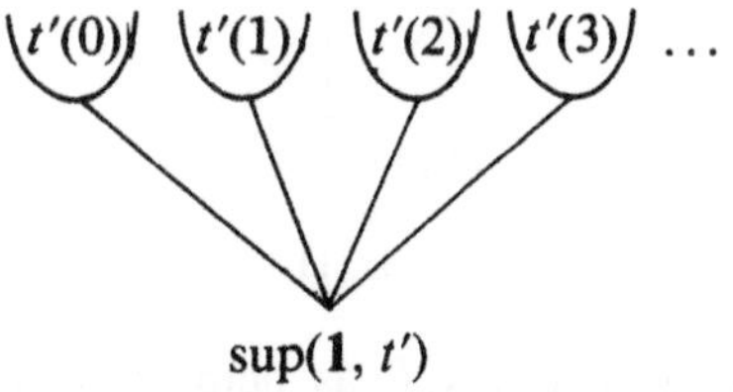

Fig. 11.2.

A countable sequence of trees in Tr is represented by an element $t' \in B[\mathbf{1}] \to \mathrm{Tr}$, that is to say a function from (a copy of) the natural numbers into Tr, and $\sup(\mathbf{1}, t')$ represents the tree in the picture above. Tr is equivalent to the class K^* discussed in 4.8.3, and WE corresponds in this case precisely to induction over K^*.

Let us now try to apply Martin-Löf's semantics to this special case. An element $t_0 \in \mathrm{Tr}$ can be evaluated to something of the form $\sup(t_1, t_1')$, with $t_1 \in N_2$; if $t_1 = \mathbf{1}$, t_1' maps any $v_1 \in B[\mathbf{1}]$ ($\cong N$) to $t_1'(v_1) \in \mathrm{Tr}$; this can in

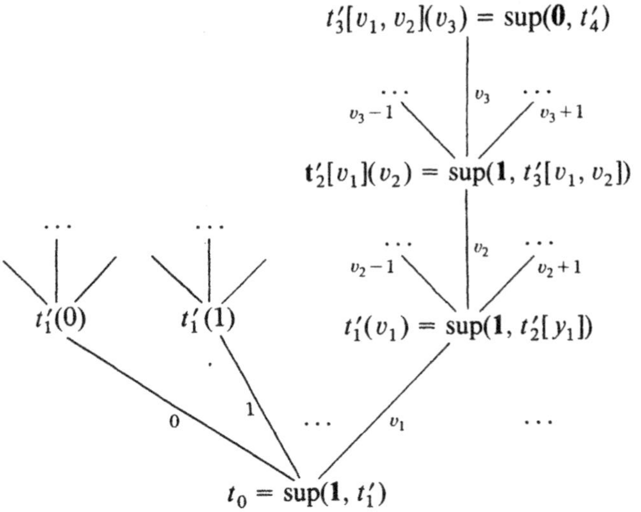

Fig. 11.3.

turn be evaluated for any v_1 as $\sup(t_2[v_1], t_2'[v_1])$ with $t_2[v_1] \in N_2$. If $t_2[v_1] = \mathbf{1}$, $t_2'[v_1]$ maps any $v_2 \in B[\mathbf{1}]$ $(\cong N)$ to $t_2'[v_1](v_2) \in \mathrm{Tr}$ which can in turn be evaluated, etc., as shown in fig. 11.3. In the picture we have identified $(n, \mathbf{e})$ with n. Thus in order to show that $t_0 \in \mathrm{Tr}$, according to the semantics we have to see that an arbitrary path obtained by successive choices of $v_1, v_2, v_3, \ldots$ terminates in an element $\sup(\mathbf{0}, t)$. On the other hand WI and WE tell us that Tr is inductively generated. Accepting this semantical interpretation as a justification of the rules WI and WE for Tr amounts to accepting bar induction (cf. the discussion in section 4.8).

The whole story can be repeated for the general case; there the acceptance of WI and WE on the basis of semantical considerations of this nature amounts to the acceptance of a generalized form of bar induction.

7.7. An important property of *W*-types is that we can always, for each element $\alpha \in Wx \in A.B$ recover the *branching-index* $\alpha^{-} \in A$ and the corresponding mapping $\alpha^{\sim} \in B[\alpha^{-}] \rightarrow Wx \in A.B$, such that $\alpha = \sup(\alpha^{-}, \alpha^{\sim})$. More precisely we have the following proposition.

PROPOSITION. *Let* α *range over* $Wx \in A.B$. *There exist functions* $^{-} \in (Wx \in A.B) \rightarrow A$ *and* $^{\sim} \in \Pi\alpha \in (Wx \in A.B).(B[\alpha^{-}] \rightarrow Wx \in A.B)$ *such that*

$$\alpha = \sup(\alpha^{-}, \alpha^{\sim})$$

and if $\alpha = \sup(t, t')$, *then* $\alpha^{-} = t$ *and* $\alpha^{\smile} = t'$.

PROOF. We take

$$\alpha^{-} := T_{x,y,z}(\alpha, x),$$

$$\alpha^{\smile} := T_{x,y,z}(\alpha, y).$$

Then, dropping the usual type premisses,

$$\frac{\alpha \in Wx \in A.B \qquad x \in A \Rightarrow x \in A}{T_{x,y,z}(\alpha, x) \in A}$$

and $T_{x,y,z}(\sup(t, t'), x) = t$; similarly

$$\frac{\alpha \in Wx \in A.B \qquad x \in A,\ y \in B \to Wx \in A.B \Rightarrow y \in B \to Wx \in A.B}{T_{x,y,z}(\alpha, y) \in B \to Wx \in A.B}$$

and $T_{x,y,z}(\sup(t, t'), y) = t' \in B[\alpha^{-}] \to Wx \in A.B$. Finally, let $g := \lambda\alpha.\sup(\alpha^{-}, \alpha^{\smile})$, then

$$\cfrac{\cfrac{\begin{array}{c}\alpha \in Wx \in A.B \\ x \in A,\ y \in B \to Wx \in A.B \Rightarrow e \in I(Wx \in A.B, g(\sup(x, y)), \sup(x, y))\end{array}}{T_{x,y,z}(\alpha, e) \in I(Wx \in A.B, g(\alpha), \alpha)}}{g(\alpha) = \alpha \in Wx \in A.B}$$

□

7.8. *The system* **MLV**^{i}. The system **MLV**^{i} is obtained from **ML**$^{\mathrm{i}}_{1}$ by adding the rules for $V := Wx \in U.x$:

$$\mathrm{VI} \qquad \frac{A \in U \quad t \in A \to V}{\sup(A, t) \in V}$$

$$\mathrm{VE} \qquad \frac{\begin{array}{c} t \in V \\ X \in U,\ y \in X \to V,\ z \in \Pi v \in X.C[w/y(v)] \Rightarrow t_1[X, y, z] \in C[w/\sup(X, y)]\end{array}}{T_{X,y,z}(t, t_1) \in C[w/t]}$$

where we used, for better readability, the capital letter X for a variable ranging over U; and we have a special equality rule corresponding to

$$\mathrm{VCONV} \qquad T_{x,y,z}(\sup(A, t), t_1) = t_1\big[A, t, \lambda v.T_{x,y,z}(tv, t_1)\big].$$

Sometimes it is slightly more convenient to take

$$T(t, t_1) := T_{x,y,z}(t, t_1xyz),$$

where

$$t_1 \in \Pi X \in U \Pi y \in X \to V$$
$$\Pi z \in (\Pi v \in X.C[w/yv]).C[w/\sup(X, y)],$$

and so

$$T(\sup(A, t), t_1) = t_1(A, t, \lambda v.T(tv, t_1)).$$

The system $\mathbf{MLV}^{\mathrm{i}}$ will be used in the next section to give an interpretation of the system **CZF** of constructive set theory.

7.9. *The extensional versions.* The addition of universes to $\mathbf{ML}_0$ proceeds just as in the case of $\mathbf{ML}_0^{\mathrm{i}}$; we do not need special equality rules. For the W-types, we do need some obvious adaptations of the rules:

$$\text{WTYP} \quad \frac{A = A' \qquad x \in A \Rightarrow B = B'}{Wx \in A.B = Wx \in A'.B'}$$

with an introduction rule

$$\text{WI} \quad \frac{t = t_1 \in A \qquad t' = t_1' \in B[x/t] \to Wx \in A.B \qquad x \in A \Rightarrow B \text{ type}}{\sup(t, t') = \sup(t_1, t_1') \in Wx \in A.B}$$

$$\text{WE} \quad \frac{\begin{array}{l} t = t' \in Wx \in A.B \\ w \in Wx \in A.B \Rightarrow C[w] \text{ type} \\ x \in A, y \in B \to Wx \in A.B, z \in \Pi v \in B.C[yv] \Rightarrow t_1 = t_1' \in C[\sup(x, y)] \end{array}}{T_{x,y,z}(t, t_1) = T_{x,y,z}(t', t_1') \in C[t]}$$

The system **ML** is obtained from $\mathbf{ML}_0$ by addition of the $\mathrm{U_nI}$, $\mathrm{U_nTYP}$ rules as before, and in addition an extra introduction rule

$$\frac{A \in U_n \qquad x \in A \Rightarrow B \in U_n}{Wx \in .B \in U_n}$$

and also the extensional W-rules as above. This is (essentially) the system of Martin-Löf (1982).

8. Constructive set theory

There exists a fair amount of literature dealing with set theories based on intuitionistic logic. In most cases these systems are not directly motivated by a constructive or intuitionistic interpretation, though some of these systems such as **IZF** (see 8.11 below) arise very naturally, e.g. by considering "Heyting-valued" interpretations of set theory (see the references in

9.7). A thorough discussion of these systems and their metamathematics falls outside the scope of this book, for this we refer the reader to the literature (9.6–7).

In this section we concentrate on **CZF**, a subsystem of **ZF** which permits a quite natural interpretation in the fragment $\mathbf{MLV}^{\mathrm{i}}$ of Martin-Löf's theory of types defined in the preceding section. This provides a constructive justification of **CZF**.

8.1. The language of **CZF** contains only $=$ and $\in$ as primitives. The logic is intuitionistic logic without equality, with operators $\perp$, $\wedge$, $\vee$, $\rightarrow$, $\forall$, and $\exists$.

DEFINITION. A *restricted* formula is a formula constructed from prime formulas using $\perp$, $\wedge$, $\vee$, $\rightarrow$, $\forall x \in y$, and $\exists x \in y$ only, where as usual

$$\forall x \in y A := \forall x(x \in y \rightarrow A),$$

$$\exists x \in y A := \exists x(x \in y \wedge A). \quad \square$$

8.2. *Axioms of* **CZF**. For any $A[x, y]$ let

$$A'[u, v] := \forall x \in u \exists y \in v A[x, y] \wedge \forall y \in v \exists x \in u A[x, y].$$

The axioms of **CZF** are:

Extensionality:

$$\text{Ext} \quad \begin{cases} x = y \leftrightarrow \forall z(z \in x \leftrightarrow z \in y), \\ x = y \wedge y \in z \rightarrow x \in z. \end{cases}$$

Set-induction:

$$\in\text{-Ind} \quad \forall y(\forall x \in y A[x] \rightarrow A[y]) \rightarrow \forall x A[x].$$

Pairing:

$$\forall xy \exists z(x \in z \wedge y \in z).$$

Union:

$$\forall x \exists z \forall y \in x \forall u \in y(u \in z).$$

Restricted Separation:

$$\text{R-Sep} \quad \forall x \exists z \forall y(y \in z \leftrightarrow y \in z \wedge A[y]) \quad (A \text{ restricted}).$$

Strong Collection:

S-Coll $\forall x \in u\, \exists y A[x, y] \to \exists v\, A'[u, v]$.

Subset Collection:

Sub-Coll $\forall uv \exists w \forall z(\forall x \in u\, \exists y \in v\, A[x, y] \to \exists w' \in w\, A'[u, w'])$,

where A may contain z free.

Infinity:

Inf $\exists x \forall y(y \in x \leftrightarrow \forall z \in y(\bot) \vee \exists u \in x(\mathrm{suc}[u, y]))$,

where $\mathrm{suc}[x, y] := \forall z \in y(z = x \vee z \in x)$.

8.3. LEMMA.
(i) $\exists! x(\forall y \in x.\, \bot)$,
(ii) $\forall xy \exists! z \forall u(u \in z \leftrightarrow u = x \vee u = y)$,
(iii) $\forall x \exists! z \forall y(y \in z \leftrightarrow \exists v \in x(y \in v))$,
(iv) $\exists! x \forall y(y \in x \leftrightarrow \forall z \in y(\bot) \vee \exists u \in x(\mathrm{suc}[u, y])$.

PROOF. (i). There exists some set x by Inf; apply to this x the axiom R-Sep with $A[x, y] \equiv \bot$. Then we find x' such that $\forall y \in x'(\bot)$. Uniqueness requires $\forall y \in x'(\bot) \wedge \forall y \in x''(\bot) \to x' = x''$, which is immediate by extensionality.

(ii), (iii): exercise.

(iv). Let $A[x] := \forall y(y \in x \leftrightarrow \forall z \in y(\bot) \vee \exists u \in x.\mathrm{suc}[u, y])$. Uniqueness of this x can be proved by $\in$-Ind applied to $B[z, x, y] := A[x] \wedge A[y] \wedge z \in x \to z \in y$; this yields $A[x] \wedge A[y] \to \forall z(z \in x \leftrightarrow z \in y)$, and hence by extensionality uniqueness follows. □

8.4. NOTATION. We write $\emptyset$, $\{x, y\}$, $\bigcup x$, and ω for the unique sets given by (i)–(iv) of 8.3, respectively. We write $\langle x, y\rangle$ for the ordered pair $\{x, \{x, y\}\}$. □

8.5. LEMMA. *ω satisfies induction*:

$$A[\emptyset] \wedge \forall x(A[x] \to A[x^+]) \to \forall x \in \omega\, A[x],$$

where x^+ is the unique y such that $\mathrm{suc}[x, y]$.

PROOF. Left to the reader, use $\in$-Ind. □

8.6. Remarks. (i). Set-induction is a contrapositive of *Foundation*

$$\exists x\, B[x] \to \exists x(B[x] \wedge \forall y \in x \neg B[y]).$$

However, Foundation implies PEM: take $B[x] \equiv (x = \emptyset \wedge A) \vee (x = \{\emptyset\})$. Then obviously $\exists x\, B[x]$; and if $B[x]$ and $\forall y \in x \neg B[y]$, we have either $x = \{\emptyset\}$ and then $\neg A$ holds, since for $\emptyset \in x$ we have $\neg B[\emptyset]$, or $x = \emptyset$ and then A holds. Foundation is therefore unsuitable as an axiom for constructive set theory.

(ii). S-Coll obviously implies *Collection*,

Coll $\quad \forall x \in u\, \exists y A[x, y] \to \exists v \forall x \in u\, \exists y \in v A[x, y],$

and this obviously implies *Replacement* in the form

Repl $\quad \forall x \in u\, \exists! y A[x, y] \to \exists v \forall x \in u\, \exists! y(\langle x, y\rangle \in v \wedge A[x, y]).$

8.7. Definition.

$$\mathrm{dom}[x] := \{y : \exists z(\langle y, z\rangle \in x\},$$

$$\mathrm{range}[x] := \{y : \exists z(\langle z, y\rangle \in x\},$$

$$\mathrm{fun}[x] := \forall y \in x\, \exists y'y''(y = \langle y', y''\rangle) \wedge \forall y \in \mathrm{dom}[x]$$
$$\exists! z(\langle y, z\rangle \in x),$$

$$\mathrm{FUN}[x, y, z] := \mathrm{fun}[x] \wedge \mathrm{dom}[x] = y \wedge \mathrm{range}[x] = z,$$

$$x \subset y := \forall z \in x(z \in y),$$

$$x \times y := \{\langle u, v\rangle : u \in x \wedge v \in y\},$$

$$\mathrm{FULL}[x, y, z] := \forall u \subset y \times z(\mathrm{dom}[u] = y \to \mathrm{range}[u] \in x)$$

(all possible ranges of relations from y to z are elements of x). If fun$[f]$, we shall use the notation $f(x)$ for f applied to y as usual, i.e.

$$f(x) := \bigcup\{y : \langle x, y\rangle \in f\}. \quad \square$$

Observe that all sets and relations defined above can be given by restricted formulas.

8.8. *Additional axioms.* We already mentioned Collection. In addition we shall consider:

Powerset:

Pow $\quad \forall x \exists y \forall z(z \in y \leftrightarrow z \subset x),$

with as a special case $\mathrm{Pow}_{\emptyset}$, stating the existence of a powerset of $\{\emptyset\}$. For the powerset of a set x we write $\mathrm{P}(x)$.

Separation (Sep):
As R-Sep, but without restriction on the A.

Exponentiation:

$$\text{Exp} \quad \forall yz \exists u \forall x (x \in u \leftrightarrow \mathrm{FUN}[x, y, z]).$$

We can show the uniqueness of this u and we will use the notation z^y for it.

Fullness:

$$\text{Full} \quad \forall yz \exists x\, \mathrm{FULL}[x, y, z].$$

Restricted excluded middle:

$$\text{REM} \quad A \vee \neg A \quad (A \text{ restricted}).$$

8.9. PROPOSITION. *Let* **CZF**$^-$ *be* **CZF** *without* Sub-Coll. *Relative to* **CZF**$^-$ *we have*:
(i) Full $\Leftrightarrow$ Sub-Coll,
(ii) Pow $\Rightarrow$ Sub-Coll $\Rightarrow$ Exp,
(iii) Pow $\Leftrightarrow$ Exp + $\mathrm{Pow}_{\emptyset}$,
(iv) Exp + REM $\Rightarrow$ Pow,
(v) Sep $\Leftrightarrow$ $\exists x(A \leftrightarrow \emptyset \in x)$ for all A.

PROOF. (i). $\Leftarrow$: assume Sub-Coll, and apply it to $A[x, y] \equiv \langle x, y \rangle \in u$. Conversely, assume Full and let $\forall x \in a\, \exists y \in b\, A[x, y]$. Define

$$B[x, z] := \exists y \in b (A[x, y] \wedge z = \langle x, y \rangle),$$

then $\forall x \in a\, \exists z\, B[x, z]$, hence by S-Coll there is a c such that

$$\forall x \in a\, \exists z \in c\, B[x, z] \wedge \forall z \in c\, \exists x \in a\, B[x, z].$$

Hence $\mathrm{dom}(c) = a$, $\forall xy(\langle x, y \rangle \in c \leftrightarrow A[x, y])$. By Full there is a d such that $\mathrm{FULL}[d, a, b]$ and thus we can find b' with $\mathrm{range}[c] = b' \in d$, and hence $\exists b' \in d\, A'[a, b']$.

(ii) Pow $\Rightarrow$ Sub-Coll is trivial and it is left to the reader. Now assume Full, and let $\mathrm{FULL}[z, x, x \times y]$, $\mathrm{FUN}[f, x, y]$. Then $f' := \{\langle u, \langle u, f(u) \rangle \rangle : u \in x\}$ is a function from x to $x \times y$, and there is a $z' \in z$ with $f =$

range $f' = z'$. With R-Sep

$$y^x := \{ f \in z : \forall u \in x\, \exists! u' \in y(\langle u, u' \rangle \in f)\}$$

is a set.

(iii) $\Leftarrow$: take $\mathrm{P}(x) := \{\{u \in x : \emptyset \in f(u)\} : f \in \mathrm{P}(\emptyset)^x\}$.

(iv) Combine (iii) with the observation that, assuming REM, $\mathrm{P}(\{\emptyset\}) = \{\emptyset, \{\emptyset\}\}$.

(v) $\Rightarrow$: let $x := \{ y \in \{\emptyset\} : A\}$, then $\emptyset \in x \leftrightarrow A$. $\Leftarrow$: given $A[y]$ and a set z, we have $\forall y \in z\, \exists x(A[y] \leftrightarrow \emptyset \in x)$; in fact we may assume $\forall y \in z\, \exists x \subset \{\emptyset\}(A[y] \leftrightarrow \emptyset \in x)$; x depends on z. Since x is uniquely determined by y, there is, by Replacement (S-Coll), an f with $\mathrm{dom}(f) = z$, such that $\forall y \in z(A[y] \leftrightarrow \emptyset \in f(y))$; with R-Sep we conclude that $\{ y \in z : \emptyset \in f(y)\} = \{y \in z : A[y]\}$ exists. □

8.10. PROPOSITION. **CZF** + PEM = **CZF** + REM + SEP = **ZF**.

PROOF. Exercise. □

8.11. **IZF** is the system **CZF** + Pow + Sep. Some of the principal relationships are indicated in the diagram of fig. 11.4 (where the arrows stand for "subsystem of").

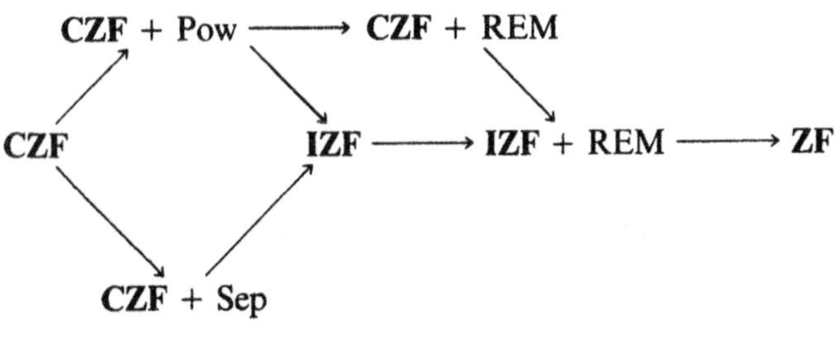

Fig. 11.4.

8.12. The remainder of this section is devoted to the interpretation of **CZF** in $\mathbf{MLV}^{\mathrm{i}}$. Our presentation is based on Aczel (1982). To simplify the notation, we shall often write $(\sup x \in A)t[x]$ for $\sup(A, \lambda x.t[x])$. Since we consider an interpretation of **CZF** in $\mathbf{MLV}^{\mathrm{i}}$, we have to distinguish carefully between $=$ and $\in$ on the one hand, which are primitive symbols in type theory, and $\dot{=}$ and $\dot{\in}$, which will be defined and are the interpretations of $=$ and $\in$ respectively in **CZF**.

The basic idea of the interpretation is easy to explain in model theoretic terms. V is to be the universe of sets in our model of **CZF**. Each element of V is of the form $\sup(A, f)$ with $A \in U$, which may be more suggestively written as $\{f(x) : x \in A\}$, where $f(x) \in V$ for each $x \in A$. That is to say, elements of V are constructed inductively as families of sets indexed by the elements of a small type. Alternatively, one may think of the elements of V as well-founded trees, where the successors of a node are always indexed by the elements of a small type; the $\in$-relation corresponds to the successor relation on such trees. So, in particular, for any element $\alpha \equiv \sup(\alpha^-, \alpha^\sim)$ of V the $\alpha^\sim(x)$ for $x \in \alpha^-$ play the role of elements. The small types play the same role here, as the ordinals do in the building of the cumulative hierarchy of sets.

If the reader keeps this idea in mind, the material that follows will appear less an exhibition of magic than a natural choice of notions.

Now we can also explain inductively the equality $\doteq$ between two such families $\{x_i : i \in A\}$ and $\{y_i : j \in B\}$ $(A, B \in U)$:

$$\{x_i : i \in A\} \doteq \{y_j : j \in B\} :=$$
$$\forall i \in A \, \exists j \in B (x_i \doteq y_j) \wedge \forall j \in B \, \exists i \in A (x_i \doteq y_j\}$$

(this is to be read as an informal explanation, not as a statement in type theory); so equality between sets is explained in terms of equality between their elements.

Set-theoretic membership $\dot\in$ is now also easily explained as

$$x \dot\in \{x_i : i \in A\} := \exists i \in A (x \doteq x_i).$$

8.13. CONVENTION. For the rest of this section we shall use Greek lower case $\alpha, \beta, \gamma, \ldots, \eta$ for elements of V. □

DEFINITION. Let $x \in V \Rightarrow B[x]$ type, then we put

$$\forall x \dot\in \alpha \, B[x] := \forall x \in \alpha^- \; B[\alpha^\sim(x)],$$
$$\exists x \dot\in \alpha \, B[x] := \exists x \in \alpha^- \; B[\alpha^\sim(x)]. \quad \square$$

NOTATION. Let T be the elimination constant for W-types introduced in the preceding section, and let f be a three-place function. We shall frequently abbreviate $T_{x,y,z}(t, f(x, y, z))$ as $T(t, f)$ and $\sup(A, \lambda x.t[x])$ as $(\sup x \in A)t[x]$. □

Observe (E11.7.5) that we have the principle of $\dot\in$-Ind

$$\forall x \in V(\forall y \dot\in x B[y] \to B[x]) \to \forall x \in V B[x].$$

8.14. THEOREM. *There is a definable type* $\doteq$ *in* $\mathbf{MLV}^{\mathrm{i}}$ *such that*

$$(\alpha \doteq \beta) = \forall x \mathrel{\dot\in} \alpha\, \exists y \mathrel{\dot\in} \beta (x \doteq y) \wedge \forall y \mathrel{\dot\in} \beta\, \exists x \mathrel{\dot\in} \alpha (x \doteq y),$$

where $\alpha \doteq \beta$ *stands for* $(\alpha, \beta) \in \doteq$, *and such that* $\alpha \doteq \beta$ *is a small proposition for all* $\alpha \in V$, $\beta \in V$, *i.e. a type in* U.

PROOF. Informally, we may reason as follows. By a double recursion over V we construct a binary propositional function $\doteq$ such that for all α, β in V $\alpha \doteq \beta$ is a small proposition, i.e. an element of U, and satisfies $(\alpha \doteq \beta) = \forall x \mathrel{\dot\in} \alpha\, \exists y \mathrel{\dot\in} \beta (x \doteq y) \wedge \forall y \mathrel{\dot\in} \beta\, \exists x \mathrel{\dot\in} \alpha (x \doteq y)$. Now assume we have already constructed $\doteq$ satisfying these requirements for all α' preceding α and all β' preceding β in the inductive generation of V. Then in particular $(\alpha^{\sim}(x) \doteq \beta^{\sim}(y))$ is a small proposition for $x \in \alpha^{-}$, $y \in \beta^{-}$, and therefore $\forall x \in \alpha^{-}\, \exists y \in \beta^{-} (\alpha^{\sim}(x) \doteq \beta^{\sim}(y))$ is also a small proposition; but this is equivalent to $\forall x \mathrel{\dot\in} \alpha\, \exists y \mathrel{\dot\in} \beta (x \doteq y)$. Similarly, $\forall y \mathrel{\dot\in} \beta\, \exists x \mathrel{\dot\in} \alpha (x \doteq y)$ is a small proposition. Thus we can recursively define $\alpha \doteq \beta$ as in the theorem.

Formally, the proof goes as follows. We define three-place functions G_1, G_2, G such that for $u \in U$, $v \in V$, $w \in u \to (V \to U)$

$$G_1(u, v, w) := \forall x \in u\, \exists y \mathrel{\dot\in} v.wxy,$$

$$G_2(u, v, w) := \forall y \mathrel{\dot\in} v\, \exists x \in u.wxy,$$

$$G(u, z, w) := \lambda v.(G_1(u, v, w) \wedge G_2(u, v, w)).$$

(So G does not actually depend on z.) Spelling it out:

$$G_1(u, v, w) \equiv \forall x \in u\, \exists y \in v^{-}.wx(v^{\sim}(y)),$$

and thus if $u \in U$, $v \in V$, $w \in u \to (V \to U)$, it follows that $wx(v^{\sim}(y)) \in U$, hence $G_1(u, v, w) \in U$. Similarly, $G_2(u, v, w) \in U$, and then also $G(u, z, w) \in V \to U$. Now $T(\alpha, G) \in V \to U$; we may thus define for $\alpha, \beta \in V$:

$$\alpha \doteq \beta := T(\alpha, G)(\beta).$$

We recall that

$$T(\sup(A, f), G) = G(A, f, \lambda u.T(fu, G))$$

for $A \in U$, $f \in A \to V$. Abbreviating $t \equiv \lambda u.T(\alpha^{\sim}(u), G)$, we find $(\alpha \doteq \beta) = T(\sup(\alpha^{-}, \alpha^{\sim}), G)(\beta) = G(\alpha^{-}, \alpha^{\sim}, t)(\beta) = G_1(\alpha^{-}, \beta, t) \wedge G_2(\alpha^{-}, \beta, t)$. However, $G_1(\alpha^{-}, \beta, t) = \forall x \in \alpha^{-}\, \exists y \mathrel{\dot\in} \beta.t(x)(y) = \forall x \in \alpha^{-}\, \exists y \mathrel{\dot\in} \beta$

$(\alpha^{\sim}(x) = y) = \forall x \dot{\in} \alpha \, \exists y \dot{\in} \beta (x \doteq y)$; similarly $G_2(\alpha^{-}, \beta, t) = \forall y \dot{\in} \beta \, \exists x \dot{\in} \alpha (x \doteq y)$, given the assertion of the theorem. □

The next lemma verifies that $\doteq$ is an equivalence relation.

8.15. LEMMA. *For* $\alpha, \beta, \gamma \in V$:
(i) $\alpha \doteq \alpha$,
(ii) $\alpha \doteq \beta \to \beta \doteq \alpha$,
(iii) $\alpha \doteq \beta \wedge \beta \doteq \gamma \to \alpha \doteq \gamma$.

PROOF. (i) Let $\alpha \in V$; then from $\forall x \dot{\in} \alpha (x \doteq x)$ it follows that $\forall x \in \alpha^{-}(\alpha^{\sim}(x) \doteq \alpha^{\sim}(x))$, hence $\forall x \in \alpha^{-} \, \exists y \in \alpha^{-}(\alpha^{\sim}(x) = \alpha^{\sim}(y))$, and so $\forall x \dot{\in} \alpha \, \exists y \dot{\in} \alpha (x \doteq y)$. Similarly $\forall x \dot{\in} \alpha (x \doteq x) \to \forall y \dot{\in} \alpha \, \exists x \dot{\in} \alpha (x \doteq y)$. Thus by $\dot{\in}$-Ind $\forall \alpha \in V (\alpha \doteq \alpha)$.

(ii) Put $F[x] := \forall y \in V (x \doteq y \to y \doteq x)$, and suppose $\forall x \dot{\in} \alpha \, F[x]$ for some $\alpha \in V$. Let $\beta \in V$, then $\alpha \doteq \beta \to \forall x \in \alpha^{-} \, \exists y \in \beta^{-}(\alpha^{\sim}(x) \doteq \beta^{\sim}(y))$, hence $\alpha \doteq \beta \to \forall x \in \alpha^{-} \, \exists y \in \beta^{-}(\beta^{\sim}(y) \doteq \alpha^{\sim}(x))$, and so $\alpha \doteq \beta \to \forall x \dot{\in} \alpha \, \exists y \dot{\in} \beta (y \doteq x)$. Similarly $\alpha \doteq \beta \to \forall y \dot{\in} \beta \, \exists x \dot{\in} \alpha (y \doteq x)$, hence $\alpha \doteq \beta \to \beta \doteq \alpha$. Thus we have proved $(\forall x \dot{\in} \alpha \, F[x]) \to F[\alpha]$, hence by $\dot{\in}$-Ind $\forall x \in V F[x]$.

(iii) Apply a similar argument to the property $F[x] := \forall y \in V \, \forall z \in V (x \doteq y \wedge y \doteq z \to x \doteq z)$ (exercise). □

8.16. DEFINITION. For $\alpha, \beta \in V$ let

$$\alpha \dot{\in} \beta := \exists y \dot{\in} \beta (\alpha \doteq y),$$

$$\alpha \subset \beta := \forall x \dot{\in} \alpha (x \dot{\in} \beta).$$

Note that $\alpha \dot{\in} \beta$ and $\alpha \subset \beta$ are small propositions (types). A predicate $F[x]$ of x is said to be $\doteq$*-extensional*, or for short, *extensional over* V if

$$\forall x \in V \, \forall y \in V (F[x] \wedge y \doteq x \to F[y]).$$

Alternatively, we can express this by saying that $\{x : F\}$ is extensional. □

8.17. LEMMA. *If* $F[x]$ *is extensional over* V *in* x, *then for* $\alpha \in V$

$$\forall x \dot{\in} \alpha \, F \leftrightarrow \forall x \in V (x \dot{\in} \alpha \to F),$$

$$\exists x \dot{\in} \alpha \, F \leftrightarrow \exists x \in V (x \dot{\in} \alpha \wedge F).$$

PROOF. We prove the first part and leave the proof of the second part as an exercise. $\forall x \in V(x \mathrel{\dot{\in}} \alpha \to F) \leftrightarrow \forall x \in V(\exists y \mathrel{\dot{\in}} \alpha(x \doteq y) \to F) \leftrightarrow \forall x \in V$ $(\exists y \in \alpha^{-}(x \doteq \alpha^{\sim}(y)) \to F) \leftrightarrow \forall y \in \alpha^{-} \forall x \in V(x \doteq \alpha^{\sim}(y) \to F) \leftrightarrow$ $\forall y \in \alpha^{-} F[x/\alpha^{\sim}(u)] \leftrightarrow \forall x \mathrel{\dot{\in}} \alpha F$. □

The following lemma verifies Extensionality for $\dot{\in}$ and $\doteq$.

8.18. LEMMA. *For* $\alpha, \beta, \gamma \in V$:
(i) $\alpha \doteq \beta \wedge \beta \mathrel{\dot{\in}} \gamma \to \alpha \mathrel{\dot{\in}} \gamma$,
(ii) $\alpha \doteq \beta \leftrightarrow \forall x \in V(x \mathrel{\dot{\in}} \alpha \leftrightarrow x \mathrel{\dot{\in}} \beta)$.

PROOF. (i) From $\alpha \doteq \beta \wedge \beta \mathrel{\dot{\in}} \gamma$ it follows that $\alpha \doteq \beta \wedge \exists z \in \gamma^{-}$ $(\beta \doteq \gamma^{\sim}(z))$, hence $\exists z \in \gamma^{-}(\alpha \doteq \gamma^{\sim}(z))$, and thus $\alpha \mathrel{\dot{\in}} \gamma$.

(ii) By (i), $x \mathrel{\dot{\in}} \alpha$ is extensional in x over V; thus, by 8.17, $\alpha \subset \beta \leftrightarrow \forall x \mathrel{\dot{\in}} \alpha$ $(x \mathrel{\dot{\in}} \beta) \leftrightarrow \forall x \in V(x \mathrel{\dot{\in}} \alpha \to x \mathrel{\dot{\in}} \beta)$; similarly $\beta \subset \alpha \leftrightarrow \forall x \in V(x \mathrel{\dot{\in}} \beta \to x \mathrel{\dot{\in}} \alpha)$, and since one easily sees that $(\alpha \subset \beta \wedge \beta \subset \alpha) \leftrightarrow \alpha \doteq \beta$, the assertion follows. □

We now verify pairing and union and a form of separation for $\doteq$ and $\dot{\in}$.

8.19. LEMMA
(i) $\forall \alpha\beta\exists\gamma\forall\eta(\eta \mathrel{\dot{\in}} \gamma \leftrightarrow \eta \doteq \alpha \vee \eta \doteq \beta)$.
(ii) $\forall \alpha\exists\gamma\forall\eta(\eta \mathrel{\dot{\in}} \gamma \leftrightarrow \exists x \mathrel{\dot{\in}} \alpha(\eta \in x))$.
(iii) *Let* $\alpha \in V$, $F \in V \to U$, *then* $\exists\gamma\forall\eta(\eta \mathrel{\dot{\in}} \gamma \leftrightarrow \exists x \mathrel{\dot{\in}} \alpha(F(x) \wedge \eta \doteq x))$.

PROOF. (i) Let $\alpha, \beta \in V$ and put $\gamma \equiv \{\alpha, \beta\} := (\sup z \in N_2)R_2(z, \alpha, \beta)$. Then $\eta \mathrel{\dot{\in}} \gamma \leftrightarrow \exists z \in N_2(\eta \doteq R_2(z, \alpha, \beta)) \leftrightarrow \eta \doteq \alpha \vee \eta \doteq \beta$ (use 4.12(iii)) and recall that $R_2(\mathbf{0}, \alpha, \beta) = \alpha$, $R_2(\mathbf{1}, \alpha, \beta) = \beta$).

(ii) Let $\alpha \in V$ and put

$$\gamma := \left(\sup z \in \Sigma x \in \alpha^{-}(\alpha^{\sim}(x))^{-}\right)\left(\alpha^{\sim}(\mathbf{p}_0 z)^{\sim}(\mathbf{p}_1 z)\right).$$

Using again lemma 4.12(i) and 7.7 we find

$$\begin{aligned}\eta \mathrel{\dot{\in}} \gamma &\leftrightarrow \exists z \in \left(\Sigma x \in \alpha^{-}(\alpha^{\sim}(x))^{-}\right)\left(\eta \doteq (\alpha^{\sim}(\mathbf{p}_0 z))^{\sim}(\mathbf{p}_1 z)\right)\\ &\leftrightarrow \exists x \in \alpha^{-} \exists y \in (\alpha^{\sim}(x))^{-}\left(\eta \doteq ((\alpha^{\sim}(x))^{\sim}(y)\right)\\ &\leftrightarrow \exists x \in \alpha^{-} \exists y \mathrel{\dot{\in}} \alpha^{\sim}(x)(\eta \doteq y)\\ &\leftrightarrow \exists x \mathrel{\dot{\in}} \alpha \exists y \mathrel{\dot{\in}} x(\eta \doteq y)\\ &\leftrightarrow \exists x \mathrel{\dot{\in}} \alpha(\eta \mathrel{\dot{\in}} x).\end{aligned}$$

(iii) Let $\alpha \in V$, $F \in V \to U$, then $A := \Sigma x \in \alpha^{-}(F(\alpha^{\sim}(x))) \in U$, and so $\gamma := (\sup z \in A)(\alpha^{\sim}(p_0 z)) \in V$. Now

$$\begin{aligned}\eta \,\dot{\in}\, \gamma &\leftrightarrow \exists z \in A(\eta \doteq \alpha^{\sim}(p_0 z))\\ &\leftrightarrow \exists x \in \alpha^{-}\, \exists y \in F(\alpha^{\sim}(x))(\eta \doteq \alpha^{\sim}x)\\ &\leftrightarrow \exists x \in \alpha^{-}(F(\alpha^{\sim}(x)) \wedge \eta \doteq \alpha^{\sim}(x))\\ &\leftrightarrow \exists x \,\dot{\in}\, \alpha(F(x) \wedge \eta \doteq x),\end{aligned}$$

again with an appeal to 4.12. □

Strong Collection and Subset-Collection for $\in$ are verified in the following lemma.

8.20. LEMMA. *Suppose* $x \in V$, $y \in V \Rightarrow F[x, y]$ type, *and put*

$$F'[x, y] := \forall u \,\dot{\in}\, x\, \exists v \,\dot{\in}\, y F[u, v] \wedge \forall v \,\dot{\in}\, y\, \exists u \,\dot{\in}\, x F[u, v].$$

Then:
(i) $\alpha^{-} = \beta^{-} \to (\forall x \in \alpha^{-} F[\alpha^{\sim}(x), \beta^{\sim}(x)] \to F'[\alpha, \beta])$,
(ii) $\forall \alpha \forall x \,\dot{\in}\, \alpha\, \exists y \in V F[x, y] \to \exists \beta \in V F'[\alpha, \beta]$,
(iii) *For* $\alpha, \beta \in V$ *there is a* $\gamma \in V$ *depending on* α *and* β *only, such that*

$$\forall x \,\dot{\in}\, \alpha\, \exists y \,\dot{\in}\, \beta F[x, y] \to \exists \delta \,\dot{\in}\, \gamma F'[\alpha, \delta].$$

PROOF. (i) Suppose $\alpha^{-} = \beta^{-}$, $\forall x \in \alpha^{-} F[\alpha^{\sim}(x), \beta^{\sim}(x)]$. Then $\forall x \in \alpha^{-}$ $\exists y \,\dot{\in}\, \beta F[\alpha^{\sim}(x), \beta^{\sim}(y)]$, hence $\forall x \,\dot{\in}\, \alpha\, \exists y \,\dot{\in}\, \beta F[x, y]$. Similarly $\forall y \,\dot{\in}\, \beta$ $\exists x \,\dot{\in}\, \alpha F[x, y]$, so $F'[\alpha, \beta]$.

(ii) Suppose $\forall x \,\dot{\in}\, \alpha\, \exists y \in V F[x, y]$; then $\forall x \in \alpha^{-}\, \forall y \in V F[\alpha^{\sim}(x), y]$. By the axiom of choice (4.10) there is an $f \in \alpha^{-} \to V$ such that $\forall x \in \alpha^{-}$ $F[\alpha^{\sim}(x), f(x)]$. Put $\beta := \sup(\alpha^{-}, f)$, then $\beta \in V$, $\alpha^{-} = \beta^{-}$, $\forall x \in \alpha^{-}$ $F[\alpha^{\sim}(x), \beta^{\sim}(x)]$ and thus by (i) $F'[\alpha, \beta]$.

(iii) Put $\gamma := (\sup z \in \alpha^{-} \to \beta^{-})(\sup x \in \alpha^{-})\beta^{\sim}(z(x))$. Clearly $\gamma \in V$. Suppose $\forall x \,\dot{\in}\, \alpha\, \exists y \,\dot{\in}\, \beta F[x, y]$, then $\forall x \in \alpha^{-}\, \exists y \in \beta^{-} F[\alpha^{\sim}(x), \beta^{\sim}(y)]$. By 4.10 there is an $f \in \alpha^{-} \to \beta^{-}$ such that $\forall x \in \alpha^{-} F[\alpha^{\sim}(x), \beta^{\sim}(f(x))]$; put $\delta := (\sup x \in \alpha^{-})\beta^{\sim}(f(x))$. Then $\delta \in V$, $\delta \doteq \gamma^{\sim}(f)$ and so $\delta \in \gamma$. Moreover $\alpha^{-} = \delta^{-}$, $\forall x \in \alpha^{-} F[\alpha^{\sim}(x), \delta^{\sim}(x)]$, and by (i) $F'[\alpha, \delta]$. □

8.21. LEMMA. *Put* $\emptyset := \sup(\bot, R_0)$. *For any* $\alpha \in V$ *let*

$$\alpha^{+} := (\sup x \in (\alpha^{-} + N_1))D_{u,v}(x, \alpha^{\sim}(u), \alpha).$$

Then:
(i) $\emptyset \in V,\ \eta \dot{\in} \emptyset \leftrightarrow \perp$,
(ii) $\eta \dot{\in} \alpha^+ \leftrightarrow \eta \dot{\in} \alpha \vee \eta \doteq \alpha$,
(iii) $\alpha^+ \doteq \emptyset \rightarrow \perp$,
(iv) $\alpha^+ \doteq \beta^+ \rightarrow \alpha \doteq \beta$.

PROOF. (i) $\emptyset \in V$ is immediate; $\eta \dot{\in} \emptyset \leftrightarrow \exists x \in \perp (\eta \doteq R_0(x)) \leftrightarrow \perp$ (by 4.12(iii)).

(ii) Let $\alpha \in V$. Then $\alpha^- \in U$, $\alpha^- + N_1 \in U$, and, for all $x \in \alpha^-$, $\alpha^-(x) \in V$. Thus $D_{u,v}(x, \alpha^-(u), \alpha) \in V$ for $x \in \alpha^- + N_1$, hence $\alpha^+ \in V$. Again by 4.12 we have $\eta \dot{\in} \alpha^+ \leftrightarrow \exists x \in (\alpha^- + N_1)(\eta \doteq D_{u,v}(x, \alpha^-(u), \alpha) \leftrightarrow \exists x \in \alpha^-(\eta \doteq \alpha^-(x)) \vee \exists y \in N_1(\eta \doteq \alpha(y)) \leftrightarrow (\eta \dot{\in} \alpha \vee \eta \doteq \alpha)$.

(iii) $\alpha^+ \doteq \emptyset \rightarrow \alpha \dot{\in} \emptyset$, by (ii), and $\alpha \dot{\in} \emptyset \rightarrow \perp$, by (i).

(iv) $\alpha^+ \doteq \beta^+ \rightarrow \alpha \dot{\in} \beta^+ \wedge \beta \dot{\in} \alpha^+$, since $\alpha^+ \doteq \beta^+ \leftrightarrow \forall x \dot{\in} \alpha^+ \exists y \dot{\in} \beta^+ (x \doteq y) \wedge \forall y \dot{\in} \beta^+ \exists x \dot{\in} \alpha^+ (x \doteq y)$, which by (ii) implies $\alpha \dot{\in} \beta^+ \wedge \beta \dot{\in} \alpha$. Hence $\alpha^+ \doteq \beta^+ \rightarrow (\alpha \dot{\in} \beta \vee \alpha \doteq \beta) \wedge (\beta \dot{\in} \alpha \vee \beta \doteq \alpha)$, and since $\alpha \dot{\in} \beta \wedge \beta \dot{\in} \alpha \rightarrow \perp$ (provable by $\dot{\in}$-Ind) we find $\alpha^+ \doteq \beta^+$. □

The next lemma establishes infinity for $\doteq$ and $\dot{\in}$.

8.22. LEMMA. *Let $n \in N$ and put*

$$\Delta := \lambda n . R_{x,y}(n, \emptyset, y^+).$$

Then $\forall n \in N(\Delta(n) \in V)$, $\omega := (\sup x \in N)\, \Delta(x) \in V$, *and*:
(i) $\emptyset \dot{\in} \omega$,
(ii) $\forall \alpha \dot{\in} \omega (\alpha^+ \dot{\in} \omega)$,
(iii) *If $F[\alpha]$ is a proposition for all $\alpha \in V$, then*

$$F[\emptyset] \wedge \forall x \dot{\in} \omega (F[x] \rightarrow F[x^+]) \rightarrow \forall x \dot{\in} \omega F[x].$$

PROOF. Observe that $\forall x \dot{\in} \omega . F[x] \leftrightarrow \forall x \in N . F[\Delta(n)]$.

(i) $\emptyset \doteq \Delta(0)$, so $\emptyset \dot{\in} \omega$.

(ii) $\forall \alpha \dot{\in} \omega (\alpha^+ \dot{\in} \omega) \leftrightarrow \forall \alpha \in N(\Delta(\alpha)^+ \dot{\in} \omega)$ which obviously holds, since $\Delta(\alpha)^+ = R_{x,y}(S\alpha, \emptyset, y^+) = \Delta(S\alpha)$.

(iii) Left as an exercise. □

8.23. LEMMA. *Let F be any formula constructed from $x \doteq y$, $x \dot{\in} y$, $\wedge$, $\vee$, $\perp$, $\rightarrow$, $\forall x \in V$, $\exists x \in V$, $\forall x \dot{\in} y$, $\exists x \dot{\in} y$. Then F is extensional over V in each of its free variables under the assumption that the other free variables are in V.*

In other words, if $F \equiv F[x_1, \ldots, x_n]$ *with all free variables shown, then* $\forall x_1 \ldots x_n \forall y_1 \ldots y_n \in V(F[x_1, \ldots, x_n] \wedge x_1 \doteq y_1 \wedge \cdots \wedge x_n \doteq y_n \to F[y_1, \ldots, y_n])$; *and this fact can be proved using only* $\alpha \doteq \beta \wedge \beta \dot{\in} \gamma \to \alpha \dot{\in} \gamma$, $\alpha \doteq \beta \leftrightarrow \forall x \in V(x \in \alpha \leftrightarrow x \in \beta)$.

PROOF. Exercise. □

8.24. Let us now formulate **CZF** in a language with $\doteq$, $\dot{\in}$, $\wedge$, $\vee$, $\perp$, $\to$, $\forall x \in V$ (for $\forall x$), $\exists x \in V$ (for $\exists x$), $\forall x \dot{\in} y, \exists x \dot{\in} y$ as primitives; we have to add to our earlier formulation two restricted quantifier axioms

$$\text{Resq} \quad \begin{cases} \forall x \dot{\in} y A[x] \leftrightarrow \forall x \in V(x \dot{\in} y \to A[x]), \\ \exists x \dot{\in} y A[x] \leftrightarrow \exists x \in V(x \dot{\in} y \wedge A[x]). \end{cases}$$

We then have the following theorem.

THEOREM. **CZF** *is true on interpretation in* $\mathbf{MLV}^{\mathrm{i}}$, *that is to say, if we interpret the primitives of* **CZF** *by the corresponding defined notions of* $\mathbf{MLV}^{\mathrm{i}}$, *each theorem A of* **CZF** *is translated into a proposition A′ of* $\mathbf{MLV}^{\mathrm{i}}$ *such that* $\mathbf{MLV}^{\mathrm{i}} \vdash t \in A'$ *for a suitable term t.*

PROOF. Since $\mathbf{MLV}^{\mathrm{i}}$ validates intuitionistic logic and since, if the restricted quantifier axioms hold, $\forall x \dot{\in} y$, $\exists x \dot{\in} y$ also behave as restricted quantifiers in intuitionistic logic, it suffices to check the axioms.

Ext is taken care of by lemma 8.18. Resq follows from 8.17 together with 8.23. $\dot{\in}$-Ind holds (end of 8.13, E11.7.5); Pairing, Union follow from 8.19(i), (ii), R-Sep from 8.19(iii) with 8.23; S-Coll and Sub-Coll from 8.20 and 8.23; Inf from 8.22. □

8.25. *Choice axioms and the presentation axiom.* The preceding theorem may be strengthened: the interpretation in $\mathbf{MLV}^{\mathrm{i}}$ also validates countable choice AC_0 and dependent choice DC (ω playing the role of N); cf. exercise 11.8.9.

Let us call a set x a *base* if choice functions over x can always be found, or more precisely, if the axiom of choice holds relative to x:

$$\forall y \in x \exists u A[y, z] \to \exists f(\mathit{fun}(f) \wedge \forall y \in x A[y, f(x)]).$$

Thus AC-N states that ω is a base.

The *presentation axiom* states that each set is the surjective image of a base. Aczel has shown that we can extend the interpretation in $\mathbf{MLV}^{\mathrm{i}}$ so as to validate the presentation axiom, either by strengthening $\mathbf{MLV}^{\mathrm{i}}$ by including an axiom of induction over U expressing that U is the least universe closed under the type forming operations (Aczel 1982), or by an inner-model construction (Aczel 1986). As noted by Aczel (1978) yet another method to validate the presentation axiom would be to use a realizability interpretation of the type discussed in section 9.5. Observe in this connection that ordinary realizability for **HA** shows that for each set $\{x : A[x]\}$ the set $\{(y, x)\} : y\mathbf{r}A[x]\}$ is a base which can be surjectively mapped onto $\{x : A[x]\}$ by j_1.

9. Notes

9.1. *Further reading on type theory.* Some papers dealing with Martin-Löf's type theories are Aczel (1977A, 1978, 1982, 1986), Jervell (1978), Diller (1980), Diller and Troelstra (1984), Troelstra (1983), Smith (A, 1984), Seely (1982, 1984) (category-theoretic equivalents to Martin-Löf's type theories), Rezus (1985).

The family of AUTOMATH-languages developed by De Bruijn and his collaborators has many features in common with Martin-Löf's type theories. These languages were developed for the automatic checking of the correctness of mathematical proofs (cf. de Bruijn 1970, 1980, Zucker 1977, van Daalen 1980, van Benthem Jutting 1977, Barendregt and Rezus 1983, Rezus 1983).

In computer science there is a rapidly growing literature on type-theory; see for example Constable et al. (1986) and the references given there.

Diller (1980) gives a version of (the arithmetical part of) the theory of Martin-Löf (1975) in which all constants and subterms of a term also carry their types along. (Diller's description of his system needs slight corrections, so as to avoid the problems signaled in 2.11 and 2.16: terms and types must be defined simultaneously with the derivations.) Seely (1984) is based on this version.

Clearly, any correct derivation in the version with typed subterms yields a correct derivation in the systems described here. That conversely a derivation in $\mathbf{ML}_0$ or $\mathbf{ML}_0^{\mathrm{i}}$ can be converted into a derivation with all subterms types seems plausible, but has not yet been proved (if we try to prove this by induction on derivations, the treatment of applications of introduction rules is straightforward, but E-rules such as ΠE present problems).

9.2. *Formulas-as-types and operational semantics.* In Martin-Löf's writings the formulas-as-types (cf. 10.9.4) concept is exploited to the full, in combination with the idea of operational semantics: the rules determine the meaning of the constants appearing in the theory.

Martin-Löf's type theory and its semantics, as illustrated in section 5, show what can be achieved along these lines. At the same time, the case of tree types (the type forming operation W of section 7) shows what has to be accepted as intuitively clear in Martin-Löf's semantical explanations, i.c. validity of bar induction.

9.3. *Intensional and extensional versions.* Martin-Löf (1975) gives an intensional version of the theory of types; its arithmetical part is quite close to $\mathbf{ML}_0^{\mathrm{i}}$, except that we have adopted the unique canonical constant $\boldsymbol{e}$ for inhabited identity types, thereby increasing the similarity to realizability. Martin-Löf (1982, 1984) present extensional versions of the theory of types. Their arithmetical part corresponds to our $\mathbf{ML}_0$. These theories can be easily modeled (classically) in **ZF**. In still more recent expositions (cf. the preface to Martin-Löf (1984) and Chapter 11 in Beeson (1985)) families of objects and families of types indexed by a type are included in the formalism; this device permits a more uniform treatment of introduction/elimination rules (cf. our remarks in 3.9.11, 4.6).

In the most recent, as yet unpublished, versions of his theory, Martin-Löf has returned to an intensional point of view, as in Martin-Löf (1975), that is to say, $t = t' \in A$ is understood as "t and t' are definitionally equal". As a consequence the rules for identity types have to be adapted:

$$\text{II} \qquad \frac{t \in A}{\boldsymbol{e}(t) \in I(A, t, t)}$$

$$\text{IE} \qquad \frac{t \in I(A, s, s') \quad x \in A \Rightarrow t' \in C[x, x, \boldsymbol{e}(x)] \quad \Delta}{J(t, t') \in C[s, s', t]}$$

where $\Delta \equiv x \in A,\ y \in A,\ z \in I(A, x, y) \Rightarrow C[x, y, z]$ type. The special equality rule becomes

$$\text{ICONV} \qquad \frac{t \in A \quad x \in A \Rightarrow t' \in C[x, x, \boldsymbol{e}(x)] \quad \Delta}{J(\boldsymbol{e}(t), t') = t'[x/t] \in C}$$

There is now no immediate connection between $t = t' \in A$ and $s \in I(A, t, t')$.

9.4. *Type theory with universes.* Early versions of Martin-Löf's type theories contained a type of all types; Girard pointed out that this gives rise to a paradox. Our presentation of the paradox in 7.4 follows Martin-Löf (1972).

In the presentation of the theory with universe in section 7 (as also in Martin-Löf (1975, 1982, 1984)) small types are themselves elements of a type. In a more cautious version of the theory one distinguishes between small types and codes for such types; the codes form the universe U. A description of such a version may be found in Smith (1984), or in Beeson (1985, ch. 11, section 21). One has to add to $\mathbf{ML}_0$ symbols U, $\boldsymbol{T}$, $\dot{N}$, $\dot{I}$, $\dot{\Pi}$, $\dot{\Sigma}$, $\dot{+}$; $\dot{N}$ is the code of type N, and $\dot{I}$, $\dot{\Pi}$, $\dot{\Sigma}$, $\dot{+}$ are the operations on codes corresponding to I, Π, Σ, $+$. $\boldsymbol{T}$ transforms a code into a small type. Some examples of the rules are

$$\frac{t \in U}{\boldsymbol{T}t} \qquad\qquad \frac{t = t' \in U}{\boldsymbol{T}t = \boldsymbol{T}t'}$$

$$\frac{t \in U \quad x \in \boldsymbol{T}t \Rightarrow t' \in U}{(\dot{\Pi}x \in t)t' \in U}$$

$$\frac{t \in U \qquad x \in \boldsymbol{T}(t) \Rightarrow t' \in U}{\boldsymbol{T}((\dot{\Pi}x \in t)t') = (\Pi x \in \boldsymbol{T}t)(\boldsymbol{T}t')}$$

etc.

9.5. Beeson (1985, ch. 13, theorem 7.5.1) and Renardel de Lavalette (1984A) contain proofs that $\mathbf{ML}_0^{\mathrm{i}}$ is conservative over **HA**, this extends the result of Goodman (1978). Even stronger results in this direction are contained in Gordeev (1988).

9.6. *Set theories based on intuitionistic logic.* The principal systems of set theory based on intuitionistic logic, "intuitionistic set theories" for short, which have been studied in the literature, are:

(1) **IZF** defined in 8.11. Of interest is also its weakening **IzF** with Replacement instead of Collection.

(2) **CZF** discussed in section 8, investigated by Aczel (1978, 1982, 1986). The system **CST** of Myhill (1975) is related to, but not identical with **CZF**.

(3) Friedman's system **B** (1977A), a weak system with the strength of first-order arithmetic. **B** contains variables for numbers and sets (regarded as disjoint domains), the first-order Peano axioms with induction restricted to sets, extensionality, existence of empty set, pairing, union, infinity (the

collection of natural numbers corresponds to a set), R-Sep, exponentiation, *Abstraction*: for a set x and restricted A

$$\{\{u \in x : A[u, y_1, \ldots, y_n]\} : y_1 \in x \wedge \cdots \wedge y_n \in x\}$$

is a set (a special case of collection), and dependent choice for restricted formulas.

9.7. *Further reading on intuitionistic set theories.* Beesons work on intuitionistic set theories (including a number of additions, and corrections of earlier proofs) is brought together in Beeson (1985, ch. 8, 9, and 12).

Gordeev (1982) investigated, among other things, set theories containing an axiom of *Completeness* contradicting Foundation (the completeness axiom states that any structure (x, R), R a binary relation on the set x, is isomorphic to a structure $(y, \in \restriction y)$ for a suitable set y). Scedrov (1985) surveys Friedman's work on intuitionistic set theories.

Friedman and Scedrov (1983) contains the interesting result that **IzF** does have the explicit definability property for sets, while **IZF** does not. So Replacement does not imply Collection.

Friedman and Scedrov (1984) is a study of the effect of large-cardinal axioms in intuitionistic set theories.

Sheaf and topos models of intuitionistic set theories are treated in, a.o., Grayson (1979), Fourman (1980), Hayashi (1981), Takeuti and Titani (1981).

Exercises

11.2.1. Complete the proof of 2.14.

11.2.2. Write down a complete set of special equality rules for the combinators introduced in 2.18.

11.2.3. Let $\mathbf{ML}_0^+$ be the system obtained from $\mathbf{ML}_0^i$ by replacing SUB and REPL 1, 2 by the following rule

$$\text{REPL*} \qquad \frac{\Gamma, \Gamma'[x/t] \Rightarrow \theta[x/t] \quad \Gamma \Rightarrow t = s \in A}{\Gamma, \Gamma'[x/s] \Rightarrow \theta[x/s]}$$

Show that $\mathbf{ML}_0 \subset \mathbf{ML}_0^+$. In particular, SUB is provable by induction on the length of deductions.

11.3.1. Complete the argument in 3.1.

11.3.2. Give a detailed proof of the proposition in 3.3.

11.3.3. Formulate the variant $\mathbf{ML}_0^{\text{comb}}$ of $\mathbf{ML}_0^{\text{i}}$ with combinators as primitives, as indicated in 3.6 (cf. also E11.2.2) and show $\mathbf{ML}_0^{\text{i}} \subset \mathbf{ML}_0^{\text{comb}}$.

11.3.4. Describe a variant $\mathbf{ML}_0'$ of $\mathbf{ML}_0^{\text{i}}$, in which not only the terms appear with a type, but also all subterms carry unequivocally a "fitting" type. *Hint.* "t" and "A" in "$t \in A$" should be treated as inseparable. Some examples of rules

$$\frac{x \in A \Rightarrow t[x \in A] \in B}{\lambda x{\in}A.t[x \in A] \in \Pi x{\in}A.B}$$

$$\frac{t \in \Pi x \in A.B \qquad t' \in A}{(t \in \Pi x{\in}A.B)(t' \in A) \in B[x{\in}A/t'{\in}A]}$$

etc.

11.3.5. Let $\mathbf{ML}_0'$ be the variant of the previous exercise. If Γ' is a context and θ' an assertion in $\mathbf{ML}_0'$, let Γ and θ be the corresponding context and assertion obtained by dropping all type indications for proper subterms. Show that if $\mathbf{ML}_0^{\text{i}} \vdash \Gamma \Rightarrow \theta$, then for some Γ', θ' we have $\mathbf{ML}_0' \vdash \Gamma' \Rightarrow \theta'$, and if $\mathbf{ML}_0' \vdash \Gamma' \Rightarrow \theta'$, then $\mathbf{ML}_0^{\text{i}} \vdash \Gamma \Rightarrow \theta$.

11.4.1. Complete the proof of 4.2.

11.4.2. Show for the definition of N_k, i_k $(i \leq k)$ and R_k as in 4.3, that N_kI, N_kE, and N_kCONV hold.

11.4.3. Complete the proof of the proposition in 4.5.

11.4.4. Give a proof of the equivalence of ΠE plus ΠCONV with ΠE* plus ΠCONV* (4.6).

11.4.5. Prove 4.7 and 4.9.

11.4.6. Prove 4.12.

11.4.7. Show in $\mathbf{ML}_0^{\text{i}}$ *without* the type-forming operation +, that for the operation +, *defined* by

$$A + B := \Sigma x \in N_2[(x = \mathbf{0} \to A) \wedge (x = \mathbf{1} \to B)],$$

one can define k_0, k_1, and an operator $D_{x,y}$, such that a weakened version of +E holds where (cf. 10.8.2) C does not contain x and y as free variables.

11.5.1. Show that in $\mathbf{ML}_0$ without primitive type forming operation + the operation + as defined in the preceding exercise satisfies +I, +E, and +CONV. *Hint.* The proof uses extensionality in an essential way: $A \to B$ is a type with a single element whenever A is empty, and for $t \in B$ there is a *unique* function in $A \to B$ with constant value t (Troelstra 1983).

11.5.2. Prove 5.6.

11.5.3. Spell out Martin-Löf's semantics for all the rules of the system $\mathbf{ML}_0$ (cf. 5.9–5.10).

11.5.4. Show that the embedding of $\mathbf{ML}_0^{\mathrm{i}}$ into **APP** also models the system $\mathbf{ML}_0^+$ of E11.2.3.

11.5.5. Show that the embedding of $\mathbf{ML}_0^{\mathrm{i}}$ into **APP** also validates ΣI with the type premiss dropped.

11.5.6. Let us abbreviate $\boldsymbol{p}_i t$ as t_i, $\boldsymbol{p}_i(\boldsymbol{p}_j t)$ as t_{ji}, etc., and let + be defined as in E11.4.7 above. Define

$$A +' B := \Sigma y \in A + B[(y_0 = \mathbf{0} \to y_{10} = \lambda x. y_{10}\boldsymbol{e} \wedge y_{11} = \lambda x'. \perp_B) \wedge$$
$$(y_0 = \mathbf{1} \to y_{10} = \lambda x. \perp_A \wedge y_{11} = \lambda x'. y_{11}\boldsymbol{e})]$$

and

$$A +'' B := \Sigma z \in A +' B[z_{10} = \lambda u.(\boldsymbol{e}, \boldsymbol{e}) \wedge z_{11} = \lambda u.(\boldsymbol{e}, \boldsymbol{e})].$$

Show in $\mathbf{ML}_0^{\mathrm{i}}$ that $+''$ satisfies +I, +E, and +CONV, for suitably defined $D_{x,y}$, and injection functions $\boldsymbol{k}_0''$ and $\boldsymbol{k}_1''$ defined by

$$\boldsymbol{k}_0'' := \lambda z.(((\mathbf{0}, (\lambda x.z, \lambda x'. \perp_B)), (\boldsymbol{e}^*, \boldsymbol{e}^*)), (\boldsymbol{e}, \boldsymbol{e})),$$

$$\boldsymbol{k}_1'' := \lambda z.(((\mathbf{1}, (\lambda x. \perp_A, \lambda x'.z)), (\boldsymbol{e}^*, \boldsymbol{e}^*)), (\boldsymbol{e}, \boldsymbol{e})),$$

where $\boldsymbol{e}^*$ abbreviates $\lambda u.(\boldsymbol{e}, \boldsymbol{e})$. *Hint*. Show that for any $t \in A +'' B$

$$t_{000} = \mathbf{0} \to t = (((\mathbf{0}, (\lambda x.t_{0010}\,\boldsymbol{e}, \lambda x'. \perp_B), (\boldsymbol{e}^*, \boldsymbol{e}^*)), (\boldsymbol{e}^*, \boldsymbol{e}^*)),$$

and similarly for $t_{000} = \mathbf{1}$.

11.6.1. Prove 6.2.

11.6.2. Prove the proposition in 6.3.

11.6.3. Construct two terms $s, t \in N \to N$ denoting primitive recursive functions such that $\mathbf{ML}_0^{\mathrm{i}} \nvdash t = s \in N \to N$, although $\mathbf{ML}_0^{\mathrm{i}} \vdash x \in N \Rightarrow t(x) = s(x) \in N$.

11.7.1. Sketch the construction of a recursive function in $\mathbf{ML}_1^{\mathrm{i}}$ which is not provably recursive in **HA** (cf. 7.3).

11.7.2. Complete the proof of the claims made in the derivation of the paradox in 7.4.

11.7.3. Intuitively it will be clear that we can only start to put elements in $Wx \in A.B$ provided that for some x $A[x]$ is empty. As a formal counterpart, establish the truth of the following two propositions:

$$Wx \in A.B[x] \to \neg\forall x \in A.B[x],$$

$$\exists x \in A \neg B[x] \to \exists y \in Wx{\in}A.B[x].$$

11.7.4. In the extensional version of the theory with W-type formation the "first tree class" Tr (cf. 7.6) contains a unique single-node tree. Why can this fail if the type theory is not extensional?

Modify the definition of Tr so as to obtain a type of countable tree ordinals with unique single-node tree, also in the intensional version of the theory.

11.7.5. Suppose that for each $x \in V\, F[x]$ is a proposition (i.e. type) and put $\forall y \mathrel{\dot\in} x A[y] := \forall y \in x^- A[x^-(y)]$ for $x \in V$. Prove transfinite induction over V in the form

$$\forall x \in V(\forall y \mathrel{\dot\in} x F[y] \to F[x]) \to \forall x \in V F[x].$$

11.8.1. Prove (ii), (iii) of 8.3 and prove 8.5.

11.8.2. Complete the proof of 8.9.

11.8.3. Prove 8.10.

11.8.4. Prove 8.15(iii).

11.8.5. Prove the second part of 8.17.

11.8.6. Prove by $\dot\in$-Induction for all $\alpha, \beta \in V \quad \alpha \mathrel{\dot\in} \beta \wedge \beta \mathrel{\dot\in} \alpha \to \bot$ (cf. end of 8.21).

11.8.7. Prove 8.22(iii).

11.8.8. Prove 8.23.

11.8.9. Show that the interpretation of **CZF** in $\mathbf{MLV}^{\mathrm{i}}$ also validates countable choice

$$\forall x \mathrel{\dot\in} \omega \exists y A[x, y] \to \exists z(Fun[z] \wedge \forall x \mathrel{\dot\in} \omega \exists y A[x, y] \wedge \langle x, y\rangle \mathrel{\dot\in} z).$$

What can you say about DC_1 ("Dependent Choices" for elements of $N \to N$)? *Hint*. Reduce countable choice in **CZF** to AC for $\mathbf{MLV}^{\mathrm{i}}$ with help of $\mathbf{MLV}^{\mathrm{i}} \vdash \Delta(n) \doteq \Delta(m) \to n =_N m$ (cf. 8.22).

CHAPTER 12

CHOICE SEQUENCES

The present chapter is rather "philosophical" in content, in particular the first two sections. Here we illustrate the possibilities and limitations of the conceptual analysis in the case of lawless sequences; familiarity with sections 6–8 from chapter 4 is presupposed.

Section 3 shows how the axioms we found for lawless sequences permit us to eliminate the lawless sequences, via a contextual definition of quantification over lawless sequences in terms of quantification over "ordinary" sequences plus the logical operations of arithmetic. Thus it is possible to regard lawless sequences as a "figure of speech".

In the final section we give a sketchy discussion of other notions of choice sequence. The reasons for concentrating on the lawless sequences is two-fold: first of all they give us the simplest representative example of a notion of choice sequence, suitable for illustrating the analysis of concepts; secondly, they are important as a link between validity in Beth models and intuitionistic validity in the intuitive sense (cf. section 13.1).

Other notions such as the GC-sequences mentioned in section 4 are perhaps directly relevant to the practice of intuitionistic mathematics, but in any case the principal interest of choice sequences lies at present in the direction of the philosophy of mathematics.

1. Introduction

1.1. *The continuum problem.* Every form of constructive mathematics has to come to terms with the continuum, i.e. $\mathbb{R}$. The standard treatments of the continuum in the context of traditional mathematics proceed by "arithmetization": the collection of reals is described via sequences or sets of rationals, or sequences of rational intervals.

From a constructivist point of view, there is a conceptual difficulty here, which led the French semi-intuitionists, and in particular Borel, to adopt the continuum as a primitive notion, to be understood as a whole; i.e. the continuum is more than the collection of its elements. This global view of the continuum is sometimes referred to as the "geometric continuum", whereas the continuum as made up of reals given by fundamental sequences is called the "arithmetical continuum".

Let us describe the difficulty in some detail. For the semi-intuitionists, any specific real should be given by a complete definition of (say) a representing fundamental sequence; and since the linguistic means at our disposal permit only countably many descriptions, there would exist only countably many reals. Cantor's diagonal argument shows that these cannot exhaust the continuum (it should be kept in mind that the semi-intuitionistics accepted classical logic; the fact that the enumeration of the reals cannot be given *effectively* is not a counterargument). Thus one could not very well conceive the continuum as being made up of its (definable) elements.

The definable elements of the continuum form what is sometimes called the "practical continuum", which is, to borrow Brouwer's terminology, "denumerably unfinished" (1907, p. 148): for every definable enumeration of elements of the continuum we can find a new element outside the enumeration (by a Cantor diagonal procedure).

Thus, the collection of definable elements of the continuum may be countable in the classical sense, but it cannot be *definably* enumerable. Restricting attention to the "practical continuum" has an unpleasant consequence: there being only countably many elements in the "practical continuum", there is a sequence of intervals with sum of length $< \varepsilon$, for any $\varepsilon > 0$, covering the practical continuum; so it looks as if the "practical continuum" has measure zero.

Brouwer occasionally discusses the "reduced continuum", which is similar to the "practical continuum", and which consists of lawlike reals, i.e. those reals which can be defined by giving a law for a fundamental sequence of rationals. The above argument also applies to the reduced continuum, which should thus get measure zero; for this reason Brouwer (1930) rejects it as inadequate for intuitionistic mathematics.

Of course, after a good deal of experience with recursive functions and recursive analysis, one nowadays realizes that the objection just mentioned is by no means conclusive. If in Borel's measure theory one would limit attention to definable coverings, and in particular definable sequences of intervals, the "measure theoretic paradox" disappears: one cannot cover the definable elements of the continuum by a *definable* sequence of intervals of arbitrarily small measure; in fact, as follows from Borel's own proof of the "Heine–Borel theorem", any *definable* countable cover of [0, 1] has a finite subcover (for if not, we can find a definable point not being covered).

As we have seen in section 6.4, in recursive analysis the apparent conflict is resolved in a different way. The recursive interval [0, 1] can indeed be covered by a recursively enumerable sequence of intervals with an arbitrarily small sum of lengths; but if one restricts attention to coverings by sequences with the sum of lengths recursively convergent, the sum of lengths is always ≥ 1.

1.2. *The introduction of choice sequences.* In Borel's thinking the adoption of the arithmetical continuum was directly connected with the acceptation of countable sequences of arbitrarily chosen objects as legitimate objects of mathematics, as is illustrated by the following illuminating quotation from his writings (1909):

> "It is necessary to say something on the notion of a continuum, the only well-known example of an uncountable set [...]. I regard this notion as obtained from geometrical intuition; one knows that the completely arithmetical concept of the continuum requires that one admits the legitimacy of a countable infinity of successive choices. This legitimacy seems to me to be highly debatable, but nevertheless one should distinguish between this legitimacy and the legitimacy of an uncountable infinity of (successive or simultaneous) choices. The latter concept seems to me, as I have remarked before, entirely meaningless [...]".

The legitimacy of an infinity of choices was, to the French semi-intuitionists, in effect what was asserted by Zermelo's "axiom of choice" (in this connection not to be viewed as an axiom schema relative to a fixed first-order language, but as an informal mathematical principle); see e.g. Borel (1914, in particular III and IV of note IV).

The quotation above already contains the germ of the idea of a choice sequence. The following quote from Borel (1912; also in 1914, VII of Note IV) is even more suggestive in this respect:

> "People will also agree on the following point: it is possible to define a decimal number of bounded length by asking thousand people to write down, arbitrarily, some digit; thus one obtains a well-defined number, if all the persons are arranged in a row, and each one writes in turn a new digit at the end of the sequence of digits already written by the people in the row preceding him. But observe where the disagreement sets in: is it possible to define a

> decimal number of unbounded length by a similar process? [...].
> On my part, I regard it as possible to ask questions of probability concerning decimal numbers obtained in this way, by choosing digits, either entirely arbitrarily, or imposing some restrictions which leave some arbitrariness, but I regard it as impossible to talk about a single individual such number, since if one denotes such a number by a, different mathematicians, in talking about a, will never be sure to be talking about the same number".

In his earlier work (before 1917), Brouwer follows the semi-intuitionists, as is clearly illustrated by the following quote from his thesis (1907, p. 150).

> "But if we introduce the logical entity: *totality* of the points of the continuum abandoning the intuition of a continuum, we shall be forced to define *the points of the continuum*, and this is only possible by *definable laws of progression for approximating dual fractions*. Now in this sense the continuum is denumerably unfinished, and so is the second number class".

Around 1917 Brouwer must have realized that from an intuitionistic point of view, there was no objection against considering sequences obtained by successive choices (instead of being completely defined in advance). What seemed "illegitimate" in the "empirical", intersubjective view of Borel, fitted very well in the Brouwerian subjectivist view of mathematics as created in the mind of an ideal mathematician. This move enabled Brouwer to reinstate the "arithmetical" account of the continuum, but now in an intuitionistic context; and this seems indeed to provide a much more satisfactory intuitive grasp of the continuum than to postulate the continuum as a primitive intuition.

Crucial, in the mathematical exploitation of choice sequences, was Brouwer's insight (in handwritten course notes on the theory of pointsets, dating from 1917) that an operation assigning completely specified objects, such as natural numbers, to all choice sequences should be continuous, i.e. the value assigned to an arbitrary choice sequence depends on an initial segment of the sequence only. A strengthening of this insight led to the "fan theorem" (Brouwer 1927) for choice sequences, an important tool in the intuitionistic reconstruction of substantial parts of traditional analysis.

Of course, right from the beginning it was clear that the process of choosing values of a choice sequence could take many different forms: we may let our choices be governed by a law, or we can refrain from making any restrictions on future choices, etc. However, initially Brouwer does not comment upon the exact nature of choice sequences, and regards them as sufficiently explained by the freedom of the "ideal mathematician" in creating them. From 1927 onwards Brouwer becomes more specific and makes remarks concerning the freedom of choice (e.g. to the effect that the choice of freedom could be restricted, but always had to conform to a spread law, i.e. the future options for choices open at any stage should lie in a spread). Brouwer apparently changed his mind more than once as to what was allowed (for a detailed account, cf. Troelstra 1982A). These attempts seem to have been motivated primarily by a desire to present an as clear as possible intuitive picture of the choice process; at least there is no indication that Brouwer's (changing) views on the precise conditions of the choice process influenced the mathematical developments.

Brouwer's new arithmetical theory of the continuum based on choice sequences could be termed "analytic" in contrast the earlier "holistic" theory where the "geometric continuum" had to be understood as a whole. In the new theory, reflection on the way an arbitrary element is to be given (in short: reflection on the notion of choice sequence, or: analysis of what it means to be given a choice sequence) justified principles such as continuity and the fan theorem.

1.3. *The analytical approach.* The method of justifying axioms for choice sequences by reflection on what it means to be given a choice sequence ("conceptual analysis of the notion") can be carried considerably farther than is done in Brouwer's writings, and leads to interesting insights and results; we shall pursue this in some detail in the next section. Choice sequences are not only a good example of the possibilities (and limitations) of conceptual analysis, but are also of interest in themselves, as demonstrating the possibilities of coherent reasoning about incomplete objects. Observe that Brouwer's early counterexamples using lawlike sequences showed that classical logic is "unreliable" in intuitionistic mathematics, while choice sequences allow one to *refute* the laws of classical logic, since continuity (WC-N) has consequences such as $\neg\forall$-PEM and $\neg$SEP (cf. 4.6.4, 4.6.5).

One source of limitations to the analytical approach is in the fact that, in building a theory, we unavoidably idealize. E.g. one thinks of choice processes carried out by an ideal mathematician, which implies that certain aspects of an actual choice process (mood, time of the day, etc.) automatically are left out (abstracted from, regarded as mathematically irrelevant). Also, we can try, in our analysis of an informal notion, to be as precise and rigorous as possible in our justification of principles valid for such a notion, but we cannot expect absolute rigour. There are, clearly, degrees of informal rigour, ranging from "it is plausible that" or "this seems to suggest that" to "we are inescapably led to the conclusion that...".

Perhaps it is needless to say that a judgement on the degree of informal rigour attained contains a subjective element. Nevertheless we can be certain that, if a plausible argument leads to the adoption of an axiom, which then by a further analysis turns out to need some restriction or modification, then the second analysis is more rigorous than the original one.

In presenting the justification for certain principles in terms of a conceptual analysis, we may have to accept that sometimes we have to make "intuitive jumps", namely when a new insight seems to be required, while we are unable (at the moment) to analyze the matter any further. In such cases, the best one can do is to show as clearly as possible what it is that we have to accept.

There is a third way of approaching the subject of choice sequences, discussed in the next section.

1.4. *Choice sequences as a "figure of speech".* In this approach one tries to *explain* what it means to quantify over choice sequences without explicit reference to individual choice sequences. In a more formal version of this approach one explains, within the context of a given language, what it means to quantify over choice sequences. In other words, one translates sentences involving choice quantifiers $\forall\alpha, \exists\alpha$ into sentences not involving such quantifiers (a "contextual definition" of $\forall\alpha, \exists\alpha$). The characteristic principles for choice sequences now hold by definition.

The "figure of speech" approach is akin to the holistic approach, inasmuch the idea of an "arbitrary individual choice sequence" does not play a role here; in contrast to the holistic approach however, the point of departure is not an intuition to be grasped as a whole, but an explanation.

The interpretation by contextual definition can in fact be used to support the insights obtained by the analytic approach; the soundness of a translation of choice quantifiers making true the principles which were obtained by concept analysis, serves as a formal test of the coherency of the analysis.

N.B. The three approaches to the subject of choice sequences (holistic, analytic, and "figure of speech") can merge gradually and almost imperceptibly into each other. (For example, it is not easy to classify Weyl's discussion (1921) of choice sequences under one of these headings. On the one hand there is a faint echo of the semi-intuitionist's holistic view of the continuum, but on the other hand some of his formulations tend towards a "figure of speech" interpretation.)

2. Lawless sequences

2.1. From the analytical point of view, there is not a single notion of choice sequence, but there are many: each notion is determined by the *type of data* which can be known about the sequences in the course of their construction. We already encountered, briefly, two such notions in 4.6.2: lawless sequences and "hesitant sequences". Perhaps the simplest among the many possible notions is that of a lawless sequence, and we shall devote this section to its analysis.

We recall that for a lawless sequence the choice of future values is at any stage completely free. Distinct lawless sequences are completely independent, in other words the values of any such sequence are not determined or restricted relative to the values of other lawless sequences.

Throughout this chapter a, b, c, d are used for lawlike sequences; in the present and in the next section α, β, γ are used for lawless sequences. LL and LS are the sets of lawlike and lawless sequences, respectively.

It is possible to skip the "critical discussions" below on a first reading, and to return to them afterwards.

2.2. *Lawless sequences; the density axiom.* We may think of a lawless sequence α as a process of generating (choosing) values $\alpha 0, \alpha 1, \alpha 2, \ldots$ in $\mathbb{N}$ without any general restriction, such that at any stage of the process only finitely many values are known (and all further choices completely left free). The complete freedom of choice suggests that we can put

LS1 $\quad \forall n \exists \alpha (\alpha \in n),$

the *density axiom*: each possible initial segment occurs as an initial segment of some lawless sequence.

A fairly good representation of this idea is that of the sequence of the casts of a die (except that now the values have to come, not from $\mathbb{N}$ but from $\{1, 2, 3, 4, 5, 6\}$); at no stage we know more than an initial segment of this sequence. However, as the values are not freely chosen, but determined by throwing the die, we cannot be certain that all initial segments will occur, no matter how many dice we take into consideration; to get a better approximation, we should permit a number of deliberate placings of the die before we start throwing; this ensures the validity of (the analogue for dice of) LS1.

N.B. The truth of a statement involving lawless sequences should not be understood in a probabilistic way; here we are only interested in what can be asserted with complete ("absolute") certainty, not in assertions with "probability 1". In a probabilistic sense we can be certain that a "6" will turn up in the sequence of casts of a die, sooner or later – but we have no absolute certainty, founded on proof.

2.3. *The density axiom; critical discussion.* There is a slight difficulty we have glossed over in the preceding subsection: if we rigorously stick to the idea that at any stage a single value is to be chosen, and no further restrictions are to be made, can we then be certain that any possible initial segment will occur as initial segment of *some* lawless sequence? How can we be certain that a particular initial segment of length 10 will occur?

As a thought experiment, we may envisage all possible sequences of casts of dice started simultaneously – we *expect*, but cannot be *certain* that all possible initial segments from $\{1, 2, \ldots, 6\}$ will occur. As to the lawless sequences themselves, the assertion that a particular initial segment will not occur (formally $\neg\exists\alpha(\alpha \in n)$) intuitively conflicts with the freedom of making choices, so we feel convinced that at least $\neg\neg\exists\alpha(\alpha \in n)$, but that is not enough to justify the density axiom).

To what extent does the picture of the casts of a die, including an initial finite number of deliberate placings, correspond to our initial description of a lawless sequence? To get a better analogy, we should for lawless sequences permit the specification of an arbitrary initial segment in advance. Then we may think of our universe of lawless sequences as containing at *least* some $\alpha_n \in n$, for each n – we may start right at the beginning to generate at least those α_n, and then determine successively a further value for α_0; for α_1, α_0; for $\alpha_2, \alpha_1, \alpha_0$, etc. Our universe may contain other lawless sequences besides the α_n. Thus the density axiom seems to be guaranteed for this slightly modified idea of lawless sequence, and it is this slight modification, permitting the stipulation of certain initial segments a priori, which we shall henceforth adopt.

The original, unmodified notion might be termed "proto-lawless sequence"; we may think of the proto-lawless sequences as forming part of the universe of lawless sequences, consisting of those lawless sequences for which no (or the empty) initial segment was specified in advance. We have no reasons to uphold LS1 for the proto-lawless sequences. Though the notion of a proto-lawless sequence is simpler than that of a lawless sequence, it does not lead to a simpler theory.

2.4. *Identity and equality between lawless sequences.* Let us write $\alpha \equiv \beta$ to indicate that α and β really refer to the *same* (identical) process. Translated in terms of "casts of a die", $\alpha \equiv \beta$ means that α, β are the same sequences of casts of the same die. Trivially, for all properties A,

$$A(\alpha) \wedge \alpha \equiv \beta \rightarrow A(\beta).$$

Also obviously

$$a \equiv \beta \rightarrow \alpha = \beta, \tag{1}$$

where as usual $\alpha = \beta := \forall x(\alpha x = \beta x)$. For lawless sequences, we also maintain

LS2* $\alpha \equiv \beta \vee \alpha \not\equiv \beta,$

since we *know* whether α and β refer to the same process or not (in terms of dice: we know whether we are referring to the same die or not).

Further it may be argued that

$$\alpha = \beta \rightarrow \alpha \equiv \beta \tag{2}$$

for the following reason: the distinctness of α and β, i.e. $\alpha \not\equiv \beta$, conflicts with $\alpha = \beta$, since $\alpha = \beta$ implies that β is completely determined by α. Combining (1), (2), and LS2* we find

LS2 $\alpha = \beta \vee \alpha \neq \beta.$

2.5. *Decidability of equality: critical discussion.* One may be tempted to try to refute LS2 by the following "counterexample". Suppose sequences α, β are given to us by means of values produced by two distinct "black boxes", one for α and one for β (a "black box" is simply a process the workings of which are unknown to us). To us (subjectively) α, β therefore appear as (proto-)lawless sequences.

But suppose we are told, after a long time, that there is a hidden connection between the boxes, such that they will keep turning out the same values; then suddenly we know that $\alpha = \beta$. We could not decide this beforehand – nor were we given, at the start, a guarantee that the decision could ever be made in the future. So we have no right to assert LS2 generally. However, this "counterexample" starts from a wrong picture of lawless sequences.

For we should look at the example in a different way: initially, α and β were given to us as distinct processes; afterwards, we learn that $\alpha = \beta$, and this is a type of information which was not permitted by our description of a lawless sequence. If, though considering α and β as distinct (i.e. $\alpha \not\equiv \beta$), we have to reckon with the possibility that later we will learn that $\alpha = \beta$, then α and β cannot be considered as lawless from the beginning; they will fall under another notion of sequence. Sequences initially given as distinct and lawless should remain independent (cf. 2.1).

In this connection, the following example is instructive. Suppose we generate a sequence of values $\gamma 0, \gamma 1, \gamma 2, \ldots$ as a lawless sequence: at no stage we make restrictions on future choices. We may *also* think of the same process as alternately choosing values for two sequences α and β: $\alpha 0, \beta 0, \alpha 1, \beta 1, \ldots$ so that $\alpha n = \gamma(2n)$, $\beta n = \gamma(2n+1)$ for all $n \in \mathbb{N}$. In this case we may think of α, β as both being lawless and independent of each other.

We cannot, however, think of γ, α, β as all three being lawless within the same context; for γ is dependent on α and β, and α and β are both dependent on γ. For example, $\forall n(\alpha n = \gamma(2n))$ represents a *general* restriction on the values of α (relative to γ). Thus, we may *decide* to regard γ as lawless, but then we cannot regard α and β as such in the same context; or we can *decide* to regard α and β as lawless (but not γ). In short, not only the choosing of values matters, but also the *decision* that we *want* to regard these as part of the construction of a certain lawless "individual".

2.6. *The principle of open data* states that for independent lawless sequences any property which can be asserted must depend on initial segments of these sequences only.

If A is a property of lawless sequences, not containing any further choice parameters, we can formulate the one-variable form of the axiom of open data:

LS3(1) $\quad A(\alpha) \to \exists n \forall \beta \in \bar{\alpha} n\, A(\beta).$

If we introduce the abbreviations

$$\neq(\alpha, \beta_0, \ldots, \beta_n) := \bigwedge_{i \leq n} \alpha \neq \beta_i,$$

$$\#(\beta_0, \ldots, \beta_n) := \bigwedge_{i<j \leq n} \beta_i \neq \beta_j,$$

we can also easily state the general form

LS3 $\quad A(\alpha, \vec{\beta}) \wedge \neq(\alpha, \vec{\beta}) \to \exists n \forall \gamma \in \bar{\alpha} n(\neq(\gamma, \vec{\beta}) \to A(\gamma, \vec{\beta}))$

(recall that $\neq(\alpha, \vec{\beta})$, in virtue of the identification of $=$ and $\equiv$, expresses that α is independent (*distinct from*) of the $\vec{\beta}$).

The condition $\neq(\alpha, \vec{\beta})$ in LS3 is really necessary. For if we take $A(\alpha, \beta) := \forall n(\alpha n = \beta \bar{n})$, and apply to this LS3 without $\neq(\alpha, \beta)$, $\neq(\gamma, \beta)$, we find $\alpha = \beta \to \exists m \forall \gamma \in \bar{\alpha} m(\gamma = \beta)$ which is clearly false (and formally refutable by the density axiom: take any $\gamma \in \bar{\alpha} m * \langle \beta m + 1 \rangle$).

2.7. *The principle of open data: critical discussion.* For the proto-lawless sequences, LS3(1) seems to be irreproachable, but as to its general validity for lawless sequences, one might raise an objection: there is one piece of information concerning a lawless sequence α which cannot be read of from an initial segment of α, namely the length of the initial segment which was fixed in advance (to ensure LS1; cf. the "deliberate placings of the die"). Specifically, let Φ_{I} be the operation assigning to each lawless α its initially specified segment. Then for any α there is an n such that $\Phi_{\mathrm{I}}(\alpha) = n$. This property gives us a counterexample to LS3(1): applying LS(1) to $\Phi_{\mathrm{I}}(\alpha) = n$, we find that, for some $m \geq n$, $\forall\beta \in \bar{\alpha}m(\Phi_{\mathrm{I}}(\beta) = n)$, and this is easily refuted, e.g. by considering a β with $\Phi_{\mathrm{I}}(\beta) = \bar{\alpha}m * \langle 0 \rangle$.

So, if we wish to maintain LS3(1), we should only consider properties A which do not refer to Φ_{I}. This seems to be enough, but the reader is perhaps assailed by a nagging doubt: is the rather "unmathematical" Φ_{I} really the only source of exceptions to LS3?

A more natural way of looking at the matter is perhaps the following. Just as in the actual process of generating values of a lawless sequence, we have abstracted from irrelevant circumstances accompanying a process of choosing in the physical world (time between the choices, the weather, etc.), it seems to be mathematically irrelevant which part of an individual lawless sequence was determined a priori, and which part generated by further free choices. Once we are past the a priori fixed values, one does not see the difference "looking backwards".

As to LS3 in general, whenever we assert at some stage $A(\alpha, \vec{\beta})$, there are three types of data regarding α on which our assertion could be based: (1) an initial segment of α, (2) Φ_{I}, which we have just excluded, (3) assertions of the form $\alpha = \beta_i$, $\alpha \neq \beta_i$, i.e. the identity or nonidentity of α with one of the other parameters. Under the assumption $\neq (\alpha, \vec{\beta})$ only the first type of data is relevant, which leads to LS3.

It is perhaps instructive to consider the following more liberal version of proto-lawless sequences: at each stage we permit the choice, not of just one, but of finitely many new values: at least one value is to be chosen. It looks as if this makes LS1 more evident: a single choice for α can guarantee any given initial segment. Nevertheless, we still have to ensure that *all* possible first choices actually *do* occur for some "liberal proto-lawless" sequence. On the other hand, at any stage there is more information than just the initial segment determined so far: we also know the stages at which the values were chosen, or equivalently, how many stages were added at each preceding stage. But just as in the case of Φ_{I} above, this ought to be regarded as irrelevant; we abstract from such details of the generation process.

The next two propositions give some easy consequences of LS1-3.

2.8. PROPOSITION. $\forall\alpha\neg\forall x(\alpha x \neq 0)$.

PROOF. Assume $\forall x(\alpha x \neq 0)$, then by LS3 $\forall\beta \in \bar{\alpha}n\, \forall x(\beta x \neq 0)$ for some n; this is refuted by taking a $\beta \in \bar{\alpha}n * \langle 0\rangle$, which is possible by LS1. □

REMARK. Note that in this derivation the use of the density axiom is essential. Intuitively, the proposition is weaker (and perhaps more evident) than the density axiom; in fact, it suffices in the proof to have $\forall n\neg\neg\exists\beta(\beta \in n)$. The essential ingredient of the proof might be paraphrased as follows: if a lawless sequence at some stage is known up to and including its nth value, but not further, we are free to choose 0 as its $(n+1)$-th value at the next stage.

LS3 also excludes the possibilities of closure of the universe of lawless sequences under any nontrivial continuous operation.

2.9. PROPOSITION. *Identity is the only lawlike operation under which the universe* LS *of lawless sequences is closed.*

PROOF. Suppose $\alpha = \Gamma\beta$, $\alpha \neq \beta$. Then $\forall\gamma \in \bar{\alpha}x(\gamma \neq \beta \to \gamma = \Gamma\beta)$ for some x, which is clearly false by LS1: choose $\gamma_0 \in \bar{\alpha}x * \langle\beta x + 1\rangle$, $\gamma_1 \in \bar{\alpha}x * \langle\beta x + 2\rangle$, and the contradiction is immediate. □

REMARK. The proposition also shows that we cannot obtain the effect of LS1 by the plausible-looking $\forall n\forall\alpha\exists\beta(n * \alpha = \beta)$, since this is false. □

2.10. *Continuity axioms.* Already LS3(1) implies WC-N without parameters

$$\forall\alpha\exists x A(\alpha, x) \to \forall\alpha\exists x\exists y\forall\beta \in \bar{\alpha}x\, A(\beta, y).$$

Here, as in the sequel, A is assumed to be extensional with respect to sequence parameters: $A(\alpha) \wedge \alpha = \beta \to A(\beta)$.

Stronger forms of continuity can be justified, however. In particular, as argued in section 4.6, we can assume that numerical values are assigned by a neighbourhood function. Let us use ξ as a variable for arbitrary sequences in $\mathbb{N}^{\mathbb{N}}$, and let $K_0 \equiv (K_0)_{\mathrm{LS}}$ be the class of neighbourhood functions with respect to LS:

$$\xi \in K_0 := \forall\alpha\exists x(\xi(\bar{\alpha}x) > 0) \wedge \forall nm(\xi n \neq 0 \rightarrow \xi n = \xi(n * m)).$$

Then, for A not containing choice parameters besides α, we can reason as in section 4.6 for choice sequences and formulate the principle

LS4*(1) $\quad \forall\alpha\exists x A(\alpha, x) \rightarrow \exists\xi \in K_0\, \forall\alpha\, A(\alpha, \xi(\alpha))$,

where $\xi(\alpha)$ is defined as before (3.7.9) by

$$\xi(\alpha) = x \leftrightarrow \exists y(\xi(\bar{\alpha}y) = x + 1).$$

From now on, we shall always assume all choice parameters to be shown. The restriction, that A in LS4*(1) does not contain other choice parameters besides α, is indeed necessary, as may be seen from the following example:

$$\forall\alpha\exists x[(\alpha = \beta \rightarrow x = 0) \wedge (\alpha \neq \beta \rightarrow x = 1)]$$

obviously holds by LS2, but the assignment of x to α cannot possibly be continuous in α.

In two directions we can strengthen LS4*(1) still further. The first strengthening is the so-called *extension principle*, stating that elements of K_0 are in fact also neighbourhood functions on all of $\mathbb{N}^{\mathbb{N}}$:

$$\xi \in K_0 \rightarrow \forall\xi'\exists x(\xi(\bar{\xi}'x) > 0), \tag{1}$$

with as an important corollary

$$\xi \in K_0 \rightarrow \forall a\exists x(\xi(\bar{a}x) > 0). \tag{2}$$

The extension principle can be motivated as follows. Suppose $\xi \in K_0$; ξ encodes some continuous functional $\Phi_\xi \in \mathrm{LS} \rightarrow \mathbb{N}$, i.e.

$$\xi(\bar{\alpha}x) = y + 1 \leftrightarrow \Phi_\xi(\alpha) = y.$$

Let now ξ' be an arbitrary element of $\mathbb{N}^{\mathbb{N}}$. We consider successively $\bar{\xi}'0, \bar{\xi}'1, \bar{\xi}'2, \ldots$; at the nth step we try to compute $\Phi_\xi(\xi')$ from $\bar{\xi}'n$. That is to say, we try to determine the value of $\Phi_\xi(\xi')$ using $\bar{\xi}'n$ as our only data, so that we may think of $\bar{\xi}'n$ as the initial segment of a lawless sequence; we abstract from all other data concerning ξ' we might possibly have. This may be slightly rephrased as follows: we try to apply Φ_ξ, taking into account more and more values of ξ': $\xi'0, \xi'1, \xi'2, \ldots$, but at each step we treat ξ' as lawless, i.e. we try to determine the value given by Φ_ξ on the assumption that we are generating a lawless sequence. The assumption is now that the method of computation associated with Φ_ξ cannot distinguish this situation from being fed the data of a lawless sequence, so ultimately Φ_ξ must come up with an answer.

Essential in this argument is the possibility of abstracting from, or forgetting extra information, as long as the remaining data may be consistently imagined to belong to a lawless sequence. Obviously the argument is not applicable with respect to global properties: we cannot take a lawlike sequence b say, decide to forget that b is lawlike, and henceforth treat it as if it were lawless.

We may paraphrase the argument as follows. The computation method for a continuous functional from LS to $\mathbb{N}$ cannot make use of the intensional information that a sequence is lawless (in contrast, a functional defined on all total recursive sequences can make use of intensional information such as gödelnumbers). In applying Φ to a lawlike a, we ask for each $n \in \mathbb{N}$ what value the computation method of Φ would yield when applied to a lawless sequence starting with initial segment $\bar{a}n$; eventually the computation method must yield a value.

A second direction in which we can strengthen LS4*(1) is by making the at first sight plausible assumption that ξ is a *lawlike* sequence; using e as a variable for lawlike elements of K_0, we obtain

$$\text{LS4}^\circ(1) \qquad \forall\alpha\exists x A(\alpha, x) \to \exists e \in K_0 \, \forall\alpha\, A(\alpha, e(\alpha)).$$

Finally we have to consider generalizations to more lawless variables. Coding p-tuples of lawless sequences into a single sequence (e.g. by $\nu^p(\alpha_1, \ldots, \alpha_p) := \lambda x.\nu^p(\alpha_1 x, \ldots, \alpha_p x)$), we may consider neighbourhood functions for p-tuples of lawless sequences. However, as one easily sees, a neighbourhood function for a p-tuple of lawless sequences is in fact a neighbourhood function with respect to each of its p arguments, and since the extension principle applies to each argument separately, it also applies to a p-tuple of arguments. All taken together this means that K_0, the class of neighbourhood functions for single lawless sequences, is the same as the class of neighbourhood functions with respect to (coded) p-tuples of lawless sequences.

Let us abbreviate $\xi(\nu^p(\vec{\alpha}))$ as $\xi(\vec{\alpha})$, then the p-variable generalization of LS4*(1) becomes

$$\text{LS4}^* \qquad \forall\vec{\alpha}(\#\vec{\alpha} \to \exists x A(\vec{\alpha}, x)) \to \exists\xi \in K_0 \, \forall\vec{\alpha}(\#\vec{\alpha} \to A(\vec{\alpha}, \xi(\vec{\alpha})).$$

An equivalent form of LS4* is

$$\forall\vec{\alpha}(\#\vec{\alpha} \to \exists x A(\vec{\alpha}, x)) \to$$
$$\exists\xi \in K_0 \, \forall n(\xi n \neq 0 \to \exists x \forall\vec{\alpha} \in n(\vec{\alpha} \to A(\vec{\alpha}, x))),$$

where $\vec{\alpha} \in n$ abbreviates $\nu^p(\vec{\alpha}) \in n$.

In this form LS4 easily generalizes to corresponding axioms for quantifier-combinations $\forall\alpha\exists a$, where a ranges over some (lawlike) domain of lawlike objects. In particular, for lawlike sequences we have

$$\text{LS4**}\quad \forall\vec{\alpha}(\#\vec{\alpha} \to \exists a A(\vec{\alpha}, a)) \to$$
$$\exists\xi \in K_0 \forall n(\xi n \neq 0 \to \exists a \forall\vec{\alpha} \in n(\#\vec{\alpha} \to A(\vec{\alpha}, a))),$$

from which LS4* can be obtained as a special case, since $\exists x A(\vec{\alpha}, x) \leftrightarrow \exists a A(\vec{\alpha}, a0)$. Combining this with the assumption of ξ being lawlike we find

$$\text{LS4}^\circ\quad \forall\vec{\alpha}(\#\vec{\alpha} \to \exists a A(\vec{\alpha}, a)) \to$$
$$\exists e \in K_0 \forall n(en \neq 0 \to \exists a \forall\vec{\alpha} \in n(\#\vec{\alpha} \to A(\vec{\alpha}, a))).$$

2.11. *Continuity axioms: critical discussion.* In particular the second strengthening mentioned in the preceding subsection, resulting in LS4°(1), is open to criticism, as we shall now show.

A consequence of LS4°(1) is "lawlike AC_{00}" for predicates without choice parameters:

$$\forall x \exists y A(x, y) \to \exists a \forall n A(n, an). \tag{1}$$

To see this, apply LS4°(1) to $\forall\alpha\exists y A(\alpha 0, y)$; we find a lawlike $e \in K_0$ such that $\forall\alpha\, A(\alpha 0, e(\alpha))$; by the extension principle, e is also a neighbourhood function for lawlike sequences, so $\forall x A((\lambda y.x)(0), e(\lambda y.x))$ and thus $\forall x A(x, ax)$ for $a := \lambda x.e(\lambda y.x)$.

It is to be noted here that the intuitive justification given for the extension principle in the preceding subsection depends on LS1: we are able to think of an *arbitrary* initial segment $\bar{\xi}'x$ as part of a lawless sequence only if LS1 is assumed. So for proto-lawless sequences our derivation of (1) from LS4°(1) is not obviously valid. Nevertheless, just as it seems plausible to strengthen LS4*(1) to LS4°(1), also for proto-lawless sequences, it looks equally plausible to strengthen

$$\forall x \exists y A(x, y) \to \exists\xi \forall x A(x, \xi x) \tag{2}$$

to (1); for what non-lawlike element could possibly enter into the ξ given by a constructive proof of $\forall x \exists y A(x, y)$, if A does not contain choice parameters? Nevertheless this argument is defective, as the following example shows (ε a variable ranging over proto-lawless sequences):

$$\forall x \exists y \exists\varepsilon(\bar{\varepsilon}x = y) \tag{3}$$

is valid; and the obvious argument for this goes as follows. Take any particular proto-lawless sequence ε, and observe that, since arbitrarily long initial segments of ε will be determined, $\forall n \exists m(\bar{\varepsilon}n = m)$ holds; therefore $\xi := \lambda n.\bar{\varepsilon}n$ satisfies $\forall n \exists \varepsilon(\bar{\varepsilon}n = \xi n)$. This yields a ξ as required by (2), but this ξ is definitely *not* lawlike, notwithstanding the fact that $\exists \varepsilon(\bar{\varepsilon}n = m)$ does not contain choice parameters.

For lawless sequences on the other hand, we can appeal to LS1 and observe that $\forall x \exists \alpha(\bar{\alpha}x = (\overline{\lambda y.0})(x))$.

The problem with (1) when applied to (3) is caused by the occurrence of statements of choice-sequence existence implicit in the property considered. LS1 in combination with LS3 permits us to show for lawless sequences

$$\exists \alpha A(\alpha) \leftrightarrow \exists n \forall \alpha \in n A(\alpha),$$

and thus in formulas of elementary analysis all existential lawless quantifiers can be removed; this argument fails for proto-lawless sequences, however.

We now proceed to show that MP fails for lawless sequences. We first need a lemma.

2.12. LEMMA. $\neg\forall\alpha\exists x(\alpha x = 0)$.

PROOF. Suppose $\forall\alpha\exists x(\alpha x = 0)$, then by continuity for some $\xi \in K_0$

$$\forall\alpha(\alpha(\xi(\alpha)) = 0).$$

By the extension principle the functional determined by ξ is defined on $\lambda x.1$. Hence for some $m = (\overline{\lambda y.1})y$ we have $\xi m \neq 0$. Choosing m such that $\mathrm{lth}(m) > \xi m \dot{-} 1$, we find $\alpha(\xi(\alpha)) = 0$ for all $\alpha \in m$, which is obviously false. □

PROPOSITION. $\neg\forall\alpha(\neg\neg\exists x(\alpha x = 0) \to \exists x(\alpha x = 0))$.

PROOF. $\neg\neg\exists x(\alpha x = 0)$, i.e. $\neg\forall x(\alpha x \neq 0)$, has been shown in 2.8. Thus $\forall\alpha(\neg\neg\exists x(\alpha x = 0) \to \exists x(\alpha x = 0))$ implies $\forall\alpha\exists x(\alpha x = 0)$, but this is false by the preceding lemma. □

2.13. *The final strengthening of continuity: bar induction* ($\mathrm{BI_M}$, $\mathrm{BI_D}$). Our continuity axioms, so far, are summarized in

LS4** $\quad \forall\vec{\alpha}(\#\vec{\alpha} \to \exists \mathfrak{a} A(\vec{\alpha}, \mathfrak{a})) \to$

$$\exists e \in K_0 \forall n(en \neq 0 \to \exists \mathfrak{a} \forall \vec{\alpha} \in n(\#\vec{\alpha} \to A(\vec{\alpha}, \mathfrak{a}))),$$

together with the extension principle

$$e \in K_0 \to \forall \xi \exists x \big(e(\bar{\xi} x) > 0 \big).$$

As a final strengthening (for choice sequences in general) we can adopt

$$K = K_0, \tag{1}$$

an assumption we have discussed at some length in section 4.8. This leads to the following form of continuity

$$\text{LS4} \quad \forall \vec{\alpha} (\# \vec{\alpha} \to \exists a A(\vec{\alpha}, a)) \to$$
$$\exists e \in K \forall n \big(en \neq 0 \to \exists a \forall \vec{\alpha} \in n (\# \vec{\alpha} \to A(\vec{\alpha}, a) \big).$$

As to the justification of (1), for lawless sequences we cannot give any further argument beyond what has already been said on this topic in section 4.8, and it seems to us to constitute the biggest single "jump" in the justification of the axioms for lawless sequences we have encountered so far.

As a consequence of LS4 we have

$$\forall \vec{\alpha} (\# \vec{\alpha} \to \exists x A(\vec{\alpha}, x)) \to \exists e \in K \forall \vec{\alpha} (\# \vec{\alpha} \to A(\vec{\alpha}, e(\vec{\alpha}))). \tag{2}$$

As in 4.8.13, or as indicated in E4.8.7, we can show that (2) implies $K = K_0$.

By arguments given already in section 4.8 we also have that $\mathrm{BI_M}$ and $\mathrm{BI_D}$ (without parameters) hold for LS, as a consequence of (2).

In the next few subsections we shall show how to derive $\mathrm{BI_M}$ *with parameters*.

2.14. Notation. We introduce the following abbreviations:

$$\dot{\forall} \alpha A(\alpha, \vec{\beta}) := \forall \alpha \big(\neq (\alpha, \vec{\beta}) \to A(\alpha, \vec{\beta}) \big),$$

$$\dot{\exists} \alpha A(\alpha, \vec{\beta}) := \exists \alpha \big(\neq (\alpha, \vec{\beta}) \wedge A(\alpha, \vec{\beta}) \big),$$

$$\dot{\forall} \vec{\alpha} A(\vec{\alpha}) := \dot{\forall} \alpha_1 \ldots \dot{\forall} \alpha_p A(\vec{\alpha}),$$

$$\dot{\exists} \vec{\alpha} A(\vec{\alpha}) := \dot{\exists} \alpha_1 \ldots \dot{\exists} \alpha_p A(\vec{\alpha}),$$

where $\vec{\alpha} \equiv (\alpha_1, \ldots, \alpha_p)$, and

$$\dot{\forall} \alpha \in n A(\alpha, \vec{\beta}) := \dot{\forall} \alpha \big(\alpha \in n \to A(\alpha, \vec{\beta}) \big),$$

$$\dot{\exists} \alpha \in n A(\alpha, \vec{\beta}) := \dot{\exists} \alpha \big(\alpha \in n \wedge A(\alpha, \vec{\beta}) \big),$$

etc., and similarly for $\forall a \in n$, $\exists a \in n$:

$$\forall a \in n A(a, \vec{\beta}) := \forall a(a \in n \to A(a, \vec{\beta})),$$

$$\exists a \in n A(a, \vec{\beta}) := \exists a(a \in n \wedge A(a, \vec{\beta})). \quad \square$$

Note that

$$\dot{\forall} \vec{\alpha} A(\vec{\alpha}) \leftrightarrow \forall \alpha_1 \cdots \alpha_p((\alpha_1, \ldots, \alpha_p) \to A(\vec{\alpha})).$$

LS3 can now be stated as

$$\dot{\forall} \alpha A(\alpha, \vec{\beta}) \to \exists n(\alpha \in n \wedge \dot{\forall} \gamma \in n A(\gamma, \vec{\beta}))$$

and LS4 as

$$\dot{\forall} \vec{\alpha} \exists a A(\vec{\alpha}, a) \to \exists e \in K \forall n(en \neq 0 \to \exists a \dot{\forall} \vec{\alpha} \in n A(\vec{\alpha}, a)).$$

2.15. Definition. Let $\vec{\beta} \equiv (\beta_1, \ldots, \beta_p)$, $e \in K$, then

$$e_{\vec{\beta}} := e[\vec{\beta}] := \lambda n.e\left(\overline{\lambda y.\nu^{p+1}((n)_y, \beta_1 y, \ldots, \beta_p y)}\, \mathrm{lth}(n)\right).$$

Thus in particular

$$e[\vec{\beta}](\bar{\alpha} x) = e\left(\overline{\lambda y.\nu^{p+1}(\alpha y, \beta_1 y, \ldots, \beta_p y)}(x)\right)$$

and

$$e[\vec{\beta}](\alpha) = e(\alpha, \vec{\beta}) \equiv e(\nu^{p+1}(\alpha, \beta_1, \ldots, \beta_p)). \quad \square$$

For notational simplicity we restrict from now on attention to the case where $p = 1$, so that we can write β for $\vec{\beta}$.

2.16. Lemma.
(i) $\forall e \in K \forall \beta (K_0(e_\beta))$,
(ii) $\forall n(e_\beta(n) \neq 0 \to Q(\beta, n)) \wedge \forall n(\forall y\, Q(\beta, n * \langle y \rangle) \to Q(\beta, n)) \to Q(\beta, \langle\ \rangle)$, *for all predicates* $Q(\beta, n)$.

Proof. (i) is proved by induction over K.
(ii) Put $P(e) \equiv \forall \beta \forall m[\forall n(e_\beta(n) \neq 0 \to Q(\beta, m * n)) \wedge \forall n(\forall y\, Q(\beta, m * n * \langle y \rangle) \to Q(\beta, m * n)) \to Q(\beta, m)]$ and prove $\forall e P(e)$ by induction over K. $\square$

2.17. THEOREM. BI_{M} *holds with lawless parameters present.*

PROOF. Let us write $P_\beta n$ for $P(\beta, n)$ and assume $\forall\alpha\exists x\, P_\beta(\bar{\alpha}x) \wedge \forall nm\, (P_\beta n \to P_\beta(n * m)) \wedge \forall n\, (\forall y P_\beta(n * \langle y\rangle) \to P_\beta n)$. We have to show $P_\beta(\langle\ \rangle)$. By LS3 we can find some n such that $\beta \in n$ and:

$$\forall\alpha\forall\beta \in n\, \exists x\, P_\beta(\bar{\alpha}x), \tag{1}$$

$$\forall\beta \in n\, \forall mm'\big(P_\beta m \to P_\beta(m * m')\big), \tag{2}$$

$$\forall\beta \in n\, \forall m\big(\forall y\, P_\beta(\beta, m * \langle y\rangle) \to P_\beta m\big). \tag{3}$$

From (1) we obtain

$$\dot\forall\alpha\dot\forall\beta\exists x\big(\beta \in n \to P_\beta(\bar{\alpha}x)\big)$$

and thus by LS4 for some $e \in K$

$$\forall n\Big(en \neq 0 \to \dot\forall\alpha\dot\forall\beta\big((\alpha, \beta) \in n \to P_\beta(\bar{\alpha}(en \dot{-} 1))\big)\Big) \tag{4}$$

Replacing, if necessary, e by $e' := \lambda n.en \cdot \mathrm{sg}(\mathrm{lth}(n) \dot{-} en)$, we can achieve that (4) holds and also

$$\forall n(en \neq 0 \to en \dot{-} 1 \leq \mathrm{lth}(n)),$$

which implies

$$\forall n\big(e_\beta n \neq 0 \to e_\beta n \dot{-} 1 \leq \mathrm{lth}(n)\big). \tag{5}$$

Combining (2), (4), and (5) we find

$$\forall n\big(e_\beta n \neq 0 \to P_\beta n\big), \tag{6}$$

and thus from (3), (6), and lemma 2.16(ii) it follows that $P_\beta\langle\ \rangle$. □

2.18. *Axioms for binary lawless sequences.* Completely similar to the argument given above we can justify axioms for lawless sequences with initial segments in a fixed fan (finitely branching tree). To fix ideas, we may think of the binary tree T_{01} of 01-sequences.

Let n, m range over codes for sequences of the tree. We can take over LS1, LS2, and LS3. The schema LS4 is replaced by a form of the *fan theorem*:

$\mathrm{LS4_F}$ $\quad \dot\forall\vec{\alpha}\exists a\, A(\vec{\alpha}, a) \to \exists z \forall n\big(\mathrm{lth}(n) = z \to \exists a \dot\forall\vec{\alpha} \in n\, A(\vec{\alpha}, a)\big).$

3. The elimination translation for the theory LS

3.1. *Lawless sequences as a "figure of speech".* The principal aim of this section is to demonstrate the possibility of interpreting speech about lawless sequences, that is to say, the use of quantification over lawless sequences, as a "figure of speech".

In fact we shall describe an axiomatic theory **LS** (based on the axioms LS1–4 discussed in the previous section) with a lawlike fragment **S** (the part of **LS** not involving lawless variables, such that the following result holds.

THEOREM (*elimination theorem*). *There is a syntactically defined translation* τ, *mapping formulas of* **LS** *without free lawless variables to formulas of* **S** *such that*:

(i) $A \equiv \tau(A)$ *for* A *without lawless variables*,

(ii) $\mathbf{LS} \vdash A \leftrightarrow \tau(A)$. □

If we simply define **S** *as those formulas not involving lawless variables which are provable in* **LS**, *we obviously also have*

(iii) $\mathbf{LS} \vdash A \Leftrightarrow \mathbf{S} \vdash \tau(A)$. □

However, we can obtain a stronger theorem in this case by giving a simpler independent characterization of **S**. For **S** one can in fact take the theory $\mathbf{IDB}_1$, based on those *axioms* in the particular axiomatization of **LS** given below which do not refer to lawless sequences. Essentially, $\mathbf{IDB}_1$ is the extension of **EL** with the constant K and axioms K1–3 for the inductively generated class of neighbourhood functions plus the countable choice axiom AC_{01}. (iii) with $\mathbf{IDB}_1 \equiv \mathbf{S}$ has an important corollary:

COROLLARY. **LS** *is conservative over* $\mathbf{IDB}_1$. □

The strengthening (iii) will not be proved here, though the idea for the proof of this result is sketched in 15.6.11.

3.2. *The theory* $\mathbf{IDB}_1$. The language of $\mathbf{IDB}_1$ is that of **EL**, with variables for numbers (x, y, z, u, v, w) and sequences (for which we shall use a, b, c, d, e, f, and not greek lower case letters as in our earlier discussions of **EL**), and in addition a constant K for a set of sequences. Intuitively we think of the sequence variables as ranging over *lawlike* sequences of natural numbers (= lawlike elements of $\mathbb{N}^{\mathbb{N}}$). K is the inductively defined class of (lawlike) neighbourhood functions. Accordingly, $\mathbf{IDB}_1$ is axiomatized by the axioms and rules of **EL**, K1–3 for K, and the countable axiom of choice

$$\mathrm{AC}_{01} \qquad \forall x \exists a A(x, a) \to \exists b \forall x A(x, (b)_x),$$

where $(b)_x := \lambda y.bj(x, y)$, as before. Note that the prime formulas are now of two kinds: $t = s$ for numerical terms t and s, and $\varphi \in K$ (or $K\varphi$) for function terms φ. K can be proved to be extensional (exercise).

3.3. *The theory* **LS**. The language of **LS** is obtained by adding lawless function variables – $\alpha, \beta, \gamma, \delta$ – to $\mathscr{L}(\mathbf{IDB}_1)$, together with lawless quantifiers $\forall\alpha, \exists\alpha$.

The only lawless terms are lawless variables; for the numerical terms we add a clause

$$t \text{ a numerical term} \Rightarrow \alpha(t) \text{ a numerical term.}$$

On the other hand, the clause for explicit definition (λ-abstraction) must be restricted so as to apply only to terms without lawless variables.

For a compact description of the axioms, we need, in addition to 2.14, the following abbreviations and notations.

NOTATION. As before, when writing $A(\vec{\alpha})$ or a formula it is intended that all free lawless variables are among the $\vec{\alpha}$.

$$\dot{\forall}\alpha A(\alpha, \vec{\beta}) := \dot{\forall}\alpha \in \langle\ \rangle A(\alpha, \vec{\beta}),$$

$$\dot{\exists}\alpha A(\alpha, \vec{\beta}) := \dot{\exists}\alpha \in \langle\ \rangle A(\alpha, \vec{\beta}).$$

It should be noted that in this definition not all $\alpha, \vec{\beta}$ need actually occur in A; so the meaning of $\dot{\forall}, \dot{\exists}$ depends on the way we have chosen to exhibit A.

Let $\vec{\alpha} \equiv (\alpha_1, \ldots, \alpha_p)$, $\vec{n} \equiv (n_1, \ldots, n_p)$, and $\vec{m} \equiv (m_1, \ldots, m_p)$. We write

$$\vec{\alpha} \in \vec{n} := \alpha_1 \in n_1 \wedge \cdots \wedge \alpha_p \in n_p,$$

$$\dot{\forall}\vec{\alpha} \in \vec{n} A := \dot{\forall}\alpha_1 \in n_1 \ldots \dot{\forall}\alpha_p \in n_p A, \quad \text{etc.},$$

$$\vec{n} \circledast \vec{m} := \left(n_1 * m_1, \ldots, n_p * m_p\right).$$

Let $k_i^m (1 \le i \le m)$ be primitive recursive functions such that

$$k_i^m \langle\ \rangle := \langle\ \rangle,$$

$$k_i^m(\langle x_0, \ldots, x_q \rangle) := \langle \boldsymbol{p}_{i-1}^m(x_0), \boldsymbol{p}_{i-1}^m(x_1), \ldots, \boldsymbol{p}_{i-1}^m(x_q) \rangle.$$

We put

$$\vec{n} \circledast m := \left(n_1 * k_1^p m, \ldots, n_p * k_p^p m\right),$$

$$\vec{\alpha} \in m := \alpha_1 \in k_1^p m \wedge \cdots \wedge \alpha_p \in k_p^p m.$$

Note that $\vec{\alpha} \in m \leftrightarrow \nu^p(\vec{\alpha}) \in m$. □

We can now compactly state the specific axioms for lawless sequences:

LS1 $\forall n \exists \alpha (\alpha \in n)$,

LS2 $\forall \alpha \forall \beta (\alpha \neq \beta \vee \alpha = \beta)$,

LS3 $\dot{\forall} \alpha \left[A(\alpha, \vec{\beta}) \to \exists n \left(\alpha \in n \wedge \dot{\forall} \gamma \in n A(\gamma, \vec{\beta}) \right) \right]$,

LS4 $\dot{\forall} \vec{\alpha} \exists a A(\vec{\alpha}, a) \to \exists e \in K\, \forall n \left(en \neq 0 \to \exists a \dot{\forall} \vec{\alpha} \in n A(\vec{\alpha}, a) \right)$.

If, for example, we spell out LS4, we get

$$\dot{\forall} \alpha_1 \ldots \dot{\forall} \alpha_p \exists a A(\alpha_1, \ldots, \alpha_p, a) \to$$
$$\exists e \in K\, \forall n \left(en \neq 0 \to \exists a \dot{\forall} \alpha_1 \in k_1^p n \ldots \dot{\forall} \alpha_p \in k_p^p n A(\alpha_1, \ldots, \alpha_p, a) \right)$$

or

$$\forall \alpha_1 \ldots \alpha_p \left(\#(\alpha_1, \ldots, \alpha_p) \to \exists a A(\alpha_1, \ldots, \alpha_p, a) \right) \to$$
$$\exists e \in K\, \forall n \left(en \neq 0 \to \right.$$
$$\left. \exists a \forall \alpha_1 \in k_1^p n \ldots \forall \alpha_p \in k_p^p n \left((\alpha_1, \ldots, \alpha_p) \to A(\alpha_1, \ldots, \alpha_p, a) \right) \right).$$

As a preliminary to the proof of (i) and (ii) of the elimination theorem, we present several lemmas.

3.4. LEMMA. **LS** $\vdash \dot{\forall} \vec{\alpha} \exists x A(\vec{\alpha}, x) \to$
$\exists e \in K\, \forall n (en \neq 0 \to \exists x \dot{\forall} \vec{\alpha} \in n A(\vec{\alpha}, x)$.

PROOF. An easy consequence of LS4. □

3.5. LEMMA

(i) **LS** $\vdash \dot{\forall} \vec{\alpha} \in \vec{n} (A(\alpha, \vec{\beta}) \to B(\vec{\alpha}, \vec{\beta})) \leftrightarrow$
$\forall m (\dot{\forall} \vec{\alpha} \in \vec{n} \circledast m A(\vec{\alpha}, \vec{\beta}) \to \dot{\forall} \alpha \in \vec{n} \circledast m B(\vec{\alpha}, \vec{\beta}))$,

(ii) **LS** $\vdash \dot{\exists} \alpha \in n B(\alpha, \vec{\beta}) \leftrightarrow \exists m \dot{\forall} \alpha \in n*m B(\alpha, \vec{\beta})$.

PROOF. (i) The direction from the left to the right is immediate. For the converse, suppose first $\vec{\alpha} \equiv \alpha$, $\vec{n} \equiv n$ and assume

$$\forall m\big(\dot{\forall}\alpha \in n * m A\big(\alpha, \vec{\beta}\big) \to \dot{\forall}\alpha \in n * m B\big(\alpha, \vec{\beta}\big)\big). \tag{1}$$

Let $\neq(\alpha, \vec{\beta})$, $\alpha \in n$, $A(\alpha, \vec{\beta})$; then, by LS3, $\alpha \in m'$ and $\dot{\forall}\gamma \in m'(\gamma \in n \wedge A(\gamma, \vec{\beta}))$ for some m'. Without loss of generality we can assume $m' = n * m$, hence $\dot{\forall}\gamma \in n * m A(\gamma, \vec{\beta})$; therefore by our assumption (1) $\dot{\forall}\gamma \in n * m B(\gamma, \vec{\beta})$, and so $B(\alpha, \vec{\beta})$. Thus $\dot{\forall}\alpha \in n(A(\alpha, \vec{\beta}) \to B(\alpha, \vec{\beta}))$. For a vector $\vec{\alpha}$ of length p in the statement of the lemma we have to apply this argument p times.

(ii) is easy and left to the reader. □

3.6. LEMMA. *Let* $\mathfrak{a}$ *be either a number variable or a variable for lawlike sequences, then*

$$\mathbf{LS} \vdash \dot{\forall}\vec{\alpha} \in \vec{n}\,\exists \mathfrak{a} A(\vec{\alpha}, \mathfrak{a}) \leftrightarrow$$
$$\exists e \in K \forall m\big(em \neq 0 \to \exists \mathfrak{a} \dot{\forall}\vec{\alpha} \in \vec{n} \circledast m A(\vec{\alpha}, \mathfrak{a})\big).$$

PROOF. The direction from the right to the left is easy, we leave it to the reader. For the converse, let $\vec{n} \equiv \langle n_1, \ldots, n_p \rangle$, and assume the left-hand side, then for some $f \in K$

$$\forall n\big(fm \neq 0 \to \exists \mathfrak{a} \dot{\forall}\vec{\alpha} \in m\big(\vec{\alpha} \in \vec{n} \to A(\vec{\alpha}, \mathfrak{a})\big)\big).$$

We define e by

$$en = y + 1 :=$$
$$\exists n'\big(n_1 * k_1^p n \succcurlyeq k_1^p n' \wedge \cdots \wedge n_p * k_p^p n \succcurlyeq k_p^p n' \wedge fn' = y + 1\big).$$

We leave it as an exercise to show that $e \in K$; it is also easy to see that e satisfies the right-hand side of the statement of the lemma. □

3.7. The following lemma shows that we can in fact interpret $\dot{\exists}\alpha \in n A(\alpha, \vec{\beta})$ as $\dot{\forall}\alpha \in n(\neq(\alpha, \vec{\beta}') \to A(\alpha, \vec{\beta}))$, where $\vec{\beta}'$ is the subsequence of $\vec{\beta}$ consisting of those lawless variables *actually* occurring free in A.

LEMMA.
(i) $\mathbf{LS} \vdash \dot{\forall}\alpha \in n A(\alpha, \vec{\beta}) \leftrightarrow \dot{\forall}\alpha \in n[A(\alpha, \vec{\beta}) \wedge \gamma = \gamma]$,
(ii) $\mathbf{LS} \vdash \dot{\exists}\alpha \in n A(\alpha, \vec{\beta}) \leftrightarrow \dot{\exists}\alpha \in n[A(\alpha, \vec{\beta}) \wedge \gamma = \gamma]$.

PROOF. We prove (i) and (ii) by simultaneous induction on the logical complexity of A, where (ii) at each step is deduced from (i). We treat some of the typical cases and leave the rest as an exercise.

Case 1. $A(\alpha, \vec{\beta}) \equiv B(\alpha, \vec{\beta}) \wedge C(\alpha, \vec{\beta})$. Then

$$\begin{aligned}
&\dot{\forall}\alpha \in n(A(\alpha, \vec{\beta}) \wedge \gamma = \gamma)\\
&\leftrightarrow \dot{\forall}\alpha \in n((B(\alpha, \vec{\beta}) \wedge \gamma = \gamma) \wedge (C(\alpha, \vec{\beta}) \wedge \gamma = \gamma))\\
&\leftrightarrow \dot{\forall}\alpha \in n(B(\alpha, \vec{\beta}) \wedge \gamma = \gamma) \wedge \dot{\forall}\alpha \in n(C(\alpha, \vec{\beta}) \wedge \gamma = \gamma)\\
&\leftrightarrow \dot{\forall}\alpha \in n\,B(\alpha, \vec{\beta}) \wedge \dot{\forall}\alpha \in n\,C(\alpha, \vec{\beta})\\
&\leftrightarrow \dot{\forall}\alpha \in n\,A(\alpha, \vec{\beta}).
\end{aligned}$$

Case 2. Let $A(\alpha, \vec{\beta}) \equiv \forall\alpha' B(\alpha, \alpha', \vec{\beta})$. Without loss of generality we can suppose $\neq (\gamma, \vec{\beta})$. Then the following are equivalent:

$$\dot{\forall}\alpha \in n(\forall\alpha' B(\alpha, \alpha', \vec{\beta}) \wedge \gamma = \gamma), \tag{1}$$

$$\dot{\forall}\alpha \in n(\gamma = \gamma \wedge \dot{\forall}\alpha' B(\alpha, \alpha', \vec{\beta}) \wedge B(\alpha, \alpha, \vec{\beta}) \wedge \cdots \wedge B(\alpha, \beta_i, \vec{\beta}) \wedge \cdots), \tag{2}$$

$$\dot{\forall}\alpha \in n[\dot{\forall}\alpha'(B(\alpha, \alpha', \vec{\beta}) \wedge \gamma = \gamma) \wedge B(\alpha, \alpha, \vec{\beta}) \wedge \cdots \wedge B(\alpha, \beta_i, \vec{\beta}) \wedge \cdots \wedge B(\alpha, \gamma, \vec{\beta})], \tag{3}$$

$$\dot{\forall}\alpha \in n\,\forall\alpha'(B(\alpha, \alpha', \vec{\beta}) \wedge \gamma = \gamma), \tag{4}$$

$$\dot{\forall}\alpha \in n\,\forall\alpha' B(\alpha, \alpha', \vec{\beta}). \tag{5}$$

(1) $\leftrightarrow$ (2) holds by the meaning of $\dot{\forall}\alpha'$ and LS2. (2) $\leftrightarrow$ (3) holds because the outer quantifier ensures $\neq (\gamma, \alpha)$ and the inner quantifier ensures $\forall\alpha'(\neq (\alpha', \alpha, \vec{\beta}) \rightarrow B(\alpha, \alpha', \vec{\beta}))$. The remaining equivalences are immediate.

Case 3. Suppose (i) has been proved; then by (ii) of lemma 3.5 $\dot{\exists}\alpha \in n\, A(\alpha, \vec{\beta}) \leftrightarrow \exists m \dot{\forall}\alpha \in n * m\, A(\alpha, \vec{\beta})$, therefore $\dot{\exists}\alpha \in n(A(\alpha, \vec{\beta}) \wedge \gamma = \gamma) \leftrightarrow \exists m \dot{\forall}\alpha \in n * m(A(\alpha, \vec{\beta}) \wedge \gamma = \gamma) \leftrightarrow \exists m \dot{\forall}\alpha \in n * m\, A(\alpha, \vec{\beta}) \leftrightarrow \dot{\exists}\alpha \in n\, A(\alpha, \vec{\beta})$. □

3.8. LEMMA. **LS** $\vdash \dot{\forall}\vec{\alpha} \in \vec{n}(t_1[\vec{\alpha}] = t_2[\vec{\alpha}]) \leftrightarrow \forall\vec{a} \in \vec{n}(t_1[\vec{a}] = t_2[\vec{a}])$, *where* $\forall\vec{a} \in \vec{n}$ *is defined in the same way as* $\dot{\forall}\vec{\alpha} \in \vec{n}$.

Proof. By induction on the complexity of a numerical term t one can prove:

$$\exists e \in K\, \forall n(en \neq 0 \to \exists x \forall \vec{\alpha} \in n\, \forall \vec{a} \in n(t[\vec{\alpha}] = t[\vec{a}] = x))$$

and from this the statement of the lemma readily follows. We leave the somewhat tedious details as an exercise. □

3.9. *Definition of* τ *and proof of the elimination theorem.*

(a) The first step in defining $\tau(A)$ consists in rewriting A in terms of quantifiers $\dot{\forall}\alpha, \dot{\exists}\alpha$ using the equivalences ($\vec{\beta} \equiv (\beta_0, \ldots, \beta_p)$)

$$\forall\alpha\, A(\alpha, \vec{\beta}) \leftrightarrow \dot{\forall}\alpha\, A(\alpha, \vec{\beta}) \wedge \bigwedge_{i \leq p} A(\beta_i, \vec{\beta}),$$

$$\exists\alpha\, A(\alpha, \vec{\beta}) \leftrightarrow \dot{\exists}\alpha\, A(\alpha, \vec{\beta}) \wedge \bigvee_{i \leq p} A(\beta_i, \vec{\beta}).$$

The result is uniquely determined if $\alpha, \vec{\beta}$ is a list of *all* the lawless variables actually free in A. However in the steps described under (b) we also need the dummy variables. This is uniquely determined modulo logical equivalence by lemma 3.7; the replacement transforms any formula into an **LS**-provably equivalent formula.

(b) We shall now define the effect of τ on formulas written with propositional operators, and $\dot{\forall}\alpha \in t$, $\dot{\exists}\alpha \in t$ (t without lawless variables); these operations are regarded as logical primitives in the syntactic definition. $\dot{\forall}\alpha, \dot{\exists}\alpha$ are regarded as synonymous with $\dot{\forall}\alpha \in \langle\ \rangle$, $\dot{\exists}\alpha \in \langle\ \rangle$.

Disjunction is treated as defined, and we take $0 = 1$ as falsum. We first eliminate all occurrences of $\dot{\exists}\alpha \in n$ by replacements

$$\dot{\exists}\alpha \in n A(\alpha, \vec{\beta}) \mapsto \exists m \dot{\forall}\alpha \in n * m A(\alpha, \vec{\beta}).$$

Next we show how to transform formulas of the form $\dot{\forall}\vec{\alpha} \in \vec{t} A(\vec{\alpha})$ into formulas with fewer logical operations within the scope of blocks $\dot{\forall}\vec{\alpha} \in \vec{t}'$. This process can be continued till we arrive at $\dot{\forall}\vec{\alpha} \in \vec{t}(P)$ for P prime. The necessary replacements are given by

$$\dot{\forall}\vec{\alpha} \in \vec{n}(A \wedge B) \mapsto \dot{\forall}\vec{\alpha} \in \vec{n} A \wedge \dot{\forall}\vec{\alpha} \in \vec{n} B,$$

$$\dot{\forall}\vec{\alpha} \in \vec{n}(A \to B) \mapsto \forall \vec{m}(\dot{\forall}\vec{\alpha} \in \vec{n} \circledast \vec{m} A \to \dot{\forall}\vec{\alpha} \in \vec{n} * \vec{m} B),$$

$$\dot{\forall}\vec{\alpha} \in \vec{n}\, \exists \mathfrak{a} A(\alpha, \mathfrak{a}) \mapsto \exists e \in K\, \forall n(en \neq 0 \to \exists \mathfrak{a} \dot{\forall}\alpha \in n \circledast m A(\vec{\alpha}, \mathfrak{a})),$$

$$\dot{\forall}\vec{\alpha} \in \vec{n}\, \forall \mathfrak{a} A(\vec{\alpha}, \mathfrak{a}) \mapsto \forall \mathfrak{a} \dot{\forall}\alpha \in \vec{n} A(\vec{\alpha}, \mathfrak{a}),$$

where $\mathfrak{a}$ is a numerical or lawlike sequence variable. Each step replaces a formula by an **LS**-provably equivalent one. (We may have to appeal to 3.7, e.g. to see that $\dot{\forall}\alpha(B(\alpha,\vec{\beta})\wedge C(\alpha,\vec{\gamma}))\leftrightarrow(\dot{\forall}\alpha\, B(\alpha,\vec{\beta})\wedge\dot{\forall}\alpha\, C(\alpha,\vec{\gamma}))$.)

Finally, we have to show how to eliminate quantifier strings $\dot{\forall}\vec{\alpha}\in\vec{t}$ in front of prime formulas, where it is assumed that $\vec{t}$ does not contain lawless variables. We use

$$\dot{\forall}\vec{\alpha}\in\vec{t}(s_1[\vec{\alpha}]=s_2[\vec{\alpha}])\mapsto\forall\vec{a}\in\vec{t}(s_1[\vec{a}]=s_2[\vec{a}])$$

and

$$\dot{\forall}\alpha\in\vec{t}(\varphi\in K)\mapsto\varphi\in K$$

if no variable of $\vec{\alpha}$ occurs in φ, and

$$\dot{\forall}\vec{\alpha}\in\vec{t}(\varphi[\vec{\alpha}]\in K)\mapsto\dot{\forall}\vec{\alpha}\in\vec{t}\,\exists e(e\in K\wedge\forall x(\varphi[\vec{\alpha}](x)=ex))$$

in all other cases. In each of these transformations **LS**-provable equivalence holds (lemma 3.8).

3.10. *The theory* $\mathbf{LS}_1$. The simplest version of a theory of binary lawless sequences has only variables for numbers and binary lawless sequences. The lawlike part of this theory is just **HA**. In a definitional extension we can add variables ranging over codes of binary finite sequences; let us reserve n, m for such codes. Then $\mathbf{LS}_1$ can be axiomatized by LS1, LS2, LS3, and

$$\text{LS4}^*_{\text{F}}\qquad \dot{\forall}\alpha\exists x A(\alpha,\vec{\beta},x)\rightarrow\exists z\forall n_z\exists x\dot{\forall}\alpha\in n_z\, A(\alpha,\vec{\beta},x),$$

where "$\forall n_z$" abbreviates "$\forall n(\text{lth}(n)=z\rightarrow\ldots$" This is equivalent to

$$\text{LS4}_{\text{F}}\qquad \dot{\forall}\vec{\alpha}\exists x A(\vec{\alpha},x)\rightarrow\exists z\forall n_z\exists x\dot{\forall}\vec{\alpha}\in n_z\, A(\vec{\alpha},x).$$

For this theory $\mathbf{LS}_1$ we can also give an elimination translation (exercise).

In precisely the same way we can formulate the theory **LS**(T) for lawless sequences ranging over a fan T (T fixed and arithmetically definable).

3.11. *Uniform forcing and elimination of lawless sequences.* Let T be a fixed definable fan. We may now consider a fragment of **LS**(T) where the only formulas with choice variables we shall consider are of the form $\forall\alpha\in n$ $A(\alpha)$, A not containing choice quantifiers. The clauses of an elimination translation for such formulas become

$$\forall\alpha \in n P\alpha \mapsto \forall m P(\lambda x.(n * m)_x) \quad \text{for } P \text{ prime},$$

$$\forall\alpha \in n(B \wedge C) \mapsto \forall\alpha \in nB \wedge \forall\alpha \in nC,$$

$$\forall\alpha \in n(B \to C) \mapsto \forall m \succcurlyeq n(\forall\alpha \in mB \to \forall\alpha \in mC),$$

$$\forall\alpha \in n \forall x B(x) \mapsto \forall x \forall\alpha \in nB(x),$$

$$\forall\alpha \in n \exists x B(x) \mapsto \exists z \forall m \succcurlyeq_z n \exists x \forall\alpha \in mB(x),$$

and if disjunction is treated as a primitive, one should add a clause

$$\forall\alpha \in n(B \vee C) \mapsto \exists z \forall m \succcurlyeq_z n(\forall\alpha \in mB \vee \forall\alpha \in mC).$$

Here "$\forall m \succcurlyeq_z n$" abbreviates "$\forall m(\text{lth}(m) = \text{lth}(n) + z \to \ldots$" Comparing the clauses for $\wedge, \to, \forall, \exists, \vee$ with the corresponding clauses for (uniform) forcing as introduced in 9.7.17, we note an obvious parallel: "$\forall\alpha \in nA$" corresponds to "$n \Vdash A$". There are differences too, caused by the fact that in the case of forcing in the theory **APP**, terms are not necessarily defined, and also because in the definition of $n \Vdash A$ the variables behave rather as ranging over "operations lawlike in α" whereas the variables $x, y, \ldots$ in **LS**(T) simply range over lawlike objects, i.e. natural numbers. These differences most notably appear in the handling of prime formulas.

However, it can be proved that for arithmetical formulas A (under the obvious embedding of **HA** in **APP**)

$$\mathbf{APP} \vdash (n \Vdash A(x_1, \ldots, x_n))[\hat{x}_1, \ldots, \hat{x}_n/\lambda\alpha.x_1, \ldots, \lambda\alpha.x_n] \leftrightarrow \tau(\forall\alpha \in nA(x_1, \ldots, x_n)),$$

where τ is the translation indicated above, followed by the embedding of **HA** in **APP**. We leave the proof as an exercise. Further understanding of this connection between forcing and the elimination translation may be gained from 13.1.9–10 and 15.6.11.

4. Other notions of choice sequence

4.1. This section briefly surveys the possibilities for other notions of choice sequence and the study of universes of sequences constructed from lawless sequences.

As we have seen in section 2, LS is not closed under any continuous operation except identity, and lawless sequences do not permit "global" restrictions (i.e. restrictions involving infinitely many values of the sequence). Thus a domain of "lawless reals" based on LS via a standard enumeration of $\mathbb{Q}$ is mathematically unsatisfactory: the elements of $\mathbb{Q}$ are not themselves represented among the lawless reals, and there are poor closure properties; cf. E12.4.1.

Therefore there is an obvious interest in the study of more "flexible" universes of sequences having better closure properties, while retaining some of the strength of C-N or FAN.

4.2. *Projections.* One possibility for getting nice universes is the method of projecting lawless sequences, that is to say the study of collections of sequences of the form

$$\left\{\Gamma\nu^p(\alpha_1,\ldots,\alpha_p) : \Gamma \in \mathscr{C}, \alpha_1,\ldots,\alpha_p \in \mathrm{LS}\right\}, \tag{1}$$

or of the form

$$\{\Gamma\alpha : \Gamma \in \mathscr{C}\} \quad \text{for fixed } \alpha \in \mathrm{LS}, \tag{2}$$

where $\mathscr{C}$ is a class of lawlike continuous functionals in $\mathbb{N}^{\mathbb{N}} \to \mathbb{N}^{\mathbb{N}}$ from functions to functions.

A particularly interesting example is obtained by taking in (1) for $\mathscr{C}$ all continuous operations representable by neighbourhood functions in K. For the theory of this example see Troelstra (1977, 4.5–10); this universe satisfies C-N! (i.e. instances of C-N with premiss strengthened to $\forall\alpha\exists!n A(\alpha, n)$).

If in (2) we take for $\mathscr{C}$ again all K-representable continuous operations, we obtain a class of models intimately connected with a topological model for intuitionistic analysis over Baire space as given in J.R. Moschovakis (1973) (adapted from a model devised by D.S. Scott (1968, 1970A)). We shall not discuss projections here; for information on topological models, see sections 15.1–3.

4.3. *Other notions of choice sequence.* In 4.6.2 we already briefly encountered an example of a notion of choice sequence, clearly distinct from that of a lawless sequence: the "hesitant sequences" HS. Another possibility is suggested by Brouwer (1925): at any stage in the construction of a choice sequence one chooses a value for the next argument, and we may impose a further overall-restriction on possible future choices, namely that future choices must be within a specified lawlike spread (i.e. a decidable tree without finite branches). Let us call the universe of such sequences BS ("*Brouwer sequences*").

There are choice sequences in BS (extensionally) equal to any lawlike sequence, since a single lawlike sequence represents a very special case of a spread, a spread with a single infinite branch. As soon as we have decided at some stage, that the sequence generated has henceforth to conform to such a "singleton" spread, the choice sequence has become equal to a lawlike sequence.

We may picture the generation of an element α of BS as a continuing process of choosing pairs $\langle \alpha(0), S_0 \rangle, \langle \alpha(1), S_1 \rangle, \langle \alpha(2), S_2 \rangle, \ldots$, where $\alpha(0), \alpha(1), \alpha(2), \ldots$ are numbers and represent the values of α, and $S_0, S_1, S_2, \ldots$ are spreads such that $\{\alpha : \alpha \in S_{n+1}\} \subset \{\alpha : \alpha \in S_n\}$ for all n, and $\bar{\alpha}(n+1) \in S_n$ for all n. For a suggestive picture see fig 12.1.

HS and BS are more satisfactory than LS, since the lawlike sequences form (extensionally) a subdomain of HS as well as of BS. On the other hand, the closure properties are still unsatisfactory: both universes are not closed under any nontrivial lawlike continuous operation.

The argument for this fact is based on the motivation of axioms for HS and BS, from which the rigidity can be derived. It would, however, carry us to far to go into details.

As to the plausibility of rigidity for BS, consider e.g. the operation of doubling all values and suppose $\alpha, \beta \in$ BS to be related by $\forall x(\alpha x = 2\beta x)$. As long as β is not yet predeterminate, α is not fixed either, but the condition that all values of α are completely determined relative to β

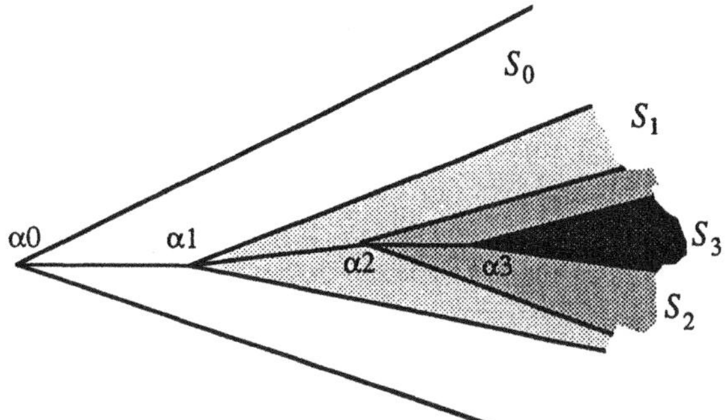

Fig. 12.1.

(namely by doubling the values) and the restriction of α to a given lawlike spread are of *different kinds*. Thus if $\beta \in \mathrm{BS}$, α cannot be in BS as well.

This suggests that we can only achieve closure under nontrivial continuous operations by building it into the notion of choice sequence right from the beginning. We sketch two such notions in the next subsection.

4.4. *General choice sequences and* GC-*sequences.* Closure under nontrivial, lawlike, continuous operations, $\Gamma, \Gamma', \Gamma'', \ldots$ can be guaranteed by permitting restrictions (to be imposed at some stage) which state that the sequence α is related to other choice sequences $\alpha_1, \ldots, \alpha_p$ via a continuous operation Γ, say: $\alpha = \Gamma\nu^p(\alpha_1, \ldots, \alpha_p)$ (provided this is compatible with the restrictions already imposed on $\alpha_1, \ldots, \alpha_p, \alpha$).

The most liberal position we can adopt, is to permit the imposition, at each stage, of any meaningful restriction (or condition), provided the possibility of continuing the sequence remains guaranteed. A meaningful restriction R in such a step-by-step process should be given by a condition R' to be satisfied by all initial segments, i.e. $R(\alpha) \leftrightarrow \forall x\, R'(\bar{\alpha}x)$; this permits us to ensure R by making sure with each choice of a value of α that R' holds for the initial segment created. Choice sequences generated in this way may be dubbed "general (choice) sequences" (GS).

If instead we go to the other extreme, and permit *only* conditions of the form $\alpha = \Gamma\alpha_1$, or $\alpha = j(\alpha_1, \alpha_2) \equiv \nu^2(\alpha_1, \alpha_2)$, we obtain the GC-sequences (GC: Generated by Continuous operations). More precisely, we may think of the generation of a GC-sequence as follows.

We start generating $\alpha 0, \alpha 1, \ldots$, until at some stage n_0 we decide to make α dependent on one or two new (fresh) GC-sequences, α_0 or α_0, α_1, via a continuous operation Γ, e.g. by imposing that $\bar{\alpha}n_0$ is to be continued with values from $\Gamma\alpha_0$ or $\Gamma(\alpha_0, \alpha_1)$, say $\alpha = \bar{\alpha}n_0 * \Gamma\alpha_0$. α_0 is generated in turn: $\alpha_0 0, \alpha_0 1, \ldots$ until at stage n_2 in the generation of α_0 we decide that $\alpha_0 = \bar{\alpha}_0 n_2 * \Gamma'(\alpha_2, \alpha_3)$ must be satisfied, etc.

4.5. *Axioms for* GS *and* GC. We now set out to formulate axioms (principles) valid for these notions, by reflecting on the way these choice sequences are given, in the same way as we did for LS in section 2. It would carry us too far to attempt to do this with the same amount of detail and critical discussion as for LS; some indications of what is involved must suffice.

For example, $\forall\alpha\exists x$-continuity (C-N) may be motivated for general sequences with the remark (already used in 4.6.3) that any method which always produces a natural number at some definite stage in the construction of any α, should do so on the basis of a finite initial segment, since for an *arbitrary* general sequence the only information which is guaranteed to be available at any stage is a finite initial segment.

This is at best a plausibility argument, but it is difficult to do better in view of the chaotic variety of possible restrictions. Also, our presentation above of the argument for $\forall\alpha\exists x$-continuity was too simple-minded, as the following example shows. At first sight, it seems that the argument should equally apply to a collection of sequences which is the union of LS and all lawlike sequences LL, since here too the only information we can assume to be *always* available for a sequence in LS ∪ LL is an initial segment. However, consider the following property (α ranging over LS ∪ LL)

$$A(\alpha, x) \equiv [(x = 0 \wedge \exists a(\alpha = a)) \vee (x = 1 \wedge \neg\exists a(\alpha = a))].$$

Then $\forall\alpha\exists!x A(\alpha, x)$ (since each α is either in LL or not); but obviously, x cannot depend continuously on α. The argument for $\forall\alpha\exists x$-continuity fails in this case, since, for any α at any stage, we do not only know an initial segment, but also know whether α belongs to LS or to LL.

Thus our argument for general sequences is incomplete; the reason that the sort of counterexample just given for LS ∪ LL is not available for general sequences is due to the fact that we cannot single out a decidable subclass $\mathscr{C}$ (at least not one which is definable in the language $\mathscr{L}$(**LS**)). Alternatively we might say that in the case of general sequences the only *definite* (encodable by a number) data which are certain to become available are initial segments, whereas in the case of LS ∪ LL also membership of either LS or LL is a piece of definite information.

We must leave it at this (cf. the intuitive argument for UP in 4.9.1).

Next, by way of an example, we shall present a plausibility argument for the validity of the "principle of analytic data", the analogue of LS3 for GC-sequences. The *principle of analytic data* may be stated as

$$A(\alpha) \to \exists\Gamma(\exists\beta(\alpha = \Gamma\beta) \wedge \forall\gamma A(\Gamma\gamma)) \tag{1}$$

(Γ ranging over lawlike continuous functionals from $\mathbb{N}^{\mathbb{N}}$ to $\mathbb{N}^{\mathbb{N}}$). Suppose α is a GC-sequence, and at some definite stage we know that $A(\alpha)$ holds. The data concerning α are then given by e.g.

$$\alpha = \bar{\alpha}(n_0) * \Gamma_0(\alpha_0),$$

$$\alpha_0 = \bar{\alpha}_0(n_1) * \Gamma_1(\alpha_1),$$

$$\dots$$

$$\alpha_{p-1} = \bar{\alpha}_{p-1}(n_p) * \Gamma_p(\alpha_p),$$

while α_p is as yet completely undetermined. As a result of these data, we can find a single Γ (at the stage considered) such that

$$\alpha = \Gamma\alpha_p. \tag{2}$$

Since α_p is completely undetermined, and $A(\alpha)$ holds on the basis of (2), we have

$$\forall\gamma A(\Gamma\gamma);$$

this justifies (1).

4.6. *The theory* **CS**. We now introduce the formal theory of the GC-sequences, based on the language $\mathscr{L}(\mathbf{LS})$. The lawlike part of the axiom system yields $\mathbf{IDB}_1$ as before. We can treat $e|\alpha = \beta$, $e(\alpha) = t$, etc., as abbreviations of, respectively,

$$\forall y(\lambda n.e(\langle y\rangle * n)(\alpha)) = \beta y, \quad \exists x(e(\bar{\alpha}x) = t + 1);$$

or we can add special variables ranging over K and introduce $e|\alpha$, $e(\alpha)$ in a definitional extension.

The axioms for GC-sequences are:

GC1 $\quad \forall e \in K\,\forall\alpha\exists\beta(e|\alpha = \beta), \qquad \forall\alpha\beta\exists\gamma(j(\alpha,\beta) = \gamma)$

(*closure under continuous operations and pairing*);

GC2 $\quad A\alpha \to \exists e \in K[\exists\beta(\alpha = e|\beta) \wedge \forall\beta A(e|\beta)]$

(*principle of analytic data*);

GC3 $\quad \forall\alpha\exists a A(\alpha, a) \to \exists e\forall n(en \neq 0 \to \exists a\forall\alpha \in n\, A(\alpha, a));$

(*continuity*);

GC4 $\quad \forall\alpha\exists\beta A(\alpha,\beta) \to \forall\alpha\exists e A(\alpha, e|\alpha)$

(*local function-continuity*).

Let k_n be an element of K such that

$$k_n|\alpha \in n \quad \text{for all } \alpha,$$
$$(k_n|\alpha)(x) = \alpha x \quad \text{for } x \geq \text{lth}(n).$$

Then it is easily seen that GC3 may be formulated as

GC3′ $\quad \forall\alpha\exists a A(\alpha, a) \to \exists e\forall n(en \neq 0 \to \exists a\forall\alpha\, A(k_n|\alpha, a)).$

The following interesting proposition expresses so-called $\forall\alpha\exists\beta$-continuity.

4.7. Proposition. CS $\vdash \forall\alpha\exists\beta A(\alpha, \beta) \to \exists e \forall\alpha A(\alpha, e|\alpha)$.

Proof. Assume $\forall\alpha\exists\beta A(\alpha, \beta)$, then by GC4 $\forall\alpha\exists e A(\alpha, e|\alpha)$, and thus by GC3

$$\exists e \forall n (en \neq 0 \to \exists f \forall\alpha \in n A(\alpha, f|\alpha)).$$

It is not hard to see that AC_{01} has as a corollary the schema

$$\forall n \exists f B(n, f) \to \exists f \forall n B(n, f_{\langle n \rangle}),$$

where $f_{\langle n \rangle} = \lambda m.f(\langle n \rangle * m)$. Therefore, for suitable e and f

$$\forall n \big(en \neq 0 \to \forall\alpha \in n A(\alpha, f_{\langle n \rangle}|\alpha)\big).$$

Now let h be given by

$$h(e, n) = \begin{cases} \text{the longest } m \preccurlyeq n \text{ such that } \forall m' \prec m(em' = 0) \text{ if } en \neq 0, \\ 0 \quad \text{otherwise,} \end{cases}$$

and put

$$f'0 = 0, \qquad f'(\langle x \rangle * n) = \text{sg}(en) \cdot f_{\langle h(e, n) \rangle}(\langle x \rangle * n).$$

Assuming $f' \in K$, we see that if $\alpha \in n$, $en \neq 0$ also $e(h(e, n)) \neq 0$, $h(e, n) \preccurlyeq n$ and $A(\alpha, f_{\langle h(e, n) \rangle}|\alpha)$, while $f_{\langle h(e, n) \rangle}|\alpha = f'|\alpha$; therefore

$$\forall n (en \neq 0 \to \forall\alpha \in n A(\alpha, f'|\alpha))$$

and so $\forall\alpha A(\alpha, f'|\alpha)$. It remains to prove the following lemma.

Lemma. *In* **CS**:
(i) $en \neq 0 \to h(e, n) = h(e, n * m)$,
(ii) $\forall e f \forall a[\forall nm(em \neq 0 \to an = a(n * m)) \to \lambda n.\text{sg}(en) \cdot f(\langle an \rangle * n) \in K]$,
(iii) *f' as defined above is in K.*

Proof. (i). Immediate.
(ii) By induction over K with respect to e (exercise).
(iii) Immediate from (i) and (ii). □

4.8. *Elimination of choice sequences for* **CS**. We can now easily describe an elimination translation for **CS**. First of all, we treat $\vee$ as a defined connective; secondly, in defining the elimination mapping τ for any formula A without choice parameters we replace subformulas $\exists\alpha B\alpha$ by $\exists a\, Ba$. The number of logical operators within the scope of a choice quantifier $\forall\alpha$ is then successively reduced by making replacements according to the following table:

$$\forall\alpha(B\alpha \wedge C\alpha) \mapsto \forall\alpha\, B\alpha \wedge \forall\alpha\, C\alpha;$$

$$\forall\alpha(B\alpha \to C\alpha) \mapsto \forall e[\forall\alpha\, B(e|\alpha) \to \forall\alpha\, C(e|\alpha)];$$

$$\forall\alpha\forall\mathfrak{a}\, A(\alpha, \mathfrak{a}) \mapsto \forall\mathfrak{a}\forall\alpha A(\alpha, \mathfrak{a});$$

$$\forall\alpha\forall\beta\, A(\alpha, \beta) \mapsto \forall\gamma\, A(j_1\gamma, j_2\gamma);$$

$$\forall\alpha\exists\mathfrak{a}\, A(\alpha, \mathfrak{a}) \mapsto \exists e\forall n(en \neq 0 \to \exists\mathfrak{a}\forall\alpha\, A(k_n|\alpha, \mathfrak{a}));$$

$$\forall\alpha\exists\beta\, A(\alpha, \beta) \mapsto \exists e\forall\alpha\, A(\alpha, e|\alpha);$$

$$\forall\alpha\, P(\alpha) \mapsto \forall a\, P(a) \quad \text{for } P \text{ prime},$$

where $\mathfrak{a}$ is any variable for numbers, lawlike sequences or elements of K. For the elimination theorem we need 4.7 and a lemma.

LEMMA
(i) $\mathbf{CS} \vdash \exists\alpha\, A(\alpha) \leftrightarrow \exists a\, Aa$,
(ii) $\mathbf{CS} \vdash \forall\alpha[A\alpha \to B\alpha] \leftrightarrow \forall e[\forall\alpha\, A(e|\alpha) \to \forall\alpha\, B(e|\alpha)]$,
(iii) $\mathbf{CS} \vdash \forall\alpha\forall\beta\, A(\alpha, \beta) \leftrightarrow \forall\gamma\, A(j_1\gamma, j_2\gamma)$,
(iv) $\mathbf{CS} \vdash \forall\alpha[A\alpha \vee \neg A\alpha] \to (\forall\alpha\, A\alpha \leftrightarrow \forall a\, Aa)$.

THEOREM (*elimination theorem for* **CS**). *For* τ *as defined above,* $\tau(A) \equiv A$ *for* A *in* $\mathscr{L}(\mathbf{IDB}_1)$, *and*

$$\mathbf{CS} \vdash \tau(A) \leftrightarrow A.$$

PROOF *of lemma and theorem.* Exercise. (iv) of the lemma is used to justify the replacement of $\forall\alpha P(\alpha)$ by $\forall a P(a)$ for P prime. □

5. Notes

5.1. *Further reading.* Taken together, the references of Troelstra (1977) and (1983A) constitute a nearly complete bibliography of choice sequences. Some recent additions are Fourman (1984), van der Hoeven and Moerdijk (1984A, 1984B), J.R. Moschovakis (1987).

Historical information on choice sequences may be found in Heyting (1981), Troelstra (1977, Appendix A; 1982A).

5.2. *Lawless sequences* appear nowhere explicitly in Brouwer's published writings, and in his unpublished writings only once, namely in a letter to A. Heyting dated 26-VI-1924 (cf. Troelstra 1982A, 6.2).

Kreisel (1958A) was the first to propose an axiomatization for lawless sequences ranging over fans; in Kreisel (1968) this was extended to lawless sequences of natural numbers (a correction to the axiomatization in the latter paper is given in Troelstra 1970A).

The informal analysis, motivating the axioms of **LS**, presented in section 2 started with Kreisel (1968) and was gradually refined and sharpened.

Van Dalen (1978) showed that LS3(2) does not follow from LS1, 2, 4 and LS3(1); in van Dalen and Lodder (1982), and van der Hoeven and Moerdijk (1983) it is shown that LS3(3) does not follow from LS1, LS2, LS4, LS3(2). Here LS3(n) is the case of LS3 where the length of $\vec{\beta}$ is at most $n - 1$.

J.R. Moschovakis (1987) shows, under the assumption of a special classical set-theoretical axiom (namely: every definable well-ordered subclass of $\mathbb{N}^{\mathbb{N}}$ is countable), that a considerable part of **LS** is compatible with classical logic.

5.3. *The elimination theorem.* Part (i) and (ii) of the theorem in 3.1 are sometimes called the "*first elimination theorem*"; the "*second elimination theorem*" is the statement

$$\mathbf{LS} \vdash A \Leftrightarrow \mathbf{IDB}_1 \vdash \tau(A). \tag{1}$$

The first elimination theorem for **LS** is proved in Kreisel (1968), this paper also contains a corresponding result for **LS** extended with relation variables. Already in (1958A) Kreisel gave a partial result for lawless sequences ranging over fans.

A complete proof of (1) is not in print, but the details are very similar to the proof of the second elimination theorem for the theory **CS** of section 4, full details of which may be found in Kreisel and Troelstra (1970).

5.4. *Projections.* The possibility of creating models for other theories of choice sequence via projections of lawless sequences is already mentioned in Kreisel (1967, 1968) and has been extensively studied (Troelstra 1969A, 1970, 1970A)); van Dalen and Troelstra (1970); Troelstra and van der Hoeven (1979; van der Hoeven (1982, 1982A); van der Hoeven and Moerdijk (1984, section 4)).

An interesting observation by van der Hoeven (in Troelstra 1983A) links the projection model of type (1) in 4.2, where Γ is the collection of all continuous mappings, to a topological permutation model considered by Krol' (1978); cf. also Grayson (1981, appendix).

5.5. BS *and* GC. The concept of BS is discussed in Myhill (1967), Troelstra (1968A, 1969B), for technical research connected with BS see in particular Troelstra (1970) (connection with projections), van der Hoeven and Moerdijk (1984A) (sheaf models).

The concept of GC-sequences was first introduced in Troelstra (1968A), and further elaborated in (1969, 1969B). Approximation of GC-sequences by means of projections is studied in van der Hoeven and Troelstra (1979), van der Hoeven (1982), van der Hoeven and Moerdijk (1984, section 4). Sheaf models for GC-sequences are also studied in van der Hoeven and Moerdijk (1984).

5.6. *Extended bar induction of type zero.* EBI_0 is a generalization of BI_{M} in which the sequences do not take arbitrary natural numbers as values, but only values in a set $A \subset \mathbb{N}$. In Troelstra (1980A) EBI_0 is discussed at length, and it is claimed that $\mathbf{EL} + \mathrm{EBI}_0$ is equal in proof-theoretic strength to the theory of the finitely iterated positive inductive definitions. The proof is defective however; in particular, an incorrect generalization of the principle of analytic data (needed to extend the second elimination theorem) has been chosen. A correct version yields the result that $\mathbf{EL}$ + $\mathrm{EBI}_0^{\mathrm{ar}}$, i.e. $\mathbf{EL}$ plus EBI_0 restricted to arithmetical A, is of the same proof-theoretic strength as $\mathbf{IDB}_1$. Refinement of the argument yields that $\mathbf{EL} + \mathrm{EBI}_0^{\mathrm{ar}}$ is conservative over $\mathbf{IDB}_1$ with respect to arithmetical sentences (Renardel de Lavalette 1984A).

It remains open to establish the proof-theoretic strength of $\mathbf{EL} + \mathrm{EBI}_0$; we think it likely that it is indeed equal to that of the theory of finitely iterated positive inductive definitions.

Exercises

12.2.1. Show that for lawless $\alpha\ \neg\forall y \exists x(\alpha y = 2x)$.

12.2.2. Show that for lawless $\alpha\ \neg\forall y(\alpha y = 0 \to \exists x > y(\alpha x = 0))$.

12.2.3. Show, parallel to 4.8.13, that (2) of 2.13 implies $K = K_0$.

12.2.4. Give the proof of 2.16 in full.

12.2.5. Show that $\forall e \in K(\lambda n.en \cdot \mathrm{sg}(\mathrm{lth}(n) \dot{-} en) \in K)$.

12.2.6. Show that (relative to the other axioms) the axiom $\mathrm{LS4}_{\mathrm{F}}$ for lawless sequences ranging over a finitely branching tree is equivalent to

$$\mathrm{LS4}_{\mathrm{F}}^{*} \quad \forall\alpha\big(\#(\alpha,\vec{\beta}) \to \exists a A(\alpha,\vec{\beta},a)\big) \to \exists z \forall n\big(\mathrm{lth}(n) = z \to \exists a \forall \alpha \in n\big(\#(\alpha,\vec{\beta}) \to A(\alpha,\vec{\beta},a)\big)\big).$$

12.3.1. Prove that K is extensional, i.e. $Ka \wedge a = b \rightarrow Kb$.

12.3.2. Give a formal proof of $\mathrm{BI_M}$ without parameters in **LS** (cf. 4.8.13).

12.3.3. Prove 3.5(ii).

12.3.4. Show that the e defined in the proof of 3.6 indeed belongs to K. *Hint.* Use K-induction with respect to f.

12.3.5. Complete the proof of 3.7.

12.3.6. Prove 3.8 in full.

12.3.7. Give a proof of the equivalence of forcing and elimination of a lawless variable (3.11).

12.3.8. Formulate and prove the elimination theorem for $\mathbf{LS}_1$ (3.10).

12.4.1. Let $\langle q_n \rangle_n$ be a standard enumeration of the rationals; show that, for lawless α, $\langle q_{\alpha n} \rangle_n$ cannot be a fundamental sequence.

12.4.2. Let x_ξ be the real defined from a sequence ξ via the representation for CSM-spaces of 7.2.3; if ξ is lawless, x_ξ is said to be a lawless real. Show that no rational can be a lawless real.

12.4.3. Prove the equivalence of GC3 and GC3′ (4.6).

12.4.4. Prove the lemma in 4.7.

12.4.5. Prove the lemma in 4.8.

12.4.6. Assume our pairing function j to be monotone in both arguments; let $(\alpha)_n :=$ $\lambda y.\alpha j(n, y)$, and put $\alpha^n = n * (\alpha)_n$. Let

$$\#(n_0, \ldots, n_p) := \bigwedge_{i<j\leq p} n_i \neq n_j.$$

We define a syntactical mapping Γ for the language of **LS** as follows. Let $v_0, v_1, v_2, \ldots$ be the list of numerical variables, $w_0, w_1, w_2, \ldots$ the list of lawless variables, and let α be a fixed lawless variable.

(i) If t is a term, Γt is the term with each w_i replaced by α^i, and each v_j replaced by v_{2j+1}.

(ii) $\Gamma(\forall w_i A) := \forall v_{2i} \Gamma(A)$, $\Gamma(\exists w_i A) := \exists v_{2i} \Gamma(A)$;
$\Gamma(\forall v_i A) := \forall v_{2i+1} \Gamma(A)$, $\Gamma(\exists v_i A) := \exists v_{2i+1} \Gamma(A)$;
Γ is a homomorphism with respect to all other logical operations.

For any $B(\beta_1, \ldots, \beta_p)$ write $B^*_\alpha(\boldsymbol{n}_1, \ldots, \boldsymbol{n}_p)$ for $\Gamma(B)$ where $\boldsymbol{n}_i \equiv v_{2j}$ if $\beta_i \equiv w_j$. Let $\pi_n(\bar{a}x)$ be the longest initial segment $\overline{a^n}(y)$ which can be determined from $\bar{a}x$. Then prove

$$\forall u \forall n_1 \cdots n_p \Big(\#(n_1, \ldots, n_p) \rightarrow$$
$$\big\{ \dot{\forall} \in u\, B^*_\alpha(\boldsymbol{n}_1, \ldots, \boldsymbol{n}_p) \leftrightarrow \forall^* \beta_1 \in \pi_{n_1} u \ldots \dot{\forall} \beta_p \in \pi_{n_p} u\, B(\beta_1, \ldots, \beta_p) \big\} \Big),$$

$u \notin \mathrm{FV}(B) \cup \mathrm{FV}(B^*_\alpha)$ (Troelstra 1970A, 2.3).

CHAPTER 13

SEMANTICAL COMPLETENESS

The present chapter discusses completeness for **IQC** for various types of semantics.

We are especially interested in two aspects here: (1) the connection between various kinds of "truth-value" semantics for intuitionistic logic and intuitionistic validity in the intuitive ("naive") sense, and (2) the possibility of giving straightforward constructive completeness proofs for the various kinds of semantics.

In chapter 2 we have already introduced Kripke semantics and presented a Henkin-style completeness proof, using classical reasoning on the meta-level.

For the greater part of this chapter familiarity with the material on Kripke semantics is not really necessary.

Section 1 introduces Beth models, discusses the transformation of Beth models into Kripke models, and the connection with intuitionistic validity.

Section 2 presents Friedman's completeness proof for Beth semantics, with an application to the theory of a single Skolem function.

In section 3 we present a result showing that we cannot expect to obtain completeness for intuitionistic validity in full generality.

The remaining sections are devoted to algebraic semantics. Section 4 contains some preliminaries on lattices, section 5 discusses algebraic semantics ("Heyting-valued" semantics) for **IPC**, and section 6 extends this to **IQC**; some applications are given. In particular, familiarity with some of the material in sections 4–6 is assumed in the next chapter. Section 7 presents validity as forcing, section 8 discusses realizability models.

1. Beth models

In this section we discuss Beth models and their connection with intuitionistic validity. Beth models were introduced by Beth (1956, 1959).

1.1. DEFINITION. A *Beth model* for the language $\mathscr{L}$ of pure predicate logic is a quadruple $\mathscr{B} \equiv (\mathrm{K}, \preccurlyeq, \mathrm{D}, \Vdash)$ such that

(i) $(\mathrm{K}, \preccurlyeq)$ is a spread, $\preccurlyeq$ the standard ordering of the nodes determined by the initial-segment relation (we shall use $k, k', k'', \ldots$ for elements of K);

(ii) the domain-function D assigns inhabited sets to elements k of K such that if $k \preccurlyeq k'$, then $\mathrm{D}(k) \subset \mathrm{D}(k')$;

(iii) the forcing relation $\Vdash$ is a binary relation between elements of K and prime sentences $P(d_1, \ldots, d_n)$ with constants from $\mathrm{D} := \bigcup\{\mathrm{D}(k) : k \in \mathrm{K}\}$ such that

B1 $$k \Vdash P(d_1, \ldots, d_n) \Leftrightarrow \forall \alpha \in k \, \exists m(\bar{\alpha}m \Vdash P(d_1, \ldots, d_n)) \quad \text{and}$$

$$d_1, \ldots, d_n \in \mathrm{D}(k); \quad k \nVdash \bot \ \text{for all } k \in \mathrm{K}.$$

Here α ranges over the infinite branches of $(\mathrm{K}, \preccurlyeq)$. Note that B1 implies monotonicity of forcing, i.e. if $k \Vdash P$ and $k' \succcurlyeq k$, then $k' \Vdash P$.

For *compound* sentences, we extend $\Vdash$ by the clauses:

B2 $$k \Vdash A \wedge B := k \Vdash A \quad \text{and} \quad k \Vdash B;$$

B3 $$k \Vdash A \vee B := \forall \alpha \in k \, \exists n(\bar{\alpha}n \Vdash A \quad \text{or} \quad \bar{\alpha}n \Vdash B)$$

B4 $$k \Vdash A \to B := \forall k' \succcurlyeq k(k' \Vdash A \Rightarrow k' \Vdash B)$$

B5 $$k \Vdash \exists x A(x) := \forall \alpha \in k \, \exists n \exists d \in \mathrm{D}(\bar{\alpha}n)(\bar{\alpha}n \Vdash A(d))$$

B6 $$k \Vdash \forall x A(x) := \forall k' \succcurlyeq k \, \forall d \in \mathrm{D}(k')(k' \Vdash A(d)). \quad \square$$

We have the following lemma.

1.2. LEMMA. *For all sentences A:*

(i) $\forall \alpha \in k \, \exists n(\bar{\alpha}n \Vdash A) \Rightarrow k \Vdash A$ (*covering property*), *which is classically equivalent to* $k \nVdash A \Rightarrow \exists \alpha \in k \, \forall n(\bar{\alpha}n \nVdash A)$;

(ii) $k' \succcurlyeq k$ *and* $k \Vdash A \Rightarrow k' \Vdash A$ (*monotonicity*).

PROOF. Induction on A. $\square$

Observe that, for a Beth model with constant domain D, B6 simplifies to

$$k \Vdash \forall x A(x) := \forall d \in \mathrm{D}(k \Vdash A(d)).$$

1.3. REMARKS. (i) Instead of regarding $k \Vdash A$ as false if $A \notin \mathscr{L}(\mathrm{D}(k))$, we may choose to regard the meaning of $k \Vdash A$ as unspecified if A is not in $\mathscr{L}(\mathrm{D}(k))$ – this does not affect the forcing relation for sentences of $\mathscr{L}$. B1 should then be stated as

If $P(d_1, \ldots, d_n) \in \mathscr{L}(\mathrm{D}(k))$, then

$$k \Vdash P(d_1, \ldots, d_n) \Leftrightarrow \forall \alpha \in k\, \exists m (\bar{\alpha} m \Vdash P(d_1, \ldots, d_n)),$$

$$k \nVdash \bot \quad \text{for all } k \in \mathrm{K}.$$

(ii) A model for a language $\mathscr{L}(\mathrm{C})$ instead of $\mathscr{L}$ is obtained by fixing an interpretation $\mathrm{c} \in \mathrm{D}(\langle\ \rangle)$, for each $c \in \mathrm{C}$.

(iii) In Beth's original definition $(\mathrm{K}, \preccurlyeq)$ is always a fan, and the domain function is constant. Van Dalen (1984) introduced Beth models with expanding domains.

(iv) Obviously the definition above also makes sense if the elements of K are finite sequences of elements of an arbitrary set.

The notion of a Beth model may also be generalized to arbitrary partially ordered $(\mathrm{K}, \leq)$, with the role of the α played by maximal linearly ordered subsets.

1.4. *Informal interpretation of forcing in Beth models.* Beth forcing permits an informal interpretation akin to the "information-content" interpretation for Kripke models described in 2.5.1. In a Kripke model, "higher" in the partial ordering corresponds to "more information". In Beth models one may think of the nodes representing points in time: a higher node corresponds to a later moment in time, and different incomparable nodes represent different possibilities for extending our information in the future. (We certainly do not claim this informal reading to be directly connected with "intuitionistic validity").

Compare e.g. the clauses for disjunction: $k \Vdash A \vee B$ in a Kripke model means: with the state of information represented by node k, I know that I can decide between A and B, hence $k \Vdash A$ or $k \Vdash B$.

In a Beth model, $k \Vdash A \vee B$ means: at stage k, I know that I can decide in the future between $A \vee B$, no matter how my information will be extended in the future: so every branch passing through k must permit at some stage the decision between A and B. The other forcing clauses can be paraphrased in a similar way.

This suggests a method of constructing a Beth model from a Kripke model; each node k of the Kripke model is repeated indefinitely in the Beth model, since we may have to move to the next moment in time without an increase in information.

Thus, to the Kripke model on the left corresponds the Beth model on the right. See fig. 13.1.

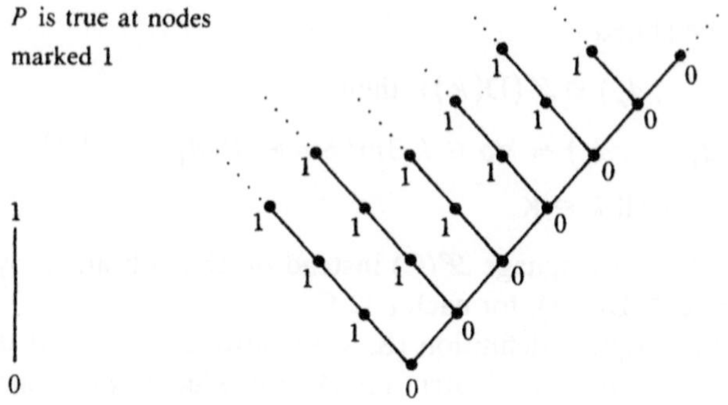

Fig. 13.1.

This suggests the formal connection described below.

1.5. *Transformation of a Kripke model into a Beth model.* Let $\mathscr{K} \equiv (\mathrm{K}, \leq_{\mathrm{K}}, \Vdash, \mathrm{D})$ be a Kripke model; we transform this into a Beth model $\mathscr{B} \equiv (\mathrm{K}', \preccurlyeq', \Vdash', \mathrm{D}')$ as follows:

(i) K' consists of all finite nondecreasing sequences in $(\mathrm{K}, \leq_{\mathrm{K}})$ of length > 0.

(ii) $\sigma \preccurlyeq' \tau$ iff σ is an initial segment of τ;

(iii) $\mathrm{D}'(\langle k_0, \ldots, k_n \rangle) := \mathrm{D}(k_n)$;

(iv) $\langle k_0, \ldots, k_n \rangle \Vdash' P := k_n \Vdash P$, for P an atomic sentence in the language $\mathscr{L}(\mathrm{D}(k_n))$. □

The last condition extends to all sentences.

THEOREM. *For all* $A \in \mathscr{L}(\mathrm{D}(k_n))$,

$$\langle k_0, \ldots, k_n \rangle \Vdash' A \quad \textit{iff} \quad k_n \Vdash A,$$

and for $A \in \mathscr{L}$

$$\mathscr{B} \Vdash' A \quad \textit{iff} \quad \mathscr{K} \Vdash A.$$

PROOF. By a straightforward formula induction. Consider for example the case of $\vee$. Let $\sigma \equiv \langle k_0, \ldots, k_p \rangle$, then

$$\sigma \Vdash' A \vee B \Leftrightarrow \forall \alpha \in \sigma\, \exists x (\bar{\alpha} x \Vdash' \quad \text{or} \quad \bar{\alpha} x \Vdash' B)$$

$$\Leftrightarrow \forall \alpha \in \sigma\, \exists x > \mathrm{lth}(\sigma) (\bar{\alpha} x \Vdash' A \quad \text{or} \quad \bar{\alpha} x \Vdash' B).$$

Applying the final assertion to $\alpha \equiv \sigma * \lambda y.k_p$ we find $\bar{\alpha}x \equiv \sigma * \langle k_p, \ldots, k_p \rangle$ with $\bar{\alpha}x \Vdash' A$ or $\bar{\alpha}x \Vdash' B$, hence by the induction hypothesis $k_p \Vdash A$ or $k_p \Vdash B$, i.e. $k_p \Vdash A \vee B$. The converse is immediate. The rest of the proof we shall leave as an exercise. □

1.6. *Transformation of a Kripke model into a Beth model with constant domain.* By a more sophisticated transformation due to Kripke (1965) we can even transform a countable Kripke model into a Beth model with domain $\mathbb{N}$ as follows. Let $\mathscr{K} \equiv (\mathrm{K}, \leq, \Vdash, \mathrm{D})$ be a Kripke model, and let $(\mathrm{K}', \preccurlyeq')$ be as above.

Let $\mathbb{N}$ be split into countably many disjoint countably infinite sets N_i, so $\mathbb{N} = \bigcup\{\mathrm{N}_i : i \in \mathbb{N}\}$ and $i \neq j \rightarrow \mathrm{N}_i \cap \mathrm{N}_j = \emptyset$. $\mathrm{M}_i = \bigcup\{\mathrm{N}_j : j \leq i\}$. We define for any $\sigma \equiv \langle k_0, \ldots, k_n \rangle \in \mathrm{K}'$ a surjection ψ_σ from M_n onto $\mathrm{D}(k_n)$ such that

$$\psi_{\sigma * \langle k \rangle}(m) = \psi_\sigma(m) \quad \text{for } m \in \mathrm{M}_{\mathrm{lth}(\sigma)-1},$$

$$\psi_{\sigma * \langle k \rangle} \text{ maps } \mathrm{N}_{\mathrm{lth}(\sigma)} \text{ onto } \mathrm{D}(k).$$

The new Beth model $(\mathrm{K}', \leq', \Vdash', \mathbb{N})$ is then specified by

$$\sigma \Vdash' P(n_0, \ldots, n_p) := k \Vdash P(\psi_\sigma n_0, \ldots, \psi_\sigma n_p)$$

and we can prove the following theorem.

THEOREM. *For the transformation described above*

$$\sigma \Vdash' A(n_0, \ldots, n_p) \Leftrightarrow k_n \Vdash A(\psi_\sigma n_0, \ldots, \psi_\sigma n_p).$$

PROOF. We consider one of the crucial inductive steps. Suppose $\sigma = \langle k_0, \ldots, k_n \rangle$, $A(n') \equiv \forall x B(x, n')$. If $\sigma \Vdash' A(n')$, then $\forall \tau \geq' \sigma \forall m (\tau \Vdash' B(m, n'))$. Choose $\tau = \sigma * \langle k \rangle$, $k \geq k_n$, then $\forall m (\sigma * \langle k \rangle \Vdash' B(m, n'))$, $n' \in \mathrm{M}_{\mathrm{lth}(\sigma)-1}$, so $\psi_\sigma(n') = \psi_{\sigma * \langle k \rangle}(n') = d \in \mathrm{D}(k_n)$ for some d. By the induction hypothesis, $\forall d' \in D(k)(k \Vdash B(d', d))$ (let m range over $\mathrm{N}_{\mathrm{lth}(\sigma)}$), hence $k_n \Vdash \forall x B(x, d)$, so $k_n \vdash \forall x B(x, \psi_\sigma n')$.

Conversely, suppose $k_n \Vdash \forall x B(x, \psi_\sigma n')$; because of the constant domain, it suffices to show $\forall n (\sigma \Vdash B(n, n'))$. So pick n and let $\tau = \sigma * \langle k_n, \ldots, k_n, k' \rangle$ be so long that $n \in \mathrm{M}_{\mathrm{lth}(\tau)-1}$. ψ_τ maps $\mathrm{M}_{\mathrm{lth}(\tau)-1}$ onto $\mathrm{D}(k')$; $\psi_\tau n' = \psi_\sigma n'$. Thus we find $\forall d \in \mathrm{D}(k')(k' \Vdash B(d, d'))$, therefore $k' \Vdash B(\psi_\tau n, \psi_\tau n')$, and thus by the induction hypothesis $\tau \Vdash' B(n, n')$. This holds for all τ with $n \in \mathrm{M}_{\mathrm{lth}(\tau)-1}$, hence $\sigma \Vdash' B(n, n')$, by the covering property of Beth models. Therefore $\sigma \Vdash' \forall x B(x, n')$.

We leave the other cases as an exercise. □

REMARK. We cannot expect a similar transformation of Beth models into Kripke models (cf. Lopez-Escobar 1981), since there are Beth models in which $\Vdash A \vee B$, while $\nVdash A$ and $\nVdash B$; this situation cannot arise in Kripke models.

1.7. *Intuitionistic validity.* We shall now study the connection between intuitionistic validity and truth in Beth models. Intuitionistic validity was already briefly discussed in 2.5.1; we recapitulate the definition. Let $F(R_1, \ldots, R_n)$ be any formula of pure predicate logic, containing the predicate symbols $R_1, \ldots, R_n$ (and no others) and suppose R_i to have n_i argument places.

We shall define "*F is intuitionistically valid*" ("valid" for short, or $\models_i F$) by

$$\models_i F(R_1, \ldots, R_n) := \forall \mathrm{D} \forall \mathrm{R}_1 \ldots \mathrm{R}_n F^{\mathrm{D}}(\mathrm{R}_1, \ldots, \mathrm{R}_n),$$

where D ranges over all intuitionistically meaningful domains, and R_i ranges over all n_i-ary relations over D; $F^{\mathrm{D}}(\mathrm{R}_1, \ldots, \mathrm{R}_n)$ results from F by relativizing all quantifiers to D, and substituting R_i for R_i.

Of course, the logical operations in the right-hand formula should be read intuitionistically. This is a definition completely similar to Tarski's definition of truth. As it stands, there seem to be two sources of vagueness: the extent of the notions of intuitionistic domain and relation, and the vagueness inherent to the intuitionistic reading of the logical operations. Notwithstanding this, a few assumptions on possible domains, relations, and in particular the admission of lawless sequences as mathematical objects, will enable us to get some quite definite results on the extent of $\Vdash_i$ in the next section. Compare the classical case: there too, for a completeness theorem we are not required to consider all possible domains and relations: $\mathbb{N}$ with Δ^0_2-definable relations suffices for countable languages! (See e.g. Kleene 1952, §72.)

1.8. *Weak and strong definitions of forcing in Beth models.* Clause B1 states that if all branches through k pass through a node where $P' \equiv P(d_1, \ldots, d_n)$ holds, i.e. the nodes $\{k' : k' \Vdash P'\}$ bar k (in the terminology of 4.8.18), then $k \Vdash P'$. Clause B3 expresses that $k \Vdash A \vee B \Leftrightarrow \{k' : k' \Vdash A$ or $k' \Vdash B\}$ bars k, and similarly for B5. These clauses can be expressed in a stronger way if we accept the neighbourhood function principle NFP

(4.6.12) and $K = K_0$; for example, the first part of B1 may then be replaced by the equivalent

$$k \Vdash P \Leftrightarrow \exists e \in K\, \forall k'(ek' \neq 0 \rightarrow \exists k'' \preccurlyeq k * k'(k'' \Vdash P) \quad \text{and}$$

$$P \in \mathscr{L}(\mathrm{D}(k)).$$

If $(\mathrm{K}, \preccurlyeq)$ is a fan, we can use the following alternative versions of B1, B3, B5, using as an abbreviation $\forall k' \succcurlyeq_z k(\ldots) := \forall k' \succcurlyeq k(\mathrm{lth}(k') - \mathrm{lth}(k) = z \rightarrow \ldots)$:

B1′ $\quad k \Vdash P \Leftrightarrow \exists z \forall k' \succcurlyeq_z k\, \exists k'' \preccurlyeq k'(k'' \Vdash P) \quad \text{and} \quad P \in \mathscr{L}(\mathrm{D}(k))$;

B3′ $\quad k \Vdash A \vee B := \exists z \forall k' \succcurlyeq_z k(k' \Vdash A \quad \text{or} \quad k' \Vdash B)$;

B5′ $\quad k \Vdash \exists x A(x) := \exists z \forall k' \succcurlyeq_z k\, \exists d \in \mathrm{D}(k')(k' \Vdash A(d))$.

B1′ may be split into two conditions:

B1a′ $\quad \exists z \forall k' \succcurlyeq_z k(k' \Vdash P) \Rightarrow k \Vdash P$,

B1b′ $\quad k \Vdash P \quad \text{and} \quad k' \succcurlyeq k \Rightarrow k' \Vdash P$.

A slightly different version of these clauses is

B1″ $\quad k \Vdash P \Leftrightarrow \exists m\, \forall \alpha \in k\, \exists n \leq m(\bar{\alpha}n \Vdash P) \quad \text{and} \quad P \in \mathscr{L}(\mathrm{D}(k))$,

B3″ $\quad k \Vdash A \vee B := \exists m\, \forall \alpha \in k(\bar{\alpha}m \Vdash A \quad \text{or} \quad \bar{\alpha}m \Vdash B)$,

B5″ $\quad k \Vdash \exists x A(x) := \exists m\, \forall \alpha \in k\, \exists d \in \mathrm{D}(\bar{\alpha}m)(\bar{\alpha}m \Vdash A(d))$.

If we assume α to range over branches *dense* in the tree $(\mathrm{K}, \preccurlyeq)$, i.e. $\forall k \exists \alpha(\alpha \in k)$, then B1″, B3″, and B5″ are equivalent to B1′, B3′, and B5′, respectively. If we also assume FAN, then B1, B3, B5 are equivalent to B1′, B3′, and B5′, respectively.

It is to be noted that B1′, B3′, B5′ do not refer to the infinite branches of the tree. We shall call the definition of Beth model and Beth forcing based on B1–B6 the *weak* definition, and the definition based on B1′, B2, B3′, B4, B5′ and B6 the *strong* definition.

1.9. *Intuitionistic validity and Beth models.* Let $\mathscr{B} \equiv (\mathrm{K}, \preccurlyeq, \Vdash, \mathrm{D})$ be a Beth model with constant domain, and let $\mathrm{LS_K}$ be the collection of lawless branches in $(\mathrm{K}, \preccurlyeq)$. Let $\langle R_n \rangle_n$ be an enumeration of all proposition and relation symbols of the language; R_n has $r(n)$ argument places. Let $\langle \mathrm{R}_n \rangle_n$ be a sequence of relations, $\mathrm{R}_n \subset \mathrm{LS_K} \times \mathrm{D}^{r(n)}$. We write $\mathrm{R}^\alpha_n(\vec{t})$ for $\mathrm{R}_n(\alpha, \vec{t})$, and we stipulate that $\mathscr{B}$ and $\langle \mathrm{R}_n \rangle_n$ are related as follows:

$$\exists x(\bar{\alpha}x \Vdash R_n(d_1,\ldots,d_{r(n)})) \quad \text{iff} \quad \mathrm{R}^\alpha_n(d_1,\ldots,d_{r(n)}). \tag{1}$$

If we take for $\mathscr{B}$ the *weak* definition of forcing, and for $\mathrm{LS_K}$ we assume "open data" and "density" (the analogues of LS3 and LS1), then (1) is equivalent to

$$k \Vdash R_n(d_1,\ldots,d_{r(n)}) \quad \text{iff} \quad \forall\alpha \in k\, \mathrm{R}^\alpha_n(d_1,\ldots,d_{r(n)}). \tag{2}$$

We leave the proof of this fact as an exercise.

REMARKS. (i) It will be clear that to each Beth model $\mathscr{B}$ with constant domain we can assign a sequence $\langle \mathrm{R}_n \rangle_n$ by (1), and conversely to each $\langle \mathrm{R}_n \rangle_n$ we can construct a Beth model such that (2) holds, and these constructions are inverse to each other. For any branch $\alpha \in \mathrm{LS_K}$ of the Beth model, $\langle \mathrm{R}^\alpha_n \rangle_n$ is an intuitionistic interpretation (model) $\mathscr{M}_\alpha$ of the language, such that for prime sentences P

$$k \Vdash P \Leftrightarrow \forall\alpha \in k(\mathscr{M}_\alpha \models_{\mathrm{i}} P).$$

(ii) Of course, the equivalence between (1) and (2) also holds if we adopt the *strong* definition of forcing of $\mathscr{B}$, and FAN (4.7.3, without additional choice parameters) for $\mathrm{LS_K}$, in addition to open data (LS3(1) in 12.2.6) and density (12.2.2).

For arbitrary formulas we have the following theorem describing the relation between intuitive validity and Beth forcing.

1.10. PROPOSITION. *Let* $\mathrm{D} = \mathbb{N}$ *and let* $(\mathrm{K}, \preccurlyeq)$ *be a fan. Assume the Beth model* $\mathscr{B} \equiv (\mathrm{K}, \preccurlyeq, \Vdash, \mathrm{D})$ *and the sequences* $\langle \mathrm{R}_n \rangle_n$ *to be related by* (1) *or* (2) *of* 1.9.

(i) *Let* $\mathscr{B}$ *be based on the strong definition, and assume the range of the variable* α *to satisfy density, open data (in a single variable), and the fan axiom (without choice parameters). Then for all* A

$$(*) \qquad \begin{cases} \exists n(\bar{\alpha}n \Vdash A) \Leftrightarrow A^\alpha, \\ k \Vdash A \Leftrightarrow \forall\alpha \in k(A^\alpha), \end{cases}$$

where A^α *is obtained from* A *by substitution of* R^α_n *for* R_n *in* A *everywhere.*

(ii) $(*)$ *also holds if* $\mathscr{B}$ *is based on the weak definition, and the range of the variable* α *is supposed to satisfy density and open data in a single variable.*

PROOF *of* (i). The forcing clauses for $k \Vdash A$ correspond literally to the clauses of the elimination translation for binary lawless sequences, restricted to formulae of the form $\forall \alpha \in k\, A^{\alpha}$; the clauses of the translation preserve logical equivalence precisely when density, open data, and the fan axiom are assumed.

The proof of (ii) is left as an exercise. □

2. Completeness for intuitionistic validity

2.1. In this section we shall present a *constructive* completeness proof for Beth semantics based on the strong definition of Beth forcing. The proof is due to Friedman and applies to the following three cases:

(i) the $\bot$ -free fragment of **IQC**,

(ii) **MQC**, and

(iii) **IQC** relative to a nonstandard notion of Beth model (*fallible Beth model*), where $\bot$ may be true at certain nodes.

As to case (ii), in a Beth model for **MQC** we permit $\bot$ to be true at certain nodes, provided $\bot$ is not true in the whole model; this corresponds to the idea that in minimal logic $\bot$ plays the role of an arbitrary unprovable proposition.

We shall only present the proof for languages without function symbols. For the case of languages with function symbols, see E13.2.2.

The proof produces in each of the cases (i)–(iii) a Beth model $\mathscr{B}$ which is universal, in the sense that for sentences A in the language and logic (**IQC** or **MQC**) under consideration, $\vdash A \Leftrightarrow \mathscr{B} \Vdash A$ (A is Beth-valid in the model). The first step in constructing $\mathscr{B}$ is the construction of a certain labelled binary tree.

2.2. *Construction of the labelled binary tree.* Let $\mathscr{L}$ be a language without function symbols and with at least one predicate symbol with arity at least one. Let $\mathrm{C} \equiv \{c_n : n \in \mathbb{N}\}$ be a collection of constants not in $\mathscr{L}$. Let $\mathscr{L}' := \mathscr{L}(\mathrm{C}) = \mathscr{L} \cup \mathrm{C}$. For any set of formulas Γ, let $\mathscr{L}(\Gamma)$ be the language containing only constants from C occurring in formulas of Γ. Let $\langle A_n \rangle_n$ be an enumeration with infinite repetitions, i.e. $\forall n \exists m > n(A_m \equiv A_n)$, such that in case (i) all $\bot$ -free sentences of $\mathscr{L}'$ are enumerated and in cases (ii) and (iii) all sentences of $\mathscr{L}'$ are enumerated.

The Beth model to be constructed is based on the binary tree of finite 01-sequences; we shall let k, k' range over such sequences. To each node k we are going to assign a finite set Γ_k of formulas of $\mathscr{L}'$, the *label* of k. Let

$\Gamma \vdash_m A$ mean that A is deducible from Γ by a deduction with code number less than m, where deducibility is interpreted in the cases (i) and (iii) as deducibility in **IQC**, and in case (ii) as deducibility in **MQC**. (Observe that $\Gamma \vdash_m A$ is decidable).

The labels are now defined by induction on the length of the nodes as follows. We put $\Gamma_{\langle\,\rangle} = \emptyset$.

Let $\mathrm{lth}(k) = u$. In the definition of $\Gamma_{k*\langle i\rangle}$ we consider four cases:

(i) $A_u \notin \mathscr{L}(\Gamma_k)$. Take $\Gamma_{k*\langle 0\rangle} \equiv \Gamma_{k*\langle 1\rangle} := \Gamma_k$.

(ii) $A_u \in \mathscr{L}(\Gamma_k)$, $A_u \equiv B \vee C$, $\Gamma_k \vdash_u B \vee C$. Take $\Gamma_{k*\langle 0\rangle} := \Gamma_k \cup \{B\}$, $\Gamma_{k*\langle 1\rangle} := \Gamma_k \cup \{C\}$.

(iii) $A_u \in \mathscr{L}(\Gamma_k)$, $A_u \equiv \exists x B(x), \Gamma_k \vdash_u \exists x B(x)$. Let c_i be the first constant of C not in $\Gamma_k \cup \{B(x)\}$ and take $\Gamma_{k*\langle 0\rangle} \equiv \Gamma_{k*\langle 1\rangle} := \Gamma_k \cup \{B(c_i)\}$.

(iv) If (i)–(iv) do not apply, take $\Gamma_{k*\langle 0\rangle} := \Gamma_k$, $\Gamma_{k*\langle 1\rangle} := \Gamma_k \cup \{A_u\}$.

REMARK. Since for some formula Bx with $\mathrm{FV}(B) = \{x\}$, $B' \equiv \exists x(Bx \to Bx)$ is provable, and this B' is repeated infinitely often in $\langle A_n \rangle_n$, ultimately each c_i will occur in $\mathscr{L}(\Gamma_k)$, provided k is sufficiently long; it is easily seen that a bound on $\mathrm{lth}(k)$ can be given, depending on i, which guarantees that $c_i \in \mathscr{L}(\Gamma_k)$.

An alternative possibility would have been to define $\mathscr{L}(\Gamma_k)$ as $\mathscr{L}_{\mathrm{lth}(k)}$, where $\mathscr{L}_0 = \mathscr{L}$, $\mathscr{L}_n = \mathscr{L} \cup \{c_0, c_1, \ldots, c_{n-1}\}$; then we can always take c_u for c_i in clause (iii).

Below we shall write $\vdash \Gamma \to A$ for $\vdash \bigwedge \Gamma \to A$, and $\forall k' \succcurlyeq_x k(\ldots)$ for $\forall k' \succcurlyeq k(\mathrm{lth}(k') = \mathrm{lth}(k) + x \to \ldots)$; the x is introduced for use in proofs by induction.

2.3. LEMMA. *For all $A \in \mathscr{L}(\Gamma_k)$ and all $x \in \mathbb{N}$*

$$(\vdash \Gamma_k \to A) \Leftrightarrow \forall k' \succcurlyeq_x k(\vdash \Gamma_{k'} \to A).$$

PROOF. By induction on x. The case $x = 0$ is trivial; the basis is the case $x = 1$, for which we have to show, for arbitrary k and A,

$$(\vdash \Gamma_k \to A) \Leftrightarrow (\vdash \Gamma_{k*\langle 0\rangle} \to A) \quad \text{and} \quad (\vdash \Gamma_{k*\langle 1\rangle} \to A).$$

Let us check the clauses of the definition of $\Gamma_{k*\langle i\rangle}$ in 2.2.

Clause (i). Obvious.

Clause (ii). Let $A_k \equiv B_1 \vee B_2, \Gamma_k \vdash B_1 \vee B_2$, so $\Gamma_{k*\langle i\rangle} = \Gamma_k \cup \{B_i\}$. If $\Gamma_k \vdash A$, then a fortiori $\vdash \Gamma_k \wedge B_i \to A$. Conversely, if $\vdash \Gamma_k \wedge B_i \to A$ for $i = 0, 1$, then by logic $\vdash \Gamma_k \wedge (B_1 \vee B_2) \to A$; and since in this case $\Gamma_k \vdash B_1 \vee B_2$, also $\vdash \Gamma_k \to A$.

Clause (iii). $\Rightarrow$ trivial. For $\Leftarrow$ assume $\vdash \Gamma_k \wedge B(c_i) \to A$. Since $A \in \mathscr{L}(\Gamma_k)$, $c_i \notin \mathscr{L}(\Gamma_k)$, we have $\vdash \Gamma_k \wedge \exists x B(x) \to A$, and therefore $\vdash \Gamma_k \to A$, since $\Gamma_k \vdash \exists x B(x)$.

Clause (iv). Trivial, use $\vdash \Gamma_{k*\langle 0\rangle} \to A$.

Induction now yields the statement of the lemma. □

Note that we immediately have $\vdash \Gamma_k \to A \Leftrightarrow \forall k' \succ k(\vdash \Gamma_{k'} \to A)$.

2.4. Definition. In the Beth model $\mathscr{B}^* \equiv (\mathrm{K}, \preccurlyeq, \mathrm{C}, \Vdash)$ $(\mathrm{K}, \preccurlyeq)$ is the binary tree of finite 01-sequences, $\mathrm{C} \equiv \{c_n : n \in \mathbb{N}\}$ is the constant domain which may be identified with $\mathbb{N}$, and $\Vdash$ for prime sentences is given by

$$k \Vdash P := \vdash \Gamma_k \to P. \quad \square$$

From lemma 2.3 we see by an easy inductive argument that this indeed defines a Beth model.

N.B. The model construction depends on the enumeration of deductions and formulas used.

2.5. Lemma. In the model $\mathscr{B}^*$ for all sentences $A \in \mathscr{L}(\Gamma_k)$

$$k \Vdash A \quad \Leftrightarrow \quad \vdash \Gamma_k \to A.$$

Proof. By induction on A.

Case 1. For A prime the assertion holds by definition.

Case 2. Let $A \equiv B \vee C$. Suppose $k \Vdash B \vee C$, then by the definition of $\Vdash$ for some $x \in \mathbb{N}$

$$\forall k' \succcurlyeq_x k(k' \Vdash B \text{ or } k' \Vdash C).$$

Hence by the induction hypothesis

$$\forall k' \succcurlyeq_x k(\vdash \Gamma_{k'} \to B \text{ or } \vdash \Gamma_{k'} \to C),$$

and thus

$$\forall k' \succcurlyeq_x k(\vdash \Gamma_{k'} \to B \vee C),$$

and therefore $\vdash \Gamma_k \to B \vee C$ by 2.3.

Conversely, let $\vdash \Gamma_k \to B \vee C$, then $\Gamma_k \vdash_x B \vee C$ for some $x \in \mathbb{N}$. Let

$$y := \min_z[z > \max(\mathrm{lth}(k), x) \wedge A_z \equiv B \vee C] \dot{-} \mathrm{lth}(k),$$

then

$$\forall k' \succcurlyeq_{y+1} k(B \in \Gamma_{k'} \text{ or } C \in \Gamma_{k'}),$$

hence

$$\forall k' \succcurlyeq_{y+1} k((\vdash \Gamma_{k'} \to B) \text{ or } (\vdash \Gamma_{k'} \to C)),$$

therefore (induction hypothesis)

$$\forall k' \succcurlyeq_{y+1} k(k' \Vdash B \text{ or } k' \Vdash C),$$

and thus $k \Vdash B \vee C$.

Case 3. Let $A \equiv \exists x B(x)$. Assume $k \Vdash \exists x B(x)$, then for some $y \in \mathbb{N}$

$$\forall k' \succcurlyeq_y k \exists c_i(k' \Vdash B(c_i)).$$

For sufficiently large y, we have $B(c_i) \in \mathscr{L}(\Gamma_{k' * k''})$ for all k'' of length y, and thus we can arrange that for some $z \geq x$

$$\forall k' \succcurlyeq_z k \exists c_i \in \mathscr{L}(\Gamma_{k'})(k' \Vdash B(c_i)),$$

and therefore by the induction hypothesis

$$\forall k' \succcurlyeq_z k \exists c_i(\vdash \Gamma_{k'} \to B(c_i)),$$

so

$$\forall k' \succcurlyeq_z k(\vdash \Gamma_{k'} \to \exists x B(x)),$$

and therefore, by lemma 2.3, $\vdash \Gamma_k \to \exists x B(x)$.

Let conversely $\vdash \Gamma_k \to \exists x B(x)$, then $\Gamma_k \vdash_y \exists x B(x)$ for some $y \in \mathbb{N}$. Let

$$z := \min_u[u > y \wedge u > \mathrm{lth}(k) \wedge (A_u \equiv \exists x B(x))] \dot{-} \mathrm{lth}(k),$$

then for $\mathrm{lth}(k') = z$

$$\Gamma_{k * k' * \langle j \rangle} = \Gamma_{k * k'} \cup \{B(c_i)\},$$

where $c_i \in \mathscr{L}(\Gamma_{k * k' * \langle j \rangle}) \backslash \mathscr{L}(\Gamma_{k * k'})$. Hence $\vdash \Gamma_{k * k' * \langle j \rangle} \to B(c_i)$, and since $c_i \in \mathscr{L}(\Gamma_{k * k' * \langle j \rangle})$ we have by the induction hypothesis

$$k * k' * \langle j \rangle \Vdash B(c_i).$$

Therefore $k * k' * \langle j \rangle \Vdash \exists x Bx$, and since this holds for all k' of length z and all $j \in \{0, 1\}$, we have $k \Vdash \exists x Bx$.

Case 4. Let $A \equiv B \to C$, and assume $k \Vdash B \to C$, i.e.

$$\forall k'(k * k' \Vdash B \Rightarrow k * k' \Vdash C).$$

Let

$$x = \min_y \left[y > \mathrm{lth}(k) \wedge \left(A_y \equiv B \vee B \right) \right] \dot{-} \mathrm{lth}(k),$$

then for any k' of length x

$$\Gamma_{k * k'} \vdash_x B \vee B \quad \text{or} \quad \Gamma_{k * k'} \nvdash_x B \vee B.$$

In the first case $B \in \Gamma_{k * k' * \langle 0 \rangle} = \Gamma_{k * k' * \langle 1 \rangle}$, in the second case $\Gamma_{k * k' * \langle 1 \rangle} \equiv \Gamma_{k * k'} \cup \{ B \vee B \}$, and therefore, for all k' of length x, $\vdash \Gamma_{k * k' * \langle 1 \rangle} \to B$. By the induction hypothesis, $k * k' * \langle 1 \rangle \Vdash B$, hence $k * k' * \langle 1 \rangle \Vdash C$; therefore, again by the induction hypothesis, $\vdash \Gamma_{k * k' * \langle 1 \rangle} \to C$. So $\vdash \Gamma_{k * k'} \to (B \to C)$ for all k' of length x, hence $\vdash \Gamma_k \to (B \to C)$.

Conversely, let $\vdash \Gamma_k \to (B \to C)$; if $k' \succcurlyeq k$, $k' \Vdash B$, then $\vdash \Gamma_{k'} \to B$, hence $\vdash \Gamma_{k'} \to C$, and thus $k' \Vdash C$; so $k \Vdash B \to C$.

Case 5. Let $A \equiv \forall x B(x)$. Assume $k \Vdash \forall x B(x)$ and suppose $\vdash_y \exists x(Cx \to Cx)$ for some formula C with $\mathrm{FV}(C) = \{x\}$, then $A_z \equiv \exists x(Cx \to Cx)$ for some $z > y, \mathrm{lth}(n)$. Then for all $k' \succcurlyeq k$ with $\mathrm{lth}(k') = z$ there is a $c_{i(k')}$ such that for all k'' of the form $k' * \langle j \rangle$ we have $B(c_{i(k')}) \in \mathscr{L}(\Gamma_{k''}) \setminus \mathscr{L}(\Gamma_{k'})$, and thus $k'' \Vdash B(c_{i(k')})$.

By the induction hypothesis $\vdash \Gamma_{k''} \to B(c_{i(k')})$, hence $\vdash \Gamma_{k'} \to B(c_{i(k')})$; it follows that $\vdash \Gamma_{k'} \to \forall x B(x)$, since $c_{i(k')} \notin \mathscr{L}(\Gamma_{k'})$. This holds for all $k' \succ k$ of length z, and therefore $\vdash \Gamma_k \to \forall x B(x)$.

Conversely, let $\vdash \Gamma_k \to \forall x B(x)$, then $\vdash \Gamma_k \to B(c_i)$ for all $c_i \in \mathrm{C}$. Fix i; choose y such that $B(c_i) \in \mathscr{L}(\Gamma_{k * k'})$ for all k' of length y, then from $\forall k' \succcurlyeq_y k(\vdash \Gamma_k \to B(c_i))$ we obtain $\forall k' \succcurlyeq_y k(k' \Vdash B(c_i))$ (induction hypothesis). Therefore $k \Vdash B(c_i)$ by lemma 2.3. This holds for all i, therefore $k \Vdash \forall x B(x)$. $\square$

N.B. we have used the strong forcing clauses for prime formulas, $\vee$ and $\exists$.

We now immediately conclude the following.

2.6. Theorem.

(i) *For the $\bot$-free fragment of* **IQC** *there is a universal Beth model $\mathscr{B}_{\mathrm{i}}$ such that for all sentences A*

$$\mathscr{B}_{\mathrm{i}} \Vdash A \quad \textit{iff} \ \vdash A. \tag{1}$$

(ii) *For minimal predicate logic* **MQC** *there is a universal Beth model* $\mathscr{B}_{\mathrm{m}}$ *such that for all sentences A*

$$\mathscr{B}_{\mathrm{m}} \Vdash A \quad \textit{iff} \ \vdash A. \tag{2}$$

In this model $\perp$ *is interpreted as an arbitrary fixed proposition letter, which is not forced in* $\langle\ \rangle$. □

REMARK. Note, that if we consider a theory in $\perp$-free **IQC** or in **MQC**, axiomatized by a recursively enumerable set of sentences Δ, no changes in the arguments for the lemmas 2.2 and 2.3 are necessary; only the meaning of $\vdash_x$ changes. This gives a ready generalization of the theorem to recursively axiomatized theories.

2.7. DEFINITION. A *fallible Beth model* is defined exactly as a Beth model, except that the condition $\forall k(k \nVdash \perp)$ is weakened to $\langle\ \rangle \nVdash \perp$; the forcing of prime sentences P must satisfy $k \Vdash \perp \Rightarrow k \Vdash P$. □

N.B. In a fallible Beth model, $k \Vdash \perp$ entails $k \Vdash A$ for all sentences A.

2.8. THEOREM. *For* **IQC** *there exists a fallible Beth model* $\mathscr{B}_{\mathrm{f}}$ *such that*

$$\mathscr{B}_{\mathrm{f}} \Vdash A \Leftrightarrow \vdash A$$

for all sentences A.

PROOF. Nothing new is involved, the construction proceeds as before, with $\vdash$ interpreted as deducibility in **MQC**, with additional axioms of the form $\perp \to P$ for P prime; cf. the remark in 2.6. □

The result on the connection between Beth validity and intuitionistic validity now immediately leads to the following theorem.

2.9. THEOREM. *Let F be* $\perp$*-free. There is a sequence of relations* $\langle \mathrm{R}_n^\alpha \rangle_n$ *depending on a parameter* α *ranging over lawless* 01*-sequences, such that*

$$\forall \alpha \left[F^{\mathbb{N}}\left(\mathrm{R}_{i(1)}^\alpha, \ldots, \mathrm{R}_{i(n)}^\alpha\right)\right] \Leftrightarrow \vdash F\left(R_{i(1)}, \ldots, R_{i(n)}\right),$$

where $R_{i(k)}$ *is a relation symbol with the same number of arguments as* $\mathrm{R}_{i(k)}^\alpha$. *The* R_n^α *are enumerable in* α, *the other arguments are in* $\mathbb{N}$; *as axioms for lawless parameters we need only* FAN, *density, and open data in a single parameter.* □

Inspection of the proof of lemma 2.5 shows that if we base ourselves on the weak definition of Beth forcing, then we have to appeal to FAN in case

2 and case 3 of the proof. In fact we only need FAN_{D}. Consider e.g. case 2, and suppose $k \Vdash A \vee B$, then

$$\forall \alpha \in k\, \exists x(\bar{\alpha}x \Vdash B \text{ or } \bar{\alpha}x \Vdash C).$$

By the induction hypothesis

$$\forall \alpha \in k\, \exists x(\vdash \Gamma_{\bar{\alpha}x} \to B \text{ or } \vdash \Gamma_{\bar{\alpha}x} \to C),$$

in fact

$$\forall \alpha \in k\, \exists y \exists x \leq y(\vdash_y \Gamma_{\bar{\alpha}x} \to B \text{ or } \vdash_y \Gamma_{\bar{\alpha}x} \to C).$$

With FAN_{D}

$$\exists z \forall \alpha \in k(\vdash \Gamma_{\bar{\alpha}z} \to B \text{ or } \vdash \Gamma_{\bar{\alpha}z} \to C),$$

hence $\exists z \forall \alpha \in k(\vdash \Gamma_{\bar{\alpha}z} \to B \vee C)$, and so with 2.3 $\vdash \Gamma_{\text{k}} \to B \vee C$.

As a result we obtain the following improvement of the preceding theorem.

COROLLARY (*to the analysis of the proof of* 2.5). *The preceding theorem holds with* FAN *replaced by* FAN_{D}. □

2.10. *Classical proof of completeness for Beth semantics.* Essentially the same construction can be used for a classical completeness proof for full **IQC**.

Either we make a simple adaptation in the construction of the model, changing clause (iv) into

(iv′) If (i)–(iii) do not apply, take $\Gamma_{k*\langle 0 \rangle} := \Gamma_k$, and let $\Gamma_{k*\langle 1 \rangle} := \Gamma_k \cup \{A_u\}$ provided $\Gamma_k \cup \{A_u\}$ is consistent, otherwise let $\Gamma_{k*\langle 1 \rangle} := \Gamma_k$,

or we observe, after having constructed the universal fallible Beth model, that we may cut out (delete) all nodes k where $k \Vdash \bot$, without affecting the forcing relation at other nodes.

If we adopt the modified clause (iv′), a slight change in the treatment of implication in the proof of 2.5 is needed; we leave this as an exercise.

2.11. *The axiomatization of the function-free fragment of the theory of a single Skolem function.* Just as in the case of Kripke models, the process of saturation in the Beth model construction described above may be varied for various special applications. As an example, we shall identify the f-free fragment of the theory of a single (unary) Skolem function f.

Let **T** be the theory **IQC** + $\forall x R(x, fx)$ in the language $\mathcal{L}$ with $=$, a single unary function symbol f, and a single binary relation R. Let $\mathcal{L}'$ be

$\mathscr{L}$ without f. **T** is not conservative over **IQC** + $\forall x \exists y R(x, y)$; in fact, we shall show the following.

THEOREM. *The $\mathscr{L}'$-fragment of* **T** *is axiomatized by the axioms*

$$\mathrm{Fun}_0 \qquad \forall x_0 \exists y_0 \ldots \forall x_n \exists y_n \Big(\bigwedge_i R(x_i, y_i) \wedge \bigwedge_{i<j} (x_i = x_j \to y_i = y_j) \Big)$$

for i, j ranging over $\{0, \ldots, n\}$, and $n \in \mathbb{N}$.

PROOF. We have to modify the construction of the fallible Beth model (2.4, 2.7) for **IQC** as follows. Instead of new constants $\{c_n : n \in \mathbb{N}\}$, we now have constants $\{c_n^m : n, m \in \mathbb{N}\}$. The c_n^n take the place of the c_n, that is to say they are used to guarantee saturation for existential formulas. The other constants serve to give uniform saturation for Fun_0, so that in the model we have $R(c_n^m, c_n^{m+1})$ for all n and m, and also $c_n^m = c_{n'}^{m'} \to c_n^{m+1} = c_{n'}^{m'+1}$.

The result will be a Beth model of **IQC** + Fun_0 in which the Skolem function f can be defined by $f(c_n^m) = c_n^{m+1}$, or more accurately

$$\big(k \Vdash f(c_n^m) = c_{n'}^{m'}\big) := k \Vdash c_{n'}^{m'} = c_n^{m+1}.$$

The essential modification is in clause (iii) of the construction of the labelled tree. There we now take $\Gamma_{k * \langle 0 \rangle} \equiv \Gamma_{k * \langle 1 \rangle} := \Gamma_k \cup \{B(c_n^0), \mathrm{Fun}_{n+1}\}$. Here n is the least n' such that $c_{n'}^0$ is not in $\Gamma_k \cup \{B(x)\}$. Fun_{n+1} is the set of all axioms of the form

$$(*) \qquad \forall x_0 \exists y_0 \ldots \forall x_m \exists y_m \Big[\bigwedge_i R(s_i, t_i) \wedge \bigwedge_{i<j} (s_i = s_j \to t_i = t_j) \Big].$$

Here i, j range over $\{0, \ldots, m'\}$ for $m' \geq m$, and (s_i, t_i) is either (x_i, y_i) (for $i \leq m$) or (c_q^p, c_q^{p+1}), where $p, q \leq n$ (and similarly for (s_j, t_j)). The new constants introduced at this stage are therefore $\{c_n^p : p \leq n+1\} \cup \{c_q^{n+1} : q \leq n\}$.

We need lemmas corresponding to 2.3 and 2.5 for our modified construction; to obtain these, we first show that for A not containing the new constants

$$\Gamma_k + \mathrm{Fun}_{n+1} \vdash A \Rightarrow \Gamma_k \vdash A. \tag{1}$$

We observe that Fun_{n+1} is closed under conjunction, in the sense that finitely many axioms $F_1, \ldots, F_r$ from Fun_{n+1} are implied by some F in Fun_{n+1}; so assume $F \vdash A$, where F has the form $(*)$.

Let p be maximal such that c_q^{p+1} occurs in F, c_q^{p+1} not in A. Then c_q^{p+1} can only occur in contexts

$$R\left(c_q^p, c_q^{p+1}\right),$$

$$s_i = c_q^p \to t_i = c_q^{p+1},$$

$$c_q^p = s_j \to c_q^{p+1} = t_j.$$

Choose a new variable y_q^p and replace c_q^{p+1} by y_q^p in F; after an $\exists$-introduction the result is a formula

$$F^* := \exists y_q^p \forall x_0 \exists y_0 \ldots \forall x_m \exists y_m \left[\bigwedge_i R(s_i t_{i'}) \wedge \bigwedge_{i<j} \left(s_i = s_j \to t_{i'} = t_{j'}\right) \right],$$

where $t_{i'}$ is obtained from t_i by replacing c_q^{p+1} by y_q^p. We have $\Gamma_k + F^* \vdash A$. c_q^p can now occur in the following contexts in F^*:

$$\text{(I)} \quad \begin{cases} R\left(c_q^p, y_q^p\right), \\ s_i = c_q^p \to t_i = y_q^p, \\ c_q^p = s_j \to y_q^p = t_j. \end{cases} \qquad \text{(II)} \quad \begin{cases} R\left(c_q^{p-1}, c_q^p\right), \\ s_i = c_q^{p-1} \to t_i = c_q^p, \\ c_q^{p-1} = s_j \to c_q^p = t_j. \end{cases}$$

We may now replace all occurrences of c_q^p of type (I) by a new variable x_q^p; F^* is implied by

$$F^{**} := \forall x_q^p \, \exists y_q^p \forall x_0 \exists y_0 \ldots \forall x_m \exists y_m \left[\bigwedge_i R(s_{i''}, t_{i''}) \wedge \bigwedge_{i<j} \left(s_{i''} = s_{j''} \to t_{i''} = t_{j''}\right) \right],$$

where $s_{i''} := s_i[c_q^p/x_q^p]$, $t_{j''} := t'_j[c_q^p/x_q^p]$. $F^{**} \in \mathrm{Fun}_{n+1}$, but does not contain c_q^{p+1} anymore, and $p-1$ is now the maximal r (for fixed q) such that c_q^r occurs in F^{**}.

We may thus repeat this process till we have eliminated all fresh constants; the resulting formula obtained from F belongs to Fun_n, and $\mathrm{Fun}_n \subset \Gamma_k$.

Now we can establish lemma 2.3 as before; the treatment of clause (ii) in the proof now requires us to show

$$\left(\vdash \Gamma_k \wedge B\left(c_n^0\right) \wedge F \to A\right) \Rightarrow \left(\vdash \Gamma_k \to A\right),$$

where F is a finite conjunction of instances of Fun_{n+1}; this follows readily with the help of (1).

There are similar slight modifications needed in the proof of lemma 2.5; we leave it to the reader to check the details. □

3. Incompleteness results

3.1. The results of the preceding section might lead us to believe that completeness for full **IQC** for the notion of intuitionistic validity is within reach. We shall show that, nevertheless, we cannot expect to achieve this.

As in the preceding section we concentrate on pure first-order predicate logic without equality and function symbols.

Warning. In this section α runs over all (choice) 01-sequences.

First we shall show how, for each primitive recursive relation $A \subset \mathbb{N} \times \{0,1\}^{\mathbb{N}}$ we can find a formula A° of **IQC** such that the validity of A° expresses that $\forall\alpha\exists n A(n, \alpha)$. (*A functional* $F \in \mathbb{N} \times \{0,1\}^{\mathbb{N}} \to \mathbb{N}$ *is primitive recursive* if $\lambda n.F(n, \alpha)$ is primitive recursive in α, i.e. can be defined from Z, S, $\boldsymbol{p}^i_n$, and α by means of recursion and composition, and a relation $A \subset \mathbb{N} \times \{0,1\}^{\mathbb{N}}$ is primitive recursive iff its characteristic functional is, cf. E3.7.3.)

3.2. *The construction of* A°. The idea is to imitate the definition of $A(n, \alpha)$. Let $=$, S be binary predicate symbols, and Z a unary predicate symbol, with intended interpretation equality for $=$, "y is successor of x" for $S(x, y)$, and "x is zero" for $Z(x)$. "Succ" is the conjunction of the universal closures of

$$x = x;$$

$$x = y \wedge x = z \to y = z;$$

$$\mathrm{Repl}(Z) \wedge \mathrm{Fun}(Z);$$

$$\mathrm{Repl}(S) \wedge \mathrm{Fun}(S);$$

$$S(x, z) \wedge S(y, z) \to x = y;$$

$$Z(x) \to \neg S(y, x);$$

where for any n-ary predicate P

$$\mathrm{Repl}(P) := \forall\vec{x}\vec{x}'(\vec{x} = \vec{x}' \wedge P(\vec{x}) \to P(\vec{x}')),$$

and

$$\mathrm{Fun}(P) := \forall\vec{x}zz'(P(\vec{x}, z) \wedge P(\vec{x}, z') \to z = z').$$

So Succ axiomatizes the theory of the successor relation. The existence of copies of $\mathbb{N}$ can be expressed by a formula "Num", where

$$\mathrm{Num} := \mathrm{Succ} \wedge G,$$

$$G := \exists x Z(x) \wedge \forall x \exists y S(x, y).$$

If $A(n, \alpha)$ is in PRIM, then its characteristic function χ_A is built up from a finite list of defining equations of the form

$$f_q(x) = \alpha(x), \tag{1'}$$

$$f_q(x) = 0, \tag{2'}$$

$$f_q(x) = Sx, \tag{3'}$$

$$f_q(x_1, \dots, x_n) = x_i \quad (1 \le i \le n), \tag{4'}$$

$$f_q(x_1, \dots, x_n) = f_s(f_{s_1}(x_1, \dots, x_n), \dots, f_{s_m}(x_1, \dots, x_n)) \quad \text{with } s < q,\ s_i < q, \tag{5'}$$

$$\begin{cases} f_q(0, \vec{x}) = f_s(\vec{x}) \quad (s < q), \\ f_q(Sy, \vec{x}) = f_t(y, \vec{x}, f_q(y, \vec{x})) \quad (t < q), \end{cases} \tag{6'}$$

$q \le k$ for all equations in the list, and $f_k(n) = \chi_A(n, \alpha)$.

For each f_q of the list we introduce a corresponding predicate P_q, with $f_q(\vec{x}) = y$ as the intended interpretation of $P_q(\vec{x}, y)$.

"Comp" is the conjunction of the closures of the following formulas (the first six groups corresponding to (1′)–(6′) above)

$$\begin{cases} Q(x) \wedge Z(y) \to P_0(x, y), \\ \neg Q(x) \wedge Z(y) \wedge S(y, z) \to P_0(x, z); \end{cases} \tag{1}$$

$$Z(y) \to P_q(x, y); \tag{2}$$

$$S(x, y) \to P_q(x, y); \tag{3}$$

$$P_q(\vec{x}, x_i); \tag{4}$$

$$P_{s_1}(\vec{x}, y_1) \wedge \cdots \wedge P_{s_m}(\vec{x}, y_m) \wedge P_s(y_1, \dots, y_m, z) \to P_q(\vec{x}, z); \tag{5}$$

$$\begin{cases} P_s(\vec{x}, y) \wedge Z(z) \to P_q(z, \vec{x}, y), \\ P_t(y, \vec{x}, z, w) \wedge P_q(y, \vec{x}, z) \wedge S(y, y') \to P_q(y', \vec{x}, w); \end{cases} \tag{6}$$

$$\mathrm{Fun}(P_i), \mathrm{Repl}(P_i) \quad \text{for } i \le k; \tag{7}$$

$$Q(x) \vee \neg Q(x). \tag{8}$$

Here $Q(x)$ corresponds to $\alpha(x) = 0$, i.e. we may think of α as the characteristic function χ_Q of Q. We define

$$A^\circ := H' \to H$$

where $H' := \text{Num} \wedge \text{Comp}$ and $H := \exists x\, \exists y(Zy \wedge P_k(x, y))$. Then we can prove the following lemma.

3.3. LEMMA. $\forall\alpha\exists n A(n, \alpha)$ *iff* $\models_i A^\circ$.

PROOF. Let $\mathscr{M}$ be any interpretation of the language of **IQC**, so in particular we have relations $Z^*, S^*, =^*, Q^*, P_i^*$ for $i \leq k$ over a domain D. Without loss of generality we may assume that $=^*$ is the identity of D (if not, take equivalence classes modulo $=^*$), and henceforth we write $=$ instead of $=^*$.

Suppose $\forall\alpha\exists n A(n, \alpha)$, and assume H' to hold in $\mathscr{M}$. Hence Num holds, so we can find a unique $0^* \in \mathrm{D}$ such that $Z^*(0^*)$. Also $\forall d \in \mathrm{D}\, \exists d' \in \mathrm{D}$ $S^*(d, d')$; using the axiom of dependent choices we find a sequence $0^*, 1^*, 2^*, \ldots$ such that $S^*(n^*, (n+1)^*)$ for all $n \in \mathbb{N}$. Thus $\mathbb{N}^* \equiv \{n^*: n \in \mathbb{N}\}$ is an isomorphic copy of $\mathbb{N}$. We also have, on assumption of H' for $\mathscr{M}$, that

$$\forall d \in \mathrm{D}(Q^*d \vee \neg Q^*d),$$

and thus

$$\forall d \in \mathrm{D}\, \exists! d' \in \mathrm{D}[(d' = 0^* \wedge Q^*d) \vee (d' = 1^* \wedge \neg Q^*d)),$$

hence using a function comprehension axiom following from dependent choices we find an $\alpha^* \in \mathbb{N}^* \to \{0^*, 1^*\}$ such that

$$\alpha^*(d) = \begin{cases} 0^* & \text{if } Q^*(d), \\ 1^* & \text{if } \neg Q^*(d). \end{cases}$$

α^* is therefore the characteristic function of Q^* on the domain D. Let α be the corresponding function $\mathbb{N} \to \{0, 1\}$. For each f_q occurring in the definition of f we can prove with the help of induction that

$$f_q(n_1, \ldots, n_t) = n_{t+1} \quad \text{iff} \quad P_q(n_1^*, \ldots, n_{t+1}^*),$$

and hence also

$$f_k(n, \alpha) = 0 \quad \text{iff} \quad P_k^*(n^*, 0^*).$$

Since $\exists n(f(n, \alpha) = 0)$ also $\exists xy(Zy \wedge P_k(x, y))$ in $\mathscr{M}$, i.e. H holds in $\mathscr{M}$. Therefore we have shown $\models_i H' \to H$, i.e. $\models_i A^\circ$.

Conversely, let $\models_i A^\circ$. Choose any $\alpha \in \mathbb{N} \to \{0, 1\}$. Since A° is valid, it is valid in particular for the intended standard model with $\mathrm{D} = \mathbb{N}$,

$Z^*(x) := x = 0$, etc. If we choose Q^* such that Q^*n iff $\alpha(n) = 0$, we conclude from $\models_{\mathrm{i}} A^\circ$ that $\exists n A(n, \alpha)$. This holds for all α, so $\forall\alpha\exists n A(n, \alpha)$. □

3.4. LEMMA. $\forall\alpha\neg\neg\exists n A(n, \alpha) \Rightarrow \models_{\mathrm{i}} \neg\neg A^\circ$.

PROOF. $\neg\neg(H' \to H) \leftrightarrow \neg(H' \wedge \neg H)$, by intuitionistic propositional logic. Now assume $H' \wedge \neg H$ to hold in some model $\mathcal{M}$. As before, because of H' we can find in $\mathcal{M}$ an isomorphic copy $\mathbb{N}^* = \{n^* : n \in \mathbb{N}\}$ of $\mathbb{N}$, and a function $\alpha^* \in \mathbb{N}^* \to \{0^*, 1^*\}$. By $\neg H$ we also have

$$\forall d \in \mathrm{D}\, \forall d' \in \mathrm{D} \neg(Z^*(d) \wedge P_k^*(d, d')),$$

hence in particular $\forall n^* \in \mathbb{N}^* \neg P_k^*(n^*, 0^*)$, and so, by the same reasoning as before, $f_k(n, \alpha) \neq 0$ for all $n \in \mathbb{N}$, where α corresponds to α^*. It follows that $\exists\alpha\forall n \neg A(n, \alpha)$, contradicting $\forall\alpha\neg\neg\exists n A(n, \alpha)$. Therefore $\neg(H' \wedge \neg H)$ must hold in $\mathcal{M}$. $\mathcal{M}$ was arbitrary, hence $\models_{\mathrm{i}} \neg\neg A^\circ$. □

3.5. LEMMA. *Let B be a prenex formula equivalent in* **IQC** *to $\neg A^\circ$, i.e. $H' \wedge \neg H$. Then*

$$(\vdash \neg B) \Rightarrow \forall\alpha\exists n A(n, \alpha).$$

PROOF. For notational simplicity we drop parentheses and commas in $S(x, y)$, etc. $H' \wedge \neg H$ can be spelt out as

$$G \wedge \mathrm{Succ} \wedge \mathrm{Comp} \wedge \forall xy \neg(Zy \wedge P_k xy);$$

also

$$G \leftrightarrow \exists x \forall y \exists z (Zx \wedge Syz).$$

This gives rise to a prenex form B

$$\exists x \forall y \exists z \forall \vec{u} \forall vw (Zx \wedge Syz \wedge T' \wedge \neg(Zv \wedge Svw))$$

with $T' := T \wedge \neg(Zv \wedge P_k wv)$, and $\forall\vec{u}(T) \leftrightarrow \mathrm{Succ} \wedge \mathrm{Comp}$.

$\neg B$ is the negation of a prenex formula, and hence by Herbrand's theorem (cf. E10.5.6) we can find q, variables $x_0, \ldots, x_q$, $y_0, \ldots, y_q$, terms $t_0, \ldots, t_q$ such that the following negated conjunction is provable:

$$\neg[(Zx_0 \wedge St_0 z_0 \wedge T'_0) \wedge (Zx_1 \wedge St_1 z_1 \wedge T'_1) \wedge \cdots \wedge (Zx_q \wedge St_q z_q \wedge T'_q)], \qquad (1)$$

where

$$t_0 \in \{x_0\},$$
$$t_1 \in \{x_0, x_1, z_0\},$$
$$t_2 \in \{x_0, x_1, x_2, z_0, z_1\}, \dots$$
$$t_q \in \{x_0, \dots, x_q, z_0, \dots, z_{q-1}\},$$

and where the T_i' for $i \leq q$ are substitution instances of T'.

If we now make substitutions

$$x_0 = 0, \qquad t_0 = x_0 = 0, \qquad z_0 = 1,$$
$$x_1 = 0, \qquad z_1 = t_1 + 1;$$
$$x_2 = 0, \qquad z_2 = t_2 + 1, \dots,$$

and interpret $Z, S, \dots$ in the standard way, and Qx as $\alpha x = 0$, then $Zx_0, \dots, Zx_q$, $St_0z_0, St_1z_1, \dots, St_qz_q$ are all made true. Also in each $T_i' \equiv T_i \wedge \neg(Zs_i \wedge P_k s_i')$ the T_i are true, since they consist of conjunctions of instances of universal formulas which are true on the standard interpretation; therefore by (1)

$$\neg(\neg(Zs_0 \wedge P_k(s_0', s_0)) \wedge \neg(Zs_1 \wedge P_k(s_1', s_1)) \wedge \cdots) \tag{2}$$

is true on this interpretation. In the standard interpretation the P_k are decidable, and so (2) is equivalent to

$$(Zs_0 \wedge P_k(s_0', s_0)) \vee \cdots \vee (Zs_q \wedge P_k(s_q', s_q)),$$

or

$$f_k(s_0', \alpha) = 0 \vee \cdots \vee f_k(s_q', \alpha) = 0$$

and thus, by checking these finitely many cases, we find an n such that $f_k(n, \alpha) = 0$; so $\forall \alpha \exists n (f_k(n, \alpha) = 0)$. □

3.6. THEOREM.

(i) *Completeness of* **IQC** *implies for each primitive recursive* $A(n, \alpha)$

$$\forall \alpha \neg\neg \exists n A(n, \alpha) \to \forall \alpha \exists n A(n, \alpha).$$

(ii) *Weak incompleteness in the form: for all* B *of* **IQC** $(\models_i B) \Rightarrow (\neg\neg \vdash_i B)$, *where* $\vdash_i$ *denotes derivability in* **IQC**, *implies*

$$\forall \alpha \neg\neg \exists n A(n, \alpha) \to \neg\neg \forall \alpha \exists n A(n, \alpha).$$

PROOF. Immediate by 3.4 and 3.5. □

COROLLARY. *Completeness for* **IQC** *implies in particular Markov's principle* $\mathrm{MP}_{\mathrm{PR}}$. □

3.7. *Discussion of the assumptions underlying theorem* 3.6. The essential assumptions are found in the proofs of lemmas 3.3, 3.4. We may think the proof carried out in a fragment of intuitionistic type theory, say. The proof uses DC:

$$\forall d \in \mathrm{D}\, \exists d' \in \mathrm{D}\, S^*(d, d') \wedge \exists d \in \mathrm{D}\, Z^*(d) \to$$
$$\exists \alpha \in \mathbb{N} \to \mathrm{D}(Z^*(\alpha 0) \wedge \forall n\, S^*(\alpha n, \alpha(n+1))).$$

If the *range* of any function $\alpha \in \mathbb{N} \to \mathrm{D}$ is again regarded as a set, α is an isomorphism from $\mathbb{N}$ onto range $(\alpha)/\approx$, where $\approx$ is the equivalence relation interpreting $=$ on D. The function comprehension axiom follows from

$$\forall n \in \mathbb{N}\, \exists! m \in \{0,1\} A(n, m) \to \exists \alpha \in \mathbb{N} \to \{0,1\} \forall n\, A(n, \alpha n)$$

and is a consequence of DC. For the rest, we need only induction.

In particular, we may think of the α as ranging over choice sequences satisfying the principles of **CS** and the relations as being lawlike relative to choice sequences. Then the argument can be formalized in an extension **CS*** of **CS** with lawlike relation variables added, and comprehension for formulas not containing relation quantifiers. In such a system we have $\forall \alpha \neg\neg \exists a(\alpha = a)$, and hence $\forall a \exists n A(n, a) \to \forall \alpha \neg\neg \exists n A(n, \alpha)$. Now let $A(n, \alpha) := R(\bar{\alpha} n)$, where R is the predicate determining the Kleene tree (cf. 4.7.6). For this tree clearly $\neg \forall \alpha \exists n R(\bar{\alpha} n)$, but assuming CT for lawlike sequences $\forall a \exists n A(n, a)$, so $\forall \alpha \neg\neg \exists n A(n, \alpha)$. Thus weak completeness implies $\neg$CT, while CT is consistent with **CS***.

3.8. CT_0 *and completeness of* **IQC**. In particular, from the preceding discussion we also see, that if we quite generally assume CT_0, **IQC** is incomplete. But in this case better results are known: assuming CT_0, the collection of valid formulas of predicate logic is not r.e. (Kreisel 1970). Leivant (1976) slightly improved this by weakening CT_0 to

$$\forall n(An \vee \neg An) \to \neg\neg \exists m \forall n(An \leftrightarrow \exists k\, T(m, n, k)),$$

(a decidable set is not not r.e.). See also 9.4.

4. Lattices, Heyting algebras and complete Heyting algebras

4.1. Definition. A *lattice* is a partially ordered set (poset) A such that for each $a,b \in A$ there is a least upper bound $a \vee b$ (the *join* of a and b) and a greatest lower bound $a \wedge b$ (the *meet* of a and b). This can be expressed by the following axioms: for all $a,b,c \in A$:

$$a \leq a \vee b, \qquad b \leq a \vee b, \qquad (a \leq c) \wedge (b \leq c) \rightarrow (a \vee b \leq c);$$

$$a \wedge b \leq a, \qquad a \wedge b \leq b, \qquad (c \leq a) \wedge (c \leq b) \rightarrow (c \leq a \wedge b).$$

A *zero* or *bottom element* (denoted by 0 or $\bot$) in a lattice satisfies $\forall a \in A(\bot \leq a)$, a *top* or *unit element* (denoted by 1 or $\top$) satisfies $\forall a \in A$ $(a \leq \top)$. If existing, top and bottom are unique. □

For the appropriate notions of sublattice and homomorphism we must regard a lattice as a structure $(A, \wedge, \vee)$. This suggests an alternative definition.

4.2. Definition. A *lattice* $(A, \wedge, \vee)$ is a set A with two binary operations $\wedge, \vee$ such that for all $a, b, c \in A$:

L1 $\quad a \wedge a = a, \qquad a \vee a = a$ (*idempotency*),

L2 $\quad a \wedge b = b \wedge a, \qquad a \vee b = b \vee a$ (*commutativity*),

L3 $\quad a \wedge (b \wedge c) = (a \wedge b) \wedge c, \qquad a \vee (b \vee c) = (a \vee b) \vee c$ (*associativity*),

L4 $\quad a \vee (a \wedge b) = a, \qquad a \wedge (a \vee b) = a$ (*absorption*).

If, starting from this definition of a lattice, we put $a \leq b := (a \wedge b = a)$ (or equivalently $a \leq b := (a \vee b = b)$) the two definitions of a lattice can be shown to be equivalent (exercise).

Two lattices $(A, \wedge, \vee)$ and $(A', \wedge', \vee')$ are said to be *isomorphic* if there is a bijection φ from A to A' which is a *homomorphism* with respect to $\wedge, \vee$, i.e. $\varphi(a \wedge b) = \varphi a \wedge' \varphi b$, $\varphi(a \vee b) = \varphi a \vee' \varphi b$ for all $a, b \in A$.

$(A', \wedge', \vee')$ is a *sublattice* of $(A, \wedge, \vee)$ if $A' \subset A$, and $\wedge', \vee'$ are the restrictions of $\wedge, \vee$ to A'. □

REMARK. Sometimes a lattice is defined as having top and bottom; in this case one must add to L1–4 the axiom

L5 $a \wedge 0 = 0, \qquad a \vee 1 = 1.$

4.3. DEFINITION. A lattice $(A, \wedge, \vee)$ is *distributive* iff for all $a,b,c \in A$:

D1 $a \wedge (b \vee c) = (a \wedge b) \vee (a \wedge c),$

D2 $a \vee (b \wedge c) = (a \vee b) \wedge (a \vee c).$

D1 implies D2 and vice versa (exercise). □

EXAMPLES. L_1, L_2 are not distributive; L_3, L_4 are distributive. See fig. 13.2.

We note in passing the following nice characterization of distributive lattices: a lattice is distributive iff it does not contain a sublattice isomorphic to L_1 or L_2 (proof e.g. in Birkhoff 1948, ch. 9).

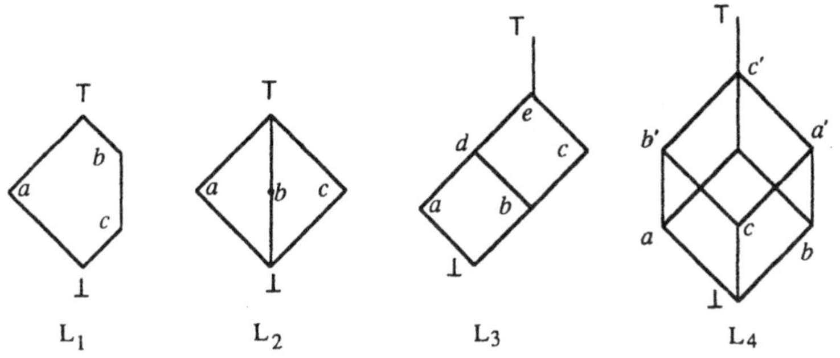

Fig. 13.2.

4.4. DEFINITION. Let $(A, \wedge, \vee)$ be a lattice, $B \subset A$. $a \in A$ is the *join* of B (notation: $\bigvee B$) iff a is a least upper bound of B, i.e.

$$\forall b \in B(b \leq a), \qquad \forall c \in A(\forall b \in B(b \leq c) \rightarrow a \leq c).$$

Similarly, $a \in A$ is the *meet* of B (notation: $\bigwedge B$) iff a is the greatest lower bound of B. In particular, if $\top$, respectively $\bot$ exist, then

$$\bigwedge \emptyset = \top, \qquad \bigvee \emptyset = \bot.$$

If $\varphi(a)$ is an expression denoting a lattice element for each $a \in A$, we sometimes use

$$\bigwedge_{a \in A} \varphi(a), \qquad \bigvee_{a \in A} \varphi(a)$$

as alternative notations for $\bigwedge\{\varphi(a) : a \in A\}$, $\bigvee\{\varphi(a) : a \in A\}$, respectively. □

4.5. Definition. A lattice $(A, \wedge, \vee)$ is *join-complete* (*meet-complete*) iff joins (meets) exist for all $B \subset A$. The lattice is *complete* iff it is join- and meet-complete. □

Example. The opens of any topological space $T \equiv (X, \mathcal{O})$ form a complete, distributive lattice with

$$U \wedge U' := U \cap U', \qquad U \vee U' := U \cup U';$$

$$\bigvee \mathscr{V} := \bigcup\{V : V \in \mathscr{V}\}, \qquad \bigwedge \mathscr{V} := \mathrm{Int} \bigcap\{V : V \in \mathscr{V}\}.$$

4.6. Proposition. *A lattice is join-complete iff it is meet-complete.*

Proof. Suppose $(A, \wedge, \vee)$ to be join-complete, let $B \subset A$ and let B^* be the set of lower bounds of B, i.e. $B^* := \{a \in A : \forall b \in B(a \le b)\}$; put $\bigwedge B := \bigvee B^*$. We leave it to the reader to verify that $\bigwedge B$ indeed is the meet of B. □

4.7. Definition. A *Heyting algebra* (Ha for short) is a structure $(A, \wedge, \vee, \perp, \to)$ such that $(A, \wedge, \vee)$ is a lattice with bottom $\perp$, and $\to$ a binary operation on A such that

$$(a \wedge b \le c) \Leftrightarrow (a \le b \to c).$$

We permit ourselves an abuse of language by using $\to$ also for this lattice operation (because of its narrow relationship with logical implication) but this will not cause confusion.

Heyting algebra's are sometimes called *pseudo-Boolean algebras*. In the literature the lattice-operation $\to$ is called *implication* or *relative pseudo-complementation* ($a \to b$ being the pseudo-complement of a relative to b; $a \to \perp$ is then the pseudo-complement of a). We write

$$\neg a := a \to \perp . \quad \square$$

4.8. PROPOSITION.

(i) *A Heyting algebra is a distributive lattice.*

(ii) *If* $(A, \wedge, \vee, \bot, \rightarrow)$ *is an* Ha, *and* $\bigvee B$ *for some* $B \subset A$ *exists, then for all* $a \in A$ *the infinitary distributive law*

$$\mathrm{D} \qquad a \wedge \bigvee B = \bigvee\{a \wedge b : b \in B\}$$

holds; in particular, the right-hand side of this equation is well-defined for all $a \in A$.

(iii) *Any complete lattice* $(A, \wedge, \vee)$ *satisfying* D *can be turned into an* Ha *by defining*

$$a \rightarrow b := \bigvee\{c : c \wedge a \leq b\}.$$

PROOF. Left as an exercise. □

4.9. DEFINITION. A *complete Heyting algebra* (cHa for short) is a structure $(A, \wedge, \vee, \bot, \rightarrow, \bigwedge, \bigvee)$ such that $(A, \wedge, \vee, \bot, \rightarrow)$ is an Ha, and which is complete as a lattice with infinite joins and meets $\bigwedge, \bigvee$. □

Of course, $\wedge, \vee, \bot$ are in fact redundant, i.e. we may think of a cHa as a structure $(A, \bigwedge, \bigvee, \rightarrow)$. (As an object, a cHa may also be given as $(A, \wedge, \bigvee, \top)$; this gives rise to a different notion of homomorphism. cHa's regarded as structures with $\wedge, \bigvee, \top$ as primitives are often called *frames*.)

4.10. EXAMPLE. The opens of a topological space $T = (X, \mathcal{O})$ form a cHa with $\vee, \wedge, \bigwedge, \bigvee$ defined as before, and

$$U \rightarrow V := \bigcup\{W : U \cap W \subset V\}.$$

It can be shown that

$$U \rightarrow V = \mathrm{Int}\{x : x \in U \rightarrow x \in V\}$$

(the $\rightarrow$ on the right-hand side is a logical implication !), and classically also

$$U \rightarrow V = \mathrm{Int}(V \cup (X \setminus U)).$$

We now turn to the notions of filter and ideal.

4.11. DEFINITION. Let $(A, \wedge, \vee)$ be a lattice. An inhabited set $F \subset A$ is a *filter* iff:

(i) $a, b \in F \Rightarrow a \wedge b \in F$ (closure under $\wedge$),

(ii) $a \in F$ and $a \leq b \Rightarrow b \in F$ (upward closure).

These two conditions may be combined into

$$a \wedge b \in F \;\Leftrightarrow\; a \in F \text{ and } b \in F.$$

The dual notion is that of an ideal; an inhabited set I is an *ideal* iff

$$a \vee b \in I \;\Leftrightarrow\; a \in I \text{ and } b \in I,$$

or equivalently, expressed in two conditions

(iii) $a \in I$ and $b \in I \Rightarrow a \vee b \in I$,

(iv) $a \in I$ and $b \leq a \Rightarrow b \in I$.

A filter (ideal) is said to be *proper* if it is a proper subset of the lattice. A filter F is *prime* iff

$$a \vee b \in F \;\Rightarrow\; a \in F \text{ or } b \in F.$$

The *filter generated by* a set X is the least filter $F \supset X$ (this exists, since the intersection of a family of filters is again a filter). □

REMARK. The filter generated by X consists of all b such that $b \geq a_1 \wedge \cdots \wedge a_n$ for some finite subset $\{a_1, \ldots, a_n\} \subset X$. Equivalently, this filter can be described as the least set Y closed under $a \in X \Rightarrow a \in Y$; $x, y \in Y \Rightarrow x \wedge y \in Y$; $x \in Y \wedge x \leq y \Rightarrow y \in Y$.

4.12. PROPOSITION. *Let F be a filter, $a \to b \notin F$. Then the filter generated by $F \cup \{a\}$ does not contain b.*

PROOF. Suppose b to be an element of the filter generated by $F \cup \{a\}$; then by the preceding remark, $b \geq c \wedge a$ for some $c \in F$, hence $c \leq a \to b$, but then $a \to b \in F$ would follow. □

The next proposition is classical and requires Zorn's lemma.

4.13. PROPOSITION. *Let F be a filter, $a \notin F$. Then there is a prime filter $F' \supset F$, $a \notin F'$.*

PROOF. The proof uses PEM and Zorn's lemma (exercise). □

5. Algebraic semantics for IPC

5.1. Algebraic semantics is the most general type of semantics for **IPC**; an algebraic model for **IPC** is nothing else but a valuation of the proposition variables with values in an Ha, and as we shall see, any theory containing **IPC** in its logical basis can be almost trivially interpreted as an algebraic model in this sense.

We shall take the following Hilbert-type axiomatization of **IPC** (the propositional part of $\mathrm{H_1}$-**IQC** in 2.4.1) as our starting point.

Axiom schemata. $A \to (B \to A)$, $[A \to (B \to C)] \to [(A \to B) \to (A \to C)]$, $A \wedge B \to A$, $A \wedge B \to B$, $A \to (B \to A \wedge B)$, $A \to A \vee B$, $B \to A \vee B$, $(A \to C) \to [(B \to C) \to (A \vee B \to C)]$, $\perp \to A$, and the following rule.

Rule. If A, $A \to B$, then B (modus ponens).

5.2. Definition. Let $\Theta \equiv (A, \wedge, \vee, \perp, \to)$ be an Ha; a *Θ-valuation* or *Θ-model* is a mapping from $\mathscr{P}$, the set of proposition letters of **IPC**, into Θ. If in a certain context a valuation φ is given, we shall often write $[\![P]\!]$ for $\varphi(P)$.

A Θ-valuation is extended to all formulas of **IPC** by

$$[\![A \circ B]\!] = [\![A]\!] \circ [\![B]\!] \quad \text{for } \circ \in \{\wedge, \vee, \to\}, \qquad [\![\perp]\!] = \perp .$$

(On the left-hand side "$\circ$" denotes a logical operation, and on the right-hand side "$\circ$" denotes an Ha-operation).

A formula A is *φ-valid* iff $[\![A]\!] = \top$; Ai is Θ-valid iff A is valid in all Θ-models. A is *Ha-valid* iff A is Θ-valid for all Ha's Θ. These notions are extended to sets of formulas in the obvious way.

More generally, we shall say that A is an *Ha-consequence* of Γ ($\Gamma \Vdash_{\mathrm{Ha}} A$, or $\Gamma \Vdash A$ if no confusion can arise) iff for each Θ-model in which all sentences of Γ hold, A also holds. □

5.3. Theorem (*soundness for Ha-validity*). *If $\Gamma \vdash A$ in* **IPC**, *then $\Gamma \Vdash A$ is Ha-valid.*

Proof. Let a Θ-model be given in which Γ is Θ-valid. For an axiom F we have to show $[\![F]\!] = \top$. First we observe that $[\![A \to B]\!] = \top$ is equivalent to $\top \leq [\![A \to B]\!]$ or $\top \leq [\![A]\!] \to [\![B]\!]$, i.e. $[\![A]\!] \leq [\![B]\!]$ (by the definition of $\to$ in an Ha).

Consider now for example the case of the schema $(A \to C) \to [(B \to C) \to (A \vee B \to C)]$, then $[\![(A \to C) \to [(B \to C) \to (A \vee B \to C)]]\!] = \top \Leftrightarrow [\![A \to C]\!] \leq [\![(B \to C) \to (A \wedge B \to C)]\!] \Leftrightarrow [\![A \to C]\!] \vee [\![B \to C]\!] \leq [\![A \vee B \to C]\!] \Leftrightarrow [\![A \to C]\!] \vee [\![B \to C]\!] \wedge [\![A \vee B]\!] \leq [\![C]\!]$. To see this, assume $a \leq [\![A \to C]\!] \wedge [\![B \to C]\!] \wedge [\![A \vee B]\!]$, i.e. $a \leq [\![A \to C]\!]$, $a \leq [\![B \to C]\!]$, $a \leq [\![A \vee B]\!]$, then $a \wedge [\![A]\!] \leq [\![C]\!]$, $a \wedge [\![B]\!] \leq [\![C]\!]$, hence by distributivity $a \wedge [\![A \vee B]\!] \leq [\![C]\!]$. We leave the verification of the other axioms as an exercise to the reader.

Modus ponens is also immediate: if $[\![A]\!] = \top$, $[\![A \to B]\!] = \top$, then $[\![A]\!] = \top$ and $[\![A]\!] \leq [\![B]\!]$, so $[\![B]\!] = \top$. □

We can also quite easily prove completeness by a so-called Lindenbaum construction.

5.4. THEOREM (*completeness for Ha-validity*). *If* $\Gamma \Vdash A$, *then* $\Gamma \vdash A$ *in* **IPC**.

PROOF. We construct a canonical model (valuation) in the so-called Lindenbaum algebra. On the formulas of **IPC** we define an equivalence relation

$$A \sim B := (\Gamma \vdash A \leftrightarrow B)$$

and we put $[A] := \{B : B \sim A\}$. On the equivalence classes we define a partial order by

$$[A] \leq [B] := \Gamma \vdash A \to B.$$

Let X be the set of all equivalence classes. X becomes an Ha Θ_Γ with

$$[A] \circ [B] := [A \circ B] \quad \text{for } \circ \in \{\wedge, \vee, \to\},$$

$[\bot]$ is the bottom of Θ_Γ.

It remains to be shown that for this definition of $\wedge, \vee, \to$ on X we indeed obtain a Heyting algebra; thus $[A] \wedge [B]$ must be the meet of $[A]$ and $[B]$, etc. All the required properties easily follow from the axioms and rule of **IPC**, and we leave this as an exercise. The canonical valuation is now defined by $[\![P]\!] := [P]$ for all proposition letters P, and trivially $[\![A]\!] = \top$ iff $\Gamma \vdash A \leftrightarrow \top$ iff $\Gamma \vdash A$. □

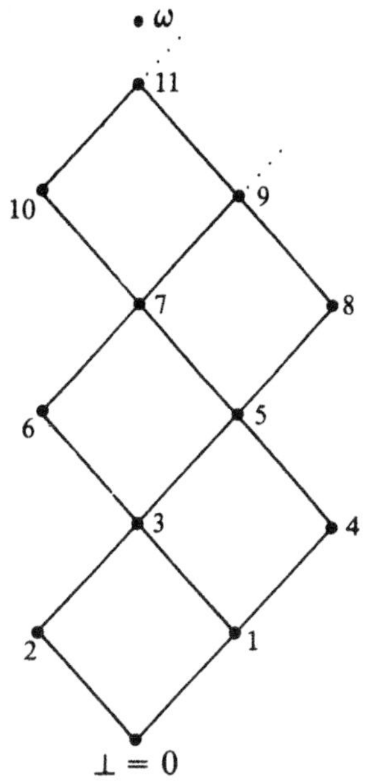

Fig. 13.3.

5.5. As an application, we consider the Rieger–Nishimura lattice (2.1.10), which is actually nothing but the Lindenbaum algebra of one-variable formulas of **IPC**. We recall the definition. Let P be a fixed proposition letter. Then

$$\mathrm{A}_0(P) := \bot\,, \qquad \mathrm{A}_1(P) := P, \qquad \mathrm{A}_2(P) := \neg P,$$

$$\mathrm{A}_{2n+3}(P) := \mathrm{A}_{2n+1}(P) \vee \mathrm{A}_{2n+2}(P)$$

$$\mathrm{A}_{2n+4}(P) := \mathrm{A}_{2n+2}(P) \to \mathrm{A}_{2n+1}(P)$$

$$\mathrm{A}_\omega(P) := P \to P.$$

In E2.1.6 it was asked to prove that all one-variable formulas in **IPC** were logically equivalent to one of the A_n for $n \in \mathbb{N} \cup \{\omega\}$; therefore their equivalence classes under logical equivalence form a Heyting algebra Θ.

On the other hand, by inspection we see that the diagram of established implications (see fig. 13.3) also yields a Heyting algebra Θ', and we want to show that Θ and Θ' coincide. We have to show that no implications hold between the A_n which are not shown in the picture.

It is lengthy, but routine to check that Θ' must be a homomorphic image of Θ; e.g. $6 \to 3 = 8$ in Θ' corresponds to $\vdash (\mathrm{A}_6 \to \mathrm{A}_3) \leftrightarrow \mathrm{A}_8$, etc. On the other hand we can see at one stroke that $\not\Vdash \mathrm{A}_n$ for all $n \in \mathbb{N}$, since for the canonical valuation given by $[\![P]\!] = 1$ we have, by the preceding remark, $[\![\mathrm{A}_n]\!] = n$. Accordingly none of the A_n is provable, nor can any two $\mathrm{A}_n, \mathrm{A}_m$ with $n \neq m$ be equivalent; so Θ is isomorphic to Θ'. □

5.6. *Relationship between Kripke models and* Ha-*models.* In one direction the correspondence is very easy. Let us call an Ha-model for an Ha Θ which consists of the opens of a topological space a *topological model.*

Every propositional Kripke-model $\mathcal{K} \equiv (\mathrm{K}, \leq, \Vdash)$ can be transformed into a topological model over $\Theta \equiv (\Omega_\mathrm{K}, \cap, \cup, \emptyset, \to)$, where Ω_K consists of all upwards monotone subsets of $(\mathrm{K}, \leq)$; i.e. $X \in \Omega_\mathrm{K}$ iff $X \subset \mathrm{K}$ and $\forall k \in X(k \leq k' \Rightarrow k' \in X)$. Ω_K is a topology on K. The valuation $[\![\ ]\!]$ determined by $\mathcal{K}$ is

$$[\![P]\!] = \{k : k \Vdash P\}.$$

We leave it as an exercise to the reader to show that for all A

$$[\![A]\!] = \{k : k \Vdash A\}.$$

We observe the following.

COROLLARY (*to* 2.6.11). **IPC** *is complete with respect to valuations in finite* Ha'*s.* □

For the converse, we have to argue classically.

5.7. THEOREM (PEM, *Zorn's lemma*). *Suppose a valuation in an* Ha $\Theta \equiv (A, \wedge, \vee, \bot, \to)$ *to be given. We construct a Kripke model* $\mathcal{K} \equiv \{Pr(\Theta), \subset, \Vdash\}$, *where* $\Pr(\Theta)$ *is the set of (proper) prime filters of* Θ, *and where* $\Vdash$ *is given by*

$$F \Vdash P := [\![P]\!] \in F.$$

Under the correspondence just described, for all formulas A of **IPC**

$$F \Vdash A \Leftrightarrow [\![A]\!] \in F.$$

PROOF. By induction on the complexity of A. The only interesting case is implication. So let $A \equiv B \to C$. If $[\![B \to C]\!] \in F$, then for all $F' \supset F$ such that $[\![B]\!] \in F'$, also $[\![B]\!] \wedge ([\![B]\!] \to [\![C]\!]) \in F'$, so $[\![C]\!] \in F'$. Hence (induction hypothesis) $\forall F' \supset F(F' \Vdash B \Rightarrow F' \Vdash C)$, i.e. $F \Vdash B \to C$. If $[\![B \to C]\!] \notin F$, then there is a prime $F' \supset F \cup \{[\![B]\!]\}$ with $[\![C]\!] \notin F'$; so $F' \Vdash B$ and $F' \nVdash C$, hence $F \nVdash B \to C$. The other cases are left to the reader. □

6. Ω-sets and structures

In this section we describe the extension of algebraic semantics to intuitionistic first-order predicate logic. In what follows $\Omega, \Omega', \ldots$ will be used for cHa's. First we describe what we shall call "global Ω-structures", representing the most direct and straightforward definition of Ω-valued models for **IQC**. Then we shall define the more general Ω-structures for **IQCE**-languages, establish completeness using constructive metamathematical reasoning only.

N.B. In the next three sections "E" (*extent*) is a constant.

6.1. *Global Ω-structures for* **IQC** *without equality.* Let $\mathscr{L}$ be an **IQC**-language, possibly with constants and function symbols, but without equality, and let $X_1, \ldots, X_n$ be sets.

An *Ω-valued* (n-ary) *relation* (Ω-relation) on $X_1, \ldots, X_n$ is a mapping $X_1 \times \cdots \times X_n \to \Omega$.

A *global Ω-model* (*global Ω-structure*) for a many-sorted language $\mathscr{L}$ assigns to each sort i a set X_i; to each relation symbol R with arguments of sorts $i_1, \ldots, i_n$ an Ω-relation $[\![R]\!]$ on $X_{i_1}, \ldots, X_{i_n}$; to each function symbol F with arguments of sorts $i_1, \ldots, i_n$ and value of sort j a function $[\![F]\!] \in X_{i_1} \times \cdots \times X_{i_n} \to X_j$, and to a constant c of sort i an element $[\![c]\!]$ of X_i.

Note that an n-ary Ω-relation is essentially the same as a unary Ω-relation (Ω-predicate) on a cartesian product of n factors.

We can now assign a value $[\![t]\!] \in X_i$ to each term of sort i, and a value $[\![A]\!] \in \Omega$ to each sentence A in the language $\mathscr{L}'$ obtained by adding constants for the elements of the X_i to $\mathscr{L}$, as follows.

We use a also as the constant symbol denoting $a \in X_i$. Then $[\![\ ]\!]$ is extended to arbitrary terms in the language $\mathscr{L}'$ ($\equiv \mathscr{L}$ extended with constants for the $a \in X_i$) by taking

$$[\![a]\!] := a \quad (a \in X_i,\ i \text{ a sort of } \mathscr{L}),$$

$$[\![F(t_1, \ldots, t_n)]\!] := [\![F([\![t_1]\!], \ldots, [\![t_n]\!])]\!] := [\![F]\!]([\![t_1]\!], \ldots, [\![t_n]\!]);$$

$[\![\ ]\!]$ is defined for sentences of $\mathcal{L}'$ by

$$[\![R(t_1,\ldots,t_n)]\!] := [\![R([\![t_1]\!],\ldots,[\![t_n]\!])]\!] := [\![R]\!]([\![t_1]\!],\ldots,[\![t_n]\!]),$$

$$[\![\perp]\!] := \perp,$$

$$[\![A \circ B]\!] := [\![A]\!] \circ [\![B]\!] \quad \text{for } \circ \in \{\wedge, \vee, \to\},$$

$$[\![\forall x A(x)]\!] := \bigwedge\{[\![A(a)]\!] : a \in X_i\} \quad (x \text{ a variable of sort } i),$$

$$[\![\exists x A(x)]\!] := \bigvee\{[\![A(a)]\!] : a \in X_i\} \quad (x \text{ a variable of sort } i).$$

Global Ω-structures with equality. In order to extend the notion of global Ω-model to languages with equality, we need, for each sort i of the language considered, a privileged Ω-relation $\mathrm{Eq}_i \in X_i \times X_i \to \Omega$ such that the equality axioms are validated if we take $[\![a = a']\!] := \mathrm{Eq}_i(a, a')$. This is where we introduce the notion of a (global) Ω-set. Instead of Eq_i we write from now on $\lambda x, y[\![x = y]\!]$ or even simply $[\![\cdot = \cdot]\!]$.

6.2. DEFINITION. A *global Ω-set* $\mathcal{M} \equiv (M, [\![\cdot = \cdot]\!])$ is a set M with a mapping $[\![\cdot = \cdot]\!] \in M \times M \to \Omega$ such that

$$[\![x = x]\!] = \top, \qquad [\![x = y]\!] = [\![y = x]\!],$$

$$[\![x = y]\!] \wedge [\![y = z]\!] \leq [\![x = z]\!].$$

Where necessary to distinguish the Ω-equality for different Ω-sets, we use subscripts: $[\![\cdot = \cdot]\!]_{\mathcal{M}}, [\![\cdot = \cdot]\!]_{\mathcal{M}'}, \ldots$.

The *product* $\mathcal{M} \equiv \mathcal{M}_1 \times \cdots \times \mathcal{M}_n$ of Ω-sets $\mathcal{M}_1, \ldots, \mathcal{M}_n$ is the Ω-set

$$(M_1 \times \cdots \times M_n, [\![\cdot = \cdot]\!]_{\mathcal{M}})$$

with

$$[\![\vec{a} = \vec{b}]\!]_{\mathcal{M}} := \bigwedge\{[\![a_i = b_i]\!]_{\mathcal{M}} : 1 \leq i \leq n\}.$$

$\mathcal{M}$ is clearly again an Ω-set (cf. the discussion in 14.2.2). □

We now adapt the notions of Ω-relation, Ω-function.

6.3. DEFINITION. Let $\mathcal{M}_1, \ldots, \mathcal{M}_n, \mathcal{M}, \mathcal{N}$ be global Ω-sets.

A unary *Ω-relation on* $\mathcal{M} \equiv (M, [\![\cdot = \cdot]\!])$ (*Ω-predicate* for short) is a mapping $P \in M \to \Omega$ such that

$$[\![a = a']\!] \wedge P(a) \leq P(a'). \tag{1}$$

An *Ω-relation* on $\mathcal{M}_1, \ldots, \mathcal{M}_n$ is the same as a unary Ω-relation on $\mathcal{M}_1 \times \cdots \times \mathcal{M}_n$.

An *Ω-function* from $\mathscr{M} \equiv (M, [\![\cdot = \cdot]\!]_{\mathscr{M}})$ to $\mathscr{N} \equiv (N, [\![\cdot = \cdot]\!]_{\mathscr{N}})$ (*Ω-function* for short) is a function $F \in M \to N$ such that

$$[\![a = a']\!] \le [\![Fa = Fa']\!].$$

F is *injective* (an *embedding*) if $[\![Fa = Fb]\!] \le [\![a = b]\!]$.

We can now define *global Ω-structure* (*global Ω-model*) as before. □

6.4. Examples *of global Ω-structures.*

Example (a). A trivial example is the *constant* Ω-set $\hat{M} \equiv (M, [\![\cdot = \cdot]\!])$ where

$$[\![a = a']\!] := \begin{cases} \top & \text{if } a = a' \\ \bot & \text{otherwise.} \end{cases}$$

The definition of $[\![\cdot = \cdot]\!]$ can be formulated more constructively as

$$[\![a = a']\!] := \bigvee\{\top : a = a'\}.$$

If we wish to distinguish between x as element of M and as element of $\hat{M}$, we write $\hat{x}$ for the latter. The global Ω-structures of 6.1 are automatically made into global Ω-structures according to 6.3 by adopting this definition of Ω-equality. Given some relation R on M^n we may associate with this the Ω-relation

$$\hat{R}(\hat{a}_1, \ldots, \hat{a}_n) := \bigvee\{\top : R(a_1, \ldots, a_n)\}. \tag{1}$$

Each function $F \in M^n \to M$ gives rise to an Ω-function $\hat{F}$ with

$$\hat{F}\hat{x} = \hat{y} \leftrightarrow Fx = y. \tag{2}$$

If, for example, we consider a language with $=$, a single n-ary predicate R, and a single function symbol F, then the interpretation of $=$ by $[\![\cdot = \cdot]\!]$, R by $\hat{R}$, and F by $\hat{F}$ results in what is essentially a model in the usual sense ($\models_i$) for this language.

Example (b) (Scott 1968). Let Ω be the cHa $\mathscr{O}(T)$ of open sets of a topological space T, and let $\mathscr{M}$ be the set of total continuous functions in $T \to \mathbb{R}$. We interpret $=, <, +, \cdot$ on $\mathscr{M}$ by

$$[\![a = b]\!] := \mathrm{Int}\{t : a(t) = b(t)\},$$

$$[\![a < b]\!] := \{t : a(t) < b(t)\},$$

$$[\![a + b]\!] := \lambda t.a(t) + b(t),$$

$$[\![a \cdot b]\!] := \lambda t.a(t) \cdot b(t);$$

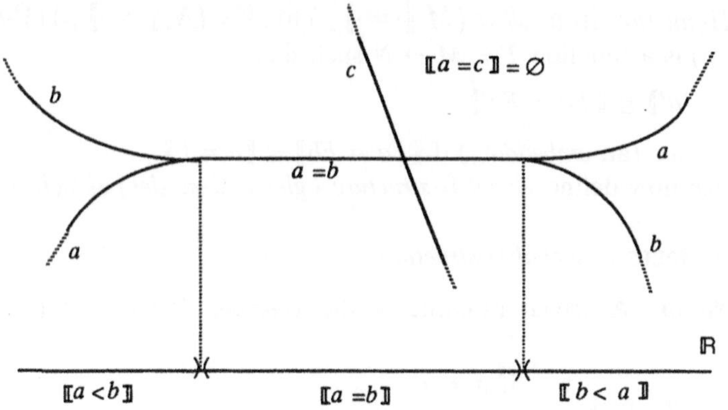

Fig. 13.4.

so $[\![\cdot < \cdot]\!]$ corresponds to $<$, and $\lambda x, y.[\![x + y]\!]$ and $\lambda x, y.[\![x \cdot y]\!]$ are the respective Ω-functions interpreting $+$ and $\cdot$. Thus we obtain an Ω-structure for the theory of real numbers in the language with $=, <, +, \cdot$. It is easy to see that for $T = \mathbb{R}$

$$[\![a < b \to a < c \vee c < b]\!] = \mathbb{R}$$

for all a, b, c in M but that

$$[\![\forall xy(x < y \vee y < x \vee x = y)]\!] = \bot$$

(exercise). Three elements a, b, and c of the model for $T \equiv \mathbb{R}$ are illustrated in fig. 13.4.

If T is a topological space, then an element $\hat{x}$ of a constant $\mathcal{O}(T)$-set $\hat{M}$ can be identified with the function $\lambda t.x$.

Example (c). Let $\mathscr{B} = (\mathrm{K}, \leq, \Vdash, \mathrm{D})$ be a Beth model with constant domain D, (K, $\leq$) a spread. $\mathscr{B}$ is regarded as a model of a (sublanguage of) **IQC**.

If (K, $\leq$) is given the obvious spread topology (the topology induced on the infinite branches of (K, $\leq$) when viewed as a subspace of Baire space), then, for this topology Ω, $\mathscr{B}$ gives rise to a global Ω-model by stipulating

$$[\![R(d_1, \ldots, d_n)]\!] := \bigcup\{V_k : k \Vdash R(d_1, \ldots, d_n)\}$$

(k ranging over K, $V_k = \{\alpha : \alpha \in k\}$ as before). One readily shows by induction on A that $[\![A]\!] := \bigcup\{V_k : k \Vdash A\}$ for all sentences A of $\mathscr{L}(\mathrm{D})$.

Similarly for a Kripke model with constant domain, if we take for Ω the topology given by the upwards monotone subsets of the partial ordering of the model.

Ω-structures for **IQCE** (6.5–7). We now consider a more general notion of structure where elements and functions may be partial. Our basis is the notion of Ω-set.

6.5. DEFINITION. An *Ω-set* $\mathcal{M} \equiv (M, [\![\cdot = \cdot]\!])$ is a set M with a binary mapping $[\![\cdot = \cdot]\!] \in M \times M \to \Omega$ such that for all $x, y, z \in M$,

$$[\![x = y]\!] = [\![y = x]\!], \qquad [\![x = y]\!] \wedge [\![y = z]\!] \leq [\![x = z]\!].$$

We put

$$E(x) := [\![x = x]\!].$$

$E(x)$ is called the *extent* of x.

With the same definition as before the *product* of Ω-sets is again an Ω-set. □

6.6. DEFINITION. Let $\mathcal{M}, \mathcal{M}_1, \mathcal{M}_2, \ldots$ be Ω-sets. An *n-ary Ω-relation* on $\mathcal{M}_1, \mathcal{M}_2, \ldots, \mathcal{M}_n$ is a unary Ω-relation (set) on the corresponding cartesian product, as above; a unary Ω-relation P on an Ω-structure $\mathcal{M}$ satisfies (1) of 6.3 and in addition *strictness for predicates*,

$$\text{STR} \qquad P(a) \leq E(a),$$

and a (*partial*) *Ω-function* F satisfies

$$E(Fa) \wedge [\![a = a']\!] \leq [\![Fa = Fa']\!] \tag{3}$$

and *strictness for functions*

$$\text{STR} \qquad E(Fa) \leq Ea.$$

A *total Ω-function* F satisfies (3) and $E(Fa) = Ea$ or equivalently STR and (2) of 6.3.

Ω-models can now be defined as before, with $[\![\cdot = \cdot]\!]$ the interpretation of $=$ on each sort, and E the interpretation of the existence predicate $\boldsymbol{E}$; we only have to adapt the notion of valuation for the quantifiers

$$[\![\forall x Ax]\!] := \bigwedge_{a \in X} (Ea \to [\![Aa]\!]) \equiv \bigwedge_{a \in X} [\![\boldsymbol{E}a \to Aa]\!],$$

$$[\![\exists x Ax]\!] := \bigvee_{a \in X} (Ea \wedge [\![Aa]\!]) \equiv \bigvee_{a \in X} [\![\boldsymbol{E}a \wedge Aa]\!],$$

where X is the domain for the sort of the variable x.

Note that global Ω-models are special cases of Ω-models where $Ex \equiv [\![x = x]\!] = \top$ for all x.

We can now prove in a routine way the following theorem.

6.7. THEOREM (*soundness for Ω-models*). *If* **IQCE** $+ \Gamma \vdash A$ *for a set of sentences* Γ *and a sentence* A, *then* $[\![A]\!] = \top$ *in each* Ω*-model for which* $[\![B]\!] = \top$ *for all* $B \in \Gamma$, *or in a short notation* $\Gamma \Vdash_{\mathrm{cHa}} A$. □

6.8. EXAMPLES *of nonglobal Ω-structures*. The examples of 6.4 are of course a fortiori examples of nonglobal Ω-structures. Some examples of properly nonglobal Ω-structures are given below.

Example (d). We consider a modification of example (b) above where for M we take $\mathbb{R}_T$, the set of all *partial* continuous functions from T to $\mathbb{R}$ with their domain open in T. The interpretation of $=, <, +, \cdot$ may be defined as before, and in addition we can put

$$Ea := \mathrm{Int}\{t : a(t) = a(t)\} = \mathrm{dom}(a).$$

We may add $^{-1}$ as a symbol for a partial function to the language and extend the interpretation by

$$[\![a^{-1}]\!] := \lambda t.a(t)^{-1},$$

where on the right-hand side $a(t)^{-1}$ is defined iff $a(t)$ is defined and $a(t)0$. As a standard example the reader should keep $\mathbb{R}_{\mathbb{R}}$ in mind.

Example (e). Kripke models with nonconstant domains are properly nonglobal Ω_{K}-structures, where

$$[\![d = d']\!] := \{k : d \sim_k d' \wedge d \in \mathrm{D}(k) \wedge d' \in \mathrm{D}(k)\}.$$

Similarly, for Beth models with nonconstant domain function

$$[\![d = d']\!] := \bigcup\{V_k : d \in \mathrm{D}(k) \wedge d \in \mathrm{D}(k) \wedge d \sim_k d'\}, \quad \text{etc.}$$

Restriction of attention to an **IQC**-language $\mathcal{L}$ with total constants and functions makes Beth and Kripke models into nonglobal structures for **IQC**-based theories only; in this connection it is to be noted, that restricting the forcing relation $k \Vdash A$ to $A \in \mathcal{L}(\mathrm{D}(k))$, has the effect that at any node k we are always dealing with elements which are total on the neighbourhood determined by k (i.e. $\{k' : k' \geq k\}$ in the case of Kripke models, V_k in the case of Beth models).

6.9. REMARK. A nonglobal Ω-structure $\mathscr{M}$ with only total functions may be regarded as a global Ω-structure $\mathscr{M}'$ if we put

$$[\![a = b]\!]_{\mathscr{M}'} := (Ea \vee Eb \to [\![a = b]\!]_{\mathscr{M}}) \equiv [\![a \simeq b]\!]_{\mathscr{M}}.$$

We shall use this observation below in 6.16. □

6.10. *Functions and functional relations.* Instead of interpreting a function symbol by a function in the Ω-structure, one may also consider the possibility of interpreting a function symbol F by an Ω-relation R_F with functional character:

$$R_F(\vec{a}, b) \wedge R_F(\vec{a}, b') \leq [\![b = b']\!].$$

This is a wider notion than the interpretation by functions, as is demonstrated by the following example: let us consider the two-point topological space $\{0,1\}$ with the discrete topology; Ω is the set of opens of this space. As on Ω-set, we take all total functions $\{0,1\} \to \mathbb{N}$ which are not constant. A function f from $\{0,1\}$ to $\mathbb{N}$ can be specified by a pair (a, b) such that $f(0) = a$, $f(1) = b$. Addition on this Ω-set exists as a functional relation

$$R_+(f, g, h) := \{x : f(x) + g(x) = h(x)\}.$$

We may think of R_+ as the interpretation of the ternary relation $\{(x, y, z) : x + y = z\}$. Clearly we expect $R_+((0,1),(1,0),(1,1))$, but $(1,1)$ is not in the Ω-set. On the other hand,

$$R_+((0,1),(1,0),(1,0)) = \{0\}, \qquad R_+((0,1),(1,0),(0,1)) = \{1\}.$$

As a result we find

$$\exists x\, R_+((0,1),(1,0), x) = \top,$$

but we cannot find an a in the Ω-set such that $R_+((0,1),(1,0), a) = \top$. (If we use the suggestive notation $[\![x + y = z]\!]$ for $R_+(x, y, z)$, we can say that $[\![(0,1) + (1,0) = (1,0)]\!] = \{0\}$, $[\![(0,1) + (1,0) = (0,1)]\!] = \{1\}$, $[\![\exists x((0,1) + (1,0) = x)]\!] = \{0,1\} = \top$.)

The distinction between functions and functional relations disappears for sheaves, which may be regarded as Ω-sets with additional properties and structure (cf. section 14.1).

Our next aim will be to extend the completeness theorem for **IPC**, relative to algebraic semantics, to **IQC** and **IQCE**. We again use a Lindenbaum-algebra construction. First we have to define a generalization of the notions of Ω-structure and global Ω-structure in which the requirement of completeness of the Heyting algebra used for the valuation is relaxed. We carry out the developments in some detail for **IQCE**; the case of global models for **IQC** is very similar.

6.11. DEFINITION. Let Θ be an Ha. A Θ-structure is defined exactly as an Ω-structure. A Θ-structure is *definitionally complete* (for the language considered) if, whenever $[\![Ba]\!] \in \Theta$ for all $a \in X_i$, we also have

$$\bigvee\{Ea \wedge [\![Ba]\!] : a \in X_i\} \in \Theta, \qquad \bigwedge\{Ea \to [\![Ba]\!] : a \in X_i\} \in \Theta$$

for all formulas $B(x)$ of the language (x of sort i). Global Θ-structures and definitional completeness of global Θ-structures, for languages based on **IQC**, may be defined similarly, dropping Ea in the definition above. □

6.12. THEOREM. *Let Γ be a theory formulated in the language $\mathscr{L}$. Then there is a Θ-structure $\mathscr{M}$ which is definitionally complete (relative to $\mathscr{L}$), such that*

$$\Gamma \vdash A \quad \textit{iff} \quad \mathscr{M} \Vdash A \quad \textit{iff} \quad [\![A]\!]_{\mathscr{M}} = \top .$$

PROOF. We give the proof for Γ in a one-sorted language. As the domain D of $\mathscr{M}$ we take the set of all terms of the language, including the free variables. For Θ we take the set of equivalence classes $\{[A] : A$ a formula of $\mathscr{L}\}$, where

$$[A] := \{B : \Gamma \vdash A \leftrightarrow B\},$$

with a partial order

$$[A] \le [B] := \Gamma \vdash A \to B.$$

The result is an Ha such that:

$$\bot_\Theta = [\bot], \tag{1}$$

$$[A \circ B] = [A] \circ [B] \quad \text{for } \circ \in \{\wedge, \vee, \to\} \tag{2}$$

and

$$[\forall x Ax] = \bigwedge\{[Ed] \to [Ad] : d \in \mathrm{D}\}, \tag{3}$$

$$[\exists x Ax] = \bigvee\{[Ed] \wedge [Ad] : d \in \mathrm{D}\}. \tag{4}$$

For example, let us check (3). Let $d \in \mathrm{D}$; d is a term of the language, so

$$\vdash \forall x Ax \wedge Ed \to Ad$$

and therefore $\vdash \forall x Ax \to (Ed \to Ad)$, i.e.

$$[\forall x Ax] \leq [Ed] \to [Ad] \quad \text{for all } d \in \mathrm{D}.$$

Conversely, let

$$[C] \leq [Ed] \to [Ad] \quad \text{for all } d \in \mathrm{D},$$

then in particular

$$[C] \leq [Ex \to Ax]$$

for some variable x not occurring free in C, hence

$$\Gamma \vdash C \to (Ex \to Ax),$$

so $\Gamma \vdash C \to \forall x Ax$, which means $[C] \leq [\forall x Ax]$. Therefore also

$$\bigwedge\{[Ed] \to [Ad] : d \in \mathrm{D}\} \leq [\forall x Ax].$$

The verification of (4) we leave to the reader. We obtain a Θ-structure if we put

$$[\![d = d']\!] := [d = d'],$$

$$[\![Ed]\!] := [Ed],$$

$$[\![R(d_1, \ldots, d_n)]\!] := [R(d_1, \ldots, d_n)],$$

while the function symbol f is interpreted by the mapping assigning to terms $t_1, \ldots, t_n$ the term $f(t_1, \ldots, t_n)$. □

For languages based on **IQC**, the proof can be given in a completely similar way, dropping all reference to E, thus yielding a definitionally complete global Θ-structure.

The following theorem will enable us to strengthen 6.12 to a completeness theorem for Ω-structures.

6.13. THEOREM. *Any* Ha *can be embedded in a* cHa *preserving* $\wedge, \vee, \to, \bot$ *and all existing meets and joins.*

PROOF. We first define the notion of a c-ideal (*complete ideal*). Let $\Theta \equiv (A, \wedge, \vee, \bot, \to)$ be an Ha. A c-ideal is a subset $I \subset A$ satisfying

$$\bot \in I;$$

$$b \in I \quad \text{and} \quad a \leq b \Rightarrow a \in I;$$

$$X \subset I \quad \text{and} \quad \bigvee X \quad \text{exists in } \Theta \Rightarrow \bigvee X \in I.$$

Now let Ω be the lattice of c-ideals of Θ ordered by inclusion; we can show that Ω has arbitrary $\bigwedge$ given by set-theoretic intersection, and $\bigvee$ by

$$\bigvee_{b \in B} I_b := \left\{ \bigvee X \colon X \subset \bigcup_{b \in B} I_b \text{ and } \bigvee X \text{ exists} \right\}$$

This is again a c-ideal, and we can moreover show the validity of the infinitary distributive law (D)

$$I \wedge \bigvee_{b \in B} I_b = \bigvee_{b \in B} (I \wedge I_b)$$

(left as an exercise). Thus Ω is a cHa. We define the embedding $i \in \Theta \to \Omega$ by

$$i(x) := \{ y \in \Theta \colon y \leq x \}.$$

i preserves $\wedge, \vee, \to, \bot$, and existing $\bigwedge, \bigvee$ (exercise). □

6.14. Remark. Ω as constructed above is also a universal completion, that is, Ω is minimal in the following sense: for all complete Ω', all $f \in \Theta \to \Omega'$ preserving $\wedge, \top$, and existing $\bigvee$, there is a unique f' preserving $\wedge, \top$, and all existing $\bigvee$ such that fig. 13.5 commutes, i.e. $f' \circ i = f$ (exercise.)

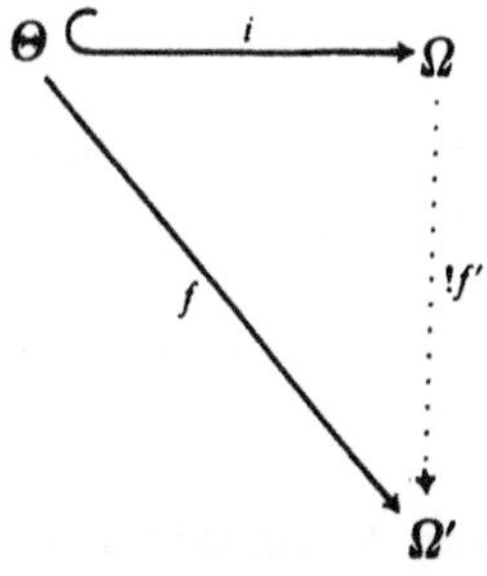

Fig. 13.5.

6.15. Corollary. *Let Γ be a theory, then there is a* cHa Ω *and an Ω-structure $\mathcal{M}$ such that $\Gamma \vdash A$ iff $[\![A]\!] = \top$ in $\mathcal{M}$.*

Proof. Immediate, combining 6.12 and 6.13. □

The corollary gives us immediately the completeness theorem.

THEOREM (*completeness theorem for Ω-models*). $\Gamma \vdash A$ *iff* $\Gamma \Vdash_{\mathrm{cHa}} A$. □

6.16. As a simple application, we can prove the theorem on the relationship between E-logic and ordinary logic described in 2.2.8 (cf. 10.5.6). We assume the E-symbol of **IQC*** to be written as E.

THEOREM.

$$\mathbf{IQCE} + \Gamma \vdash A \Leftrightarrow \mathbf{IQC}^* + \Gamma^E \vdash A^E.$$

PROOF. $\Rightarrow$ is proved by induction on the length of derivations.
$\Leftarrow$. Assume $\mathbf{IQC}^* + \Gamma^E \vdash A^E$; construct an Ω-structure $\mathcal{M}$ relative **IQCE**, validating Γ as in 6.15. $\mathcal{M}$ may at the same time be regarded as a global Ω-structure $\mathcal{M}^*$ relative to **IQC*** if we put

$$[\![x = y]\!]_{\mathcal{M}^*} := [\![x \simeq y]\!]_{\mathcal{M}},$$

retain the interpretation of terms, and put for all other relations of the language

$$[\![R(t_0, \ldots, t_{n-1})]\!]_{\mathcal{M}^*} := [\![R(t_0, \ldots, t_{n-1})]\!]_{\mathcal{M}}.$$

Now prove by induction on the complexity of A

$$[\![A^E]\!]_{\mathcal{M}^*} := [\![A]\!]_{\mathcal{M}}.$$

Consider e.g. $A := t \simeq s$; then $[\![(t \simeq s)^E]\!]_{\mathcal{M}^*} = [\![Et \vee Es \to t = s]\!]_{\mathcal{M}^*} = [\![Et \vee Es \to t \simeq s]\!]_{\mathcal{M}} = [\![t \simeq s]\!]_{\mathcal{M}}$. Since Γ holds in $\mathcal{M}$, Γ^E holds in $\mathcal{M}^*$, therefore (by soundness) A^E holds in $\mathcal{M}^*$; hence A holds in $\mathcal{M}$, so $\mathbf{IQCE} + \Gamma \vdash A$. □

7. Validity as forcing

Kripke and Beth models, originally introduced via a relation "k forces A", were seen to be special cases of Ω-structures. We now show that, conversely, we can extend the forcing approach to arbitrary Ω-structures.

7.1. DEFINITION. Let $\Theta \subset \Omega$ be a *basis* for Ω, i.e.

$$\forall p \in \Omega(p = \bigvee\{q \in \Theta : q \leq p\}).$$

Below we shall use q, q', q'' for elements of Θ. For any Ω-model $\mathcal{M}$ (relative **IQCE**) with an Ω-set $\mathcal{X}$ as domain we can define forcing as a relation between elements of Θ and atomic sentences A (in the language with constants for all elements of $\mathcal{M}$) by

$$q \Vdash A := q \leq [\![A]\!].$$

We define the *domain of* $\mathcal{M}$ *at* q, or the *set of sections of* $\mathcal{X}$ *over* q by

$$\mathcal{X}(q) := \{a \in M : q \Vdash Ea\}.$$

N.B. $q \Vdash A$ is *monotone* in q:

$$q \Vdash A \quad \text{and} \quad q' \leq q \Rightarrow q' \Vdash A$$

(observe that the ordering is the converse of the ordering in the usual Kripke/Beth forcing), and has the *covering property*

$$\forall q \in I(q \Vdash A) \Rightarrow \forall q' \leq \bigvee I(q' \Vdash A).$$

(If $\bigvee I$ is again in Θ the conclusion becomes $\bigvee I \Vdash A$).

$\mathcal{X} \subset \Omega$ is said to *cover* p if $p = \bigvee\{q : q \in \mathcal{X}\}$; let us write $\mathscr{C}(p)$ for the collection of covers of p contained in Θ. We can now extend the definition of $\Vdash$ to arbitrary sentences by the clauses

$$q \Vdash A \wedge B := q \Vdash A \quad \text{and} \quad q \Vdash B;$$

$$q \Vdash A \to B := \forall q' \leq q(q' \Vdash A \Rightarrow q' \Vdash B);$$

$$q \Vdash A \vee B := \exists I \in \mathscr{C}(q)\forall q' \in I(q' \Vdash A \quad \text{or} \quad q' \Vdash B);$$

$$q \Vdash \forall x A(x) := \forall q' \leq q \forall a \in \mathcal{X}(q')(q' \Vdash A(a));$$

$$q \Vdash \exists x A(x) := \exists I \in \mathscr{C}(q)\forall q' \in I\, \exists a \in \mathcal{X}(q')(q' \Vdash A(a)). \quad \square$$

Then we have the following proposition.

7.2. PROPOSITION.

(i) *For all sentences A, $q \Vdash A$, as defined above, is monotone in q and has the covering property, and satisfies*

$$q \Vdash A \quad \textit{iff} \quad q \leq [\![A]\!].$$

(ii) *If we define $q \Vdash A := q \leq [\![A]\!]$, then the clauses above are fulfilled.*

(iii) *A forcing which is monotone and has the covering property for atomic sentences determines an Ω-model such that $q \Vdash A$ iff $q \leq [\![A]\!]$.*

PROOF. Left to the reader. $\square$

7.3. Remark. As a result of the assumption of strictness for the basic relations and functions in the language, forcing for sentences with all constants existing locally (*d exists locally at p* iff $d \in \mathscr{X}(p)$ iff $p \Vdash Ed$) completely determines forcing for all sentences.

Beth and Kripke forcing are now obviously special cases of forcing as defined in 7.1. □

7.4. Remark. The constructive completeness proof for Beth models with nonstandard semantics for $\perp$ (case (iii) in 2.1) is now quite easily transformed into an alternative constructive completeness proof for validity in (special) Ω-models.

The completeness proof of section 2 yields completeness with respect to a generalized Ω-model $\mathscr{M}$, for $\Omega = 2^{\mathbb{N}}$, in the sense that $[\![\perp]\!]_{\mathscr{M}}$ need not coincide with $\perp_\Omega = \emptyset$, and that $[\![\perp]\!]_{\mathscr{M}} \leq [\![A]\!]_{\mathscr{M}}$ for all A. Let $\varphi : p \mapsto p \vee [\![\perp]\!]_{\mathscr{M}}$, then φ maps Ω homomorphically onto $\Omega^* := \{ p \geq [\![\perp_{\mathscr{M}}]\!] : p \in \Omega\}$, with $\perp_{\Omega^*} := [\![\perp]\!]_{\mathscr{M}}$. φ transforms the generalized Ω-model $\mathscr{M}$ into a standard Ω^*-model, and for all sentences A

$$[\![A]\!]_{\mathscr{M}^*} = \varphi[\![A]\!]_{\mathscr{M}}.$$

Finally we discuss the relationship between the "Aczel slash" introduced in 3.5.7. and "connectification" of Ω-models; as an application we give a model-theoretic proof of the existence property for **HA**. For simplicity of exposition, we give a classical treatment. In the exercises is indicated how connectification can be treated constructively.

7.5. Let Ω be a cHa, and let Ω^* be obtained from Ω by adding a new top $*$ ($= \top_{\Omega^*}$) to Ω *above* all elements of Ω, that is to say the elements of Ω^* are the elements of Ω plus $*$, and $\leq^*$ is given by

$$p \leq^* q \quad \text{iff } p \leq q, \quad \text{for } p, q \in \Omega;$$

$$* \leq^* *; \qquad p \leq^* * \quad \text{for all } p \in \Omega.$$

Ω^* has the property that every cover of $*$ contains $*$: Ω^* is *indecomposable* (this property is sometimes called strong compactness, but this term may cause confusion). We shall call Ω^* the *connectification* of Ω. See fig. 13.6.

Let $\mathscr{M}$ be an Ω-structure relative to **IQCE**, with an Ω-set $\mathscr{X} = (X, [\![\cdot = \cdot]\!])$ as domain, and let C be a set of constants in the language with parameters from $\mathscr{M}$, such that $\top_\Omega \Vdash Ec$ for all $c \in$ C. Without loss of generality we may assume $X \cap \mathrm{C} = \emptyset$.

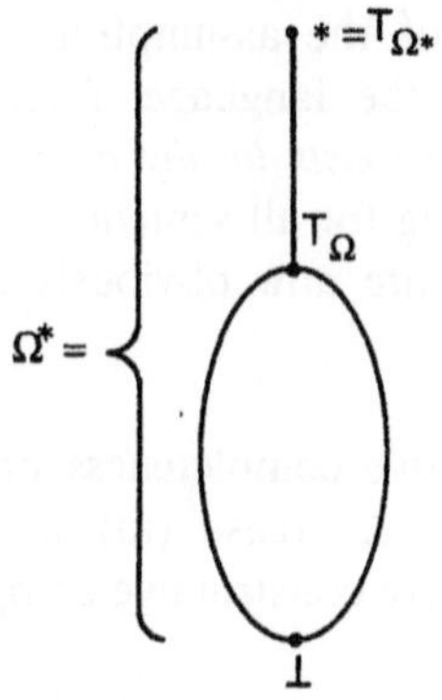

Fig. 13.6.

Now we construct an Ω^*-structure $\mathcal{M}^*$ with domain $\mathcal{X}^* = (X \cup \mathrm{C}, [\![\cdot = \cdot]\!]^*)$, as follows.

$$[\![c = c']\!]^* := [\![c = c']\!] \cup \{* : c = c'\} \quad \text{for } c, c' \in \mathrm{C},$$

$$[\![d = d']\!]^* := [\![d = d']\!] \quad \text{for } d, d' \in X,$$

$$[\![c = d]\!]^* = [\![d = c]\!]^* := [\![[\![c]\!] = d]\!] \quad \text{for } c \in \mathrm{C},\ d \in X,$$

where $[\![c]\!]$ is the interpretation of $c \in \mathrm{C}$ in $\mathcal{M}$. Thus $\mathcal{X}^*(*) = \mathrm{C}$. We put

$$* \Vdash^* P \ \text{ iff } \ \top_\Omega \Vdash P,$$

for all prime sentences P with parameters in $\mathcal{L}(\mathrm{C})$, and

$$\forall p \in \Omega (p \Vdash^* P \text{ iff } p \Vdash P)$$

for all prime sentences P with parameters locally existing in p. $\mathcal{M}^*$ is called the *connectification* of $\mathcal{M}$. We now easily prove the following proposition.

7.6. PROPOSITION. *For all sentences A with parameters from $\mathcal{M}$ and all $p \in \Omega$:*

$$p \Vdash A \ \text{ iff } p \Vdash^* A.$$

Forcing in $*$ satisfies:
(i) $* \Vdash^* P \Leftrightarrow \top_\Omega \Vdash^* P \Leftrightarrow \top_\Omega \Vdash P$;
(ii) $* \Vdash^* A \wedge B \Leftrightarrow * \Vdash^* A$ *and* $* \Vdash^* B$;
(iii) $* \Vdash^* A \vee B \Leftrightarrow * \Vdash^* A$ *or* $* \Vdash^* B$;

(iv) $* \Vdash^* A \to B \Leftrightarrow \top_\Omega \Vdash A \to B$ *and* $(* \Vdash^* A \Rightarrow * \Vdash^* B)$;
(v) $* \Vdash^* \exists x Ax \Leftrightarrow \exists c \in \mathrm{C}(* \Vdash^* Ac)$;
(vi) $* \Vdash^* \forall x Ax \Leftrightarrow \forall c \in \mathrm{C}(* \Vdash^* Ac)$, *and* $\top_\Omega \Vdash \forall x Ax$.

PROOF. Completely straightforward; clauses (iii) and (v) use the (classical) fact every cover of $*$ contains $*$. □

7.7. REMARK. If we now take for $\mathscr{M}$ the model produced by the Lindenbaum construction, we have at $\top_\Omega$: $\top_\Omega \Vdash A$ iff $\Gamma \vdash A$, or $\vdash_\Gamma A$. Reading $* \Vdash A$ as "$|A$" we see that in this case the clauses above have the same form as the definition of the Aczel slash in 3.5.7.

This suggests an alternative, model-theoretic version of the proof of DP and EDN, namely by means of the connectification.

Take e.g. **HA**, formulated on the basis of **IQCE**. As we know this yields the same provable sentences as in the usual language on the basis of **IQC**. Construct $\mathscr{M}$ by the Lindenbaum construction, and $\mathscr{M}^*$ as before, with C the set of *numerals* $0, 1, 2, 3, \ldots$.

Now we have to verify that $\mathscr{M}^*$ is again a model of **HA**. The crucial axioms are the instances of induction. We have to show (writing $\Vdash$ for $\Vdash^*$)

$$* \Vdash A(0) \wedge \forall x[A(x) \to A(Sx)] \to \forall x A(x).$$

Since induction is already valid in $\mathscr{M}^*$ below $\top_\Omega$, we only have to show

$$* \Vdash A(0) \wedge \forall x[A(x) \to A(Sx)] \Rightarrow * \Vdash \forall x A(x).$$

Assume therefore

$$* \Vdash A(0), \qquad * \Vdash \forall x[A(x) \to A(Sx)];$$

The second assertion means, since $\mathscr{X}^*(*) = \mathrm{C} = \{0, 1, 2, \ldots\}$, that $* \Vdash A(\bar{n}) \to A(\bar{n} + 1)$, hence by induction $\forall n(* \Vdash A(\bar{n}))$; and since $\top_\Omega \Vdash \forall x A(x)$, indeed $* \Vdash \forall x Ax$.

Now let $\mathbf{HA} \vdash \exists x A(x)$; it follows that $* \Vdash \exists x A(x)$ (since $\mathscr{M}^*$ is a model of **HA**), and thus $* \Vdash A(\bar{n})$ for some $n \in \mathbb{N}$; but this entails $\top_\Omega \Vdash A(\bar{n})$, i.e. $\mathbf{HA} \vdash A(\bar{n})$.

This model-theoretic proof could already have been given for Kripke models, and also has a variant for Beth models; see 9.7.

For another application of this argument see E13.7.3.

8. Postscript on realizability

This section does not deal with completeness questions, but supplements the earlier discussions of realizability (sections 4.4, 9.5) by sketching how realizability can also be viewed as a semantics of the truth-value type, along similar lines as the Ω-valued semantics described in the preceding sections.

8.1. *Logical operators.* For any sentence A in $\mathscr{L}(\mathbf{HA})$ let $[\![A]\!]$ be the set of its realizing numbers for numerical $\mathbf{r}$-realizability (4.4.2), i.e. $[\![A]\!] := \{x : x\,\mathbf{r}\,A\}$. The propositional operators $\wedge$, $\vee$, and $\to$ then correspond to certain operations on subsets of $\mathbb{N}$, for example

$$[\![A \to B]\!] := \{z : \forall x \in [\![A]\!](\{z\}(x) \in [\![B]\!])\}.$$

We may think of the set $[\![A]\!]$ as the "truth-value" of A under the realizability interpretation, and $\to, \wedge, \vee$ then correspond to certain "truth-functions". Thus for arbitrary subsets $X, Y \subset \mathbb{N}$ we may define

$$X \wedge Y := \{x : j_1 x \in X \wedge j_2 x \in Y\};$$

$$X \vee Y := \{(x,0) : x \in X\} \cup \{(y, Sz) : y \in Y, z \in \mathbb{N}\};$$

$$X \to Y := \{x : \forall y \in X(\{x\}(y) \in Y)\}.$$

If we identify $\perp$ with $\emptyset$, then

$$\neg X := X \to \emptyset.$$

We also put

$$X \leftrightarrow Y := (X \to Y) \wedge (Y \to X).$$

We have permitted ourselves an abuse of language in using the symbols for the propositional operators, but this will not cause confusion.

It is not immediately clear how to define truth-function analogues to the quantifiers. In particular, if we look at realizability for **HAS** (cf. E4.4.11) we see that the set-quantifiers are handled very differently from the numerical quantifiers:

$$x\,\mathbf{r}\,\forall X A := \forall X(x\,\mathbf{r}\,A),$$

$$x\,\mathbf{r}\,\exists X A := \exists X(x\,\mathbf{r}\,A)$$

(i.e. in the definition in E4.4.11 we have identified X^* with X for notational simplicity). Below we shall show how quantification over such different domains can nevertheless for realizability be uniformly represented.

As analogues of quantification we take unions and intersections, but use the suggestive notation of join and meet: that is to say, if $\mathscr{X} \subset \mathrm{P}(\mathrm{P}(\mathbb{N}))$, we put

$$\bigvee \mathscr{X} := \bigcup \mathscr{X}, \qquad \bigwedge \mathscr{X} := \bigcap \mathscr{X}.$$

Note that this corresponds to the treatment of set-quantifiers in **HAS**: $x \in [\![\forall X A]\!]$ iff $\forall X(x \in [\![A]\!])$, etc. It does not correspond to the treatment of numerical quantifiers.

The definition of the operations $\wedge$, $\vee$, and $\rightarrow$ on subsets of $\mathbb{N}$ may be generalized to any model $\mathscr{A} \equiv (V, \bullet, k, s, p, p_0, p_1, d, 0, S, P)$ of **APP** in the obvious way: for $X, Y \subset \mathrm{P}(V)$

$$X \wedge Y := \{p \bullet x \bullet y : x \in X \wedge y \in Y\};$$

$$X \vee Y := \{p \bullet x \bullet 0 : x \in X\} \cup \{p \bullet y \bullet (S \bullet 0) : y \in Y\};$$

$$X \rightarrow Y := \{x : \forall y \in X(x \bullet y \in Y)\}.$$

8.2. *Realizability sets and relations.* Let $\mathscr{A}$ be a model of **APP** as above; we let $\mathrm{P}(V)$, the powerset of V, play the role of a collection of truth-values. To emphasize this we use Ω for $\mathrm{P}(V)$ below (this use of Ω for realizability truth values, is strictly local; elsewhere in this book Ω is used for a cHa). There is, however, not a single element of Ω representing truth, but a whole collection: all inhabited elements of Ω denote truth ("a sentence is realizable iff there is a realizing element for it"). We use Ω^* for the collection of inhabited elements of Ω.

An *Ω-set* will now be a pair $(X, [\![\cdot = \cdot]\!])$ with $\lambda x, y.[\![x = y]\!] \in X \times X \rightarrow \Omega$ such that the axioms for equality hold

$$\bigwedge_{x, y \in X} ([\![x = y]\!] \rightarrow [\![y = x]\!]) \in \Omega^*,$$

$$\bigwedge_{x, y, z \in X} ([\![x = y]\!] \wedge [\![y = z]\!] \rightarrow [\![x = z]\!]) \in \Omega^*;$$

$\bigwedge$ plays the role of a universal quantifier. Spelling it out, this means that we must have elements $a, b \in V$ such that for all $u, u_0, u_1 \in V$

$$u \in [\![x = y]\!] \rightarrow a \bullet u \in [\![y = x]\!],$$

$$(u_0 \in [\![x = y]\!]) \wedge (u_1 \in [\![y = z]\!]) \rightarrow b \bullet (p \bullet u_0 \bullet u_1) \in [\![x = z]\!].$$

As an example, choose PRO as the model for **APP**, and let $\hat{\mathbb{N}}$ (the "standard" interpretation of $\mathbb{N}$) be the Ω-set $(\mathbb{N}, [\![\cdot = \cdot]\!])$ with

$$[\![n = m]\!] := \{n : n = m\}.$$

Let $\mathscr{X}_i \equiv (X_i, [\![\cdot = \cdot]\!]_i)$ for $i < n$ be Ω-sets, then the *product* $\mathscr{X}_0 \times \cdots \times \mathscr{X}_{n-1}$ is the Ω-set $(X_0 \times \cdots \times X_{n-1}, [\![\cdot = \cdot]\!])$ with equality given by

$$[\![(x_0, \ldots, x_{n-1}) = (y_0, \ldots, y_{n-1})]\!] :=$$
$$[\![x_0 = y_0]\!]_0 \wedge \cdots \wedge [\![x_{n-1} = y_{n-1}]\!]_{n-1}.$$

In order to be certain that this definition is correct for $n > 2$, we have to prove associativity, that is to say we must show

$$((X \wedge Y) \wedge Z \leftrightarrow X \wedge (Y \wedge Z)) \in \Omega^*$$

for arbitrary $X, Y, Z \in \Omega$; we leave this as an exercise.

A (strict) *Ω-predicate* (*unary Ω-relation*) on an Ω-set $(X, [\![\cdot = \cdot]\!])$ is a mapping $R \in X \to \Omega$ such that, if we put

$$Ex := [\![x = x]\!],$$

$$[\![\text{strict}(R)]\!] := \bigwedge\{(Rx \to Ex) : x \in X\},$$

$$[\![\text{repl}(R)]\!] := \bigwedge\{(Rx \wedge [\![x = y]\!] \to Ry) : x, y \in X\},$$

then $[\![\text{strict}(R)]\!] \in \Omega^*$ (R is *strict*) and $[\![\text{repl}(R)]\!] \in \Omega^*$ (R satisfies *replacement*).

Ω-relations on $\mathscr{X}_0, \ldots, \mathscr{X}_{n-1}$ may be defined as Ω-predicates on $\mathscr{X}_0 \times \cdots \times \mathscr{X}_{n-1}$.

Returning to the example $\hat{\mathbb{N}}$ of an Ω-set, we see that $R \in \mathbb{N} \to \Omega$ represents an Ω-predicate if there are n_0, n_1 such that

$$\forall nm(m \in R(n) \to \{n_0\}(m) = n), \tag{1}$$

$$\forall nn'm(j_1 m \in R(n) \wedge j_2 m \in [\![n = n']\!] \to \{n_1\}(m) \in R(n')). \tag{2}$$

The second condition is trivially fulfilled, since if $j_2 m \in [\![n = n']\!]$, then $n = n'$, and hence $n_1 = \Lambda m. j_1 m$ satisfies (2).

8.3. *Realizability morphisms.* A (strict) *Ω-morphism* from $\mathscr{X}_0$ to $\mathscr{X}_1$ is a strict and functional relation R on $\mathscr{X}_0$, $\mathscr{X}_1$, i.e. besides the requirements for relations we must also satisfy ($\mathscr{X}_i \equiv (X_i, [\![\cdot = \cdot]\!]_i)$ for $i \leq 1$):

$$\bigwedge_{x \in X_0} \bigwedge_{y,z \in X_1} \{R(x, y) \wedge R(x, z) \to [\![y = z]\!]_1\} \in \Omega^*, \tag{1}$$

$$\bigwedge_{x \in X_0} \Big(Ex \to \bigvee_{y \in X_1} R(x, y) \Big) \in \Omega^*. \tag{2}$$

Condition (2) expresses that the function represented by the relation is total.

We observe that in our example $\hat{\mathbb{N}}$, there corresponds to $f \in \mathbb{N} \to \mathbb{N}$ a morphism F from $\hat{\mathbb{N}}$ to $\hat{\mathbb{N}}$ given by

$$F(n, m) := \{0 : f(n) = m)\}.$$

8.4. *Realizability models.* It should now be clear how to define the notion of an Ω-model for any first-order language $\mathscr{L}$, namely by specifying an Ω-set $\mathscr{X}$ together with certain elements c of $\mathscr{X}$ with $Ec \in \Omega^*$ as interpretation of the constants, and certain Ω-relations and Ω-morphisms for the interpretation of the relations and functions of the language.

To keep the presentation simple, let us assume the Ω-morphisms to be representable by actual functions $f \in X^n \to X$. It is then obvious how to interpret terms in the language $\mathscr{L}(X)$ ($\equiv \mathscr{L}$ extended with constants for all elements of X).

Sentences A in $\mathscr{L}(X)$ get assigned a truth value $[\![A]\!]$ as follows. For prime formulas $[\![t = t']\!]$ has already been specified, for $R(t_1, \ldots, t_n)$, R a predicate symbol of the language, we take

$$[\![R(t_1, \ldots, t_n)]\!] := [\![R]\!](t_1, \ldots, t_n),$$

where $[\![R]\!]$ is the Ω-relation interpreting R, and for compound formulas we let

$$[\![\bot]\!] := \emptyset,$$

$$[\![A \circ B]\!] := [\![A]\!] \circ [\![B]\!] \quad \text{for } \circ \in \{\wedge, \vee, \to\},$$

$$[\![\forall x A]\!] := \bigwedge\{Ex \to [\![A]\!] : x \in X\},$$

$$[\![\exists x A]\!] := \bigvee\{Ex \wedge [\![A]\!] : x \in X\}.$$

If we take $\hat{\mathbb{N}}$ for $\mathscr{X}$, and for $\mathscr{L}$ the language of **HA**, we have the following theorem.

8.5. THEOREM. *For* PRO *as* **APP**-*model, and* $\Omega = \mathrm{P}(\mathbb{N})$, *we have for all sentences* A *of* $\mathscr{L}(\mathbf{HA})$

$$[\![A]\!] \in \Omega^* \Leftrightarrow \exists x(x\,\mathbf{r}\,A).$$

In fact there are recursive φ_1, φ_2 *such that*

$$(x\,\mathbf{r}\,A) \to \varphi_1 x \in [\![A]\!], \qquad y \in [\![A]\!] \to (\varphi_2 y\,\mathbf{r}\,A).$$

PROOF. Exercise. It is to be noted that a numerical quantifier $\forall n A$ is interpreted as $[\![\forall n A(n)]\!] \equiv \bigcap\{[\![n = n]\!] \to [\![A(n)]\!] : n \in \mathbb{N}\}$, that is to say $x \in [\![\forall n A(n)]\!]$ requires $\forall y(y \in [\![n = n]\!] \to \{x\}(y) \in [\![A(n)]\!])$ which is equivalent to $\forall n(\{x\}(n) \in [\![A(n)]\!])$. The dependency on n for the element in $[\![A(n)]\!]$ is here introduced via the conditional $Ex \to$. □

8.6. *Higher-order logic.* We restrict attention to PRO as model for **APP**. If to each Ω-set $\mathscr{X}$ we can associate an Ω-*powerset* $\mathscr{P}(\mathscr{X})$, and to each pair of Ω-sets $\mathscr{X}$ and $\mathscr{Y}$ an Ω-*function set* $\mathscr{Y}^{\mathscr{X}}$, we can interpret higher-order logic. Thus in order to interpret **HAS** we need an Ω-set $\mathscr{P}(\hat{\mathbb{N}})$ which can serve as the interpretation of the range of the subsets of $\mathbb{N}$ (the reader should compare the developments here with the treatment of power- and function-sheaves in section 14.5).

For $\mathscr{P}(\mathscr{X})$, we can take the collection of mappings $X \to \Omega$. We put

$$[\![x \in P]\!] := P(x),$$

$$[\![P = Q]\!] := \bigwedge\{P(x) \leftrightarrow Q(x) : x \in X\} \wedge [\![\mathrm{strict}(P)]\!] \wedge [\![\mathrm{repl}(P)]\!]$$

for $P, Q \in X \to \Omega$. It is to be noted that if

$$Y := [\![\forall x(x \in P \leftrightarrow x \in Q)]\!] \wedge [\![\mathrm{strict}(P)]\!] \wedge [\![\mathrm{repl}(P)]\!],$$

then $Y \leftrightarrow [\![P = Q]\!] \in \Omega^*$ (exercise).

In a similar way we can take for $\mathscr{Y}^{\mathscr{X}}$ the Ω-set consisting of all Ω-morphisms from $\mathscr{X} \equiv (X, [\![\cdot = \cdot]\!]_X)$ to $\mathscr{Y} \equiv (Y, [\![\cdot = \cdot]\!]_Y)$ with equality

$$[\![R = R']\!] := \bigwedge\{(R(x, y) \leftrightarrow R'(x, y)) \wedge [\![\mathrm{strict}(R)]\!] \wedge [\![\mathrm{repl}(R)]\!] : x \in X, y \in Y\}.$$

Then $\mathscr{P}(\hat{\mathbb{N}})$ consists of mappings $P \in \mathbb{N} \to \Omega$ with a number realizing strictness:

$$\exists k \forall nm(m \in P(n) \to \{k\}(m) = n).$$

8.7. THEOREM. *For* $\mathbf{r}$*-realizability as defined for* **HAS** (E4.4.11)

$$\mathbf{HAS} \vdash \exists x(x \mathbf{r} A) \Leftrightarrow [\![A]\!] \in \Omega^*.$$

PROOF. We shall leave most of the details to the reader. Crucial in the proof is the following observation. Each P of $\mathscr{P}(\mathbb{N})$ with $[\![\text{strict}(P)]\!] \in \Omega^*$ "is equal to" a standard P' with $\Lambda n. j_2 n \in [\![\text{strict}(P')]\!]$, i.e. $[\![P = P']\!] \in \Omega^*$. To see this, one has only to take for $P'(n) := \{j(m, n) : m \in P(n)\}$.

For standard elements Q of $\mathscr{P}(\mathbb{N})$ there is a fixed element in $[\![Q = Q]\!]$, namely $j(j(j(\Lambda n.n, \Lambda n.n), \Lambda n. j_2 n), \Lambda n. j_1 n)$. Quantification over $\mathscr{P}(\mathbb{N})$ may now be replaced by quantification over standard elements only; the fixed element in $[\![Q = Q]\!]$ makes the realizability of a quantifier $\forall X$ uniform. □

For more information see Hyland (1982).

9. Notes

9.1. *Beth models.* Beth models with constant domain over finitely branching trees were introduced by Beth (1956). Beth gave a classical completeness proof for **IQC** for his semantics, and presented further arguments purporting to show that completeness for his semantics held also intuitionistically. Gödel, Heyting, and Kreisel raised objections to this claim; the echo of this discussion may be seen in Beth (1959, 1959A).

Beth's completeness proof was very thoroughly investigated in Dyson and Kreisel (1961). In his book (1959) Beth pointed out that the root of the difficulty lay in the interpretation of completeness: for an intuitionistically valid argument for "A valid $\Rightarrow A$ provable" one needed validity of A on a fan of "potential models" (the ad-hoc terminology is ours) not all of which were (proper) Beth models, and the subset of (proper) Beth models of this fan is not necessarily a fan itself.

Veldman (1976, as preprint 1974) discovered that a liberalization of the notion of Kripke model, where it is permitted that certain nodes validate $\perp$, made it possible to give an intuitionistic Henkin-style completeness proof (cf. also Lopez-Escobar and Veldman 1975). De Swart (1976) adapted this idea to Beth models and a Gödel-style completeness proof (for exposition of De Swart's proof, see Troelstra (1977, ch. 7)). The generalization of Beth models considered by De Swart is a special case of the fallible

Beth models in our section 2; De Swart's models satisfy $k \Vdash \bot \Rightarrow \forall k' \in \mathrm{K}\ (k' \Vdash \bot)$.

An extensive discussion of the semantics proposed by Veldman and De Swart, with variants, is given in Dummett (1977, section 5.7); cf. also Troelstra (1981, Note C). In a sense Veldman (1976) and De Swart (1976) may be regarded as a rediscovery and continuation of Beth's ideas as found in his book (1959, section 145).

Dummett also observed that, since $\bot$ is the only logical constant for which the semantics of Veldman and De Swart are nonstandard, their proofs give in fact intuitionistically acceptable arguments for the completeness of the $\bot$-free fragment of **IQC** relative to ordinary Beth and Kripke semantics.

9.2. *Intuitionistic validity and validity in Beth models.* Independently of Dummett, De Swart, and Veldman, Friedman (1975A, 1977B, 1977C) gave a completeness proof for the $\bot$-free fragment; a version of this proof is presented in section 2. Friedman formulated his result not for Beth models, but for intuitionistic validity for predicates depending on a lawless parameter (cf. 1.9–10).

Kreisel (1958A) noted the connection between validity in topological models over a fan and intuitive validity for predicates depending on a lawless parameter. Since Beth models with constant domains are easily seen to be topological models over a spread, this can also be formulated as a connection between intuitionistic validity ($\models_i$) and Beth models (cf. also Kripke 1965, Kreisel 1970).

9.3. *The function-free fragment of the theory of a single Skolem function.* The characterization of the f-free part of the theory of a single Skolem function f is due to Mints (Smorynski 1977, 540–541); Smorynski (1977, 1978) gave classical proofs using Kripke semantics.

The proof in 2.11 was obtained by transferring the ideas of Smorynski (1977) to Friedman's completeness proof, thereby providing a straightforwardly constructive argument for the theorem.

A proof-theoretical (constructive) argument is in Schwichtenberg (1979); for generalizations see Motohashi (1982).

9.4. *The incompleteness result* in section 3 is taken from Kreisel (1962A), and is essentially due to Gödel. We already mentioned (3.8) the strong incompleteness of **IQC**, if CT_0 is assumed. Recently McCarty (1986B) showed that

$$T + \mathrm{WCT} + \mathrm{MP} \vdash (\{\ulcorner A \urcorner : \models_i A\} \text{ is nonarithmetic}),$$

where T is a suitable fragment of type theory with function comprehension, in which we can talk about domains and which permits the formulation of the standard definition of satisfaction for arithmetic (e.g. **CZF** of 11.8.2), MP is Markov's principle and WCT is a weakening of Church's thesis:

$$\mathrm{WCT} \quad \forall n(An \vee \neg An) \rightarrow$$
$$\neg\neg\exists m(\{m\} \text{ is total and } \forall n(\{m\}(n) = 0 \leftrightarrow An))$$

Leivant (1985A) mentions further unpublished incompleteness results by Friedman.

9.5. *Algebraic and topological semantics.* Heyting algebras were *not* (contrary to Johnstone 1982, p. 36) introduced by Heyting. In the literature Heyting algebras appear under a variety of other names, such as "Brouwerian lattices", "pseudo-complemented lattices", "pseudo-Boolean algebras". Jaskowski (1936) used some very special finite Heyting algebras for a completeness proof for **IPC**. Stone (1937) implicitly establishes completeness relative to valuations in a particular topological space; there is no *explicit* formulation of a completeness theorem. Tarski (1938), independently of Stone, explicitly obtains a completeness proof for **IPC** relative to valuations in suitable topological spaces (including $\mathbb{R}^n$). At the end of his paper, Tarski briefly sketches an extension of the topological semantics to valuations in a suitable class of lattices. Ogasawara (1939) elaborated this and noted that the appropriate class of lattices corresponded to the so-called residuated lattices with bottom element. Birkhoff (1940), inspired by Stone (1937), independently observed the correspondence between intuitionistic logic and residuated lattices; there is no explicit statement of a completeness theorem. McKinsey and Tarski (1946) contains an extensive study of "Brouwerian algebras", the dual of a Heyting algebra. Mostowski (1949) was the first to apply topological valuations to **IQC**, thereby obtaining underivability results. Rasiowa (1951) proved completeness of **IQC** for algebraic semantics. A standard source incorporating much of the work on algebraic semantics before 1963 is Rasiowa and Sikorski (1963).

The proof of 5.7 follows Fitting (1969).

9.6. *Ω-sets and structures.* Section 6 is mainly based on Fourman and Scott (1979). We have emphasized the total/partial distinction (corresponding to global Ω-sets versus Ω-sets) since the global structures are more directly analogous to models of classical predicate logic with values in Ω instead of the two-element Boolean algebra. Ω-sets were also studied by Higgs (1973), cf. Fourman and Scott (1979, bottom of p. 304). The study of Ω-sets is a natural historical sequel to the exploration of Boolean-valued models for classical set theory.

The global case of 6.12 is treated in Rasiowa and Sikorski (1963, X 3.2) going back to earlier papers by the same authors. A result similar to 6.13, but with a classical proof may be found in Rasiowa and Sikorski (1963, IV 10.1) going back to Sikorski (1958). The constructive argument for 6.13 has been noted by several people (see e.g. Grayson 1984).

9.7. *Slashing and glueing.* The essential idea of the "connectification" of 7.5, 7.6 appears in many guises in the literature, and applies not only to Ω-models, but also to Kripke models (cf. E2.6.6–7) and with suitable modifications also to Beth models. The idea is used in the algebraic case already in McKinsey and Tarski (1946, 1948). Smorynski (1973) makes extensive use of connectification for Kripke models.

The construction also has a syntactic variant ("slashing"), in particular the Aczel slash, Kleene slash, and its extensions. Smorynski explicitly noted the connection between his version of connectification and the Aczel slash (Smorynski 1973, 5.1.12–21).

For higher-order logic, Freyd described a model-theoretic construction applicable to elementary toposes (the Freyd-cover of a topos) which is the counterpart of Friedman's syntactically defined extension of the Kleene slash (cf. Scedrov and P. Scott 1982). As to variations on the construction of the Freyd cover, see e.g. Moerdijk (1982).

Van Dalen (1984, 1986A) applied glueing techniques for Beth models to obtain DP and ED for various theories in the language of **EL** (e.g. $\mathbf{EL} + \mathrm{BI}_{\mathrm{M}} + \mathrm{KS} + \mathrm{AC}_{01} + \mathrm{WC\text{-}N}$); the second paper shows how in some cases intuitionistic (nonclassical) principles can be used to prove their own preservation under the glueing operation.

9.8. *Realizability as a truth-value semantics.* The account in section 8 is based on Hyland (1982), where the abstract framework of Hyland et al. (1980) is applied to the specific case of recursive realizability. Hyland (1982) mentions unpublished work of W. Powell as a source of inspiration, and attributes the idea of realizability as a kind of truth-value semantics to D.S. Scott. Independently this idea was formulated for first-order theories by

Dragalin (1979). For our exposition we have also profited from some unpublished notes by Grayson.

Exercises

13.1.1. Complete the proof of the theorem in 1.5.

13.1.2. Complete the proof of the theorem in 1.6.

13.1.3. Prove (ii) of 1.10.

13.1.4. Extend the correspondence between intuitionistic validity and Beth models described in 1.8–1.9 to languages where equality and function symbols are present.

13.2.1. Make the necessary modifications in the completeness proof when clause (iv) in the definition of the labels (2.2) is replaced by (iv′) (2.10).

13.2.2. Extend the completeness proof to the case where function symbols are present. *Hint:* take care that, for any function symbol f, $\exists x(fc_i = x)$ is realized *uniformly* by some c_j at all relevant nodes of the model.

13.2.3. Show intuitionistically that there is a universal Beth model for **IPC**. *Hint:* Recall that we proved constructively the decidability of **IPC** in section 10.4.

13.2.4. Show, that for the following modification of the original construction, analogues of 2.3 and 2.5 can be proved for *all* sentences A (not just those in $\mathscr{L}(\Gamma_k)$). The labels are defined as follows. $\Gamma_{\langle\,\rangle} := \emptyset$.

If $\mathrm{lth}(n) = u$ there are three cases corresponding to (ii)–(iv) in 3.2:

(ii′) If $A_u \equiv B \vee C$, $\Gamma_n \vdash_u B \vee C$, put $\Gamma_{n*\langle 0\rangle} := \Gamma_n \cup \{B\}$, $\Gamma_{n*\langle 1\rangle} := \Gamma_n \cup \{C\}$.
(iii′) Let $A_u \equiv \exists x Bx$, $\Gamma_n \vdash_u \exists x Bx$, and suppose $\langle c_i^n\rangle_i$ enumerates all parameters not in $\Gamma_n \cup \{\exists x Bx\}$. Put $\Gamma_{n*\langle i\rangle} = \Gamma_n \cup \{Bc_i^n\}$.
(iv′) $\Gamma_{n*\langle 0\rangle} := \Gamma_n$, $\Gamma_{n*\langle 1\rangle} := \Gamma_n \cup \{A_u\}$ in all other cases.

Let $\mathrm{Fan}_{n,A}$ be the collection of subfans F of the tree, containing all predecessors of n, and such that for any $m \in F$, at any application of (iii′), the $\Gamma_{m*\langle i\rangle}$ permitted are such that $c_i^m \notin \mathscr{L}(A)$; at (i′), (ii′) all successors are taken. Show:

LEMMA. *For any* A, $\forall F \in \mathrm{Fan}_{n,A} \forall x[(\Gamma_n \vdash A) \Leftrightarrow \forall m \in F(\mathrm{lth}(m) = x \Rightarrow (\Gamma_{n*m} \vdash A))]$.

LEMMA. *Putting* $n \Vdash P := \Gamma_n \vdash P$ *for* P *prime, we find* $n \Vdash A \Leftrightarrow \Gamma_n \vdash A$.

13.2.5. Show that the construction in 2.2–5 also yields a model for any r.e. theory Γ.

13.2.6. Complete the proof in 2.11.

13.2.7. Adapt the completeness proof to **IQCE**.

13.4.1. Show the equivalence of the two definitions in 4.1 and 4.2.

13.4.2. Show that D1 implies D2 and vice versa (4.3).

13.4.3. Complete the proof of 4.6.

13.4.4. Prove 4.8.

13.4.5. Prove the assertion in 4.10.

13.4.6. Give a proof of 4.13.

13.5.1. Complete the proof of the soundness theorem 5.3.

13.5.2. Complete the proof of 5.4.

13.5.3. Show (5.6) $[\![A]\!] = \{k : k \Vdash A\}$, for all A.

13.5.4. Complete the proof of 5.7.

13.5.5. Give a complete list of all homomorphic images of the Rieger–Nishimura lattice. *Hint:* Show that a homomorphism φ can be characterized by $\mathrm{Ker}_\varphi := \{a : \varphi a = \top\}$; for the Rieger–Nishimura lattice, Ker_φ always has a minimal element.

13.5.6. If $(A, \wedge, \vee)$ is a lattice, then $a \in A$ is *join-irreducible* iff $a \neq \bigvee\{b : b < a\}$. For any lattice A let A^0 be the set of join-irreducibles of A. A is *join-representable* iff for all $a \in A$ $a = \bigvee\{a' : a' \in A^0\}$.

If $(P, \leq)$ is any poset, Ω_P is the lattice of upwards monotone subsets with $\cap, \cup$ as lattice operations. Show that $(\Omega_P, \cap, \cup)$ is join-representable, and that $a \in \Omega_P$ is join-irreducible iff it is of the form $\{q : q \geq p\}$ for some $p \in P$.

Show also that for complete, join-representable lattices L satisfying the infinitary distributive law (D) L is isomorphic to Ω_{L^0}.

13.5.7. Show that the Rieger–Nishimura lattice is join-representable; what does the corresponding poset look like? How can one transform a valuation in a join-representable Ha into a Kripke model? Construct the posets corresponding to some finite Heyting algebras.

13.6.1. Prove the soundness theorem 6.7.

13.6.2. Show that in example 6.4(b) with $T = \mathbb{R}$:

$$[\![a < b \to a < c \vee c < b]\!] = T, \qquad [\![\forall xy(x < y \vee y < x)]\!] = \emptyset.$$

What about $[\![\forall xy(x \leq y \vee y \leq x)]\!]$?

13.6.3. Show in example 6.4(c) that, for all sentences A of $\mathscr{L}(\mathrm{D})$, $[\![A]\!] := \bigcup\{V_k : k \Vdash A\}$.

13.6.4. Show that (i) the c-ideals are closed under infinite joins and satisfy (D), and (ii) the embedding i preserves $\wedge, \vee, \to, \bot$ and $\bigwedge, \bigvee$ where existing (theorem 6.13). Prove also the remark 6.14.

13.6.5. Check $\Rightarrow$ in theorem 6.16.

13.7.1. Prove monotonicity and the covering property (7.1).

13.7.2. Prove 7.2.

13.7.3. Extend the method of 7.7 to prove the numerical existence property for **HAS** as follows. Treat **HAS** as a two-sorted theory, let $\mathcal{M}$ be the canonical structure obtained for **HAS** by the Lindenbaum construction. In the formulation of **HAS** comprehension constants C_A with axioms $x \in C_A \leftrightarrow A(x)$ should be added.

Construct the connectification $\mathcal{M}^*$, taking for domain of the number variables at $*$ the numerals (as for **HA**), and for the domain of the sets the collection of set constants C_A, for each formula $A(x)$ in the language of **HAS** with only x free, and put

$$* \Vdash^* (\bar{n} \in C_A) \quad \text{iff} \quad \top_\Omega \Vdash A(\bar{n}),$$

$$p \Vdash^* (d \in C_A) \quad \text{iff} \quad p \Vdash A(d) \quad \text{for } p \neq *.$$

13.7.4. Let $\{\mathcal{M}_i : i \in I\}$ be a set of models of a theory **T**, and let validity in these models be specified as forcing $p \Vdash_i A$, $p \in \Omega_i$, Ω_i the cHa underlying $\mathcal{M}_i$. The product model $\mathcal{M} = \prod\{\mathcal{M}_i : i \in I\}$ has as forcing $\Vdash$, defined by $\langle p_i \rangle_{i \in I} \Vdash A$ iff $p_i \Vdash_i A$ for all $i \in I$. $\mathcal{M}$ is again a model of **T**. Note that for **T** $\equiv$ **HA** we can identify the interpretations of $\bar{n}$ in the various $\mathcal{M}_i$.

13.7.5. Combine completeness (without appealing to the specific Lindenbaum construction) with connectification to obtain EDN and DP for **HA**. *Hint:* Suppose **HA** $\vdash \exists x Ax$, but for all $\bar{n}$, **HA** $\nvdash A\bar{n}$.

Consider a product of models $\mathcal{M}_n$ of **HA** such that $\mathcal{M}_n \nVdash A\bar{n}$. Use connectification and derive a contradiction. (This is a "local" variant of the argument in 7.5–7.7.)

13.7.6. Let Ω be any cHa and define Ω^* as the collection of all $U \subset \{*\} \cup \Omega$ $(* \notin \Omega)$ satisfying:
(i) $p, q \in \Omega$, $q \in U$, $p \le q \Rightarrow p \in U$,
(ii) $* \in U \Rightarrow p \in U$ for all $p \in \Omega$,
(iii) the supremum in Ω of $U \cap \Omega$ belongs to U, i.e.

$$\bigvee_\Omega\{p : p \in U \cap \Omega\} \in U.$$

Then show
(iv) $(\Omega^*, \subset)$ is a cHa;
(v) $\bigvee_{\Omega^*}\{U_i : i \in I\} = \{* : * \in \bigcup\{U_i : i \in I\}\} \cup \{p \in \Omega : p \le \bigvee_\Omega(\bigcup\{U_i \cap \Omega : i \in I\})\}$;
(vi) $(\Omega^*, \subset)$ is classically isomorphic to the Ω^* as defined in 7.5.
Finally use Ω^* with its properties (i) and (ii) for a constructive version of connectification and a constructive proof of EDN for **HA**.

13.8.1. Show that for all $X, Y \in \Omega$, Ω an **APP**-model:

$$(X \wedge Y) \leftrightarrow \bigwedge\{(X \to (Y \to Z)) : Z \in \Omega\} \in \Omega^*,$$

$$(X \vee Y) \leftrightarrow \bigwedge\{((X \to Z) \wedge (Y \to Z) \to Z) : Z \in \Omega\} \in \Omega^*,$$

and for any $\mathscr{X} \subset \Omega$

$$\bigvee\mathscr{X} \leftrightarrow \bigwedge\{(\bigwedge\{X \to Z : X \in \mathscr{X}\}) \to Z : Z \in \Omega\} \in \Omega^*$$

(the proof requires impredicative comprehension.)

13.8.2. Show that the product of finitely many Ω-sets is again an Ω-set.

13.8.3. Prove theorem 8.5 and verify the statements in 8.6.

13.8.4. Establish a soundness theorem for **IQC**, for Ω-models as outlined in 8.4. Can you extend this to E-logic?

13.8.5. Describe explicitly the composition of Ω-morphisms.

13.8.6. Prove theorem 8.7.

13.8.7. $\mathbf{q}$-realizability (E4.4.7) can also be treated as a truth-value semantics. Take again an **APP**-model, say PRO (9.1.9), and take for Ω a set of pairs

$$\{(X, p) : X \subset \mathbb{N} \wedge p \subset \{0\} \wedge \forall x \in X(0 \in p)\}.$$

Thus p ranges over P({0})). Define

$$\bot' := (\emptyset, \emptyset),$$

$$(X, p) \wedge' (Y, q) := (X \wedge Y, p \cap q),$$

$$(X, p) \vee' (Y, q) := (X \vee Y, p \cup q),$$

$$(X, p) \to' (Y, q) := (\{z : (z \in X \to Y) \wedge (0 \in p \to 0 \in q)\}, \{0 : 0 \in p \to 0 \in q\}),$$

and for $\mathscr{X} \equiv \{(X_i, p_i) : i \in I\}$

$$\bigwedge\mathscr{X} := (\bigcap\{X_i : i \in I\}, \bigcap\{p_i : i \in I\}),$$

and similarly for $\bigvee\mathscr{X}$. Let

$$\Omega^* := \{(X, \{0\}) : \exists x(x \in X)\}.$$

Models for first order theories may now be defined as before. Let $[\![\mathbb{N}]\!] = \hat{\mathbb{N}}$ be the set $(\mathbb{N}, [\![\cdot = \cdot]\!])$ with $[\![n = m]\!] := (\{n : n = m\}, \{0 : n = m\})$.

Show for sentences A in $\mathscr{L}(\mathbf{HA})$ that $[\![A]\!] \in \Omega^* \leftrightarrow \exists x(x\,\mathbf{q}\,A)$.

13.8.8. Ω itself can be made into an Ω-set, if we put $[\![x = y]\!]_\Omega := X \leftrightarrow Y$. Assuming the categorical notions of section 14.3 known, prove that the Ω-sets form a category with products and exponentials, and that $\mathscr{P}(\mathscr{X})$ is isomorphic to $\Omega^{\mathscr{X}}$ in this category.

CHAPTER 14

SHEAVES, SITES AND HIGHER-ORDER LOGIC

The present chapter continues the discussion of semantics for intuitionistic logic. It may be regarded as a sequel to the preceding chapter; its principal aim, however, is to lay the basis for the applications of sheaf-semantics to specific intuitionistic theories in the next chapter. Only a small part of the preceding chapter is being presupposed, in particular 13.6.1–13, 13.7.1–3.

Perhaps the reader is inclined to wonder why we still continue to introduce different kinds of semantics, beyond the quite general notion of Ω-model of the preceding chapter, with its constructive completeness theorem. The reason is that in dealing with specific formal theories the general notion of Ω-model is not always very manageable, as must have become clear from the example discussed in 13.6.10. For that reason we shall introduce Ω-sheaf models which may be regarded as a special class of Ω-models with pleasant properties as regards the existence of functions; these properties permit us to think of sheaf models in a more "geometric" way than is possible for Ω-sets in general. Sheaf models are particularly suited to the study of higher-order systems.

More flexibility in constructing models is obtained by sheaf models over *sites*.

A still more general notion of model of intuitionistic higher-order logic is provided by the notion of "elementary topos", but this falls outside the scope of this book.

For information about elementary toposes see e.g. Lambek and Scott (1986). Just as any theory in **IPC** can be regarded as an algebraic structure by considering its Lindenbaum algebra, so a theory based on intuitionistic higher-order logic can be regarded as an elementary topos.

Contents of the chapter. The first section introduces presheaves and sheaves for cHa's, and describes their relationship to the Ω-sets of chapter 13. It is shown how Ω-sets can be transformed into Ω-presheaves, and Ω-presheaves into Ω-sheaves without affecting logical validity. Section 2 describes Ω-sheaf models as a generalization of Ω-structures.

Section 3 contains category-theoretic preliminaries for the next two sections. In section 4, the notion of "sheaf over a cHa" is considerably generalized to "sheaf over a site"; a site is a category provided with a covering system (= Grothendieck topology). The forcing version of validity in a Ω-structure is now extended to forcing over a site. Section 5 discusses in particular sheaf models over a site for intuitionistic higher-order logic, often called Grothendieck toposes.

1. Presheaves, sheaves and sheaf-completion

1.1. *Introduction.* Let us return to the example discussed in 13.6.10, where we exhibited a functional relation not corresponding to a function, in an Ω-model where Ω is the topology of the discrete two-element space $\{0,1\}$. One can expand the model described there in two directions:

(i) By adding, for any f in the model and any open $U \subset \{0,1\}$ its restriction $f \upharpoonright U$ as an element. This produces the function $\lambda x.n$ on $\{0\}$ and on $\{1\}$, as well as the empty function.

(ii) By glueing together compatible collections. $\{f_i : i \in I\}$ is *compatible* if $f_i \upharpoonright \mathrm{dom}(f_i) \cap \mathrm{dom}(f_j) = f_j \upharpoonright \mathrm{dom}(f_i) \cap \mathrm{dom}(f_j)$, or equivalently, $f_i \upharpoonright \mathrm{dom}(f_j) = f_j \upharpoonright \mathrm{dom}(f_i)$. $f = \bigcup\{f_i : i \in I\}$ then is the compatible collection glued together. This produces the constant total functions $\lambda x.n$ on $\{0,1\}$.

In the model thus expanded we have addition as a function: $(f+g)(x) = f(x) + g(x)$ for $x \in \mathrm{dom}(f) \cap \mathrm{dom}(g)$.

It is instructive to carry out the same operations on a slightly more interesting model, a variant of 13.6.4(b) consisting of all *total* uniformly continuous functions on $\mathbb{R}$; we leave this to the reader.

The addition of restrictions and closure under glueing makes our models into sheaves, over which functional relations can be interpreted as functions. Moreover, the validity of formulas with parameters from the original model is not affected, as we shall see below. Thus the completion of Ω-sets to sheaves will give us another model-theoretic proof of the conservativity of adding descriptions. As a first step we introduce structures with restrictions.

1.2. DEFINITION. A *presheaf* $\mathcal{M}$ *over* Ω, or an *Ω-presheaf* is a triple $(M, E, \upharpoonright)$ consisting of a set M, a map of *extent* $E \in M \to \Omega$, and a

restriction map $\upharpoonleft \in M \times \Omega \to M$ such that for all $a \in M$ and all $p,q \in \Omega$:
(i) $a \upharpoonleft Ea = a$,
(ii) $(a \upharpoonleft p) \upharpoonleft q = a \upharpoonleft (p \wedge q)$,
(iii) $E(a \upharpoonleft p) = (Ea) \wedge p$.
If $\mathcal{M} \equiv (M, E, \upharpoonleft)$, we also write $|\mathcal{M}|$ for M. □

N.B. From the viewpoint of presheaves on sites, to be treated in section 4, it would have been more natural to define $a \upharpoonleft p$ only for $p \leq Ea$. In any case (i) and (ii) imply $a \upharpoonleft p = a$ for $p \geq Ea$.

1.3. Examples.
(a) The collection of all *partial*, continuous a from $\mathbb{R}$ to $\mathbb{R}$ with *open* domain is a presheaf with $Ea := \mathrm{dom}(a)$, and $\upharpoonleft$ is ordinary restriction, i.e. $a \upharpoonleft U := a \restriction U$. Observe that the equality is the old one (cf. 13.6.8): $[\![a = b]\!] = \mathrm{Int}\{t: a(t) = b(t)\}$. The reader will do well to keep this in mind as the prime example of a presheaf; all notions to be introduced should be visualized in terms of this particular presheaf.

(b) Similarly for the collection of all bounded, partial, continuous functions from $\mathbb{R}$ to $\mathbb{R}$.

(c) If we consider the Ω_{K}-model corresponding to a Kripke model $(\mathrm{K}, \leq, \Vdash, \mathrm{D})$ (standard presentation with increasing domains, as in 2.5.7), we may define, for elements d of $\mathbf{D} = \bigcup\{\mathrm{D}(k): k \in \mathrm{K}\}$, $Ed := \{k: d \in \mathrm{D}(k)\}$. The result is in general not a presheaf. Closure under restrictions is easily achieved however: for each $U \in \Omega_{\mathrm{K}}$ and each $d \in \mathbf{D}$ we add d_U with $E(d_U) = Ed \cap U$, $k \Vdash d = d_U$ for each $k \in U$. Putting $d = d_{Ed}$, we can take $d_U \upharpoonleft U' = d_{U \cap U'}$.

(d) The Ω-set of total continuous functions in $\mathbb{R} \to \mathbb{R}$ is a subset of (a) which is not a presheaf, since it is not closed under restrictions.

Ω-sets and Ω-presheaves are not all that different. In an Ω-presheaf we can measure the extent to which elements coincide, by means of the following definition (think of example (a) above).

1.4. Definition (*of equality on a presheaf*). If $\mathcal{M} \equiv (M, E, \upharpoonleft)$ is an Ω-presheaf, we define $[\![\cdot = \cdot]\!] \in M \times M \to \Omega$ by

$$[\![a = b]\!] := \bigvee\{p \leq Ea \wedge Eb : a \upharpoonleft p = b \upharpoonleft p\}. \quad \square$$

Observe that with this definition the equality in example (a) above becomes $[\![a = b]\!] := \mathrm{Int}\{t: a(t) = b(t)\}$, which is the old definition of 13.6.8(d).

1.5. PROPOSITION. *If* $\mathcal{M}$ *be an* Ω*-presheaf, with* $[\![\cdot = \cdot]\!]$ *as in* 1.4, *then* $(M, [\![\cdot = \cdot]\!])$ *is an* Ω*-set satisfying*

$$[\![a \upharpoonright q = b]\!] = [\![a = b]\!] \wedge q.$$

PROOF. Symmetry, i.e. $[\![a = b]\!] = [\![b = a]\!]$, is obvious. As to transitivity, we can evaluate $[\![a = b]\!] \wedge [\![b = c]\!]$ by repeated use of the infinitary distributive law for cHa's:

$$\bigvee\{p \le Ea \wedge Eb : a \upharpoonright p = b \upharpoonright p\} \wedge \bigvee\{q \le Eb \wedge Ec : b \upharpoonright q = c \upharpoonright p\}$$
$$= \bigvee\{p \wedge q : p \le Ea \wedge Eb,\ q \le Eb \wedge Ec,\ a \upharpoonright p = b \upharpoonright p,\ b \upharpoonright q = c \upharpoonright q\}.$$

Now, if $p \le Ea \wedge Eb$ and $q \le Eb \wedge Ec$, then $p \wedge q \le Ea \wedge Ec$. Also $a \upharpoonright p = b \upharpoonright p \Rightarrow a \upharpoonright (p \wedge q) = b \upharpoonright (p \wedge q)$, $b \upharpoonright q = c \upharpoonright q \Rightarrow b \upharpoonright (p \wedge q) = c \upharpoonright (p \wedge q)$, hence the above supremum is less than

$$\bigvee\{p \wedge q : p \wedge q \le Ea \wedge Ec,\ a \upharpoonright (p \wedge q) = c \upharpoonright (p \wedge q)\},$$

which is $[\![a = c]\!]$.

Finally, $[\![a \upharpoonright q = b]\!] = \bigvee\{p \le Ea \wedge Eb \wedge q : a \upharpoonright p = b \upharpoonright p\} \le [\![a = b]\!] \wedge q$; on the other hand, if $p \le Ea \wedge Eb$ and $a \upharpoonright p = b \upharpoonright p$, then $p \wedge q \le Ea \wedge Eb \wedge q$ and $a \upharpoonright (p \wedge q) = b \upharpoonright (p \wedge q)$; thus by the distributive law $[\![a = b]\!] \wedge q \le [\![a \upharpoonright q = b]\!]$. □

1.6. DEFINITION. Let $\mathcal{M}$ be an Ω-presheaf, $a, b \in |\mathcal{M}| = M$. a is a *restriction* of b (or b an *extension* of a), iff $a = b \upharpoonright Ea$ (notation: $a \le b$).

Let $X \subset M$; X is *compatible* in $\mathcal{M}$ iff $\forall b, b' \in X(b \upharpoonright Eb' = b' \upharpoonright Eb)$. A *join* a of X is a minimal upper bound for X (i.e. $\forall b \in X(b \le a)$, $\forall a'(\forall b \in X(b \le a') \wedge a' \le a \rightarrow a = a')$).

A subset $X \subset M$ is *bounded* iff $\exists a \in M\, \forall b \in X(b \le a)$. □

N.B. The verification that $\le$ is a partial order is an easy exercise: lemma 1.7(i) below.

1.7. LEMMA. *Let* $\mathcal{M} \equiv (M, E, \upharpoonright)$ *be an* Ω*-presheaf, and let* $a, b \in M$. *Then*:

(i) $\le$ *partially orders* M;

(ii) $b \le a \Rightarrow Eb \le Ea$;

(iii) *if* $X \subset M$, *then* $a \in M$ *is a join for* X *iff*

$$Ea = \bigvee\{Eb : b \in X\} \quad \textit{and} \quad \forall b \in X(b \le a);$$

(iv) *$a \restriction \bigvee\{p_i : i \in I\}$ is a join of $\{a \restriction p_i : i \in I\}$ for $\{p_i : i \in I\} \subset \Omega$;*
(v) *every bounded subset of M is compatible and has a join.*

PROOF. Exercise. □

1.8. DEFINITION. An Ω-set is *separated* iff $[\![a \simeq b]\!] = \top \Rightarrow a = b$; an Ω-presheaf is *separated* iff joins of compatible subsets, when existing, are unique. □

The notions of separated Ω-set and separated Ω-presheaf are connected in the obvious way.

PROPOSITION. *An Ω-presheaf is separated as an Ω-set iff it is a separated presheaf.*

PROOF. Exercise. □

1.9. EXAMPLES.

(e) The examples (a) and (b) in 1.3 are also separated presheaves and hence separated Ω-sets.

(f) A Kripke model in standard presentation (i.e. with increasing domains, cf. 1.3(c)) containing two domain elements d, d' such that $d \neq d'$ but $\forall k \in K(d \in \mathrm{D}(k) \leftrightarrow d' \in \mathrm{D}(k))$ is in general not separated *as an* Ω_K-set. Adding restrictions as in 1.3(c) makes it into a presheaf which is also not separated in general.

(g) Let Ω be the collection of opens of a Hausdorff space (X, Ω) with at least two points. The constant functions in $X \to \mathbb{N}$ and the functions $U \to \{0\}$ for opens $U \subset X$ constitute the elements of a presheaf $\mathcal{M}$; restriction is defined by $a \restriction X := a$, $a \restriction U :=$ the unique constant function in $U \to \{0\}$ for $U \neq X$. The extent of an element f is simply the domain of f. $\mathcal{M}$ is not separated (exercise). N.B. Our definition is classical.

The following proposition lists conditions on a presheaf equivalent to being separated, plus two additional properties of separated presheaves.

1.10. PROPOSITION. *Let $\mathcal{M} \equiv (M, E, \restriction)$ be an Ω-presheaf. Then the following conditions are equivalent:*
(i) *$\mathcal{M}$ is separated;*
(ii) $\forall i \in I(a \restriction p_i = b \restriction p_i) \Rightarrow a \restriction \bigvee\{p_i : i \in I\} = b \restriction \bigvee\{p_i : i \in I\}$;
(iii) *$a \restriction [\![a = b]\!] = b \restriction [\![a = b]\!]$ for all $a, b \in M$;*

(iv) $a \upharpoonright [\![a \simeq b]\!] = b \upharpoonright [\![a \simeq b]\!]$ *for all* $a, b \in M$;
(v) *for all* $a, b \in M$, $p \in \Omega$: $p \leq [\![a \simeq b]\!] \Leftrightarrow a \upharpoonright p = b \upharpoonright p$;
(vi) *for all* $a, b \in M$: $a \leq b \Leftrightarrow Ea \leq [\![a = b]\!]$.
In addition, for separated presheaves:
(vii) $[\![a \simeq b]\!] = \bigvee\{p \in \Omega : a \upharpoonright p = b \upharpoonright p\}$,
(viii) $a \upharpoonright Eb = b \upharpoonright Ea \Leftrightarrow [\![a = b]\!] = Ea \wedge Eb$.

PROOF. Lengthy, but straightforward; we leave the proof as an exercise. □

1.11. DEFINITION.
(i) Let $\mathcal{M}, \mathcal{M}'$ be Ω-sets. $\mathcal{M}$ is an *Ω-subset* of $\mathcal{M}'$ (notation $\mathcal{M} \subset \mathcal{M}'$) if $M \subset M'$ and $[\![\cdot = \cdot]\!]_{\mathcal{M}}$ is the restriction of $[\![\cdot = \cdot]\!]_{\mathcal{M}'}$ to $M \times M$. $\mathcal{M}$ *generates* $\mathcal{M}'$ iff $Eb = \bigvee\{[\![a = b]\!] : a \in M\}$ for all $b \in M'$.
(ii) If $\mathcal{M}, \mathcal{M}'$ are Ω-presheaves, $\mathcal{M}$ is said to be a *sub-presheaf* of $\mathcal{M}'$ iff $M \subset M'$, and $E_{\mathcal{M}}$, $\upharpoonright_{\mathcal{M}}$ are the restrictions of $E_{\mathcal{M}'}$, $\upharpoonright_{\mathcal{M}'}$ to M. $\mathcal{M}$ *generates* $\mathcal{M}'$ iff for all $b \in M'$

$$Eb = \bigvee\{p \leq Ea \wedge Eb : a \in M, a \upharpoonright p = b \upharpoonright p, p \in \Omega\}. \quad \square$$

1.12. PROPOSITION. *Let $\mathcal{M} \subset \mathcal{M}'$, $\mathcal{M}$ an Ω-set, $\mathcal{M}'$ a separated presheaf.*
(i) *$\mathcal{M}$ generates $\mathcal{M}'$ iff each $b \in M'$ is a join of restrictions of elements from $\mathcal{M}$.*
(ii) *Let Θ be a basis for Ω, i.e. $\forall p \in \Omega(\bigvee\{q \leq p : q \in \Theta\} = p)$, and suppose $M = \{a \in M' : Ea \in \Theta\}$; then $\mathcal{M}$ generates $\mathcal{M}'$.*

PROOF. (i) Suppose $\mathcal{M}$ generates $\mathcal{M}'$, then for any $b \in M'$ $b = b \upharpoonright Eb = b \upharpoonright \bigvee\{[\![a = b]\!] : a \in M\} =$ (1.7(iv)) $\bigvee\{b \upharpoonright [\![a = b]\!] : a \in M\} =$ (by separatedness) $\bigvee\{a \upharpoonright [\![a = b]\!] : a \in M\}$.

Conversely, suppose that for each $b \in M'$

$$b = \bigvee\{a_i \upharpoonright p_i : i \in I\} \quad (p_i \in \Omega, a_i \in M).$$

Then $a_i \upharpoonright p_i \leq b$, so $E(a_i \upharpoonright p_i) = [\![a_i \upharpoonright p_i = b]\!]$, and thus $Eb = \bigvee\{E(a_i \upharpoonright p_i) : i \in I\} = \bigvee\{[\![a_i \upharpoonright p_i = b]\!] : i \in I\} = \bigvee\{[\![a_i = b]\!] \wedge p_i : i \in I\} \leq \bigvee\{[\![a = b]\!] : a \in M\}$. Also $[\![a = b]\!] \leq Eb$, so $\bigvee\{[\![a = b]\!] : a \in M\} \leq Eb$ and thus $Eb = \bigvee\{[\![a = b]\!] : a \in M\}$.

(ii) Clearly, for $b \in M'$, $b = b \upharpoonright \bigvee\{p : p \in \Theta, p \leq Eb\} = \bigvee\{b \upharpoonright p : p \in \Theta, p \leq Eb\}$, and thus the assertion follows from (i). □

Note that if $\mathcal{M}$ generates $\mathcal{M}'$ and $\mathcal{M}'$ generates $\mathcal{M}''$, then $\mathcal{M}$ generates $\mathcal{M}''$.

1.13. DEFINITIONS.
(i) A presheaf is a *sheaf* iff every compatible subset has a unique join.
(ii) $\mathscr{N}$ is a *subsheaf relative to the presheaf* $\mathscr{M}$ if $\mathscr{N}$ is a sub-presheaf of $\mathscr{M}$ and for compatible collections $\mathscr{X} \subset \mathscr{N}$

$$x = \bigvee \mathscr{X} \in \mathscr{M} \Rightarrow x \in \mathscr{N}.$$

(iii) If $\mathscr{N}$ is a subpresheaf of $\mathscr{M}$, and $\mathscr{M}$ is a sheaf, then $\mathscr{N}$ is a *subsheaf* of $\mathscr{M}$. □

N.B. A sheaf is therefore automatically a *separated* presheaf! Example 1.3(a) is a sheaf, 1.3(b) not; why? A subsheaf relative to a sheaf $\mathscr{N}$ is a subsheaf of $\mathscr{N}$.

1.14. *Embedding an Ω-set into a presheaf.* Let $\mathscr{M}$ be an Ω-set $(M, [\![\cdot = \cdot]\!])$. We can transform $\mathscr{M}$ into an Ω-presheaf $\mathscr{M}^*$ by formally adding restrictions of elements of M.

DEFINITION. $\mathscr{M}^*$ has as elements equivalence classes modulo $\approx$ of

$$\{(a, p) : a \in M,\ p \in \Omega,\ p \leq Ea\},$$

where

$$(a, p) \approx (b, q) := (p = q) \text{ and } p \leq [\![a = b]\!],$$

$$E(a, p)_{\approx} := p, \qquad (a, p)_{\approx} \upharpoonright q := (a, p \wedge q)_{\approx}.$$

$\mathscr{M}$ is mapped into $\mathscr{M}^*$ by

$$\varphi : a \mapsto (a, Ea)_{\approx}. \ \square$$

1.15. PROPOSITION. *The mapping φ is an embedding, i.e. it preserves* $[\![\cdot = \cdot]\!]$, $\mathscr{M}^*$ *is separated, and* $\mathscr{M}$ *generates* $\mathscr{M}^*$.

PROOF. Let us write $[\![\cdot = \cdot]\!]^*$ for the defined equality in $\mathscr{M}^*$. The we have to show $[\![\varphi a = \varphi b]\!]^* = [\![a = b]\!]$ (exercise). Observe that $\mathscr{M}^*$ may be obtained by first identifying those a and b for which $[\![a \simeq b]\!] = \top$, then adding restrictions. □

Note that if we apply φ to a Kripke model as an Ω_K-set (13.6.8(e)) we obtain the Kripke model as an Ω_K-presheaf (1.3(c)).

1.16. *Completing a presheaf to a sheaf.* We shall now show how to embed an Ω-presheaf into an Ω-sheaf, preserving E, $\upharpoonright$, and hence $[\![\cdot = \cdot]\!]$.

We achieve this by defining a suitable operation $^+$ on presheaves; $^+$ transforms a presheaf into a separated presheaf, and a separated presheaf into a sheaf. Thus any presheaf is transformed into a sheaf by applying $^+$ twice. We shall give an example (1.20) to show that applying $^+$ only once is, in general, not enough. The idea is to construct $\mathcal{M}^+$ as the compatible set of collections of $\mathcal{M}$ under an appropriate equivalence relation.

1.17. Definition. Let $\mathcal{M}$ be an Ω-presheaf; $X \subset \mathcal{M}$ is said to be *closed* iff

$$\forall a \in X \forall p \in \Omega (a \upharpoonright p \in X).$$

Let Y be closed and compatible; put

$$E(Y) := \bigvee\{Ec : c \in Y\}, \qquad Y \upharpoonright p := \{c \upharpoonright p : c \in Y\},$$

$$Y \cong Y' := (E(Y) = E(Y')) \text{ and for some cover } \mathcal{D} \text{ of } E(Y)$$

$$\forall p \in \mathcal{D}(Y \upharpoonright p \cup Y' \upharpoonright p \text{ is compatible}) \text{ (cf. 13.7.1)}.$$

$\cong$ is an equivalence relation; we write $Y_{\cong}$ for the equivalence class, and put

$$E(Y_{\cong}) := E(Y), \qquad (Y_{\cong}) \upharpoonright p := (Y \upharpoonright p)_{\cong}.$$

Then $\mathcal{M}^+$ consists of the $\cong$-classes with extent E, restriction $\upharpoonright$. $\mathcal{M}$ is embedded into $\mathcal{M}^+$ by

$$\varphi' : a \mapsto \{a \upharpoonright p : p \in \Omega\}_{\cong}. \quad \square$$

1.18. Proposition. *$\mathcal{M}^+$ is a separated presheaf, φ' preserves E and $\upharpoonright$, and φ' is an embedding (in the sense of Ω-sets). Moreover $\varphi[\mathcal{M}]$ generates $\mathcal{M}^+$.*

Proof. Exercise. $\square$

1.19. Proposition. *If $\mathcal{M}$ is a separated presheaf, then $\mathcal{M}^+$ is a sheaf.*

Proof. Let $\mathcal{C}$ be a family of closed compatible subsets of $\mathcal{M}$. Writing EY for $E(Y_{\cong})$ if Y is compatible, and $Y \upharpoonright p$ for $\{c \upharpoonright p : c \in Y\}$, the compatibility of the $\cong$-equivalence classes of $\mathcal{C}$ amounts to

$$\forall Y, Y' \in \mathcal{C}(Y \upharpoonright EY' \cong Y' \upharpoonright EY). \tag{1}$$

By definition, for $Y, Y' \in \mathscr{C}$,

$$Y \restriction EY' = \{c \restriction EY' : c \in Y\} = \{c \restriction \bigvee\{Ed : d \in Y'\} : c \in Y'\},$$

$$Y' \restriction EY = \{c' \restriction EY : c' \in Y'\} = \{c' \restriction \bigvee\{Ed : d \in Y\} : c' \in Y'\}.$$

Assume (1) for $Y, Y' \in \mathscr{C}$. That is to say, for a cover $\{p_i : i \in I\}$ of $E(Y \restriction EY') = \bigvee\{Ec \wedge Ec' : c \in Y,\ c' \in Y'\}$, and for all $i \in I$ and all $c \in Y$, $c' \in Y'$

$$\{c \restriction \bigvee\{Ed' : d' \in Y'\} \wedge p_i\} \cup \{c' \restriction \bigvee\{Ed : d \in Y\} \wedge p_i\}$$

is compatible, hence for all $i \in I$

$$c \restriction \bigvee\{Ed \wedge Ed' \wedge p_i : d \in Y, d' \in Y'\}$$
$$= c' \restriction \bigvee\{Ed \wedge Ed' \wedge p_i : d \in Y, d' \in Y'\},$$

and therefore $c \restriction (Ec' \wedge p_i) = c' \restriction (Ec \wedge p_i)$.

$\{c \restriction (p_i \wedge Ec') : i \in I\}$ is bounded, and thus has a join which is unique (since $\mathscr{M}$ is separated), and equal to the join of $\{c' \restriction (p_i \wedge Ec) : i \in I\}$; therefore $c \restriction Ec' = c' \restriction Ec$, and so $Y^* = \bigcup \mathscr{C}$ is compatible; we shall show that $Y^*_{\cong}$ is the join of the $Y_{\cong}$ for $Y \in \mathscr{C}$. Thus it has to be shown that $Y^* \restriction EY \cong Y$, and that $EY^* = \bigvee\{EY : Y \in \mathscr{C}\}$. The latter assertion is obvious. As to the first assertion, we note that

$$Y^* \restriction EY = \{c' \restriction \bigvee\{Ec : c \in Y\} : c' \in Y^*\};$$

clearly $E(Y^* \restriction EY) = EY$, and $(c' \restriction \bigvee\{Ec : c \in Y\}) \restriction Ec'' = c' \restriction Ec'' = c'' \restriction (Ec' \wedge Ec'')$ for $c'' \in Y$. $c'' \restriction (Ec' \wedge \bigvee\{Ec : c \in Y\}) = c'' \restriction (Ec' \wedge Ec'')$, i.e. $(Y^* \restriction EY) \cup Y$ is compatible, which implies $Y^* \restriction EY \cong Y$. Thus we have verified the sheaf property. □

Notation. By the observation at the end of 1.12 any presheaf $\mathscr{M}$ generates a sheaf $\mathscr{M}^{++}$, which we shall call $\mathscr{M}^{\mathrm{sh}}$, the *sheaf completion* of $\mathscr{M}$. □

Remark. If $\mathscr{M}$ is an Ω-set, then $\mathscr{M}^{\mathrm{sh}}$ and $\mathscr{M}$ are isomorphic Ω-sets in the sense that there is a binary Ω-relation R on $\mathscr{M} \times \mathscr{M}^{\mathrm{sh}}$ which is functional and bijective, i.e.

$$[\![\forall x \in \mathscr{M}\ \exists! y \in \mathscr{M}^{\mathrm{sh}}\ R(x, y) \wedge \forall y \in \mathscr{M}^{\mathrm{sh}}\ \exists! x \in \mathscr{M}\ R(x, y)]\!] = \top_\Omega$$

(exercise, cf. also 3.16).

The following example shows that in general we have to apply $^+$ twice to obtain a sheaf from a presheaf.

1.20. EXAMPLE. We will use classical reasoning. Let us take for Ω the opens of the two-point topological space $X \equiv \{0,1\}$ with the discrete topology. As presheaf $\mathscr{M}$ we take the set

$$\{a_0, b_0, a_1, b_1, a, b\}$$

with

$$Ea_0 = Eb_0 = \{0\}, \qquad Ea_1 = Eb_1 = \{1\}, \qquad Ea = Eb = \emptyset,$$

and an obvious restriction map

$$a_i \restriction \emptyset = a, \qquad b_i \restriction \emptyset = b,$$

$$a_i \restriction \{j\} = a_i \restriction \{i\} \cap \{j\} \quad (i, j \in \{0,1\}).$$

The compatible closed families are

$$\{a_0, a_1, a\}, \qquad \{b_0, b_1, b\}, \qquad \{a_0, a\}, \qquad \{b_0, b\},$$

$$\{a_1, a\}, \qquad \{b_1, b\}, \qquad \{a\}, \qquad \{b\}, \qquad \emptyset.$$

Over the empty cover of $\emptyset$, everything has to coincide: $\{a\} \cong \{b\} \cong \emptyset$. Call this element of $\mathscr{M}^+$ $*$. No other families are identified under $\cong$.

If we now apply $^+$ again, all closed compatible families are generated from the following compatible families Y (by closure under restrictions)

$$\{\{a_0, a_1, a\}\}; \qquad \{\{b_0, b_1, b\}\}; \qquad \{\{a_0, a\}, \{b_1, b\}\},$$
$$\{\{a_1, a\}, \{b_0, b\}\} \quad (EY = X);$$

$$\{\{a_0, a\}\}; \{\{b_0, b\}\} \quad (EY = \{0\});$$

$$\{\{a_1, a\}\}; \{\{b_1, b\}\} \quad (EY = \{1\});$$

$$\{*\} \quad (EY = \emptyset).$$

None of these are to be identified.

The elements f_Y of $\mathscr{M}^+$ may be identified with certain partial mappings from X into $\{a^*, b^*\}$; if $Y_{\cong} \in \mathscr{M}^+$, and $a_i \in Y$, then $f_Y(i) = a^*$ for the corresponding partial function; similarly $b_i \in Y \Rightarrow f_Y(i) = b^*$. Restriction becomes restriction of functions; the extent of f_Y is its domain.

Collections $\mathscr{C}$ of partial functions are compatible, if for $f, f' \in \mathscr{C}$: $f \restriction \mathrm{dom}(f') = f' \restriction \mathrm{dom}(f)$. $\mathscr{M}^+$ contains as total functions only f_0 and f_1 satisfying

$$f_0(0) = f_0(1) = a^* \quad \text{and} \quad f_1(0) = f_1(1) = b^*.$$

$\mathcal{M}^{++}$ contains new elements with extent X obtained by joining compatible functions from $\mathcal{M}^{+}$ such as the partial functions f' with domain $\{0\}$ and f'' with domain $\{1\}$ such that $f'(0) = a^*$ and $f''(1) = b^*$; together they yield f_2 defined on $\{0, 1\}$ with $f_2(0) = a^*$, $f_2(1) = b^*$.

1.21. *Sheaves over a basis (digression).* One can also study sheaves and sheaf-completion with respect to a basis Θ of a cHa Ω. We briefly describe this here; we shall encounter the same approach in a much more general context in section 4. We assume Θ to be closed under $\wedge$.

Let $p, p', p_i, \ldots$ range over Θ. A *presheaf* $\mathcal{D}$ *over* Θ is specified by giving a collection $\{D(p) : p \in \Theta\}$ such that $a \in D(p) \leftrightarrow p \leq Ea$, with a restriction map $\varphi^p_{p'} \in D(p) \to D(p')$ for each $p' \leq p$; for $\varphi^p_{p'}(a)$ we write $a \upharpoonright p'$ and the a at $D(p)$ are now thought of as having domain p. For restrictions we should always have $(a \upharpoonright p) \upharpoonright q = a \upharpoonright (p \wedge q)$, and $a \upharpoonright p = p$ if $a \in D(p)$.

A compatible collection $\{a_i \in D(p_i) : i \in I\}$ is such that for all $i, j \in I$ $a_i \upharpoonright p_j = a_j \upharpoonright p_i$, or equivalently, $a_i \upharpoonright p_i \wedge p_j = a_j \upharpoonright p_i \wedge p_j$.

A presheaf over Θ is a *sheaf over* Θ if for each compatible $\{a_i \in D(p_i) : i \in I\}$ and each $p \leq \bigvee\{p_i : i \in I\}$, there is a unique $a \in D(p)$ such that $\forall i \in I(a \upharpoonright p_i = a_i)$.

In this formulation, Kripke models defined with restriction mappings (cf. 2.5.12) are already sheaves with respect to the basis consisting of all $\{k' : k' \geq k\}$, $k \in \mathrm{K}$ (minimal neighbourhoods of points) for Ω_{K}. But in general, Kripke models, when viewed as Ω_{K}-models, are not sheaves, though the difference is clearly inessential (exercise).

2. Ω-presheaf and Ω-sheaf structures

This section will be devoted to particular Ω-structures, where all domains are interpreted by presheaves or sheaves (instead of just Ω-sets). As a result of the sheaf-completion described in the previous section, we shall also obtain a constructive model-theoretic proof of the conservativity of adding a description operator. As a preliminary, we discuss products of presheaves.

2.1. *Products of presheaves.* Let $\mathcal{M}, \mathcal{N}$ be presheaves. A presheaf $\mathcal{M} \times \mathcal{N}$ is obtained as the set

$$\{(a, b) : a \in M \wedge b \in N \wedge Ea = Eb\},$$

with

$$E(a,b) := Ea = Eb,$$
$$(a,b) \upharpoonright p := (a \upharpoonright p, b \upharpoonright p). \quad \Box$$

2.2. Proposition. *If $\mathcal{M}, \mathcal{N}$ are separated, then so is $\mathcal{M} \times \mathcal{N}$; if $\mathcal{M}, \mathcal{N}$ are sheaves, then so is $\mathcal{M} \times \mathcal{N}$.*

Proof. For the second assertion, let Y be any compatible subset of $\mathcal{M} \times \mathcal{N}$, and take

$$\bigvee Y := (\bigvee\{a : (a,b) \in Y\}, \bigvee\{b : (a,b) \in Y\}). \quad \Box$$

Remark. The definition of "product of presheaves" at this stage seems ad hoc. However, in the appropriate category of Ω-presheaves the product as defined is indeed the categorical product which is uniquely defined up to isomorphism (exercise; cf. section 3).

For Ω-sets we defined a product (cf. 13.6.2, 13.6.5) $\mathcal{M} \times' \mathcal{N}$ by taking as underlying set $M \times N$ and putting

$$[\![(a,b) = (a',b')]\!]_{\mathcal{M}\times'\mathcal{N}} := [\![a = a']\!]_{\mathcal{M}} \wedge [\![b = b']\!]_{\mathcal{N}}.$$

This definition is of no use here, since, for presheaves $\mathcal{M}$ and $\mathcal{N}$, $\mathcal{M} \times' \mathcal{N}$ need not be a presheaf. It should be noted however, that when both $\mathcal{M} \times \mathcal{N}$ and $\mathcal{M} \times' \mathcal{N}$ are conceived as Ω-sets, then the first generates the second; as a result, because of lemma 2.5 below, both definitions of product yield the same result in evaluating quantifiers.

2.3. Definition. *An Ω-presheaf structure (Ω-sheaf structure) is an Ω-structure where all domains are interpreted by presheaves (sheaves).* $\Box$

After the discussion of products it is obvious how n-ary relations should be interpreted as sub-presheaves (sub-sheaves) of the product.

2.4. Proposition. *Let $\mathcal{M} \subset \mathcal{M}'$, $\mathcal{M}'$ a sheaf; any structure on $\mathcal{M}$ can be extended to a structure on $\mathcal{M}'$. Moreover, if $\mathcal{M}$ generates $\mathcal{M}'$, any strict relation or function on $\mathcal{M}$ can be extended to a unique strict relation or function on $\mathcal{M}'$. Similarly for many sorted structures, where $\mathcal{M}_i \subset \mathcal{M}'_i$ for all sorts i, etc.*

Proof. A relation R can be extended by

$$R(a) := \bigvee\{R(b) \wedge [\![a \simeq b]\!] : b \in M\}.$$

The extended R again satisfies replacement (exercise). An operation F is extended by

$$F(a) = \bigvee\{F(b) : b \leq a \wedge b \in M\}.$$

If $\mathcal{M}$ generates $\mathcal{M}'$, and R is strict, we can take

$$R'(a) = \bigvee\{R(b) \wedge [\![a = b]\!] : b \in M\}$$

for the extension R' of R. Let R'' be any other strict extension. Then $R''(a) \wedge [\![a = b]\!] = R''(b) \wedge [\![a = b]\!]$, so $R''(a) = R''(a) \wedge Ea = \bigvee\{[\![a = b]\!] : b \in M\}) \wedge R''(a) = \bigvee\{R''(a) \wedge [\![a = b]\!] : b \in M\} = \bigvee\{R''(b) \wedge [\![a = b]\!] : b \in M\} = \bigvee\{R(b) \wedge [\![a = b]\!] : b \in M\} = R'(a)$. The rest is an exercise. □

2.5. LEMMA. *Let $\mathcal{M}' \subset \mathcal{M}$, and suppose that $\mathcal{M}'$ generates $\mathcal{M}$. Then in any structure over $\mathcal{M}$:*

(i) $$\bigwedge_{a \in M} (Ea \rightarrow [\![A(a)]\!]) = \bigwedge_{b \in M'} (Eb \rightarrow [\![A(b)]\!]);$$

(ii) $$\bigvee_{a \in M} (Ea \wedge [\![A(a)]\!]) = \bigvee_{b \in M'} (Eb \wedge [\![A(b)]\!]).$$

PROOF. We verify (i) and leave the verification of (ii) as an exercise.

$$\bigwedge_{a \in M} (Ea \rightarrow [\![A(a)]\!]) \leq \bigwedge_{b \in M'} (Eb \rightarrow [\![A(b)]\!])$$

is trivial, since $M' \subset M$.

On the other hand, $[\![A(b)]\!] \wedge [\![a = b]\!] = (Eb \rightarrow [\![A(b)]\!]) \wedge [\![a = b]\!]$, since $[\![a = b]\!] \leq Eb$; also $[\![A(b)]\!] \wedge [\![a \simeq b]\!] \leq [\![A(a)]\!]$, therefore a fortiori $[\![A(b)]\!] \wedge [\![a = b]\!] \leq [\![A(a)]\!]$; thus

$$(Eb \rightarrow [\![A(b)]\!]) \wedge [\![a = b]\!] \leq [\![A(a)]\!]$$

for all $b \in M'$, hence

$$\bigwedge_{d \in M'} (Ed \rightarrow [\![Ad]\!]) \wedge \bigvee_{b \in M'} [\![a = b]\!] \leq [\![A(a)]\!], \quad \text{so}$$

$$\bigwedge_{d \in M'} (Ed \rightarrow [\![Ad]\!]) \wedge Ea \leq [\![A(a)]\!],$$

hence

$$\bigwedge_{d \in M'} (Ed \rightarrow [\![Ad]\!]) \leq (Ea \rightarrow [\![A(a)]\!]),$$

therefore

$$\bigwedge_{d \in M'} (Ed \to [\![Ad]\!]) \leq \bigwedge_{a \in M} (Ea \to [\![Aa]\!]). \quad \square$$

2.6. Theorem. *Let $\mathcal{M}$ be an Ω-structure, $\mathcal{M}' \subset \mathcal{M}$, $\mathcal{M}'$ generates $\mathcal{M}$, t any closed term in the language $\mathcal{L}(\mathcal{M}')$, A a sentence of $\mathcal{L}(\mathcal{M}')$, then*

$$[\![t]\!]_{\mathcal{M}'} = [\![t]\!]_{\mathcal{M}}, \qquad [\![A]\!]_{\mathcal{M}'} = [\![A]\!]_{\mathcal{M}}.$$

Proof. For terms this is immediate; for formulas we apply formula-induction, where the quantifier-case is taken care of by lemma 2.5, and the case of prime formulas by 2.4. □

Example. Let Y be a set and let Ω be the powerset $P(Y)$ (i.e. Y is provided with the discrete topology). An Ω-sheaf $\mathcal{N}$ consists of all partial functions from Y into a set X, and $\mathcal{N}$ is generated by the Ω-set of constant elements $\hat{X} = \{\hat{x} : x \in X\}$, $\hat{x} := \lambda y \in Y.x$. In other words, $\mathcal{N} \equiv (\hat{X})^{\mathrm{sh}}$.

Let $\mathcal{L} \equiv (R_0, R_1, \ldots, F_0, F_1, \ldots, \{c_i : i \in I\})$ be a first-order language and let $\mathcal{M} \equiv (X, R_0, R_1, \ldots, F_0, F_1, \ldots, \{c_i : i \in I\})$ be a model for $\mathcal{L}$. $\mathcal{M}_\Omega$ is the corresponding Ω-structure with domain $\hat{X}^{\mathrm{sh}}$, and Ω-relations, Ω-functions, and Ω-constants interpreting the relations, functions, and constants of $\mathcal{L}$ given by

$$[\![R_i]\!](f_1, \ldots, f_n) := \{y : \mathrm{R}_i(f_1(y), \ldots, f_n(y))\},$$

$$[\![F_i]\!](f_1, \ldots, f_n)(y) := \mathrm{F}_i(f_1(y), \ldots, f_n(y)),$$

$$[\![c_i]\!] := \lambda y \in Y.\mathrm{c}_i.$$

Then, for arbitrary formulas A of $\mathcal{L}(X)$, $x_1 \in X, \ldots, x_n \in X$

$$[\![A(\hat{x}_1, \ldots, \hat{x}_n)]\!] = Y \Leftrightarrow \mathcal{M} \Vdash_i A(x_1, \ldots, x_n),$$

$$y \in [\![A(f_1, \ldots, f_n)]\!] \Leftrightarrow \mathcal{M} \Vdash_i A(f_1(y), \ldots, f_n(y)).$$

We leave the verification of these facts as an exercise.

2.7. Proposition. *Let $\mathcal{M}$ be a sheaf and let $A(x)$ be any formula of the language $\mathcal{L}(\mathcal{M})$. If we put*

$$\Phi(a) := [\![Ea \wedge \forall x(A(x) \leftrightarrow x = a)]\!],$$

then $\{a \upharpoonright \Phi(a) : a \in M\}$ is a compatible collection; we denote its join by

$[\![Ix.A(x)]\!]$. *For this interpretation of the description operator the axiom*

$$\text{DESCR} \qquad \forall y[y = Ix.A(x) \leftrightarrow \forall x(A(x) \leftrightarrow y = x)]$$

holds.

PROOF.

$$[\![a = Ix.A(x)]\!] = [\![Ea \wedge \forall x(A(x) \leftrightarrow x = a)]\!]$$

holds by definition for all a; since $[\![a = Ix.A(x)]\!] \leq Ea$, we have for all a

$$Ea \rightarrow [\![a = Ix.A(x) \leftrightarrow \forall x(A(x) \leftrightarrow x = a)]\!] = \top ,$$

and therefore

$$\forall y[y = Ix.A(x) \leftrightarrow \forall x(A(x) \leftrightarrow x = y)]$$

holds. □

2.8. COROLLARY. *Many-sorted* **IQCE** + DESCR *is conservative over many-sorted* **IQC**.

PROOF. Combine 1.15, 1.18, 2.6, and 2.7: if we extend an Ω-set $\mathcal{M}$ to $\mathcal{M}^{\text{sh}}$, the same sentences from $\mathcal{L}(\mathcal{M})$ remain valid, while at the same time DESCR holds in $\mathcal{M}^{\text{sh}}$. □

2.9. CONVENTION. For future use we introduce the following convention. Let $\mathcal{M}$ be any Ω-set, and let $\mathcal{M}'$ be an Ω-structure in which sort i is interpreted by $\mathcal{M}$; let x be a variable of sort i. Then for $[\![\forall x A(x)]\!]$ and $[\![\exists x A(x)]\!]$ we shall often write $[\![\forall a \in \mathcal{M}\ A(a)]\!]$ and $[\![\exists a \in \mathcal{M}\ A(a)]\!]$, respectively. Similarly, if a set X is interpreted by an Ω-set $\mathcal{X}$, $[\![\forall x \in X\ A(x)]\!] = [\![\forall a \in \mathcal{X} A(a)]\!]$, $[\![\exists x \in X A(x)]\!] = [\![\exists a \in \mathcal{X} A(a)]\!]$. □

3. Some notions from category theory

In this section we collect some basic notions from category theory which play a role in the sequel. We do not pretend to give an introduction to category theory, we only have included a *minimum* needed to follow the developments in later sections. Thus, for example, we shall not discuss adjointness. A standard reference for notions from category theory is MacLane (1971). We omit proofs, these can be found in the textbooks or may be supplied as exercises.

3.1. DEFINITION. A *category* $\mathscr{C}$ consists of a collection of *objects* Ob($\mathscr{C}$) and a collection of *morphisms* (or *arrows*) Hom($\mathscr{C}$); to each arrow f is associated a *domain*, dom(f) $\in$ Ob($\mathscr{C}$), and a *codomain* cod(f) $\in$ Ob($\mathscr{C}$). We write

$$A \to_f B \quad \text{or} \quad A \xrightarrow{f} B$$

for an arrow f with $A = \text{dom}(f)$, $B = \text{cod}(f)$. For the set of morphisms from A to B, i.e. morphisms with domain A, codomain B, we write $\mathscr{C}(A, B)$ or $\text{Hom}_{\mathscr{C}}(A, B)$. The morphisms of a category must satisfy

(i) If $f \in \mathscr{C}(A, B)$, $g \in \mathscr{C}(B, C)$, then the *composition* $g \circ f \in \mathscr{C}(A, C)$ is defined, satisfying associativity

$$h \circ (g \circ f) = (h \circ g) \circ f;$$

(ii) for each $A = \text{Ob}(\mathscr{C})$ there is an *identity morphism* $1_A \in \mathscr{C}(A, A)$ such that for $g \in \mathscr{C}(A, B)$

$$1_B \circ g = g, \qquad g \circ 1_A = g.$$

In this section we use $\mathscr{C}$, $\mathscr{D}$ for categories. □

REMARK. In our examples (a)–(f) below the categories are "locally small", that is to say, thinking in terms of classical set theory, $\mathscr{C}(A, B)$ is always a set, even though the collection of objects may be a proper class. In examples (h) and (i) locally small categories are constructed from locally small ones. For our applications to intuitionistic metamathematics in the next chapter all this is irrelevant: there we have no need to consider categories $\mathscr{C}$ with Ob($\mathscr{C}$) a proper class.

3.2. EXAMPLES *of categories*.

(a) The category Set has as objects all sets, and Set(x, y) consists of all (set-theoretic) mappings from x into y.

(b) Top is the category with as objects all topological spaces; Top(X, Y) consists of the continuous maps from X to Y.

(c) Let $(P, \leq)$ be a partially ordered set. This can be made into a category $\mathscr{P}$ with $\text{Ob}(\mathscr{P}) \equiv P$, and $\mathscr{P}(p, q) = \{i_{p,q} : p \leq q\}$, that is to say $\mathscr{P}(p, q)$ contains a single element, say $i_{p,q}$, iff $p \leq q$. As important special cases we have cHa's ordered by $\leq$, or topologies $\mathscr{O}(X)$ ordered by inclusion.

(d) A *monoid* is a category with a single object $*$. In section 15.5 we meet with an example of a monoid where $*$ is Baire space $\mathbb{B}$ and $\mathrm{Hom}(\mathbb{B}, \mathbb{B})$ consists of all continuous $f \in \mathbb{B} \to \mathbb{B}$. (A monoid from algebra is just the set of morphisms of a categorical monoid.)

(e) The category Ha of Heyting algebras has Heyting algebras as objects, and $\mathrm{Ha}(C, D)$ consists of all homomorphisms from C to D.

(f) The category Grp of groups with $\mathrm{Grp}(C, D)$ consisting of all group homomorphisms.

(h) If $\mathscr{C}$ is a category, then the *opposite* (or *dual*) category $\mathscr{C}^{\mathrm{op}}$ is given by

$$\mathrm{Ob}(\mathscr{C}^{\mathrm{op}}) := \mathrm{Ob}(\mathscr{C}), \qquad \mathscr{C}^{\mathrm{op}}(C, D) := \mathscr{C}(D, C),$$

where $f \circ g$ in $\mathscr{C}^{\mathrm{op}}$ corresponds to $g \circ f$ in $\mathscr{C}$. It is obvious that $\mathscr{C}^{\mathrm{op}}$ is again a category.

(i) Let $\mathscr{C}$ be a category, $C \in \mathrm{Ob}(\mathscr{C})$. The category $\mathscr{C} \downarrow C$ ($\mathscr{C}$ *over* C, often called *comma category*, *slice category*, or *localization of* $\mathscr{C}$ *over* C) has as objects all arrows $D \to_f C$ in $\mathscr{C}$, and as arrows from $D \to_f C$ to $D' \to_{f'} C$ all commuting triangles (i.e. $f' \circ g = f$): see fig. 14.1. (A *diagram* is a directed graph where the edges represent morphisms. A diagram *commutes* if for every two points all morphisms associated with directed paths between these points coincide.)

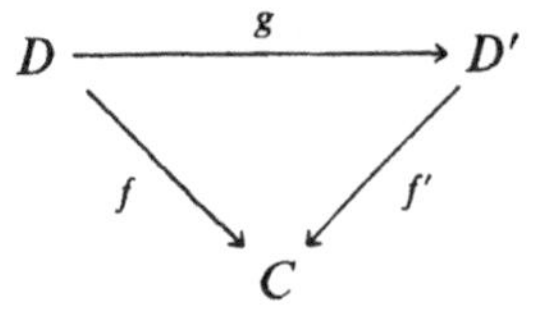

Fig. 14.1.

3.3. The category Set may be regarded as a standard example: many constructions/notions familiar from Set may be lifted to other categories. Thus there are analogues to the notions of injection, surjection, bijection: monomorphism, epimorphism, invertible morphism.

DEFINITION. Let $\mathscr{C}$ be a category, $A, B, C, D \in \mathrm{Ob}(\mathscr{C})$. $B \to_f C$ is a *mono-arrow* (*monomorphism*, the arrow is *mono*) iff for any pair $A \to_g B$, $A \to_h B$ we have

$$f \circ g = f \circ h \Rightarrow g = h,$$

and an *epi-arrow* (*epimorphism*, the arrow is *epi*) when for any two

$C \to_g D$, $C \to_h D$

$$g \circ f = h \circ f \Rightarrow g = h.$$

$B \to_f C$ is *invertible* (the arrow is *iso*, or an *isomorphism*) iff there is a $C \to_g B$ such that

$$f \circ g = 1_C, \qquad g \circ f = 1_B.$$

Two objects B, C are *isomorphic* iff there is an invertible $B \to_f C$. □

REMARK. In Set mono + epi is equivalent to invertible, but this is not generally the case.

If a concept is formulated for an arbitrary category $\mathscr{C}$, and we apply it to $\mathscr{C}^{\mathrm{op}}$, then translate the definition back in terms of $\mathscr{C}$, we obtain the dual concept. Note that monic and epi are duals, and that invertible is self-dual.

"Universal constructions" are a typical feature of category theory. We first define terminal objects and products.

3.4. DEFINITION. $\mathbb{1}$ is a *terminal object* in a category $\mathscr{C}$ if for each $A \in \mathscr{C}$ there is a unique arrow $A \to \mathbb{1}$. □

It is easy to see that terminal objects are unique up to isomorphism. In Set each singleton set is terminal object; in Grp it is a trivial group consisting of a unit element only. Dual to the notion of terminal object is *initial object*; an initial object $\mathbb{0}$ has a unique arrow $\mathbb{0} \to A$ for each object A. Set has also an initial object, namely the empty set.

3.5. DEFINITION. A *product* $A \times B$ of two objects A, B in a category $\mathscr{C}$ is an object with arrows $A \times B \to_{p_0} A$, $A \times B \to_{p_1} B$ such that for each pair of arrows $C \to_f A$, $C \to_g B$ there is a unique arrow h so that fig. 14.2 commutes. □

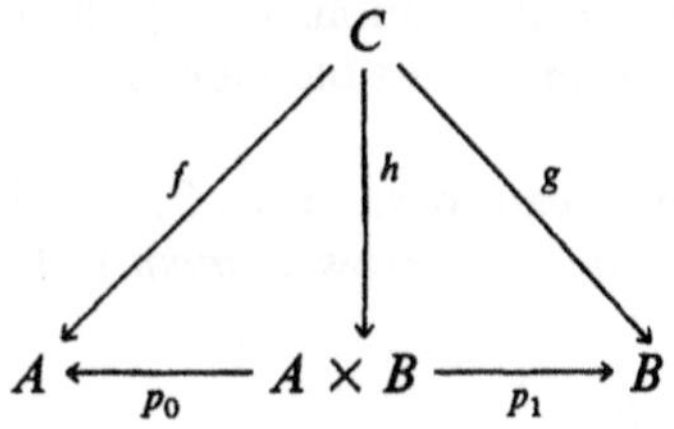

Fig. 14.2.

Proposition. *Products are unique up to isomorphism.*

Remark. The dual of "product" is called "sum". In Set products are cartesian products, in Top topological products. What is the product in the category $\mathscr{P}$ of a poset $(P, \leq)$?

Products are special cases of the more general notion of a pullback.

3.6. Definition. Let $A \to_f C$, $B \to_g C$ be arrows in a category $\mathscr{C}$. A *pullback* of f and g is an object P with arrows $P \to_{p_0} A$, $P \to_{p_1} B$ such that $f \circ p_0 = g \circ p_1$, i.e. fig. 14.3. commutes, and such that for each D with $D \to_{g'} A$, $D \to_{f'} B$ such that $f \circ g' = g \circ f'$, there is a *unique* arrow $D \to_h P$ such that $g' = p_0 \circ h$, $f' = p_1 \circ h$, i.e. fig. 14.4 commutes. □

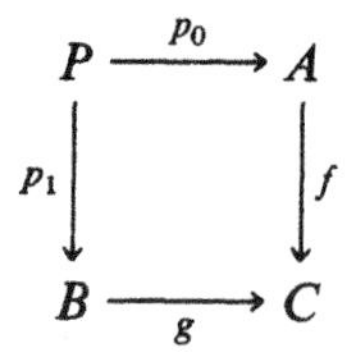

Fig. 14.3.

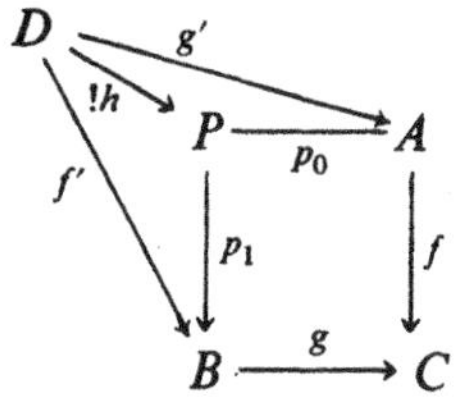

Fig. 14.4.

3.7. Proposition. *Pullbacks are unique up to isomorphism.* □

For the pullback of $B \to_g C$, $A \to_f C$ one often write $A \times_C B$; note however that a pullback is determined not only by the objects involved but by the *arrows* f, g, as shown by the following example.

Example. In Set and Top pullbacks always exist. In Set the pullback of $A \to_f C$, $B \to_g C$ is the subset of $A \times B$ defined by

$$\{(a, b) : a \in A \wedge b \in B \wedge f(a) = g(b)\}.$$

In Top it is the same set with as topology the relative topology inherited from the topological product $A \times B$ (exercise).

Products are special cases of pullbacks: a product of A and B is the pullback of $A \to \mathbf{1}$, $B \to \mathbf{1}$.

We now turn to exponentials.

3.8. Definition. Let $A, B \in \mathrm{Ob}(\mathscr{C})$, $\mathscr{C}$ a category. An *exponential* A^B in $\mathscr{C}$ is an object such that there is a fixed arrow (*evaluation*) $A^B \times B \to_{\mathrm{ev}} A$ and to each arrow $C \times B \to_f A$ a unique arrow $C \to_{\hat{f}} A^B$ such that

$$\mathrm{ev} \circ (\hat{f} \times 1) = f,$$

i.e. fig. 14.5 commutes. Here $f \times g$ for $C \to_f A$, $D \to_g B$ is the unique arrow making the diagram in fig. 14.6 commute, i.e. $f \circ p_0 = p_0' \circ (f \times g)$, $g \circ p_1 = p_1' \circ (f \times g)$. □

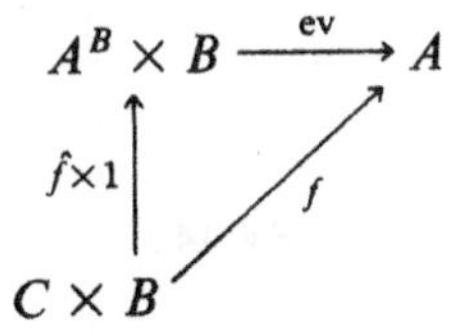

Fig. 14.5.

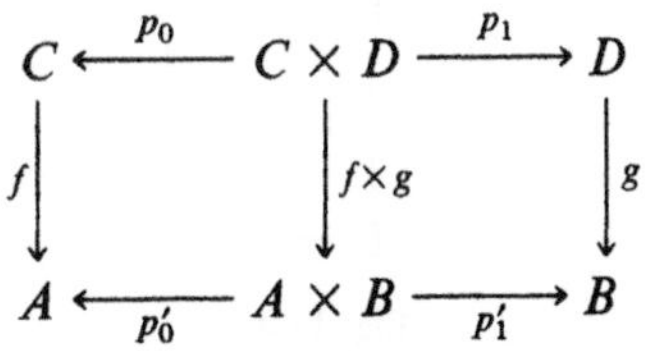

Fig. 14.6.

3.9. Proposition. *Exponentials are unique up to isomorphism.* □

We now turn to mappings between categories, namely functors.

3.10. Definition. A *functor* F from a category $\mathscr{C}$ into a category $\mathscr{D}$ (notation: $F: \mathscr{C} \to \mathscr{D}$) maps $\mathrm{Ob}(\mathscr{C})$ into $\mathrm{Ob}(\mathscr{D})$, and each $\mathscr{C}(A, B)$ into $\mathscr{D}(F(A), F(B))$ and $F(1_A) = 1_{F(A)}$, $F(f \circ g) = F(f) \circ F(g)$. □

3.11. Examples.

(a) The "*forgetful functor*" G maps Top to Set: $G(X, \mathcal{O}) = X$, $G(f) = f$. For many other categories where the objects are sets plus extra structure, and the morphisms set-theoretic mappings with extra properties, we have similar forgetful functors.

(b) The *powerset functor* P mapping Set to Set, defined by $\mathrm{P}(X) :=$ the powerset of X, and for $X \to_f Y$, $\mathrm{P}(f) = f'$ where $f' : Y \mapsto f[Y]$ for $Y \subset X$.

(c) Let $\mathscr{C}$ be any category, $A \in \mathrm{Ob}(\mathscr{C})$. We define a functor F from $\mathscr{C}$ to Set by

$$B \in \mathrm{Ob}(\mathscr{C}) \mapsto \mathscr{C}(A, B),$$

$$B \to_f C \in \mathscr{C}(B, C) \mapsto F(f) \in \mathscr{C}(A, B) \to \mathscr{C}(A, C),$$

where $F(f)(g) = f \circ g$ for all $g \in \mathscr{C}(A, B)$. This is called the *representable functor* $\mathscr{C}(A, -)$.

We also need the notion of natural transformation between functors.

3.12. Definition. A *natural transformation* τ from a functor $F: \mathscr{C} \to \mathscr{D}$ to a functor $G: \mathscr{C} \to \mathscr{D}$ is a collection of arrows $\{\tau_C : C \in \mathrm{Ob}(\mathscr{C})\} \subset \mathrm{Hom}(\mathscr{D})$ such that for all arrows $C \to_f C'$ in $\mathrm{Hom}(\mathscr{C})$ the diagram in fig. 14.7 commutes. Notation: $\tau : F \dot{\to} G$.

A *natural isomorphism* $\tau : F \cong G$ is a natural transformation τ such that each τ_C is an isomorphism for each $C \in \mathrm{Ob}(\mathscr{C})$. □

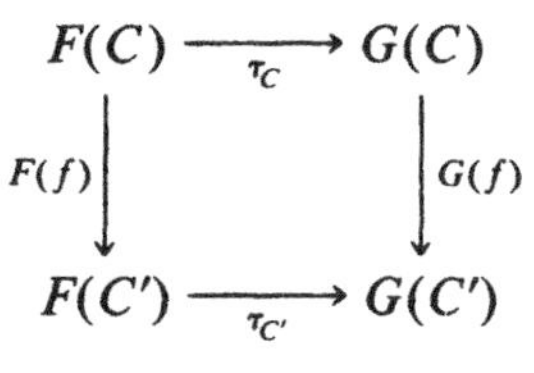

Fig. 14.7.

Example. Let H be a fixed group; then the mapping $G \mapsto H \times G$ for $G \in \mathrm{Ob}(\mathrm{Grp})$ determines a functor F_H (usually denoted by $H \times -$); this functor assigns to $G \to_g G'$ the arrow $H \times G \to_{1_H \times g} H \times G'$. With any homomorphism in $H \to_h H'$ corresponds a natural transformation from F_H to $F_{H'}$ (or from $H \times -$ to $H' \times -$).

We can now formulate a notion of equivalence between categories.

3.13. DEFINITION. Two categories $\mathscr{C}, \mathscr{D}$ are *equivalent* if there are $F: \mathscr{C} \to \mathscr{D}$, $G: \mathscr{D} \to \mathscr{C}$ with natural isomorphisms $F \circ G \cong I_{\mathscr{C}}$, $G \circ F \cong I_{\mathscr{D}}$, where $I_{\mathscr{D}}, I_{\mathscr{C}}$ are the obvious identity-functors. □

3.14. DEFINITION. Let $F: \mathscr{C} \to \mathscr{D}$ be a functor. F is *full* if to each arrow $F(C) \to_g F(D)$ there is an arrow $C \to_f D$ with $F(f) = g$ (surjectivity). F is *faithful* if for each pair $C \to_f D$, $C \to_g D$ with $F(f) = F(g)$ it follows that $f = g$ (injectivity). □

3.15. PROPOSITION. Let $F: \mathscr{C} \to \mathscr{D}$ be full and faithful and suppose that for each $D \in \mathrm{Ob}(\mathscr{D})$ there is a $C \in \mathrm{Ob}(\mathscr{C})$ such that $F(C)$ is isomorphic to D. Then the categories $\mathscr{C}$ and $\mathscr{D}$ are equivalent.

PROOF. Exercise. The proof requires an appeal to the axiom of choice. □

REMARK. It follows from the proof that, if F is full and faithful, and there is a mapping $G: \mathrm{Ob}(\mathscr{D}) \to \mathrm{Ob}(\mathscr{C})$ such that $F(G(D))$ is isomorphic to D for all $D \in \mathrm{Ob}(\mathscr{D})$, then no appeal to the axiom of choice is necessary.

3.16. *The categories* $\mathrm{Set}(\Omega)$ *and* $\mathrm{Sh}(\Omega)$. We illustrate some of the preceding notions for $\mathrm{Set}(\Omega)$ and $\mathrm{Sh}(\Omega)$.

The Ω-sheaves become a category by taking as morphisms from $\mathscr{M}$ to $\mathscr{N}$ all strict and total mappings from $\mathscr{M}$ to $\mathscr{N}$ (cf. 13.6.6).

The Ω-sets can be made into a category $\mathrm{Set}(\Omega)$ by taking as $\mathrm{Hom}(\mathscr{M}, \mathscr{M}')$ all Ω-predicates R on $\mathscr{M} \times \mathscr{M}'$ which internally represent total functions, i.e.

$$R(a, b) \wedge R(a, b') \leq [\![b = b']\!],$$

$$\bigvee\{b: R(a, b)\} = \top\,;$$

we leave it to the reader to verify that $\mathrm{Set}(\Omega)$ and $\mathrm{Sh}(\Omega)$ are categories. We have the following proposition.

PROPOSITION. $\mathrm{Set}(\Omega)$ *and* $\mathrm{Sh}(\Omega)$ *are equivalent categories.*

PROOF. Cf. 1.19, remark. We define a full and faithful functor $F: \mathrm{Sh}(\Omega) \to \mathrm{Set}(\Omega)$ as follows: on the objects of $\mathrm{Sh}(\Omega)$ we let F (essentially) be the identity. A morphism in $\mathrm{Sh}(\Omega)$, i.e. a strict and total function $f \in \mathscr{M} \to \mathscr{M}'$ is mapped by F to the corresponding functional relation

$$R_f(a, b) := [\![f(a) = b]\!]_{\mathscr{M}}.$$

It is easy to see that F is full and faithful; F also satisfies the condition of 3.15, remark: for an arbitrary Ω-set $\mathcal{M}$, $\mathcal{M}^{\text{sh}}$ is isomorphic to $\mathcal{M}$ in Set(Ω). □

3.17. *Presheaves.* The notion of a presheaf over a category $\mathscr{C}$ may be formulated as follows: a presheaf is a functor $F: \mathscr{C}^{\text{op}} \to \text{Set}$. If we take for $\mathscr{C}$ the partially ordered set $(\Omega, \leq)$, Ω a cHa, we obtain indeed the notion of an Ω-presheaf: $F(p)$ is the collection of all a with $Ea = p$, and if $p \leq q$, then the restriction map $\upharpoonright$ is $F(i_{p,q}) \in \text{Set}(F(q), F(p))$.

The presheaves over Ω form themselves also a category Psh(Ω), with as morphisms total and strict functional relations. In Set(Ω) and Psh(Ω) there are in general other morphisms besides the strict and total functions between two objects.

3.18. *Kripke models.* Let $(P, \leq)$ be a poset with associated category $\mathscr{P}$, and let D be a functor from the category $\mathscr{P}$ to Set. Then $\text{D}(i_{p,q}) \in \text{D}(p) \to \text{D}(q)$ if $p \leq q$; $(P, \leq, \text{D})$ represents a Kripke model for pure logic with equality if we interpret $=$ at each node k by equality in $\text{D}(k)$, and φ^p_q is the restriction map from p to q if $p \leq q$ (cf. 1.3(c)).

Structures of a fixed similarity type can be made into a category. For example, consider structures for a language with a single binary relation symbol R and a single unary function symbol f. As morphisms from the structure (M, R, f) to the structure $(M', \text{R}', \text{f}')$ we take mappings φ from M to M' such that for all $a, b \in M$

$$\text{R}(a, b) \Rightarrow \text{R}'(\varphi a, \varphi b), \qquad \varphi(\text{f}a) = \text{f}'\varphi(a).$$

Let us call the resulting category $\mathscr{C}$. A functor from $\mathscr{P}$ to $\mathscr{C}$ now represents a Kripke model for the language (R, f), with $(P, \leq)$ as the underlying poset.

4. Forcing over sites

In this section we consider a further generalization of the notion of an Ω-model, which will be needed for certain applications in the next chapter. Specifically, we generalize the forcing-variant of Ω-set semantics (cf. 13.7.1) to "forcing over a site".

4.1. Definition. Let $C \in \text{Ob}(\mathscr{C})$. A *sieve* on C is a collection of morphisms S with codomain C and closed under right composition, i.e. if $f \in S$ and $D' \to_g D \to_f C$, then $f \circ g \in S$. □

4.2. DEFINITION. A *covering system* (usually called a *Grothendieck topology*) on a category $\mathscr{C}$ is a function J which associates to each $C \in \mathrm{Ob}(\mathscr{C})$ a family $J(C)$ of sieves on C (the *covering sieves of* C) such that:

(i) (*Trivial cover*). For each $C \in \mathrm{Ob}(\mathscr{C})$ the maximal sieve $\{f : \mathrm{cod}(f) = C\} \in J(C)$.

(ii) (*Stability*). If $S \in J(C)$, $f \in \mathscr{C}(D, C)$, then $f^*(S) := \{E \rightarrow_g D : f \circ g \in S\} \in J(D)$.

(iii) (*Transitivity*). If $R \in J(C)$, S a sieve on C such that $f^*(S) \in J(D)$ for all $D \rightarrow_f C \in R$, then $S \in J(C)$. □

REMARK. Grothendieck topologies generalize the notion of covering, not of topology as given by a collection of open sets; therefore we choose the term "covering system".

We now illustrate (and motivate) the notion of a "covering system" in a topological space $(X, \mathscr{O})$. As our category $\mathscr{C}$ we take the partially ordered set $(\mathscr{O}, \subset)$. A sieve is (or may be identified with) a collection $\mathscr{U} \subset \mathscr{O}$ iff $\forall U \in \mathscr{U}\, \forall U' \in \mathscr{O}(U' \subset U \rightarrow U' \in \mathscr{U})$. $\mathscr{U}$ is a sieve on U_0 if $\forall U \in \mathscr{U}(U \subset U_0)$.

$J(U)$ for $U \in \mathscr{O}$ consists of all sieves covering U. One easily sees in this example: (i) expresses that U covers itself, (ii) corresponds to the fact that if $\mathscr{U}$ covers $U \in \mathscr{O}$ and $U' \subset U$ then $\{U'' \cap U' : U'' \in \mathscr{U}\}$ covers U', and (iii) expresses the following: if $\{U_i : i \in I\}$ covers $U \in \mathscr{O}$, $\{U_{i,j} : j \in I'(i)\}$ covers U_i, then $\{U_{i,j} : i \in I,\ j \in I'(i)\}$ covers U.

4.3. DEFINITION. A *site* is a pair $(\mathscr{C}, J)$, $\mathscr{C}$ a category, J a covering system on $\mathscr{C}$. A site is *consistent* if $\exists C \in \mathscr{C}\, \forall U \in J(C)$ (U inhabited). □

4.4. EXAMPLES.

(a) Let $\mathscr{C}$ be the category obtained from a cHa Ω with partial ordering $\leq$. A collection of morphisms $S \equiv \{p_i \rightarrow_{f_i} p : i \in I\}$ is in this case entirely determined by the p_i since for $p_i \leq p$ there is only a single morphism from p_i to p. If we put $S \in J(p) := (p = \bigvee\{p_i : i \in I\} \wedge S$ a sieve), then $S \in J(p)$ iff it is a sieve and a cover of p in the usual sense for cHa's. So this example straightforwardly generalizes the case of topological spaces in 4.2, remark.

(b) Take as category the monoid Mon($\mathbb{B}$) with single object $\mathbb{B}$ (Baire space), and as Hom($\mathbb{B}, \mathbb{B}$) all local homeomorphisms from $\mathbb{B}$ to $\mathbb{B}$. ($f \in X \rightarrow Y$ is a *local homeomorphism* from the topological space $(X, \mathscr{O}(X))$ to the topological space $(Y, \mathscr{O}(Y))$ iff f is continuous and $\forall x \in X\, \exists U \in \mathscr{O}(X)\, \exists U' \in \mathscr{O}(Y)\, (x \in U \wedge f(x) \in U'$ and f a homeomorphism from U to U').

A local homeomorphism is always an open mapping (i.e. maps open sets to open sets). We put

$$S \in J(\mathbb{B}) := (\bigcup\{f[\mathbb{B}] : f \in S\} = \mathbb{B}).$$

We leave it as an exercise to show that this is indeed a site.

(c) For any category $\mathscr{C}$ we may take the trivial notion of a covering system

$$S \in J(C) := (S = \{f : \mathrm{cod}(f) = C\}) \equiv 1_c \in S$$

(S consists of all arrows with codomain C).

4.5. Remark. The intersection of a family of covering systems on $\mathscr{C}$ is again a covering system on $\mathscr{C}$, that is to say, if $\{J_i : i \in I\}$ is a family of covering systems, then J defined by $J(C) := \bigcap\{J_i(C) : i \in I\}$ is also a covering system.

It is not always convenient to work with covering systems of sieves directly. Therefore we introduce the following definition.

Definition. Let $\mathscr{C}$ be a category, and let for each $C \in \mathrm{Ob}(\mathscr{C})$ a family $\mathscr{Y}(C)$ of sets of arrows with codomain C be given. Then the covering system *generated* by $\mathscr{Y}$ is the least covering system J such that for each sieve S

$$S \in J(C) := \exists X \in \mathscr{Y}(C)(X \subset S).$$

(A least covering system exists, by the preceding remark.) □

In particular we define the following.

4.6. Definition. Let $\mathscr{C}$ be a category; a function K assigning to each $C \in \mathrm{Ob}(\mathscr{C})$ families of morphisms with codomain C is called a *covering basis* if (i)–(iii) below hold:

(i) (*Trivial cover*). $\{C \to_1 C\} \in K(C)$.

(ii) (*Stability*). If $\{C_i \to_{f_i} C\}_{i \in I} \in K(C)$ and $D \to_g C$, then there is a family $\{D_j \to_{h_j} D\}_{j \in I'} \in K(D)$ such that for each $j \in I'$ there is an $i \in I$ and a morphism $D_j \to_{g'} C_i$ with $f_i \circ g' = g \circ h_j$, i.e. fig. 14.8 commutes.

(iii) (*Transitivity*). If $\{C_i \to_{f_i} C\}_{i \in I} \in K(C)$ and for each $i \in I$ there is a $(C_{i,j} \to_{g_{i,j}} C_i\}_{j \in I(i)} \in K(C_i)$, then there is a family $\{D_k \to_{g_k} C\}_{k \in I'} \in$

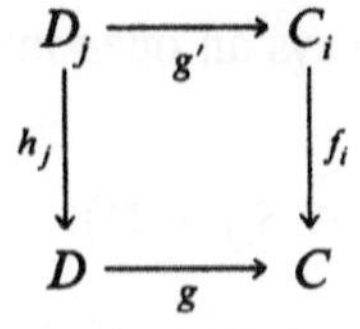

Fig. 14.8.

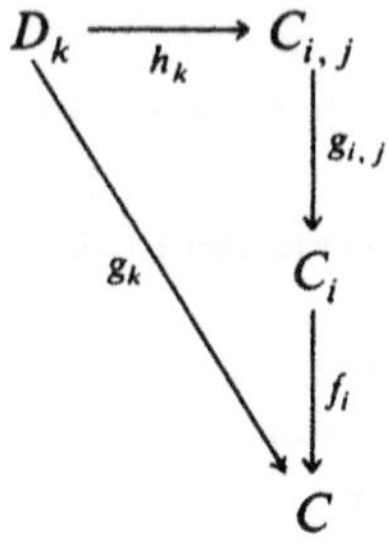

Fig. 14.9.

$K(C)$ such that for each $k \in I'$ there are $i \in I$, $j \in I(i)$ with morphisms $D_k \to_{h_k} C_{i,j}$ such that fig. 14.9 commutes, i.e. $g_k = f_i \circ g_{i,j} \circ h_k$. □

The reader should check the meaning of (i), (ii), and (iii) for the example of covers in a topological space.

PROPOSITION. *The covering basis K determines a covering system J satisfying*

$$R \in J(C) \Leftrightarrow R \text{ a sieve on } C \text{ and } \exists S \in K(C)(S \subset R).$$

PROOF. Exercise. □

Note that J is consistent iff for some $C \in \mathrm{Ob}(\mathscr{C})$ $\forall S \in K(C)$ (S is inhabited).

4.7. DEFINITION. A *domain* $\mathscr{M}$ on a category $\mathscr{C}$, or $\mathscr{C}$*-presheaf*, is a functor from $\mathscr{C}^{\mathrm{op}}$ to Set, i.e. a collection of sets $\{\mathscr{M}(C) : C \in \mathrm{Ob}(\mathscr{C})\}$ with *restriction maps* $\mathscr{M}(f) \in \mathscr{M}(D) \to \mathscr{M}(C)$ for each $f \in \mathscr{C}(C, D)$; we write $x \upharpoonright f$ for $\mathscr{M}(f)(x)$. The restriction maps must satisfy

$$x \upharpoonright 1 = x, \qquad (x \upharpoonright f) \upharpoonright g = x \upharpoonright (f \circ g).$$

A *domain* on a site $(\mathscr{C}, J)$ (or $(\mathscr{C}, J)$*-presheaf*) is simply a domain on $\mathscr{C}$. □

4.8. Examples.

(a) An Ω-presheaf is a domain on the category $(\Omega, \leq)$.

(b) A $\mathscr{C}$-presheaf in the case of example 3.2(d) is a set with a restriction map $\upharpoonright$ such that

$$(x \upharpoonright f) \upharpoonright g = x \upharpoonright (f \circ g), \quad x \upharpoonright 1 = x.$$

4.9. Definition. Let $\mathscr{M}$ be a $(\mathscr{C}, J)$-presheaf, then:

(i) $\mathscr{M}$ is *J-separated* iff $\forall x, y \in \mathscr{M}(C)(\exists R \in J(C) \forall f \in R(x \upharpoonright f = y \upharpoonright f) \to x = y)$ for all $C \in \mathrm{Ob}(\mathscr{C})$;

(ii) a collection $\{x_f \in \mathscr{M}(\mathrm{dom}(f)) : f \in R\}$, where $R \in J(C)$ is *compatible* iff $\forall f \in R\, \forall g(\mathrm{cod}(g) = \mathrm{dom}(f) \to x_f \upharpoonright g = x_{f \circ g})$;

(iii) $\mathscr{M}$ is a *J-sheaf* in $(\mathscr{C}, J)$, or a $(\mathscr{C}, J)$*-sheaf*, where $(\mathscr{C}, J)$ is a site, iff for each compatible $\{x_f : f \in R\}$, $R \in J(C)$, there is a unique $x \in \mathscr{M}(C)$ such that $x \upharpoonright f = x_f$ for all $x \in R$; x is called the *join* of $\{x_f : f \in R\}$.

(iv) Let $\mathscr{M}' \subset \mathscr{M}$, $\mathscr{M}$ a $(\mathscr{C}, J)$-sheaf. $\mathscr{M}'$ *generates* $\mathscr{M}$ iff each element of $\mathscr{M}$ is a join of restrictions of elements of $\mathscr{M}'$ (cf. 1.12(i)). □

Remark. We can also define the notion of a compatible collection with respect to a covering basis K: $\{x_f : f \in R\}$ for $R \in K(C)$ is *compatible* iff for all $f, g \in R$ and all h, k such that $f \circ h = g \circ k$ we have $x_f \upharpoonright h = x_g \upharpoonright k$.

4.10. *Sheaf-completions.* Here we can proceed in complete analogy to the method for sheaf completion for Ω-presheaves.

For a $(\mathscr{C}, J)$-presheaf $\mathscr{M}$ we define the equivalence relation $\cong$ on compatible families as:

$$\{x_f : f \in S\} \cong \{y_g : g \in T\}$$
$$:= \exists R \in J(C)\big(R \subset S \cap T \wedge \forall f \in R(x_f = y_f)\big),$$

where $S, T \in J(C)$.

Restriction on the equivalence classes is defined for $S \in J(C)$, $D \to_g C$ by

$$\big(\{x_f : f \in S\}/\cong\big) \upharpoonright g := \{x_h : h \in g^*(S)\}/\cong .$$

The result is a J-presheaf $\mathscr{M}^+$ (exercise).

4.11. PROPOSITION. *Let $\mathcal{M}$ be a $\mathscr{C}$-presheaf, J a covering system on $\mathscr{C}$; $\mathcal{M}^+$ is a J-separated presheaf, and if $\mathcal{M}$ is J-separated, $\mathcal{M}^+$ is a J-sheaf generated by $\mathcal{M}$.*

PROOF. Similar to the case of Ω-presheaves, and left as an exercise. □

NOTATION. $\mathcal{M}^{\mathrm{sh}}$ is the sheaf-completion of a presheaf $\mathcal{M}$. □

4.12. DEFINITION. Let $\mathcal{M}$ be a $\mathscr{C}$-presheaf. A *$\mathscr{C}$-constant* c in $\mathcal{M}$ is a mapping defined on $\mathrm{Ob}(\mathscr{C})$ such that $c(C) \in \mathcal{M}(C)$ for all $C \in \mathrm{Ob}(\mathscr{C})$, and $c(D) \upharpoonright f = c(D')$ for each $D' \to_f D$. A *$(\mathscr{C}, J)$-constant* is the same as a $\mathscr{C}$-constant.

We can formulate this also in terms of natural transformations; let $\mathbb{1}_{\mathscr{C}}$ be the trivial $(\mathscr{C}, J)$-sheaf with $\mathbb{1}_{\mathscr{C}}(C) = \{*\}$ for all $C \in \mathrm{Ob}(\mathscr{C})$, then a constant is a natural transformation

$$c : \mathbb{1}_{\mathscr{C}} \dot{\to} \mathcal{M}. \quad \square$$

4.13. PROPOSITION (*product of presheaves*). *If $\mathcal{M}, \mathcal{M}'$ are $\mathscr{C}$-presheaves, then so is $\mathcal{M} \times \mathcal{M}'$, given by*

$$(\mathcal{M} \times \mathcal{M}')(C) := \{(x, y) : x \in \mathcal{M}(C) \wedge y \in \mathcal{M}'(C)\}$$

with restriction map

$$(x, y) \upharpoonright f := (x \upharpoonright f, y \upharpoonright f).$$

PROOF. Left to the reader. □

4.14. DEFINITION. Let $\mathcal{M}_1, \ldots, \mathcal{M}_n$ be $\mathscr{C}$-presheaves, $(\mathscr{C}, J)$ a site. A *$(\mathscr{C}, J)$-relation* R on $\mathcal{M}_1, \ldots, \mathcal{M}_n$ is a mapping assigning to each $C \in \mathrm{Ob}(\mathscr{C})$

$$R(C) \subset \mathcal{M}_1(C) \times \cdots \times \mathcal{M}_n(C),$$

such that for all $D \to_g C$

$$(x_1, \ldots, x_n) \in R(C) \to (x_1 \upharpoonright g, \ldots, x_n \upharpoonright g) \in R(D), \quad \text{and}$$

$$\forall S \in J(C)[\forall f \in S(x_1 \upharpoonright f, \ldots, x_n \upharpoonright f) \in R(\mathrm{dom}(f)) \to (x_1, \ldots, x_n) \in R(C)]. \quad \square$$

REMARK. An n-ary $(\mathscr{C}, J)$-relation over $\mathcal{M}_1, \ldots, \mathcal{M}_n$ is the same as a unary $(\mathscr{C}, J)$-relation ($(\mathscr{C}, J)$-predicate) over $\mathcal{M}_1 \times \cdots \times \mathcal{M}_n$.

4.15. Definition. Let $\mathcal{M}_1 \dots \mathcal{M}_n, \mathcal{M}_{n+1}$ be $\mathscr{C}$-presheaves. A *$\mathscr{C}$-function* or *$(\mathscr{C}, J)$-function* from $\mathcal{M}_1 \times \cdots \times \mathcal{M}_n$ to $\mathcal{M}_{n+1}$ is a natural transformation

$$f : \mathcal{M}_1 \times \cdots \times \mathcal{M}_n \dot{\to} \mathcal{M}_{n+1} .$$

That is to say, to each $C \in \mathrm{Ob}(\mathscr{C})$ we have an $f_C : \mathcal{M}_1(C) \times \cdots \times \mathcal{M}_n(C) \to \mathcal{M}_{n+1}(C)$, such that if $D \to_g C$, then

$$f_D(x_1 \upharpoonright g, \dots, x_n \upharpoonright g) = f_C(x_1, \dots, x_n) \upharpoonright g . \quad \square$$

Remark. Presheaves over $(\mathscr{C}, J)$ form a category $\mathrm{Psh}(\mathscr{C}, J)$ with as morphisms total functional $(\mathscr{C}, J)$-relations; so here there are in general fewer $\mathscr{C}$-functions than morphisms from $\mathcal{M}$ to $\mathcal{M}'$. In the subcategory $\mathrm{Sh}(\mathscr{C}, J)$ of sheaves over $(\mathscr{C}, J)$ morphisms correspond to $\mathscr{C}$-functions, as in the special case of $\mathrm{Sh}(\Omega)$ (cf. 3.17).

4.16. Definition. A *structure $\mathcal{M}$ over a site* $(\mathscr{C}, J)$ consists of a family of $\mathscr{C}$-presheaves $\{\mathcal{M}_i : i \in I\}$ over $\mathscr{C}$ (I is the collection of *sorts*), together with a collection of $\mathscr{C}$-constants, $\mathscr{C}$-functions, and $(\mathscr{C}, J)$-relations over the $\mathcal{M}_i$.

We adopt the conventions of 2.9 also for structures over sites. $\square$

4.17. Definition (*forcing over sites*). Let $\mathscr{C}$ be a many-sorted language with sorts in I, and let $\mathcal{M}$ be a $(\mathscr{C}, J)$-structure with family of presheaves $\{\mathcal{M}_i : i \in I\}$. To each constant symbol c of sort i, function symbol f, and relation symbol R of $\mathscr{L}$ we assign a corresponding $(\mathscr{C}, J)$-constant $[\![c]\!]$ in $\mathcal{M}_i$, a $(\mathscr{C}, J)$-function $[\![f]\!]$, a $(\mathscr{C}, J)$-relation $[\![R]\!]$ of $\mathcal{M}$, respectively.

We now define forcing $C \Vdash A$ for all $C \in \mathrm{Ob}(\mathscr{C})$, and all sentences A in $\mathscr{L}(C)$, i.e. sentences in $\mathscr{L}$ with parameters from the $\mathcal{M}_i(C)$, $i \in I$, relative to a site $(\mathscr{C}, J)$.

We first assign to each term t of sort j with parameters from the collection $\{\mathcal{M}_i(C) : i \in I\}$ an element $[\![t]\!]_C$ of $\mathcal{M}_j(C)$ as follows:

(a) if $a \in \mathcal{M}_i(C)$, $[\![a]\!]_C := a$;
(b) $[\![c]\!]_C := [\![c]\!](C)$;
(c) $([\![f(t_1, \dots, t_0)]\!]_C := [\![f]\!]_C([\![t_1]\!]_C, \dots, [\![t_n]\!]_C)$.

The forcing clauses now become:

(i) $C \Vdash t = t' := \exists S \in J(C) \forall f \in S\, ([\![t]\!]_C \upharpoonright f = [\![t']\!]_C \upharpoonright f)$; if the presheaves in $\mathcal{M}$ are separated, this simplifies to $C \Vdash t = t' := [\![t]\!]_C = [\![t']\!]_C$;
(ii) $C \Vdash R(t_1, \dots, t_n) := ([\![t_1]\!]_C, \dots, [\![t_n]\!]_C) \in [\![R]\!](C)$;
(iii) $C \Vdash \bot := \emptyset \in J(C)$;
(iv) $C \Vdash A \wedge B := (C \Vdash A) \wedge (C \Vdash B)$;

(v) $C \Vdash A \vee B := \exists S \in J(C) \forall f \in S(\text{dom}(f) \Vdash A \upharpoonright f \vee \text{dom}(f) \Vdash B \upharpoonright f)$;

(vi) $C \Vdash A \to B =: \forall D \forall f \in \mathscr{C}(D, C)((D \Vdash A \upharpoonright f) \to (D \Vdash B \upharpoonright f))$;

(vii) $C \Vdash \exists x^i A(x) := \exists S \in J(C) \forall f \in S \exists d \in \mathscr{M}_i(\text{dom}(f))(\text{dom}(f) \Vdash (A \upharpoonright f)(d))$;

(viii) $C \Vdash \forall x^i A(x) := \forall D \forall f \in \mathscr{C}(D, C) \forall d \in \mathscr{M}_i(D)(D \Vdash (A \upharpoonright f)(d))$.

Here $A \upharpoonright f$ for $D \to_f C$ is obtained from the formula A by replacing all parameters a^i from $\mathscr{M}_i(C)$ in A by $a^i \upharpoonright f$. □

4.18. LEMMA.

(i) *If* $D \to_f C$ *and* $C \Vdash A$, *then* $D \Vdash A \upharpoonright f$ *(monotonicity).*

(ii) *If* $S \in J(C)$, *and* $D \Vdash A \upharpoonright f$ *for all* $D \to_f C \in S$, *then* $C \Vdash A$ *(covering property).*

PROOF. By induction on the complexity of A (exercise). □

4.19. DEFINITION. Let $\mathscr{M}$ be a $\mathscr{C}$-structure, $(\mathscr{C}, J)$ a site. $\mathscr{M} \Vdash A$ iff $C \Vdash A$ for all $C \in \text{Ob}(\mathscr{C})$. $\Gamma \Vdash A$ means: for all sites $(\mathscr{C}, J)$, and all (C, J)-structures $\mathscr{M}$, $\forall B \in \Gamma(\mathscr{M} \Vdash B) \Leftrightarrow \mathscr{M} \Vdash A$. □

4.20. THEOREM (*soundness and completeness for forcing over sites*). *For derivability in many-sorted intuitionistic predicate logic* **IQC**:

$$\Gamma \vdash A \quad \Leftrightarrow \quad \Gamma \Vdash A.$$

PROOF. ⇒, the soundness, we leave as an exercise. ⇐, the completeness, is immediate since we have completeness for Ω-valued structures, Ω a cHa, as shown in the preceding chapter, and these can be viewed as special cases of structures over sites. □

We did not discuss the interpretation of partial functions in sites; therefore we have formulated the theorem for **IQC** only. The extension to **IQCE** is not difficult and left to the reader. The following lemma corresponding to 2.5 is often useful in evaluating formulas over sites.

4.21. LEMMA. *Let* $\mathscr{M}'$ *be a* $\mathscr{C}$*-presheaf*, $\mathscr{M} = (\mathscr{M}')^{\text{sh}}$ *the* $(\mathscr{C}, J)$*-sheaf completion. Then*

$$(C \Vdash \forall a \in \mathscr{M}' A(a)) \Rightarrow (C \Vdash \forall a \in \mathscr{M} A(a)),$$

$$(C \Vdash \exists a \in \mathscr{M}' A(a)) \Rightarrow (C \Vdash \exists a \in \mathscr{M} A(a)).$$

Proof. Left to the reader. □

5. Sheaf models for higher-order logic

The interpretation of higher-order logic requires that we give a standard meaning to the power sheaf of a given sheaf and to the sheaf of functions from sheaf $\mathcal{M}$ to sheaf $\mathcal{N}$.

We shall now first describe power sheaves, and then sheaves of functions; the power sheaves are special cases of sheaves of functions. We start with Ω-sheaves, and afterwards consider the generalization to sheaves over sites.

In this section C and D are objects of a category, A and B are formulas.

Power sheaves over cHa's (5.1–5.5).

5.1. Definition. Let $\mathcal{N}$ be an Ω-set. The *power sheaf* $\mathcal{P}(\mathcal{N})$ has as elements pairs (P, p) with $p \in \Omega$, P a strict Ω-predicate on $\mathcal{N}$, i.e. P satisfies, for all a, b in $\mathcal{N}$

$$Pa \wedge [\![a = b]\!] \le P(b),$$

$$P(a) \le Ea, \qquad P(a) \le p.$$

Extent and restriction are defined by

$$E(P, p) := p,$$

$$(P, p) \upharpoonright q := (P \upharpoonright q, p \wedge q),$$

where $P \upharpoonright q = \lambda a.(P(a) \wedge q)$. The membership relation is given by

$$[\![a \in P]\!] := P(a). \quad \square$$

5.2. Proposition. *$\mathcal{P}(\mathcal{N})$ is always a sheaf and each element is the restriction of a global element (i.e. an element with extent* $\top$*). Moreover, for* $N = |\mathcal{N}|$

$$[\![(P, p) = (Q, q)]\!] = p \wedge q \wedge \bigwedge\{P(a) \leftrightarrow Q(a) : a \in N\}.$$

Proof. The presheaf properties are straightforward. Obviously, $(P, p) = (P, \top) \upharpoonright p$. Suppose $\mathcal{Y}$ is any compatible collection in $\mathcal{P}(\mathcal{N})$, i.e.

$$(P_i, p_i) \in \mathcal{Y} \wedge (P_j, p_j) \in \mathcal{Y} \Rightarrow (P_i \upharpoonright p_j, p_i \wedge p_j) = (P_j \upharpoonright p_i, p_i \wedge p_j).$$

Take for $\bigvee\mathcal{Y}$ the pair (P, p) with

$$P(a) := \bigvee\{P_i(a) : (P_i, p_i) \in \mathcal{Y}\},\ p := \bigvee\{p_i : i \in I\}.$$

For equality we find

$$[\![(P, p)=(Q, q)]\!] = \bigvee\{r \leq E(P, p) \wedge E(Q, q) :$$
$$(P, p) \upharpoonright r = (Q, q) \upharpoonright r\} = \bigvee\{r \leq p \wedge q : (P \upharpoonright r) = (Q \upharpoonright r)\}.$$

Now observe $(P \upharpoonright r) = (Q \upharpoonright r) \Leftrightarrow \forall a \in N(P(a) \wedge r = Q(a) \wedge r) \Leftrightarrow r \leq \bigwedge\{P(a) \leftrightarrow Q(a) : a \in \mathcal{N}\}$, and so $[\![(P, p) = (Q, q)]\!] = \bigvee\{r : r \leq p \wedge q \wedge \bigwedge\{P(a) \leftrightarrow Q(a) : a \in \mathcal{N}\}\} = p \wedge q \wedge \bigwedge\{P(a) \leftrightarrow Q(a) : a \in N\}$. □

Next we show that $\mathcal{P}(\mathcal{N})$ has the usual properties, namely extensionality (replacement) and comprehension.

5.3. Proposition. *If $\mathcal{N}$ is any Ω-set, then*
(i) $[\![\forall P, Q \in \mathcal{P}(\mathcal{N})\,(P = Q \leftrightarrow \forall a \in \mathcal{N}(a \in P \leftrightarrow a \in Q)]\!] = \top$,
(ii) *For all $A(x)$,* $[\![\exists P \in \mathcal{P}(\mathcal{N})\,\forall a \in \mathcal{N}(a \in P \leftrightarrow A(a))]\!] = \top$.

Proof. (i) Each element of $\mathcal{P}(\mathcal{N})$ is the restriction of a global element; for global elements $(P, \top)$, $(Q, \top)$ we have by 5.2

$$[\![(P, \top) = (Q, \top)]\!] = \bigwedge\{P(a) \leftrightarrow Q(a) : a \in N\}.$$

Now P is strict, so $P(a) \wedge Ea = P(a)$ and hence $P(a) \to Q(a) = Ea \wedge P(a) \to Q(a) = Ea \to (P(a) \to Q(a))$. Similarly $Q(a) \to P(a) = Ea \wedge Q(a) \to P(a) = Ea \to (P(a) \to Q(a))$. Thus $[\![(P, \top) = (Q, \top)]\!] = \bigwedge\{Ea \to (P(a) \leftrightarrow Q(a)) : a \in N\} = [\![\forall a \in \mathcal{N}(P(a) \leftrightarrow Q(a))]\!]$, and since global elements generate the powersheaf, (i) holds.

(ii) Let $A(x)$ be any formula with a single free variable, and let P be given by

$$P(a) := [\![Ea \wedge A(a)]\!].$$

P is a strict predicate on $\mathcal{N}$ (use the fact that for formulas we have replacement), and thus represents a global element of $\mathcal{P}(\mathcal{N})$; for all $a \in N$

$$Ea \leq [\![a \in P \leftrightarrow A(a)]\!]$$

and thus (ii) holds. □

5.4. Remarks. (i) Ω is itself an Ω-set, with

$$[\![p = q]\!] = p \wedge q, \quad \text{and hence } Ep = p.$$

We may think of Ω as consisting of the (trivially strict) Ω-predicates over a singleton, say $\{0\}$.

In order to obtain a sheaf, we consider the Ω-set $\mathbf{\Omega} = \mathscr{P}(\{0\})$, with $|\mathbf{\Omega}| = \{(p, q) : p \leq q\}$. Writing $a = (|a|, Ea)$ for $a \in |\mathbf{\Omega}|$, we have, in keeping with definition 5.1,

$$a \upharpoonright p := (|a| \wedge p, Ea \wedge p).$$

In this case 5.2 implies

$$[\![a = b]\!] := (|a| \leftrightarrow |b|) \wedge Ea \wedge Eb.$$

(ii) The *global* elements of $\mathscr{P}(\mathscr{N})$ correspond precisely to the total strict operations from $\mathscr{N}$ to $\mathbf{\Omega}$: if $(P, \top)$ is a global element, we can associate to it f_P given by $f_P(a) := (P(a), Ea)$; and conversely, given $f \equiv \lambda a.(P(a), Ea)$ from $\mathscr{N}$ to $\mathbf{\Omega}$ we obtain the global element p_f as $(\lambda a.P(a), \top)$.

5.5. Definition. Let $\mathscr{M}$ be any presheaf, $p \in \Omega$. $\mathscr{M} \downarrow p$ ("$\mathscr{M}$ over p", $\mathscr{M}$ *restricted* to p) has elements

$$|\mathscr{M} \downarrow p| \equiv M \upharpoonright p := \{a \upharpoonright p : a \in M\},$$

and the extent E and restriction $\upharpoonright$ on $\mathscr{M} \downarrow p$ are simply the restrictions of $\mathrm{E}_{\mathscr{M}}$, $\upharpoonright_{\mathscr{M}}$ to $M \upharpoonright p$. If $\mathscr{M}$ is a sheaf, then so is $\mathscr{M} \downarrow p$. □

Remark. If $\mathscr{M}$ is a presheaf, $\mathscr{N}$ a sheaf, and $f \in M \to N$ a total and strict operation from $\mathscr{M}$ to $\mathscr{N}$, then f also maps $M \upharpoonright p$ to $N \upharpoonright p$ for any $p \in \Omega$. Thus $f \restriction (M \upharpoonright p)$ is a mapping from $\mathscr{M} \downarrow p$ to $\mathscr{N} \downarrow p$ which is strict and total.

Function sheaves over cHa's (5.6–8).

5.6. Definition. Let $\mathscr{M}$ be a presheaf, $\mathscr{N}$ a sheaf. Then $\mathscr{N}^{\mathscr{M}}$, *the function sheaf from* $\mathscr{M}$ *to* $\mathscr{N}$, is a sheaf with elements $\boldsymbol{f} \equiv (f, p_f)$, where f is strict and total from $\mathscr{M} \downarrow p_f$ to $\mathscr{N} \downarrow p_f$. That is to say

$$E(fa) = Ea \quad \text{for all } a \text{ with } Ea \leq p_f,$$

$$[\![a = b]\!] \leq [\![fa = fb]\!] \quad \text{for all } a \text{ and } b \text{ with } Ea, Eb \leq p_f.$$

Extent E and restriction $\upharpoonright$ on $\mathcal{N}^{\mathcal{M}}$ are defined by

$$Ef := p_f,$$

$$\boldsymbol{f} \upharpoonright q := (f \upharpoonright q, p_f \wedge q),$$

where $f \upharpoonright q := \lambda a.f(a) \upharpoonright q$. We interpret application for $\boldsymbol{f} = (f, p_f)$, $a \in \mathcal{M}$ as

$$\boldsymbol{f}(a) := f(a \upharpoonright p_f). \quad \square$$

5.7. Remarks. (i) We may also think of the elements (f, p_f) of $\mathcal{N}^{\mathcal{M}}$ as strict operations from $\mathcal{M}$ to $\mathcal{N}$ which are strict and total precisely on $\mathcal{M} \downarrow p_f$. That is to say, given a strict and total $f \in \mathcal{M} \downarrow p_f \to \mathcal{N} \downarrow p_f$, we can think of this as obtained from an f' given by

$$f' \equiv \lambda a.f(a \upharpoonright p_f).$$

Clearly $\forall a \in \mathcal{M}(E(f'a) \leq p_f)$.

(ii) Suppose that $\mathcal{N}'$, $\mathcal{M}'$ generate the sheaf $\mathcal{N}$ and the presheaf $\mathcal{M}$, respectively. Then we can also represent $\mathcal{N}^{\mathcal{M}}$ as a subsheaf of $\mathcal{P}(\mathcal{N}' \times \mathcal{M}')$, namely

$$\{(R, p_R) : R \text{ a strict } \Omega\text{-relation on } \mathcal{N}' \times \mathcal{M}',\ p_R = [\![\mathrm{fun}(R)]\!]\},$$

where "fun(R)" expresses "R is functional and total", i.e. $\mathrm{fun}(R) := \forall x\, \boldsymbol{E}(\boldsymbol{I}y.R(x, y))$, or $\mathrm{fun}(R) := \forall x\, \exists! y R(x, y)$.

With (R, p_R) corresponds $\boldsymbol{f}_R \equiv (f_R, p_R)$ with $f_R(a) = \bigvee\{b \upharpoonright R(a, b) \wedge p_R : b \in N'\}$ for $Ea \leq p_R$; conversely, given $\boldsymbol{f} \equiv (f, p_f)$ we find the corresponding (R, p_R) as $(\lambda a,b.[\![fa = b]\!], p_f)$. Restriction corresponds to

$$(R, p_R) \upharpoonright q := (R \upharpoonright q, p_R \wedge q),$$

where $(R \upharpoonright q)(a, b) := R(a \upharpoonright q, b \upharpoonright q)$. The equality relation is inherited from $\mathcal{P}(\mathcal{N}' \times \mathcal{M}')$, namely

$$(R, p_R) = (R', p_{R'}) :=$$
$$p_R \wedge p_{R'} \wedge \bigwedge\{R(a, b) \leftrightarrow R'(a, b) : a \in N' \wedge b \in M'\}. \quad \square$$

5.8. Proposition. *Let $\mathcal{N}$ be a sheaf, $\mathcal{M}$ a presheaf. In $\mathcal{N}^{\mathcal{M}}$*

$$[\![\boldsymbol{f} = \boldsymbol{g}]\!] = E\boldsymbol{f} \wedge E\boldsymbol{g} \wedge [\![\forall x \in \mathcal{M}(\boldsymbol{f}(x) = \boldsymbol{g}(x))]\!]$$
$$= p_f \wedge p_g \wedge \bigwedge\{ Ea \to [\![f(a) = g(a)]\!] : a \in M \}.$$

Proof. Exercise. $\square$

5.9. PROPOSITION. *Let $\mathcal{N}$ be a sheaf, $\mathcal{M}$ a presheaf. Then $\mathcal{N}^{\mathcal{M}}$ validates function comprehension i.e.*

$$[\![\forall x \in \mathcal{M}\, \exists! y \in \mathcal{N} A(x, y) \to \exists f \in \mathcal{N}^{\mathcal{M}}\; \forall x A(x, f(x))]\!] = \top .$$

PROOF. Exercise. □

5.10. Let **HAH** be a version of intuitionistic higher-order arithmetic based on intuitionistic higher-order logic, with for each sort σ a power-sort $[\sigma]$, and for each pair of sorts σ, τ a product sort $\sigma \times \tau$ and a function sort $\sigma \to \tau$.

THEOREM. *Let* **HAH** *be interpreted in* $\mathrm{Sh}(\Omega)$ *as follows. Assign to* $\mathbb{N}$ *as interpretation the simple Ω-sheaf* $(\hat{\mathbb{N}})^{\mathrm{sh}}$, *the simple sheaf obtained by completion of the constant Ω-set* $\hat{\mathbb{N}}$ *(defined as in* 13.6.4). $\hat{0} \equiv [\![0]\!]$ *is the interpretation of* 0; *the interpretation of the successor S,* $[\![S]\!] \equiv \hat{S}$, *is determined by its action on* $\hat{\mathbb{N}}$: $\hat{S}\hat{n} = (n+1)^{\wedge}$. *The domains assigned to products, power sorts, and function sorts are determined according to* 2.1, 5.1, 5.6 *above. Then* $\mathrm{Sh}(\Omega)$ *is a model of* **HAH**.

PROOF. Straightforward with 5.3, 5.8, and 5.9. □

We may assume that the logical basis is in fact **IQCE** + DESCR.

5.11. *Power sheaves and function sheaves over sites.* For the applications in the next chapter (in particular sections 15.2–3) soundness of cHa-sheaf models for intuitionistic higher-order arithmetic **HAH** suffices. Nevertheless it is of interest to see how the construction of power sheaves and function sheaves can also be carried out for sites. This also removes the ad hoc character of the interpretations of $\mathbb{B}, \mathbb{N}^{\mathbb{B}}, \mathbb{B} \to \mathbb{B}$ in the models of sections 15.5–6, since it shows that the corresponding function sheaves are constructed according to a uniform recipe.

5.12. DEFINITION. Let $(\mathscr{C}, J)$ be a site and $\mathcal{M}$ a $(\mathscr{C}, J)$-presheaf; the functor $\mathscr{X}$ is a *sub-presheaf* of $\mathcal{M}$ if $\mathscr{X}$ is a $(\mathscr{C}, J)$-presheaf such that:

(i) $\forall C \in \mathscr{C}(\mathscr{X}(C) \subset \mathcal{M}(C))$,

(ii) $\forall x \in \mathscr{X}(C)(x \restriction_{\mathscr{X}} f = x \restriction_{\mathcal{M}} f)$ i.e. restriction in $\mathscr{X}$ is the same as restriction in $\mathcal{M}$ for x in $\mathscr{X}(C)$, $C \in \mathscr{C}$.

$\mathscr{X}$ is a *subsheaf relative to* $\mathcal{M}$ if in addition

(iii) $\forall S \in J(C)\, \forall x \in \mathcal{M}(C)[\forall f \in S(x \restriction f \in \mathscr{X}(\mathrm{dom}(f)) \to x \in \mathscr{X}(C)]$ (cf. the corresponding condition on relations in 4.14). □

N.B. A subsheaf relative to a sheaf is again a sheaf.

The notion of subsheaf is global; in order to define power sheaves we need a localized version.

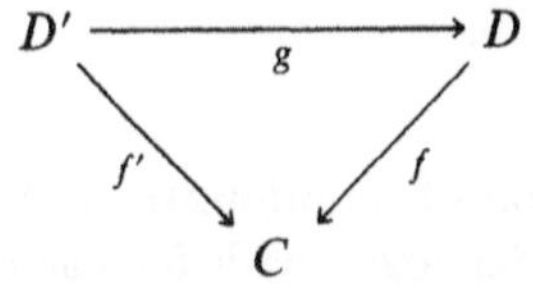

Fig. 14.10.

Let us write (g, f) for an arrow in $\mathscr{C} \downarrow C$, as in fig. 14.10. We note that if $(\mathscr{C}, J)$ is a site and $C \in \mathrm{Ob}(\mathscr{C})$, then J induces a covering system J_C on $\mathscr{C} \downarrow C$ defined by

$$S \equiv \{(g_i, f) : i \in I\} \in J_C(f) := \{g_i : i \in I\} \in J(\mathrm{dom}(f)).$$

A $\mathscr{C}$-presheaf $\mathscr{M}$ induces a presheaf $\mathscr{M} \downarrow C$ in $\mathscr{C} \downarrow C$ given by

$$(\mathscr{M} \downarrow C)(f) := \mathscr{M}(\mathrm{dom}(f))$$

$$[(\mathscr{M} \downarrow C)(g, f)](x) := x \upharpoonright_{\mathscr{M}} g \equiv \mathscr{M}(g)(x) \quad \text{for } x \in \mathscr{M}(\mathrm{dom}(f)).$$

So, essentially, $\mathscr{M} \downarrow C$ is obtained by restricting $\mathscr{M}$ to objects D such that $D \rightarrow_f C$ for some f, and restricting $\upharpoonright_{\mathscr{M}}$ to g such that $(g, f) \in \mathrm{Hom}(\mathscr{C} \downarrow C)$. If $\mathscr{M}$ is a $(\mathscr{C}, J)$-sheaf, then $\mathscr{M} \downarrow C$ is a $(\mathscr{C} \downarrow C, J_C)$-sheaf (exercise).

5.13. Definition. Let $\mathscr{M}$ be a $(\mathscr{C}, J)$-presheaf. Then $\mathscr{P}(\mathscr{M})$, the *power sheaf* of $\mathscr{M}$ is given by:

(i) $\mathscr{P}(\mathscr{M})(C)$ is the collection of subsheaves relative to $\mathscr{M}$ in $\mathscr{C} \downarrow C$;

(ii) Let $\mathscr{X} \in \mathscr{P}(\mathscr{M})(C)$, $D \rightarrow_f C \in \mathscr{C}$, then for all $D' \rightarrow_g D$ *restriction* in $\mathscr{P}(\mathscr{M})$ is defined by

$$(\mathscr{X} \upharpoonright f)(g) := \mathscr{X}(f \circ g).$$

Membership can now be interpreted by

$$C \Vdash a \in \mathscr{X} := a \in \mathscr{X}(C) \quad \text{for } \mathscr{X} \in \mathscr{P}(\mathscr{M})(C). \quad \square$$

We now turn to function sheaves.

$$\begin{array}{ccc} \mathcal{M} \downarrow C(f) & \xrightarrow[\tau_f]{} & \mathcal{N} \downarrow C(f) \\ {\scriptstyle \mathcal{M} \downarrow C(g,f)} \Big\downarrow & & \Big\downarrow {\scriptstyle \mathcal{N} \downarrow C(g,f)} \\ \mathcal{M} \downarrow C(f \circ g) & \xrightarrow[\tau_{f \circ g}]{} & \mathcal{N} \downarrow C(f \circ g) \end{array}$$

Fig. 14.11.

5.14. Definition. Let $(\mathscr{C}, J)$ be a site, $\mathcal{M}$ a presheaf over $(\mathscr{C}, J)$, $\mathcal{N}$ a sheaf over $(\mathscr{C}, J)$. The *function sheaf* $\mathcal{N}^{\mathcal{M}}$ is defined as follows.

(i) $\mathcal{N}^{\mathcal{M}}(C)$ consists of all natural transformations $\tau\colon\ \mathcal{M} \downarrow C \dot{\to} \mathcal{N} \downarrow C$. That is to say, $\tau = \{\tau_f : f \in \mathcal{M} \downarrow C\}$ such that for $(g, f) \in \mathrm{Hom}(\mathscr{C} \downarrow C)$ the square in fig. 14.11 commutes.

(ii) *Restrictions* are defined for $D \to_f C$ by

$$(\tau \upharpoonright f)_g := \tau_{f \circ g}.$$

(iii) *Evaluation*: if $\tau \in \mathcal{N}^{\mathcal{M}}(C)$, $x \in \mathcal{M}(C)$, then

$$\mathrm{ev}_C(\tau, x) \equiv \tau(x) := \tau_1(x). \quad \Box$$

5.15. Theorem (*soundness for sites*). *Let $(\mathscr{C}, J)$ be a site, and let $\mathbb{N}$ be interpreted by the sheaf generated by the constant presheaf $\mathcal{N}$ with $\mathcal{N}(C) = \mathbb{N}$ for all $C \in \mathrm{Ob}(\mathscr{C})$, 0 by 0 in $\mathcal{N}(C)$, S by S on each $\mathcal{N}(C)$, and let $\upharpoonright$ be the identity on $\mathcal{N}(C)$. Then with the standard interpretation of product, power, and function types,* $\mathrm{Sh}(\mathscr{C}, J)$ *is a model for* **HAH**.

Proof. Let $\{\mathcal{M}_i : i \in I\}$ be a collection of $(\mathscr{C}, J)$-sheaves. We shall check the most complicated axiom schema, namely function comprehension. That is to say, we have to show that for all $C \in \mathrm{Ob}(\mathscr{C})$ and all formulas A with $\mathrm{FV}(A) \subset \{x, y\}$ and parameters from the $\mathcal{M}_i(C)$:

$$\forall x \in \mathcal{M}\ \exists! y \in \mathcal{N} A(x, y) \to \exists f \in \mathcal{N}^{\mathcal{M}}\ \forall x \in \mathcal{M} A(x, fx)$$

holds at C. This amounts to showing that for all $D \in \mathrm{Ob}(\mathscr{C})$ and all $g \in \mathscr{C}(D, C)$

$$D \Vdash (\forall x \in \mathcal{M}\ \exists! y \in \mathcal{N} A(x, y)) \upharpoonright g \Rightarrow$$

$$D \Vdash (\exists f \in \mathcal{N}^{\mathcal{M}}\ \forall x \in \mathcal{M} A(x, fx)) \upharpoonright g.$$

Equivalently, we must show for arbitrary $C \in \mathrm{Ob}(\mathscr{C})$ and arbitrary $A(x, y)$

with $\mathrm{FV}(A) \subset \{x, y\}$ and parameters from the $\mathscr{M}_i(C)$

$$C \Vdash \forall x \in \mathscr{M}\, \exists! y \in \mathscr{N} A(x, y) \Rightarrow C \Vdash \exists f \in \mathscr{N}^{\mathscr{M}}\, \forall x \in \mathscr{M} A(x, fx).$$

For notational simplicity we shall assume that $A(x, y)$ does not contain parameters from the $\mathscr{M}_i(C)$. Suppose

$$C \Vdash \forall x \in \mathscr{M}\, \exists! y \in \mathscr{N} A(x, y), \text{ i.e.} \tag{1}$$

$$\forall f \in \mathscr{C}(D, C)\, \forall a \in \mathscr{M}(D) (D \Vdash \exists! y \in \mathscr{N} A(a, y)),$$

that is

$$\forall f \in \mathscr{C}(D, C) \forall a \in \mathscr{M}(D) \exists S \in J(D) \forall g \in S\, \exists b \in \mathscr{N}(\mathrm{dom}(g))$$
$$(\mathrm{dom}(g) \Vdash A(a \upharpoonright g, b)), \tag{2}$$

and also

$$\forall f \in \mathscr{C}(D, C) \forall a \in \mathscr{M}(D) \forall g \in \mathscr{C}(D', D) \forall b \in \mathscr{N}(D')$$

$$\forall g' \in \mathscr{C}(D'', D') \forall b' \in \mathscr{N}(D'')[D'' \Vdash A(a \upharpoonright g \circ g', b \upharpoonright g') \wedge$$
$$A(a \upharpoonright g \circ g', b') \rightarrow b \upharpoonright g' = b']. \tag{3}$$

By (2), for given $f \in \mathscr{C}(D, C)$ and $a \in \mathscr{M}(D)$ there is an $S \in J(D)$ such that for all $g \in S$ there is a suitable b_g such that

$$\mathrm{dom}(g) \Vdash A(a \upharpoonright g, b_g).$$

If $\mathrm{cod}(g') = \mathrm{dom}(g)$, we also have

$$\mathrm{dom}(g \circ g') \Vdash A(a \upharpoonright g \circ g', b_{g \circ g'}),$$

$$\mathrm{dom}(g \circ g') \Vdash A(a \upharpoonright g \circ g', b_g \upharpoonright g'),$$

and by (3) it follows that

$$b_g \upharpoonright g' = b_{g \circ g'}.$$

It follows that $\{b_g : g \in S\}$ is compatible, so there is a unique b such that

$$\forall g \in S(\mathrm{dom}(g) \Vdash A(a \upharpoonright g, b \upharpoonright g)),$$

and therefore by the covering property of forcing

$$D \Vdash A(a, b).$$

We have to show

$$C \Vdash \exists f \in \mathcal{N}^{\mathcal{M}} \forall x \in \mathcal{M} A(x, fx), \quad \text{i.e.}$$

$$\exists S' \in J(C) \forall f' \in S' \exists f'' \in \mathcal{N}^{\mathcal{M}}(\mathrm{dom}(f')) \forall g \in \mathscr{C}(D, \mathrm{dom}(f'))$$
$$\forall a \in \mathcal{M}(D)[D \Vdash A(a, (f'' \upharpoonright g)(a))].$$

Clearly it suffices to show for suitable $f'' \in \mathcal{N}^{\mathcal{M}}(C)$

$$C \Vdash \forall x \in \mathcal{M} A(x, f''x).$$

An $f'' \in \mathcal{N}^{\mathcal{M}}(C)$ is in fact a natural transformation $\tau : \mathcal{M} \downarrow C \dot{\to} \mathcal{N} \downarrow C$. For any $D \to_f C$ such that $D \Vdash A(a, b)$, we take $\tau_f(a) = b$. We must show that this defines a natural transformation, that is to say for $D' \to_g D$

$$\tau_f(a) \upharpoonright g = \tau_{f \circ g}(a \upharpoonright g). \tag{5}$$

To see this, note that by definition

$$D \Vdash A\big(a, \tau_f(a)\big), \qquad D' \Vdash A\big(a \upharpoonright g, \tau_{f \circ g}(a \upharpoonright g)\big);$$

from the first we obtain by monotonicity

$$D' \Vdash A\big(a \upharpoonright g, \tau_f(a) \upharpoonright g\big).$$

(5) follows because of (3). The rest of the proof is left to the reader. □

6. Notes

6.1. The connection between sheaves, categories, and logic is a rather recent topic. Moreover, many results have not been published and belong to "folklore". Therefore we shall not attempt to give detailed credits and history. A survey of the development of sheaf theory is found in Gray (1979). Johnstone's monograph on topos theory (1977) also contains an extensive bibliography and many historical notes. For the presentation of the material in this chapter we are indebted to Fourman and Scott (1979) and handwritten notes by Moerdijk.

6.2. The pioneer in the field has undoubtedly been Lawvere (see e.g. his survey 1971, and 1976); he observed that the notion of elementary topos, a generalization of the notion of a Grothendieck topos, could be regarded as a category-theoretic generalization of set-theory. Particularly important in this connection was Lawvere's discovery of the role of subobject-classifiers (see below) in toposes.

An (*elementary*) *topos* may be described as a category with finite limits (in fact, terminal objects and pullbacks suffice), exponentials and a *subobject-classifier*, i.e. a truth-value object Ω with a mono arrow $\mathbb{1} \to_{\text{true}} \Omega$ such that for every mono arrow $S \rightarrowtail_m X$ there is exactly one arrow $X \to_\varphi \Omega$ such that the pair (m, s) is the pullback of (φ, true), and vice versa (each φ derives from some m), see fig. 14.12.

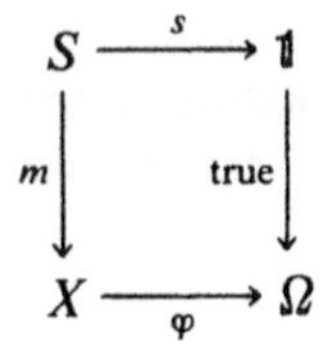

Fig. 14.12.

In Set S is isomorphic to a subset of X, Ω is $\mathrm{P}(\{0\})$, and φ is the characteristic function of S. As to the truth-value object in a topos $\mathrm{Sh}(\mathscr{C}, J)$, see E14.5.5. For the theory of elementary topoi we refer the reader to Johnstone (1977), Barr and Wells (1985), Lambek and Scott (1986); Goldblatt (1979) is suitable for a first orientation. A concise treatment of the connection with intuitionistic higher-order logic is presented in Fourman (1977), Boileau and Joyal (1981).

6.3. Mitchell (1972) was presumably the first to associate an explicit formal language $\mathscr{L}_{\mathscr{E}}$ and an interpretation (*functorial semantics*) with any topos $\mathscr{E}$; Johnstone (1977, section 5.4) calls this the Mitchell–Benabou language.

An alternative, but equivalent, semantics has been formulated by Joyal and is widely known under the name of *Kripke–Joyal semantics* (Beth–Joyal semantics might have been more appropriate). This semantics is given as a forcing relation between the objects of a topos and the formulas of a language; for the case of a Grothendieck topos, i.e. a topos of the form $\mathrm{Sh}(\mathscr{C}, J)$, this may also be formulated as forcing between the objects of $\mathscr{C}$ and the formulas of the language. It is the latter version which has been used in this book (cf. Scedrov 1984, section 0.6).

For the connection between first-order (infinitary) logic and categories, see Reyes (1974) and the monograph by Makkai and Reyes (1977).

An alternative to the introduction of an explicit formal language with its interpretation, more directly in line with Lawvere's original approach, is the use of category-theoretic equivalents or analogues to logical concepts; see e.g. Barr and Wells (1985).

Exercises

14.1.1. Check the examples in 1.3.

14.1.2. Prove 1.7.

14.1.3. Prove the proposition in 1.8.

14.1.4. Check the assertions in 1.9.

14.1.5. Prove 1.10.

14.1.6. Explain why the example in 1.3(b) is not a sheaf.

14.1.7. Complete the proofs of 1.15 and 1.18.

14.1.8. Check the assertions in 1.21.

14.1.9. Let $\pi \in Y \to X$ be a continuous map between topological spaces Y and X. We define an $\mathcal{O}(X)$-set π_X as consisting of all continuous $f \in U \to Y$ with $U \in \mathcal{O}(X)$, $\pi \circ f$ the identity on U, $Ef = U$, and

$$[\![f = g]\!] := \mathrm{Int}\{t \in Ef \cap Eg : f(t) = g(t)\};$$

show that this is a sheaf.

14.1.10. We call a pair (Y, π) as in E14.1.9 a sheaf space if π is a local homeomorphism. The following construction associates to any presheaf $\mathcal{M}$ over $\mathcal{O}(X)$ a sheaf space for the sheaf generated by $\mathcal{M}$. Let

$$\mathcal{M}(U) := \{a \in \mathcal{M} : Ea = U\},$$

$$a \sim_x b := \exists U(x \in U \wedge a \restriction U = b \restriction U).$$

$\sim_x$ is readily seen to be an equivalence relation. Put

$$[a]_x := \{b : b \sim_x a\};$$

(the equivalence classes are called germs). Take for Y the space with as set of points

$$\{(x, [a]_x) : x \in X, a \in \mathcal{M}(U), U \in \mathcal{O}(X)\}$$

and with a topology generated by the basis consisting of all sets

$$a_U := \{(x, [a]_x) : x \in U\},$$

where $a \in \mathcal{M}(U)$. π is defined by

$$\pi((x, [a]_x)) = x.$$

π is not only continuous, but in fact a local homeomorphism. If $\mathcal{M}$ is a sheaf, the sheaf π_X (E14.1.9) is isomorphic to $\mathcal{M}$.

14.2.1. Prove proposition 2.2 and the assertion following it.

14.2.2. Complete the proof of 2.4.

14.2.3. On a separated presheaf $\mathscr{M}$, a strict relation is given as an $R \in M^n \to \Omega$ such that

$$\begin{cases} R(a_1 \restriction p,\ldots,a_n \restriction p) = R(a_1,\ldots,a_n) \wedge p, \\ R(a_1,\ldots,a_n) \leq \bigwedge\{Ea_i : 1 \leq i \leq n\}, \end{cases} \tag{1}$$

and a strict function by an $F \in M^n \to M$ such that

$$\begin{cases} F(a_1 \restriction p,\ldots,a_n \restriction p) = F(a_1,\ldots,a_n) \restriction p, \\ E(F(a_1,\ldots,a_n)) \leq \bigwedge\{Ea_i : 1 \leq i \leq n\}. \end{cases} \tag{2}$$

Prove this fact. Show that these characterizations do not necessarily hold on a presheaf which is not separated. That is to say, (1) and (2) determine relations and functions, but not every relation or function on the presheaf satisfies (1) or (2); cf. the example in 1.20.

14.2.4. Let $\Omega = \mathcal{O}(\mathbb{C})$ be the open sets in the complex plane, and let us consider the structure $\mathscr{M}$ of holomorphic functions $f \in U \to \mathbb{C}$ with $U \in \mathcal{O}(\mathbb{C})$, and equality $[\![f = g]\!] := \mathrm{Int}\{z : f(z) = g(z)\}$. Show that with $[\![f \# g]\!] := \{z : f(z) \# g(z)\}$ we have an apartness relation such that (classically) $\mathscr{M} \Vdash \forall f(f = 0 \vee f \# 0)$. *Hint*: use the fact that two holomorphic functions agreeing on some open set agree everywhere on their common domain.

14.3.1. Verify the examples in 3.2.

14.3.2. Give an example of a category in which mono + epi is not the same as invertible (3.3).

14.3.3. Prove the uniqueness of terminal objects, products and pullbacks (3.4-6). What are the product and the sum in the category $\mathscr{P}$ of a poset $(P, \leq)$?

14.3.4. If $\mathscr{C}$ has pullbacks, then so has $\mathscr{C} \downarrow C$.

14.3.5. Check the example in 3.7.

14.3.6. Prove the uniqueness of exponentials (3.9). Show that there is a biunique correspondence between $\mathscr{C}(A \times B, C)$ and $\mathscr{C}(A, C^B)$.

14.3.7. Check examples in 3.11 and 3.12.

14.3.8. Prove 3.15.

14.3.9. Check the assertions in 3.16.

14.3.10. Let Ω be any cHa and let $\mathscr{N}$ be an Ω-presheaf. $\mathrm{Kr}(\Omega)$ is the cHa of all downwards closed subsets of Ω. $\mathscr{N}(p)$ is the set of sections over p, i.e. the collection $\{x \in \mathscr{N} : Ex = p\}$. Let $\mathscr{N}^*$ be the $\mathrm{Kr}(\Omega)$-presheaf with sets of sections given by

$$\mathscr{N}^*(K) = \varprojlim\{\mathscr{N}(q) : q \in K\}.$$

That is to say, $\mathcal{N}^*(K)$ consists of all maps $f \in K \to N$ such that $f(q) \in N$, $E(f(q)) = q$, $f(q') = f(q) \restriction q'$ for all $q, q' \in K$. Show that $\mathcal{N}^*$ is a Kr(Ω)-sheaf with $\mathcal{N}^*(\{q : q \leq p\}) = \mathcal{N}(p)$, and that each Kr($\Omega$)-sheaf $\mathcal{N}$ is obtained from the Ω-presheaf $\mathcal{N}^-$ with $\mathcal{N}^-(p) := \mathcal{N}(\{q\colon q \leq p\})$. Show that this amounts to an equivalence between the categories Psh(Ω) and Sh(Kr(Ω)).

14.4.1. Check that example (b) in 4.4 yields a covering system.

14.4.2. Prove the proposition in 4.6.

14.4.3. Show that a compatible collection $\{x_f\colon f \in R\}$ with respect to a covering basis K, $R \in K(C)$, $C \in \mathrm{Ob}(\mathscr{C})$ gives rise to a compatible collection $\{x_f \restriction g\colon g \in \mathscr{C}(D, C), f \in R\}$ with respect to $J(C)$ (J the covering system generated by K) with the same join.

14.4.4. Prove 4.11.

14.4.5. Prove 4.18.

14.4.6. Prove soundness for forcing in sites (4.20).

14.4.7. Let A be a distributive lattice viewed as a poset-category $\mathscr{A}$. Let J_{fc} be given by $R \in J_{\mathrm{fc}}(a) :=$ R a sieve and $\bigvee F = a$ for some finite $F \subset R$. Prove that $(\mathscr{A}, J_{\mathrm{fc}})$ is a site (the *finite-cover site*).

14.4.8. Let $(\mathscr{A}, J_{\mathrm{fc}})$ be as in E14.4.7, and let $A_a := \{b \in A\colon b \leq a\}$. A presheaf on $(\mathscr{A}, J_{\mathrm{fc}})$ is obtained by considering the assignment $a \in A \mapsto \mathrm{Hom}(L, A_a)$ (L a distributive lattice, $\mathrm{Hom}(L, A_a)$ the collection of lattice homomorphisms from L to A_a). Show that this presheaf is in fact a sheaf.

14.4.9. Let $\mathscr{C}$ be any category, $C \in \mathrm{Ob}(\mathscr{C})$. Let $Z(C)$ be the collection of sieves on C partially ordered by inclusion. Then $Z(C)$ can be made into a cHa, and for any $f \in \mathscr{C}(D, C)$ the mapping $f^* \in Z(C) \to Z(D)$, where $*$ is defined as in 4.2(ii), preserves $\wedge, \bigwedge, \vee, \to$.

14.4.10. Let $(\mathscr{C}, J)$ be a site. A sieve R on $C \in \mathrm{Ob}(\mathscr{C})$ is called *J-closed* if for each $f \in \mathscr{C}(D, C)$ we have $f^*(R) \in J(D) \Rightarrow f \in R$. If R is J-closed on C and $f \in \mathscr{C}(D, C)$, then $f^*(R)$ is J-closed ($*$ defined as in 4.2(ii)). Then the functor $\Omega\colon \mathscr{C}^{\mathrm{op}} \to \mathrm{Set}$ with $\Omega(C)$ the collection of J-closed sieves on C, with restriction given by $R \restriction f = f^*(R)$, is a $(\mathscr{C}, J)$-sheaf. Show that $\Omega(C)$ is a cHa for each C and that $\restriction$ preserves $\wedge, \bigwedge, \vee, \to$.

14.5.1. Check the details of 5.7.

14.5.2. Prove 5.8 and 5.9.

14.5.3. Prove the fact mentioned at the end of 5.12.

14.5.4. Prove the soundness theorem of **HAH** (5.15) in full.

14.5.5. With notation as in E14.4.10, show that $\mathscr{P}(\mathcal{N})$ is isomorphic to $\Omega^{\mathcal{N}}$ (i.e. Ω is the truth-value object of Sh($\mathscr{C}$, J)).

CHAPTER 15

APPLICATIONS OF SHEAF MODELS

In this chapter we shall give some illustrations of the use of sheaf-semantics in obtaining metamathematical results: independence, consistency, and derived rules. The models we shall consider also have an independent interest as mathematical structures.

We first investigate what $\mathbb{N}$, $\mathbb{Z}$, $\mathbb{Q}$, $\mathbb{R}$, and $\mathbb{N}^{\mathbb{N}}$ "look like" as sheaves over $\mathcal{O}(T)$. The interpretation of $\mathbb{N}$ in $\mathrm{Sh}(\mathcal{O}(T))$ is fixed up to an isomorphism; we can take as interpretation of $\mathbb{N}$ the sheaf generated by an "isomorphic copy" $\hat{\mathbb{N}}$ of $\mathbb{N}$ itself and then, on the standard interpretation of exponentials, products, etc., described in section 14.5, $\mathbb{Q}$, $\mathbb{R}$, and $\mathbb{N}^{\mathbb{N}}$ are fixed as well. However, we shall spend some effort on finding more concrete representations for the sheaves $\mathbb{R}$ and $\mathbb{N}^{\mathbb{N}}$ given abstractly by the general recipe of section 14.5.

In section 2 we show that for sheaf models over topological spaces, so-called *spatial models*, $\mathbb{R}^{\mathrm{c}} = \mathbb{R}^{\mathrm{d}}$ does not always hold, and that $\mathbb{R}^{\mathrm{c}} = \mathbb{R}^{\mathrm{d}}$ does not imply AC_0 in general.

Section 3 is devoted to internal topologies, it illustrates the failure of the intermediate value theorem, and shows that $\mathbb{C}$ is not, in general, algebraically closed in sheaf models. We use a nonspatial model $\mathrm{Sh}(\Omega)$ to give an example in which $[0,1]$ is not compact. In section 4 a derived rule of continuity is established.

Section 5 presents a monoid model for **CS**, and section 6 a site model for **LS**.

Our methods are sometimes ad hoc, since in certain cases we have preferred to verify special instances of a general result directly, instead of presenting the general result. It should be kept in mind that this chapter is not intended as a systematic development of mathematics in sheaves, but as an illustration of the methods only.

1. Interpretation of $\mathbb{N}, \mathbb{Q}, \mathbb{Z}, \mathbb{R}, \mathbb{N}^{\mathbb{N}}$ in $\mathrm{Sh}(\mathcal{O}(T))$

1.1. *Simple sheaves.* A *simple sheaf* in $\mathrm{Sh}(\Omega)$ is a sheaf of the form $\hat{X}^{\mathrm{sh}}$, $\hat{X}$ a constant Ω-set (cf. 13.6.4(a), 14.2.6, example) obtained from a set X. The element in $\hat{X}$ corresponding to $x \in X$ is denoted by $\hat{x}$, if it is necessary to make a distinction. Over a topology ($\Omega = \mathcal{O}(T)$) $\hat{x}$ is identified with $\lambda t.x$.

It is easy to see that an element a of $\hat{X}^{\mathrm{sh}}$ can be given as a compatible collection

$$\{(\hat{x}_i, p_i) : x_i \in X,\ i \in I,\ p_i \in \Omega\}.$$

Such a collection represents the join of the $(\hat{x}_i, p_i)$, the restrictions of the $\hat{x}_i$ to p_i. Compatibility for such a collection means that $\bigvee\{p_i \wedge p_j : x_i = x_j\} = p_i \wedge p_j$ (exercise). If $b \equiv \{(\hat{x}'_i, p'_i) : x'_j \in X,\ i \in I',\ p'_i \in \Omega\}$, then

$$[\![a = b]\!] \equiv \bigvee\{p_i \wedge p'_j : x_i = x'_j,\ i \in I,\ j \in I'\},$$

$$Ea \equiv \bigvee\{p_i : i \in I\},$$

$$a \upharpoonright p \equiv \{(\hat{x}_i, p_i \wedge p) : x_i \in X,\ i \in I,\ p_i \in \Omega\}.$$

Constructively, if equality on X is decidable, the compatibility condition reduces to $\forall i, j \in I(x_i \neq x_j \to p_i \wedge p_j = \bot)$. Using classical logic, compatibility is equivalent to $\forall i, j \in I(i \neq j \leftrightarrow p_i \wedge p_j = \bot)$.

If Ω is of the form $\mathcal{O}(T)$, then $\hat{X}^{\mathrm{sh}}$ can be described as consisting of all locally constant functions from U to X with $U \in \mathcal{O}(T)$.

However, we do not actually need this more concrete description of $\hat{X}^{\mathrm{sh}}$ for the evaluation of (the truth-value of) formulas, since $\hat{X}$ *generates* $\hat{X}^{\mathrm{sh}}$.

We recall that (13.6.4), for any set X, relations R can be automatically interpreted as Ω-relations $\hat{R}$ over $\hat{X}$, the constant set (or simple sheaf), by

$$\hat{R}(\hat{x}_1, \ldots, \hat{x}_n) = [\![R(\hat{x}_1, \ldots, \hat{x}_n)]\!] = \bigvee\{\top : R(x_1, \ldots, x_n)\},$$

where the $x_i \in X$. For the simple sheaf $\hat{X}^{\mathrm{sh}}$ we can extend this

$$\hat{R}(x_1, \ldots, x_n) = \bigvee\{p_{1,i_1} \wedge \cdots \wedge p_{n,i_n} : R(\hat{x}_{i_1}, \ldots, \hat{x}_{i_n})\},$$

where $x_k \equiv \{(\hat{x}_{i_k}, p_{k,i_k}) : \hat{x}_{i_k} \in X,\ p_{k,i_k} \in \Omega,\ i_k \in I_k\}$.

Functions F from X to Y can be treated in the same way: they give rise automatically to functional relations between $\hat{X}$ and $\hat{Y}$, representable by functions $\hat{F} \in \hat{X} \to \hat{Y}$.

1.2. *$\mathcal{P}_{\mathrm{fin}}(\mathcal{N})$ and $\mathcal{N}^*$ for a sheaf $\mathcal{N}$.* For a sheaf $\mathcal{N}$ we can interpret the collection of finitely indexed subsets of $\mathcal{N}$ by a sheaf $\mathcal{P}_{\mathrm{fin}}(\mathcal{N})$, generated by an Ω-set

$$\{X \subset A : X \text{ finite} \wedge \forall a, b \in X(Ea = Eb)\},$$

with equality

$$[\![X = Y]\!] = \bigwedge_{a \in X} \bigvee_{b \in Y} [\![a = b]\!] \wedge \bigwedge_{b \in X} \bigvee_{a \in Y} [\![a = b]\!].$$

Similarly, the sheaf of finite sequences $\mathcal{N}^*$ is generated by

$$\{\langle a_0, \ldots, a_{n-1}\rangle : n \in \mathbb{N}, \forall i < n(a_i \in A), \forall i, j < n(Ea_i = Ea_j)\}$$

with equality

$$[\![\langle a_0, \ldots, a_{n-1}\rangle = \langle b_0, \ldots, b_{m-1}\rangle]\!] :=$$
$$\bigwedge\{[\![a_i = b_i]\!] : i < n\} \quad \text{if } n = m, \quad \bot \text{ otherwise.}$$

If $\mathcal{N}$ is a simple sheaf, then so are $\mathcal{P}_{\mathrm{fin}}(\mathcal{N})$ and $\mathcal{N}^*$ (exercise).

1.3. *Arithmetic and the representation of* $\mathbb{N}$. As our standard interpretation of $\mathbb{N}$ (cf. 14.5.10) we take the sheaf completion $\hat{\mathbb{N}}^{\mathrm{sh}}$ of the constant Ω-set $\hat{\mathbb{N}} \equiv (\hat{\mathbb{N}}, [\![\cdot = \cdot]\!])_{\mathbb{N}}$ with $[\![\hat{n} = \hat{m}]\!]_{\mathbb{N}} = \bigvee\{\top : n = m\}$. The successor S corresponds on $\hat{\mathbb{N}}$ to $\hat{S}$ satisfying

$$\hat{S}\hat{n} = (Sn)\hat{}.$$

We have for all arithmetical statements A

$$[\![A(\hat{m}_1, \ldots, \hat{m}_n]\!] = \top \Leftrightarrow A(m_1, \ldots, m_n),$$

and in particular

$$[\![\hat{0} \neq S\hat{0}]\!] = \top_{\Omega},$$

$$[\![A(\hat{0}) \wedge \forall n \in \mathbb{N}(A(n) \to A(Sn)) \to \forall n \in \mathbb{N} A(n)]\!] = \top_{\Omega}.$$

Thus in $\mathrm{Sh}(\Omega)$ $\hat{\mathbb{N}}^{\mathrm{sh}}$ is a Peano-structure. Since full second-order logic is validated in $\mathrm{Sh}(\Omega)$, the interpretation of $\mathbb{N}$ is essentially canonical: any two objects satisfying full induction with respect to an injective successor function are isomorphic (cf. 3.8.6).

Via fixed (primitive recursive) enumerations, the structures of $\mathbb{Z}$ and $\mathbb{Q}$, finite sets of natural numbers, finite sequences of natural numbers, etc., can be encoded in arithmetic over $\mathbb{N}$; and thus $\mathbb{Z}$ and $\mathbb{Q}$ behave in the model just as $\mathbb{Z}$ and $\mathbb{Q}$ itself. In fact, we can represent $\mathbb{Z}$ and $\mathbb{Q}$ in the model by $\hat{\mathbb{Z}}^{\text{sh}}$ and $\hat{\mathbb{Q}}^{\text{sh}}$.

1.4. *The interpretation of* $\mathbb{R}^{\text{d}}$ *in* $\text{Sh}(\mathcal{O}(T))$. In intuitionistic higher-order logic the Dedekind reals as an ordered structure are determined up to isomorphism by the fact that $\mathbb{R}^{\text{d}}$ is the order-completion of $\mathbb{Q}$ with respect to bounded located sets (cf. 5.5.12(iii)). So the interpretation of $\mathbb{R}^{\text{d}}$ in $\text{Sh}(\Omega)$ is also determined up to isomorphism. We shall show that the Dedekind reals in $\text{Sh}(\Omega)$ correspond externally to the sheaf $(\mathbb{R}^{\text{d}})_T$. Our starting point is the symmetric version of the notion of Dedekind cut (cf. E5.5.3).

A *Dedekind real* is a pair (L, R) with $L \subset \mathbb{Q}$, $R \subset \mathbb{Q}$ such that for p, q ranging over $\mathbb{Q}$:

(i) $\exists p(p \in L), \exists q(q \in R)$;
(ii) $L \cap R = \emptyset$;
(iii) $\forall pq(p < q \wedge q \in L \to p \in L), \forall pq(p < q \wedge p \in R \to q \in R)$;
(iv) $\forall p \in L\, \exists q \in L(p < q), \forall p \in R\, \exists q \in R(q < p)$;
(v) $\forall pq(p < q \to p \in L \vee q \in R)$.

$\mathbb{R}^{\text{d}}$ is then the collection of such reals with equality

$$(L, R) = (L', R') := L = L' \wedge R = R'$$

and strict order $<$

$$(L, R) < (L', R') := \exists p(p \in R \cap L').$$

In the definition of equality "$\wedge R = R'$" is in fact redundant. For the conjunction of (i)–(v) we write $\text{Cut}(L, R)$.

1.5. PROPOSITION. *In the standard model* $\text{Sh}(\mathcal{O}(T))$ *the Dedekind reals are represented by the sheaf* $\mathbb{R}^{\text{d}}_T$.

PROOF. The predicate "Cut" introduced above determines (by comprehension, cf. 14.5.3) a subsheaf of $\mathcal{P}(\hat{\mathbb{Q}}) \times \mathcal{P}(\hat{\mathbb{Q}})$, namely the subsheaf generated by the $\mathcal{O}(T)$-set of pairs (L, R) of strict $\mathcal{O}(T)$-predicates on $\hat{\mathbb{Q}}$, restricted to $[\![\text{Cut}(L, R)]\!]$. We may think of these restricted pairs as

$$((L, R), [\![\text{Cut}(L, R)]\!]),$$

or we may put

$$E(L, R)_{\text{Cut}} := [\![\text{Cut}(L, R)]\!],$$

where the subscript "Cut" means that we are interested in the pair only as elements of the subsheaf determined by "Cut".

Equality is inherited from $\mathscr{P}(\hat{\mathbb{Q}}) \times \mathscr{P}(\hat{\mathbb{Q}})$ relativized to "Cut".

(a) Given any $a \in \mathbb{R}^{\mathrm{d}}_T$, we define a pair (L_a, R_a) of $\mathcal{O}(T)$-predicates by

$$\begin{cases} [\![\hat{r} \in L_a]\!] := [\![\hat{r} < a]\!] = \{t : r < a(t)\}, \\ [\![\hat{r} \in R_a]\!] := [\![a < \hat{r}]\!] = \{t : a(t) < r\}. \end{cases} \tag{1}$$

Then

$$Ea = [\![\mathrm{Cut}(L_a, R_a)]\!] = [\![E(L_a, R_a)]\!].$$

To see this, we have to consider the conjuncts of $\mathrm{Cut}(L_a, R_a)$. Take for example openness; we have to show

$$Ea \leq [\![\forall r \in L_a \, \exists q \in L_a(q > r)]\!].$$

Now if $t \in \mathrm{dom}(a) = Ea$, then $t \in [\![\hat{r} \in L_a]\!] \Leftrightarrow r < a(t) \Leftrightarrow \exists q(r < q < a(t)) \Leftrightarrow \exists q(t \in [\![\hat{q} \in L_a \wedge \hat{r} < \hat{q}]\!]) \Leftrightarrow t \in [\![\exists q \in L_a(\hat{r} < q)]\!]$, so $t \in [\![\forall r \in L_a \, \exists q \in L_a(r < q)]\!]$. We leave the other clauses to the reader.

Also, if $t \in [\![\mathrm{Cut}(L_a, R_a)]\!]$, then $t \in [\![\exists r \in \mathbb{Q}(r \in L_a)]\!]$, hence $t \in [\![\hat{r} \in L_a]\!])$ for some r, i.e. $r < a(t)$. Therefore $t \in \mathrm{dom}(a) = Ea$.

(b) Suppose now that L, R be strict predicates on $\hat{\mathbb{Q}}$; we shall associate an element $a \in \mathbb{R}_T$ with this pair such that $[\![\mathrm{Cut}(L, R) = \mathrm{Cut}(L_a, R_a)]\!]$. Put

$$L_t := \{r : t \in [\![\hat{r} \in L]\!]\}, \qquad R_t := \{r : t \in [\![\hat{r} \in R]\!]\}.$$

Claim: if $t \in [\![\mathrm{Cut}(L, R)]\!]$, then (L_t, R_t) is a Dedekind real. For example, $t \in [\![\forall r \in L \, \exists s \in L(r < s)]\!]$ holds by the assumption $t \in [\![\mathrm{Cut}(L, R)]\!]$; hence

$$t \in \bigwedge\{[\![\hat{r} \in L]\!] \to [\![\exists s \in L(\hat{r} \in s)]\!] : r \in \mathbb{Q}\}.$$

So if $r \in L_t$, i.e. $t \in [\![\hat{r} \in L]\!]$, we also have

$$t \in [\![\exists s \in L(\hat{r} < s)]\!] = \bigvee\{[\![\hat{s} \in L]\!] \wedge [\![\hat{r} < s]\!] : s \in \mathbb{Q}\},$$

hence $t \in [\![\hat{s} \in L]\!]$, $t \in [\![\hat{r} < \hat{s}]\!]$, and thus $r < s$, $s \in L_t$, etc. We leave the verification of the other clauses to the reader.

We now define

$$a(t) := (L_t, R_t); \tag{2}$$

this defines a real-valued a with domain $Ea = [\![\mathrm{Cut}(L, R)]\!]$. This shows

$[\![\mathrm{Cut}(L, R)]\!] = [\![\mathrm{Cut}(L_a, R_a)]\!]$. The continuity of a is immediate by the fact that

$$\{t : r < a(t) < s\} = [\![\hat{r} \in L]\!] \cap [\![\hat{s} \in R]\!]$$

is open. We observe that if a and (L, R) are linked via (1) or (2), then $[\![\hat{r} \in L_a]\!] = [\![\hat{r} < a]\!] = \{t : r < a(t)\} = \{t : r \in L_t\} = [\![\hat{r} \in L]\!]$, so the two constructions are inverse to each other.

Note that the proof also shows that in fact the $\mathcal{O}(T)$-set we used to generate the sheaf interpreting the reals is already a sheaf. □

1.6. PROPOSITION. *In* $\mathrm{Sh}(\mathcal{O}(T))$ *Baire space* $\mathbb{B} \equiv \mathbb{N}^{\mathbb{N}}$ *is represented by* $\mathbb{B}_T$, *the sheaf of partial continuous functions from* T *into* $\mathbb{B}$.

PROOF. Since $\mathbb{B}$ consists of all functions $\alpha \in \mathbb{N} \to \mathbb{N}$, it corresponds in $\mathrm{Sh}(\Omega)$ to the subsheaf of $\mathcal{P}(\hat{\mathbb{N}} \times \hat{\mathbb{N}})$ generated by elements

$$(R, E(R)) \quad \text{with } E(R) = [\![\mathrm{fun}(R)]\!], \quad R \text{ a strict } \mathcal{O}(T)\text{-relation.}$$

As before, $\mathrm{fun}(R)$ expresses that R is a function: $\mathrm{fun}(R) \equiv \forall x \exists! y\, R(x, y)$. Any $a \in \mathbb{B}_T$ determines an $\mathcal{O}(T)$-predicate R_a by

$$[\![R_a(\hat{m}, \hat{n})]\!] := \{t : a(t)(m) = n\}$$

such that

$$Ea = [\![\mathrm{fun}(R_a)]\!].$$

Conversely, given any $\mathcal{O}(T)$-relation R on $\hat{\mathbb{N}} \times \hat{\mathbb{N}}$, we can determine $a \in \mathbb{B}_T$ with

$$\mathrm{dom}(a) = Ea = [\![\mathrm{fun}(R)]\!]$$

by

$$a(t)(m) = n := t \in [\![R(\hat{m}, \hat{n})]\!];$$

a is continuous by definition. For arbitrary $x \in \hat{\mathbb{N}}^{\mathrm{sh}}$ and f in $\mathbb{N}^{\mathrm{sh}} \to \mathbb{N}^{\mathrm{sh}}$, represented by an $a \in \mathbb{B}_T$, we have

$$\mathrm{ev}(f, x)(t) = n \Leftrightarrow a(t)(x(t)) = n.$$

As before, we see that the generating $\mathcal{O}(T)$-set is itself already a sheaf. □

2. The axiom of countable choice

In this section we shall illustrate the use of sheaf models in independence proofs in the case of the axiom of countable choice.

2.1. Proposition. $\mathrm{Sh}(\mathcal{O}(\mathbb{B})) \Vdash \mathrm{AC}_0$.

Proof. We have to show, for $\mathcal{N} \equiv \mathcal{M}^{\hat{\mathbb{N}}}$,

$$U \equiv [\![\forall n \in \mathbb{N}\, \exists y \in \mathcal{M} A(n, y)]\!] \leq [\![\exists f \in \mathcal{N}\, \forall n \in \mathbb{N}\, A(n, fn)]\!].$$

Clearly

$$U \subset \bigcap_{n \in \mathbb{N}} \bigcup_{a \in \mathcal{M}} ([\![A(\hat{n}, a)]\!] \wedge Ea),$$

hence for all $n \in \mathbb{N}$

$$U \subset \bigcup_{a \in \mathcal{M}} ([\![A(\hat{n}, a)]\!] \wedge E\breve{a}),$$

therefore $\{[\![A(n, a)]\!] \wedge Ea : a \in \mathcal{M}\}$ is an open cover of U.

Let $n \in \mathbb{N}$ and $a \in \mathcal{M}$ be fixed, and let $V_p := \{\alpha : \alpha \in p\} \subset U$. Put

$$V(a) := [\![A(\hat{n}, a)]\!] \wedge Ea,$$

then

$$V_p \subset \bigcup\{V(a) : a \in \mathcal{M}\}.$$

With the help of the principle of bar induction we construct a countable disjoint refinement $\{U_k : k \in \mathbb{N}\}$ of $\{V(a) \cap V_p : a \in \mathcal{M}\}$, that is to say

$$\forall kk'(k \neq k' \to U_k \cap U_{k'} = \emptyset),$$

$$\forall k \exists a \in \mathcal{M}\big(U_k \subset V(a) \cap V_p\big).$$

First of all, if

$$P(n) := \exists a \in \mathcal{M}\big(V_{p*n} \subset V(a)\big),$$

we have, because of $\forall \beta \in p\, \exists a \in \mathcal{M}\, \exists n (V_{\bar{\beta}n} \subset V(a))$,

$$\forall \alpha \exists n\, P(\bar{\alpha}n). \tag{1}$$

If we put

$Q(n) :=$ V_{p*n} is covered by a countable disjoint refinement of

$$\{V_p \cap V(a) : a \in \mathcal{M}\},$$

it follows that

$$\forall n(Pn \to Qn), \tag{2}$$

$$\forall nm(Qn \to Q(n*m)), \tag{3}$$

$$\forall n(\forall k\, Q(n*\langle k\rangle) \to Qn). \tag{4}$$

Bar induction in the form $\mathrm{BI}'_{\mathrm{M}}$ (cf. 4.8.11) applied to (1), (2), (3), and (4) yields $Q(\langle\ \rangle)$, which is what we wanted. Let $\{U_k : k \in \mathbb{N}\}$ be the countable disjoint refinement of $\{V(a) \cap V_p : a \in \mathscr{M}\}$; apply AC_0 externally to find $\langle a_k\rangle_k$ such that

$$\forall k\big(U_k \subset V(a_k) \cap V_p\big).$$

Now $\{a_k \upharpoonright U_k : k \in \mathbb{N}\}$ is a compatible collection because of the disjointness of the U_k; let $\varphi(\hat{n})$ be the join, then

$$E\varphi(\hat{n}) = \bigcup\{U_k : k \in \mathbb{N}\} = V_p.$$

Now $U_k \subset [\![A(\hat{n}, a_k \upharpoonright U_k)]\!] \cap U_k = [\![A(\hat{n}, a_k \upharpoonright U_k) \wedge a_k \upharpoonright U_k = \varphi(\hat{n})]\!] \subset \bigvee\{[\![A(\hat{n}, d) \wedge d = \varphi(\hat{n})]\!] : d \in \mathscr{M}\} = [\![\exists d \in \mathscr{M}(A(\hat{n}, d) \wedge d = \varphi(\hat{n}))]\!] = [\![A(\hat{n}, \varphi(\hat{n})]\!]$. Hence $V_p \subset [\![A(\hat{n}, \varphi(\hat{n}))]\!]$. Since $V_p \subset U$ is arbitrary, we have found $U \subset [\![\exists f \in \mathscr{N}\, \forall n \in \mathbb{N}\, A(n, fn)]\!]$. □

2.2. Proposition. *Let* $\mathbb{R}^{\mathrm{c}}$ *be the set of Cauchy reals, then*

$$\mathrm{Sh}(\mathscr{O}(\mathbb{R})) \Vdash \mathbb{R}^{\mathrm{c}} \neq \mathbb{R}^{\mathrm{d}}, \quad \textit{hence } \mathrm{Sh}(\mathscr{O}(\mathbb{R})) \nVdash \mathrm{AC}_0.$$

Proof. Via the coding of $\mathbb{Q}$ in $\mathbb{N}$ we see that a sequence $\mathbb{N} \to \mathbb{Q}$ in a model $\mathrm{Sh}(\mathscr{O}(T))$ corresponds to a partial continuous $f \in T \to (\mathbb{N} \to \mathbb{Q})$ (cf. the result in 1.5). Let U be any *connected* open contained in $\mathrm{dom}(f)$. For fixed n, $f(n)$ is a (locally) continuous map over U which is therefore necessarily constant over U. Thus, *on* U, f is a sequence of constant total functions with values in $\mathbb{Q}$. (This argument can be generalized, cf. E15.2.6.)

It is easy to see that f represents a Cauchy sequence on U, iff $\langle f(t)(n)\rangle_n$, for any $t \in U$, is a Cauchy sequence; and if f is a Cauchy sequence, its limit is a constant function over U with as value a Cauchy real.

On the other hand, $\lambda x.x$ is a global element of $\mathbb{R}_{\mathbb{R}}$ which is nowhere locally constant; hence for any f representing a Cauchy real in the model $[\![f = \lambda x.x]\!] = \emptyset$. Therefore $\mathbb{R}^{\mathrm{c}} \neq \mathbb{R}^{\mathrm{d}}$, and since $\mathrm{AC}_0 \Rightarrow \mathbb{R}^{\mathrm{c}} = \mathbb{R}^{\mathrm{d}}$, we also have $\mathrm{Sh}(\mathscr{O}(\mathbb{R})) \nVdash \mathrm{AC}_0$. □

However, in general $\mathbb{R}^{\mathrm{c}} = \mathbb{R}^{\mathrm{d}}$ is weaker than AC_0:

2.3. PROPOSITION. *Let T be the space with points $\mathbb{R} \cup \{*\}$ $(* \notin \mathbb{R})$, and open sets*

$$\{U : U \subset \mathbb{R} \cup \{*\}, U \cap \mathbb{R} \in \mathcal{O}(\mathbb{R}), \forall x \in \mathbb{R}(x \in U \to * \in U)\}.$$

(Classically, $\mathcal{O}(T)$ consists of $\emptyset$ and all $U \cup \{*\}$ with $U \in \mathcal{O}(\mathbb{R})$.) For this space T

$$\mathrm{Sh}(T) \Vdash \mathbb{R}^{\mathrm{c}} = \mathbb{R}^{\mathrm{d}}, \qquad \mathrm{Sh}(T) \nVdash \mathrm{AC}_0.$$

PROOF. Suppose $U \in \mathcal{O}(T)$; a continuous $f \in U \to \mathbb{R}$ (i.e. an element of $\mathbb{R}$ in $\mathrm{Sh}(\mathcal{O}(T))$ must be constant. To see this, let $p < f(t) < q$ for some $t \in U$ and $p, q \in \mathbb{Q}$; then, since $[\![\hat{p} < f < \hat{q}]\!]$ is open, in particular $p < f(*) < q$. Since p, q were arbitrary, we see that $f(t) = f(*)$. From this we easily obtain $\mathrm{Sh}(\mathcal{O}(T)) \Vdash \mathbb{R}^{\mathrm{c}} = \mathbb{R}^{\mathrm{d}}$ (exercise).

On the other hand, AC_0 is not generally true. We define $\mathcal{O}(T)$-predicates A_n on $\hat{\mathbb{Q}}$ by

$$[\![\hat{q} \in A_n]\!] := \{t : |q - t| < 2^{-n}\} \cup \{*\}.$$

Clearly

$$[\![\forall n \exists q(q \in A_n)]\!] = T,$$

since the sets $\{[\![\hat{q} \in A_n]\!] : q \in \mathbb{Q}\}$ cover T for each n. Via the coding of $\mathbb{Q}$ in $\mathbb{N}$ we see, using 1.6, that $\mathbb{N} \to \mathbb{Q}$ is represented by the set of partial continuous mappings in $T \to (\mathbb{N} \to \mathbb{Q})$; as before we can show that for each fixed $n \in \mathbb{N}$ the resulting mapping $T \to \mathbb{Q}$ is constant.

Suppose $t \in [\![\exists f \in \mathbb{N} \to \mathbb{Q}\, \forall n \in \mathbb{N}(f(n) \in A_n)]\!] \cap \mathbb{R}$, then $t \in U = [\![\forall n \in \mathbb{N}(\varphi(n) \in A_n)]\!] \cap \mathbb{R}$ for some φ (assigning a constant $\hat{q}$ to each $\hat{n}$) and some interval $U \subset \mathbb{R}$. Hence $U \subset [\![\varphi(\hat{n}) \in A_n]\!]$ for all n, but this is impossible since the diameter of $[\![\varphi(\hat{n}) \in A_n]\!] \cap \mathbb{R}$ tends to zero. So AC_0 fails in the model. □

3. Topologies in sheaves over a cHa

In this section we shall investigate topological spaces and the continuity of functions between topological spaces in sheaf models. From now on $\mathbb{R}$ will be $\mathbb{R}^{\mathrm{d}}$.

3.1. DEFINITION. As before, let X_T be the sheaf of partial continuous functions from T into a topological space X. $T \times X$ is a topological space equipped with the product topology. For any $W \in \mathcal{O}(T \times X)$ we define a strict $\mathcal{O}(T)$-predicate $\underline{W}$ on X_T by

$$[\![a \in \underline{W}]\!] := \{t : (t, a(t)) \in W\}$$

(this set is open, because of the continuity of a). $\underline{\mathcal{O}}(X_T)$ is the $\mathcal{O}(T)$-set consisting of all such predicates, with equality

$$[\![\underline{W} = \underline{W}']\!] := [\![\forall a(a \in \underline{W} \leftrightarrow a \in \underline{W}')]\!].$$

$E\underline{W} = T$ for all $\underline{W}$ on this definition, so $\underline{\mathcal{O}}(X_T)$ consists of global elements of $\mathcal{P}(X_T)$, generating a subsheaf of $\mathcal{P}(X_T)$. □

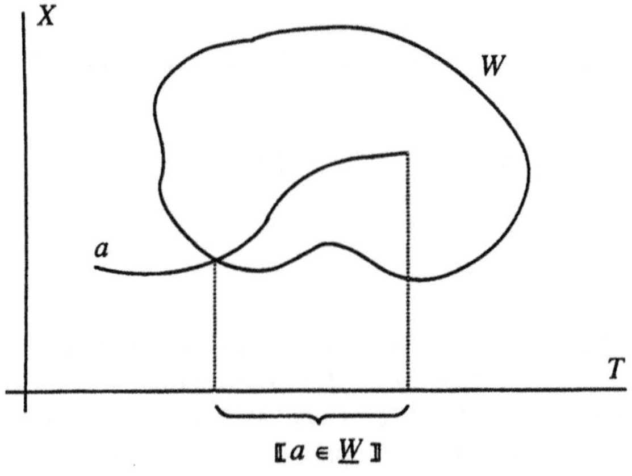

Fig. 15.1.

The predicate $[\![a \in \underline{W}]\!]$ is illustrated by fig. 15.1.

The property of being a topology on X, for a subset of $\mathcal{P}(X)$, is clearly expressible in higher-order logic. We have the following proposition.

3.2. PROPOSITION. $\mathrm{Sh}(\mathcal{O}(T)) \Vdash (\underline{\mathcal{O}}(X_T)^{\mathrm{sh}}$ *is a topology*).

PROOF. It suffices to show that in $\mathrm{Sh}(\mathcal{O}(T))$ the elements of $\underline{\mathcal{O}}(X_T)$ are closed under finite intersections and arbitrary unions.

(i) Closure under finite intersections is verified as follows:

$$[\![a \in \underline{W} \cap \underline{W'}]\!] = [\![a \in \underline{W}]\!] \wedge [\![a \in \underline{W'}]\!];$$

$$t \in [\![a \in \underline{W \cap W'}]\!] \leftrightarrow (t, a(t)) \in W \cap W'$$
$$\leftrightarrow (t, a(t)) \in W \wedge (t, a(t)) \in W'$$
$$\leftrightarrow t \in [\![a \in \underline{W}]\!] \wedge t \in [\![a \in \underline{W'}]\!].$$

Hence

$$[\![a \in \underline{W} \cap \underline{W'}]\!] = [\![a \in \underline{W \cap W'}]\!].$$

(ii) Arbitrary unions. Let Φ be any $\mathcal{O}(T)$-predicate on $\underline{\mathcal{O}}(X_T)$, and put

$$W_0 = \{(t, x) : \exists W(t \in \Phi(\underline{W}) \cap [\![\hat{x} \in \underline{W}]\!])\}.$$

Then $\underline{W_0} = \bigcup\{\underline{W} : \Phi(\underline{W})\}$ is valid in Sh($\mathcal{O}(T)$), since

$$t \in [\![a \in \bigcup\{\underline{W} : \Phi(\underline{W})\}]\!]$$
$$\leftrightarrow t \in [\![\exists W \in \Phi(a \in W)]\!]$$
$$\leftrightarrow t \in \bigcup\{\Phi(\underline{W}) \cap [\![a \in \underline{W}]\!] : W \in \mathcal{O}(T \times X)\}$$
$$\leftrightarrow \exists W(t \in \Phi(\underline{W}) \wedge t \in [\![a \in \underline{W}]\!]) \leftrightarrow \exists W(t \in \Phi(\underline{W}) \wedge (t, a(t)) \in W)$$
$$\leftrightarrow t \in [\![a \in W_0]\!]. \quad \square$$

In the case of $X = \mathbb{R}$ or $\mathbb{B}$, it is not difficult to show that $\underline{\mathcal{O}}(\mathbb{R}_T)^{\text{sh}}$, $\underline{\mathcal{O}}(\mathbb{B}_T)^{\text{sh}}$ indeed represent $\mathcal{O}(\mathbb{R})$, $\mathcal{O}(\mathbb{B})$ *within* Sh($\mathcal{O}(T)$). In fact, we have the following proposition.

3.3. Proposition. *If $\mathscr{B}$ is a basis for X, then $\hat{\mathscr{B}} \equiv \{\hat{V} : V \in \mathscr{B}\}$, where $\hat{V} = \underline{T \times V}$, internally generates a basis for the internal topology $\underline{\mathcal{O}}(X_T)^{\text{sh}}$.*

Proof. We have to show (cf. fig. 15.2)

$$[\![\forall W \in \underline{\mathcal{O}}(X_T) \forall a \in W \exists V \in \hat{\mathscr{B}}(x \in V \subset W]\!] = T.$$

So let $\underline{W} \in \underline{\mathcal{O}}(X_T)$, $t \in Ea$, $t \in [\![a \in \underline{W}]\!]$, i.e. $(t, a(t)) \in W$. We show that $t \in [\![a \in \hat{V} \subset \underline{W}]\!]$ for a suitable $V \in \mathscr{B}$. We first note that for $U \in \mathcal{O}(T)$

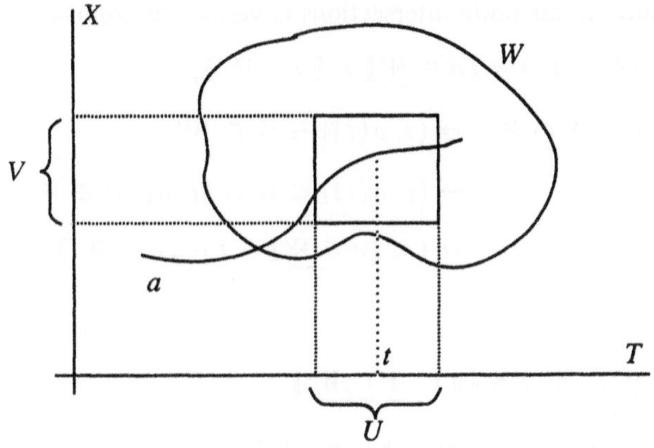

Fig. 15.2.

and $V \in \mathcal{O}(X)$, $U \subset [\![\underline{U \times V} = \hat{V}]\!]$, hence for U and V such that $U \times V \subset W$:

$$U \subset [\![\underline{U \times V} \subset \underline{W}]\!] \wedge [\![\underline{U \times V} = \hat{V}]\!] \subset [\![\hat{V} \subset \underline{W}]\!].$$

Now $(t, a(t)) \in W$, so there exists a $V \in \mathcal{B}$ and $U \in \mathcal{O}(T)$ such that $(t, a(t)) \in U \times V \subset W$, therefore $t \in [\![a \in \hat{V}]\!]$ and $t \in [\![\hat{V} \subset \underline{W}]\!]$. So $t \in [\![\exists V \in \hat{\mathcal{B}}(a \in V \subset \underline{W})]\!]$. This finishes the proof. Observe that, by logic, $[\![\underline{W} = \bigcup\{V \in \mathcal{B} : V \subset \underline{W}\}]\!] = T$. □

This argument shows in particular that $\underline{\mathcal{O}}(\underline{\mathbb{R}}_T)^{\mathrm{sh}}$ coincides internally in $\mathrm{Sh}(\mathcal{O}(T))$ with the topology generated by the rational open intervals, and that the topology $\underline{\mathcal{O}}(\underline{\mathbb{B}}_T)^{\mathrm{sh}}$ coincides internally with the topology generated by the basic opens $V_n = \{\alpha : \exists x(\bar{\alpha}x = n)\}$.

3.4. Proposition. *Let T, X, Y be topological spaces, Y a T_0-space, i.e. $\forall x, y \in Y(\forall U \in \mathcal{O}(Y)\ (x \in U \leftrightarrow y \in U) \to x = y)$. In $\mathrm{Sh}(\mathcal{O}(T))$ a continuous F from X_T to Y_T, with $E(F) = U \in \mathcal{O}(T)$ is externally (i.e. can be represented externally by) a continuous $f \in U \times X \to Y$ such that*

$$F(a)(t) = f(t, a(t)) \tag{1}$$

and vice versa.

Proof. Given F, define $f(t, y) := F(\hat{y})(t)$. We have to show, assuming $a(t) = b(t)$, that $F(a)(t) = F(b)(t)$. Let $V \in \mathcal{O}(Y)$. For any

$W \in \mathcal{O}(T \times X)$, $\lambda a.[\![F(a) \in \underline{W}]\!]$ determines a predicate $F^{-1}[\underline{W}] \equiv [\![F^{-1}[\underline{W}]]\!]$. This is a global element of $\mathcal{P}(X_T)$, and since F is internally open, this must actually be an element $\underline{W'}$ of $\mathcal{O}(X_T)$ such that $[\![\underline{W'} = F^{-1}[\underline{W}]]\!] = T$. Therefore for a suitable $W \in \mathcal{O}(T \times X)$ also

$$T = [\![\underline{W} = F^{-1}[\underline{U \times V}]]\!],$$

so $F(a)(t) \in V \Leftrightarrow t \in [\![F(a) \in \underline{U \times V}]\!] \Leftrightarrow t \in [\![a \in F^{-1}[\underline{U \times V}]]\!] \Leftrightarrow t \in [\![a \in \underline{W}]\!] \Leftrightarrow (t, a(t)) \in W$. Therefore $(t, a(t)) \in W$.

Similarly $(t, b(t)) \in W \Leftrightarrow F(b)(t) \in V$. Now this holds for arbitrary $V \in \mathcal{O}(Y)$, and since Y is a T_0-space, this means that $F(a)(t) = F(b)(t)$.

To show that f is continuous, let $V \in \mathcal{O}(Y)$ and let as before $W \in \mathcal{O}(T \times X)$ be such that $T = [\![\underline{W} = F^{-1}[\underline{U \times V}]]\!]$. Then $(t, x) \in W \Leftrightarrow t \in [\![\hat{x} \in F^{-1}(\underline{U \times V})]\!] \Leftrightarrow t \in [\![F(\hat{x}) \in \underline{U \times V}]\!] \Leftrightarrow f(t, x) = F(\hat{x})(t) \in V$, so $f^{-1}(V) = W$ is open.

Conversely, suppose f to be given, and define F by (1); $F(a)$ is continuous for each a. Let $W \in \mathcal{O}(T \times Y)$, then by the continuity of f

$$W' := \{(t, x) : (t, f(t, x)) \in W\} \in \mathcal{O}(T \times X).$$

Also $t \in [\![a \in F^{-1}[\underline{W}]]\!] \Leftrightarrow t \in [\![F(a) \in \underline{W})]\!] \Leftrightarrow (t, F(a)(t)) \in W \Leftrightarrow (t, f(t, a(t))) \in W \Leftrightarrow (t, a(t)) \in W' \Leftrightarrow t \in [\![a \in \underline{W'}]\!]$, and therefore $T = [\![F^{-1}[\underline{W}] = \underline{W'}]\!]$. F is therefore internally continuous. Clearly, for the F defined from f, $E(F) = U$. □

Internally continuous functions from $\mathbb{R}$ to $\mathbb{R}$ can therefore be represented by continuous functions in $\mathbb{R} \times \mathbb{R} \to \mathbb{R}$, i.e. surfaces over the plane.

For example, the internal identity function is representable by $f : (t, x) \mapsto x$. Another example is the interpretation of $(\mathbb{R}, +, \cdot, 0, 1)$ as an internal ring; the operations $+, \cdot$ correspond to mappings $f_+, f_\cdot \in \mathbb{R}^3 \to \mathbb{R}$ given by

$$f_+(t, x, y) := x + y,$$

$$f_\cdot(t, x, y) := x \cdot y.$$

representing $+, \cdot$ satisfying

$$a + b = \lambda t.a(t) + b(t),$$

$$a \cdot b = \lambda t.a(t) \cdot b(t), \quad \text{etc.}$$

3.5. *Failure of the intermediate value theorem in sheaves over* $\mathbb{R}$. We will construct a continuous function F from $\mathbb{R}_{\mathbb{R}}$ to $\mathbb{R}_{\mathbb{R}}$ such that internally

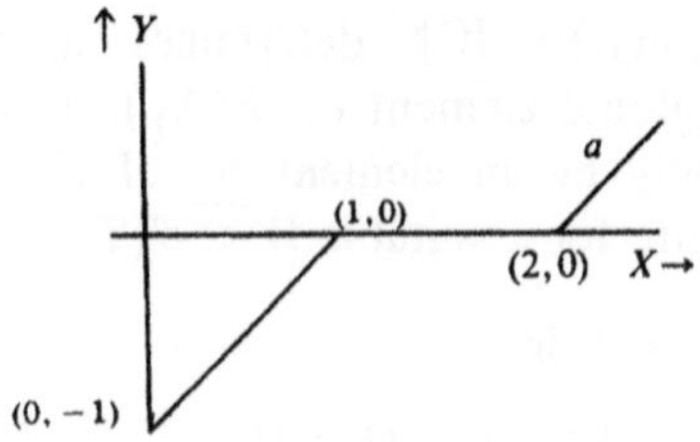

Fig. 15.3.

$F(\hat{0}) \le \hat{0}$ and $F(\hat{3}) > \hat{0}$, but such that F has no zero between $\hat{0}$ and $\hat{3}$. Let $d \in \mathbb{R}$, $0 < d < 1$. We define F by means of its representation f:

$$f(t, x) := a(x + d(t-1)) + d(1-t),$$

where a is the function given by the graph in fig. 15.3 (we encountered a before, as f in 6.1.2).

The resulting surface over $\mathbb{R} \times \mathbb{R}$ is shown in figs. 15.4 and 15.5 (one global, the other more detailed) below. The surface is obtained by shifting the graph of a. Observe that the plane $t = 1$ cuts the surface along a. It is easy to see that

$$[\![F(\hat{0}) < \hat{0}]\!] = \{t : a(x + d(t-1)) + d(1-t) < 0\} = \mathbb{R},$$

$$[\![F(\hat{3}) > \hat{0}]\!] = \{t : a(x + d(t-1)) + d(1-t) > 0\} = \mathbb{R}.$$

The bold line represents the intersection of the surface $\{(t, x, f(t, x)) : x, t \in \mathbb{R}\}$ with the OTX-plane. The solution $x_0(t)$ such that $f(t, x_0(t)) = 0$ exists locally as a continuous function everywhere except in $t = 1$. As a result, $[\![\exists y(F(y) = 0)]\!] = \mathbb{R} \setminus \{1\}$.

It should be noted that this is quite close to the reasoning behind the weak counterexample (6.1.2) to the intermediate value theorem: $f(x) := a(x) + d = 0$ does not have a solution depending continuously on d.

We can do even better in $\mathrm{Sh}(\mathcal{O}(\mathbb{R}^2))$.

3.6. Theorem. $\mathbb{R}$ *is not real-closed in* $\mathrm{Sh}(\mathcal{O}(\mathbb{R}^2))$.

Proof. Consider the polynomial

$$p(x) := x^3 + ax + b$$

and interpret it internally as a continuous function from $\mathbb{R}$ to $\mathbb{R}$, taking for a and b

$$a := \lambda uv.u, \qquad b := \lambda uv.v.$$

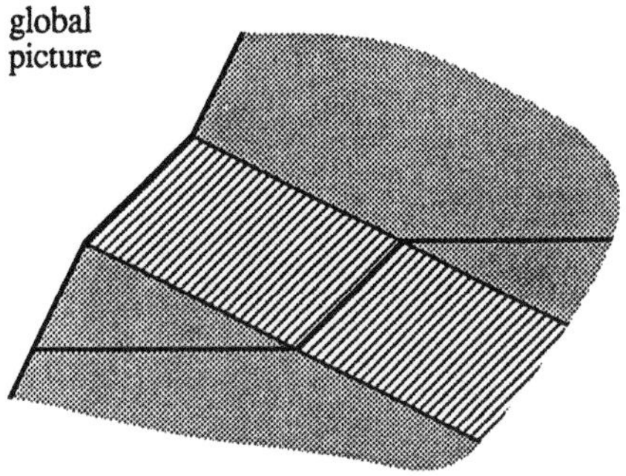

Fig. 15.4.

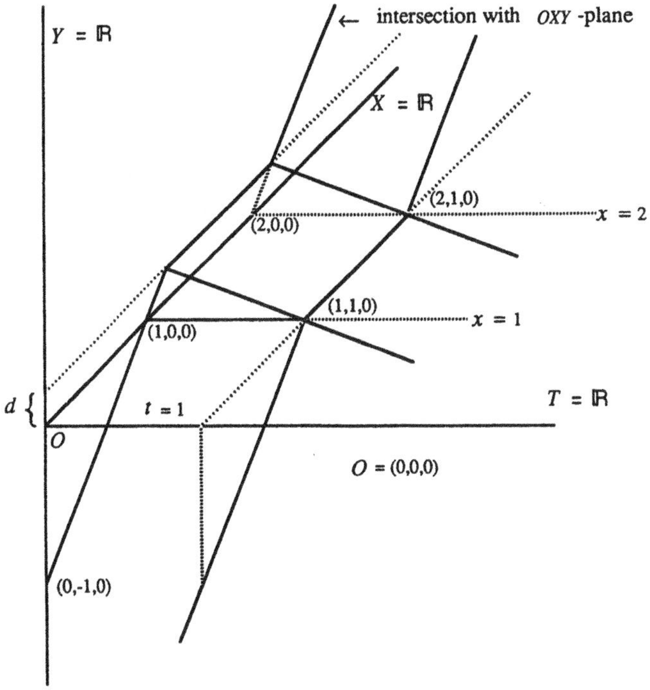

Fig. 15.5.

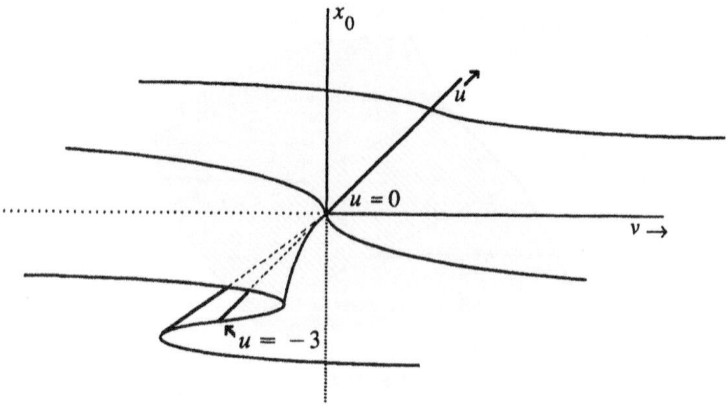

Fig. 15.6.

Externally p corresponds to the function f_p defined as

$$f_p((u, v), x) := x^3 + ux + v.$$

We can now draw a picture of the values $x_0(u, v)$ which yield solutions of $x_3 + ux + v = 0$, see fig. 15.6. We leave the verification of the details to the reader.

In a neighbourhood of $(0, 0)$ we cannot find a root, for this should be a continuous a such that, for all (u, v) in some open U with $(0, 0) \in U$, $a(u, v)$ lies on the surface of roots. Obviously, such a U does not exist! □

3.7. Theorem. $\mathbb{C}$ *is not algebraically closed over* $\mathrm{Sh}(\mathcal{O}(\mathbb{C}))$.

Proof. Left as an exercise. □

Certain topological properties automatically "lift" from X to X_T; for example, we have the following theorem.

3.8. Theorem. X_T *is compact in* $\mathrm{Sh}(\mathcal{O}(T))$ *iff* X *is compact.*

Proof. It clearly suffices to consider coverings by elements of a basis.

Let now $\hat{\mathscr{B}}$ be the internal basis generated by $\{\hat{U} : U \in \mathcal{O}(X)\}$, and suppose $\mathrm{Sh}(\mathcal{O}(T)) \Vdash \mathscr{U} \subset \hat{\mathscr{B}}$ and

$$t \in [\![\forall x \in X_T \exists U \in \mathscr{U}(x \in U)]\!]$$

(i.e. $\mathcal{U}$ covers X_T at t), and thus for each $x \in X_T$ with $t \in Ex$

$$t \in \bigcup\{[\![x \in \hat{U} \wedge \hat{U} \in \mathcal{U}]\!] : U \in \mathcal{O}(X)\}.$$

Applying this to $\hat{x}$, we find

$$t \in \bigcup\{[\![\hat{x} \in \hat{U} \wedge \hat{U} \in \mathcal{U}]\!] : U \in \mathcal{O}(X)\},$$

so for all $x \in X$

$$x \in \bigcup\{U : t \in [\![\hat{U} \in \mathcal{U}]\!] \wedge U \in \mathcal{O}(X)\}.$$

Since X is compact, we can find a finite subcover $\{U_i : i < n\}$ of the cover $\{U : t \in [\![\hat{U} \in \mathcal{U}]\!]\}$, therefore

$$t \in [\![\bigwedge_{i<n} (\hat{U}_i \in \mathcal{U}) \wedge \forall x \in X_T \Big(\bigvee_{i<n} (x \in \hat{U}_i)\Big)]\!]. \quad \square$$

3.9. Corollary. $[0,1]$ *is compact in all spatial models* $\mathrm{Sh}(\mathcal{O}(T))$. $\square$

This corollary motivates the study of sheaf models $\mathrm{Sh}(\Omega)$, where Ω is not of the form $\mathcal{O}(T)$ for some topological space T; as we shall see below, for a suitable Ω we find that $[0,1]$ is not compact in $\mathrm{Sh}(\Omega)$.

3.10–19. *Noncompactness of* $[0,1]$.

3.10. Recall that a topological space T is *separated* (7.1.11) if there is an apartness $\#$ on T such that $\forall x \in T(\{y : y \# x\} \in \mathcal{O}(T))$, and $\forall U \in \mathcal{O}(T)$ $\forall y \in T \forall x \in U(y \in U \vee x \# y)$. $\mathbb{R}$ and $[0,1]$ are obviously separated.

Definition. Let T be separated. We define: $U \in \mathcal{O}(T)$ is *coperfect* iff

$$\forall V \in \mathcal{O}(T) \forall n \forall t \in T\big(\mathrm{Int}(V \setminus \{t\}) \subset U \;\Rightarrow\; V \subset U\big). \quad \square$$

Classically, the coperfect sets are just the complements of perfect sets (a set A is *perfect* iff it is closed and has no isolated points; exercise).

3.11. Proposition. The coperfect opens $K(T)$ of a separated topological space form a cHa, obtainable as the lattice of fixed points of the operator $F \in \mathcal{O}(T) \to \mathcal{O}(T)$ given by

$$F(U) = \bigcup\{V \in \mathcal{O}(T) : \exists n \exists t_1 \dots t_n(\mathrm{Int}(V \setminus \{t_1, \dots, t_n\}) \subset U)\};$$

$F(U)$ is in fact the least coperfect open set containing U. $\square$

For the proof we need several lemmas.

3.12. LEMMA. *Let* Λ *be a complete lattice and let* $f \in \Lambda \to \Lambda$ *be order-preserving. Then the set of fixed points of* f, $\Lambda_f := \{p \in \Lambda : p = fp\}$ *is a complete lattice with join operation* $\bigvee'$ *given by*

$$\bigvee'\Theta := \bigwedge Q \quad \textit{where } Q \equiv \{q : fq \leq q \wedge \forall p \in \Theta(p \leq q)\},$$

for all $\Theta \subset \Lambda_f$.

PROOF. First observe that

$$q \in Q \Rightarrow fq \in Q. \tag{1}$$

Let $\Theta \subset \Lambda_f$, then, $\forall p \in \Theta(p \leq \bigvee'\Theta)$; since f preserves $\leq$ and $p \in \Lambda_f$, it follows that $p \leq f(\bigvee'\Theta)$. $\forall q \in Q(\bigwedge Q \leq q)$, hence again by the preservation of $\leq$ under f, $\forall q \in Q(f(\bigwedge Q) \leq fq \leq q)$, so $f(\bigwedge Q) \leq \bigwedge Q$, and therefore $\bigwedge Q \in Q$. By (1) $f(\bigwedge Q) \in Q$, hence $\bigwedge Q \leq f(\bigwedge Q)$, so $\bigwedge Q \in \Lambda_f$. Any $r \in \Lambda_f$ which is an upper bound for Θ is in Q, hence $\bigwedge Q \leq r$.

It follows from 13.4.6 that Λ_f is complete, since the meet of a set can be defined as the joins of the collection of lower bounds for the given set. □

COROLLARY (*to the proof*). *The argument shows that we can put*

$$\bigvee'\Theta := \bigwedge\{q : fq = q \wedge \forall p \in \Theta(p \leq q)\}. \quad \square$$

3.13. LEMMA. *Let* Ω *be a* cHa *and let* $f \in \Omega \to \Omega$ *be a multiplicative operator, i.e.*

$$f(p \wedge q) = f(p) \wedge f(q),$$

then Ω_f *is a* cHa.

PROOF. The preservation of $\wedge$ implies that f is order-preserving. By the preceding lemma the set of fixed points Ω_f of Ω is a complete lattice. It remains to establish the infinitary distributive law. Let $\Theta \subset \Omega_f$, $q \in \Omega_f$. Then $\forall p \in \Theta(p \leq \bigvee'\Theta)$, so $\forall p \in \Theta(q \wedge p \leq q \wedge \bigvee'\Theta \in \Omega_f)$, hence

$$\bigvee'\{q \wedge p : p \in \Theta\} \leq q \wedge \bigvee'\Theta.$$

Next we observe that

$$\forall p,q \in \Omega_f(f(p \to q) \leq p \to q). \tag{1}$$

(To see this note that $p \wedge (p \to q) \leq q$, so $fp \wedge f(p \to q) \leq fq$, and therefore $f(p \to q) \leq fp \to fq = p \to q$). Now

$$\forall p \in \Theta(p \wedge q \leq \mathsf{V}'\{q \wedge p' : p' \in \Theta\}), \quad \text{so}$$

$$\forall p \in \Theta(p \leq q \to \mathsf{V}'\{q \wedge p' : p' \in \Theta\}); \quad \text{also}$$

$$f(q \to \mathsf{V}'\{q \wedge p' : p' \in \Theta\}) \leq (q \to \mathsf{V}'\{q \wedge p' : p' \in \Theta\}),$$

and therefore $\mathsf{V}'\Theta \leq q \to \mathsf{V}'\{q \wedge p' : p' \in \Theta\}$ by the definition of V'. Thus

$$q \wedge \mathsf{V}'\Theta \leq \mathsf{V}'\{q \wedge p' : p' \in \Theta\}. \quad \square$$

3.14. DEFINITION. Let Ω be a cHa. $J \in \Omega \to \Omega$ is a called a *j-operator* or *nucleus* if:
(i) $p \leq Jp = JJp$,
(ii) $J(p \wedge q) = J(p) \wedge J(q)$
for all $p, q \in Q$. $\square$

3.15. LEMMA. *The range Ω_J of a j-operator J is a* cHa *with V' as defined in* 3.12. *Moreover*

$$\mathsf{V}'\Theta = J(\mathsf{V}\Theta),$$

and J, as a mapping from Ω to Ω_J, preserves V.

PROOF. From the preceding lemma; we leave the proof as an exercise. $\square$

3.16. The proposition 3.11 is a corollary of the following lemma.

LEMMA. *Let Ω be a* cHa, *and let f satisfy for all $p, q \in \Omega$*

$$f(p \wedge q) = f(p) \wedge f(q), \quad p \leq f(p).$$

Then J with $J(p) := \bigwedge\{q : p \leq q = fq\}$ is a j-operator with $\Omega_J = \Omega_f$ and $J(p)$ is the least fixed point of f above p.

PROOF. By the corollary of 3.12 $J(p) = \mathsf{V}'\{p\}$, so $Jp \in \Omega_f$ and actually Jp is the least fixed point of f above p. This shows

$$p \leq Jp, \tag{1}$$

$$fJp = Jp, \tag{2}$$

$$JJp = Jp, \tag{3}$$

$$p \leq q \Rightarrow Jp \leq Jq. \tag{4}$$

It remains to show that J preserves $\wedge$. Obviously $J(p \wedge q) \leq Jp \wedge Jq$. We have $p \wedge q \leq J(p \wedge q)$ (1), so

$$q \leq p \to J(p \wedge q). \tag{5}$$

From $\forall pr(p \wedge (p \to r) \leq r)$ we conclude

$$fp \wedge f(p \to J(p \wedge q)) \leq fJ(p \wedge q) = J(p \wedge q).$$

Using $p \leq fp$, it follows that $p \wedge f(p \to J(p \wedge q)) \leq J(p \wedge q)$, hence

$$f(p \to J(p \wedge q)) \leq p \to J(p \wedge q) \leq f(p \to J(p \wedge q)),$$

i.e. $p \to J(p \wedge q)$ is a fixed point of f. From (4) and (5) we get

$$Jq \leq J(p \to J(p \wedge q)) = f(p \to J(p \wedge q)) = p \to J(p \wedge q),$$

and therefore

$$p \leq Jq \to J(p \wedge q). \tag{6}$$

Repeating the argument above with (5) instead of (6) we find that with $Jq \to J(p \wedge q)$ is a fixed point of f, and so

$$Jq \leq Jp \to J(p \wedge q)$$

and $Jp \wedge Jq \leq J(p \wedge q)$. □

3.17. Definition. Let T be separated. $K(T)^*$ is the collection of all $U \subset T \cup \{*\}$ $(* \notin T)$ such that

$$U \cap T \in K(T), \qquad t \in U \cap T \;\Rightarrow\; * \in U. \;\square$$

Proposition. *$K(T)^*$ is again a* cHa *with* $\subset$ *as partial order; in* $K(T)^*$ *and* $K(T)$ $\wedge$ *coincides with* $\cap$.

Proof. Exercise. □

As an example consider $(-\infty, t) \cup (t, \infty) = \{s \in \mathbb{R} : s \# t\}$. $(-\infty, t)$ and (t, ∞) are both in $K(T)$, and in $K(T)$ $(-\infty, t) \vee (t, \infty) = T$: joins add points. Similarly in $K(T)^*$ $((-\infty, t) \cup \{*\}) \vee (\{*\} \cup (t, \infty)) = T \cup \{*\}$.

3.18. Lemma. *Suppose T is coperfect open in $\mathcal{O}(T)$ (classically this is the same as: T has no isolated points). Then reals in* $\mathrm{Sh}(K(T)^*)$ *are locally constant, i.e. for any real ξ in the model there is an $x \in \mathbb{R}^d$ such that $E\xi \subset [\![\hat{x} = \xi]\!]$.*

PROOF. For any proposition P

$$\bigcup\{T: P\}$$

is coperfect open, and

$$\hat{P} = \bigcup\{T \cup \{*\} : P\} \in K(T)^*.$$

Clearly

$$* \in \hat{P} \leftrightarrow P.$$

Also for $U \in K(T)^*$ and $X \subset T$

$$U \subset (* \in U)^\wedge, \qquad U \subset X \Rightarrow (* \in U)^\wedge \subset (* \in X)^\wedge.$$

Therefore, if $\mathscr{A} \subset K(T)^*$, we have for each $U \in \mathscr{A} : U \subset (* \in U)^\wedge \subset (* \in \bigcup\mathscr{A})^\wedge$, and since $(* \in \bigcup\mathscr{A})^\wedge \in K(T)^*$, also $\bigvee\mathscr{A} \subset (* \in \bigcup\mathscr{A})^\wedge$. From this we see that

$$* \in \bigvee\mathscr{A} \Rightarrow * \in \bigcup\mathscr{A}.$$

Now let ξ be an element of $\mathbb{R}^{\mathrm{d}}$ in $\mathrm{Sh}(K(T)^*)$, and $t \in E\xi$; then also $* \in E\xi$, and

$$L_\xi = \{r \in \mathbb{Q} : * \in [\![\hat{r} < \xi]\!]\}, \qquad U_\xi = \{r \in \mathbb{Q} : * \in [\![\xi < \hat{r}]\!]\}$$

yield an external Dedekind real $(L_\xi, U_\xi) \equiv x$. To see this, suppose for example $r \in L_\xi$. Then $* \in [\![\hat{r} < \xi]\!]$, and since ξ is a real in $\mathrm{Sh}(K(T)^*)$,

$$* \in [\![\exists r' \in \mathbb{Q}(\hat{r} < r' \wedge r' < \xi)]\!];$$

that is to say for some $r' \in \mathbb{Q}$

$$* \in \bigvee\{[\![\hat{r} < (r')^\wedge \wedge (r')^\wedge < \xi]\!] : r' \in \mathbb{Q}\},$$

therefore $* \in \bigcup\{[\![\hat{r} < (r')^\wedge]\!] \wedge [\![(r')^\wedge < \xi]\!] : r' \in \mathbb{Q}\}$, so $* \in [\![\hat{r} < (r')^\wedge]\!]$, $* \in [\![(r')^\wedge < \xi]\!]$ for some r', that is to say $r < r' \wedge r' \in L_\xi$, and so on. Claim:

$$E\xi \subset [\![\hat{x} = \xi]\!].$$

For suppose $t \in E\xi \cap [\![\xi < \hat{x}]\!]$, then $(* \in E\xi) \wedge (* \in [\![\xi < \hat{x}]\!])$. From the latter, $* \in [\![\exists r \in \mathbb{Q}(\xi < r < \hat{x})]\!]$, i.e. $* \in \bigvee\{[\![\xi < \hat{r} < \hat{x}]\!] : r \in \mathbb{Q}\}$, so for some $r \in \mathbb{Q}$, $* \in [\![\xi < \hat{r} < \hat{x}]\!]$. Then $* \in [\![\xi < \hat{r}]\!]$, i.e. $r \in U_\xi$, and also $* \in [\![\hat{r} < \hat{x}]\!]$, i.e. $r < x$, hence $r \in L_\xi$. This contradicts $(L_\xi, U_\xi) \in \mathbb{R}^{\mathrm{d}}$. Therefore $E\xi \cap [\![\xi < \hat{x}]\!] = \emptyset$, and similarly $E\xi \cap [\![\hat{x} < \xi]\!] = \emptyset$; hence $t \in E\xi \Rightarrow t \in [\![\hat{x} \le \xi]\!] \wedge [\![\xi \le \hat{x}]\!]$, i.e. $t \in [\![\xi = \hat{x}]\!]$. □

3.19. THEOREM. $[0,1]$ *is not compact over* $K([0,1])^*$.

PROOF. We define internal opens Φ_n by

$$[\![\hat{x} \in \Phi_n]\!] = \{*\} \cup \{t \in [0,1] : |x - t| > 2^{-n}\}.$$

This correctly defines a predicate in view of the fact that $\{t \in [0,1]: |x - t| > 2^{-n}\}$ is coperfect open, so $[\![\hat{x} \in \Phi_n]\!]$ is an element of $K([0,1])^*$. It is also not difficult to see that

$$t \in [\![\hat{x} \in \Phi_n]\!] \to \exists p,q \in \mathbb{Q}(t \in [\![\hat{p} < \hat{x} < \hat{q}]\!] \wedge t \in [\![(\hat{p},\hat{q}) \subset \Phi_n]\!]),$$

so the Φ_n are indeed open in the topology generated by the rational open intervals (exercise).

In $\mathrm{Sh}(K([0,1]^*)$, $\{\Phi_n : n \in \mathbb{N}\}$ covers $[\hat{0},\hat{1}]$, since for all x

$$[\![\hat{x} \in \bigcup\{\Phi_n : n \in \mathbb{N}\}]\!] = \bigvee\{[\![\hat{x} \in \Phi_n]\!] : n \in \mathbb{N}\} \supset$$
$$\bigcup\{[\![\hat{x} \in \Phi_n]\!] : n \in \mathbb{N}\} = \{*\} \cup [0,x) \cup (x,1].$$

Since $\bigvee\{[\![\hat{x} \in \Phi_n]\!] : n \in \mathbb{N}\}$ must be the *least* element of $K([0,1])^*$ containing $\{*\} \cup [0,x) \cup (x,1]$, we find $\bigvee\{[\![\hat{x} \in \Phi_n]\!] : n \in \mathbb{N}\} = \{*\} \cup [0,1]$. Clearly, the Φ_n increase with increasing n; on the other hand, for all $x \in [0,1]$, $k \in \mathbb{N}$, we find $x \notin [\![\hat{x} \in \Phi_k]\!]$, and therefore

$$[\![[\hat{0},\hat{1}] \subset \bigcup\{\Phi_n : n \leq k\}]\!] = \{*\},$$

so $[\hat{0},\hat{1}]$ is not compact in the model. □

REMARK. The treatment of the topology in this case is ad hoc. It follows from the theorem that $K([0,1])^*$ is not spatial, i.e. it is not of the form $\mathcal{O}(X)$. This can also be proved directly (exercise). □

3.20–24. *The continuity of functions in* $\mathbb{R} \to \mathbb{R}$. We shall now give a *classical* argument showing that under suitable conditions on T, the functions in $\mathbb{R} \to \mathbb{R}$ are continuous in the model $\mathrm{Sh}(\mathcal{O}(T))$. It is an open problem whether a constructive proof is possible.

3.20. In what follows T is a topological space without isolated points, *first-countable* (each point has a countable neighbourhood basis) and which is *completely regular*, i.e. $\forall U \in \mathcal{O}(T) \forall t \in U \exists f \in T \to [0,1](f \text{ continuous} \wedge f(t) = 0 \wedge \forall s \notin U(f(s) = 1))$.

Let F be a function from $\mathbb{R}_T$ to $\mathbb{R}_T$, total over U, i.e. $U \cap Ea \le E(Fa)$ for all $a \in \mathbb{R}_T$. We introduce $f \in U \times \mathbb{R} \to \mathbb{R}$ by

$$f(t, x) := F(\hat{x})(t).$$

We have to show that F satisfies $f(t, a(t)) = F(a)(t)$.

3.21. Lemma. *Let T be as above, $U \in \mathcal{O}(T)$, $t \in U$. Then there is a sequence $\langle t_n \rangle_n$ with limit t and a continuous function $g \in T \to [0,1]$ such that*

$$g(t) = 0, \quad \forall n(g(t_n) > g(t_{n+1})), \quad \forall s \notin U(g(s) = 1).$$

Proof. We can find neighbourhoods U_n and points t'_n ($n \in \mathbb{N}$) such that

$$U_0 = U, \quad \forall n(U_n \supset U_{n+1} \wedge t'_n \in U_n \setminus U_{n+1}), \quad \bigcap\{U'_n : n \in \mathbb{N}\} = \{t\}.$$

By complete regularity there are continuous $g_n \in T \to [0,1]$ such that $g_n(t) = 0$ and $\forall s \notin U_n(g_n(s) = 1)$; put

$$g(s) := \sum_{m=0}^{\infty} g_m(s) \cdot 2^{-m}.$$

It follows that

$$g(t'_n) \ge \sum_{m>n} g_m(t'_n) \cdot 2^{-m} = \sum_{m>n} 2^{-m} = 2^{-n},$$

thus $g(t'_n) > 0$ for all n, $g(t) = 0$. Since g is continuous, there is a subsequence $t_m := t'_{n(m)}$, $m \in \mathbb{N}$, such that for all m $g(t_{m+1}) < g(t_m)$. □

3.22. Lemma. *Let T, U be as before, $t \in U$, a and $b \in \mathbb{R}_T$. If $a(t) = b(t)$, then $F(a)(t) = F(b)(t)$.*

Proof. Assume $a(t) = b(t)$. We want to find a $c \in \mathbb{R}_T$ such that

$$t \in \mathrm{dom}(c), \qquad t \in [\![a = c]\!]^- \cap [\![b = c]\!]^-, \tag{1}$$

where $^-$ is the closure operator of T. For if we have (1), it follows that $t \in [\![F(a) = F(c)]\!]^- \cap [\![F(b) = F(c)]\!]^-$ and $F(a)(t) = F(c)(t) = F(b)(t)$, since $F(a)$, $F(b)$, and $F(c)$ are continuous. So let $U' = U \cap Ea \cap Eb$, and let $g, \langle t_n \rangle_n$ be given by the preceding lemma (with U' for U). Assume $g(t_0) < 1$, and choose y_n such that

$$y_0 = 1, \qquad g(t_n) < y_{n+1} < g(t_{n+1}).$$

Define c on U' such that if $y_n \le g(s) \le y_{n+1}$, then

$$n = 0\ (\mathrm{mod}\ 4) \Rightarrow c(s) = a(s);$$

$$n = 2\ (\mathrm{mod}\ 4) \Rightarrow c(s) = b(s);$$

$$n = 1\ (\mathrm{mod}\ 4) \Rightarrow c(s) = \left[b(s)\cdot(y_n - g(s)) + a(s)\cdot(g(s) - y_{n+1})\right]\cdot(y_n - y_{n+1})^{-1};$$

$$n = 3\ (\mathrm{mod}\ 4) \Rightarrow c(s) = \left[a(s)\cdot(y_n - g(s)) + b(s)\cdot(g(s) - y_{n+1})\right]\cdot(y_n - y_{n+1})^{-1}.$$

Continuity follows from the continuity of a, b, and g, and for all n $t_{4n} \in [\![a = c]\!]$, $t_{4n+2} \in [\![b = c]\!]$. □

3.23. LEMMA. *f is continuous.*

PROOF. Continuity in the first argument is immediate, since $f(t, x) = F(a)(t)$ for $a = \hat{x}$, and $F(a)$ is continuous on U. We prove sequential continuity of f. Let $\langle x_m\rangle_m$ converge to x, and let $t \in T$; by lemma 3.21 we can find g and $\langle t_n\rangle_n$ as specified there. For each m $\langle f(t_n, x_m)\rangle_n$ converges to $f(t, x_m)$, so there is a strictly increasing sequence α such that

$$\forall n \ge \alpha(m)(|f(t, x_m) - f(t_n, x_m)| < 2^{-m}).$$

Now let $y_m = g(t_{\alpha(m)})$ and let φ be a continuous function from $[0,1]$ to $\mathbb{R}$ such that $\varphi(0) = x$, $\varphi(y_m) = x_m$ for all m (take e.g. φ piecewise linear). If $\psi = \varphi \circ g$, then ψ is a continuous map from T to $\mathbb{R}$ such that $\psi(t) = x$ and $\psi(t_{\alpha(m)}) = x_m$ for all m.

Now $F(\psi)$ is continuous and $F(\psi)(t) = f(t, \psi(t)) = f(t, x)$, $F(\psi)(t_{\alpha(m)}) = f(t_{\alpha(m)}, x_m)$. Therefore $f(t_{\alpha(m)}, x_m)$ converges to $f(t, x)$; it readily follows that $f(t_m, x_m)$ converges to $f(t, x)$. □

Combining the three preceding lemmas with 3.4 we have the following theorem.

3.24. THEOREM (*classical*). *If T is completely regular, first-countable, and has no isolated points, then* $\mathrm{Sh}(\mathcal{O}(T)) \Vdash$ *all $f \in \mathbb{R} \to \mathbb{R}$ are continuous.* □

COROLLARY. *In combination with 3.9 we see that in* $\mathrm{Sh}(\mathcal{O}(T))$ *all functions in* $[0,1] \to \mathbb{R}$ *are uniformly continuous.* □

4. A derived rule of local continuity

4.1. On the one hand we know that in intuitionistic higher-order logic **HAH** we can show that $\mathbb{R}^c$ is real-closed. If we give the fundamental sequences of rationals with standard rate of convergence a Baire space topology (via the encoding of $\mathbb{Q}$ in $\mathbb{N}$), then the solutions x of $p(x) = 0$, where p is a polynomial of odd degree, are found continuously in the fundamental sequences determining the coefficients of p (cf. 6.2.7). On the other hand we cannot prove in **HAH** that $\mathbb{R}^d$ is real-closed (cf. 3.6). This impossibility is connected with the fact that we cannot find the solution to a polynomial equation continuously in the coefficients for the usual topology on $\mathbb{R}^d$. We also know that, in the presence of the axiom of countable choice, $\mathbb{R}^c = \mathbb{R}^d$.

Turning things the other way round, *if* we can show in **HAH**, in the *absence* of AC_0, that $\forall y \in \mathbb{R}^d\, \exists x \in \mathbb{R}^d(p(x, y) = 0)$ for a polynomial p, then perhaps we can find at least locally a solution $f(y)$ for x which is continuous in the usual topology on $\mathbb{R}$. This is indeed the case; below we present a proof, due to Joyal, for a rule of local continuous choice.

4.2. As we have seen (soundness), the theory **HAH** is valid in all models $\mathrm{Sh}(\mathcal{O}(T))$.

By induction on the formula complexity of A one can show that, for any topological space T definable in **HAH**, "$\mathrm{Sh}(\mathcal{O}(T)) \Vdash A$" can be expressed by a formula of **HAH**.

Secondly, by an induction on the length of derivations in **HAH**, one can prove a formalized version of the soundness theorem: for any sentence A,

$$\mathbf{HAH} \vdash A \;\Rightarrow\; \mathbf{HAH} \vdash (\mathrm{Sh}(\mathcal{O}(T)) \Vdash A).$$

Finally, the basic results, such as the representation of the internal reals as the sheaf $\mathbb{R}_T$, can be established within **HAH**.

4.3. *The space T.* Joyal's argument now depends on a clever choice of the space to which the formalized soundness theorem can be applied. Let T_i be $\mathbb{R}^d$ with the interval topology, and T_d the set $\mathbb{R}^d$ provided with the discrete topology: every subset of $\mathbb{R}^d$ counts as open.

We use the following notation for disjoint unions:

$$Z \dot{\cup} Z' := (\{0\} \times Z) \cup (\{1\} \times Z').$$

As the points of our space T we take the points of the disjoint union $T_i \dot{\cup} T_d \equiv \mathbb{R}^d \dot{\cup} \mathbb{R}^d$. For each subset $Y \subset \mathbb{R}^d \dot{\cup} \mathbb{R}^d$ we have $Y = Y_i \dot{\cup} Y_d$,

$Y_i = \{x : (0, x) \in Y\}$ and $Y_d = \{x : (1, x) \in Y\}$. As opens we take

$$\mathcal{O}(T) := \{U \subset T_i \dot{\cup} T_d : U_d \supset U_i \in \mathcal{O}(\mathbb{R}^d)\}.$$

We observe the following lemma.

4.4. Lemma.

(i) $\{1\} \times \mathbb{R}^d$ *is open in* T, *with the relativized discrete topology.*

(ii) *If* φ *is a partial continuous mapping from* T *into* Y, Y *a separated space* (15.3.10), *then* $\forall x \in \mathbb{R}(\varphi(0, x) = \varphi(1, x))$, *and* $\varphi_0 := \lambda x.\varphi(0, x)$ *is continuous.*

(iii) $\mathbb{R}_T := \{\varphi : \varphi$ *continuous from* U *into* $\mathbb{R}$, $U \in \mathcal{O}(T)\}$ *represents* $\mathbb{R}$ *in* $\mathrm{Sh}(\mathcal{O}(T))$. *The mapping* π *with*

$$\pi(0, x) = \pi(1, x) = x$$

is continuous, and hence a global element of $\mathbb{R}_T$.

Proof. (i) and (iii) are obvious. As to (ii), suppose $y_0 = \varphi(0, x) \# \varphi(1, x) = y_1$ and take $U = \{y : y \# y_1\}$, then $\varphi^{-1}[U] = U_i \dot{\cup} U_d$ with U_i open in $\mathbb{R}$, $U_d \supset U_i$. It follows that $(1, x) \in \varphi^{-1}[U]$, so $y_1 \in U$; contradiction, hence $y_0 = y_1$.

4.5. Lemma. *For any space* Y *and* $U \in \mathcal{O}(Y)$, *let* F *be the functor from* $\mathrm{Sh}(\mathcal{O}(Y))$ *to* $\mathrm{Sh}(\mathcal{O}(U))$, ($\mathcal{O}(U)$ *the relative topology on* U *induced by* $\mathcal{O}(Y)$) *given by*

$$F(a) = a \upharpoonright U.$$

Then one has for all formulas $A(\vec{a})$

$$[\![A(\vec{a})]\!]_Y \cap U = [\![A(F\vec{a})]\!]_U,$$

where $[\![A]\!]_Y$ *and* $[\![A]\!]_U$ *are the values of* A *in* $\mathrm{Sh}(\mathcal{O}(Y))$ *and* $\mathrm{Sh}((\mathcal{O}(U))$, *respectively.*

Proof. By formula induction. □

Below we shall apply this lemma with $Y = T$, $U = \{1\} \times T_d \equiv T'_d$.

4.6. Theorem. *The derived rule of local continuity holds in* **HAH**:

$$\vdash \forall x \in \mathbb{R}\, \exists y \in \mathbb{R}\, A(x, y) \Rightarrow$$
$$\vdash \forall x \in \mathbb{R}\, \exists U \exists f \in U \to \mathbb{R}(x \in U \wedge f \text{ continuous } \wedge \forall y \in U A(y, fy)).$$

PROOF. Assume $\vdash \forall x \in \mathbb{R}\, \exists y \in \mathbb{R}\, A(x, y)$, then also

$$\vdash [\mathrm{Sh}(T) \Vdash \forall x \in \mathbb{R}\, \exists y \in \mathbb{R}\, A(x, y)],$$

hence

$$\vdash [\![\exists y \in \mathbb{R}\, A(\pi, y)]\!]_T = T, \quad \pi \text{ as in 4.4(iii)}.$$

We reason informally in **HAH**. Let $x \in \mathbb{R}$; then there is a $U \in \mathcal{O}(T)$ with $(0, x) \in U$, and a continuous $\varphi \in U \to \mathbb{R}$ (i.e. $\varphi \in \mathbb{R}_T$) such that

$$U \subset [\![A(\pi, \varphi)]\!]_T.$$

From this we obtain with the lemma

$$U^* \equiv U \cap T'_{\mathrm{d}} = \{1\} \times U_{\mathrm{d}} \subset [\![A(F\pi, F\varphi)]\!]_{T'_{\mathrm{d}}} = [\![A(1', \varphi')]\!]_{T'_{\mathrm{d}}},$$

where F is the functor from $\mathrm{Sh}(\mathcal{O}(T))$ to $\mathrm{Sh}(\mathcal{O}(T'_{\mathrm{d}}))$ (cf. 4.5), and where $1'$ is the mapping $(1, z) \mapsto z$, $\varphi' = \varphi \restriction T'_{\mathrm{d}}$. We can evaluate pointwise over T'_{d}, since the topology is discrete (14.2.6, example). Thus $U^* \subset [\![A(1', \varphi')]\!]_{T'_{\mathrm{d}}}$ is equivalent to $\forall y \in U^*\, A(1'(y), \varphi(y))$, i.e.

$$\forall z \in U_{\mathrm{d}} A(z, \varphi(1, z)).$$

Since $\varphi(0, z) = \varphi(1, z)$ (4.4(ii)) and $U_{\mathrm{i}} \supset U_{\mathrm{d}}$ it follows that U_{i} is a neighbourhood of x in T_{i}, and

$$\forall z \in U_{\mathrm{i}} A(z, \varphi(0, z)).$$

Because of the continuity of $\lambda z.\varphi(0, z)$ (4.4(ii)) we have proved

$$\mathbf{HAH} \vdash \forall x \in \mathbb{R}\, \exists U\, \exists f \in U \to \mathbb{R}$$
$$(x \in U \wedge f \text{ continuous} \wedge \forall y \in U A(y, fy)). \quad \square$$

REMARK. The proof is a sketch inasmuch we have not carried out the formalization of the soundness theorem in **HAH**. If we assume the correctness of **HAH** on the informal (intuitive) level, the proof yields *without* formalization of the soundness theorem the result:

If $\vdash \forall x \in \mathbb{R}\, \exists y \in \mathbb{R}\, A(x, y)$, then $\forall x \in \mathbb{R}\, \exists U\, \exists f \in U \to \mathbb{R}$
$(x \in U \wedge f \text{ continuous} \wedge \forall y \in U A(y, fy))$ is true.

5. The monoid model for CS

5.1. We now turn to models over sites. In this section we shall discuss a model for the theory **CS** introduced in 12.4.6. The underlying site has a particularly simple structure: the category has only a single object, say $*$, with a set $M = \mathrm{Hom}(*, *)$. Below we use f, g, h for elements of M.

A presheaf or M-set is now simply a pair $\mathscr{X} \equiv (X, \upharpoonright)$, X a set, $\upharpoonright \in M \times X \to X$ such that

$$x \upharpoonright 1 = x,$$

$$(x \upharpoonright f) \upharpoonright g = x \upharpoonright f \circ g.$$

A morphism from $\mathscr{X}$ to $\mathscr{Y}$ is given by a function $\alpha \in X \to Y$ commuting with $\upharpoonright$:

$$\alpha(x \upharpoonright f) = \alpha(x) \upharpoonright f.$$

Given a covering system, we can formulate the notion of a sheaf. If $\mathscr{X}$ is a presheaf, $\mathscr{Y}$ a sheaf, we can describe $\mathscr{Y}^{\mathscr{X}}$ as the set of morphisms

$$\alpha \in \mathscr{M} \times \mathscr{X} \to \mathscr{Y},$$

where $\mathscr{M} := (M, \upharpoonright)$, $f \upharpoonright g := f \circ g$, and

$$(\alpha \upharpoonright f)(g, x) = \alpha(f \circ g, x).$$

We leave it as an exercise to show that this indeed coincides with our earlier definition of exponential when specialized to the present case; α corresponds to a natural transformation τ with $\tau_g(x) = \alpha(g, x)$. Evaluation is given by

$$\mathrm{ev}(\alpha, x) := \alpha(1, x).$$

5.2. *Description of the monoid model for* **CS**. The monoid M is the set $\mathrm{Cont}(\mathbb{B}, \mathbb{B})$ of all continuous $f \in \mathbb{B} \to \mathbb{B}$ with function composition as the operation $\circ$. The covering system J is defined by: a *sieve* $S \in J$ iff there is a collection of open embeddings $\{f_i : i \in I\} \subset S$ such that $\bigcup\{f_i[\mathbb{B}] : i \in I\} = \mathbb{B}$. ($f_i$ is an open embedding means: $f_i[\mathbb{B}]$ is open in $\mathbb{B}$ and f_i is a homeomorphism from $\mathbb{B}$ to $f_i[\mathbb{B}]$).

REMARK. Let h_n be the standard homeomorphic embedding

$$h_n : \mathbb{B} \xrightarrow{\cong} V_n \hookrightarrow \mathbb{B}$$

given by $h_n(\alpha) = n * \alpha$. Then, if f is an open embedding with $U \equiv f[\mathbb{B}]$, we can find, to each $V_n \subset U$, an element g of the monoid such that $f \upharpoonright g = f \circ g = h_n$. For since $f^{-1}[V_n] \cong \mathbb{B}$, we can take for g a homeomorphism from $\mathbb{B}$ to $f^{-1}[V_n]$ with $f \circ g = h_n$.

So we know that all h_n with $V_n \subset f[\mathbb{B}]$ for some $f \in S$ belong to S. Thus $S \in J$ iff there is an open cover $\{V_n : i \in I\}$ such that $\{h_n : i \in I\} \subset S$. In other words, it suffices to look at covers consisting of basic neighbourhoods. Alternatively, we might have defined J from the covering basis of the h_n, where $\{h_n : n \in I\}$ is a cover iff $\bigcup\{V_n : n \in I\} = \mathbb{B}$ (cf. 14.4.6).

Now $\mathrm{Sh}(M, J)$ is our model.

5.3. *Characteristic function of a cover.* Let K_0 be the set of all neighbourhood functions on $\mathbb{B}$. If we assume either continuity C-N (cf. 4.6.8) or classical logic, we can associate to each $S \in J$ a characteristic neighbourhood function $e_S \in K_0$ such that

$$\forall n \in \mathbb{N}(e_S(n) \neq 0 \leftrightarrow h_n \in S), \tag{1}$$

where h_n is as above. In fact, the following special consequence of C-N

$$\forall \alpha \exists x \forall y \geq x A(\bar{\alpha} y) \to \exists \gamma \in K_0 \forall n(\gamma n \neq 0 \to A(n)),$$

applied to $\forall \alpha \exists x (h_{\bar{\alpha} x} \in S)$ suffices.

Conversely, given $e \in K_0$ we can associate with f a cover

$$S_e := \{g \in \mathrm{Cont}(\mathbb{B}, \mathbb{B}) : \exists n(e(n) \neq 0 \wedge g[\mathbb{B}] \subset V_n)\}. \tag{2}$$

Clearly $S_e \in J$, and (1), (2) are inverse to each other.

The next few subsections are devoted to concrete representations of the sheaves interpreting $\mathbb{N}, \mathbb{B}$, K_0, the lawlike elements of $\mathbb{B}$, and the lawlike elements of K_0.

5.4. NOTATION. As before, α, β, γ range over $\mathbb{B}$. h_n is the standard embedding of 5.2, $\mathbf{1}$ is the identity in $\mathrm{Cont}(\mathbb{B}, \mathbb{B})$.

For arbitrary x, $\hat{x}$, or $x^\wedge$ is an abbreviation of the constant function $\lambda \alpha \in \mathbb{B}.x$.

Let N be the sheaf generated by the constant M-set $\hat{\mathbb{N}} \equiv \{\hat{n} : n \in \mathbb{N}\}$ with restriction $\hat{n} \upharpoonright f := \hat{n} = \hat{n} \circ f$. Henceforth $\boldsymbol{n}, \boldsymbol{m}, \boldsymbol{k}, \boldsymbol{l}$ range over N. □

We choose N as our interpretation of $\mathbb{N}$; the following lemma provides an explicit representation.

LEMMA. *N can be represented as* $\mathrm{Cont}(\mathbb{B}, \mathbb{N})$ *with* $\boldsymbol{n} \upharpoonright f = \boldsymbol{n} \circ f$ *for any* $\boldsymbol{n} \in N$.

PROOF. $\{\hat{n}_f : f \in S\}$, $n_f \in \mathbb{N}$, $S \in J$, is compatible (cf. 14.4.9) iff $\hat{n}_f \upharpoonright g = \hat{n}_{f \circ g}$ for all $g \in \mathrm{Cont}(\mathbb{B}, \mathbb{B})$. S is a cover, so there is a collection of open embeddings $\{f_i : i \in I\} \subset S$ such that $\bigcup\{f_i[\mathbb{B}] : i \in I\} = \mathbb{B}$; let $U_i \equiv f_i[\mathbb{B}]$. Define now $\boldsymbol{n} \in \mathbb{B} \to \mathbb{N}$ by

$$\boldsymbol{n}(x) = n_{f_i} \quad \text{iff } x \in U_i .$$

$\boldsymbol{n}$ is well-defined, for if $x \in U_i \cap U_j$, we can find $V_n \subset U_i \cap U_j$, $x \in V_n$, and h_n is then a restriction of f_i as well as of f_j, hence $n_{h_n} = n_{f_i} = n_{f_j}$. Clearly, $\boldsymbol{n}$ is continuous. We leave it to the reader to check that equivalent compatible collections yield the same function in $\mathbb{B} \to \mathbb{N}$.

We have to check that restriction is indeed correctly defined by $\boldsymbol{n} \upharpoonright g = \boldsymbol{n} \circ g$. Restriction along g of a compatible collection modulo $\cong$ is defined as

$$(\{x_f : f \in S\}/\cong) \upharpoonright g := \{x_{g \circ f'} : f' \in g^*(S)\}/\cong ,$$

where $S \in J$, and $g^*(S) = \{g' : g \circ g' \in S\}$. Hence in particular

$$(\{\hat{n}_f : f \in S\}/\cong) \upharpoonright g = \{\hat{n}_{g \circ f'} : g \circ f' \in S\}/\cong . \tag{1}$$

Let $\boldsymbol{n}, \boldsymbol{n}'$ be defined from the collections $\{\hat{n}_f : f \in S\}$ and $\{\hat{n}_{g \circ f'} : g \circ f' \in S\}$, respectively, as indicated above. Then

$$\boldsymbol{n}'(\alpha) = n_{g \circ f'} \quad \text{for some } f' \text{ such that } \alpha \in f'[\mathbb{B}],\ g \circ f' \in S,$$
$$f' \text{ an open embedding},$$
$$(\boldsymbol{n} \circ g)(\alpha) = \boldsymbol{n}(g(\alpha)) = n_f \quad \text{for some } f \in S \text{ such that } g(\alpha) \in f[\mathbb{B}],$$
$$f \text{ an open embedding}.$$

Let $f'' \equiv (f'^{-1}(\alpha))\hat{}$, $f''' \equiv (f^{-1}(g(\alpha)))\hat{}$, then $g \circ f' \circ f'' = (g(\alpha))\hat{} = f \circ f'''$ and hence $n_{g \circ f'} = n_{g \circ f' \circ f''} = n_{f \circ f'''} = n_f$. Therefore for all $\alpha \in \mathbb{B}$, $\boldsymbol{n}'(\alpha) = \boldsymbol{n}(\alpha)$.

Conversely, from $\boldsymbol{n} \in \mathrm{Cont}(\mathbb{B}, \mathbb{N})$ we can construct an open cover $\{\boldsymbol{n}^{-1}(\{m\}) : m \in \mathbb{N}\}$, and hence an $S_{\boldsymbol{n}} \in J$ given by

$$S_{\boldsymbol{n}} := \{f : \exists m (f^{-1}[\mathbb{B}] \subset \boldsymbol{n}^{-1}(\{m\}))\},$$

together with a compatible collection $\{\hat{n}_f : f \in S_{\boldsymbol{n}}\}$ such that

$$n_f = m \leftrightarrow f[\mathbb{B}] \subset \boldsymbol{n}^{-1}(\{m\}).$$

We leave it as an exercise to show that these constructions are indeed inverse to each other. $\square$

5.5. NOTATION. $\boldsymbol{B}$ is the sheaf N^N; $\boldsymbol{\alpha}, \boldsymbol{\beta}, \gamma$ range over $\boldsymbol{B}$. □

We have the following representation for $\boldsymbol{B}$.

LEMMA. *The sheaf $\boldsymbol{B}$ interpreting $\mathbb{B}$ in the model can (externally) be represented by* Cont($\mathbb{B}, \mathbb{B}$); *if $\boldsymbol{\alpha} \in \boldsymbol{B}$ restrictions are given by*

$$\boldsymbol{\alpha} \upharpoonright g := \boldsymbol{\alpha} \circ g.$$

Evaluation, i.e. function application, becomes

$$\mathrm{ev}(\boldsymbol{\alpha}, \boldsymbol{n}) := \lambda\beta \in \mathbb{B}.(\boldsymbol{\alpha}(\beta))(\boldsymbol{n}(\beta))$$

for $\boldsymbol{\alpha} \in \boldsymbol{B}$, $\boldsymbol{n} \in N$.

PROOF. $\boldsymbol{B} = N^N$ is an exponential in the model, i.e. the sheaf corresponding to $\mathbb{B}$ consists of mappings

$$\tau \in \mathrm{Cont}(\mathbb{B}, \mathbb{B}) \times \mathrm{Cont}(\mathbb{B}, \mathbb{N}) \to \mathrm{Cont}(\mathbb{B}, \mathbb{N})$$

with

$$(\tau \upharpoonright f)(g, \boldsymbol{n}) = \tau(f \circ g, \boldsymbol{n}) \quad \text{and} \quad \tau(g \upharpoonright f, \boldsymbol{n} \upharpoonright f) = \tau(g, \boldsymbol{n}) \upharpoonright f;$$

the restrictions on Cont($\mathbb{B}, \mathbb{B}$), Cont($\mathbb{B}, \mathbb{N}$) are simply given by composition to the right, as follows from the preceding discussions.

Such a τ can be represented by $\boldsymbol{\alpha}_\tau \in$ Cont($\mathbb{B}, \mathbb{B}$) given by

$$\boldsymbol{\alpha}_\tau(x) := \tau(1, \hat{n})(x);$$

conversely, from $\boldsymbol{\alpha} \in$ Cont($\mathbb{B}, \mathbb{B}$) we find a corresponding $\tau^{\boldsymbol{\alpha}}$ by

$$\tau^{\boldsymbol{\alpha}}(g, \boldsymbol{n})(\beta) := \boldsymbol{\alpha}(g(\beta))(\boldsymbol{n}(\beta)).$$

It remains to show that these constructions are inverse to each other.

For example, for $\boldsymbol{\beta} \equiv \boldsymbol{\alpha}_\tau$, $\tau^{\boldsymbol{\beta}} = \tau$ can be seen as follows. Take any γ, $\boldsymbol{n}$, g; then $\boldsymbol{n}$ is constant on some neighbourhood V_p of γ. Let $f \in$ Cont($\mathbb{B}, \mathbb{B}$) be such that $f[\mathbb{B}] = V_p$, $\forall\beta \in V_p(f(\beta) = \beta)$; then $f \circ f = f$.

$$\begin{aligned}
\tau^{\boldsymbol{\beta}}(g, \boldsymbol{n})(\gamma) &= \boldsymbol{\alpha}_\tau(g(\gamma))(\boldsymbol{n}(\gamma)) \text{ (by definition)} \\
&= \tau(1, \boldsymbol{n}(\gamma)^\wedge)(g(\gamma)) \quad \text{(by definition)} \\
&= \big(\tau(1, \boldsymbol{n}(\gamma)^\wedge) \upharpoonright g\big)(\gamma) = \tau(g, \boldsymbol{n}(\gamma)^\wedge \circ g)(\gamma) \\
&= \tau(g, \boldsymbol{n}(\gamma)^\wedge)(\gamma) = \tau(g, \boldsymbol{n}(\gamma)^\wedge)(f(\gamma)) \\
&= \tau(g \circ f, \boldsymbol{n}(\gamma)^\wedge \circ f)(\gamma) \\
&= \tau(g \circ f, \boldsymbol{n} \circ f)(\gamma) = \tau(g, \boldsymbol{n})(f(\gamma)) \\
&= \tau(g, \boldsymbol{n})(\gamma), \quad \text{etc.}
\end{aligned}$$

We leave it to the reader to check that for $\boldsymbol{\alpha} \in \boldsymbol{B}$, $\boldsymbol{n}, \boldsymbol{m} \in \boldsymbol{N}$ we have $\mathrm{ev}(\boldsymbol{\alpha}, \boldsymbol{n}) := \lambda\beta \in \mathbb{B}(\boldsymbol{\alpha}(\beta))(\boldsymbol{n}(\beta))$, so $\Vdash \boldsymbol{\alpha n} = \boldsymbol{m} \Leftrightarrow \forall\beta \in \mathbb{B}(\boldsymbol{\alpha}(\beta)(\boldsymbol{n}(\beta)) = \boldsymbol{m}(\beta))$. □

REMARK. The lemma shows that $\mathrm{Cont}(\mathbb{B}, \mathbb{B}) = M = \boldsymbol{B}$. We use boldface greek letters for elements of $\mathrm{Cont}(\mathbb{B}, \mathbb{B})$ in order to emphasize that we are thinking of them as elements of $\boldsymbol{B}$.

5.6. *Representation of* $\boldsymbol{N}^{<\boldsymbol{N}}$, *the interpretation of* $\mathbb{N}^{<\mathbb{N}}$. Since, arithmetically, $\mathbb{N}^{<\mathbb{N}}$ can be coded onto $\mathbb{N}$, we may identify $\mathbb{N}^{<\mathbb{N}}$ and $\mathbb{N}$, not only externally, but also inside the model. Thus $\boldsymbol{N}^{<\boldsymbol{N}} := \mathrm{Cont}(\mathbb{B}, \mathbb{N}^{<\mathbb{N}}) = \mathrm{Cont}(\mathbb{B}, \mathbb{N})$. For $\boldsymbol{\alpha} \in \boldsymbol{B}$, $\boldsymbol{n} \in \boldsymbol{N}$ one has that $\bar{\boldsymbol{\alpha}}(\boldsymbol{n})$ as an internal object corresponds to $\lambda\beta \in \mathbb{B}.\overline{\boldsymbol{\alpha}(\beta)}(\boldsymbol{n}(\beta))$ as an external object, and so

$$(\Vdash \boldsymbol{\alpha} \in \boldsymbol{n}) \quad \Leftrightarrow \quad \forall\beta \in \mathbb{B}(\boldsymbol{\alpha}(\beta) \in \boldsymbol{n}(\beta)).$$

5.7. *Lawlike domains.* In our interpretation of **CS**, $\boldsymbol{B}$ will be our interpretation of the domain of choice sequences. However, we also need certain domains of lawlike objects, in particular lawlike sequences, for our interpretation of **CS**.

The obvious solution is to interpret lawlike objects as objects on which our information is complete, that is to say by restriction we do not get more information. This is the case, for example, for an $\boldsymbol{\alpha} \in \boldsymbol{B}$ which is constant. We define lawlike elements as follows.

DEFINITION. Let $\mathscr{X}$ be a sheaf; the subsheaf $\mathscr{X}_{\mathrm{L}}$ of *lawlike elements* of $\mathscr{X}$ is the sheaf generated by

$$\{x \in \mathscr{X} : \forall f \in M(x \upharpoonright f = x)\}. \quad \square$$

Thus $x \in \mathscr{X}_{\mathrm{L}}$ iff $x \in \mathscr{X}$ and for some $S \in J$

$$\forall f \in S\, \forall g \in M(x \upharpoonright f = x \upharpoonright f \circ g).$$

Note that

$$(\boldsymbol{N})_{\mathrm{L}} = \boldsymbol{N}, \qquad (\boldsymbol{N}^{<\boldsymbol{N}})_{\mathrm{L}} = \boldsymbol{N}^{<\boldsymbol{N}}.$$

An element $\boldsymbol{\alpha} \in \boldsymbol{B}$ is invariant under restrictions iff $\boldsymbol{\alpha}$ is a constant function; hence $\boldsymbol{B}_{\mathrm{L}}$ is the sheaf of locally constant functions from $\mathbb{B}$ to $\mathbb{B}$.

5.8. *Internal neighbourhood functions.* $\boldsymbol{N}^{<\boldsymbol{N}} \to \boldsymbol{N}$ may be identified with $\boldsymbol{N} \to \boldsymbol{N}$, i.e. $\boldsymbol{B}$, so, in the model, $\mathbb{N}^{<\mathbb{N}} \to \mathbb{N}$ is represented by $\gamma \in$

Cont($\mathbb{B}, \mathbb{B}$) with restrictions $\gamma \upharpoonright f := \gamma \circ f$ for $f \in M$. γ represents an element of K_0 in the model, i.e. $\gamma \in \boldsymbol{K}_0$ iff

$$\Vdash \forall\alpha\exists n(\gamma(\bar{\alpha}(n)) > 0 \wedge \forall nm(n \preccurlyeq m \wedge \gamma(n) > 0 \rightarrow \gamma(n) = \gamma(m))).$$

NOTATION. For internally lawlike elements, i.e. elements of $(\boldsymbol{B})_{\mathrm{L}}$, we use $\boldsymbol{a}, \boldsymbol{b}, \boldsymbol{c}$. □

Elements of $\boldsymbol{N}^{<\boldsymbol{N}}$ invariant under restrictions correspond to constant mappings in Cont($\mathbb{B}, \mathbb{B}$). The following lemma is readily verified.

5.9. LEMMA. *For each $e \in K_0$ the element $\tilde{e} \in \boldsymbol{K}_0$ given by*

$$\tilde{e}(\alpha)(u) := e(u), \quad \textit{where } \alpha \in \mathbb{B},\ u \in \mathbb{N}^{<\mathbb{N}},$$

is invariant under restrictions, and conversely, if $\gamma \in \boldsymbol{K}_0$ is invariant under restrictions, then $\gamma = \tilde{e}$ for some $e \in K_0$. The elements of $(\boldsymbol{K}_0)_{\mathrm{L}}$ are therefore locally of the form $\tilde{e}$ for some $e \in K_0$.

PROOF. Left as an exercise. □

We now turn to the characterization of $(\boldsymbol{N}^{\boldsymbol{N}} \rightarrow \boldsymbol{N}^{\boldsymbol{N}})_{\mathrm{L}} = (\boldsymbol{B}^{\boldsymbol{B}})_{\mathrm{L}}$.

5.10. *Lawlike elements of $\boldsymbol{B}^{\boldsymbol{B}}$.* By our definition of exponentials for the monoid, $\boldsymbol{N}^{\boldsymbol{N}} \rightarrow \boldsymbol{N}^{\boldsymbol{N}}$ consists of morphisms

$$F \in \mathrm{Cont}(\mathbb{B},\mathbb{B}) \times \mathrm{Cont}(\mathbb{B},\mathbb{B}) \rightarrow \mathrm{Cont}(\mathbb{B},\mathbb{B})$$

such that

$$(F \upharpoonright f)(g, \boldsymbol{\alpha}) = F(f \circ g, \boldsymbol{\alpha}),$$

$$F(f \upharpoonright h, \boldsymbol{\alpha} \upharpoonright h) = F(f, \boldsymbol{\alpha}) \upharpoonright h = F(f, \boldsymbol{\alpha}) \circ h.$$

Note that $\mathrm{ev}(F, \boldsymbol{\alpha}) = F(\mathbf{1}, \boldsymbol{\alpha})$. F is invariant under restrictions iff for all f $F(f, \boldsymbol{\alpha}) = (F \upharpoonright f)(\mathbf{1}, \boldsymbol{\alpha})$, so F can be represented by an $F^* \in \mathrm{Cont}(\mathbb{B},\mathbb{B}) \rightarrow \mathrm{Cont}(\mathbb{B},\mathbb{B})$ with $F^*(\boldsymbol{\alpha}) := F(\mathbf{1}, \boldsymbol{\alpha})$; in this case $\mathrm{ev}(F^*, \boldsymbol{\alpha}) = F^*(\boldsymbol{\alpha})$. Moreover, we have

$$F^*(g) = F(1, g) = F(g, g) = F(\mathbf{1}, \mathbf{1}) \upharpoonright g = F^*(\mathbf{1}) \upharpoonright g = F^*(\mathbf{1}) \circ g,$$

so that for $\xi := F^*(\mathbf{1})$ we have

$$F^*(\boldsymbol{\alpha}) = \xi \circ \boldsymbol{\alpha}.$$

For an arbitrary F let π be some fixed homeomorphism in $\mathbb{B} \to \mathbb{B} \times \mathbb{B}$, e.g. $\pi(\alpha) := (\lambda n. j_1(\alpha n), \lambda n. j_2(\alpha n))$, and let us write π_i for $\lambda\alpha\lambda n. j_i(\alpha n)$. Then

$$F(f, g) = F\big((\pi_1 \circ \pi \circ \pi^{-1})(f, g), (\pi_2 \circ \pi \circ \pi^{-1})(f, g)\big)$$
$$= F(\pi_1 \circ \pi, \pi_2 \circ \pi) \restriction (\pi^{-1}(f, g)),$$

so $F(f, \alpha) = (\xi \circ \pi^{-1})(f, \alpha)$ with $\xi := F(\pi_1 \circ \pi, \pi_2 \circ \pi)$.

Given a $\xi \in \mathrm{Cont}(\mathbb{B}, \mathbb{B})$, there is (externally) a (nonunique) $e_\xi \in K_0$ such that

$$\forall \beta\big(\xi(\beta)(n) = m \leftrightarrow \exists k\big(e_\xi(\langle n\rangle * \bar{\beta}(k)) = m + 1\big)\big).$$

Let $\tilde{e}_\xi$ correspond to e_ξ as in 5.9, then $\tilde{e}_\xi$ is invariant under restrictions. Now we prove the following lemma.

5.11. LEMMA. *With* $F_\xi = \lambda g. \xi \circ g$ *we have*

$\Vdash \tilde{e}_\xi$ *is a neighbourhood function for* F_ξ.

PROOF. Suppose $\Vdash \tilde{e}_\xi((\langle m\rangle * u)\hat{\ }) = (n + 1)\hat{\ }$. Then $e_\xi((\langle m\rangle * u)\hat{\ }) = n + 1$. Suppose furthermore $\Vdash \boldsymbol{\alpha} \in \hat{u}$ for an element $\boldsymbol{\alpha}$ of $\boldsymbol{B}$, then $\forall\beta$ $\forall y < \mathrm{lth}(u)(\alpha(\beta)(y) = (u)_y)$, and $\Vdash F_\xi(\boldsymbol{\alpha})(\hat{m}) = \hat{n} \Leftrightarrow \forall\beta(\xi \circ \alpha(\beta)(m) = n + 1) \Leftrightarrow \forall\beta(\xi(\alpha(\beta))(m) = n) \Leftrightarrow \exists k(e_\xi(\langle m\rangle * \overline{\alpha(\beta)}(k)) = n)$; the last assertion is true, hence $\Vdash F(\boldsymbol{\alpha})(\hat{m}) = \hat{n}$. We leave it to the reader to complete the argument. □

5.12. COROLLARY. *For* $F \in (\boldsymbol{B}^{\boldsymbol{B}})_{\mathrm{L}}$: $\Vdash F$ *is continuous.*

PROOF. Immediate. □

5.13. The next few subsections will be devoted to the validity of the axioms of **CS** in the monoid model.

The monotonicity and covering properties for forcing over sites specialize in this case to

$$\Vdash A(a_1, \ldots, a_n) \;\Rightarrow\; \Vdash A(a_1 \restriction f, \ldots, a_n \restriction f)$$

and if $S \in J$

$$\forall f \in S(\Vdash A(a_1 \restriction f, \ldots, a_n \restriction f)) \;\Rightarrow\; \Vdash A(a_1, \ldots, a_n).$$

Here A is a sentence with parameters $a_1, \dots, a_n$ from sheaves in the model. We note in addition, that if we accept classical logic externally, then either $\Vdash A$ or $\Vdash \neg A$ for each closed A (exercise).

5.14. LEMMA. *Let $\mathscr{X}$ be an (M, J)-sheaf, and let $A(a_1, \dots, a_n, x)$ be a formula with only x free, $a_1, \dots, a_n$ parameters from the model. Then*

$$\Vdash \exists x \in \mathscr{X} A(a_1, \dots, a_n, x) \Leftrightarrow \exists p \in \mathscr{X}(\Vdash A(a_1, \dots, a_n, p)).$$

PROOF. $\Leftarrow$ is immediate. As to $\Rightarrow$, assume $\Vdash \exists x \in \mathscr{X} A(\vec{a}, x)$; then there is an $S \in J$, and a p_f for each $f \in S$ such that

$$\forall f \in S(\Vdash A(\vec{a} \upharpoonright f, p_f)).$$

The cover S has a characteristic function $e \in K_0$ (cf. our remark in 5.3) such that

$$en \neq 0 \to h_n \in S,$$

where h_n is the standard homeomorphic embedding of V_n into $\mathbb{B}$. Let $I := \{n : en \neq 0 \wedge \forall m \prec n(em = 0)\}$. From $\{p_{h_n} : n \in I\}$ we obtain a compatible subset of $\mathscr{X}$ over the cover $S^* := \{h_n \circ f : n \in I \wedge f \in M\}$ by taking simply $p_{h_n \circ f} := p_{h_n} \upharpoonright f$ (observe that h_n and f in the elements of S^* are uniquely determined); $S^* \subset S$.

Hence, by the sheaf property of $\mathscr{X}$, there is a unique $p \in \mathscr{X}$ such that $p \upharpoonright h_n = p_{h_n}$. Therefore we have $\forall f \in S^*(\Vdash A(\vec{a} \upharpoonright f, p \upharpoonright f))$ and thus with the covering property $\Vdash A(\vec{a}, p)$. □

5.15. THEOREM. *The schema of lawlike countable choice holds in the model ($\mathscr{X}$ any sheaf; A contains only lawlike parameters besides x):*

$$\Vdash \forall n \exists x \in \mathscr{X} A(n, x) \to \exists F \in (\mathscr{X}^N)_L \forall n A(n, Fn).$$

PROOF. Assume $\Vdash \forall n \exists x A(n, x)$. We now use the following facts, which are easily verified:

(i) each element of a sheaf has a lawlike restriction,

(ii) $\Vdash \exists x \in \mathscr{X} B(x, \vec{a}) \to \exists x \in (\mathscr{X})_L B(x, \vec{a})$, where the parameters $\vec{a}$ are all lawlike (the so-called *specialization property*), and we conclude that to each $n \in \mathbb{N}$ there is an $x_n \in \mathscr{X}_L$ such that

$$\Vdash A(\hat{n}, x_n) \quad \text{with } x_n \upharpoonright f = x_n \quad \text{for all } f \in M.$$

We now define a morphism

$$F : \text{Cont}(\mathbb{B}, \mathbb{B}) \times \text{Cont}(\mathbb{B}, \mathbb{N}) \to \mathscr{X}$$

via

$$F(\mathbf{1}, \hat{n}) := x_n \quad \text{for all } n \in \mathbb{N};$$

then

$$F(f, \hat{n}) = F(\mathbf{1}\upharpoonright f, \hat{n}\upharpoonright f) = F(\mathbf{1}, \hat{n})\upharpoonright f = x_n\upharpoonright f = x_n,$$

$$(F\upharpoonright f)(g, \hat{n}) = F(f\circ g, \hat{n}) = x_n = F(g, \hat{n}),$$

so F is invariant under restrictions, hence lawlike. Clearly, we also have $\Vdash A(\hat{n}, F\hat{n})$, and thus $\Vdash \forall n A(n, Fn)$. □

5.16. THEOREM. *The schema* C-N *holds in the model*

$$\Vdash \forall\alpha\exists n A(\alpha, n) \to \exists e \in (K_0)_{\mathrm{L}}\forall m(em > 0 \to \exists n\forall\alpha \in mA(\alpha, n)).$$

PROOF. Suppose $\Vdash \forall\alpha\exists n A(\alpha, n)$, and choose $\boldsymbol{\alpha} = \mathbf{1}$, then $\Vdash \exists n \in \mathbb{N}$ $A(\mathbf{1}, n)$ and therefore by lemma 5.14 for some $\boldsymbol{n} \in N$

$$\Vdash A(\mathbf{1}, \boldsymbol{n}).$$

$\boldsymbol{n} \in \mathrm{Cont}(\mathbb{B}, \mathbb{N})$, and thus there exists an $e \in K_0$ such that

$$\forall mk(em = k + 1 \leftrightarrow \forall\beta \in m(\boldsymbol{n}(\beta) = k)).$$

We define $\boldsymbol{e} \in \mathrm{Cont}(\mathbb{B}, \mathbb{B}) = \mathrm{Cont}(\mathbb{B}, \mathbb{N}^{<\mathbb{N}} \to \mathbb{N})$ by

$$\boldsymbol{e}(\beta)(m) = em,$$

then $\boldsymbol{e} \in (\boldsymbol{K}_0)_{\mathrm{L}}$ (cf. 5.10). Now suppose

$$\Vdash \boldsymbol{e}(\hat{m}) > 0,$$

where $\hat{m} = \lambda\beta.m \in \mathrm{Cont}(\mathbb{B}, \mathbb{N}^{<\mathbb{N}}) = N^{<N}$; then $\boldsymbol{n}$ is constant on V_m, with value p, say; also

$$\Vdash A(\mathbf{1}, \boldsymbol{n})\upharpoonright\boldsymbol{\alpha}, \quad \text{i.e.} \quad \Vdash A(\boldsymbol{\alpha}, \boldsymbol{n}\upharpoonright\boldsymbol{\alpha}),$$

so $\Vdash \forall\alpha \in \hat{m} A(\alpha, \hat{p})$; hence $\Vdash \boldsymbol{e}(\hat{m}) > 0 \to \exists k \in \mathbb{N}\,\forall\alpha \in \hat{m} A(\alpha, k)$, etc. □

5.17. THEOREM. *If all parameters in A are lawlike, and* $\mathrm{FV}(A) \subset \{\alpha, \beta\}$, *then*

$$\Vdash \forall\alpha\exists\beta A(\alpha, \beta) \to \exists F \in (\boldsymbol{B}^{\boldsymbol{B}})_{\mathrm{L}}\forall\alpha A(\alpha, F\alpha).$$

PROOF. Suppose $\Vdash \forall\alpha\exists\beta A(\alpha, \beta)$, and choose $\mathbf{1}$ for $\boldsymbol{\alpha}$, then for some $\boldsymbol{\beta}$ (by lemma 5.14)

$$\Vdash A(\mathbf{1}, \boldsymbol{\beta}).$$

Therefore by the monotonicity of forcing

$$\Vdash A(\mathbf{1} \upharpoonright f, \boldsymbol{\beta} \upharpoonright f) \quad \text{for all } f \in M.$$

Now put $F(g, h) = \boldsymbol{\beta} \circ h$; this clearly is an element of $(\boldsymbol{B}^{\boldsymbol{B}})_{\mathrm{L}}$, since it is invariant under restrictions. Note that $\mathrm{ev}(F, \boldsymbol{\alpha}) = F(\mathbf{1}, \boldsymbol{\alpha}) = \boldsymbol{\beta} \circ \boldsymbol{\alpha}$. So

$$\Vdash \exists F \in (\boldsymbol{B}^{\boldsymbol{B}})_{\mathrm{L}} \forall\alpha A(\alpha, F\alpha). \quad \square$$

5.18. THEOREM. *The schema of analytic data holds in the model: for $A(\alpha)$ a formula with all parameters other than α lawlike,*

$$\Vdash \forall\alpha\big(A(\alpha) \to \exists F \in (\boldsymbol{B}^{\boldsymbol{B}})_{\mathrm{L}}(\exists\beta(\alpha = F\beta) \wedge \forall\gamma A(F\gamma))\big).$$

PROOF. Suppose $\Vdash A(\boldsymbol{\alpha})$ for $\boldsymbol{\alpha} \in \boldsymbol{B}$. Then F defined by $F(f, g) := \boldsymbol{\alpha} \upharpoonright g$, for $g \in M$, is in fact an element of $(\boldsymbol{B}^{\boldsymbol{B}})_{\mathrm{L}}$ since it is invariant under restrictions; clearly $\Vdash F(\mathbf{1}) = \alpha$, and $\Vdash \forall\gamma A(F\gamma)$, by the monotonicity of forcing. $\square$

Observe that by 5.12 we have continuous choice in 5.17, and a continuous operator in 5.18 (hence the "analytic").

5.19. PROPOSITION. *Assuming $K_0 = K$ externally, we have internally $\Vdash (\boldsymbol{K}_0)_{\mathrm{L}} = (\boldsymbol{K})_{\mathrm{L}}$, that is to say we can prove internally that $(\boldsymbol{K}_0)_{\mathrm{L}}$ satisfies the axioms* K1–3 *for $\boldsymbol{K}$, where the schema* K3 *is taken to refer to predicates with lawlike parameters only.*

PROOF. It is sufficient to prove K1–3 with respect to the constant elements of $(\boldsymbol{K}_0)_{\mathrm{L}}$, i.e. the elements which are invariant under restrictions; and for those elements K1–3 follow straightforwardly from the external assumption $K_0 = K$. Instead of checking K3 we may also check the equivalent principle IUS (cf. 4.8.12), induction over insecured sequences. We leave the details to the reader. $\square$

Summarizing the results in 5.15–5.19 we have the following theorem.

5.20. THEOREM. *Under the indicated interpretations of* $\mathbb{N}$, *the domain of choice sequences, the domain of lawlike sequences and lawlike* K_0, $\mathrm{Sh}(M, J)$ *is a model for* **CS**.

REMARK. The actual verification of this fact does not need the theory of sheaf-completions for sites; it suffices to verify ad hoc that the interpretations of the various domains make the axioms for them valid. However, the fact that $\boldsymbol{B}$ as chosen *is* indeed the exponential N^N, teaches us that the interpretation chosen is not ad hoc, but that our choice of interpretation of $\mathbb{B}$ is "the canonical one."

On the other hand, we frequently appealed to the fact that certain sets generate the sheaves and that it is sufficient to compute with generating sets. (cf. section 14.4). □

5.21. *Formalizing the metamathematics of the model.* It can be shown that the treatment of the fragment of the model $\mathrm{Sh}(M, J)$ needed for the interpretation of **CS** can be carried out completely in $\mathbf{IDB}_1$.

In order to show this, some extra work has to be done; below we indicate the principal points which are involved. First of all, in $\mathbf{IDB}_1$ we have no variables for elements of $\mathrm{Cont}(\mathbb{B}, \mathbb{N})$ and $\mathrm{Cont}(\mathbb{B}, \mathbb{B})$. The solution is to interpret $\mathrm{Cont}(\mathbb{B}, \mathbb{N})$ and $\mathrm{Cont}(\mathbb{B}, \mathbb{B})$ by the neighbourhood functions in K; each $e \in K$ represents a $\Phi_e \in \mathrm{Cont}(\mathbb{B}, \mathbb{N})$ and a $\Psi_e \in \mathrm{Cont}(\mathbb{B}, \mathbb{B})$ as indicated in 4.6.8, 4.6.15. Composition of the Ψ_e's can be represented by a certain primitive recursive operation: such that $\Psi_{e:f} = \Psi_e \circ \Psi_f$. Thus we have a representation of the monoid.

The covers in J can be handled via their characteristic functions (5.3 above); one has to show in $\mathbf{IDB}_1$ that they constitute a covering system.

It will be clear from the results earlier in this section that we can define representations for $\boldsymbol{N}$, $\boldsymbol{B}$, $\boldsymbol{N}^{<N}$, $\boldsymbol{K}_0, (\boldsymbol{K}_0)_{\mathrm{L}}, (\boldsymbol{B})_{\mathrm{L}}$ within $\mathbf{IDB}_1$.

Next we have to show that forcing in the monoid model, for formulas of **CS**, can be expressed in $\mathbf{IDB}_1$. Since the lawlike sheaves are generated by their constant elements, quantification over $\boldsymbol{N}, (\boldsymbol{K}_0)_{\mathrm{L}}, (\boldsymbol{B})_{\mathrm{L}}$ can be expressed by quantification over $\mathbb{N}$, K_0, $\mathbb{B}$, etc. Thus we find, for example,

$$(\Vdash \alpha\hat{n} = \hat{m}) := \forall \beta \in \mathbb{B}(\alpha(\beta)(n) = m).$$

Here α is an element of $\mathrm{Cont}(\mathbb{B}, \mathbb{B})$ (representing an element of $\mathbb{B}$ in the model); within $\mathbf{IDB}_1$, therefore, we can use an element of $K_0 = K$ to represent $\boldsymbol{\alpha}$. That is to say, if we think of $e \in K$ as representing $\boldsymbol{\alpha}$, we have $\Vdash \alpha\hat{n} = \hat{m} := \forall \beta \in \mathbb{B}((e|\beta)(n) = m)$.

The constant element $\boldsymbol{\alpha} := \lambda\beta.\alpha$ (a lawlike element of $\mathbb{B}$ in the model) satisfies

$$(\Vdash \boldsymbol{\alpha}\hat{n} = \hat{m}) \Leftrightarrow (\alpha n = m).$$

Quite generally, functional constants $F \in \mathbb{B}^n \to \mathbb{B}$ are simply interpreted as $F(\boldsymbol{\alpha}_1, \ldots, \boldsymbol{\alpha}_n) := \lambda\beta \in \mathbb{B}.F(\boldsymbol{\alpha}_1(\beta), \ldots, \boldsymbol{\alpha}_n(\beta))$.

Assume now that variables $e_1, \ldots, e_n \in K$ are associated to parameters $\alpha_1, \ldots, \alpha_n$, then to any term $t[\alpha_1, \ldots, \alpha_n]$ of **CS** we can, in an obvious way, associate $t^*_\beta[e_1, \ldots, e_n]$, obtained by replacing each occurrence of α_i in t by $e_i|\beta$. Then

$$\Vdash t[\alpha_1, \ldots, \alpha_n] = s[\alpha_1, \ldots, \alpha_n] \Leftrightarrow$$

$$\forall\beta \in \mathbb{B}\big(t^*_\beta[e_1, \ldots, e_n] = s^*_\beta[e_1, \ldots, e_n]\big).$$

Let us adopt the following convention: if $A[\alpha_1, \ldots, \alpha_n]$ has its choice variables among $\{\alpha_1, \ldots, \alpha_n\}$, we write $\Vdash A[e_1, \ldots, e_n]$ for $\Vdash A[\alpha_1, \ldots, \alpha_n]$; and $\Vdash A \upharpoonright e := \Vdash A[e_1 : e, \ldots, e_n : e]$.

The other forcing clauses may now be expressed as follows:

$$\Vdash A \wedge B := (\Vdash A) \wedge (\Vdash B),$$

$$\Vdash A \vee B := \exists e \in K\, \forall n(en \neq 0 \rightarrow (\Vdash A \upharpoonright h_n) \vee (\Vdash B \upharpoonright h_n)),$$

where h_n is the continuous mapping of 5.2; for a lawlike domain X interpreted as $\mathscr{X}_{\mathrm{L}}$ (e.g. $\mathbb{N}, (\mathbb{B})_{\mathrm{L}}, (\boldsymbol{K}_0)_{\mathrm{L}}$)

$$\Vdash \forall x \in \mathscr{X}_{\mathrm{L}}\, A(x) := \forall x \in X(\Vdash A(\hat{x})).$$

$$\Vdash \forall\alpha\, A(\alpha) := \forall f(\Vdash A(f)),$$

$$\Vdash \exists x \in \mathscr{X}_{\mathrm{L}}\, A(x) := \exists e \in K\, \forall n(en \neq 0 \rightarrow \exists x \in X(\Vdash (A \upharpoonright h_n)(\hat{x}))),$$

$$\Vdash \exists\alpha\, A(\alpha) := \exists e \in K(\Vdash A(e)).$$

The final clause is justified by an appeal to lemma 5.14.

It is not hard to show that for formulas $A(x_1, \ldots, a_1, \ldots, e_1, \ldots)$ of $\mathbf{IDB}_1$ we have

$$(\Vdash A(\hat{x}_1, \ldots, \hat{a}_1, \ldots, \hat{e}_1, \ldots)) \Leftrightarrow A(x_1, \ldots, a_1, \ldots, e_1, \ldots).$$

Finally one obtains the following result.

THEOREM. **CS** *is conservative over* $\mathbf{IDB}_1$. □

6. A site model for LS

In this section we describe the construction of a site model for the theory **LS** of lawless sequences. The model is more complicated in one sense; the underlying category has many objects, but simpler in another sense: there is a very limited supply of morphisms. The covering system is similar. We also adopt the notational conventions of 12.2.14; in particular the dotted quantifiers

$$\dot\forall \varepsilon_1 \ldots \dot\forall \varepsilon_n A(\varepsilon_1,\ldots,\varepsilon_n) := \forall \varepsilon_1 \ldots \varepsilon_n (\#(\varepsilon_1,\ldots,\varepsilon_n) \to A(\varepsilon_1,\ldots,\varepsilon_n)).$$

6.1. *Description of the site.* The underlying category $\mathscr{C}$ has as objects all finite products

$$V_{n_0} \times \cdots \times V_{n_p}$$

of elementary neighbourhoods of Baire space, with $\emptyset$ as initial object (a product of zero factors). Any morphism

$$V_{n_0} \times \cdots \times V_{n_p} \xrightarrow{\varphi} V_{m_0} \times \cdots \times V_{m_q}$$

is specified by an injection φ' from $\{0,\ldots,q\}$ into $\{0,\ldots,p\}$ such that $n_{\varphi'(i)} \succcurlyeq m_i$ for $i \leq q$; φ is then given by

$$\varphi(\alpha_0,\ldots,\alpha_p) := (\alpha_{\varphi'0},\ldots,\alpha_{\varphi'q}).$$

Note that projections (from Cartesian products) are special cases of morphisms, as well as inclusions

$$V_n \hookrightarrow V_m \quad \text{for } n \succcurlyeq m.$$

Below we shall not always notationally distinguish between φ and φ'.

The covering system J on $\mathscr{C}$ is defined in the same way as for the monoid model:

$$S \in J(V) \quad \text{iff } \{\varphi[U] : \varphi \in S\} \text{ covers } V.$$

It should be noted that each inhabited object $U \in \mathrm{Ob}(\mathscr{C})$ is in fact homeomorphic to $\mathbb{B}$ (though the homeomorphism is not necessarily a morphism of the site).

6.2. *Representing* $\mathbb{N}, \mathbb{N}^{<\mathbb{N}}, \mathbb{B}$. Many details are similar to the ones of the monoid model. Thus we take for the interpretation N of $\mathbb{N}$ the sheaf generated by the constant functions $\hat{n}$. More precisely, $\hat{n}$ is at object U

represented by $(\hat{n})_U = \lambda\alpha \in U.n$, and $(\hat{n})_U \upharpoonright f = (\hat{n})_V$ for $V \to_f U$. We mostly drop the subscript $_U$.

LEMMA. $\mathbb{N}$ *is represented by the sheaf* $\boldsymbol{N}$ *with*

$$\boldsymbol{N}(U) = \mathrm{Cont}(U, \mathbb{N})$$

and for $V \to_\varphi U$, $\boldsymbol{n} \in \boldsymbol{N}(U)$ *we have*

$$\boldsymbol{n} \upharpoonright \varphi := \boldsymbol{n} \circ \varphi.$$

PROOF. Left as an exercise. □

Via the standard encoding of finite sequences we find the same result for $\boldsymbol{N}^{<\boldsymbol{N}} = \boldsymbol{N}$.

6.3. LEMMA. $\mathbb{B}$ *is represented by the sheaf* $\boldsymbol{B}$ *with*

$$\boldsymbol{B}(U) = \mathrm{Cont}(U, \mathbb{B})$$

and for $V \to_\varphi U$, $\boldsymbol{\alpha} \in \boldsymbol{B}(U)$

$$\boldsymbol{\alpha} \upharpoonright \varphi = \boldsymbol{\alpha} \circ \varphi.$$

Application is defined locally: *if* $\boldsymbol{\alpha} \in \boldsymbol{B}(U)$, $\boldsymbol{n} \in \boldsymbol{N}(U)$, *then*

$$\mathrm{ev}(\boldsymbol{\alpha}, \boldsymbol{n}) = \lambda\beta \in U.(\boldsymbol{\alpha}(\beta))(\boldsymbol{n}(\beta)) \in \boldsymbol{N}(U),$$

and

$$(U \Vdash \boldsymbol{\alpha n} = \boldsymbol{m}) \Leftrightarrow \forall\beta \in U(\boldsymbol{\alpha}(\beta))(\boldsymbol{n}(\beta)) = \boldsymbol{m}(\beta)).$$

PROOF. $\boldsymbol{B}$ must be an exponential, i.e. an element of $\boldsymbol{B}(U)$ is a collection τ of the form

$$\{\tau_\varphi \in \mathrm{Cont}(U, \mathbb{N}) \to \mathrm{Cont}(V, \mathbb{N}) : V \to_\varphi U\}$$

for which we define, for $V' \to_\psi V$:

$$\tau_\varphi \upharpoonright \psi := \tau_{\varphi \circ \psi},$$

and such that for $\boldsymbol{n} \in \boldsymbol{N}(V) = \mathrm{Cont}(V, \mathbb{N})$

$$\tau_\varphi(\boldsymbol{n}) \upharpoonright \psi = \tau_{\varphi \circ \psi}(\boldsymbol{n} \upharpoonright \psi).$$

τ is now completely determined from $f_\tau \in \mathrm{Cont}(U, \mathbb{B})$ given by

$$(\hat{n} = (\hat{n})_U = \lambda\beta \in U.n, \xi \in U) \quad \text{and}$$

$$f_\tau(\xi)(n) := \tau_1(\hat{n})(\xi),$$

where 1 is the identity $U \to_1 U$; conversely, given $f \in \mathrm{Cont}(V, \mathbb{B})$ we find a τ by taking

$$\tau_\varphi(\boldsymbol{n})(\xi) = (f \circ \varphi(\xi))(\boldsymbol{n}(\xi)). \quad \square$$

6.4. *Lawlike sheaves* $(\boldsymbol{B})_{\mathrm{L}}, (\boldsymbol{B}^{\boldsymbol{B}})_{\mathrm{L}}$. As before, we can define, for any sheaf $\mathscr{X}$, the subsheaf $(\mathscr{X})_{\mathrm{L}}$ as the sheaf generated by those elements of $\mathscr{X}$ which are invariant under restrictions. In particular, we can prove the following lemma.

LEMMA. *The elements of* $\boldsymbol{B}^{\boldsymbol{B}}$ *which are invariant under restrictions are represented by functions* $F \in \mathrm{Cont}(\mathbb{B}, \mathbb{B})$, *such that for* $\alpha \in \mathrm{Cont}(U, \mathbb{B})$, *we have* $\mathrm{ev}(F, \alpha) := F \circ \alpha$ *in the model.*

PROOF. Exercise. □

LEMMA. *The internal lawlike neighbourhood functions* $(\boldsymbol{K}_0)_{\mathrm{L}}$ *are generated by internal neighbourhood functions invariant under restrictions, and those are represented by the elements of* K_0; *if* $e \in K_0$, $\boldsymbol{n} \in \boldsymbol{N}(U)$, *then* $\mathrm{ev}(e, \boldsymbol{n}) = \lambda\xi \in U.e(\boldsymbol{n}(\xi))$ *in the model.*

PROOF. Exercise. □

6.5. *Interpretation of* LS. The universe of lawless sequences LS is interpreted by (the sheaf generated by) the presheaf $(\boldsymbol{B})_{\mathrm{LS}}$ of $\boldsymbol{B}$, consisting of the projections π_i with $\pi_i(\xi_1, \ldots, \xi_n) = \xi_i$, so

$$(\boldsymbol{B})_{\mathrm{LS}}(U) = \mathscr{C}(U, \mathbb{B}).$$

Note that the restriction of a projection is again a projection. From now on we let $\boldsymbol{\varepsilon}, \boldsymbol{\eta}$ range over elements of $(\boldsymbol{B})_{\mathrm{LS}}$.

6.6. PROPOSITION. $U \Vdash \forall \varepsilon\eta(\varepsilon = \eta \vee \neg\varepsilon = \eta)$.

PROOF. Suppose at U $\boldsymbol{\varepsilon} = \pi_i$, $\boldsymbol{\eta} = \pi_j$. If $i = j$, then $U \Vdash \boldsymbol{\varepsilon} = \boldsymbol{\eta}$. But if $i \neq j$, then $U \Vdash \neg\boldsymbol{\varepsilon} = \boldsymbol{\eta}$; for suppose $V \to_\varphi U$ to be a morphism in $\mathscr{C}$, and

$$U = V_{n_0} \times \cdots \times V_{n_p}, \qquad V = V_{m_0} \times \cdots \times V_{m_q},$$

and φ coming from an injection $\varphi' \in \{0, \ldots, p\} \to \{0, \ldots, q\}$, then $\pi_i \upharpoonright \varphi = \pi_{\varphi'(i)}$, $\pi_j \upharpoonright \varphi = \pi_{\varphi'(j)}$, and $\varphi'(i) \neq \varphi'(j)$, therefore also $V \nVdash \pi_i \upharpoonright \varphi = \pi_j \upharpoonright \varphi$. □

6.7. PROPOSITION. $U \Vdash \forall n \exists \varepsilon(\varepsilon \in n)$.

Proof. It suffices to verify for constant $\hat{n}$ that $U \Vdash \exists\varepsilon(\varepsilon \in \hat{n})$. The obvious projection

$$U \times V_n \to_\pi U$$

is a cover of $U \equiv V_{n_0} \times \cdots \times V_{n_{p-1}}$; clearly $U \times V_n \Vdash \pi_p \in \hat{n}$, hence $U \Vdash \exists\varepsilon(\varepsilon \in \hat{n})$. $\square$

6.8. Proposition. *The axiom schema of open data is valid in the model.*

Proof. We have to show, for any $U = V_{u_1} \times \cdots \times V_{u_p}$ and $\boldsymbol{\varepsilon}_1, \ldots, \boldsymbol{\varepsilon}_m$, that

$$U \Vdash A(\boldsymbol{\varepsilon}_1, \ldots, \boldsymbol{\varepsilon}_m) \wedge \#(\boldsymbol{\varepsilon}_1, \ldots, \boldsymbol{\varepsilon}_m) \to$$
$$\exists n_1 \ldots \exists n_m(\boldsymbol{\varepsilon}_1 \in n_1 \wedge \cdots \wedge \boldsymbol{\varepsilon}_m \in n_m$$
$$\wedge \dot{\forall}\eta_1 \in n_1, \ldots, \dot{\forall}\eta_m \in n_m A(\eta_1, \ldots, \eta_m)).$$

It actually suffices to show that if the premiss is forced at U, then so is the conclusion, for arbitrary A with all parameters lawlike. Suppose

$$U \Vdash A(\boldsymbol{\varepsilon}_1, \ldots, \boldsymbol{\varepsilon}_m) \wedge \#(\boldsymbol{\varepsilon}_1, \ldots, \boldsymbol{\varepsilon}_m).$$

Then, since $U \Vdash \#(\boldsymbol{\varepsilon}_1, \ldots, \boldsymbol{\varepsilon}_m)$, $\boldsymbol{\varepsilon}_1, \ldots, \boldsymbol{\varepsilon}_m$ are represented by distinct projections $\pi_{i_1}, \ldots, \pi_{i_m}$ in $\mathscr{C}(U, \mathbb{B})$, and $m \le p$. Take $\boldsymbol{n}_k = \hat{u}_{i_k}$, then

$$U \Vdash \pi_{i_k} \in \boldsymbol{n}_k, \quad 1 \le k \le m.$$

For the second half, suppose $W \to_\varphi U$ to be a morphism, $W = V_{w_1} \times \cdots \times V_{w_q}$, and let $\pi_{j_1}, \ldots, \pi_{j_m}$ be m distinct projections at W such that

$$W \Vdash \pi_{j_k} \in \boldsymbol{n}_k, \quad 1 \le k \le m; \tag{1}$$

it remains to show that

$$W \Vdash A(\pi_{i_1}, \ldots, \pi_{i_m}).$$

By (1), w_{j_k} extends n_k. Suppose $\{1, \ldots, p\} \setminus \{i_1, \ldots, i_m\} = \{i'_1, \ldots, i'_{p-m}\}$, and let for $t(k) = u_{i'_k}$

$$W' := W \times V_{t(1)} \times \cdots \times V_{t(p-m)},$$

a product of $q + p - m$ factors. The injection $\psi \in \{1, \ldots, p\} \to \{q + p - m\}$ given by

$$\psi(i_k) = j_k \quad (i \le k \le n),$$
$$\psi(i'_r) = q + i'_r \quad (1 \le r \le p - m),$$

induces a morphism $W' \to_\varphi U$, since $w_{j_k} \succcurlyeq n_k = u_{i_k}$. Thus from $U \Vdash A(\pi_{i_1}, \ldots, \pi_{i_m})$ we obtain by restriction

$$W' \Vdash A(\pi_{i_1} \upharpoonright \psi, \ldots, \pi_{i_m} \upharpoonright \psi), \quad \text{i.e.}$$

$$W' \Vdash A(\pi_{j_1}, \ldots, \pi_{j_m}).$$

On the other hand, the obvious projection $W' \to_\pi W$ is a cover, so

$$W \Vdash A(\pi_{j_1}, \ldots, \pi_{j_m}). \quad \square$$

6.9. PROPOSITION. *The principle of continuous choice holds in the model:*

$$U \Vdash \dot{\forall} \varepsilon_1 \ldots \dot{\forall} \varepsilon_m \exists n \in \mathbb{N} A(\varepsilon_1, \ldots, \varepsilon_m, n) \to$$
$$\exists F \in (\boldsymbol{B}^m \to \boldsymbol{N})_{\mathrm{L}} \dot{\forall} \varepsilon_1 \ldots \dot{\forall} \varepsilon_m A(\varepsilon_1, \ldots, \varepsilon_m, F(\varepsilon_1, \ldots, \varepsilon_m)).$$

PROOF. Suppose U forces the premiss, and let $\mathbb{B}^m \times U \to_\varphi U$ be a projection. At $\mathbb{B}^m \times U$ we take $\pi_1, \ldots, \pi_m$ for $\varepsilon_1, \ldots, \varepsilon_m$, then

$$\mathbb{B}^m \times U \Vdash \exists n A(\varepsilon_1, \ldots, \varepsilon_m, n).$$

Therefore there is a cover $\{\varphi_i \in \mathscr{C}(W_i, \mathbb{B}^m \times U) : i \in I\}$ with elements $\boldsymbol{n}_i$ such that

$$W_i \Vdash A(\pi_1 \upharpoonright \varphi_i, \ldots, \pi_n \upharpoonright \varphi_i, \boldsymbol{n}_i);$$

without loss of generality we may suppose $\boldsymbol{n}_i = \hat{n}_i$. Therefore also for $W_i' = \varphi_i[W_i] \subset \mathbb{B}^m \times U$

$$W_i' \Vdash A(\pi_1, \ldots, \pi_m, \hat{n}_i).$$

Without loss of generality we may assume the W_i' to be disjoint; as a result this defines an $F \in \mathrm{Cont}(\mathbb{B}^m \times U, \mathbb{N})$ by

$$F(\alpha) = n_i \quad \text{iff } \alpha \in W_i.$$

F can be chosen so as to factorize as

$$\mathbb{B}^m \times U \xrightarrow{\psi} \mathbb{B}^n \xrightarrow{F'} \mathbb{N}$$

with $F = F' \circ \psi$.

To see this, observe that for any open $U' \subset \mathbb{B}^m \times U$, say $U' = V_{u_1} \times \cdots \times V_{u_m} \times U''$ with U'' open, $U'' \subset U$, and $U' \Vdash A(\pi_1, \ldots, \pi_m, \hat{n})$, we also have

$$V_{u_1} \times \cdots \times V_{u_m} \times U \Vdash A(\pi_1, \ldots, \pi_m, \hat{n}) \tag{1}$$

since, obviously, by restriction along the projection χ

$$V_{u_1} \times \cdots \times V_{u_m} \times U'' \times U' = U' \times U \xrightarrow{\chi} U'$$

we find $U' \times U \Vdash A(\pi_1, \ldots, \pi_m, \hat{n})$, while

$$U' \times U \xrightarrow{\chi'} V_{u_1} \times \cdots \times V_{u_m} \times U,$$

given by $\chi'(\xi_1, \ldots, \xi_m, \chi_1, \chi_2) = (\xi_1, \ldots, \xi_m, \chi_2)$, covers $V_{u_1} \times \cdots \times V_{u_m} \times U$, hence (1). Thus we can assume the W_i' to be of the form $W_i'' \times U$ and from this we find F'. Now

$$\mathbb{B}^m \times U \Vdash A(\pi_1, \ldots, \pi_m, F'(\pi_1, \ldots, \pi_m)).$$

From this in turn we can straightforwardly show

$$U \Vdash \dot{\forall} \varepsilon_1 \ldots \dot{\forall} \varepsilon_m A(\varepsilon_1, \ldots, \varepsilon_m, F'(\varepsilon_1, \ldots, \varepsilon_m)). \quad \square$$

6.10. THEOREM. Sh($\mathscr{C}$, J) *is a model for* **LS**.

PROOF. Immediate from 6.6–9, combined with the observation that if $K_0 \dot{=} K$ externally, then within the model $(\boldsymbol{K}_0)_{\mathrm{L}} = (\boldsymbol{K})_{\mathrm{L}}$, that is to say K1–3 can be proved to hold for $(\boldsymbol{K}_0)_{\mathrm{L}}$ internally. □

6.11. *Formalization of the metamathematics of the model* can be carried out along the same lines as in 5.21, yielding the following theorem.

THEOREM. **LS** *is conservative over* $\mathbf{IDB}_1$. □

7. Notes

7.1. In our exposition in sections 1–3 we have in particular drawn upon Grayson (1980, 1981, 1984) and Fourman and Scott (1979).

7.2. *First applications to intuitionistic analysis.* The first to apply a particular topological model to the study of intuitionistic real analysis was D.S. Scott (1968, 1970A). Scott's model used only total elements, but may be regarded as a sheaf model "avant la lettre". In particular Scott (1970A) proved the continuity of all real-valued functions in his model (3.24 for the case $T \equiv \mathbb{B}$). J.R. Moschovakis (1973) gave a similar model for $\mathbb{B}$, making a direct comparison with the treatment of intuitionistic analysis in Kleene and Vesley (1965) possible. Van Dalen (1978) similarly treated various theories of choice sequences in terms of Beth models.

Krol' (1978) constructed a topological model invariant under a suitable group of permutations ("permutation model") in which elementary analysis plus KS, C-N, RDC, and BI_M are satisfied. This work was extended to **IZF** by Scedrov (1981).

7.3. *Axiom of choice, compactness of the interval, continuity of real-valued functions.* Already right at the beginning of the development of "categorical logic", it was realized that the axiom of choice was not generally valid in a topos (e.g. Lawvere 1971). It was Diaconescu (1975) who showed that the general validity of AC in a topos was equivalent to the topos being Boolean (i.e. obeying classical logic); Goodman and Myhill (1978) gave a simple logical version of this result (cf. E4.2.1). The more special failure of the axiom of countable choice implicit in $\mathbb{R}^c \neq \mathbb{R}^d$ is mentioned, for example, in Lawvere (1976, p. 117). The proofs given in section 2 follow Fourman and Hyland (1979).

The noncompactness of $[0,1]$ over $K([0,1])^*$ is proved in the same paper; in the proofs of the lemmas 3.12–16 we follow Fourman and Scott (1979). Generalizations of 3.2 and 3.4 may be found, e.g., in Fourman and Scott (1979, 8.13, 8.17, 8.23).

Fourman and Hyland's argument for the underivability of compactness of $[0,1]$ was extended to set theory by Scedrov (1982, 1984).

The proof of 3.24, which generalizes the main result of D.S. Scott (1970A), is due to Grayson (1981). Other generalizations of Scott's result are given in Hyland (1979).

7.4. *The continuity rule.* The proof of the continuity rule in section 4 is due to Joyal (Grayson 1984). Hayashi (1980) established the same rule by proof-theoretic methods, using normalization in natural deduction calculi for **HAH**. Joyal's proof is reminiscent of glueing techniques (cf. 13.9.7).

7.5. *Models for* **CS**. Section 5 is based on Van der Hoeven and Moerdijk (1984). The monoid model for **CS** was independently discovered by Fourman and by Grayson. Van der Hoeven and Moerdijk (1984, section 3) observed that interpretation in the monoid model is in fact a model-theoretic version of the elimination translation (12.4.8); this fact was also noted by Grayson.

7.6. *Models for* **LS**. Van der Hoeven and Moerdijk (1984, subsection 5.2), inspired by Fourman's talk at the Brouwer Centenary Symposium (considerably different from the published paper) also gave a monoid model for the theory of lawless sequences **LS**. Later Moerdijk recast it in the form of a model over a site as presented in section 6; our exposition has made use of handwritten notes by Moerdijk. An earlier model for **LS**, based on Cohen forcing, was presented in Van Dalen (1978; see also the appendix to the paper).

7.7. Further work along the lines of sections 5 and 6 may be found in Van der Hoeven and Moerdijk (1983, 1984A, 1984B; cf. also Fourman 1982, 1984).

The soundness theorem for sheaf models requires full comprehension externally in order to show the validity of full comprehension internally. Grayson (1983) shows how to avoid the use of full comprehension, thereby permitting the application of sheaf-model techniques to theories much weaker than **HAH**.

Of the many other interesting results obtained by sheaf-model techniques we mention only a single example, namely the independence of FAN from the Heine–Borel theorem, i.e. the compactness of $[0,1]$ (Moerdijk 1984 using techniques beyond the scope of this book).

Exercises

15.1.1. Check the correctness of the explicit definition of simple sheaves in 1.1.

15.1.2. If $\mathscr{A}$ is a simple sheaf, then so are $\mathscr{A}^*$ and $\mathscr{P}_{\mathrm{fin}}(\mathscr{A})$ (cf. 1.2).

15.1.3. Complete the proof of 1.5.

15.1.4. Carry out the proof of 1.5 for the definition of $\mathbf{R}^{\mathrm{d}}$ as the collection of (located) left cuts (as in chapter 5).

15.1.5. Let T be a topological space. A partial function f from T to $\mathbf{R}$ is *lower semi-continuous* (*lsc*) if $f^{-1}[(x,\infty)]$ is open for all $x \in \mathbf{R}$, i.e. $\varphi(x) = \{y : x < f(y)\}$ is open for all x. If $\{f_i : i \in I\}$ is a set of lsc functions, then so is $g(x) := \sup\{f_i(x) : i \in I\}$ with $\mathrm{dom}(g) = \bigcup\{\mathrm{dom}(f_i) : i \in I\}$.

A *pre-cut* is defined as a weak left cut except that strong monotonicity is weakened to monotonicity

$$r \in S \wedge r' < r \to r' \in S.$$

Show that the pre-cuts in $\mathrm{Sh}(\mathscr{O}(\mathrm{T}))$ correspond to lsc functions from T to $\mathbf{R}$ (classical reasoning permitted).

15.1.6. Let f and g be lsc functions (cf. the preceding exercise) from T to $\mathbf{R}$. Define $[\![f < g]\!] := \mathrm{Int}\{t : f(t) < g(t)\}$, $[\![g \le f]\!] := \mathrm{Int}([\![f > g]\!]^{\mathrm{c}} \cap Ef \cap Eg)$, $[\![f = g]\!] := [\![f \le g \wedge$

$g \leq f]\!]$. If $f^\circ(x) := \sup\{h(x) : x \in [\![f = h]\!],\ h$ an lsc function from T to $\mathbb{R}\}$, then $[\![f^\circ = f]\!] = \mathrm{dom}(f)$ and f° corresponds internally in $\mathrm{Sh}(\mathcal{O}(T))$ to a unique element of $\mathbb{R}^{\mathrm{be}}$. Prove this (classical reasoning permitted; Troelstra 1980).

15.2.1. Show that the schema of dependent choices DC holds in $\mathrm{Sh}(\mathbb{B})$.

15.2.2. $p \in \Omega$ is said to be *countably compact*, if for each $\langle q_n \rangle_n$ such that $\bigvee\{q_n : n \in \mathbb{N}\} \geq p$ there is an m such that $q_0 \vee \cdots \vee q_m \geq p$. Ω is *countably compact*, iff $\top_\Omega$ is countably compact. Ω is *locally countably compact iff* $\top_\Omega = \bigvee\{p_i : i \in I\}$, p_i countably compact. Show that if Ω is locally countably compact, then BI holds in $\mathrm{Sh}(\Omega)$ (*Hint*: use BI externally.)

15.2.3. $r \in \Omega$ is *weakly connected* iff $\forall p,q \in \Omega((p \wedge q) = \bot \wedge (p \vee q \geq r) \rightarrow (p \geq r) \vee (q \geq r))$. Ω is *locally weakly connected* if $\top_\Omega = \bigvee\{p_i : i \in I\}$ with each p_i weakly connected. Show that if Ω is locally weakly connected, then $\mathrm{BI_D}$ holds over $\mathrm{Sh}(\Omega)$, assuming $\mathrm{BI_D}$ externally. *Hint*. Let $\mathrm{BI_D}(P,Q)$ be an instance of $\mathrm{BI_D}$; without essential loss of generality suppose that the value of the hypothesis of $\mathrm{BI_D}(P,Q)$ is $\top_\Omega$. Apply $\mathrm{BI_D}$ externally to $P^*n := ([\![P\bar{n}]\!] = \top_\Omega)$, $Q^*n := ([\![Q\bar{n}]\!] = \top_\Omega)$ (Fourman and Hyland 1979, 3.7).

15.2.4. Show that BI holds in $\mathrm{Sh}(\mathbb{B})$, assuming BI externally. *Hint*. Use the schema of double bar induction DBI (cf. E4.8.13). Let $\mathrm{BI}(R)$ be an instance of BI, and assume, without loss of generality, the premisses to have value $\mathbb{B}$. Apply DBI to $P(n,m) := \forall\alpha\beta\exists nm([\![R(\bar{\alpha}n)]\!] \supset V_{\bar{\beta}m})$ (Fourman and Hyland 1979, 3.5, 3.6).

15.2.5. Construct a space T such that $\neg\mathrm{BI_D}(P,Q)$ holds for a suitable choice of P, Q. *Hint*. Topologize $\mathbb{N}^{<\mathbb{N}}$ by $U \in \mathcal{O}(T) \Leftrightarrow \forall n \in U \exists m \forall k > m(n * \langle k \rangle \in U)$, and take $[\![P(\bar{n})]\!] = [\![Q(\bar{n})]\!] := \mathbb{N}^{<\mathbb{N}} \setminus \{m : m \succcurlyeq n\}$ (Grayson; Fourman and Hyland 1979, 3.8).

15.2.6. Show that in $\mathrm{Sh}(\mathcal{O}(T))$, where T is locally weakly connected (cf. E15.2.3), for simple sheaves $(\hat{A})^{\mathrm{sh}}$, $(\hat{B})^{\mathrm{sh}}$ the exponential is isomorphic to $((A^B)^\wedge)^{\mathrm{sh}}$.

15.3.1. Give a detailed proof of 3.6.

15.3.2. Prove 3.7.

15.3.3. Show that, classically, the coperfect sets are just the complements of perfect (closed) sets.

15.3.4. Prove 3.15.

15.3.5. Prove the proposition in 3.17.

15.3.6. Show that the Φ_n in 3.19 are internally open.

15.3.7. Show that the cHa $K(\mathbb{R}^2)$ is locally weakly connected, and that each open rectangle in $K(\mathbb{R}^2)$ is weakly connected.

15.3.8. Show that $\neg(\mathcal{O}(\mathbb{R}^{\mathrm{d}}) \cong \mathcal{O}(\mathbb{R}^{\mathrm{c}}))$ over $K(\mathbb{R}^2)$ (i.e. the Dedekind reals and the Cauchy reals have different topologies). *Hint*: Use E15.3.7 (Grayson 1981, 5.3).

15.3.9. Show that separatedness of X does not in general entail internal separatedness of X_T. *Hint:* take T to be $\mathbb{C}$, and let $X = \mathbb{Q} \cup \{*\}$ with a basis for $\mathbb{Q}$ together with all cofinite sets containing $*$ as neighbourhood basis.

15.3.10. A (separated) space X is *strongly Hausdorff* iff $\forall x, y \in X \forall U \in \mathcal{O}(X)[x \in U \to (y \in U \vee \exists V, W \in \mathcal{O}(X)(x \in V \wedge y \in W \wedge V \cap W = \emptyset))]$, and X is *regular* iff $\forall U \forall x \in U \exists V(X = (\mathrm{Int}(V) \cup U^c) \wedge (x \in V))$. Show that the properties of being strongly Hausdorff and of being regular are preserved, i.e. if X possesses them, so does X_T (Grayson 1981, 4.2).

15.3.11. Show that compact, strongly Hausdorff implies *normality*;

$$\forall UV[X = U \cup V \to \exists W(X = U \cup W = (V \cup \neg W)]$$

($\neg W = \mathrm{Int}(W^c)$). *Hint.* Show that strongly Hausdorff and compact implies regularity. Combine this with the following lemma to prove normality.

LEMMA. *X is normal iff every finite open cover has a finite strong refinement* (*Grayson* 1981, 6.2, *proposition* 2), *where, for covers $\mathcal{U}, \mathcal{V}$, we say that $\mathcal{V}$ strongly refines $\mathcal{U}$ if $\forall V \in \mathcal{V} \exists U \in \mathcal{V}$ $(Int(V^c) \cup U = X)$.*

15.3.12. Show that $[0,1]$ is not normal over $K([0,1])^*$, and that $\mathbb{R}$ is not normal over $K(\mathbb{R})^*$ (Grayson 1981, 6.2, proposition 2).

15.3.13. Give a simplified version of the proof of 3.24 for $T \equiv$ Baire space or $T \equiv \mathbb{R}$ (Scott 1970A).

15.3.14. Show that the cHa $K(X)$ of coperfect opens of a separated space X is *pointless*, i.e. there are no mappings from $K(X)$ to $\mathrm{P}(\{0\})$ preserving $\wedge, \mathrm{V}, \top$ (cf. 7.6.7; Fourman and Scott 1979, 3.5(ii)).

15.3.15. Show, assuming classical logic, that Kripke's schema KS (4.9.3) holds in $\mathrm{Sh}(\mathcal{O}(\mathbb{B}))$ (J.R. Moschovakis 1973).

15.4.1. Show in some detail that $\mathrm{Sh}(\mathcal{O}(T)) \Vdash A$ can be expressed by a formula of **HAH**, provided T is definable in **HAH**. Show also that the soundness theorem for **HAH** can be formalized (end of 4.2).

15.4.2. Prove lemma 4.5.

15.5.1. Show that the definition given for an exponential in 5.1 is a specialization of the earlier general definition of exponentials over sites (14.5.14).

15.5.2. Show that the construction of $f_S \in K_0$ from $S \in J$, and of S_f from $f \in K_0$ are inverse to each other (5.3).

15.5.3. Complete the proof of lemmas 5.4 and 5.5.

15.5.4. Prove lemma 5.9.

15.5.5. Complete the proof of 5.11.

15.5.6. Prove facts (i) and (ii) in 5.15.

15.5.7. Give a detailed proof of 5.19.

15.5.8. Prove in $\mathbf{IDB}_1$ that the covers in J, given by their characteristic functions as in 5.3, form a covering system.

15.5.9. Show in $\mathbf{IDB}_1$ that for all terms of the language of **CS**

$$\mathbf{IDB}_1 \vdash (\Vdash \forall\alpha(t[\alpha] = s[\alpha])) \leftrightarrow \forall a(t[a] = s[a]).$$

15.5.10. Show, within $\mathbf{IDB}_1$, that the formulas of $\mathbf{IDB}_1$ are preserved under forcing (end of 5.21).

15.6.1. Prove the lemma in 6.2.

15.6.2. Prove the lemma in 6.4.

15.6.3. Investigate the connection between the elimination translation for **LS** in 12.3.9 and the forcing definition for the model $\mathrm{Sh}(\mathscr{C}, J)$.

CHAPTER 16

EPILOGUE

In this chapter we present, tentatively, some general views on the foundations of constructive mathematics against the background of the technical developments in the earlier chapters; we also discuss some controversial issues. The discussion is far from exhaustive. In particular we refrain from any attempt to survey the extensive literature on the foundational and philosophical aspects of constructivism.

In section 1 we discuss the role of language and informal rigour; section 2 discusses the semantics of intuitionistic logic and formalization of intuitionistic and constructive mathematics from a slightly more technical side.

Section 3 reviews the theory of the idealized mathematician, and section 4 presents a summary and discussion of Dummett's argument that intuitionistic logic is better suited to an adequate "theory of meaning".

1. The role of language and "informal rigour"

1.1. L.E.J. Brouwer regarded intuitionistic mathematics as the only legitimate form of mathematics, that is to say, the only kind of mathematics which subjectively could be described as "certain". Mathematics, in his view, consists of mental constructions, created by the individual in free action, independent of experience. Language is considered as a secondary phenomenon, e.g. in Brouwer (1907, p. 127):

"The words of your mathematical demonstration merely accompany a mathematical *construction* that is effected without words".

Even in logic and mathematics, Brouwer says, "no two persons will think the same thing in case of the fundamental notions" (Brouwer 1905, p. 37).

The key problem for mathematics in relation to language is that of exactness. The following passage from Brouwer (1933) succinctly summarizes Brouwer's view:

> "Now, if on the basis of rational reflection the exactness of mathematics, in the sense of exclusion of error and misunderstand-

> ing, cannot be assured by any linguistic means, the question arises whether this assurance can come forth by any other means. The answer to this question is that the languageless constructions originating by the self-unfolding of the primordial intuition are, by virtue of their presence in memory alone, exact and correct; that the human power of memory, however, which has to survey these constructions, even when it summons the assistance of linguistic signs, by its very nature is limited and fallible. For a human mind equipped with an unlimited memory, a pure mathematics which is practised in solitude and without the use of linguistic signs would be exact; this exactness, however, would again be lost in an exchange *between* human beings with unlimited memory, since they remain committed to language as a means of communication".

Thus, the certainty of Brouwerian intuitionistic mathematics is *a certainty for the ideal mathematician only*. Intersubjective certainty is a notion which does not even make sense on Brouwer's solipsist view. The certainty of intuitionist mathematics, firmly embedded in Brouwer's solipsist philosophy, is therefore an idealization that one has to accept or reject.

1.2. Given the fact that the view of intuitionistic mathematics as a languageless mental activity undeniably is a (strong) idealization of the actual practice, we have to recognize that intuitionism can only be a *theory* about our mathematical experience, or, to borrow a phrase from Bernays (1970), it corresponds at best in a schematic way to our experience. And since we ourselves are already quite soon forced to use symbol manipulation in our mathematics, it may well be that we have no proper understanding what mathematics according to Brouwer's principles really looks like, since the divergence between mathematics conforming to the theoretical postulate of *languageless mental activity* and our actual (intuitionistic) practice is too great.

The idealization, i.e. the *theoretical* aspect is met at a very early stage, namely in the picture we form of the natural numbers: all constructivists, except the ultra-finitists, treat all natural numbers as objects "of the same kind", even if some of them have been introduced by powerful operations such as exponentiation.

There are considerable difficulties to be overcome in developing the ultra-finitist point of view into a viable and coherent theory; certainly we have much less difficulty managing the idealized concept of the natural numbers, even though it is a highly sophisticated one. Nevertheless the ultra-finitist critique should be taken seriously, the conclusion being not so much that the ultra-finitist approach is the "correct" one, as well as that, right from the beginning, the intuitionist picture of mathematics diverges from our actual experience, and that language and more abstract notions (such as exponentiation) cannot be avoided even in developing the theory of $\mathbb{N}$.

The use of hypothetical reasoning is another aspect of intuitionistic mathematical reasoning, which presents itself early on, and seems in practice to make the use of language inescapable (although we do not regard hypothetical reasoning without language as impossible in principle); cf. the discussion in 1.5.2.

1.3. Since the *perfect* introspection postulated for Brouwer's ideal mathematician is not accessible to us, the question arises: what takes its place? The answer seems to be: informal, rigorous concept analysis ("informal rigour" in the terminology of Kreisel (1967)). In sections 12.1 and 12.2 we had an opportunity to see informal rigour at work in formulating the axioms for lawless sequences. We expand a bit on the notion of informal rigour in the following points (a)–(e), partly a recapitulation of earlier remarks in section 12.1. "Informal rigour" as explained here is akin to a phenomenological analysis as expounded, for example, in Tragesser (1977), but we hesitate to equate the two approaches.

(a) Informal rigour is not an absolute notion; today's rigour may be tomorrow's vagueness. Nevertheless it makes sense to argue about degrees of informal rigour: one analysis may be regarded as more rigorous than another.

(b) Degrees of informal rigour may vary from barely plausible to completely convincing. Heyting (1958A), in reaction to Griss' criticism of the use of negation in intuitionistic mathematics, distinguishes between degrees of intuititive evidence in intuitionistic mathematics (cf. 1.5.2).

(c) Nearly always we have to take certain intuitive "jumps" or "leaps", if we arrive at a point where we do not know how to push our informal concept analysis any further. We can only try to make the jumps as small as possible.

(d) Concept analysis may reach a point of diminishing returns, in particular if we do not know how to incorporate the results of a more refined analysis in a formal mathematical theory.

(e) We are practically always left with a margin of uncertainty whether our formal rendering of the results of an informal conceptual analysis is adequate, precisely because of the "hazy edges" of informal notions.

The use of informal rigour is not restricted to intuitionistic and constructive mathematics. It plays a role in mathematics in general; see for example the history of the notion of a polyhedron (Lakatos 1976), and the development of a notion of area of pointsets in the plane, starting from a few simple intuitively clear properties (finite additivity, monotonicity, agreement with the standard definition of area for a rectangle or a triangle).

Neither is rigorous analysis of informal concepts the only source of mathematics; for example, analogy, generalization, and the development of theories prompted by external motivations (e.g. from physics) play a role as well. Nevertheless, for intuitionistic mathematics in the tradition of Brouwer informal rigour is the principal, perhaps the only source of mathematical knowledge. (The use of generalization or analogy sometimes appears as a limiting case of the use of informal rigour, with mere plausibility as a motivation.)

Historically, the informal concept fitting a theory does not necessarily precede the theory, it may also be discovered afterwards as a result of the mathematical experience gained with the theory (cf. again the history of the polyhedron-concept; possibly the history of the concept of set is another example, see e.g. Hallett (1984)).

The theory of the creative subject, to be discussed in section 3, is an example of an aspect of Brouwerian intuitionism which is in an unsatisfactory state as to concept analysis.

The admissibility of impredicative comprehension and "full" power-set is another case which is in an unsatisfactory state from the viewpoint of informal rigour. On a liberal approach one might defend impredicative comprehension as follows: we have a good grasp of what it means to be a subset of some already understood set; hence we understand quantification over arbitrary subsets, and we can define sets by means of such quantification. It is not easy to give a convincing argument against this defense, but on the other hand there is, we think, no doubt that a more predicative approach is, at least in principle, conceptually clearer. (In passing we note that the assumption of quantification over functions, together with KS, or more generally a parametrization principle, automatically justifies impredicative set comprehension, cf. section 4.9 and E4.9.5.) Hence we think it only prudent to avoid full impredicative comprehension where possible.

It is of historical interest to point out that the predicative-impredicative controversy was ignored by Brouwer and traditional intuitionism.

1.4. *The role of language.* As argued above, we cannot develop constructive mathematics, as we know it, without the help of language. Because of our limitations, formalization of our concept analysis serves as a check on the coherence and viability of our informal analysis.

Finite combinatorial manipulations represent a stable intersubjective basis for communication; in practice (i.e. putting aside the vagaries of "private language") we are able to communicate the rules for such manipulations without fear of being misunderstood. Deductions in explicitly presented axiom systems in a specified formal language are of a finite combinatorial nature; the formal rendering of the results of our informal analysis thus carries them into the intersubjective domain.

However, it seems arbitrary to restrict mathematics, in a narrow formalist sense, to this intersubjective distillate, since this leaves unexplained how we arrive at (and even use) our formalisms. We certainly tend to have more confidence in a formalism if we have a better intuitive picture of the informal notions (partially) described by the formalism. On the other hand, developing formal consequences may shape or modify our informal picture; sometimes we are led to adapt our informal notions to the properties of the formalism.

2. Intuitionistic logic, formalisms, and equality

The present section contains some loosely connected remarks on the two basic types of semantics for intuitionistic logic (2.1–4), the merits of different formalizations (2.5–6), proof-theoretic strength and ontological commitments (2.7), freedom in the interpretation of logic (2.6), equality and choice principles (2.9–10).

2.1. *Two types of semantics.* Two distinct semantical ideas dominate the subject of intuitionistic logic.

The first idea is the one embodied in the BHK-interpretation; formal versions of this type of semantics are for example Kleene's realizability, the propositions-as-types idea as elaborated in constructive type theory, abstract realizability in **APP** (cf. section 9.5). In combination with restrictive assumptions about the universe of operations, in particular versions of Church's thesis, a conflict with classical logic results.

The second idea is represented by the introduction of choice sequences. Here we encounter not a restrictive assumption on the universe of mathematical objects, but rather the demand that our logic should also be applicable to a new kind of objects, namely incompletely given sequences.

The simplest illustration is provided by the universe of lawless sequences. As we have seen in 13.1.10, being uniformly valid in a lawless parameter corresponds in a rather straightforward way to validity in a Beth model, a notion which can be understood without reference to lawless sequences. Beth models on the other hand are nothing but topological models over (a subspace of) Baire space. The generalization of this semantical idea ultimately leads to sheaf models over a site. Though there exist semantics of sufficient generality so as to cover both realizability interpretations and sheaf models over a site, these generalizations are perhaps too general to be very informative on the intuitive level (notwithstanding the technical usefulness of such generalizations), in other words there is at present no direct connection on the conceptual level between the two "trends" in the semantics of intuitionistic logic. This is not to say that the two ideas cannot peacably coexist in the interpretation of a single intuitionistic formal system. There are at least two instances of such a coexistence.

2.2. The first instance of such a coexistence is offered by the theory of **LS** discussed in chapter 12. The choice-free part of **LS** is the theory $\mathbf{IDB}_1$, which, for example, can be interpreted in constructive type theory and thus has a realizability-type semantics, and is also correct for numerical realizability.

The addition of lawless variables may be thought of as the introduction of sequences of numbers from "outside", i.e. we have no information on the working of the process and we cannot influence the outcome. The elimination process shows how we can interpret quantification over such sequences relative to $\mathbf{IDB}_1$ (in $\mathbf{IDB}_1$ we can control the universe of lawless sequences as a whole, but not an individual lawless sequence).

2.3. The second example of coexistence of ideas derived from realizability and choice sequences is provided by Kleene's realizability by functions. This is defined in the same way as numerical realizability, except that now the realizing objects are functions, and partial recursive function application is replaced by partial continuous function application. Thus for example

$$\beta \mathbf{r} \forall\alpha A(\alpha) := \forall\alpha(E(\beta|\alpha) \wedge (\beta|\alpha) \mathbf{r} A(\alpha)).$$

This interpretation is essentially different from the elimination translation for **CS** as described in section 12.3; function realizability validates in elementary analysis the schema GC (9.6.9). To see the difference, note that function realizability reduces $\mathbf{EL} + \mathrm{AC}_{01} + \mathrm{BI}_{\mathrm{D}} + \text{C-N}$ to $\mathbf{EL} + \mathrm{AC}_{01} +$

BI_D, i.e. a system incompatible with CT, whereas the elimination translation reduces **EL** + AC_{01} + BI_D + C-N to $\mathbf{IDB}_1$, which is compatible with CT.

Whereas in the elimination of lawless sequences, combined with realizability-type semantics, the functions realizing implications and universal quantifiers may be thought of as "lawlike in the lawless parameters", in function realizability every realizing object is itself a choice sequence.

2.4. There is a link between constructive type theory and the theory of choice sequences which we already mentioned in 11.7.6: Martin-Löf's justification of the elimination rule for W-types presupposes an intuition akin to bar induction. On the other hand, we observed in our discussion of the axioms for lawless sequences that, although we had a lot of "informal rigour" arguments for "open data" and "continuity axioms", we had very little which could serve as a justification of bar induction.

2.5. *Formalisms for constructive mathematics.* The formalisms in the metamathematics of constructive mathematics are of three types.

(a) *Systems based on typed languages*, such as elementary analysis **EL** (3.6.2), finite-type arithmetic $\mathbf{HA}^\omega$ (9.1.5–7), second-order arithmetic with sets **HAS** (3.8.4), intuitionistic higher-order arithmetic **HAH** (3.9.4), and the systems introduced by Martin-Löf (chapter 11). These typed systems may be further subdivided into three groups: (a1) systems with variables for numbers and functionals, with function application as a primitive, (a2) systems with variables for sets, but without variables for functions or functionals, and (a3) systems with variables for sets *and* for functionals.

(b) *Applicative systems*, i.e. systems based on a primitive concept of operation with a partial application operation (e.g. **APP**, 9.3.2–3) which may be combined with the idea of a classification or set (e.g. $\mathbf{EM}_0$, 9.8.2).

(c) *Systems based on the language of set-theory*, such as **CZF** (11.8.1–2) and **IZF** (11.8.11).

We wish to make a few comments on the relative merits of these three types. The basic intuitive concept underlying a theory such as **APP** is that of a rule or operator, with the partial recursive functions as a paradigm. In **APP** the essential aspects of partial operations are expressed without commitment to the precise notion of partial recursive function. Theories based on **APP** and its extensions have already proved their value in metamathematical research. Thus **APP** is, as we have seen, the natural setting for an abstract version of Kleene's realizability. The absence of type distinctions makes for formal simplicity, when compared with the more complicated systems of type (a); on the other hand, we have to pay a price if we want to retain the partial recursive functions as our basic model: we must base the theory on logic with existence predicate.

In working with systems of type (c) one can to some extent profit from the extensive experience gained with classical set theory. Also the formalisms are simple and elegant. On the other hand it looks as if the intuitive motivation (justification) for such systems is either formal analogy (**IZF** as an obvious intuitionistic analogue of **ZF**) or indirect, via interpretation in systems of type (a) or (b) (Aczel's interpretation of **CZF** in **MLV** in section 11.8 is a good example).

From a conceptual point of view typed systems seem to us to provide the simplest framework, although they are not always optimal for the actual formalization of constructive mathematics. But a good deal of flexibility may be obtained by having a liberal supply of type-forming operators and by the use of the variable types.

For the reasons just described we decided to concentrate in this book on systems of type (a) and (b).

2.6. We remarked in section 1 that "informal rigour" is the main source of mathematical knowledge in intuitionistic mathematics. We further note that the usefulness of intuitionistic logic in mathematics in general, in particular topos theory, consists in its wide applicability; it applies in many situations where classical logic fails. To illustrate the point, there are many phenomena in traditional classical mathematics that find their coherent or elegant formulation in an intuitionistic formalism. For example "continuity in parameters" (cf. Hyland, 1982A), and "differentiable objects" (synthetic differential geometry, cf. Kock (1981), and the forthcoming monograph by Moerdijk and Reyes).

But it is to be noted that such applications also lead us to consider mathematics in structures which in certain respects considerably differ from traditional constructive mathematics, for example by failure of the axiom of countable choice (cf. 15.3.7). Thus the search for applications of intuitionistic logic introduces "external motivations" for the acceptation and study of certain principles, and leads to the study of axiomatizations divergent from the traditional ones.

2.7. *Proof-theoretic strength and "ontological commitment".* The lack of proof-theoretic strength has often been invoked as an argument against intuitionistic mathematics. Recent research has made this argument lose much, if not all, of its force. Much can be achieved by choosing systems based on languages with strong *expressive power*, but weak in proof-theoretic strength (cf. Sieg 1985, Mints 1976, and the work flowing from Friedman's idea of reverse mathematics, such as Friedman et al. 1983, Simpson 1984). Before this, the work in classical predicative analysis had already shown that examples of theorems that really require impredicative comprehension are scarce. Though this aspect has not been investigated on the same scale in constructive mathematics, there seems to be little doubt that the same general conclusion holds there as well. See also the illuminating discussion in Feferman(A).

2.8. *Freedom in the interpretation of logic.* Our experience with translations of one system into another, and the semantical counterpart, models which can be shown to validate certain principles internally without having to assume them externally, shows that we have a good deal of freedom also in our interpretation of intuitionistic logic.

Thus the Dialectica interpretation (Gödel 1958, cf. 4.5.1) validates Markov's principle, numerical realizability (section 4.4) justifies Church's thesis, and elimination of choice sequences (section 12.3) validates continuity axioms. It should be kept in mind however, that all these justifications do not have an absolute character, but hold *relative to a specific language* (in other words, language is essential in this respect). Of course, a direct intuitive grasp of the validity of a principle is as a rule more satisfactory then an indirect justification based on reinterpretation (in the vein of the familiar relative consistency proofs).

In Martin-Löf's type theories "logic" is even completely absorbed by the domains of mathematical objects (types).

These observations tend to support Brouwer's thesis that logic is not prior to mathematics, but inextricably bound to it.

2.9. *Equality in constructive mathematics.* Is equality to be regarded as a "logical primitive" in constructive mathematics, or should it be regarded as defined? In the treatment of classical set theory one sometimes presents equality between sets as defined, but there it seems more natural to regard it as a logical primitive; the principle $x = y \leftrightarrow \forall z(z \in x \leftrightarrow z \in y)$ just spells out what the basis notion of equality means in this particular case.

In a constructive setting it is not so clear that equality should be regarded as a logical primitive. First of all, we have seen that "logical" and "mathematical" cannot be separated from the constructive point of view; cf. the discussion of the BHK-interpretation and Martin-Löf's type theories.

Let us consider some examples of domains:

(a) $\mathbb{N}$, the natural numbers;
(b) LS, the lawless sequences;
(c) LL, the lawlike sequences;
(d) TREC, the collection of total recursive functions;
(e) $\mathbb{R}$, the Cauchy reals.

For $\mathbb{N}$ it is clear what is meant by equality. In the case of LS the obvious equality relation is $\equiv$, introduced in 12.2.4, which, by sheer luck, happens to coincide with the ordinary *defined* $=$ because of the special properties of LS.

For LL we may be in some doubt as to what equality ($=_{\mathrm{LL}}$) means. Of course, $a =_{\mathrm{LL}} b :=$ a and b are (given by) the same law. However, what is equality between laws? It seems that we have to accept this notion as primitive.

Elements $f, g \in$ TREC are given to us by algorithms from $\mathbb{N}$ to $\mathbb{N}$ together with proofs that these algorithms are total. Should $f =_{\mathrm{TREC}} g$ be taken to mean that the algorithms (or their code numbers) are identical, *and* that the *proofs* of the algorithms being total on $\mathbb{N}$ are also the same? If we accept this as an explanation of equality for TREC, the problem of equality reduces to the question of equality between proofs.

Finally, $x \in \mathbb{R}$ and $y \in \mathbb{R}$ are given by fundamental sequences $\langle r_n \rangle_n$ and $\langle s_n \rangle_n$, respectively, and equality is defined via the equivalence relation $\approx$ (5.2.2). Note that constructively we can present a Cauchy real *only* via a fundamental sequence and therefore the equality *has* to be specified by a definition.

So it looks as if, in constructive mathematics, the slogan of *equality as a logical primitive* can only mean that understanding a domain means at the same time understanding equality between elements of the domain. This certainly permits us to think of a domain as given as a collection with an equivalence relation on it, provided we do not think of these two components as necessarily independently specified.

Our doubts, as to what exactly $a =_{\mathrm{LL}} b$ or $f =_{\mathrm{TREC}} g$ means, may be interpreted as an indication that we do not fully understand LL and TREC as domains. It is to be noted that this uncertainty does not prevent us from using these notions in mathematics. The uncertainties in the case of LL: what exactly is an arbitrary law, and when are two laws one and the same, are really of the same kind, or might even be said to coincide.

The examples given above suggest that for certain "basic" domains such as $\mathbb{N}$ and LL understanding equality is immediately connected with our understanding the domains as such; other domains such as $\mathbb{R}$ have to be introduced in a more indirect way, requiring a definition of equality, namely via an equivalence relation on the, more basic, underlying domain of fundamental sequences of rationals. As a rule the basic domains pose genuine philosophical problems, which in our approach have been discussed in the framework of informal rigour.

2.10. *Equality, choice principles, and the presentation axiom.* There is a connection between equality on a domain D and the validity of choice principles over D.

If a domain $(D, =_D)$ is *basic*, in the sense that two elements are D-equal $(d =_D d')$ iff they are given to us as the same object, it is natural to assume

$$\text{AC-D} \qquad \forall d \in D\, \exists y \in D'\, A(d, y) \to \exists f \in D \to D'\, \forall d \in D\, A(d, fd),$$

since the constructive truth of the premiss presupposes a method for finding y from $d \in D$ (cf. the discussion in 4.2.1). Clearly AC-D need not hold (think of the example of $\mathbb{R}$, cf. (1) in 4.2.1) if $=_D$ is an equivalence relation between "presentations" of elements of D.

Reflecting on the possible ways to create domains we arrive at the following plausible principle: *each domain* $(D, =_D)$ *is obtained from a domain* $(D', =_{D'})$ *such that* $(D', =_{D'})$ *is basic and the elements of D are equivalence classes with respect to some defined equivalence relation on* D'. Combination of this principle with AC-D$'$ gives the presentation axiom (cf. 4.2.2).

There is no *direct* connection between the validity of the axiom of choice over a basic domain, and the assumption that equality on a basic domain is decidable. Decidability of equality on a basic domain seems intuitionistically quite plausible: we know whether two objects are given to us as the same object or not. The usual argument for this is closely bound up with assumptions about constructive proofs, and runs as follows. We can always transform a domain $(D, =_D)$ in a basic domain D' consisting of pairs (p, d), where p is a proof that $d \in D$. Assuming moreover that d can actually be extracted (read off) from p, and that equality between proofs is decidable, D' has a decidable equality. (For certain domains such as $\mathbb{N}$ this yields nothing new: $n \in \mathbb{N}$ can be taken to be its own "proof of $n \in \mathbb{N}$".)

However, there are two problems here. First of all, equality between proofs is not completely understood. Secondly, it has been argued, for example by Sundholm (1983), that a proof p of a statement A really contains two components: a mathematical construction p' witnessing the truth of A, and an argument p'' convincing us of the fact that p' witnesses A. p'' cannot itself be regarded as a mathematical object, if we want to avoid an infinite regress. On this view the intuitive decidability of the predicate "p proves A" does not entail the decidability of "p' is a witness for A". This situation is illustrated by realizability: "p' realizes A" need not be decidable, and the corresponding p'' is so to speak mathematically invisible. The view just outlined also enters into Martin-Löf's motivation for his theory of types (in keeping with our observation that the interpretation of logic in Martin-Löf's theories is a form of realizability).

If Sundholm's arguments are accepted, the construction of a basic domain D' from $(D, =_D)$ can only mean taking pairs (p', d), where p' is the *mathematical* witness of a proof that $d \in D$; but now the decidability of equality on the basic domain is not plausible any more. On the other hand the argument for AC-D′ over such a domain D' does not lose its force, since p' contains all *mathematical* information contained in an argument showing that $d \in D$, and any method transforming, for arbitrary $d \in D$, a proof of $d \in D$ into an element y, can only make use of p' and d itself.

3. Brouwer's theory of the creative subject

3.1. The theory of the idealized mathematician or "creative subject" has already been briefly surveyed in section 4.9; its origin is to be found in Brouwer (1948) (according to Brouwer, the idea dates back to 1927), where he uses explicit reference to the activity of the idealized mathematician (IM for short) in order to give an example of a real number a such that we can prove $a \neq 0$, but not $a \# 0$ (thus Markov's principle, cf. 4.5.4, is refuted). By means of this example Brouwer wanted to show that negation could not always be eliminated from mathematical statements, in opposition to Griss' claim (1946) that negation ought not to be admitted in intuitionistic mathematics.

3.2. Slightly simplified, Brouwer's argument runs as follows. Let A be an assertion for which $\neg A \vee \neg\neg A$ is not known. Then the IM can create, in connection with A, a sequence of rationals $\langle r_n^A \rangle_n$ according to the following recipe:

(1) as long as, in the course of choosing the r_n^A the IM has neither experienced the truth, nor the falsity of A, then r_n^A is chosen to be 0;

(2) if, between the choice of r_n^A and r_{n+1}^A the IM has proved A or $\neg A$, r_{n+k+1}^A is taken to be 2^{-n} for all k.

Clearly, $\langle r_n^A \rangle_n$ is a fundamental sequence and thus defines a real number a. If the IM would know a priori that $a \# 0$, then $a < 0$ or $a > 0$; the first is excluded by the construction, the second would mean $a > 2^{-k}$ for some k, hence the IM a priori knows that he shall have established $A \vee \neg A$ before choosing r_{k+1}^A; but the IM cannot know this in advance, since neither $\neg A$ nor $\neg\neg A$ is known to hold.

On the other hand, $a \neq 0$ holds; for if $a = 0$ were *known* it would mean that at no stage (2) applies, hence the IM *knows* that he will *never* prove A *nor ever* prove $\neg A$ (i.e. refute A), which means that he would *know* that $\neg A$ as well as $\neg\neg A$, which is contradictory. Hence $a \neq 0$.

In this argument the solipsistic element in Brouwer's view of mathematics enters in the assumption that the certainty that the IM cannot prove A means that $\neg A$ has been established.

3.3. The first proposal for a formalization of the theory of the IM has been made by Kreisel (1967), using a primitive $\Box_{\Sigma, m}A$: the IM Σ has found evidence for A at stage m of his activities.

In later presentations, such as Myhill (1976), the dependence on Σ is dropped since the assumption of more than one IM clashes with the basic solipsistic tenets of Brouwer's philosophy: for the IM only his *own* experiences represent truth, the existence of other creative subjects must remain conjectural (Brouwer 1949).

What has also been dropped in Kreisel's and many later presentations of the theory, is the fact that Brouwer does not postulate numbered stages in the activity of the IM in an absolute sense, but only *relative* to the proposition A and the sequence $\langle r_n^A \rangle_n$ to be constructed (certain authors have tried to take this into account, e.g. Posy 1976).

Heyting (1956) presents the argument above in the first person, i.e. "I (myself)" is identified with the IM. The dependence of the stages on the problem A considered is not stressed by him.

3.4. *Weak and strong versions of the theory.* Let us consider, for the time being, the version of the theory where all activity of the IM is supposed to proceed in ω discrete stages. The solipsistic point of view then suggests that for all propositions A

$$A \leftrightarrow \exists n(\Box_n A). \tag{1}$$

However, Brouwer in his argument seems to assume only

$$(\neg\exists n(\Box_n A) \rightarrow \neg A) \wedge (\exists n(\Box_n A) \rightarrow A). \tag{2}$$

Only in Brouwer (1954, p. 4, penultimate paragraph) something like (1) seems to be asserted, though not used.

Some elaboration might be helpful here. The solipsistic view identifies "true" with "provable by the IM", which suggests a principle $A \leftrightarrow \Box A$, where $\Box$ is a modal operator "provable by the IM". This operator is obviously redundant; nonetheless it may serve to underline that (1) acts to structure the (global) proving activity of the IM in *stages*.

One may also consider ω-sequences of stages, not for the whole mathematical activity of the IM, but only relative to a proposition B, that is to say we only distinguish ω stages with respect to the IM's attempts to prove (or refute) B. This means that the IM does not necessarily meet, in the ω-sequence relative to B, all proofs accessible to him, but given his freedom of activity, it is absurd to assume, if A can be proved at all, that he shall never encounter a proof of A in this ω-sequence. This makes it plausible that one can assert $\neg\exists n(\Box_{n,B}A) \to \neg A$ without asserting $A \leftrightarrow \exists n(\Box_{n,B}A)$.

In another respect (1) may be thought to have paradoxical consequences as well. For example, if $\Box_n \forall m A(m)$ holds, it follows that

$$\forall m \exists n (\Box_n A(\overline{m})),$$

i.e. the IM is forced to conclude $A(\overline{m})$ at some stage, for each numeral $\overline{m}$. However, as Dummett (1977, section 6.3) observes, this problem can be avoided if we realize that the correct intuitive reading of $\exists nC$ is: the IM knows how to find an n such that C holds. For then $\exists n \Box_n A(m)$ means that the IM can arrange his mathematical activity so as to have a proof of $A(m)$ at some stage. On this reading (1) is plausible, and there seems to be no particular reason to assert only (2). Dummett's way of reading agrees with our usual way of accepting $A(10^{10^{10}}) \vee \neg A(10^{10^{10}})$ for a decidable A; that we have a proof of this statement does not mean that we either have a proof of $A(10^{10^{10}})$ or a proof of $\neg A(10^{10^{10}})$, but only that we know how to arrive at a proof of the one or the other. (2) has also been defended in connection with a theory of several creative subjects, but this seems to be rather alien to Brouwer's intentions.

3.5. We now sketch the difficulties associated with the development of a coherent version of the theory based on a single sequence of ω stages. The crucial problem seems to be: what does it mean to have "evidence for A at stage n", or "to have established A at stage n", in particular if we also require that this notion is decidable, i.e.

$$\Box_n A \vee \neg \Box_n A.$$

If we can assume that we can draw only finitely many conclusions at each stage, there seems to be no problem in assuming that at each stage n we draw at most one conclusion, say $A^{(n)}$ ("the one-conclusion-per-stage hypothesis"), so

$$\Box_n A^{(n)}, \qquad m < n \to \neg \Box_m A^{(n)}.$$

But this easily leads us into paradoxes. For consider statements of the form "α is a sequence in $\mathbb{N} \to \mathbb{N}$ fixed by a recipe (possibly by explicit reference to the stages of the IM's activity)". By inspection the IM can see whether an $A^{(n)}$ is of this form or not, so we may enumerate all sequences appearing in this manner in some $\Box_n A^{(n)}$; let β_n be the nth such sequence. Then

$$\gamma := \lambda n.\beta_n(n) + 1$$

is a sequence fixed by a recipe which therefore should appear at some stage as β_m, say; but then $\beta_m(m) = \beta_m(m) + 1$ gives a contradiction.

This suggests that we should give up the idea of "one-conclusion-per-stage". But then another problem crops up: if $\Box_n A$, what else may be regarded as established at stage n besides A? For example, we might suppose that

$$\Box_n(\forall m A(m)) \to \forall m(\Box_n A(m)),$$

$$(\Box_n(A \to B) \wedge \Box_n A) \to \Box_n B.$$

On the other hand it is not very plausible that

$$\Box_n A \wedge (\mathbf{IQC} + A \vdash B) \to \Box_n B,$$

since deducibility in **IQC** is not decidable.

3.6. All proposed solutions to these difficulties seem to us to suffer from one of the following defects: either the proposed revised theory replaces formulations of the theory discussed above by formulations which are at least as problematic (this applies, for example, to the hierarchy of "levels of self-reflection" considered in Troelstra 1969), or by theories which seem to deviate rather far from Brouwer's original intentions. The first defect is unacceptable; for the second type of solution we must require that at least Brouwer's counterexamples can be recovered. In this connection the solution proposed by Niekus (1987) deserves further investigation.

Since already for other reasons (cf. section 1) we came to reject Brouwer's solipsistic and language-free picture of mathematics as inadequate, it is presumably necessary to deviate from Brouwer's original theory, even if we wish to save the essence of his counterexamples based on the IM.

Summing up, we can say that the attempts to formalize the theory of the IM as envisaged by Brouwer cannot be said to be satisfactory examples of "informal rigour".

3.7. Finally we want to point out that Kreisel's theory of the creative subject, with axioms (4.9.3):

IM1 $\square_n A \vee \neg \square_n A$,

IM2 $\square_n A \to \square_{n+m} A$,

IM3 $A \leftrightarrow \exists n(\square_n A)$,

has a surprisingly simple model (van Dalen 1978): consider a Beth model and define $k \Vdash \square_n A := \forall k' \succcurlyeq k(\mathrm{lth}(k') \geq n \to (k' \Vdash A))$. One easily verifies the axioms. Observe however, that this interpretation obliterates many distinctions.

4. Dummett's anti-realist argument

4.1. Dummett has presented a general philosophical argument for rejecting classical logic in favour of intuitionistic logic (1975, 1976, 1977, sections 7.1–2). Prawitz (1977) summarizes the main points of Dummett (1975) and presents supplementary discussion. Another account is found in Sundholm (1986) (the "theory of sense$_D$" in Sundholm's paper on the whole coincides with the "theory of meaning" as used in this section). It is not possible to present Dummett's arguments here in all their ramifications, but we shall make an attempt to sketch the principal points in 4.2–5 below; for details we must refer the reader to the sources just mentioned. For our presentation we have drawn mainly on Dummett (1977) and Prawitz (1977).

4.2. *On the principle that meaning is determined by use*. "Meaning" is very much a philosophically determined notion. For classical/platonist mathematics (philosophy) the meaning of a sentence is determined by its truth conditions. Instead of the meaning based on truth conditions, Dummett, following Wittgenstein, has proposed another theory of meaning based on the principle "meaning is determined by use". Here "use" means total use in all its aspects, "determined" means: if two expressions are used in the same way, then they have the same meaning, or: the meaning of a sentence must be fully manifest in its use. Dummett's notion of meaning is *epistemic* in nature, i.e. meaning must be recognizable by *us*. Hence truth-conditionally determined meaning (in the essence of two-valued logic) is not acceptable, as it escapes recognition—since truth in general is not decidable (cf. 4.4). Thus Dummett's theory can be termed *anti-realistic*, since it rejects the realist notion of truth (as determined, and existing independently of us) as an acceptable point of departure. The principle "meaning is use" is supported by the following arguments:

(1) Meaning has to be communicable and communication has to be observable. A notion of meaning not satisfying these conditions would be irrelevant to mathematics as a social enterprise.

(2) To understand a language is to have knowledge of meaning.

(3) Knowledge of meaning must ultimately be *implicit*: even if we define the meaning of an expression in terms of other expressions, this requires knowledge of the meaning of the latter. If we are to avoid an infinite regress we must eventually arrive at expressions for which the meaning is implicit.

(4) To learn a language is to learn to use it systematically. For, the meaning involved in the knowledge of the language, being implicit, must be manifested in some way, in particular in the (observable) use of language.

Meaning is closely tied up with correctness: when we know the meaning of A, we know under what conditions A may be correctly asserted. It is in this sense that "use" has to be understood.

4.3. *The rejection of holism.* On the holistic view, nothing less than the total use of language determines its meaning; as a consequence no other principles of meaning can be given. In particular it is impossible to provide the meaning of a single statement by itself, or even that of a particular (say mathematical) theory, since such a theory can only be understood in the wider framework of the whole language.

In a restricted version of holism one may admit a certain class of statements as possessing direct content and meaning, e.g. the numerical equations in arithmetic, whereas the meaning of statements at large remains determined by the total use of the language. In such a restricted holism there should be a certain harmony between the privileged class of statements and the language at large, in the sense that it should be impossible to deduce from (correct) statements outside the privileged class correct privileged statements not originally given as correct. For, the meaning of the privileged statements is immediately given, and one should not be able to arrive at correct privileged statements by employing a larger class of statements. In the case of arithmetic this is Hilbert's requirement that arithmetic should be conservative over its closed atomic statements. Thus the foundational practice of formalism may be regarded as a restricted form of holism.

On Dummett's view holism and restricted holism are to be rejected, since they violate the requirement that meaning is recognizably determined by use: "total use of the language" patently escapes recognizable use. Therefore Dummett, following Frege, adopts a molecular view, where the understanding of a given sentence depends on the understanding of its constituent parts. The molecular view obviously requires that it is possible to rank sentences in a hierarchy according to their complexity.

4.4. *The rejection of platonism.* If meaning is to be determined by use, could we not just take a pragmatic (sociological) view and say that since mathematicians largely use bivalent logic and Tarskian truth, meaning is (at least in mathematics) determined by platonistic truth conditions?

This defense of platonistic logic is rejected by Dummett in his observation that this particular usage is recognition-transcendent, as certain instances of the truth-conditions turn out to be undecidable, thus bringing us into conflict with (4) in 4.2.

In particular, a platonist "theory of meaning" requires us to have an understanding of quantification over an infinite domain not relating to our own restricted means of recognizing such sentences as true. If we assume that every statement is determinately either true or false we cannot explain the truth of a universal statement in terms of the procedure available to us for recognizing such a statement as true, since already quite elementary mathematical theories contain many sentences which are not decidable, and the condition for whose truth cannot be stated without circularity. For a universal arithmetical statement there will be no guarantee that, if it is true, we shall be able to recognize its truth condition as fulfilled: for all we know, the condition may be such that no human being will ever be able to recognize it as being fulfilled. In this case the notion of "knowledge of the condition for truth" seems without substance.

In other words, on a platonistic meaning theory we cannot give substance to the conception of having implicit knowledge of what the condition for the truth of a mathematical statement is, since nothing we can do can amount to a manifestation of such knowledge.

Dummett (1977) considers three possible defenses of the platonist view. We mention here only the one which has the greatest appeal, and would presumably be the line of reasoning followed by most mathematicians if called upon to defend a platonistic position.

On this view, we understand the condition for the truth of complex sentences in terms of what it would be to be able to recognize effectively their truth or falsity in a direct manner, an ability which we do not ourselves possess, but of which we can form a conception by *analogy* with those abilities we do have.

Dummett rejects this view with the remark that the language that we use, when we are engaged in mathematics as well as in other activities, is *our language*, and its meaning must be connected with our own capacities (here we interpret Dummett's "connected" in a "computational" sense).

Of course, we should distinguish between the rejection of a platonist theory of meaning, and the rejection of the use of classical logic in mathematics. As we have seen from the results reported in this book, large parts of classical mathematics permit at least an indirect constructive justification, via the negative translation and other devices (such translations, however, change the meaning).

4.5. *Consequences of "meaning is use".* The use of "use" as stipulated in 4.2 rules out mere "use by convention", as found in some specific mathematical practice. Since the subject matter of this book is mathematics, let us illustrate "meaning is use" for the logical operators in mathematical statements.

The meaning and use of a statement A are determined by "the conditions under which A may be correctly asserted" – in other words, by what it means to give a proof of A. Proofs are constructed via inferences, and a valid rule of inference should transform correct proofs into correct proofs; we therefore have to explain "proof", "meaning" and "inference" simultaneously.

A schema on which such an explanation can be based is provided by the BHK-interpretation, since this interpretation explains the proof conditions of a compound statement in terms of the proof conditions for its components, and thus conforms to the molecular view mentioned above. In order to prevent confusion, let us call the notion of proof as intended in the BHK-interpretation "*canonical proof*". One should distinguish between canonical proofs and (formal) demonstrations, which correspond to proofs in the usual sense of the word, i.e. arguments which show us how we can arrive at a canonical proof.

The introduction rules correspond precisely to the clauses of the BHK-interpretation describing the form of canonical proofs. For example, the rule $\wedge$I tells us that a canonical proof of $A \wedge B$ is given by presenting a canonical proof of A and a canonical proof of B. The elimination rules are justified indirectly; thus the rules $\wedge$E are justified precisely because all canonical proofs of $A \wedge B$ must have the form as specified by the $\wedge$I-rule.

A canonical proof of an implication is a function; in a demonstration we may use modus ponens $A \to B, A/B$; applied to canonical proofs d, d' of $A \to B$ and A, respectively, this gives a demonstration of B which is not, as it stands, a canonical proof of B, but a canonical proof can be extracted from the demonstration: one only has to apply the function d to the canonical proof d' and find the canonical proof which is the value of $d(d')$.

Thus the system of rules consists of rules which directly give us the form of canonical proofs, and a second group of rules which is justified indirectly; there exists a harmony between these two groups of rules since the second group does not lead to the acceptance of new forms of canonical proofs and in particular does not permit new proofs of atomic statements (i.e. their addition is conservative).

Martin-Löf's type theories discussed in chapter 11 can be said to implement this idea on a large scale. (It will have become clear from this discussion in chapter 11 that the canonical elements (proofs) describe, in actual fact, only part of what we call a proof or argument in the usual sense.)

In practice we do not just give (methods for finding) canonical proofs of statements, but we also need to convince ourselves that the description of a canonical proof of A actually leads to such a proof; but as has been argued at length, by e.g. Sundholm (1983), this element of "acquiring the insight that the given description of a canonical proof does the job" is not itself to be regarded as a mathematical object (in contrast to the canonical proofs themselves). Cf. the discussion in 1.5.3, concerning Kreisel's variant of the BHK-interpretation.

According to Dummett, the *general notion* of a canonical proof is as yet not clear, and he mentions the possibility that the notion for statements of a particular form is time-dependent, i.e. new forms of canonical proof might be discovered in the future.

4.6. *Some anti-anti-realist remarks.* Undoubtedly Dummett's argument has a certain appeal, though there remain problems, in particular when confronted with mathematical practice.

One objection concerns the rejection of the defense of the platonist mentioned above. If we understand Dummett correctly, the platonists idealization of human cognitive processes is really one step too far. On the other hand, intuitionistically we do accept that "in principle" we can view $10^{10^{10}}$ as a sequence of units (i.e. we reject the ultrafinitist objection), and the authors are not sure that this is really less serious than the platonists extrapolation. At least, it needs argument. (A similar objection is raised in Wright 1982.)

Another defect of both Dummett's and Prawitz's discussion seems to us to be a rather too exclusive concentration on the meaning of the *logical* operators; the authors do not think that logic, as it functions *in mathematics*, can be separated from mathematical objects and operations. However, this objection is met by Martin-Löf's work, in the sense that his theories show how, and to what extent, for an already powerful piece of mathematics, a theory of meaning along the lines of Dummett's thinking is viable. In these theories the logical operators appear in fact as special cases of more general mathematical operations.

Nevertheless it is not clear how the interplay between the nature of mathematical objects and logic, which leads to the formulation of principles via the process of "informal rigour", fits into Dummett's theory.

One may well feel some doubt as to whether mathematical practice, and in particular the way in which mathematicians arrive at axioms, and keep on adding or revising axioms, is in keeping with a "molecular" approach to the meaning of mathematical statements. Perhaps the mathematicians way of assigning meaning is neither all-out holistic, nor purely "molecular". It should be added, of course, that here "meaning" is not the purified theoretical notion of Dummett's discussion. What does it mean, that we add a new axiom schema involving compound logical statements, e.g. Markov's principle, or the $\forall\alpha\exists x$-continuity axiom in the theory of choice sequences, after having convinced ourselves (by "informal rigour") that it is valid? The authors are inclined to say that this constitutes a revision of the meaning of the certain logical operations (or a making precise of the meaning within a given language), in particular, in the example just mentioned, of the quantifier combination $\forall\alpha\exists x$.

The objections raised might be met by applying the Procrustus method of denying the legitimacy of anything that does not fit the framework, but that does not seem to be fruitful. Different attitudes are possible, more in keeping with Dummett's ideas:

(i) Adopt the "hard-headed" point of view that mathematics based on such extra axioms is in reality nothing but the making of logical deductions from such axioms as hypotheses. "Logical" may be relaxed to "by means of principles justified in the style of Martin-Löf's type theories".

Or, what amounts to the same, one regards such additional axioms as being "in the waiting room" until they have been justified in the same manner as other principles in constructive type theory. Such a solution is formalistic in nature and therefore mildly holistic.

(ii) One may attempt to treat the extra axioms as constituting a revision of the notion of canonical proof (cf. Dummett's remarks mentioned above, to the effect that the notion of canonical proof is perhaps time-dependent). In the authors' opinion this solution also approaches holism, rejected by Dummett.

(iii) Provide an indirect justification for the new axioms (e.g. by elimination of choice sequences in the case of the $\forall\alpha\exists x$-continuity axiom mentioned above). This solution avoids the issue and does not attack the problem of meaning.

Therefore none of these solutions seems to be quite satisfactory from Dummett's point of view.

BIBLIOGRAPHY

Below we have only listed those items which have been referred to in this book. A more complete bibliography of constructivism (in particular the items classified under F50–65) is in Müller (1987).

Full titles of journals can easily be removed from Müller (1987) or in the Mathematical Reviews. The exceptions concern certain frequently cited journals for which we use the abbreviations below:

JSL	the Journal of Symbolic Logic,
AML	Annals of Mathematical Logic, continued as
APAL	Annals of Pure and Applied Logic,
ZMLGM	Zeitschrift für mathematische Logik und Grundlagen der Mathematik,
Arch.ML	Archiv für mathematische Logik und Grundlagenforschung.

In addition we use in the title of certain proceedings the abbreviation

LMPS Logic, Methodology and Philosophy of Science.

Congress proceedings and the papers in such proceedings appear under the year of publication, not the year of the congress. If a volume of a journal appeared over more than one calendar year, papers in the volume are listed under the year appearing on the title page of the completed volume, which is as a rule also the year in which the last issue of the volume appears (occasionally there are awkward exceptions!).

Items with more than one author are listed in full under the author which is named first in the title of the official publication; for the co-authors there is an indirect reference. E.g. under "Bernays" we find "Bernays, P., see *Hilbert and Bernays (1934, 1939)*". We have refrained from giving indirect references in the case of multiple editors.

A few items we only know from reviews or references by others. In such cases the reference is followed by "(n.v.)".

Aberth, O.
(1980) *Computable Analysis* (McGraw-Hill, New York).

Aczel, P.H.G.
(1968) Saturated intuitionistic theories, in: *Schmidt et al. (1968) 1–11.*
(1977) An introduction to inductive definitions, in: *Barwise (1977) 739–782.*
(1977A) The strength of Martin-Löf's intuitionistic type theory with one universe, in: S. Miettinen, S. Väänänen, eds. *Proceedings of Symposia in Mathematical Logic, Oulu, 1974, and Helsinki, 1975. Report No. 2* (University of Helsinki, Department of Philosophy, 1977) pp. 1–32.
(1978) The type-theoretic interpretation of constructive set theory, in: A. Macintyre, L. Pacholski, J. Paris, eds., *Logic Colloquium '77* (North-Holland, Amsterdam) pp. 55–66.
(1982) The type-theoretic interpretation of constructive set theory: choice principles, in: *Troelstra and Van Dalen (1982) 1–40.*
(1986) The type-theoretic interpretation of constructive set theory: inductive definitions, in: *Marcus et al. (1986) 17–49.*

Allen, S.F., see *Constable et al. (1986).*

Barendregt, H.P.
(1974) Pairing without conventional restraints, *ZMLGM* 20, 289–306.
(1981) *The Lambda Calculus, Its Syntax and Semantics* (North-Holland, Amsterdam), 2nd rev. ed. (1984).

Barendregt, H.P. and A. Rezus
(1983) Semantics for classical AUTOMATH and related systems, *Information and Control* 59, 127–147.

Barr, M. and C. Wells
(1985) *Toposes, Triples and Theories* (Springer, Berlin).

Barwise, K.J.
(1977) ed., *Handbook of Mathematical Logic* (North-Holland, Amsterdam).

Barwise, K.J., H.J. Keisler, and K. Kunen
(1980) eds., *The Kleene Symposium* (North-Holland, Amsterdam).

Beeson, M.J.
(1975) The nonderivability in intuitionistic formal systems of theorems on the continuity of effective operations, *JSL* 40, 321–346.
(1976) Half a note on Riemann's theorem about absolute convergence. Unpublished note.
(1977) Principles of continuous choice and continuity of functions in formal systems for constructive mathematics, *AML* 12, 249–322.
(1979) Goodman's theorem and beyond, *Pac. J. Math.* 84, 1–16.
(1980) Extensionality and choice in constructive mathematics, *Pac. J. Math.* 88, 1–28.
(1982) Recursive models for constructive set theories, *AML* 23, 127–178.
(1985) *Foundations of Constructive Mathematics* (Springer, Berlin).
(1986) Proving programs and programming proofs, in: *Marcus et al. (1986) 51–82.*

Benthem Jutting, L.S. van
(1977) Checking Landau's "Grundlagen" in the AUTOMATH system, PhD thesis, Technische Hogeschool, Eindhoven.

Berg, G., H. Cheng, R. Mines and F. Richman
(1976) Constructive dimension theory, *Compos. Math.* 33, 161–177.

Berg, G., W. Julian, R. Mines and F. Richman

(1975) The constructive Jordan curve theorem, *Rocky Mt. J. Math.* 5, 225–236.

(1977) The constructive equivalence of covering and inductive dimensions, *General Topology Appl.* 7, 99–108.

Bergstra, J.A.

(1976) Computability and continuity in finite types, PhD thesis, Rijksuniversiteit Utrecht.

Bernays, P.

(1970) Die schematische Korrespondenz und die idealisierten Strukturen, *Dialectica* 24, 53–66; reprinted in P. Bernays, *Abhandlungen zur Philosophic der Mathematik* (Wissenschaftliche Buchgesellschaft, Darmstadt), (1976) pp. 176–188.

Bernays, P., see *Hilbert and Bernays (1934, 1939).*

Beth, E.W.

(1956) Semantic construction of intuitionistic logic, *Kon. Nederl. Akad. Wetensch. Afd. Let. Med., Nieuwe Serie* 19/11, 357–388.

(1959) *The Foundations of Mathematics* (North-Holland, Amsterdam) 2nd ed. (1965).

(1959A) Remarks on intuitionistic logic, in: *Heyting (1959) 15–21.*

Bethke, I.

(1986) How to construct extensional combinatory algebras, *Indagationes Math.* 48, 243–257.

Birkhoff, G.

(1940) *Lattice Theory* (American Mathematical Society, Providence, RI) (n.v.)

(1948) *Lattice Theory* (American Mathematical Society, Providence, RI) 2nd ed.

Bishop, E.

(1967) *Foundations of Constructive Analysis* (McGraw-Hill, New York).

Bishop, E. and D.S. Bridges

(1985) *Constructive Analysis* (Springer, Berlin); second, thoroughly rewritten edition of *Bishop (1967).*

Boffa, M., D. van Dalen and K. McAloon

(1979) eds., *Logic Colloquium '78* (North-Holland, Amsterdam).

Boileau, A. and A. Joyal

(1981) La logique des topos, *JSL* 46, 6–16.

Borceux, F. and G. van den Bossche

(1983) *Algebra in a Localic Topos with Applications to Ring Theory* (Springer, Berlin).

Borel, E.

(1909) Sur les principes de la théorie des ensembles, in: G. Castelnuovo, ed., *Atti del IV Congresso Internazionale dei Matematici, Roma, 6–11 Aprile 1908*, (Academia dei Lincei, Roma) Vol. II, pp. 15–17; also in: *Borel (1914) as V of note IV, 160–162.*

(1912) La philosophie mathématique et l'infini, *Revue du Mois* 14, 218–227; also in: *Borel (1914) as VII of Note IV, 166–174.*

(1914) *Leçons sur la Théorie des Fonctions* (Gauthier-Villars, Paris); second edition, unchanged with an additional note VII in the third edition of 1928; the fourth edition of 1950 contains a further note VIII.

(1947) Sur l'illusion des définitions numériques, *C.R. Acad. Sc.* 224, 765–767.

Bossche, G. van den, see *Borceux and van den Bossche (1983).*

Bridges, D.S.

(1978) On the connectivity of convex sets, *Bull. Lond. Math. Soc.* 10, 86–90.

(1978A) More on the connectivity of convex sets, *Proc. Amer. Math. Soc.* 68, 214–216.

(1979) *Constructive Functional Analysis* (Pitman, London).

Bridges, D.S., see also *Bishop and Bridges (1985)*.
Bromley, H.M., see *Constable et al. (1986)*.
Brouwer, L.E.J.

(1905) *Leven Kunst. Mystiek* (Dutch) (Waltman, Delft).

(1907) *Over de Grondslagen der Wiskunde* (Dutch), Ph.D. thesis, Universiteit van Amsterdam (Maas and van Suchtelen, Amsterdam); reprinted with additional material, ed. D. van Dalen (Mathematisch Centrum, Amsterdam, 1981).

(1908) Over de onbetrouwbaarheid der logische principes (Dutch), *Tijdschrift voor Wijsbegeerte* 2, 152–158.

(1912) *Intuïtionisme en Formalisme* (Dutch), Inaugural address (Noordhoff, Groningen); also *Wiskundig Tijdschrift* 9, 180–211; translation: Intuitionism and formalism, *Bull. Amer. Math. Soc.* 20 (1914), 81–96.

(1914) Review of Schoenflies's "Die Entwicklung der Mengenlehre und ihre Anwendungen, erste Hälfte", *Jahresber. Dtsch. Math.-Ver.* 23(2), 78–83 italics.

(1918) Begründung der Mengenlehre unabhängig vom logischen Satz vom ausgeschlossenen Dritten I: Allgemeine Mengenlehre, *Nederl. Akad. Wetensch. Verh. Tweede Afd. Nat.* 12/5.

(1919) Begründung der Mengenlehre unabhängig vom logischen Satz vom ausgeschlossenen Dritten II: Theorie der Punktmengen, *Nederl. Akad. Wetensch. Verh. Tweede Afd. Nat.* 12/7.

(1921) Besitzt jede reelle Zahl eine Dezimalbruchentwickelung? *Nederl. Akad. Wetensch. Verslagen* 29, 803–812. Also in *Math. Ann.* 83, 201–210.

(1923) Intuïtionistische splitsing van mathematische grondbegrippen (Dutch), *Nederl. Akad. Wetensch. Verslagen* 32, 877–880; German translation: *Jahresber. Dtsch. Math.-Ver.* 33 (1925), 251–256.

(1923A) Begründung der Funktionenlehre unabhängig vom logischen Satz vom ausgeschlossenen Dritten, *Nederl. Akad. Wetensch. Verh. Tweede Afd. Nat.* 13/2.

(1924) Beweis dass jede volle Funktion gleichmässig stetig ist, *Nederl. Akad. Wetensch. Proc.* 27, 189–193.

(1924A) Bemerkungen zum Beweise der gleichmässigen Stetigkeit voller Funktionen, *Nederl. Akad. Wetensch. Proc.* 27, 644–646.

(1924B) Intuitionistische Ergänzung des Fundamentalsatzes der Algebra, *Nederl. Akad. Wetensch. Proc.* 27, 631–634.

(1925) Zur Begründung der intuitionistischen Mathematik I, *Math. Ann.* 93, 244–257.

(1925A) Intuitionistischer Beweis des Jordanschen Kurvensatzes, *Nederl. Akad. Wetensch. Proc.* 28, 503–508.

(1926) Zur Begründung der intuitionistischen Mathematik II, *Math. Ann.* 95, 453–472.

(1926A) Die intuitionistische Form des Heine–Borelschen Theorems, *Nederl. Akad. Wetensch. Proc.* 29, 866–867.

(1926B) Intuitionistische Einführung des Dimensionsbegriffes, *Nederl. Akad. Wetensch. Proc.* 29, 855–863.

(1927) Über Definitionsbereiche von Funktionen, *Math. Ann.* 97, 60–75.

(1927A) Virtuelle Ordnung und unerweiterbare Ordnung, *J. Reine Angew. Math.* 157, 255–257.

(1929) Mathematik, Wissenschaft und Sprache, *Monatshefte für Mathematik und Physik* 36, 153–164; a slightly different and expanded version appeared in Dutch, Brouwer (1933): Willen, weten, spreken, *Euclides* 9 (1933) 177–193; in *Brouwer (1975)* there is a translation of section III of the 1933 version; in Heyting's notes to *Brouwer (1929)* the principal differences with section I and II of the 1933 version are indicated.

(1930) *Die Struktur des Kontinuums* (Gottlieb Gistel, Vienna) 12 pp.

(1933) Weten, willen, spreken (Dutch), *Euclides* 9, 177–193.

(1942) Beweis dass der Begriff der Menge höherer Ordnung nicht als Grundbegriff der intuitionistischen Mathematik in Betracht kommt, *Indagationes Math.* 4, 154–156.

(1948) Essentieel-negatieve eigenschappen (Dutch), *Indagationes Math.* 10, 322–323.

(1949) Consciousness, philosophy and mathematics, in: E.W. Beth, H.J. Pos, and H.J.A. Hollak, eds., *Library of the Tenth International Congress of Philosophy, August 1948, Amsterdam* (North-Holland, Amsterdam) Vol. 1, pp. 1235–1249.

(1952) Over accumulatiekernen van oneindige kernsoorten (Dutch), *Indagationes Math.* 14, 439–441.

(1954) Points and spaces, *Can. J. Math.* 6, 1–17.

(1975) *Collected Works I,* ed. A. Heyting (North-Holland, Amsterdam).

(1981) *Brouwer's Cambridge Lectures on Intuitionism,* ed. D. van Dalen (Cambridge University Press, Cambridge).

Brouwer, L.E.J. and B. de Loor

(1924) Intuitionistischer Beweis des Fundamentalsatzes der Algebra, *Nederl. Akad. Wetensch. Proc.* 27, 186–188.

Bruijn, N.G. de

(1970) The mathematical language AUTOMATH, its usage, and some of its extensions, in: *Laudet et al. (1970) 29–61.*

(1980) A survey of the project AUTOMATH, in: *Seldin and Hindley (1980) 579–606.*

Buchholz, W., S. Feferman, W. Pohlers and W. Sieg

(1981) *Iterated Inductive Definitions and Subsystems of Analysis: Recent Proof-Theoretical Studies* (Springer, Berlin).

Cellucci, C.

(1969) Un' osservazione sul theorema di Minc-Orevkov (english summary), *Boll. Un. Mat. Ital.* 1, 1–8.

Cheng, H., see *Berg et al. (1976).*

Church, A.

(1932) A set of postulates for the foundation of logic, *Ann. of Math.* 33, 346–366.

(1936) An unsolvable problem of elementary number theory, *Amer. J. Math.* 58, 345–363.

Cleaveland, W.R., see *Constable et al. (1986).*

Cohen, L.J., J. Łos, H. Pfeiffer and K.-P. Podewski

(1982) eds., *LMPS VI* (North-Holland, Amsterdam).

Constable, R.L., S.F. Allen, H.M. Bromley, W.R. Cleaveland, J.F. Cremer, R.W. Harper, D.J. Howe, T.B. Knoblock, N.P. Mendler, P. Panangaden, J.T. Sasaki, and S.F. Smith

(1986) *Implementing Mathematics with the Nuprl Proof Development System* (Prentice-Hall, Englewood Cliffs, New Jersey).

Cremer, J.F., see *Constable et al. (1986).*

Crossley, J.

(1975) ed., *Algebra and Logic* (Springer, Berlin).

(1981) *Aspects of Effective Algebra* (Upside Down A Book Company, Steel's Creek, Victoria, Australia) (n.v.).

Curry, H.B.

(1930) Grundlagen der kombinatorischen Logik, *Amer. J. Math.* 52, 509–536, 789–834.

(1941) A formalisation of recursive arithmetic, *Amer. J. Math.* 63, 263–282.

Curry, H.B. and R. Feys

(1958) *Combinatory Logic I* (North-Holland, Amsterdam) 2nd ed. (1968).

Daalen, D.T. van

(1980) The language theory of AUTOMATH, PhD thesis, Technische Hogeschool, Eindhoven.

Dalen, D. van

(1977) The use of Kripke's schema as a reduction principle, *JSL* 42, 238–240.

(1978) An interpretation of intuitionistic analysis, *AML* 13, 1–43.

(1982) Singleton reals, in: *van Dalen et al. (1982) 83–94.*

(1984) How to glue analysis models, *JSL* 49, 1339–1349.

(1986) Intuitionistic logic, in: *Gabbay and Guenthner (1986) 225–339.*

(1986A) Glueing of analysis models in an intuitionistic setting, *Studia Logica* 45, 181–186.

Dalen, D. van, D. Lascar and T.J. Smiley

(1982) eds., *Logic Colloquium '80* (North-Holland, Amsterdam).

Dalen, D. van, and J.S. Lodder

(1982) Lawlessness and independence, in: *Troelstra and van Dalen (1982) 279–309.*

Dalen, D. van, and A.S. Troelstra

(1970) Projections of lawless sequences, in: *Myhill et al. (1970) 163–186.*

Dantzig, D. van

(1956) Is $10^{10^{10}}$ a finite number? *Dialectica* 9, 273–277.

Davis, M.

(1965) ed., *The Undecidable* (Raven Press, New York).

Demuth, O.

(1969) The differentiability of constructive functions (Russian), *Commentat. Math. Univ. Carolinae* 10, 357–390.

Diaconescu, R.

(1975) Axiom of choice and complementation, *Proc. Amer. Math. Soc.* 51, 176–178.

Dijkman, J.G.

(1952) Convergentie en divergentie in de intuïtionistische wiskunde (Dutch), PhD thesis, University of Amsterdam.

Diller, J.

(1980) Modified realization and the formulae-as-types notion, in: *Seldin and Hindley (1980) 491–501.*

Diller, J. and G.H. Müller

(1975) eds., *I.S.I.L.C. Proof Theory Symposium Kiel 1974. Dedicated to Kurt Schütte on the Occasion of His 65th Birthday* (Springer, Berlin).

Diller, J. and A.S. Troelstra

(1984) Realizability and intuitionistic logic, *Synthese* 60, 253–282.

Dragalin, A.G.

(1969) Transfinite completions of constructive arithmetical calculus (Russian), *Dokl. Akad. Nauk SSSR* 189, 10–12; translated in *Sov. Math. Dokl.* 10, 1417–1420.

(1979) *Mathematical Intuitionism. Introduction to Proof Theory* (Russian) (Nauka, Moskva).

(1980) New forms of realizability and Markov's rule (Russian), *Dokl. Akad. Nauk SSSR* 251, 534–537; translated in *Sov. Math. Dokl.* 21, 461–464.

Du Bois Reymond, P.
(1882) *Die allgemeine Funktionentheorie I* (Verlag der H. Laupp'schen Buchhandlung, Tübingen).

Dummett, M.A.E.
(1975) The philosophical basis of intuitionistic logic, in: *Rose and Shepherdson (1975) 5–40.*
(1976) What is a theory of meaning? (II), in: G. Evans and J. McDowell, eds., *Truth and Meaning* (Oxford University Press, Oxford) pp. 67–137.
(1977) *Elements of Intuitionism* (Clarendon Press, Oxford).

Dyson, V.H. and G. Kreisel
(1961) Analysis of Beth's semantic construction of intuitionistic logic, Technical report no. 3, Applied Mathematics and Statistics Laboratories, Stanford University.

Engelking, R.
(1968) *Outline of general topology* (North-Holland, Amsterdam; PWN, Warszawa; John Wiley and Sons, New York).

Ershov, Ju.L.
(1977) Theorie der Numerierungen III, *ZMLGM* 23, 289–371; translation of a Russian original from 1974.

Esenin-Vol'pin, A.S.
(1961) Le programme ultra-intuitionniste des fondements des mathématiques, in: *Infinitistic methods. Proceedings of the Symposium on Foundations of Mathematics, September 1959 Warsaw* (PWN, Warszawa) pp. 201–223.
(1970) The ultra-intuitionistic criticism and the anti-traditional program for foundations of mathematics. in: *Myhill et al. (1970) 3–45.*
(1981) About infinity, finiteness and finitization, in: *Richman (1981) 274–313.*

Feferman, S.
(1964) Systems of predicative analysis, *JSL* 29, 1–30.
(1975) A language and axioms for explicit mathematics, in: *Crossley (1975) 87–139.*
(1978) A more perspicuous system for predicativity, in: K. Lorenz, ed., *Spezielle Wissenschaftstheorie, Vol. 1: Konstruktionen versus Positionen. Beiträge zur Diskussion um die konstruktive Wissenschaftstheorie* (W. de Gruyter, Berlin) pp. 58–93.
(1979) Constructive theories of functions and classes, in: *Boffa et al. (1979) 59–224.*
(A) Infinity in mathematics: Is Cantor necessary? To appear.

Feferman, S., see also *Buchholz et al. (1981).*

Felscher, W.
(1985) Dialogues, strategies and intuitionistic provability, *APAL* 28, 217–254.
(1986) Dialogues as a foundation for intuitionistic logic, in: *Gabbay and Guenthner (1986) 341–372.*

Fenstad, J.E.
(1971) ed., *Proceedings of the Second Scandinavian Logic Symposium* (North-Holland, Amsterdam).
(1980) *General Recursion Theory* (Springer, Berlin).

Feys, R., see *Curry and Feys (1958).*

Fitting, M.C.
(1969) *Intuitionistic Logic, Model Theory and Forcing* (North-Holland, Amsterdam).

Flagg, R.C.
(1986) Integrating classical and intuitionistic type theory, *APAL* 32, 27–51.

Flagg, R.C. and H.M. Friedman
(1986) Epistemic and intuitionistic formal systems, *APAL* 32, 53–60.

Flannagan, T.B.
(1976) A new finitary proof of a theorem of Mostowski, in: *Müller (1976) 257–275.*

Fourman, M.P.
(1977) The logic of topoi, in: *Barwise (1977) 1053–1090.*
(1980) Sheaf models for set theory, *J. Pure Applied Algebra* 19, 91–101.
(1982) Notions of choice sequence, in: *Troelstra and van Dalen 1982, 91–105.*
(1984) Continuous truth I. Nonconstructive objects, in: *Lolli et al. (1984)* 161–180.

Fourman, M.P. and R.J. Grayson
(1982) Formal spaces, in: *Troelstra and van Dalen (1982) 107–122.*

Fourman, M.P. and J.M.E. Hyland
(1979) Sheaf models for analysis, in: *Fourman et al. (1979) 280–301.*

Fourman, M.P., C.J. Mulvey and D.S Scott
(1979) eds., *Applications of sheaves* (Springer, Berlin).

Fourman, M.P. and D.S. Scott
(1979) Sheaves and logic, in: *Fourman et al. (1979) 302–401.*

Fraenkel, A.A., Y. Bar-Hillel and A. Levy
(1973) *Foundations of Set Theory* (North-Holland, Amsterdam).

Freudenthal, H.
(1937) Zur intuitionistischen Deutung logischer Formeln, *Compos. Math.* 4, 112–116.
(1937A) Zum intuitionistischen Raumbegriff, *Compos. Math.* 4, 82–111.

Friedman, H.M.
(1971) Axiomatic recursive function theory, in: R.O. Gandy and C.M.E. Yates, eds., *Logic Colloquium '69* (North-Holland, Amsterdam), pp. 113–137.
(1973) Some applications of Kleene's methods for intuitionistic systems, in : *Mathias and Rogers (1973) 113–170.*
(1973A) The consistency of classical set theory relative to a set theory with intuitionistic logic, *JSL* 38, 315–319.
(1975) The disjunction property implies the numerical existence property, *Proc. Nat. Acad. Sci. USA* 72, 2877–2878.
(1975A) Intuitionistic completeness of Heyting's predicate calculus, *Notices Amer. Math. Soc.* 22, A–648.
(1977) On the derivability of instantiation properties, *JSL* 42, 506–514.
(1977A) Set theoretic foundations for constructive analysis, *Ann. of Math.* 105, 1–28.
(1977B) The intuitionistic completeness of intuitionistic logic under Tarskian semantics, unpublished abstract, State University of New York at Buffalo.
(1977C) New and old results on completeness of HPC, unpublished abstract, State University of New York at Buffalo.
(1978) Classically and intuitionistically provably recursive functions, in: *Müller and Scott (1978) 21–27.*
(1980) A strong conservative extension of Peano arithmetic, in: *Barwise et al. (1980) 113–122.*

Friedman, H.M. and A. Scedrov
(1983) Set existence property for intuitionistic theories with dependent choice *APAL* 25, 129–140 (corrections ibidem 26, 101).
(1984) Large sets in intuitionistic set theory, *APAL* 27, 1–24.

Friedman, H.M., S.G. Simpson and R.L. Smith
(1983) Countable algebra and set existence axioms, *APAL* 25, 141–181 (errata ibidem 28, 319–320).
Friedman, H.M., see also *Flagg and Friedman (1986)*.
Fröhlich, A. and J.E. Shepherdson
(1955) Effective procedures in field theory, *Philos. Trans. Roy. Soc. London (Ser. A)* 284, 407–432.
Gabbay, D.
(1981) *Semantical Investigations in Heyting's Intuitionistic Logic* (Reidel, Dordrecht).
Gabbay, D. and F. Guenthner
(1986) eds., *Handbook of Philosophical Logic III* (Reidel, Dordrecht).
Gandy, R.O.
(1962) Effective operations and recursive functionals (abstract), *JSL* 39, 81–87.
(1980) Proofs of strong normalization, in: *Seldin and Hindley (1980) 457–477*.
(1980A) An early proof of normalization by A.M. Turing, in: *Seldin and Hindley (1980) 453–455*.
Geiser, J.
(1974) A formalization of Essenin–Volpin's proof theoretical studies by means of nonstandard analysis, *JSL* 39, 81–87.
Gelfond, A.O.
(1960) *Transcendental and Algebraic Numbers* (Dover Publications, New York). Translation from the first Russian edition by Leo F. Boron.
Gentzen, G.
(1933) Über das Verhältnis zwischen intuitionistischer und klassischer Logik. Originally to appear in the Mathematische Annalen, reached the stage of galley proofs but was withdrawn. It was finally published in *Arch. ML* 16 (1974), 119–132; a translation is in *Gentzen (1969) 53–67*.
(1935) Untersuchungen über das logische Schliessen, I, II, *Math. Z.* 39, 176–210, 405–431.
(1969) *Collected Papers*, ed. M.E. Szabo (North-Holland, Amsterdam).
Giaretta, P., see *Martino and Giaretto (1981)*
Girard, J.-Y.
(1971) Une extension de l'interprétation de Gödel à l'analyse, et son application à l'élimination des coupures dans l'analyse et la théorie des types, in: *Fenstad (1971) 63–92*.
(1973) Quelques résultats sur les interpretations fonctionelles, in: *Mathias, Rogers, 1983, 232–252*.
Glivenko, V.I.
(1928) Sur la logique de M. Brouwer, *Acad. Roy. Belg. Bull, Cl. Sci. (5)* 14, 225–228.
(1929) Sur quelques points de la logique de M. Brouwer, *Bull. Soc. Math. Belg.* 15, 183–188.
Gödel, K.
(1931) Über formal unentscheidbare Sätze der Principia Mathematica und verwandte Systeme I, *Monatshefte für Mathematik und Physik* 38, 173–198.
(1932) Zum intuitionistischen Aussagenkalkül, *Anzeiger der Akademie der Wissenschaften in Wien* 69, 65–66.
(1933) Zur intuitionistischen Arithmetik und Zahlentheorie, *Ergebnisse eines mathematischen Kolloquiums* 4, 34–38.

(1933A) Eine Interpretation des intuitionistischen Aussagenkalküls, *Ergebnisse eines mathematischen Kolloquiums 4*, 39–40.
(1934) On undecidable propositions of formal mathematical systems, mimeographed lecture notes taken by S.C. Kleene and J. Barkley Rosser, reprinted with revisions in *Davis (1965)* 39–74; also in: *Gödel (1986)* 346–371.
(1958) Über eine bisher noch nicht benützte Erweiterung des finiten Standpunktes, *Dialectica* 12, 280–287.
(1986) *Collected Works I*, eds. S. Feferman, J.W. Dawson, jr, S.C. Kleene, G.H. Moore, R.M. Solovay and J. van Heijenoort (Oxford University Press, Oxford).
(1987) *Collected works II*, eds. S. Feferman, J.W. Dawson, jr, S.C. Kleene, G.H. Moore, R.M. Solovay and J. van Heijenoort (Oxford University Press, Oxford). Added in proof: publication delayed.

Goldblatt, R.
(1979) *Topoi* (North-Holland, Amsterdam) 2nd rev. ed. (1984).

Goodman, N.D.
(1970) A theory of constructions equivalent to arithmetic, in: *Myhill et al. (1970) 101–120.*
(1978) Relativized realizability in intuitionistic arithmetic of all finite types, *JSL* 43, 23–44.

Goodman, N.D. and J. Myhill
(1978) Choice implies excluded middle, *ZMLGM* 24, 461.

Goodstein, R.L.
(1945) Function theory in an axiom-free equation calculus, *Proc. London Math. Soc.* 52, 81–106.
(1957) *Recursive Number Theory* (North-Holland, Amsterdam).
(1961) *Recursive Analysis* (North-Holland, Amsterdam).

Gordeev, L.
(1982) Constructive models for set theory with extensionality, in: *Troelstra and van Dalen (1982) 123–147.*
(1988) Proof-theoretical analysis; weak systems of functions and classes, to appear in *APAL* 37.

Gray, J.W.
(1979) Fragments of the history of sheaf theory, in: *Fourman et al. (1979) 1–79.*

Grayson, R.J.
(1979) Heyting-valued models for intuitionistic set theory, in: *Fourman et al. (1979) 402–414.*
(1980) Introduction to topological models, xeroxed handwritten notes of a course at the Department of Mathematics, University of Amsterdam, 1979–1980.
(1981) Concepts of general topology in constructive mathematics and in sheaves, *AML* 20, 1–41 (corrections ibidem 23, 99).
(1982) Concepts of general topology in constructive mathematics and in sheaves, II, *AML* 23, 55–98.
(1982A) Constructive well-orderings, *ZMLGM* 28, 495–504.
(1983) On closed subsets of the intuitionistic reals, *ZMLGM* 29, 7–9.
(1983A) Forcing in intuitionistic systems without power-set, *JSL* 48, 670–682.
(1984) Heyting-valued semantics, in: *Lolli et al. (1984) 181–208.*

Griss, G.F.C.

(1946) Negationless intuitionistic mathematics I, *Indagationes Math.* 8, 675–681.

(1955) La mathématique intuitionniste sans negation, *Nieuw Archief voor Wiskunde (3)*, 3, 134–142.

Hallett, M.

(1984) *Cantorian Set Theory and Limitation of Size* (Clarendon Press, Oxford).

Harper, R.W., see *Constable et al. (1986).*

Harrington, L.W., M.D. Morley, A. Scedrov and S.G. Simpson

(1985) eds., *Harvey Friedman's Research on the Foundations of Mathematics* (North-Holland, Amsterdam).

Harrop, R.

(1956) On disjunctions and existential statements in intuitionistic systems of logic, *Math. Ann.* 132, 347–361.

(1960) Concerning formulas of the type $A \rightarrow B \vee C$, $A \rightarrow (Ex)B(x)$ in intuitionistic formal systems, *JSL* 25, 27–32.

Hayashi, S.

(1980) Derived rules related to a constructive theory of metric spaces in intuitionistic higher order arithmetic without countable choice, *AML* 19, 33–65.

(1981) On set theory in toposes, in: *Müller et al. (1981) 23–29.*

(1982) A note on the bar induction rule, in: *Troelstra and van Dalen (1982) 149–163.*

Heijenoort, J. van

(1967) ed., *From Frege to Gödel. A Source Book in Mathematical Logic 1879–1931* (Harvard University Press, Cambridge, MA), reprinted (1970).

Henkin, L.

(1960) On mathematical induction, *American Mathematical Monthly* 67, 323–338.

Herbrand, J.

(1931) Unsigned note on Herbrand's thesis written by Herbrand himself, published in J. Herbrand, *Écrits Logiques* (Presses Universitaires de France, Paris, 1968) pp. 209–214; also: J. Herbrand, *Logical Writings,* ed. W.D. Goldfarb (Harvard University Press, Cambridge, MA, 1971) pp. 271–276.

Hermann, G.

(1926) Die Frage der endlich vielen Schritte in der Theorie der Polynomideale, *Math. Ann.* 95, 736–788.

Heyting, A.

(1925) *Intuïtionistische Axiomatiek der Projectieve Meetkunde* (Dutch), PhD. thesis, Universiteit van Amsterdam (P. Noordhoff, Groningen).

(1927) Die Theorie der linearen Gleichungen in einer Zahlenspezies mit nichtkommutativer Multiplikation, *Math. Ann.* 98, 465–490.

(1930) Die formalen Regeln der intuitionistischen Logik, *Sitzungsberichte der Preussischen Akademie von Wissenschaften. Physikalisch-mathematische Klasse*, 42–56.

(1930A) Die formalen Regeln der intuitionistischen Mathematik II, *Sitzungsberichte der Preussischen Akademie von Wissenschaften. Physikalisch-mathematische Klasse*, 57–71.

(1930B) Die formalen Regeln der intuitionistischen Mathematik III, *Sitzungsberichte der Preussischen Akademie von Wissenschaften. Physikalisch-mathematische Klasse*, 158–169.

(1930C) Sur la logique intuitionniste, *Acad. Roy. Belg. Bull. Cl. Sci.* (5) 16, 957–963.

(1931) Die intuitionistische Grundlegung der Mathematik, *Erkenntnis*, 2, 106–115.

(1934) *Mathematische Grundlagenforschung. Intuitionismus. Beweistheorie.* (Springer, Berlin); reprinted 1974; considerably expanded and updated French translation *Heyting (1955)*.

(1937) Bemerkungen zu dem Aufsatz von Herrn Freudenthal "Zur intuitionistischen Deutung logischer Formeln", *Compos. Math.* 4, 117–118.

(1941) Untersuchungen über intuitionistische Algebra, Nederl. Akad. Wetensch. Verh. Tweede Afd. Nat. 18/2.

(1955) *Les fondements des Mathématiques. Intuitionnisme. Théorie de la Démonstration* (Gauthier-Villars, Paris, E. Nauwelaerts, Louvain); an expanded translation of *Heyting (1934)*.

(1956) *Intuitionism, an Introduction* (North-Holland, Amsterdam); 2nd rev. ed. (1966), 3rd rev. ed. (1971). There exist copies with on the cover "third edition", but the text is that of the second edition; in these copies the foreword to the third edition is missing.

(1958) Intuitionism in mathematics, in: R. Klibansky, ed., *Philosophy in the Mid-century. A Survey* (La Nuova Italia Editrice, Firenze) pp. 101–115.

(1958A) Blick von der intuitionistischen Warte, *Dialectica* 12, 332–345.

(1959) ed., *Constructivity in Mathematics* (North-Holland, Amsterdam).

(1981) Continuum en keuzenrij bij Brouwer (Dutch), *Nieuw Archief voor Wiskunde (3)* 29, 125–139.

Higgs, D.

(1983) A category approach to Boolean-valued set theory, Preprint, University of Waterloo, Canada (n.v.).

Hilbert, D.

(1905) Über die Grundlagen der Mathematik, in: *Verhandlungen des Dritten Internationalen Mathematiker-Kongresses in Heidelberg vom 8. bis 13. August 1904* (Teubner Verlag, Leipzig) pp. 174–185; translated in: *van Heijenoort (1967) 130–185*.

(1922) Neubegründung der Mathematik (Erste Mitteilung), *Abhandlungen aus dem mathematischen Seminar der Hamburgischen Universität* 1, 157–177.

Hilbert, D. and P. Bernays

(1934) *Grundlagen der Mathematik I* (Springer, Berlin); 2nd ed. (1968).

(1939) *Grundlagen der Mathematik II* (Springer, Berlin); 2nd ed. (1970).

Hölder, O.

(1924) *Die mathematische Methode* (Springer, Berlin).

Hoeven, G.F. van der

(1982) *Projections of Lawless Sequences*, Mathematical Centre Tracts (Mathematisch Centrum, Amsterdam).

(1982A) An application of projections of lawless sequences, in: *Troelstra and van Dalen (1982) 487–503*.

Hoeven, G.F. van der and I. Moerdijk

(1983) On an independence result in the theory of lawless sequences, *Indagationes Math.* 45, 185–191.

(1984) Sheaf models for choice sequences, *APAL* 27, 63–107.

(1984A) On choice sequences determined by spreads, *JSL* 49, 908–916.

(1984B) Constructing choice sequences from lawless sequences of neighbourhood functions, in: G.H. Müller and M.M. Richter, eds., *Models and Sets. Proceedings, Logic Colloquium Aachen 1983. Part I* (Springer, Berlin) pp. 207–234.

Hoeven, G.F. van der and A.S. Troelstra
(1979) Projections of lawless sequences II, in: *Boffa et al. (1979) 265–298.*
Howard, W.A.
(1980) The formulae-as-types notion of construction, in: *Seldin and Hindley (1980) 480–490;* circulated as preprint since 1969.
Howard, W.A. and G. Kreisel
(1966) Transfinite induction and bar induction of types zero and one, and the role of continuity in intuitionistic analysis, *JSL* 31, 325–358.
Howe, D.J. see *Constable et al. (1986).*
Hyland, J.M.E.
(1978) The intrinsic recursion theory on the countable or continuous functionals, in: J.E. Fenstad, R.O. Gandy and G.E. Sacks, eds., *Generalized Recursion Theory II* (North-Holland, Amsterdam) p. 135–145.
(1979) Continuity in spatial toposes, in: *Fourman et al. (1979) 442–465.*
(1982) The effective topos, in: *Troelstra and van Dalen (1982) 165–216.*
(1982A) Applications of constructivity, in: *Cohen et al. (1982) 145–152.*
Hyland, J.M.E., P.T. Johnstone and A.M. Pitts
(1980) Tripos theory, *Math. Proc. Camb. Philos. Soc. 88*, 205–232.
Hyland, J.M.E., see also *Fourman and Hyland (1979).*
Jaskowski, S.
(1936) Recherches sur le système de la logique intuitionniste, in: *Actes du Congrès International de Philosophie Scientifique, Septembre 1935 Paris* (Hermann, Paris), Vol. VI, pp. 58–61; translation: *Studia Logica* 34 (1975), 117–120.
Jervell, H.R.
(1971) A normal form in first-order arithmetic, in: *Fenstad (1971) 93–108.*
(1978) Constructive universes, in: *Müller and Scott (1978) 73–98.*
Johansson, I.
(1937) Der Minimalkalkül, ein reduzierter intuitionistischer Formalismus, *Compos. Math.* 4, 119–136.
Jongh, D.H.J. de and C.A. Smorynski
(1976) Kripke models and the intuitionistic theory of species, *AML* 9, 157–186.
Johnstone, P.T.
(1977) *Topos Theory* (Academic Press, London).
(1982) *Stone Spaces* (Cambridge University Press, Cambridge).
Johnstone, P.T. see also *Hyland et al. (1980).*
Joyal, A., see *Boileau and Joyal (1981).*
Julian, W., and F. Richman
(1984) A uniformly continuous function on [0, 1] that is everywhere different from its infimum, *Pac. J. Math.* 111, 333–340.
Julian, W., see also *Berg et al. (1975).*
Kleene, S.C.
(1936) λ-definability and recursiveness, *Duke Math. J.* 2, 340–353 (errata *JSL* 2, 39).
(1945) On the interpretation of intuitionistic number theory, *JSL* 10, 109–124.
(1952) *Introduction to Metamathematics* (North-Holland, Amsterdam).
(1952A) Recursive functions and intuitionistic mathematics, in: L.M. Graves, E. Hille, P.A. Smith and O. Zariski, eds. *Proceedings of the International Congress of Mathematicians, August 1950, Cambridge, MA* (American Mathematical Society, Providence, RI) pp. 679–685.

(1957) Realizability, in: *Summaries of Talks Presented at the Summer Institute for Symbolic Logic, July 1957, Ithaca, NY* (Institute for Defense Analyses, Communications Research Division) pp. 100–104; reprinted in *Heyting (1959) 285–289.*

(1958) Extension of an effectively generated class of functions by enumeration, *Colloquium Mathematicum* 6, 67–78.

(1959) Countable functionals, in: *Heyting (1959) 81–100.*

(1960) Realizability and Shanin's algorithm for the constructive deciphering of mathematical sentences, *Logique et Analyse* 3, 154–165.

(1962) Disjunction and existence under implication in elementary intuitionistic formalisms, *JSL* 27, 11–18 (addendum ibidem 28, 154–156).

(1969) Formalized recursive functionals and formalized realizability, *Mem. Amer. Math. Soc.* 89.

(1973) Realizability: a retrospective survey, in: *Mathias and Rogers (1973) 95–112.*

Kleene, S.C. and R.E. Vesley

(1965) *The Foundations of Intuitionistic Mathematics, Especially in Relation to Recursive Functions* (North-Holland, Amsterdam).

Kneser, H.

(1940) Der Fundamentalsatz der Algebra und der Intuitionismus, *Math. Z.* 46, 287–302.

Kneser, M.

(1981) Ergänzung zu einer Arbeit von Helmut Kneser über den Fundamentalsatz der Algebra, *Math. Z.* 177, 285–287.

Knoblock, T.B., see *Constable et al. (1986).*

Kock, A.

(1976) Universal projective geometry via topos theory. *J. Pure Appl. Algebra* 9, 1–24.

(1981) *Synthetic differential geometry* (Cambridge University Press, Cambridge).

Kolmogorov, A.N.

(1925) On the principle of the excluded middle (Russian), *Mat. Sb.* 32, 646–667; translated in *van Heijenoort (1967) 414–437.*

(1932) Zur Deutung der intuitionistischen Logik, *Math. Z.* 35, 58–65.

Koymans, C.P.J.

(1982) Models of the lambda calculus, *Inf. Control* 52, 306–332.

(1984) Models of the lambda calculus, PhD thesis, Rijksuniversiteit Utrecht.

Krabbe, E.C.W.

(1985) Formal systems of dialogue rules, *Synthese* 63, 295–328.

Kreisel, G.

(1958) Mathematical significance of consistency proofs, *JSL* 23, 155–182.

(1958A) A remark on free choice sequences and the topological completeness proofs, *JSL* 23, 369–388.

(1959) Interpretation of analysis by means of constructive functionals of finite types, in: *Heyting (1959) 101–128.*

(1960) La prédicativité, *Bull. Soc. Math. Fr.* 88, 371–379.

(1962) Foundations of intuitionistic logic, in: E. Nagel, P. Suppes and A. Tarski, eds., *LMPS I* (Stanford University Press, Stanford, CA) pp. 198–210.

(1962A) On weak completeness of intuitionistic predicate logic, *JSL* 27, 139–158.

(1963) Reports of the seminar on the foundations of analysis, mimeographed report in two parts, Stanford University.

(1965) Mathematical logic, in: T.L. Saaty, ed., *Lectures on Modern Mathematics III* (Wiley and Sons, New York) pp. 95–195).

(1967) Informal rigour and completeness proofs, in: I. Lakatos, ed., *Problems in the Philosophy of Mathematics* (North-Holland, Amsterdam) pp. 138–186.

(1968) Lawless sequences of natural numbers, *Compos. Math.* 20, 222–248.

(1968A) Functions, ordinals, species, in: *van Rootselaar and Staal (1968) 145–159.*

(1970) Church's thesis: a kind of reducibility axiom for constructive mathematics, in: *Myhill et al. (1970) 121–150.*

(1971) A survey of proof theory II, in: *Fenstad (1971) 109–170.*

(1972) Which number-theoretic problems can be solved in recursive progressions on Π_1^1-paths through $\mathcal{O}$?, *JSL* 37, 311–334.

Kreisel, G. and D. Lacombe

(1957) Ensembles récursivements mésurables et ensembles récursivements ouverts ou fermés, *C.R. Acad. Sci. Paris, Ser. A–B* 245, 1106–1109.

Kreisel, G., D. Lacombe and J. Shoenfield

(1957) Fonctionelles récursivement définissables et fonctionelles récursives, *C.R. Acad. Sci. Paris, Ser. A–B* 245, 399–402.

(1959) Partial recursive functionals and effective operations, in: *Heyting 1959, 290–297.*

Kreisel, G. and A.S. Troelstra

(1970) Formal systems for some branches of intuitionistic analysis, *AML* 1, 229–387.

Kreisel, G., see also *Dyson and Kreisel (1961), Howard and Kreisel (1966).*

Kreisel, G., see also *Howard and Kreisel (1966).*

Kripke, S.A.

(1963) Semantical considerations on modal and intuitionistic logic, *Acta Philosophica Fennica* 16, 83–94.

(1965) Semantical analysis of intuitionistic logic, in: J. Crossley and M.A.E. Dummett, eds., *Formal Systems and Recursive Functions* (North-Holland, Amsterdam) pp. 92–130.

Krol', M.D.

(1978) A topological model for intuitionistic analysis with Kripke's schema, *ZMLGM* 24, 427–436.

(1978A) Distinct variants of Kripke's schema in intuitionistic analysis, *Dokl. Akad. Nauk SSSR* 271, 33–36; translation: *Sov. Math., Dokl.* 28, 27–30.

Kronecker, L.

(1887) Über den Zahlbegriff, *J. Reine Angew. Math.* 101, 337–355.

(1901) *Vorlesungen über Zahlentheorie I*, ed. K. Hensel (Teubner, Leipzig).

(1930) *Werke*, ed. K. Hensel (Teubner, Leipzig); the collected works appeared over the period 1895–1930.

Kuroda, S.

(1951) Intuitionistische Untersuchungen der formalistischen Logik, *Nagoya Math. J.* 2, 35–47.

Kushner, B.A.

(1973) *Lectures on Constructive Mathematical Analysis* (Russian) (Nauka, Moskva). Translation by E. Mendelson (American Mathematical Society, Providence, RI, 1984).

(1981) Behaviour of the general term of a Specker series (Russian), in: M. Gladkij, ed., *Mathematical Logic and Mathematical Linguistics* (Kalinin Gos. Univ., Kalinin) pp. 112–116. Math. Reviews 84i:03111 (n.v.).

(1983) A class of Specker sequences (Russian), in: *Mathematical Logic, Mathematical Linguistics and Theory of Algorithms* (Kalinin Gos. Univ., Kalinin) pp. 62–65. Math. Reviews 85e: 03152 (n.v.).

Lacombe, D., see *Kreisel and Lacombe (1957), Kreisel et al. (1957, 1959).*

Lakatos, I.
(1976) *Proofs and Refutations*, eds. J. Worrall and E. Zahar (Cambridge University Press, Cambridge).

Lambek, J. and P.J. Scott
(1986) *Introduction to Higher-Order Categorical Logic* (Cambridge University Press, Cambridge).

Lang, S.
(1965) *Algebra* (Addison-Wesley, Reading, MA).

Läuchli, H.
(1970) An abstract notion of realizability for which the predicate calculus is complete, in: *Myhill et al. (1970) 227–234.*

Laudet, M., D. Lacombe, D. Nolin and M. Schützenberger
(1970) eds., *Symposium on Automatic Demonstration, December 1980, Rocquencourt* (Springer, Berlin).

Lawvere, W.
(1971) Quantifiers and sheaves, in: *Actes du Congrès International des Mathématiciens, 1–10 Septembre 1970, Nice, France* (Gauthier-Villars, Paris) Vol. I, pp. 329–334.
(1976) Variable quantities and variable structures in topoi, in: A. Heller and M. Tierney, eds., *Algebra, Topology and Category Theory* (Academic press, New York) pp. 101–131.

Leivant, D.M.E.
(1971) A note on translations from *C* into *I*, Report ZW5/71, Mathematisch Centrum, Amsterdam.
(1976) Failure of completeness properties for intuitionistic predicate logic for constructive models, *Ann. Fac. Sci. Univ. Clermont* 13, 93–107.
(1979) Assumption classes in natural deduction, *ZMLGM* 25, 1–4.
(1985) Syntactic translations and provably recursive functions, *JSL* 50, 682–688.
(1985A) Intuitionistic formal systems, in: *Harrington et al. (1985) 231–255.*

Levy, A.
(1973) Ch. II in: *Fraenkel et al. (1973)*; also in: *Müller (1976) 173–215.*

Lifschitz, V.A.
(1979) CT_0 is stronger than $CT_0!$, *Proc. Amer. Math. Soc.* 73, 101–106.

Lodder, J.S., see *van Dalen and Lodder (1982).*

Lolli, G., G. Longo and G. Marcja
(1984) eds., *Logic Colloquium '82* (North-Holland, Amsterdam).

Loor, B. de, see *Brouwer and de Loor (1924).*

Lopez-Escobar, E.G.K.
(1981) Equivalence between semantics for intuitionism, *JSL* 46, 773–780.

Lopez-Escobar, E.G.K. and W. Veldman
(1975) Intuitionistic completeness of a restricted second-order logic, in: *Diller and Müller (1975) 198–232.*

Lorenzen, P.
(1955) *Einführung in die operative Logik und Mathematik* (Springer, Berlin); 2nd ed. (1969).

(1960) Logik und Agon, in: *Atti del XII Congresso Internazionale di Filosofia (Venezia, 12–18 Settembre 1958)* (Sansoni, Firenze) Vol. IV, pp. 187–194; also in: *Lorenzen and Lorenz (1978) 9–16.*

(1965) *Differential und Integral* (Akademische Verlagsgesellschaft, Frankfurt a.M.). Translated as: *Differential and integral*, by J. Bacon (University of Texas Press, Austin, TX, 1970).

Lorenzen, P. and K. Lorenz

(1978) eds., *Dialogische Logik* (Wissenschaftliche Buchgesellschaft, Darmstadt).

Luckhardt, H.

(1970) Ein Henkin-Vollständigkeitsbeweis für die intuitionistische Prädikatenlogik bezüglich der Kripke-Semantik, *Arch. ML* 13, 55–59.

(1973) *Extensional Gödel functional interpretation. A consistency proof of classical analysis* (Springer, Berlin).

(1977) Über das Markov-Prinzip, *Arch. ML* 18, 73–80.

(1977A) Über das Markov-Prinzip II, *Arch. ML* 18, 147–157.

Lusin, N.

(1930) *Leçons sur les Ensembles Projectives et leur Applications* (Gauthier-Villars, Paris).

Maclane, S.

(1971) *Categories for the Working Mathematician* (Springer, Berlin).

Makkai, M. and G. Reyes

(1977) *First-order Categorical Logic* (Springer, Berlin).

Malmnås, P.E. and D. Prawitz

(1968) A survey of some connections between classical, intuitionistic and minimal logic, in: *H.A. Schmidt et al. (1968) 215–229.*

Mandelkern, M.

(1976) Connectivity of an interval, *Proc. Amer. Math. Soc.* 54, 170–172.

(1982) Continuity of monotone functions, *Pac. J. Math.* 99, 413–418.

Mannoury, G.

(1909) *Methodologisches und Philosophisches zur Elementar-Mathematik* (P. Visser, Haarlem).

Marcus, R.B., G.J.W. Dorn and P. Weingartner

(1986) eds., *LMPS VII* (North-Holland, Amsterdam).

Markov, A.A.

(1954) On the continuity of constructive functions (Russian), *Usp. Math. Nauk* 9/3 61, 226–230.

(1954A) *Theory of algorithms* (Russian), Trudy Matematiceskog Istituta imeni V.A. Steklov., Vol. 42 (Izdatel'stvo Akademii Nauk SSSR, Moskva). English translation by J.J. Schoor-Kan and PST staff, Israel Program for Scientific Translations, Jerusalem (1961).

(1962) On constructive mathematics (Russian), *Tr. Mat. Inst. Steklov.* 67, 8–14; translated in *Amer. Math. Soc. Transl., II Ser.* 98 (1971), 1–9.

Martin-Löf, P.

(1972) An intuitionistic theory of types, Report, Mathematical Institute, University of Stockholm, 86 pp.

(1975) An intuitionistic theory of types: predicative part, in: *Rose and Shepherdson (1975) 73–118.*

(1982) Constructive mathematics and computer programming, in: *Cohen et al. (1982) 153–175.*

(1984) *Intuitionistic Type Theory. Notes by Giovanni Sambin of a Series of Lectures given in Padua, June 1980* (Bibliopolis, Napoli).

Martino, E. and P. Giaretta

(1981) Brouwer, Dummett and the bar theorem, in: S. Bernini, ed., *Atti del Congresso Nazionale di Logica, Montecatini Terme, 1–5 Ottobre 1979* (Bibliopolis, Napoli) 541–558.

Mathias, A.R.D. and H. Rogers

(1973) eds., *Cambridge Summer School in Mathematical Logic* (Springer, Berlin).

McCarty, D.C.

(1984) Realizability and recursive mathematics, PhD thesis, University of Oxford; also: report CMU-CS-84-131, Department of Computer Science, Carnegie-Mellon University, Pittsburgh, PA.

(1986) Subcountability under realizability, *Notre Dame Journal of Formal Logic* 27, 210–220.

(1986A) Realizability and recursive set theory, *APAL* 32, 11–194.

(1986B) Constructive validity is non-arithmetic, Report CSR-203-86, Department of Computer Science, University of Edinburgh, to appear in *JSL*.

McKinsey, J.C.C. and A. Tarski

(1946) On closed elements in closure algebras, *Ann. of Math.* 47, 122–162.

(1948) Some theorems on the sentential calculi of Lewis and Heyting, *JSL* 13, 1–15.

Mendelson, E.

(1964) *Introduction to Mathematical Logic* (Van Nostrand, New York; 2nd ed., 1979).

Mendler, N.P., see *Constable et al. (1986)*.

Metakides, G. and A. Nerode

(1975) Recursion theory and algebra, in: *Crossley (1975) 209–219*.

(1982) The introduction of non-recursive methods into mathematics, in: *Troelstra and van Dalen (1982) 319–335*.

Mines, R., F. Richman and W. Ruitenburg

(1988) *A course in Constructive Algebra* (Springer, Berlin).

Mines, R., see also *Berg et al. (1975, 1976, 1977)*.

Mints, G.E.

(1976) What can be done with PRA (Russian), *Zap. Nauchn. Semin. Leningr. Otd. Mat. Inst. Steklov.* 60, 93–102; translation: *J. Sov. Math.* 14, 1487–1492.

Mints, G.E. and V.P. Orevkov

(1967) On embedding operators (Russian), *Zap. Nauchn. Semin. Leningr. Otd. Mat. Inst. Steklov.* 4, 160–167; translation: *Sem. Math. V.A. Steklov* 4, 64–66.

Mitchell, W.

(1972) Boolean topoi and the theory of sets, *J. Pure App. Alg.* 2, 261–274.

Moerdijk, I.

(1982) Glueing topoi and higher order disjunction and existence, in: *Troelstra and van Dalen (1982) 359–375*.

(1984) Heine–Borel does not imply the fan theorem, *JSL* 49, 514–519.

(1986) Continuous fibrations and inverse limits of toposes, *Compos. Math.* 58, 45–72.

Moerdijk, I., see also *van der Hoeven and Moerdijk (1983, 1984, 1984A, 1984B)*.

Molk, J.

(1885) Sur une notion qui comprend celle de la divisibilité et sur la théorie de l'élimination, *Acta Mathematica* 6, 1–166.

Mooij, J.J.A.

(1966) *La Philosophie des Mathématiques de Henri Poincaré* (Gauthier-Villars Paris; E. Nauwelaerts, Louvain).

Moschovakis, J.R.

(1973) A topological interpretation of second-order intuitionistic arithmetic, *Compos. Math.* 26, 261–275.

(1981) A disjunctive decomposition theorem for classical theories, in: *Richman (1981) 250–259.*

(1987) Relative lawlessness in intuitionistic analysis, *JSL* 52, 68–88.

Moschovakis, Y.N.

(1964) Recursive metric spaces, *Fundam. Math.* 55, 215–238.

Mostowski, A.W.

(1948) Proofs of non-deducibility in intuitionistic functional calculus, *JSL* 13, 204–207.

Motohashi, N.

(1982) An elimination theorem of uniqueness conditions in the intuitionistic predicate calculus, *Nagoya Math. J.* 85, 223–230.

Müller, G.H.

(1976) ed., *Sets and Classes. On the Work of Paul Bernays* (North-Holland, Amsterdam).

(1987) ed., *Ω-bibliography of Mathematical Logic*, edited by G.H. Müller in collaboration with W. Lenski, *Volume 6: Proof Theory and Constructive Mathematics*, eds. D. van Dalen, J.E. Kister and A.S. Troelstra (Springer, Berlin).

Müller, G.H. and D.S. Scott

(1978) eds., *Higher Set Theory* (Springer, Berlin).

Müller, G.H., G. Takeuti and T. Tugué

(1981) eds., *Logic Symposia, Hakone 1979–1980* (Springer, Berlin).

Myhill, J.

(1963) The invalidity of Markov's schema, *ZMLGM* 9, 359–360.

(1967) Notes towards an axiomatization of intuitionistic analysis, *Logique et Analyse* 9, 280–297.

(1971) A recursive function, defined on a compact interval and having a continuous derivative that is not recursive, *Michigan Math. J.* 18, 97–98.

(1974) "Embedding classical type theory in 'intuitionistic' type theory", a correction, in: T. Jech, ed., *Axiomatic Set Theory, part II* (American Mathematical Society, Providence, RI) pp. 185–188.

(1975) Constructive set theory, *JSL* 40, 347–382.

Myhill, J., see also *Goodman and Myhill (1978).*

Myhill, J., A. Kino and R.E. Vesley

(1970) eds., *Intuitionism and Proof Theory* (North-Holland, Amsterdam).

Nelson, D.

(1947) Recursive functions and intuitionistic number theory, *Trans. Amer. Math. Soc.* 61, 307–368, 556.

Nerode, A., see *Metakides and Nerode (1975, 1982).*

Niekus, J.M.

(1987) The method of the creative subject, *Indagationes Math.* 90, 431–443.

Nishimura, I.

(1960) On formulas of one variable in intuitionistic propositional calculus, *JSL* 25, 327–331.

Normann, D.
(1980) *Recursion on the Countable Functionals* (Springer, Berlin).

Ogasawara, T.
(1939) Relation between intuitionistic logic and lattices, *Hiroshima Univ. Sci. Ser. A* 9, 157–164.

Orevkov, V.P.
(1971) Equivalence of two definitions of continuity (Russian), *Zap. Nauchn. Semin. Leningr. Otd. Mat. Inst. Steklov.* 20, 145–159, 286; translation: *J. Sov. Math.* 1 (1973), 92–99.

Orevkov, V.P., see *Mints and Orevkov (1967).*

Panangaden, P., see *Constable et al. (1986).*

Parikh, R.J.
(1971) Existence and feasibility in arithmetic, *JSL* 36, 494–508.

Peter, R.
(1940) Review of Skolem 1939, *JSL* 5, 34–35.

Plotkin, G.D.
(1972) A set-theoretical definition of application, Memo MIR-R-95, School of Artificial Intelligence, University of Edinburgh (n.v.).

Pitts, A.M. see *Hyland et al. (1980).*

Pohlers, W., see *Buchholz et al. (1981).*

Poincaré, H.
(1902) *Science et Hypothèse* (Flammarion, Paris).
(1905) *La Valeur de la Science* (Flammarion, Paris).
(1908) *Science et Méthode* (Flammarion, Paris).
(1913) *Dernières Pensées* (Flammarion, Paris).

Posy, C.J.
(1976) Varieties of indeterminacy in the theory of general choice sequences, *J. Phil. Logic* 5, 91–132.

Pottinger, G.
(1977) Normalization as a homomorphic image of cut-elimination, *AML* 12, 323–357.

Prawitz, D.
(1965) *Natural Deduction A Proof-theoretical Study* (Almquist and Wiksell, Stockholm).
(1971) Ideas and results in proof theory, in: *Fenstad (1971) 237–309.*
(1977) Meaning and proofs: on the conflict between classical and intuitionistic logic, *Theoria* 43, 1–40.

Prawitz, D., see also *Malmnås and Prawitz (1968).*

Rasiowa, H.
(1951) Algebraic treatment of the functional calculi of Heyting and Lewis, *Fund. Math.* 38, 101–116.
(1954) Constructive theories, *Bull. Acad. Polon. Sci. Ser. Sci. Math. Astron. Phys.* 1, 229–231.
(1955) Algebraic models of axiomatic theories, *Fundam. Math.* 41, 291–310.

Rasiowa, H. and R. Sikorski
(1963) *The Mathematics of Metamathematics* (PWN, Warszawa).

Rath, P.
(1978) Eine verallgemeinerte Funktionalinterpretation der Heyting Arithmetik endlicher Typen, PhD thesis, Westfälische Wilhelms Universität, Münster i.Wf.

Renardel de Lavalette, G.R.

(1984) Descriptions in mathematical logic, *Studia Logica* 43, 281–294.

(1984A) Theories with type-free application and extended bar induction, PhD thesis, University of Amsterdam.

Reyes, G.E.

(1974) From sheaves to logic, in: A. Daigneault, ed., *Studies in Algebraic Logic* (The Mathematical Association of America (Inc.)) 143–204.

Reyes, G.E. see also *Makkai and Reyes (1977).*

Rezus, A.

(1983) Abstract AUTOMATH, Mathematisch Centrum, Amsterdam.

(1985) Semantics of constructive type theory, report no. 70, Informatics Department, Katholieke Universiteit Nijmegen.

Rezus, A., see also *Barendregt and Rezus (1983).*

Rice, H.G.

(1954) Recursive real numbers, *Proc. Amer. Math. Soc.* 5, 784–791.

Richman, F.

(1981) ed., *Constructive Mathematics. Proceedings of the New Mexico State Conference* (Springer, Berlin).

(1983) Church's thesis without tears, *JSL* 48, 797–803.

Richman, F., see also *Berg et al. (1975, 1976, 1977), Julian and Richmann (1984), Mines et al. (1988).*

Rieger, L.

(1949) On the lattice theory of Brouwerian propositional logic, *Acta Univ. Carol. Math. Phys.* 189.

Rootselaar, B. van

(1952) Un problème de M. Dijkman, *Indagationes Math.* 14, 405–407.

Rootselaar, B. van and J. F. Staal

(1968) eds., *LMPS III* (North-Holland, Amsterdam).

Rose, H.E. and J. Shepherdson

(1975) eds., *Logic Colloquium '73* (North-Holland, Amsterdam).

Rosolini, G.

(1986) Continuity and effectiveness in topoi, PhD thesis, University of Oxford; also report CMU-CS-86-123, Department of Computer Science, Carnegie-Mellon University, Pittsburgh, PA.

Ruitenburg, W.

(1982) Intuitionistic Algebra, PhD thesis, Rijksuniversiteit Utrecht.

(1982A) Primality and invertibility of polynomials, in: *Troelstra and van Dalen (1982)*, 413–434.

Ruitenburg, W., see also *Mines et al. (1988).*

Sasaki, J.T., see *Constable et al. (1986).*

Scedrov, A.

(1981) Consistency and independence results in intuitionistic set theory, in: *Richman (1981) 54–86.*

(1982) Independence of the fan theorem in the presence of continuity principles, in: *Troelstra and van Dalen (1982) 435–442.*

(1984) Forcing and classifying topoi, *Mem. Amer. Math. Soc.* 295.

(1985) Intuitionistic set theory, in: *Harrington et al. (1985) 257–284.*

(1986) Embedding sheaf models for set theory into boolean-valued permutation models with an interior operator, *APAL* 32, 103–109.

Scedrov, A. and P.J. Scott
(1982) A note on the Friedman slash and Freyd covers, in: *Troelstra and van Dalen (1982) 443–452.*

Scedrov, A., see also *Friedman and Scedrov (1983, 1984).*

Schmidt, H.A., K. Schütte and H.-J. Thiele
(1968) eds., *Contributions to Mathematical Logic* (North-Holland, Amsterdam).

Schroeder-Heister, P.
(1984) A natural extension of natural deduction, *JSL* 49, 1284–1300.

Schütte, K.
(1960) *Beweistheorie* (Springer, Berlin).
(1962) Der Interpolationssatz der intuitionistischen Prädikatenlogik, *Math. Ann.* 148, 192–200.
(1968) *Vollständige Systeme modaler und intuitionistischer Logik* (Springer, Berlin).
(1977) *Proof Theory* (Springer, Berlin).

Schwichtenberg, H.
(1979) Logic and the axiom of choice, in: *Boffa et al. (1979) 351–356.*

Scott, D.S.
(1968) Extending the topological interpretation to intuitionistic analysis I, *Compos. Math.* 20, 194–210.
(1970) Constructive validity in: *Laudet et al. (1970) 237–275.*
(1970A) Extending the topological interpretation to intuitionistic analysis II, in: *Myhill et al. (1970) 235–255.*
(1975) Lambda calculus and recursion theory, in: S. Kanger, ed., *Proceedings of the Third Scandinavian Logic Symposium* (North-Holland, Amsterdam) pp. 154–193.
(1975A) Data types as lattices, in: G.H. Müller, A. Oberschelp and K. Potthoff, eds., *ISILC Logic Conference* (Springer, Berlin) pp. 579–651; also in: *SIAM J. Comput.* 5 (1976) 522–587.
(1979) Identity and existence in intuitionistic logic, in: *Fourman et al. (1979) 660–696.*

Scott, D.S., see also *Fourman and Scott (1979).*

Scott, P.J., see *Lambek and Scott (1986), Scedrov and Scott (1982).*

Seely, R.A.G.
(1982) Locally cartesian categories and type theory, *C.R. Math. Rep. Acad. Sci. Can.* 4, 271–275.
(1984) Locally cartesian categories and type theory, *Math. Proc. Camb. Phil. Soc.* 95, 33–48.

Segerberg, K.
(1974) Proof of a conjecture of McKay, *Fundam. Math.* 81, 267–270.

Seidenberg, A.
(1974) Constructions in algebra, *Trans. Amer. Math. Soc.* 197, 273–313.
(1978) Constructions in a polynomial ring over the ring of integers, *Amer. J. Math.* 100, 685–706.

Seldin, J.P. and J.R. Hindley
(1980) eds., *To H.B. Curry: Essays on Combinatory Logic, Lambda Calculus and Formalism* (Academic Press, New York).

Shanin, N.A.

(1958) On the constructive interpretation of mathematical judgments (Russian), *Tr. Mat. Inst. Steklov.* 52, 226–311. Translated in *Amer. Math. Soc. Transl., II Ser.* 23 (1963), 109–189.

(1958A) Über einen Algorithmus zur konstruktiver Dechiffrierung mathematischer Urteile (Russian), *ZMLGM* 4, 293–303.

(1962) Constructive real numbers and constructive function spaces (Russian), *Tr. Mat. Inst. Steklov.* 67, 15–294. Translation: (American Mathematical Society, Providence, RI, 1968).

Shapiro, S.

(1981) Understanding Church's thesis, *J. Philos. Logic* 10, 353–365.

(1985) ed., *Intensional Mathematics* (North-Holland, Amsterdam).

Shoenfield, J.R., see *Kreisel et al. (1957, 1959).*

Sieg, W.

(1985) Fragments of arithmetic, *APAL*, 28, 33–71.

Sieg, W., see also *Buchholz et al. (1981).*

Sikorski, R.

(1958) Some applications of interior mappings, *Fundam. Math.* 45, 200–212.

Sikorski, R., see also *Rasiowa and Sikorski (1963).*

Simpson, S.G.

(1984) Which set existence axioms are needed to prove the Cauchy/Peano theorem for ordinary differential equations?, *JSL* 49, 783–802.

Simpson, S.G., see also *Friedman et al. (1983).*

Skolem, T.

(1923) Begründung der elementaren Arithmetik durch die rekurrierende Denkweise ohne Anwendung scheinbarer Veränderlichen mit unendlichen Ausdehnungsbereich, Videnskaps Selskapet i Kristiana, Skrifter Utgit (1) 6, 1–38. Translated in *van Heijenoort (1967) 303–333.*

(1939) Eine Bemerkung über die Induktionsschemata in der rekursiven Zahlentheorie, *Monatshefte für Mathematik und Physik* 48, 268–276.

Smith, J.

(1984) An interpretation of Martin-Löf's type theory in a type-free theory of propositions, *JSL* 49, 730–753.

(A) The independence of Peano's fourth axiom from Martin-Löf's type theory without universes, *JSL*, to appear.

Smith, R.L., see *Friedman et al. (1983).*

Smith, S.F., see *Constable et al. (1986).*

Smorynski, C.A.

(1973) Applications of Kripke models, in: *Troelstra (1973) 324–391.*

(1973A) Investigations of intuitionistic formal sytems by means of Kripke models, PhD thesis, University of Chicago.

(1977) On axiomatizing fragments, *JSL* 42, 530–544.

(1978) The axiomatization problem for fragments, *AML* 14, 193–221.

(1982) Nonstandard models and constructivity, in: *Troelstra and van Dalen (1982) 459–464.*

Smorynski, C.A., see also *de Jongh and Smorynski (1976).*

Specker, E.

(1949) Nicht konstruktiv beweisbare Sätze der Analysis, *JSL* 14, 145–158.

Spector, C.
(1962) Provably recursive functionals of analysis: a consistency proof of analysis by an extension of principles formulated in current intuitionistic mathematics, in: J.C.E. Dekker, ed., *Recursive Function Theory*, Proceedings of Symposia in Pure Mathematics V (American Mathematical Society, Providence, RI) pp. 1–27.

Staples, J.
(1971) On constructive fields, *Proc. London Math. Soc.* 23, 753–768.

Stone, M.H.
(1937) Topological representations of distributive lattices and Brouwerian logics, *Casopis Pro Pestvování Matematiky a Fysiki Cast Matematická* 67, 1–25.

Strong, H.
(1968) Algebraically generalized recursive function theory, *IBM J. Research Develop.* 12, 465–475.

Sundholm, G.
(1983) Constructions, proofs and the meaning of logical constants, *J. Philos. Logic* 12, 151–172.
(1986) Proof theory and meaning, in: *Gabbay and Guenthner (1986) 471–506.*

Swart, H.C.M. de
(1976) Another intuitionistic completeness proof, *JSL* 41, 644–662.

Tait, W.W.
(1967) Intensional interpretation of functionals of finite type I, *JSL* 32, 198–212.

Takeuti, G.
(1978) *Two Applications of Logic to Mathematics* (Princeton University Press, Princeton).

Takeuti, G. and S. Titani
(1981) Heyting-valued universes of intuitionistic set theory, in: *Müller et al. (1981) 189–206.*

Tarski, A.
(1938) Der Aussagenkalkül und die Topologie, *Fundam. Math.* 31, 103–134.

Tarski, A., see also *McKinsey and Tarski (1946, 1948).*

Thomason, R.H.
(1968) On the strong semantical completeness of the intuitionistic predicate calculus, *JSL* 33, 1–7.

Titani, S., see *Takeuti and Titani (1981).*

Tragesser, R.S.
(1977) *Phenomenology and Logic* (Cornell University Press, Ithaca, N.Y.).

Troelstra, A.S.
(1966) Intuitionistic general topology, PhD thesis, Universiteit van Amsterdam.
(1967) Intuitionistic continuity, *Nieuw Archief voor Wiskunde (3)* 15, 2–6.
(1967A) Intuitionistic connectedness, *Indagationes Math.* 29, 96–105.
(1968) The use of "Brouwer's principle" in intuitionistic topology, in: *H.A. Schmidt et al. (1968) 289–298.*
(1968A) The theory of choice sequences, in: *van Rootselaar and Staal (1968) 201–223.*
(1968B) New sets of postulates for intuitionistic topology, *Compos. Math.* 20, 211–221.
(1968C) One-point compactifications of intuitionistic locally compact spaces, *Fundam. Math.* 62, 75–93.
(1969) *Principles of Intuitionism* (Springer, Berlin).
(1969A) Notes on the intuitionistic theory of sequences I, *Indagationes Math. 31*, 430–440.
(1969B) Informal theory of choice sequences, *Studia Logica* 25, 31–54.

(1970) Notes on the intuitionistic theory of sequences II, *Indagationes Math*. 32, 99–109.

(1970A) Notes on the intuitionistic theory of sequences III, *Indagationes Math*. 32, 245–252.

(1971) Notions of realizability for intuitionistic arithmetic and intuitionistic arithmetic in all finite types, in: *Fenstad (1971) 369–405*.

(1973) *Metamathematical Investigation of Intuitionistic Arithmetic and Analysis* (Springer, Berlin); for a list of errata see *Troelstra (1974)*.

(1973A) Notes on intuitionistic second-order arithmetic; in: *Mathias and Rogers (1973) 171–205*.

(1974) Corrections and additions to "Metamathematical Investigation of Intuitionistic Arithmatic and Analysis", report 74-76, Department of Mathematics, Universiteit van Amsterdam.

(1977) *Choice Sequences, a Chapter of Intuitionistic Mathematics* (Clarendon Press, Oxford); errata in: *Troelstra (1983A) 205–206*.

(1977A) Axioms for intuitionistic mathematics incompatible with classical logic, in: R.E. Butts and J. Hintikka, eds., *Logic, Foundations of Mathematics and Computability Theory. Part One of LMPS V* (Reidel, Dordrecht) pp. 59–84.

(1977B) Some models for intuitionistic finite type arithmetic with fan functional, *JSL* 42, 194–202.

(1977C) A note on non-extensional operations in connection with continuity and recursiveness, *Indagationes Math*. 39, 455–462.

(1978) A. Heyting on the formalization of intuitionistic mathematics, in: E.M.J. Bertin, H.J.M. Bos and A.W. Grootendorst, eds., *Two Decades of Mathematics in the Netherlands 1920–1940. A Retrospection on the Occasion of the Bicentennial of the Wiskundig Genootschap* (Mathematisch Centrum, Amsterdam) pp. 153–175.

(1978A) Some remarks on the complexity of Henkin–Kripke models, *Indagationes Math*. 40, 296–302.

(1980) Intuitionistic extensions of the reals, *Nieuw Archief voor Wiskunde (3)* 28, 63–113.

(1980A) Extended bar induction of type zero, in: *Barwise et al. (1980) 277–316*.

(1981) The interplay between logic and mathematics: intuitionism, in: E. Agazzi, ed., *Modern Logic–A Survey* (Reidel, Dordrecht) pp. 197–221.

(1981A) Arend Heyting and his contribution to intuitionism, *Nieuw Archief voor Wiskunde (3)* 29, 1–23.

(1982) Intuitionistic extensions of the reals II, in: *van Dalen et al. (1982) 279–310*.

(1982A) On the origin and development of Brouwer's concept of choice sequence, in: *Troelstra and van Dalen (1982) 465–486*.

(1983) Definability of finite sum types in Martin-Löf's type theories, *Indagationes Math*. 45, 475–481.

(1983A) Analyzing choice sequences, *J. Philos. Logic* 12, 197–260.

(1983B) Logic in the writings of Brouwer and Heyting, in: V.M. Abrusci, E. Casari and M. Mugnai, eds., *Atti del Convegno Internazionale di Storia della Logica, San Gimignano, 4–8 Dicembre 1982* (Cooperativa Libraria Universitaria Editrice Bologna, Bologna) pp. 193–210.

Troelstra, A.S., see also *van Dalen and Troelstra (1970), Diller and Troelstra (1984), van der Hoeven and Troelstra (1979), Kreisel and Troelstra (1970)*.

Troelstra, A.S. and D. van Dalen

(1982) eds., *The L.E.J. Brouwer Centenary Symposium* (North-Holland, Amsterdam).

Tsejtin, G.S.
(1959) Algorithmic operators in constructive complete separable metric spaces (Russian), *Dokl. Akad. Nauk SSSR* 128, 49–52.
(1962) Algorithmic operators in constructive metric spaces (Russian), *Tr. Mat. Inst. Steklov.* 67, 295–361. Translated in *Amer. Math. Soc. Transl., II Ser.* 64 (1967), 1–80.

Tsejtin, G.S. and I.D. Zaslavskij
(1962) On singular coverings and properties of constructive functions connected with them, *Tr. Mat. Inst. Steklov.* 67, 458–502. Translated in *Amer. Mat. Soc. Transl., II Ser.* 98 (1971), 41–89.

Turing, A.M.
(1937) On computable numbers, with an application to the Entscheidungsproblem, *Proc. London Math. Soc.* 42, 230–265; corrections ibidem 43, 544–546; reprinted in *Davis (1965) 116–154.*

Unterhalt, M.
(1986) Kripke-Semantik mit partieller Existenz, Dissertation, Westfälische Wilhelms-Universität, Münster i. Wf.

Veldman, W.
(1976) An intuitionistic completeness theorem for intuitionistic predicate logic, *JSL* 41, 159–166.
(1981) Investigations in intuitionistic hierarchy theory, PhD thesis, Katholieke Universiteit Nijmegen, 235 pp.

Vesley, R.E. see also *Kleene and Vesley (1965).*

Visser, A.
(1981) Aspects of diagonalization and provability, PhD thesis, Rijksuniversiteit Utrecht.

Vrijer, R.C. de
(1987) Surjective pairing and normalization: two themes in the lambda calculus, PhD thesis, Universiteit van Amsterdam.

Waerden, B. van der
(1930) Eine Bemerkung über die Unzerlegbarkeit von Polynomen, *Math. Ann.* 102, 738–739.

Wagner, E.
(1969) Uniform reflexive structures: on the nature of Gödelizations and relative computability, *Trans. Amer. Math. Soc.* 144, 1–41.

Walton, D.N.
(1985) ed., The logic of dialogue, *Synthese* 63, no. 3, 259–388.

Webb, J.C.
(1980) *Mechanism, mentalism and metamathematics*, Reidel, Dordrecht.

Weinstein, S.
(1983) The intended interpretation of intuitionistic logic, *J. Philos. Logic* 12, 261–270.

Wells, C., see *Barr and Wells (1985).*

Weyl, H.
(1918) *Das Kontinuum. Kritische Untersuchungen über die Grundlagen der Analysis* (Veit, Leipzig); reprinted in: *Das Kontinuum und andere Monographien* (Chelsea Publ. Co., New York, 1966).
(1921) Über die neue Grundlagenkrise der Mathematik, *Math. Z.* 10, 39–70.
(1924) Randbemerkungen zur Hauptproblemen der Mathematik. *Math. Z.* 20, 131–150.

Wraith, G.
(1979) Generic Galois theory of local rings, in: *Fourman et al. (1979) 739–767.*

Wright, C.
(1982) Strict finitism, *Synthese* 51, 203–282.

Zaslavskij, I.D.
(1962) Some properties of constructive real numbers and constructive functions (Russian), *Tr. Mat. Inst. Steklov.* 67, 385–457. Translated in *Amer. Math. Soc. Transl., II. Ser.* 57, 1–84.

Zaslavskij, I.D., see also *Tsejtin and Zaslavskij (1962).*

Zucker, J.I.
(1974) The correspondence between cut-elimination and normalization I, II, *AML* 7, 1–156.
(1977) Formalization of classical mathematics in AUTOMATH, in: *Colloque International de Logique, Clermont-Ferrand, France* (Editions du Centre National de Recherche, Paris) pp. 135–145.

INDEX

Below we have listed notions and terminology from both volumes. Bold-face page numbers refer to places where a definition or a description of some form is given; if more definitions in varying contexts are given they are all shown in bold-face. Pages with roman numbers refer to both volumes; the first number refers to Volume II and the second number (in brackets) refers to Volume I.

INDEX OF NAMES

The bibliography has not been indexed.

LIST OF SYMBOLS

Part I lists the abbreviations designating formal systems, part II the abbreviations for axioms or axiom schemas. The listing is alphabetical, disregarding mathematical symbols which are not roman letters. Part III lists other notations in order of appearance. Notations which are used only locally ('ad hoc') are not listed as a rule.

I. Formal systems

APP, 472	applicative theory of operations
CPC, 48	classical propositional logic
CQC, 48	classical predicate logic
CS, 670	theory of choice sequences
CZF, 620	constructive set theory
$\mathbf{E\text{-}HA}^{\omega}$, 451	extensional intuitionistic finite-type arithmetic
$\mathbf{E\text{-}HA}^{\omega}_{\rightarrow}$, 458	$\mathbf{E\text{-}HA}^{\omega}$ without product types
EL, 144	elementary analysis
$\mathbf{EM}_0$, $\mathbf{EM}_0\restriction$, 512	elementary mathematics
HA, 126	intuitionistic (first-order) arithmetic, Heyting arithmetic
$\mathbf{HA}^{\omega}$, 447	intuitionistic finite-type arithmetic
$\mathbf{HA}^{\omega}_{\rightarrow}$, 456	$\mathbf{HA}^{\omega}$ without products
$\mathbf{HA}^{\omega}_{0}$, 452	variant of $\mathbf{HA}^{\omega}$
$\lambda\text{-}\mathbf{HA}^{\omega}$, 467	$\mathbf{HA}^{\omega}$ with λ-abstraction
$\lambda\text{-}\mathbf{HA}^{\omega}_{\rightarrow}$, 468	$\lambda\text{-}\mathbf{HA}^{\omega}$ without products
HAH, 170	higher-order Heyting arithmetic, higher-order intuitionistic arithmetic
HAS, 164	second-order Heyting arithmetic, second-order intuitionistic arithmetic
$\mathbf{HAS}_0$, 167	weak **HAS**

H-**IQC** etc., 68	Hilbert-type variant of **IQC**
H-**IQCE**, etc., 73	Hilbert-type variant of **IQCE**
$\mathbf{IDB}_1$, 658	theory of inductively defined neighbourhood functions
$\mathbf{I\text{-}HA}^{\omega}$, 452	intensional intuitionistic finite-type arithmetic
IPC, 48	intuitionistic propositional logic
IQC, 48	intuitionistic predicate logic
$\mathbf{IQC}^2$, 160	intuitionistic second-order logic
IQCE, 52	E-logic (**IQC** with existence predicate)
$\mathbf{IQCE}^+$, 52	E^+-logic (variant of **IQCE**)
IZF, 620	intuitionistic (**ZF**-)set theory
IzF, 634	**IZF** with Replacement instead of Collection
LS, 659	theory of lawless sequences
$\mathbf{LS}_1$, 664	theory of binary lawless sequences
ML, 619	Martin-Löf's theory of types
$\mathbf{ML}_0$, 603	arithmetical fragment of **ML** (extensional)
$\mathbf{ML}_0^i$, 575	arithmetical fragment of **ML** (intensional)
$\mathbf{ML}_1^i$, 612	Martin-Löf type theory with a single universe
$\mathbf{MLV}^i$, 618	$\mathbf{ML}_1^i$ with iterative sets
MQC, 57	minimal predicate logic
N-**CQC** etc., 49	natural-deduction version of **CQC**
N-**IQC** etc., 49	natural-deduction version of **IQC**
N-**MQC**, 544	natural-deduction version of **MQC**
N-**MQCE**, 544	natural-deduction version of **MQCE**
PA, 128	Peano arithmetic, classical first-order arithmetic
PRA, 120	primitive recursive arithmetic
TAPP, 475	applicative theory of total operations
TYP, 169	intuitionistic simple type theory

II. Axioms and rules

AC, 189, 498	axiom of choice
AC!, 99, 498	AC with uniqueness, function comprehension
AC_0, 189	AC for numbers
AC_{00}, 189	AC from numbers to numbers
$AC_{00}!$, 188	AC_{00} with uniqueness

∈-Ind, 620	set induction
Inf, 621	infinity axiom
IP, 176	independence-of-premiss schema
IP_0, 204	weak IP
IPR, 138	independence-of-premiss rule
ITYP, 579, 603	I-type rule
IUS, 230	induction over unsecured sequences
K1, 2, 3, 226	axioms for inductively generated neighbourhood functions
KS, 236	Kripke's schema
LNP, 129	least-number principle
¬¬LNP, 130	double negation of LNP
LS1, 645	density axiom
LS2, 647	decidability of equality for LS
LS3, 648	axiom of open data
LS4, 655	continuity axiom for LS
MOD, 501	existence of modulus-of-continuity
MP, 203	Markov's principle
MP_{PR}, 204	MP for primitive recursive predicates
MR, 137	Markov's rule
MR_{PR}, 136	MR for primitive recursive predicates
NCONV, 582, 604	special equality rules for N
NE, 580, 603	N-elimination
NFP, 215	neighbourhood function principle
NI, 579, 603	N-introduction
N_kCONV, 594	special equality rules for N_k
N_kE, 594	N_k-elimination
N_kI, 594	N_k-introduction
NTYP, 579, 603	N-type rule
PCT, 244	partial Church's thesis
PEM, 10	principle of the excluded middle
∀-PEM, 15	PEM for purely universal formulas
¬∀-PEM, 15	negation of ∀-PEM
∃-PEM, 194	PEM for purely existential formulas
PF_A, 235	parametrization of functions via A
Pow, 622	powerset axiom
PT, 205	Post's theorem
QF-AC, 144	quantifier-free axiom of choice
RDC-D, 191	relativized dependent choice for domain D
REC, 144	definition by recursion

REFL, 48, 578	reflexivity
REM, 623	restricted excluded middle
REPL, 48, 578	replacement schema, axioms
Repl, 622	replacement in constructive set theory
Resq, 631	restricted-quantifier axioms
R-Sep, 620	restricted separation
SCDS, 241	subcountability of discrete sets
SC-N, 215	special continuity for numbers
S-Coll, 621	collection axiom
SEP, 195	separation of enumerable sets
¬SEP, 210	negation of SEP
Sep, 623	separation axiom
SHP, 245	Shanin's principle
SP, 478	surjectivity of pairing
STR, 52, 73	strictness axioms
SUB, 51, 120, 578	substitution axioms
Sub-Coll, 621	subset collection
SYM, 578	symmetry
THIN, 577	thinning
THIN*, 591	variant of THIN
TI_ω, 129	transfinite induction over ℕ
TRANS, 48, 578	transitivity
ΠTYP, ΣTYP, +TYP, 579, 603	Π-, Σ-, +-type rule
UI, 611	introduction rule for a single universe
UI′, 613	addition to UI
U_nI, 613	U_n-introduction
U_nTYP, 613	U_n-type rules
UP, 234	uniformity principle (for numbers)
UP!, 235	UP with uniqueness
UP_1, 235	UP for functions
UTYP, 612	U-type rule
VCONV, 618	special equality rule for iterative sets
VE, 618	V-elimination
VI, 618	V-introduction
WCONV, 616	special equality rule for W-types
WC-N, 209	weak continuity for numbers
WC-N!, 211	WC-N with uniqueness
WC-N(*T*), 216	WC-N for a spread *T*
WC-N$^\vee$, 209	WC-N for disjunctions

III. Other notations

Below we give a list of notations of more than local significance in their order of appearance. n, m are used for natural numbers, α for a function from natural numbers to natural numbers, A, A', A'' for formulas, X, Y for sets, t for a term, $\mathfrak{a}$ for arbitrary expressions.

Since the preliminaries appear in both volumes with a different pagination in lower-case roman numerals, we use "xv(xvii)" to indicate that a notation appears on page xv in volume 2, and on page xvii in volume 1, etc.

$\{x_1, \ldots, x_n\}$, xv(xvii)	finite set
$\{x : A\}$, $\{f(x) : A\}$, xv(xvii)	set notations
X^c, xv(xvii)	complement of X
$X \times Y$, (a, a'), xv(xvii)	cartesian product, pair notation
$\prod$, xv(xvii)	arbitrary cartesian products
$X/\sim$, xv(xvii)	quotient of X by equivalence relation $\sim$
$x/\sim$, $x_\sim$, $(x)_\sim$, $[x]_\sim$, xv(xvii)	equivalence class of x relative to $\sim$
$f\|X$, $f \upharpoonright X$, xv(xvii)	f restricted to X
$\mathrm{P}(X)$, xv(xvii)	powerset of X
$X \to Y$, Y^X, xv(xvii)	set of functions from X to Y
$f \in X \to Y$, xv(xvii)	f is a function from X to Y
$\lambda x.t$, $\mapsto$, xv(xvii)	lambda-abstraction notation
$\rightarrowtail$, $\twoheadrightarrow$, $\rightarrowtail\!\!\!\!\!\twoheadrightarrow$, $\hookrightarrow$, xv(xvii)	injection, surjection, bijection, inclusion mapping
χ_R, xv(xvii)	characteristic function of a relation R
(t, t'), xvi(xviii)	(code of) the pair t, t'
$\langle x_0, \ldots, x_n \rangle$, xvi(xviii)	(code of) the finite sequence $x_0, \ldots, x_n$
$\vec{t} = \vec{s}$, xvi(xviii)	equality of finite sequences
$\lambda x_1 \ldots x_n.t$, xvi(xviii)	iterated lambda-abstraction
$\lambda x_1, x_2, \ldots, x_n.t$, xiv(xviii)	simultaneous lambda-abstraction
graph(f), dom(f), xvi(xviii)	graph and domain of f
range(f), xvi(xviii)	range of f
$\mathbb{N}$, $\mathbb{Z}$, $\mathbb{Q}$, $\mathbb{B}$, $\mathbb{R}$, $\mathbb{C}$, xvi(xviii)	natural numbers, integers, rationals, Baire space, reals, complex numbers
n, m, i, j, k, xvi(xviii)	variables over $\mathbb{N}$
α, β, γ, δ, xvi(xviii)	variables over $\mathbb{N}^{\mathbb{N}}$
$\bar{n}$, $\bar{m}$, xvi(xviii)	numerals
$\langle a_n \rangle_n$, xvi(xviii)	notation for infinite sequences $a_0, a_1, a_2, \ldots$
$\mathscr{L}(\mathbf{H})$, xvi(xviii)	language for system $\mathbf{H}$
$\vdash$, xvii(xix)	deducibility
E, 13	existence predicate, E-predicate
$\mathrm{A}_n(P)$, 49	nth element of the Rieger–Nishimura lattice
$\simeq$, 52, 54	equal-and-equally-defined (partial equality)
$Ix.A$, 55	descriptor, the unique x such that A
$\vdash_c$, $\vdash_i$, $\vdash_m$, 57	deducibility in classical, intuitionistic, minimal predicate logic
A^g, 57	Gödel–Gentzen negative translation of A

A^{k}, 59	Kolmogorov's negative translation of A
A^{q}, 59	Kuroda's negative translation of A
$\mathscr{CON}$, 60	set of contexts
$\mathscr{POS}$, $\mathscr{NEG}$, 61	positive and negative contexts
$\mathscr{SPOS}$, 66	strictly positive contexts
$\models$, $\models_{\mathrm{c}}$, $\models_{\mathrm{i}}$, xix, 75	validity in all structures (classical, intuitionistic)
$\Vdash$, 77, 78, 80, 81	Kripke forcing
$\varphi_{kk'}$, 85	restriction maps for Kripke models
PRIM, 115	the class of primitive recursive functions
Z, S, 115	zero-function, successor function
p_i^n, 115	projection functions
prd, $\dot{-}$, 116	predecessor function, cut-off subtraction
sg, sg$'$, 116	signum function, inverse signum function
$\|x-y\|$, 116	absolute difference of x and y
Π, 116	finite product (for PRIM)
Σ, 117	finite sum (for PRIM)
$\min_{n \le m}$, 117	bounded minimum operator
$n\|m$, $n!$, 117	n divides m, factorial of n
j, j_1, j_2, 118	pairing with inverses for $\mathbb{N}$
ν^p, j_i^p, 118	p-tupling with inverses for $\mathbb{N}$
$\langle m_0, \ldots, m_u \rangle$, 119	code of finite sequence $m_0, \ldots, m_u$
$*$, lth, 119	concatenation, length function
$(n)_m$, 119	m-th component coded by n
$\preccurlyeq$, $\succcurlyeq$, $\prec$, $\succ$, 119	partial order of finite sequences
$\bar{\alpha}(n)$, 119	initial segment of α of length n
TREC, 131	total recursive functions
REC, 131	(partial) recursive functions
$\min_n$, 132, 135	minimum operator
T, U, 133, 155	Kleene's T-predicate, result-extracting function
$\{n\}(m_0, \ldots, m_p)$, 133	Kleene-bracket notation
$\Gamma\|A$, 139	Aczel slash; Γ slashes A
$\mathscr{RH}$, 143	class of Rasiowa–Harrop formulas
Ap, **r**, λ, 144	application, recursor, lambda-abstraction in **EL**
$\alpha(\beta_0, \ldots, \beta_p)$, 157, 158	$\mathbb{N}$-valued partial continuous function application
$\alpha\|(\beta_0, \ldots, \beta_p)$, 157, 158	$\mathbb{N}^{\mathbb{N}}$-valued partial continuous function application

$\Lambda^0 n$, $\Lambda^1 n$, $\Lambda^0\alpha$, $\Lambda^1\alpha$, 159	
Λn, $\Lambda\alpha$, 160	
X^n, Y^n, Z^n, 160	variables for n-place relations
$[\sigma]$, $\mathrm{P}(\sigma)$, 169	power type of type σ
p_n^i, 169	projection functions from product types
$\alpha \in n$, 186	α has initial segment with code n
$n * \alpha$, 186	concatenation
$\forall n_T$, $\exists n_T$, $\forall\alpha_T$, $\exists\alpha_T$, 187	quantifiers restricted to trees
T_{01}, T_U, 187	binary tree, universal tree
V_n, 187	basic neighbourhood in a tree determined by node n
$\mathrm{Int}(X)$, 188, 266	interior of X
$(\alpha)_n$, 189	n-th component of α
$\mathbf{r}$, 196, 203, 244, 245	symbol for realizability
K_0, 212	set of neighbourhood functions
K, 226	set of inductively defined neighbourhood functions
$\Box_n A$, 236	A is evident at stage n
$\mathbf{q}$, 243, 244, 502	symbol for $\mathbf{q}$-realizability
$\mathbb{R}$, 253	Cauchy reals
$\approx$, 253	equivalence on fundamental sequences
$<$, $\le$, $>$, $\ge$, 255, 272	order on $\mathbb{R}$
$\#$, 255, 280	apartness
$+$, $-$, $\lvert x\rvert$, 258	addition, subtraction, absolute value
M_f, 260	modulus-of-continuity predicate
x^{-1}, 264	inverse of x
$U(x, p)$, 265	open ball with centre p and radius x
X^-, 265	closure of X
lim, 266	limit
$\mathbb{R}^{\mathrm{d}}$, 270	Dedekind reals
$\mathbb{R}^{\mathrm{be}}$, $\mathbb{R}^{\mathrm{e}}$, 270	bounded extended reals, extended reals
(x, y), $[x, y]$, 291	open interval, closed interval or segment
$(x, y]$, $[x, y)$, 292	halfopen intervals
sup, inf, 294	supremum, infimum
f', $f^{(n)}$, 296	derivative of f, n-th derivative of f
V_n', 303	(for the standard representation of $\mathbb{R}$)
$\lvert I\rvert$, 307	length of an interval I
U, V, W, 346	variables for open sets
X, Y, Z, 346	variables for arbitrary pointsets
U_x, V_x, 346	variables for neighbourhoods of x

$U_\varepsilon(y)$, 346	open ball with centre y and radius ε
l^2, 346	separable hilbert-space
Y^-, 348	closure of Y (in topological spaces)
V_n', 352	(for standard representations of metric spaces)
$[x]_M$, 356	point with code x in metric space M
$\mathrm{ML}(Y)$, $\mathrm{ML}_M(Y)$, 359	Y is metrically located (in M)
$\mathrm{L}(Y)$, $\mathrm{L}_\Gamma(Y)$, 359	Y is topologically located (in Γ)
$\subset\subset$, 361	strong inclusion
$\rho(x, Y)$, 367	distance of x from Y
diam, 367	diameter
$\underline{\cup}$, 371	closure of union
$\#$, 384	apartness (in algebraic structures)
$\sigma : G_1 \to G_2$, 392	homomorphism σ from G_1 to G_2
$G_1 \to_\sigma G_2$, 392	homomorphism σ from G_1 to G_2
$G_1 \overset{\sigma}{\to} G_2$, 392	homomorphism σ from G_1 to G_2
A_σ, 403	anti-ideal determined by σ
F_R, 405	quotient-field of a ring R
dim, $\dim_w$, 409	dimension, weak dimension
$\dim(V) \le n$, 409	dimension not greater than n
$\dim(V) \ge n$, 409	dimension not less than n
$\dim(V) = n$, 409	dimension equal to n
$\mathrm{rank}(A)$, 410	rank of a matrix A
$\det(A)$, 412	determinant of A
$R[X]$, 415	ring of polynomials in X over R
$R[X_1, \ldots, X_n]$, 416	ring of polynomials in $X_1, \ldots, X_n$ over R
(f), 421	principal ideal generated by f
A_f, 421	anti-ideal generated by f
$F[\alpha]$, 423	factor ring over a polynomial
$g(\alpha)$, 423	$(\alpha = X + (f))$
$\mathcal{T}$, 444	set of finite-type symbols
$\mathcal{T}_\to$, 444	set of finite-type symbols without products
$(\sigma\tau)$, 0, n, 444	notations for types
$\sigma_0 \sigma_1 \ldots \sigma_n$, 444	notations for types
$(\sigma \to \tau)$, 445	$= (\sigma\tau)$
HRO, HRO_σ, 445	hereditarily recursive operations (of type σ)
HEO, HEO_σ, 446	hereditarily effective operations (of type σ)
$\boldsymbol{p}^{\sigma,\tau}$, $\boldsymbol{p}_0^{\sigma,\tau}$, $\boldsymbol{p}_1^{\sigma,\tau}$, 447	pairing and unpairing in $\mathbf{HA}^\omega$
$\boldsymbol{k}^{\sigma,\tau}$, $\boldsymbol{s}^{\rho,\sigma,\tau}$, $\boldsymbol{r}^\sigma$, 447	combinators and recursor in $\mathbf{HA}^\omega$
Ap, 447	application (in $\mathbf{HA}^\omega$)

$t_0t_1\ldots t_n$, $t[x]$, $t[x, y]$, 448	
$\forall\vec{x}$, $\exists\vec{x}$, $\forall\vec{x}\vec{y}$, $\exists\vec{x}\vec{y}$, 448	notations for quantifier strings
$A(\overrightarrow{ts})$, 448	
(t, s), 448	the pair t, s
$\lambda x.t[x]$, 449	defined lambda-abstraction in $\mathbf{HA}^{\omega}$
$\lambda\vec{x}.(t_1, \ldots, t_n)$, $\lambda\vec{x}.\vec{t}$, 449	lambda abstraction for a sequence
$[\![\ ]\!]_{\mathrm{HRO}}$, 450, 453	interpretation in HRO
$=_{e,\sigma}$, 451	defined extensional equality at type σ
ICF*, $\mathrm{ICF}_{\sigma}^{*}$, 454	variant of ICF (crude version)
ICF, ICF_{σ}, 454	intensional continuous functionals (of type σ)
$[\![\ ]\!]_{\mathrm{ICF}}$, $[\![\ ]\!]_{\mathrm{ICF}^*}$, 455	interpretation in ICF, ICF*
ECF, 456	extensional continuous functionals
ICF($\mathcal{U}$), ECF($\mathcal{U}$), 456	ICF and ECF relative to a universe $\mathcal{U}$
conv, $\succcurlyeq$, $\succ_1$, 459, 461, 466, 472	conversion, reduction, one-step reduction
NF, NF_{σ}, 459, 469	terms in normal form (of type σ)
CNF, CNF_{σ}, 459, 469	closed terms in normal form (of type σ)
T_t, $h(t)$, 461	reduction tree of t, height of T_t.
SN(t), 462	t is strongly normalizable
SC(t), 463	t is strongly computable
WCR, CR, 469	weak Church–Rosser and Church–Rosser property
UN(t), 469	t has a unique normal form
CNF^{α}, 472	CNF with a function parameter
$\boldsymbol{k}$, $\boldsymbol{s}$, $\boldsymbol{p}$, $\boldsymbol{p}_0$, $\boldsymbol{p}_1$, $\boldsymbol{d}$, P, Ap, N, 473	constants of **APP**
(s, t), $(s)_i$, 473	$= \boldsymbol{p}st$, $= \boldsymbol{p}_i s$
n, m, $\forall n$, $\exists n$, 473	number variables, number quantifiers in **APP**
PRO, 474	(model of) partial recursive operations
$\lambda x.t[x]$, 475	defined lambda-abstraction in **APP**
fix, 475	fixed-point operator in **APP**
$\boldsymbol{r}$, 476	recursor in **APP**
IT, IT_{σ}, 479	intensional (type σ) structure
ET, ET_{σ}, 480	extensional (type σ) structure
$N \to N$, N^N, 480	type 1 in **APP**
PCO, 481	(model of) partial continuous operations
$\alpha\|\beta$, $\alpha(\beta)$, 481	$\mathbb{N}$- and $\mathbb{N}^{\mathbb{N}}$-valued partial continuous application
$[\![\ ]\!]_{\mathrm{PCO}}$, 481	interpretation in PCO

$\mathrm{P}(\omega)$, 484	graph model
$[\![\ ]\!]_{\mathrm{P}(\omega)}$, 485	interpretation in the graph model
graph, fun, 486	graph and function operator in $\mathrm{P}(\omega)$
$\Sigma(\omega)$, $\mathrm{RE}(\omega)$, 489	submodels of $\mathrm{P}(\omega)$
CNFS, CNFS^{α}, 489, 490	closed normal term models for **APP**
CTS, 489, 491	closed term model for **APP**
conv, $\succcurlyeq$, $\succ_1$, 489	conversion, reduction, one-step reduction
r, 491	abstract realizability for **APP**
τ_A, 492	canonical realizing terms
q, 502	**q**-realizability for **APP**
VAL^{α}, VAL, 503	valuation predicate
SRED, SRED^{α}, 505	relation of strict reduction
Cl, 512	the class predicate
c_n, 513	class constant
conv, 532, 558	conversion
$\succcurlyeq$, $\succ_1$, 536	reduction, one-step reduction
$\Gamma \Rightarrow \theta$, 576	assertion θ under assumptions Γ
A type, 576	A is a type
$t \in A$, 576	t is element of type A
$t = s \in A$, 576	t and s are equal elements of A
$A = B$, 576	A and B are equal types
x, y, z, u, v, w, 577	(meta-)variables for elements of types
t, s, 577	terms for elements of types
A, B, C, D, 577	(meta-)variables for types
θ, θ', θ'', 577	metavariables for assertions
Γ, Γ', Γ'', 577	contexts
N, 579	type of the natural numbers
$\Pi x \in A.B$, 579	product type
$\Sigma x \in A.B$, 579	general sum type
$A + B$, 579	plus-type, finite sum type
$I(A, t, s)$, 579	identity type
O, S, $R_{x,y}$, 576, 580	zero, successor, recursor
Ap, $\lambda x.t$, 576, 580	application, lambda-abstraction in type theory
$\boldsymbol{p}$, $\boldsymbol{p}_0$, $\boldsymbol{p}_1$, 576, 580	pairing with inverses in type theory
(t, t'), 576, 580	$= \boldsymbol{p}(t, t')$
$\boldsymbol{k}_0$, $\boldsymbol{k}_1$, $D_{x,y}$, 576, 580	introduction and elimination constants for $+$-types
$\boldsymbol{e}$, 576, 580	constant for I-types

$\forall x \in A, \exists x \in A$, 580	quantifiers in type theory
$\to, \wedge, \vee, \perp$, 580	propositional operators in type theory
$\boldsymbol{k}, \boldsymbol{s}, \boldsymbol{p}', \boldsymbol{r}, \boldsymbol{f}$, 587	combinators in type theory
$\perp_A$, 592	defined absurdity in type theory
N_k, 594	k-element type
$0_k, 1_k, 2_k, \ldots$, 594	elements in N_k
R_k, 594	elimination constant for N_k
$\mathbf{0}, \mathbf{1}$, 595	$= 0_2, 1_2$
$E_{x,y}(t, t')$, 595	Σ-elimination operator
J, 598	elimination constant for I-types
U, 611	universe of small types
U_n, 613	sequence of universes
$Wx \in A.B$, 615	type of well-founded trees, W-type
$\sup(t, t')$	introduction of a W-type
$T_{x,y,z}(t, t')$	elimination of a W-type
$\bar{\ }, \tilde{\ }$, 617	branching index and branching function
V, 618	universe of iterative sets
$\in, =$, 620	elementhood and equality in set theory
$A'[\ ,\ ]$, 620	
suc, 621	successor in set theory
$\emptyset, \{x, y\}, \bigcup x, \omega$, 621	empty set, pair set, union, $\mathbb{N}$ in set theory
$\langle x, y \rangle$, 621	ordered pair in set theory
dom, range, 622	domain and range in set theory
fun, FUN, 622	functionality predicates in set theory
$\subset, \times$, FULL, 622	inclusion, product, fullness in set theory
$\mathrm{P}(x)$, 623	powerset of x (in set theory)
$\doteq, \dot{\in}$, 625	defined equality and elementhood of iterative sets
$\alpha, \beta, \gamma, \ldots, \eta$, 625	elements of V (in chapter 11)
$\forall x \dot{\in} \alpha, \exists x \dot{\in} \alpha$, 625	
$T(t, f)$, 625	abbreviation of $T_{x,y,z}(t, f(x, y, z))$
a, b, c, d, 645	variables for lawlike sequences
α, β, γ, 645	variables for lawless sequences (in 12.2–3)
$\neq (\alpha, \alpha', \alpha'', \ldots)$, 648	α differs from $\alpha', \alpha'', \ldots$
$\#(\alpha, \alpha', \alpha'', \ldots)$, 648	$\alpha, \alpha', \alpha'', \ldots$ are all distinct
$K_0, (K_0)_{\mathrm{LS}}$, 650	neighbourhood functions w.r.t. LS
$\dot{\forall}, \dot{\exists}$, 655	
$\forall a \in n, \exists a \in n$, 656	
$\vec{\alpha} \in \vec{n}, \vec{\alpha} \in n$, 659	
$\dot{\forall} \vec{\alpha} \in \vec{n}$, 659	

Printed and bound by CPI Group (UK) Ltd, Croydon, CR0 4YY
03/10/2024
01040428-0005